23rd Canadian Hydrotechnical Conference

Held as Part of the Canadian Society for Civil Engineering Annual Conference and General Meeting 2017

Vancouver, Canada
31 May – 3 June 2017

ISBN: 978-1-5108-7840-2

Printed from e-media with permission by:

Curran Associates, Inc.
57 Morehouse Lane
Red Hook, NY 12571

Some format issues inherent in the e-media version may also appear in this print version.

Printed by Curran Associates, Inc. (2019)

For permission requests, please contact Canadian Society for Civil Engineering at the address below.

Canadian Society for Civil Engineering
521-300 rue St-Sacrement
Montréal, QC
H2Y 1x4

Phone: (514) 933-2634
Fax: (514) 933-3504

info@csce.ca

Additional copies of this publication are available from:

Curran Associates, Inc.
57 Morehouse Lane
Red Hook, NY 12571 USA
Phone: 845-758-0400
Fax: 845-758-2633
Email: curran@proceedings.com
Web: www.proceedings.com

TABLE OF CONTENTS

Leadership in Sustainable Infrastructure

Leadership en Infrastructures Durables

Vancouver, Canada

May 31 – June 3, 2017/ *Mai 31 – Juin 3, 2017*

3D PHYSICAL MODELLING OF WIND AND WAVE OVERTOPPING AT THE BILLY BISHOP AIRPORT REVETMENT

Scott Baker [1, 4], Andrew Cornett [2], and Chris Glodowski [3]
[1, 2] National Research Council Canada
[3] WSP Canada Inc.
[4] scott.baker@nrc.ca

ABSTRACT

As part of a planned development for Billy Bishop Toronto City Airport, new land reclamation and related marine works are required to extend both ends of the main runway into Lake Ontario. Wave conditions at the eastern end of the runway are relatively mild, as the site is sheltered inside Toronto Harbour. However, an extension at the western end of the runway will be directly exposed to strong winds, energetic wave conditions, high water levels, and winter ice conditions. An engineered revetment is required to prevent erosion, preserve stability, and protect the exposed western reclamation from attack by waves and ice. The design challenge was compounded by the requirement to minimize the frequency and extent of wave run-up and overtopping during storms as much as possible in order to avoid frequent inundation of the runway. At the same time, the crest elevation of the perimeter revetment was limited by the elevation of the existing runway and various other requirements for safe operations at the airport.

A large-scale 3D physical modelling study was undertaken which played a crucial role in developing, testing, and optimizing the design of the perimeter revetment and for costing the marine works. In order to accommodate the stringent design requirements, several innovative revetment design concepts (featuring lower crest elevations, milder slopes, thicker filter layers, and thicker armour layers than normally seen in conventional designs) were tested in a wide range of harsh wave, wind, and water level conditions. With regards to wave overtopping of the perimeter revetment, specifically at the location of the navigational antenna array and jet blast deflector near the end of the runway, this study collected valuable quantitative measurements and qualitative observations.

Keywords: - physical modelling, revetment, wave overtopping, wind

1. INTRODUCTION

1.1 Project Background

Billy Bishop Toronto City Airport (BBTCA) was purpose-built on reclaimed land on the Toronto Islands chain and opened as Toronto's first commercial airport in 1939. It was recently ranked Canada's 9th-busiest airport by passenger numbers. In 2013 Porter Airlines, whose home base of operations is BBTCA, announced their desire to purchase a fleet of new larger aircraft with greater range (the CS100 by Bombardier) and expand operations from a regional airline serving cities in central and eastern Canada and the north-eastern US, to a continental airline serving destinations throughout North America. Introduction of the CS100 would allow Porter Airlines to vastly increase the number of domestic and trans-border destinations served from BBTCA (Porter Airlines, 2013).

Airport Infrastructure upgrades would be required in order to safely and efficiently operate the CS100 at BBTCA. PortsToronto retained WSP Canada Inc. (herein WSP) to identify operating requirements for the CS100 at BBTCA. The major element was the need to extend the main runway by about 400 m (200 m at either end) to accommodate the longer landing and take-off requirements of the larger CS100 aircraft. This extension would include a 150 m Runway End Safety Area. PortsToronto would also need to obtain an exemption to an existing tripartite agreement restricting operations by jet aircraft.

1.2 Marine Works

The proposed extension of Runway 08-26 requires land reclamation work into Lake Ontario on the west end and into the Inner Harbour on the east end (see Figure 1). Wave conditions at the eastern end of the runway are mild, as the site is well protected by the Toronto Islands chain. However, the western runway extension will be directly exposed to harsh wind, wave, and ice conditions from Lake Ontario. The western reclamation must be protected by an engineered revetment to prevent erosion and preserve stability under wave attack. In addition, to ensure that the runway is not frequently inundated by water and ice, the revetment bordering the reclamation must be designed to minimize the volume of water passing over the crest as spray or green-water flows during storm events. Moreover, the crest of the revetment must be maintained at a relatively low elevation to ensure safe landing and take-off operations.

Figure 1. BBTCA Runway 08-26 extension plan (source: Porter Airlines)

1.3 Site Conditions

The city of Toronto and BBTCA are located on the north-western shore of Lake Ontario. WSP conducted a detailed study documenting the existing coastal environment to support the design of the new marine works and the physical model study. This study produced up-to-date information on local bathymetry, lake water levels, wind climate, as well as offshore and nearshore wave climates. A wave hindcast model was used to estimate the offshore wave climate at BBTCA, as no wave measurements over a long period were available close to the project location. The offshore wave climate was computed for a point located 5.2 km south of the airport at an approximate water depth of 70 m, providing a representative offshore wave exposure with maximum fetch in all directions.

As they propagate to shore, various physical phenomena transform the waves as they start to interact with the lakebed. The SWAN numerical wave model was used to transform the offshore wave climate to the tip of the proposed western runway extension at a depth of 8.2 m. An hourly nearshore wave climate was predicted for the period from 1971-2014. In general, maximum wave heights ranging between 2.0 and 2.5 m were predicted, with the exception during the Southern Ontario Blizzard of 1978 where a maximum wave height of 3.4 m was predicted. The incident wave direction is a relatively narrow band between SE and WSW. Waves approaching from southwest are the most frequent and are expected to have the largest wave heights. Prevailing winds are from the southwest, west, and east directions, with maximum speeds on the order of 80 km/h (WSP Canada Inc. 2015).

1.4 Study Objectives

The National Research Council of Canada (NRC) was commissioned to conduct physical modelling studies to support the efficient, cost-effective, and reliable engineering design for the proposed extension to Runway 08-26. The main objectives of the study were to: verify that the proposed revetment designs would remain stable under the extreme design conditions forecasted at the site, and optimize the proposed designs to improve stability, reduce wave overtopping to acceptable safe levels, and reduce construction and maintenance costs where possible.

2. REVETMENT DESIGN CRITERIA

The design requirements for the western runway extension were unique in that the coastal protection works needed to remain relatively low-crested as to not interfere with the Obstacle Limitation Surface (OLS, a government regulation restricting the height of objects located near runways), but also limit wave overtopping and associated spray from reaching the runway or taxiway surfaces such that safe operation of the airport was maintained, especially during winter. Desktop analysis indicated that a conventional revetment design would require armour stone in the order to 10 tonne, with crest heights reaching an elevation of +6.0 to +6.5 m. While this conventional design would likely be stable under extreme wave attack, it did not comply with the OLS regulation. Furthermore, an increased crest width and/or height would be required to control wave overtopping and ensure stability on the leeward side. Information from an aggregates survey indicated that large armour stone would be difficult and/or expensive to source, especially in the quantities required. Hence, an alternative non-conventional design based on mean armour stone weights of lighter than 6 tonne was developed. The design concept was to minimize overtopping and maximize stability by increasing the volume of porous armour and filter materials. It was anticipated that increasing the filter layer thickness would improve the stability of the outer armour layer while also reducing the levels of overtopping. Damage criteria for the rock armour relates to the displacement or loss of stone that would diminish the function of the revetment to withstand extreme wave attack and to control wave overtopping. For this study, little or no displacement of stone along the leeside of the revetment crest was permitted during the principle design conditions (once yearly event); while under overload conditions there should be little or no displacement of armour stone on the front face and crest of the revetment.

The issue of wave overtopping at BBTCA was considered to be an important factor in the rubble-mound revetment design. Criteria for tolerable wave overtopping rates were primarily based on engineering guidelines (TAW, 2002; EurOtop, 2007; Krom, 2012) and were influenced by specific infrastructure to be placed at the end of the western runway extension. It is noted that these guidelines provide only limited information on dealing with wind effects on overtopping. For this study, maximum mean rates of wave overtopping in the area at the end of the runway extension were set as 0.1 L/s/m for servicing of the equipment and 1.0 L/s/m for the infrastructure itself. Although the area immediately behind the revetment crest would not normally have pedestrian or vehicular traffic, a localizer antenna array (LOC) and a jet blast deflector fence will be located near the end of the runway. The LOC is critical to the operation of the airport since it is part of the primary navigation guidance system for aircraft landings and missed approaches. The LOC array can operate with an ice covering of up to 5 cm, but must remain in perfect alignment at all times. It must also be accessible for servicing under almost all conditions. Excessive wave overtopping under freezing conditions could render the LOC inoperable. Overtopping reaching the runway and taxiway surfaces at the western extension was to be minimized as much as possible (ideally <0.1 L/s/m).

Apart from the space requirement on the crest of the revetment for the LOC placement and service access, the LOC cannot be too close to reflecting surfaces due to backscatter considerations. In particular, the rock slopes on the leeside of the revetment crest had to be at least 15 m away from the LOC across a flat area, or be at least 5 m away across a flat area if the top of the revetment crest is at least 1.5 m below the elevation of the antenna heads.

3. PHYSICAL MODEL DESIGN AND CONSTRUCTION

3.1 Model Layout and Construction

The model study was conducted in NRC's 26 m by 36 m Multidirectional Wave Basin (MWB), a state-of-the-art facility for engineering research and performance verification of maritime structures in the ocean environment. The MWB can accommodate water depths from 0 to 2.5 m and features a powerful 30 m long directional wave generator capable of generating a broad range of realistic wave conditions with significant wave heights exceeding 0.5 m.

A three-dimensional physical model of the western runway extension and surrounding lakebed bathymetry was constructed at a length scale of 1:28. This scale represents a reasonable compromise between minimizing scale and boundary effects without exceeding the limits of the MWB facility. All aspects of the model were designed using Froude scaling criterion, which assumes that gravitational and inertial forces are dominant, compared to viscous forces. At this scale, the functional area of the wave basin represented a 0.47 km^2 rectangular footprint of prototype terrain (see Figure 2). The model was oriented within the basin such that the runway extension could be tested in waves approaching from a ~45° range of directions between SE and SW. A side-wall reflection technique was used in this study in order to increase the physical area within the wave basin over which homogeneous wave conditions were generated. The technique involves purposely reflecting waves off the solid guide walls on either side of the wave machine. This side-wall reflection technique was a crucial factor in being able to generate sea-states with mean wave directions of 22.5° relative to the wave machine that were reasonably homogeneous across the test site. Highly efficient expanded-metal wave absorbers featuring reflection coefficients less than 5% were installed around the perimeter of the basin to control unwanted wave reflections.

A faithful representation of the existing lakebed, down to approximately the 9 m depth contour was constructed to ensure that local nearshore wave transformations would be properly simulated in the model. The model bathymetry was formed by placing a thin skin of cement mortar over a moulded bed of compacted gravel. Templates spaced at 1 m (model scale) intervals were used to control the final elevation of the bathymetry surface. Once the cement mortar had cured, several lakebed contours were painted onto the concrete floor to enhance the visual perception of the lakebed.

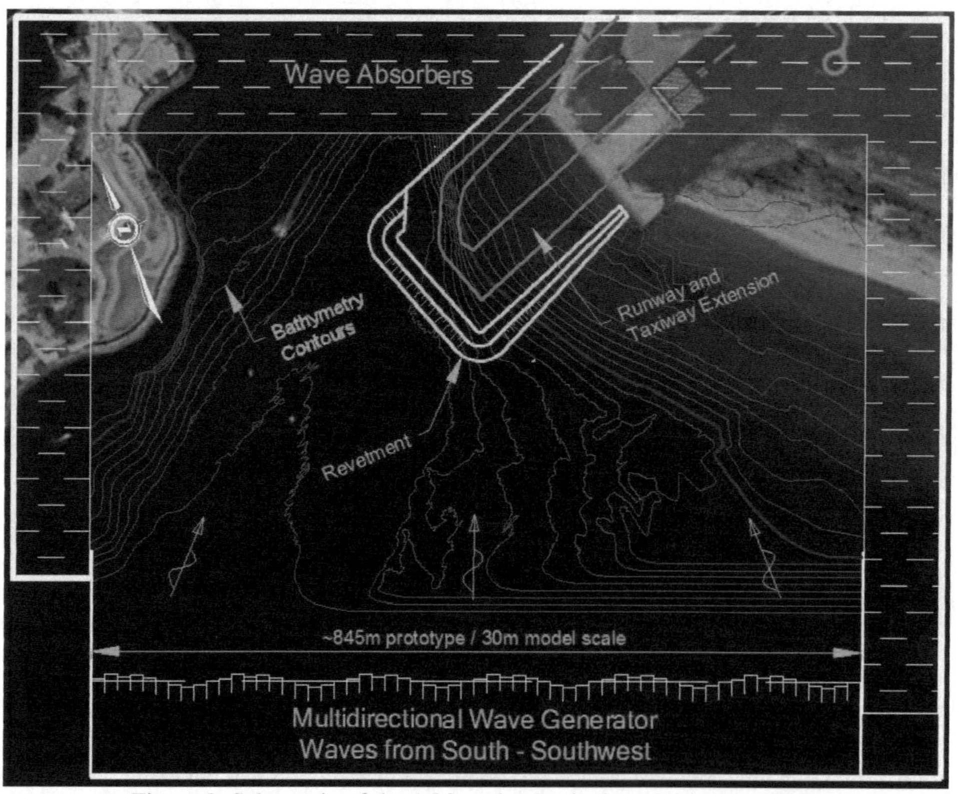

Figure 2. Schematic of the 1:28 scale physical model in the MWB

Careful attention was given to the location, dimensions, composition, and methods of construction of the model structures to ensure that they replicated the proposed designs accurately and faithfully. A total station was used to guide the position and alignment of all structures and ensure conformity with designs. Structure alignment was precisely laid out using a total station, ensuring the overall positioning and curvature of the prototype designs was very well replicated.

A small portion of the adjacent Ontario Place island (see Figures 1 & 2) was included in the model domain. This region, which was not a focus of the study, was backfilled with fine gravel and armoured with a revetment of oversized stone to ensure stability throughout the entire study. It was incorporated to ensure that any effects of waves reflecting from this island would be included in the model. It also provided a visual reminder of its proximity to the runway and the narrowing of the Western Gap harbour channel.

The north side of the proposed runway extension was defined as a vertical retaining wall, and was modelled using marine-grade plywood along the designated alignment. A plywood trench was constructed between the edge of the runway and the edge of land mass along the south and west sides of the runway extension. This trench was built to accommodate overtopping measurement systems at any location behind the revetment and to permit relocation of these systems between tests with minimal effort.

The vertical wall at the existing edge of land mass was modelled using concrete blocks levelled to the runway centre elevation of +2.4 m (all elevations refer to chart datum lake level). Portions of the runway extension and taxiway areas were modelled and simple mock-ups of the CS100 and Q400 aircraft were created to provide additional realism and sense of scale (see Figure 3).

Figure 3. Overview of the 1:28 scale physical model

Rock materials and gradations were carefully prepared to replicate the characteristics of the prototype materials under consideration for use in the prototype construction. Five different gradations of rock material were defined, including two different sizes of filter and armour stone, and one size of core material. Equivalent armour and filter materials at model scale were obtained by sieving locally available crushed limestone aggregate, effectively separating it into different sizes. Measured gradations for the filter and armour rock classes were obtained by weighing a representative sample. In some cases one of the sieved materials matched the design gradation, while in other cases two or more different stone sizes were blended together in order to produce a gradation in better agreement with specifications. In this paper, the various rock gradations are referred to by their M_{50} values. Core materials were scaled by particle diameter and checked via sieve analysis.

A large number of fibreboard templates were prepared and used to guide the placement of the core, filter, and armour materials for each rebuild. Elevations were carefully controlled by surveying the templates using an optical level, and were verified after construction for quality assurance. This technique allows the design profile (shape and elevation) to be constructed to a high degree of accuracy. Once construction was nearly complete, the templates were removed and the locally disturbed areas were patched prior to testing. Core and filter layers were bulk-placed by shovel and then gently hand-packed and shaped to match the desired profile using wooden trowels. The methods used to place the model armour stone attempted to mimic those to be used in the prototype, so that rock behaved similarly and displayed the same kind of interlocking characteristics. Armour stones were placed by hand, unit-by-unit to simulate placement by crane/excavator (see Figure 4a). The outer surface of the filter and armour layers were each painted with different colours to assist in visualizing stone motion and surface damage (see Figures 3 & 4b).

Figure 4. Revetment construction

Several banks of variable-speed fans were set up and operated in the model to generate a local onshore wind and improve the simulation of overtopping by wind-driven spray (see Figure 3). At 1:28 scale, the model is expected to provide reliable predictions of armour stone stability at full scale under equivalent conditions. However, model scale effects relating to fluid surface tension mean that physical processes that are sensitive to surface tension were not simulated correctly in the model. Surface tension is an important parameter affecting aeration and bubble content in breaking waves, the formation and character of spray, and other details of the wave breaking process. For example, the excessively large surface tension will reduce the level of aeration in the model overtopping flows, and cause spray droplets to over-sized in the model. Nevertheless, wind-driven overtopping was included in the model in an attempt to capture the approximate contribution to the total amount of overtopping.

3.2 Instrumentation

Sixteen capacitance wave gauges were deployed throughout the domain, including one gauge for performing directional measurements. Two ultrasonic anemometers were used to measure wind speeds. The anemometers were mounted on a pipe stand and levelled to +10 m elevation. One anemometer was located directly behind the SW corner while the other anemometer was situated in line with the taxiway/runway intersection.

Three overtopping collection systems were deployed, each consisting of a water storage reservoir, a wave gauge to measure the level of water in the reservoir, and a metal tray to convey the overtopping flow from the crest of the revetment into the reservoir (see Figures 3 & 4b). For this study, the collection trays were positioned along the 'edge of land mass' line, several meters (prototype) inland from the crest of the revetment (where overtopping is often measured). Much of the overtopping volumes were absorbed into the filter and armour layers on the crest of the revetment, and only the 'excess' overtopping not already absorbed into the rock reached the collection systems. Hence, the overtopping rates at the revetment crest were significantly higher than at the measurement locations. The overtopping collection systems were better suited for measuring moderate to severe overtopping flows, so measuring very small volumes of overtopping (e.g. <0.05 L/s/m) was less precise.

A photographic damage analysis system comprising eight remotely-operated digital cameras was used to monitor the movement of armour stone on the surface of the revetment. Since each camera remained fixed throughout a test series, the movement of individual stones could be detected by comparing photographs taken at different times. Four digital video cameras were also used to record all tests: two cameras provided oblique overhead views while the other two provided close-up views of wave-structure interactions and overtopping at ground level.

4. WAVE AND WIND CALIBRATIONS

Following construction of the bathymetry (but before building any revetment structures), a series of undisturbed wave calibrations were conducted to tune the wave generator command signals and verify the incident wave conditions to be used in the model study. A range of mild, moderate, and extreme conditions were tuned based on available wave climate data. The design wave condition was defined as the annual maximum wave height at three different water levels (corresponding to different months/seasons during the year), with significant wave heights ranging from 2.5 – 3.3 m. An overload condition with significant wave height of ~4.0 m corresponded to an unlikely event with a large return period. Peak wave periods varied from 8 – 11 s, and incident directions ranged from SSW to SW. The scaled wave conditions generated and measured in the model agreed very well with specifications.

Local wind fields featuring a mean wind with gustiness were generated using three banks of five fans each (see Figure 3). Each fan bank was independently controlled by electronic meters, and the resulting wind speed could be varied over a wide range by adjusting the fan RPM. These electronic meters allowed the local winds at the site to be reproduced in the model in a controlled and repeatable manner. Following construction of the runway extension and surrounding revetment, a series of wind calibrations were conducted to tune the fan control settings and verify that the desired wind conditions at the test site were being achieved. Target wind conditions were developed based on available wind climate data. Three wind conditions from the SW with mean speeds ranging from 55 – 75 km/h were modelled. These conditions were considered to be the maximum likely wind speeds at three different water levels (corresponding to different seasons within a typical year).

5. TESTING AND RESULTS

Six distinct test series were conducted during this study. Each test series involved the construction of different revetment cross-sections in various locations, followed by a test program involving exposure to increasingly harsh water level, wave, and wind conditions. The local wave conditions at the toe of the revetment varied due to the variations in local water depth and exposure to the incident waves. During the model study the revetment designs for each area were optimized to accommodate the local wave conditions. For the purposes of this paper, the revetment is divided into four main segments, namely: south face, southwest corner, west face, and northwest corner (see Figure 5). Not all segments were reconstructed for each test series. The southwest corner and southern portion of the west face experienced the most energetic wave attack and the highest amount of wave overtopping and were required to provide protection to the nearby localizer antenna array. A more detailed analysis of the revetment designs is discussed in (Baker et al. 2017).

Accurate measurement of wave overtopping in a 3D model is quite challenging. Overtopping is a complex process, and while it is possible to reliably and precisely reproduce waves in a physical model, the interactions of those waves with the rock-armoured revetment is such that overtopping rates during a repeated test may vary considerably despite being measured in the same way and at the same locations. In some cases, overtopping flows might fall between two collection systems and not be measured on either.

The first overtopping measurement system (OT1) was located directly behind the SW corner and remained unmoved throughout the entire test program. The second and third systems were shifted on several occasions to learn about overtopping rates at different sections behind the west face. Therefore, OT1 remains the best candidate for best comparing the overtopping rates against various cross-section designs. The SW corner crest elevation was lowest during Test Series 2 and 5, which predictably resulted in larger overtopping flowrates. Crest elevations during Test Series 3, 4, and 6 were nearly identical; however, the crest width was notably wider for Test Series 6. Comparing the various designs of the SW corner in terms of stability and control of overtopping, the optimized design featured in Test Series 6 had the best overall performance.

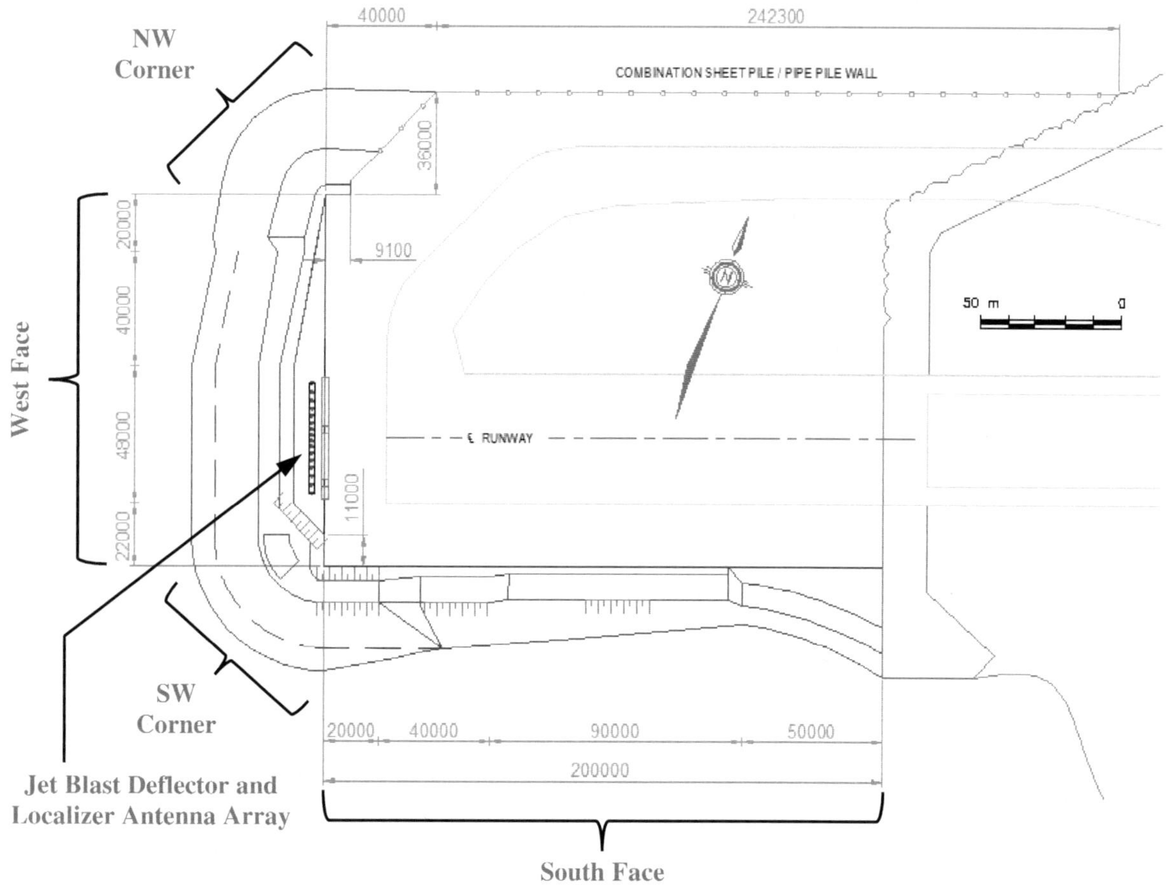

Figure 5. Plan view

On several occasions during each test series, the same seastate was repeated with and without wind, to determine its relative effect on overtopping (see Figure 6). It was absolutely clear, based on qualitative observations and quantitative measurements, that wind caused a noticeable increase in overtopping volume and extent. Visually it was apparent that, once a wave crest ran up to the front crest of the revetment, the wind helped to carry and drive the overtopping flow further inland. Less dissipation occurred on the crest of the structure, resulting in larger volumes of overtopping reaching the leeside of the crest and beyond. Some overtopping spray was observed to reach the end of the runway during severe conditions.

Efforts were made to setup and generate wind in as realistic a manner as possible; however, more rigorous testing (not within the scope of this study) would be required to more conclusively define the effect of wind. For instance, although three different wind speeds were used, they always corresponded with the same water level each time; therefore it is difficult to conclude how the speed of wind affects the overtopping. Regardless, overtopping measurements made throughout the study demonstrated that, compared to a seastate without wind which had relatively low overtopping (e.g. less than 1.0 L/s/m), the same seastate with wind typically saw a three-fold increase in overtopping volumes. The wind had a less noticeable effect when overtopping volumes were significantly larger.

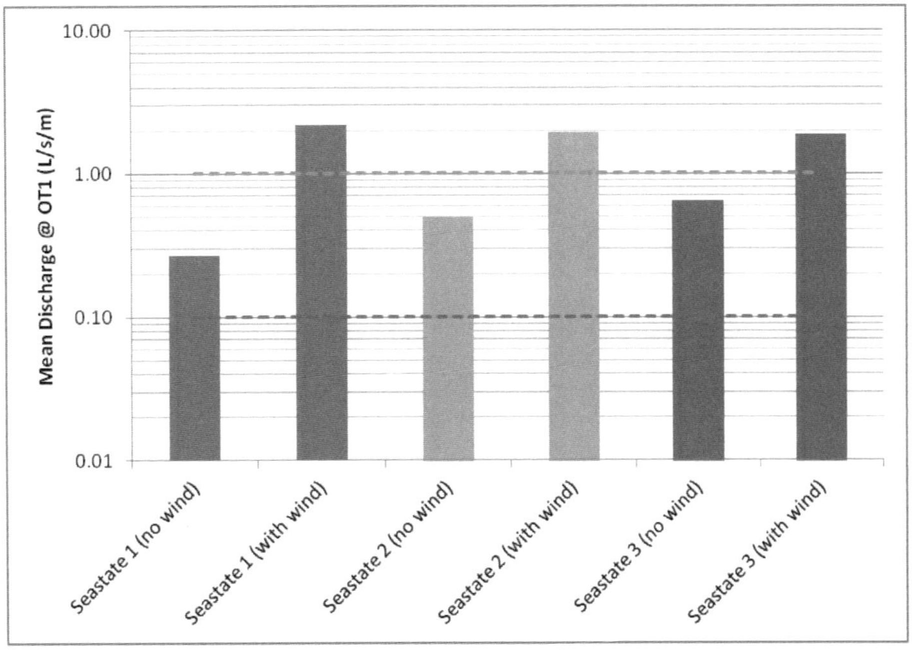

Figure 6. Effect of wind on overtopping measurements

6. CONCLUSION

With regards to a proposed runway extension at Billy Bishop Toronto City Airport, the study objective was to develop an efficient revetment design with low crest elevations that remained stable when exposed to extreme wave and water level conditions, while limiting the frequency and extent of wave overtopping that could impact airport operations. For the principle design wave conditions, the final optimized revetment design saw overtopping volumes less than 1.0 L/s/m. The development and validation of an optimized non-conventional revetment design was guided by a 3D physical model constructed at a geometric scale of 1:28 that included scaled reproductions of: the western end of the existing runway; the proposed 200 m long reclamation; new coastal protection works around the perimeter of the runway extension; the surrounding lakebed bathymetry; and the local water levels, waves, and wind.

Several rubble-mound revetment designs were modelled and assessed in a wide range of wave, wind, and water level conditions. In order to reduce wave runup and overtopping, many of the revetment designs were unconventional, featuring lower crest elevations, milder slopes, thicker filter and armour layers, and lighter armour stone than normally seen in conventional designs. Although there are significant model scale challenges associated with it, wind-driven overtopping was included in the model in an attempt to capture the approximate contribution to the total amount of overtopping. With the optimized designs, no green-water overtopping and only small amounts of spray are expected to reach the runway or taxiway surfaces at any location during once yearly storm events. Moderate amounts of spray may reach the runway or taxiway surfaces during more extreme events, especially with strong winds blowing from SW. At the location of the localizer array and blast fence, overtopping and spray were limited to low rates during extreme conditions; however, published limits for safe serviceability of this equipment are not well defined. Although more rigorous testing (beyond the scope of this study) would be required to better define the effects of wind on wave overtopping, the study demonstrated that strong winds typically caused a minimum three-fold increase in overtopping volumes (when overtopping rates are relatively low).

The physical modelling study played a crucial role in assessing structure performance and optimizing the design of the marine works. The longer runway will allow Porter Airlines to expand operations from BBTCA, reaching destinations across North America. The project will provide options to the citizens of Toronto and help boost the local economy. The longer runway will also improve safety by reducing the risk of a runway excursion.

7. REFERENCES

Baker, S. 2015. *3D Physical Modelling Studies Supporting the Extension of Runway 08-26 at Billy Bishop Toronto City Airport*, NRC Technical Report, Ottawa, Canada.

Baker, S., Cornett, A., Glodowski, C. 2017. 3D Physical Modelling and Revetment Design for the Billy Bishop Toronto City Airport Extension. *Coasts, Marine Structures and Breakwaters*, Institution of Civil Engineers, Liverpool, UK.

EurOtop. 2007. *Wave Overtopping of Sea Defences and Related Structures: Assessment Manual*. Guidance manual developed in conjunction with Environment Agency (UK), Expertise Netwerk Waterkeren (NL), and Kuratorium für Forschung im Küsteningenieurwesen(DE).

Krom, J.C. 2012. *Wave overtopping at rubble mound breakwaters with a non-reshaping berm*, M.Sc. Thesis, Delft University of Technology, Rotterdam, The Netherlands.

LPS Aviation Inc. 2013. *Porter Airlines Runway 08-26 Extension Study*, Ottawa, Canada.

Porter Airlines. 2013. *Runway Extension*, public presentation, Toronto, Canada.

TAW. 2002. *Technical report on wave run-up and wave overtopping at dikes*, Technical Advisory Committee on Water Defences, The Netherlands.

WSP Canada Inc. 2015. *Coastal Environment Study Supporting the Expansion of Billy Bishop Airport*, Québec, Canada.

WSP Canada Inc. 2015. *Billy Bishop Toronto City Airport – Preliminary Runway Design – Coastal Engineering Summary Report*, Ottawa, Canada.

Leadership in Sustainable Infrastructure
Leadership en Infrastructures Durables

Vancouver, Canada
May 31 – June 3, 2017/ *Mai 31 – Juin 3, 2017*

A SIMPLE RIVER MEANDERING MODEL

Bertrand Massé [1, 2]
[1] SNC-Lavalin Inc., Montreal, Canada
[2] bertrand.masse@snclavalin.com

ABSTRACT

Principles of mechanics are used to derive a mathematical formulation of river meandering. In its most simple formulation the model generates symmetrical meander loops whose shapes approach those produced by the sine-generated curve. For the more general case the model generates regular upstream-skewed meander loops like those commonly observed in nature. The model proposes explicit equations for evaluating meander characteristics such as loop length or wave length, based on commonly used floodplain and river parameters like the Froude number, transverse slope and the river width.

RÉSUMÉ

Les principes de mécanique sont utilisés pour développer une formulation mathématique des méandres de rivières. Selon la formulation la plus simple le modèle génère des boucles de méandres symétriques. La formulation complète permet de reproduire des boucles méandres asymétriques dirigées vers l'amont. Le modèle permet d'évaluer les caractéristiques de méandres telles que la longueur de la trajectoire, la longueur d'onde et l'amplitude sur la base des caractéristiques physiques de la plaine alluviale et de la rivière elle-même.

Keywords: - Meanders, river, sine-generated curve

1 INTRODUCTION

River meanders have always intrigued scientists. Regular-shaped meander loops are often observed but a succession of regular loops is less frequent. Most of the times the pattern is broken after a few loops only. Changes in shape or in direction occur suddenly without evident physical reason. These irregularities seem to imply that meander loops are sensitive to any modification in the physical properties of the floodplain in the flowing river direction. It is clear that a change in the composition of the material composing the river banks will most certainly disturb the meandering pattern. The fact that regularly shaped meanders are more frequently observed in small rather than large rivers indicates that the variation in surface elevation of the flood plain also plays a major role.

The first serious attempt to mathematically represent river meanders is due to Langbein and Leopold (1966) who have proposed the sine-generated curve, which reproduces fairly well the fatness or roundness of the meander loops observed in nature. However, the sine-generated curve gives no information on the characteristics of the river. Since then, more involved river meander models have been proposed by various researchers (Ikeda *et al.* 1981; Parker *et al.* 1982; Odgaard 1989). These models generally combine dynamic equations for flow in bends with a kinematical description of bank erosion to describe channel migration. Parker *et al.* (1983) have developed a model reproducing the upstream skew commonly encountered in meandering streams.

The regularly-shaped free meander model development presented herein is based on the equations of mechanics and does not require any parameter adjustment. The model allows the derivation of simple explicit equations for meander loop length, meander wave length and meander amplitude.

1. MODEL ASSUMPTIONS

Fig. 1 shows a plan view of a meandering river assumed to be in a quasi-equilibrium state. This condition means that the amplitude and shape of the meanders change slowly with time and can be considered constant during a given time frame (from a few days to several months). The x-axis coincides with the general longitudinal down valley direction.

A regularly-shaped meander is a meander developed in ideal conditions and these ideal conditions are defined by the following set of assumptions:

a) The plan form geometry is defined by the bankfull discharge

b) The river width is constant

c) The water depth is constant

d) The river banks are erodible

e) The longitudinal valley slope is constant

f) The flood plain is gently and uniformly sloping on both sides toward the main longitudinal river channel axis

The transverse flood plain slopes allow surface runoff to drain toward the river channel. This last assumption has rarely, if ever, been considered in meandering models. Nevertheless, it is considered to be an important condition for developing meanders because it forces the river to flow within a limited flood plain band on each side of the main longitudinal river axis. In order to generate symmetrical meanders, it is further assumed that this transverse slope is constant and is the same on both sides of the longitudinal axis. Unequal transverse slopes are possible but would result in asymmetrical meander loops.

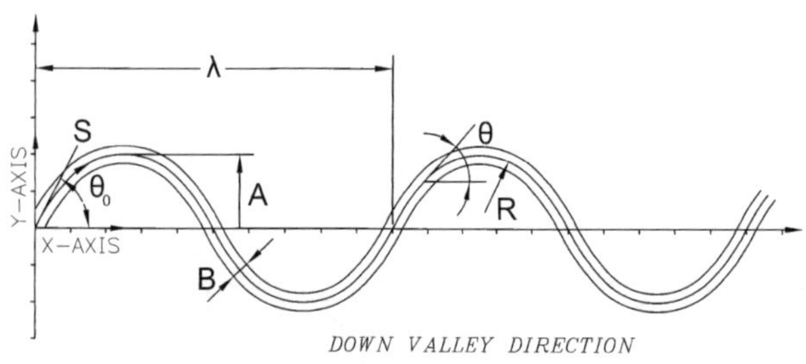

Figure 1. Definition of variables

2 MODEL DEVELOPMENT – BASE CASE

The model development first addresses the simple base case in which it is assumed that the down valley slope, denoted S_X, is much smaller than the transverse flood plain slope, denoted S_Y or, expressed differently, $S_X << S_Y$.

Let us consider an infinitesimal river segment of length Δs. Using the definition of variables shown in Fig. 2, the normal force exerted on the river bank by the curved river segment can be derived using the momentum equation.

[1] $2\rho Q V \sin\left(\frac{\Delta\theta}{2}\right) = w\,\Delta s$

For infinitesimally small Δs the equation becomes

[2] $\frac{\gamma}{g} Q V \frac{d\theta}{ds} = w$

in which γ = unit weight of water; g = acceleration due to gravity; Q = discharge; V = flow velocity; θ = deflection angle relative to x axis; s = distance along river path; and w = force per unit length applied in the river bend.

In order to achieve quasi-equilibrium this force must be resisted by the river banks. The total force exerted by the river banks in a direction normal to the flow is equal to the force on the outer bank minus the force on the inner bank of the bend. Due to the curved flow, a superelevation of the water surface is observed on the outer bank which is counterbalanced by an equivalent drop on the inner bank, as shown in Fig. 3. Let h be the average water depth, assumed to be uniform over a meander loop length. Neglecting secondary currents, the total water depth will be $h + \Delta h$ on the outer river bank and $h - \Delta h$ on the inner bank. The resulting force per unit length on the segment of length *ds* is then

[3] $w = \frac{1}{2}\gamma(h+\Delta h)^2 - \frac{1}{2}\gamma(h-\Delta h)^2 = 2\gamma h \Delta h$

in which it is assumed that $\Delta h \ll h$. Combining Eqs. (2) and (3) it follows that

[4] $\frac{\gamma}{g} Q V \frac{d\theta}{ds} = 2\gamma h \Delta h$

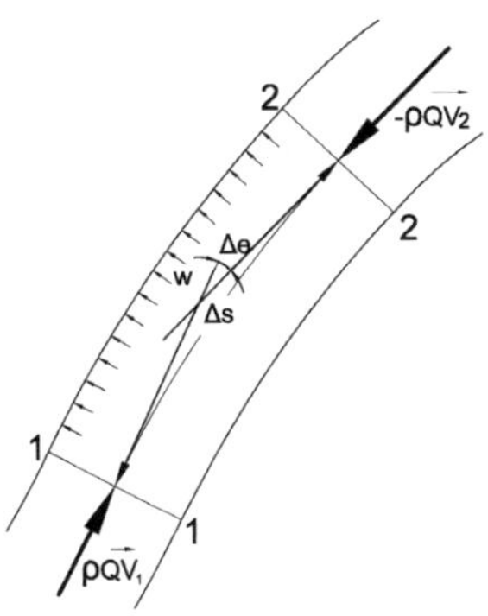

Figure 2. Momentum equation applied to a river bend

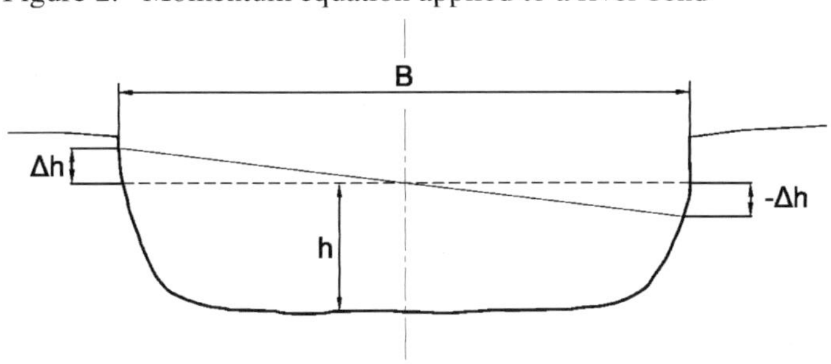

Figure 3. Superelevation of water surface in a river bend

Considering that Q = V B h, where B is the river width, and that $d\theta/ds$ is the inverse of the radius of curvature (R), Eq. (4) is equivalent to the well-known expression for estimating the superelevation of the water surface in the outer side of a river bend (Chow 1959).

[5] $\Delta h = \frac{V^2 B}{2 g R}$

Let us now make use of the assumption according to which the floodplain is not flat but is rather gently sloping, on both sides, towards the longitudinal axis, as shown in Fig. 4. The transverse floodplain slope is denoted S_Y.

Assuming a constant transverse slope, the point in the floodplain where the river bank height is exactly equal to the river depth plus the superelevation is at distance *y* from the x-axis.

It follows that Eq. (5) can be rewritten as:

$$[6] \quad \frac{\gamma}{g} Q V \frac{d\theta}{ds} = 2\gamma h \Delta h = -2\gamma h S_Y y$$

in which y is the distance from any given point of the river path to the x-axis. The minus sign is needed because, as shown in Fig. 1, the value of dθ/ds is negative when *y* is positive and vice versa.

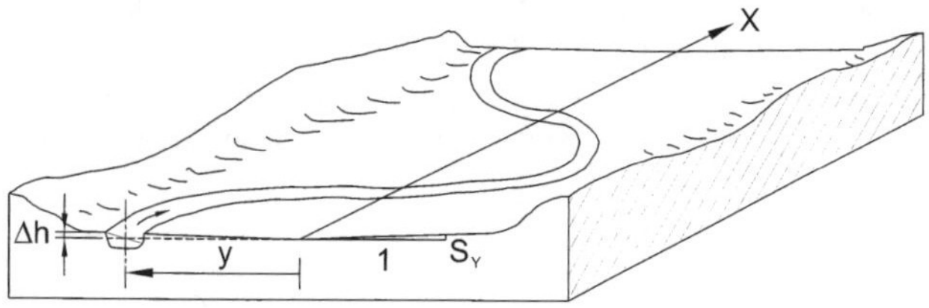

Figure 4. Definition of floodplain transverse slope

Distances *x* and *y* can be expressed in curvilinear coordinates as follows:

$$[7] \quad \begin{aligned} x &= \int \cos\theta \, ds \\ y &= \int \sin\theta \, ds \end{aligned}$$

We can therefore write

$$[8] \quad \frac{\gamma}{g} Q V \frac{d\theta}{ds} = -2\gamma h S_Y y = -2\gamma h S_Y \int \sin\theta \, ds$$

or, after differentiating with respect to *s* and rearranging,

$$[9] \quad \frac{\gamma}{g} Q V \frac{d^2\theta}{ds^2} + 2\gamma h S_Y \sin\theta = 0$$

Equation 9 is similar to other well-known equations in engineering mechanics. For example, the motion of a large amplitude pendulum is described by an equation similar to Eq. 9. The buckling of a column beam subject to large deformations is also defined by an equation of the same form as Eq. 9.

Solving Eq. 9 requires some mathematical treatment which the interested reader can find in Timoshenko and Gere (2009). After some manipulations and changes of variables the trajectory, s, of the river can be expressed in terms of elliptic integrals as follows:

$$[10] \quad s = \mu \left[F(k) - F(k/\varphi) \right]$$

where F(k) and F(k/φ) are respectively the complete and incomplete elliptic integral of the first kind, k = sin (θ_0/2) and

$$[11] \quad \varphi = \sin^{-1}\left(\frac{\sin(\theta/2)}{\sin(\theta_0/2)} \right)$$

F(k) is equivalent to F(k/(π/2)). The constant μ is equal to

[12] $\mu = \sqrt{\dfrac{QV}{2ghS_Y}} = 4F_rB\sqrt{\dfrac{1}{2\xi S_Y}}$

where F_r is the Froude number and ξ is the ratio B/h. Parameter μ has the dimension of a length and is characteristic of the meander size. It will hereafter be called the meander scale parameter.

Eq. (10) can be solved numerically (Abramowitz and Stegun 1980). For one quarter period $\varphi = 0$ and for a full period the total meander path length is

[13] $M_L = 4\mu F(k)$

It is also possible to derive mathematical expressions for both x and y which can then be used to derive some characteristics of the meander plan form.

From Eqs. (7) and (10), it can be shown that :

[14] $x = 2\mu[E(k) - E(k/\varphi)] - \mu[F(k) - F(k/\varphi)]$

where E(k) and E(k/φ) are the complete and incomplete elliptic integrals of the second kind and E(k) = E(k/(π/2)). For a full period, x corresponds to the meander wave length, λ, and is given by

[15] $\lambda = 4\mu[2E(k) - F(k)]$

Finally, the amplitude of the meander – half the meander width - is estimated using Eqs. (7) and (10) which results in

[16] $y = 2\mu\sqrt{\sin^2\left(\theta_0/2\right) - \sin^2\left(\theta/2\right)}$

and

[17] $A = y(\theta = 0) = 2\mu k$

3 CHARACTERISTICS OF THE MEANDER PATTERNS

Eqs. (10) and (14) can be solved numerically and, in combination with Eq. (16), can be used to draw the meander plan form. Fig 5 illustrates some dimensionless meander patterns for different values of the maximum deflection angle, θ_0. The coordinates are made dimensionless by dividing both x and y by the meander scale parameter (μ). The curves actually represent the channel centerlines of the meandering streams. It can be seen that the pattern increasingly departs from a sine curve as the maximum deflection angle increases. Theoretically, the maximum value of the starting deflection angle for the centerline of the stream cannot exceed about 117.5° otherwise the meander loops come into contact with each other and a cutoff is formed which destroys the meandering pattern. In practice, the maximum deflection angle would be slightly less than 117.5° because of the finite river width.

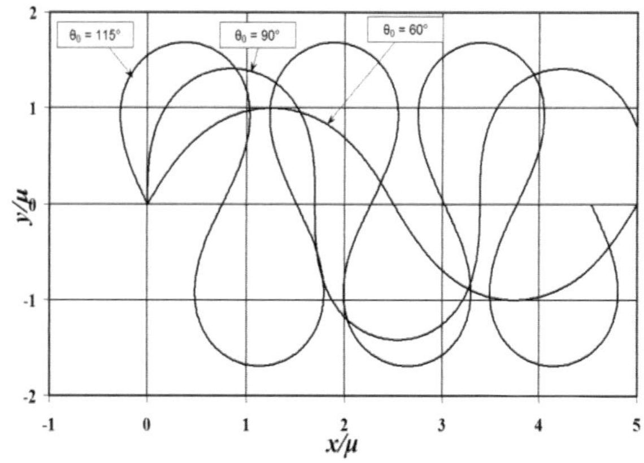

Figure 5. Some typical river meander patterns

It is interesting to visualize how parameter k and the complete elliptic integrals vary with the maximum deflection angle. Fig. 6 shows the variation of parameters k, F(k) and E(k), obtained by numerical integration, as a function of the maximum deflection angle, θ_0. It can be seen that F(k) varies from about 1.57 ($\theta_0 = 0°$) to 2.13 ($\theta_0 = 118°$) and, for the same value of parameter μ, the meander path length (Eq. 13) can vary by a factor of 1.36 at most.

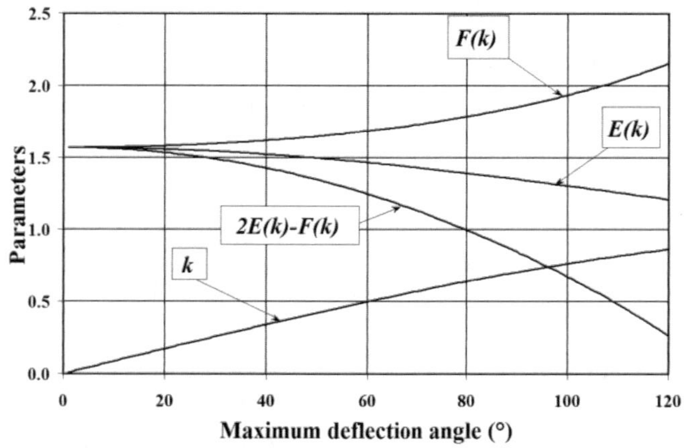

Figure 6. Variation of some general parameter as a function of the maximum deflection angle

Some empirical relationships have been proposed for expressing the wave length as a function of the river width. All these equations have the form $\lambda = \alpha B^{\beta}$, where α and β are constants. Different authors have proposed different values for α which vary from 6 to 12 and β is generally close to 1.0. However, Eq. 15 shows that the wave length is not only a function of the river width, *B*, but also depends on the Froude number, the flood plain transverse slope and the maximum deflection angle. For the same value of the length parameter, μ, figure 5 shows, for example, that the meander wave length for a deflection angle of 115° is about 3.5 times less than when the deflection angle is 60°. That ratio decreases as the deflection angle approaches 0°. Figure 7 shows the variation of the meander wave length as a function of the maximum deflection angle for a constant value of the length parameter. One must remember that the maximum deflection angle theoretically does not exceed 117.5°.

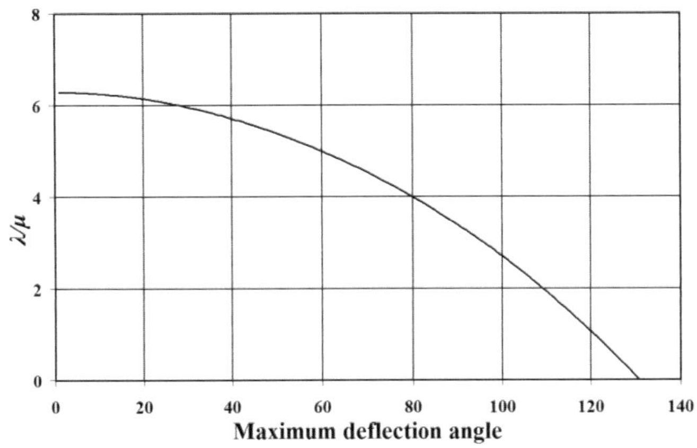

Figure 7. Variation of meander wave length as a function of the maximum deflection angle

The sinuosity, σ, of a meander is defined as the ratio of the path length to the wave length ($\sigma = M_L/\lambda$). Using Eqs. (13) and (15), we can write

$$[18] \quad \sigma = {M_L}/{\lambda} = \frac{F(k)}{2E(k) - F(k)}$$

from which it appears that the sinuosity depends solely on parameter k which is a function of the maximum deflection angle. Fig. 8 shows the variation of the sinuosity.

Hey (1976) and Hey and Thorne (1986) recognized that the maximum deflection angle had an influence on the wave length and they proposed to express the full path length as a function of the river width. The equation proposed is

$$[19] \quad M_L = 4\pi B$$

but Eq. (13) can also be written as

$$[20] \quad M_L = 4 F_r B \sqrt{\frac{1}{2\xi S_Y}} F(k)$$

which shows that M_L is still a function of the Froude number, the transverse flood plain slope and the maximum deflection angle. However, the range of variation of factor F(k) (from 1.57 to 2.12) is much smaller than the range of variation of factor 2E(k)-F(k) (from 1.57 to 0.34), which is associated with the meander wave length as shown in Fig. 6. This explains why estimates of meander path length is generally more accurate than estimates of meander wave length. Eqs. (19) and (20) can be used to obtain rough estimates of the floodplain transverse slope. By taking a value of the Froude number equal to 0.50 and an average value of the factor F(k) equal to 1.8, the product $\xi\, S_Y$ is found to be approximately equal to 0.040. The factor ξ, which is equal to the ratio B/h, normally varies, for meandering rivers, between 10 and about 60 (Odgaard 1989). This shows that the flood plain transverse slope can then be expected to vary between about 0.0040 for narrow rivers and 0.0007 for wide rivers.

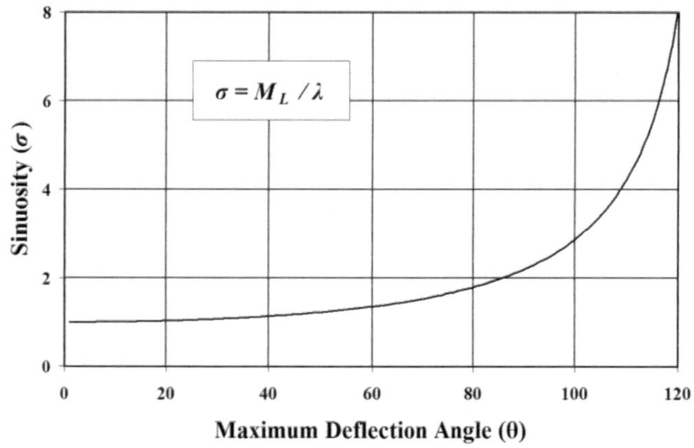

Figure 8. Variation of river meander sinuosity with the maximum deflection angle

4 MODEL DEVELOPMENT – GENERAL CASE

Upstream-skewed meander loops are frequently met in nature as can be seen in Fig. 10. The following paragraphs show the meandering model presented above can also generate upstream-skewed meander loops.

Figure 10. Meander loops in the St. Louis River near Beauharnois (Quebec) (Water flowing from southwest to northeast)

The model developed above generates symmetrical meanders with respect to the down valley axis. This was obtained because it was assumed that the longitudinal valley slope was much smaller than the floodplain transverse slope. The following development addresses the case where the longitudinal slope is not negligible with respect to the transverse slope.

Starting with Eq. (4), which remains valid for the general case, the point in the flood plain whose elevation matches the superelevation caused by the curvature of the flow is defined, according to Fig. 9, by the expression $S_Y\, y - S_X\, x$ and Eq. (6) now becomes:

$$[21]\quad \frac{\gamma}{g} Q V \frac{d\theta}{ds} = 2\gamma h \Delta h = 2\gamma h (\Delta h_Y + \Delta h_X) = -2\gamma h (S_Y\, y - S_X\, x)$$

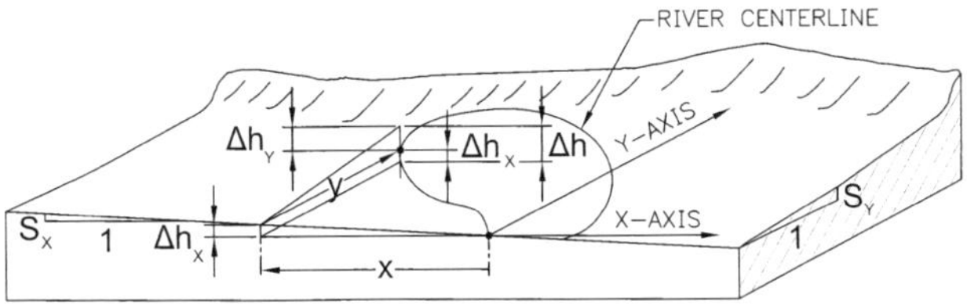

Figure 9. Mathematical representation for the general case

After a few transformations Eq. (21) becomes

$$[22]\quad \frac{\gamma}{g} Q V \frac{d^2(\theta - \delta)}{ds^2} = -2\gamma h Z \sin(\theta - \delta)$$

where $Z = (S_X^2 + S_Y^2)^{1/2}$ and $\delta = \tan^{-1}(S_X / S_Y)$.

Since δ is a constant, Eq. (22) can be rewritten

$$[23]\quad \frac{\gamma}{g} Q V \frac{d^2\Theta}{ds^2} + 2\gamma h Z \sin\Theta = 0$$

where $\Theta = \theta - \delta$. Eq. (23) has the same form as Eq. (9) and will therefore produce similar solutions. The only difference is that Eq. (23) generates meander loops making an angle δ with respect to the x-axis. A representation of the tilted loops is shown in Fig. 11 where the tilting is due to the introduction of the angle δ.

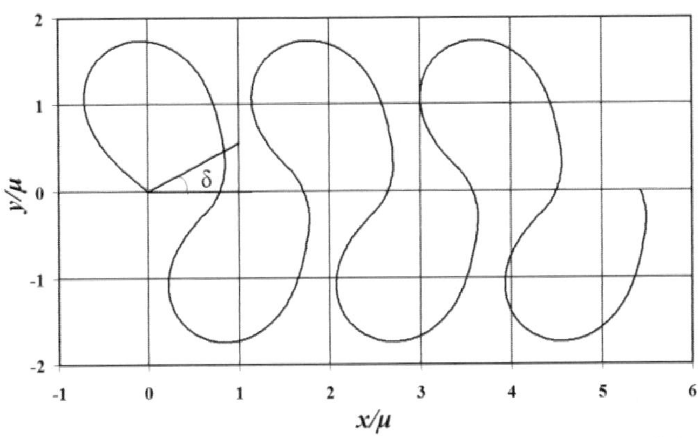

Figure 11. Meander pattern showing upstream skewing

Figure 11 shows the new meander patterns obtained when $\delta = 30°$, which is equivalent to assuming that the ratio of the longitudinal slope to the flood plain transverse slope is about 0.58. The larger the ratio of the slopes the more tilted the patterns. The above development demonstrates that the commonly observed upstream-skewed meander loops result from a non-negligible ratio of the longitudinal slope to the transverse slope.

5 SINE-GENERATED CURVE

Langbein and Leopold (1966) proposed the theory of minimum variance to describe regular meander patterns. They proposed the following equation, also called the sine-generated curve,

$$[24]\quad \theta = \theta_0 \sin 2\pi \frac{s}{M_L}$$

where M_L is the total path length along a full meander cycle. The equation includes a scale parameter, M_L, and a shape parameter, θ_0. By selecting appropriate values for these parameters it is possible to represent the shape of real meanders, including the relatively round curve shown by meanders of large amplitude. However, Eq. 24 provides no information about the variables characterizing the river (discharge, river width, river depth, etc.). The same remark applies to the meander models proposed by Ikeda *et al.* (1981), Parker *et al.* (1982 and 1983) and Odgaard (1989).

As Langbein and Leopold explained, the sine-generated curve is actually an approximation to an equation developed by von Schelling (1951) to describe the most probable path taken by a particle moving between two fixed points in a plane when the path length is fixed. According to von Schelling, the problem of the most frequent path is equivalent to finding the curve which minimizes the curvature along the loop, or

$$[25] \quad \int \left(\frac{d\theta}{ds} \right)^2 ds = \text{a minimum}$$

It is remarkable that the solutions to Eq. 25 found by von Schelling for the path length and for the x coordinate also involve elliptic integrals, similar to Eqs. 10 and 15. His solution for the y coordinate is also similar to Eq. 16. This is a further argument showing that the present model generates meanders loops having the minimum variance.

6 DISCUSSION

The introduction of constant floodplain transverse slope in this model is crucial for the development of regular-shaped minimum variance meander loops. In the simpler form of the model, this condition ensures that the flow curvature is directly proportional to the distance from the main longitudinal river axis (Eq. 6). The assumption is equivalent to stating that the rise in flood plain elevation due to the transverse slope must match the superelevation of the river flow in bends at bankfull conditions. Although it may appear surprising that this condition suffices to generate regular meander loops, it actually ensures that the rate of bank erosion is proportional to the flow curvature. In effect, when the floodplain transverse slope is constant, the difference in level between the water surface in the outer side of a river curve and the flood plain is constant over the whole meander loop. All other conditions being the same, it may then be inferred that the rate of bank erosion is proportional to the flow curvature. However, a discontinuity in the flood plain transverse slope such as a depression will cause the river flow to inundate part of the flood plain, disturbing the flow pattern in the river bank near the depression. This disturbance will be sufficient to break the meander pattern.

Irregularities in the flood plain transverse slope are much more likely to be encountered in very wide flood plains and this may explain why large rivers, although they may be sinuous or braided, rarely exhibit regularly shaped meander loops. Conversely, it is quite frequent to observe small or medium-sized nicely meandering rivers discharging into a sinuous but not regularly meandering large river.

The shear force exerted by the flow on the river banks is not considered in this model, which means that no equation for bank erosion is included. However, it is implicitly assumed that the rate of bank erosion is directly proportional to the curvature of the flow. Eqs. (10), (14) and (16), which define the shape of the meander loops, all involve a shape parameter, k, which is a function of the maximum deflection angle (θ_0). This implies that there is an infinite number of solutions to these equations, depending on the value of parameter k. In practice, the value of k should be small when the river starts meandering in its newly formed floodplain. With time, the meanders expand laterally as the value of k increases, due to erosion of the river banks. As long as the rate of erosion of the river banks at any given point is directly proportional to the curvature at that point, the shape of the loop will comply with the present model.

River meanders are known to slowly migrate in a downstream direction. Downstream migration is possible when the rate of erosion is larger downstream of the apex of the loop than upstream and this asymmetry is believed to be caused by secondary currents. Nevertheless, it may be supposed that the present model would apply, if all ideal conditions are met, for short term or medium term – from a few days to a few months - modeling of meander loops but secondary currents are to be taken into account for long term modeling.

7 CONCLUSIONS

The model presented in this paper is based on a set of assumptions which may be approached on some river reaches, thereby resulting in regularly-shaped meander loops. In its simplest form, the model generates symmetrical meander loops having minimum variance. It also provides simple explicit equations defining meander path length, meander wave length and meander amplitude. These equations clearly demonstrate that the path length and wave length are not unique functions of river width but also depend on other river and flood plain characteristics. One of the parameters included in the model is the floodplain transverse slope which, apparently, has never been considered before in other meandering models. The model development for the general case results in the generation of upstream-skewed meander loops which are commonly observed in nature. It is shown that the upstream skew is due a non-negligible ratio of the longitudinal valley slope to the flood plain transverse slope,

8 REFERENCES

Chow, V.T., 1959. *Open-Channel Hydraulics*, McGraw-Hill, New York, NY.

Hey, R.D. 1976. Geometry of river meanders. *Nature*, 262, 482-484.

Hey, R.D. and Thorne, C.R. 1986. Stable channels with mobile gravel-beds. J. *Hydraul., Eng*., ASCE, 112(8).

Ikeda, S., Parker, G., and Sawai, K. 1981. Bend theory of river meanders. Part 1. Linear development. *J. Fluid Mech.*, 112, 363-377.

Langbein, W.B., and Leopold, L.B. 1966. River meanders – Theory of minimum variance, *Geological Survey Professional Paper 422-H.*

Leopold, L.B., and Wolman, M.G. 1960. River meanders, *Bull. Geol. Soc. of America*, 71, 769-794.

Odgaard, J. 1989. River-meander model. I: Development. *J. Hydraul. Eng.*, 115(11), 1433-1450.

Parker, G., Sawai, K., and Ikeda, S. 1982. Bend Theory of River Meanders. Part 2. Nonlinear deformation of finite-amplitude bends. *J. Fluid Mech.*, 115, 303-314.

Parker, G., Diplas, P., and Akiyama, J. 1983. Meander bends of high amplitude. *J. Hydraul. Eng.*, 109(10).

Timoshenko, S.P. and Gere, J.M., 2009. *Theory of Elastic Stability*. Dover Publications. Mineola, NY.

Von Schelling, H. 1951. Most frequent particle paths in a plane., *Transactions of the American Geophysical Union*, Vol. 32, Number 2.

Leadership in Sustainable Infrastructure
Leadership en Infrastructures Durables

Vancouver, Canada
May 31 – June 3, 2017/ *Mai 31 – Juin 3, 2017*

A STOCHASTIC APPROACH TO DOWNSCALING OF MULTISITE DAILY TEMPERATURE SERIES IN THE CONTEXT OF CLIMATE CHANGE

Rahman, Mahzabeen[1,2], Nguyen, Van-Thanh-Van[1]
[1] McGill University, Canada
[2] mahzabeen.rahman@mail.mcgill.ca

Abstract: Downscaling techniques have been commonly used in many climate change impact assessment studies. In particular, statistical downscaling (SD) methods are widely employed because of their simplicity of implementation. However, most SD methods were developed for a single site without considering the observed spatial dependence of the hydrologic processes at different locations; this could significantly affect the accuracy of impact study results at the catchment scale. Therefore, in the present study an improved multivariate multisite SD approach was proposed for downscaling daily temperature series at many sites concurrently. This approach is based on multiple regression models for describing the linkages between large scale atmospheric predictors and local scale daily maximum and minimum temperatures, and on the use of the Singular Value Decomposition technique and multivariate autoregressive model for capturing the observed statistical properties of the stochastic components of these regression models. The feasibility of the proposed approach has been assessed using the available NCEP/NCAR re-analysis data and the daily extreme temperature data available in two regions with different climatic conditions: the southern Quebec-Ontario region in Canada and the Bangladesh region. Results of this illustrative application have indicated the ability of the proposed multivariate multisite SD to reproduce accurately the observed statistical properties of the daily extreme temperature series, and the spatial-temporal dependence of the underlying temperature processes at different locations.

1 INTRODUCTION

Climate change has been recognized as having a profound impact on the hydrologic cycle and therefore, many studies have been carried out to investigate this impact using outputs from Global Climate Models (GCMs). However, the GCM outputs are inadequate to represent local or regional scale features and variability due to their coarse spatial resolution (Nguyen et al. 2006). Hence, downscaling techniques have been proposed for assessing the climate change effects on hydrologic processes at the regional or local scale (Nguyen and Nguyen, 2008). Among two fundamental categories of downscaling techniques (dynamic and statistical), the statistical downscaling (SD) methods are widely used because of their simplicity of implementation and use (Wilby et al. 2002). However, most existing SD methods were developed for a single site without considering the observed spatial dependence of the hydrologic processes at different locations, which could significantly affect the accuracy of impact study results at the catchment scale. For example, Zareie et al. (2015) showed that the timing of streamflow during peak flow season was simulated more accurately by the use multisite downscaling as compared to the single site downscaling approach.

In view of the above issues, multi-site downscaling approaches have been a subject of intense interest in many recent studies (see, e.g., Khalili and Nguyen, 2016). In particular, for multisite downscaling of temperature series at different locations different methods have been proposed by Khalili et al. (2011),

Khalili and Nguyen (2012), Rahman et al. (2013), Jeong et al. (2013), Srinivas et al. (2014). In the present study, a new multivariate multisite SD approach is proposed to provide further improvement in the accuracy of multisite downscaling of daily maximum and minimum temperature (Tmax and Tmin) series. The feasibility and accuracy of the proposed SD procedure was evaluated using daily temperature data available from two networks of weather stations located in two regions with completely different climatic conditions: cold-climate southern Quebec-Ontario region in Canada and tropical-climate region in Bangladesh.

2 METHODOLGY

The proposed multisite multivariate SD method is based on a combination of a multiple linear regression model to describe the direct linkages between global climate predictors and local Tmax and Tmin series, and a multivariate autoregressive-singular value decomposition (SVD-MAR) model to reproduce accurately the observed spatial-temporal statistical properties of the stochastic component of the regression model. Equation [1] represents the regression model for daily Tmax:

$$[1] \quad \begin{pmatrix} Tmax_{i,m,1} \\ Tmax_{i,m,2} \\ \vdots \\ \vdots \\ Tmax_{i,m,s} \end{pmatrix} = \begin{pmatrix} \alpha_{Tmax_{0,m,1}} \\ \alpha_{Tmax_{0,m,2}} \\ \vdots \\ \vdots \\ \alpha_{Tmax_{0,m,s}} \end{pmatrix} + \begin{pmatrix} \sum_{j=1}^{q} \alpha_{Tmax_{j,m,1}} \, p_{Tmax_{ij,m,1}} \\ \sum_{j=1}^{q} \alpha_{Tmax_{j,m,2}} \, p_{Tmax_{ij,m,2}} \\ \vdots \\ \vdots \\ \sum_{j=1}^{q} \alpha_{Tmax_{j,m,s}} \, p_{Tmax_{ij,m,s}} \end{pmatrix} + \begin{pmatrix} E_{Tmax_{i,m,1}} \\ E_{Tmax_{i,m,2}} \\ \vdots \\ \vdots \\ E_{Tmax_{i,m,s}} \end{pmatrix}$$

where, $Tmax_{i,m,k}$ is the maximum temperature on day i in month m and at station k from a network of s stations; $\alpha_{Tmax_{j,m,k}}$ is the jth regression parameter for month m and at station k; $p_{Tmax_{ij,m,k}}$ is the value of the jth predictor, from a total of q monthly significant predictors, on day i in month m and at station k; $E_{Tmax_{i,m,k}}$ is the residual value on day i in month m and at station k. In Eq. [1] the significant predictors were selected mainly based on partial correlation coefficient, although subjective selection has been made in many cases. The expression of Eq. [1] is only for *Tmax* but the similar representation is applicable for *Tmin* as well.

The residuals of Tmax and Tmin, which are the stochastic components of the regression models have been generated simultaneously by the SVD technique (Green, 1978). As the proposed methodology considers the simultaneous downscaling of Tmax and Tmin, the general matrix of the residual values E is obtained by the concatenation of the matrices E_{Tmax} and E_{Tmin} as described in Eq. [2], in which the elements of E_{Tmax} and E_{Tmin} are the residuals obtained for daily Tmax and Tmin, respectively.

$$[2] \quad E = [E_{Tmax} \; E_{Tmin}]$$

The general idea behind using the SVD technique is that several series of the residual values need to be produced. The SVD technique is able to transform a matrix of several correlated variables into a matrix of a small number of principal components that are uncorrelated variables. These principal components are linear combinations of the standardized observed variables and are closer to the normal distribution than the observed variables. These principal components can therefore be generated independently, several times, following a standard normal distribution with zero mean and unit variance.

The SVD-MAR technique was used by Chaleeraktrakoon (1995), where a multivariate seasonal streamflow model was proposed to simulate streamflow at a single location as well as at many locations concurrently. The present study involves the combined use of SVD and MAR technique for generating simultaneously the residuals of daily Tmax and Tmin in the proposed SD model. Detailed description of SVD-MAR technique can be found in Chaleeraktrakoon (1995).

3 NUMERICAL APPLICATION

3.1 Data and Study Area

In the following, to assess the feasibility and accuracy of the proposed multivariate multisite SD procedure two sets of daily extreme temperature data representing two completely different climatic regions were selected. The first dataset is obtained from a network of seven weather stations located in the cold-climate southern Quebec-Ontario region in Canada, and the second dataset is from a network of four stations located in the tropical climate region in Bangladesh (Table1). In addition, significant climate predictors given by the NCEP reanalysis data (Kistler et al. 2001, Kalnay et al. 1996) for these two regions have been selected for this study using the same screening method proposed by Khalili et al. (2011).

Table 1: Name and coordinates of selected stations used in this study.

	Quebec-Ontario region			Bangladesh		
	Station Name	Latitude	Longitude	Station Name	Latitude	Longitude
Station 1	L'Assomption	45.81^0 N	73.43^0 W	Bogra	24.85^0 N	89.37^0 E
Station 2	Dorval	45.47^0 N	73.75^0 W	Jessore	23.2 ^{0} N	89.33^0 E
Station 3	Chenaux	45.58^0 N	76.68^0 W	Dhaka	23.78 ^{0} N	90.38^0 E
Station 4	Maniwaki UA	46.30^0 N	76^0 W	Satkhira	22.72^0 N	89.08^0 E
Station 5	Cornwall	45.02^0 N	74.75^0 W			
Station 6	Ottawa CDA	45.38^0 N	75.72^0 W			
Station 7	Morrisburg	44.92^0 N	75.19^{0}W			

3.2 Results

In this paper, both graphical and numerical criteria were used to demonstrate the accuracy of the proposed multisite SD method. More specifically, a total of 100 simulations of daily Tmax and Tmin were generated and the agreement between the observed and simulated temperature extremes was assessed using the box plots of means of Tmax and Tmin and based on the monthly average values of Mean Absolute Error (MAE) (Eq. [3]) and Root Mean Square Error (RMSE) (Eq. [4]) of daily Tmax and Tmin for each station. The mathematical expressions of MAE and RMSE are given as follows:

$$[3]\ \mathrm{MAE} = \frac{1}{t}\sum_{t=1}^{t}\left|X_{Obs,i} - X_{Sim,i}\right|$$

$$[4]\ \mathrm{MSE} = \sqrt{\frac{1}{t}\sum_{i=1}^{t}\left(X_{Obs,i} - X_{Sim,i}\right)^2}$$

where, t, $\overline{X}_{Obs}$, $X_{Obs,i}$ and $X_{Sim,i}$ represent the length of series, average of observed values, observed and simulated values for i^{th} observation respectively.

3.2.1 Results for Southern Quebec-Ontario Region

Figure 1 shows the boxplots of the mean of Tmax for each month. It can be seen that the observed means were accurately reproduced by the proposed SD procedure for almost all the months for both calibration (1961-1975) and validation (1976-1990) periods. Similar results were also found for Tmin. Table 2 presents the MAE and RMSE values for Tmax and Tmin for all stations. It is observed that very low values of MAE (RMSE) were produced ranging from 0.02~0.04 (0.03~0.06) and 0.05~0.10 (0.06~0.13) respectively for

Tmax and Tmin during calibration period. During validation period, those values were between 0.27~0.54 (0.34~0.69) and 0.29~0.59 (0.47~0.79) for Tmax and Tmin respectively.

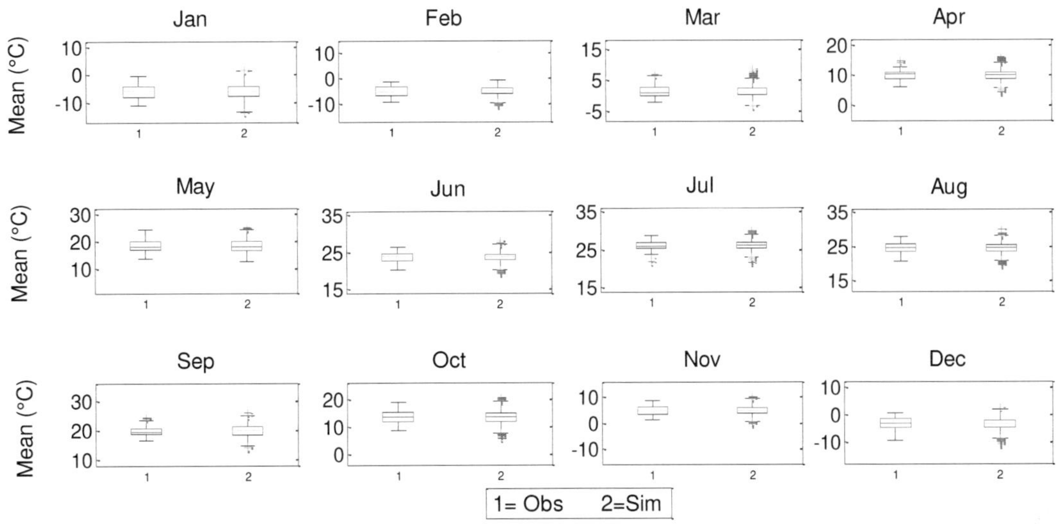

(a)

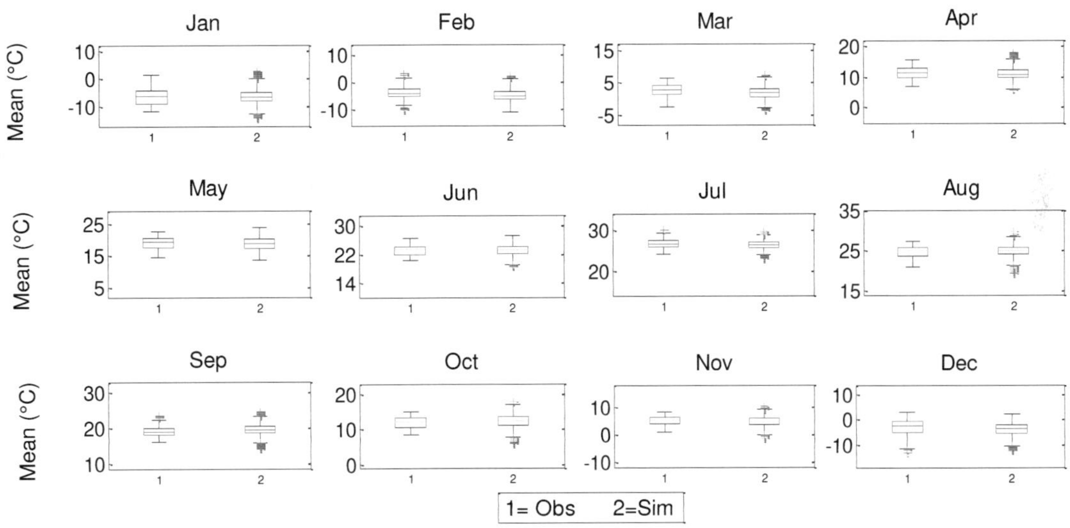

(b)

Figure 1: Boxplots of the mean of Tmax for (a) calibration and (b) validation periods for seven stations in southern Quebec-Ontario region. The band in the middle of each box denotes the median value, the boxes and whiskers denote the inter-quartile range (IQR) and 1.5×IQR respectively. Crosses beyond the whiskers are the outliers.

Table 2: MAE and RMSE values of Tmax and Tmin for seven stations in southern Quebec-Ontario region.

	Calibration Period				Validation Period			
	Tmax		Tmin		Tmax		Tmin	
	MAE	RMSE	MAE	RMSE	MAE	RMSE	MAE	RMSE
Station 1	0.03	0.03	0.07	0.09	0.44	0.54	0.54	0.77
Station 2	0.02	0.03	0.05	0.06	0.41	0.50	0.35	0.55
Station 3	0.03	0.03	0.10	0.12	0.44	0.55	0.5	0.78
Station 4	0.02	0.03	0.05	0.06	0.27	0.34	0.43	0.73
Station 5	0.04	0.06	0.08	0.10	0.54	0.69	0.59	0.79
Station 6	0.03	0.03	0.08	0.10	0.38	0.54	0.50	0.73
Station 7	0.03	0.03	0.10	0.13	0.54	0.63	0.29	0.47

Figure 2 shows the accuracy of the proposed SD method in the description of the observed lag-0 inter-station correlation for Tmin as indicated by the high value of R^2 for calibration period (R^2 = 0.94) and for validation period (R^2 = 0.84). Similar results were found for Tmax as well (not presented here).

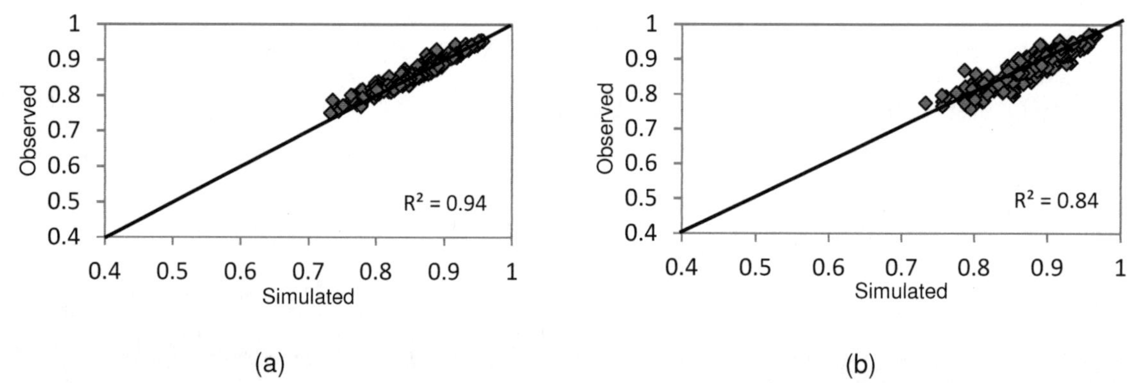

Figure 2: Lag-0 interstation correlations of daily Tmin during (a) calibration and (b) validation periods for the stations in southern Quebec-Ontario region.

The proposed SD procedure can also reproduce very well the observed autocorrelations for all stations for both Tmax and Tmin. For purposes of illustration, Figure 3 shows the very good agreement between observed and simulated values of the autocorrelations for lag one to lag three for Cornwall station for Tmax. This very good fit for Tmax and Tmin and for all stations has indicated the capability of the proposed SD method to reproduce accurately the temporal dependence of extreme temperature series at a given location of interest.

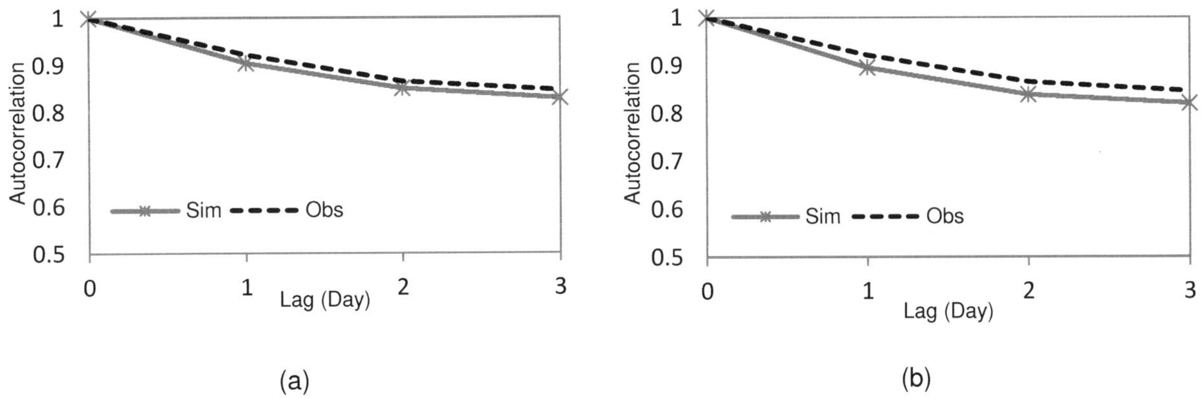

Figure 3: Observed and simulated autocorrelation coefficients of Tmax at Cornwall station during (a) calibration and (b) validation periods.

3.2.2 Results for Bangladesh

Figure 4 shows the very good agreement between observed and simulated values of monthly means of daily Tmin for all four stations in Bangladesh for both calibration (1981-1990) and validation (1991-2000) periods. The values of MAE (RMSE) for these stations ranged from 0.01 (0.01~0.02) and 0.04~0.07 (0.05~0.1) for Tmax and Tmin respectively (Table 3) during calibration period. For validation period those values are within 0.27~0.38 (0.35~0.46) and 0.16~0.38 (0.19~0.44).

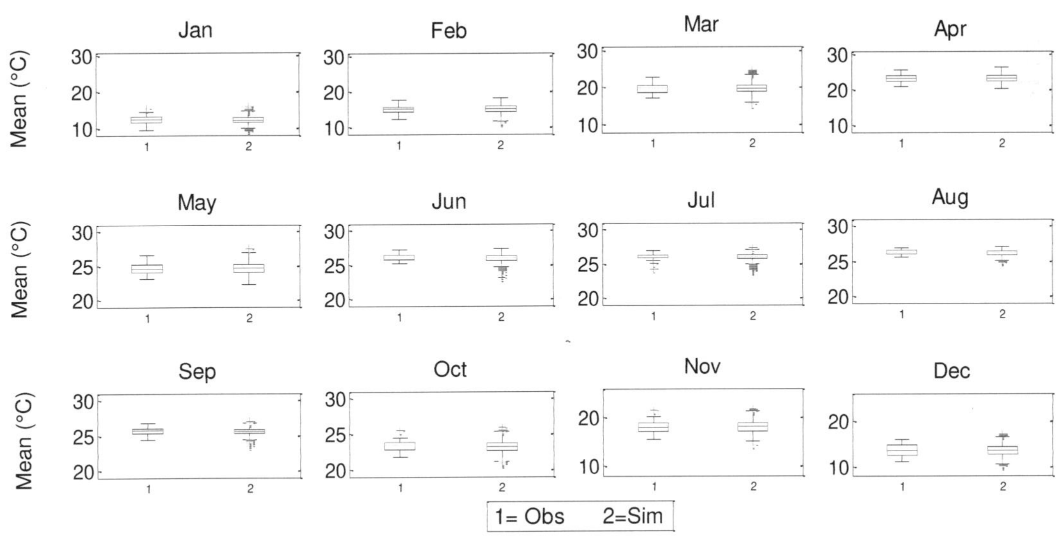

(a)

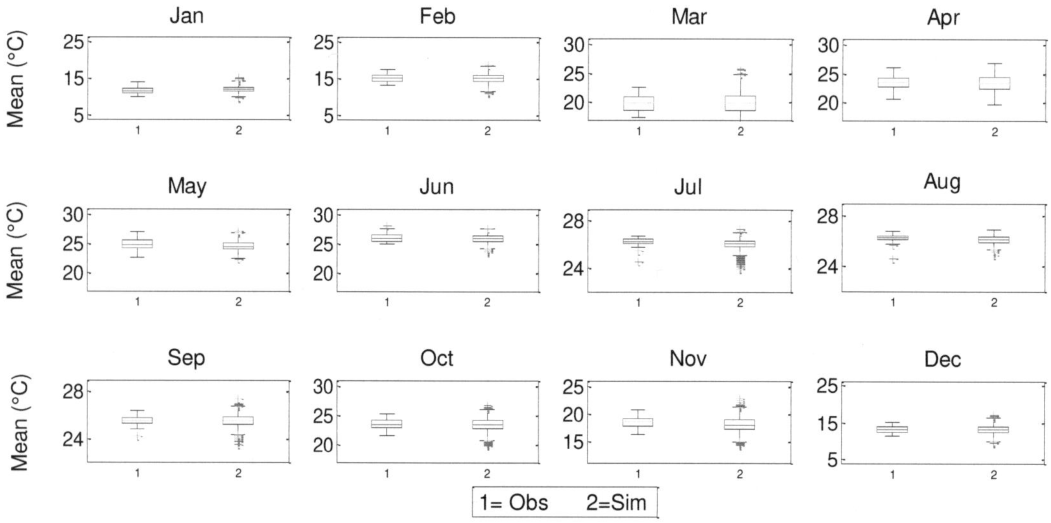

(b)

Figure 4: Same as Figure 1 but for Tmin for four stations in Bangladesh.

Table 3: Same as Table 2 but for four stations in Bangladesh.

	Calibration Period				Validation Period			
	Tmax		Tmin		Tmax		Tmin	
	MAE	RMSE	MAE	RMSE	MAE	RMSE	MAE	RMSE
Station 1	0.01	0.02	0.05	0.07	0.38	0.46	0.38	0.44
Station 2	0.01	0.02	0.04	0.05	0.34	0.40	0.23	0.28
Station 3	0.01	0.02	0.07	0.10	0.27	0.35	0.16	0.19
Station 4	0.01	0.01	0.04	0.06	0.28	0.36	0.20	0.28

Figure 5 shows the accuracy in the reproduction of the lag-0 inter-station correlations for Tmax for calibration period ($R^2 = 0.98$) and for validation period ($R^2 = 0.73$). Similar results were obtained for Tmin, which are not given here. In addition to these results, the temporal dependence at a given location was reproduced adequately for all four stations in Bangladesh. For illustration purposes, the autocorrelations for Tmin during calibration and validation periods at Dhaka station is presented in Figure 6.

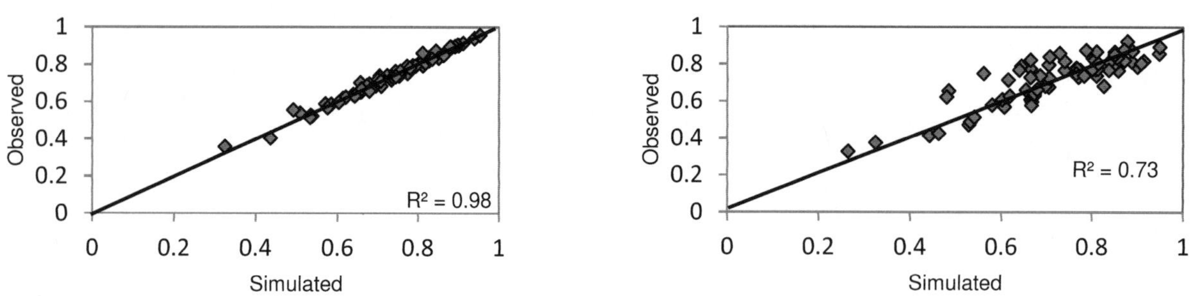

(a) (b)

Figure 5: Same as Figure 2 but for Tmax for four stations in Bangladesh.

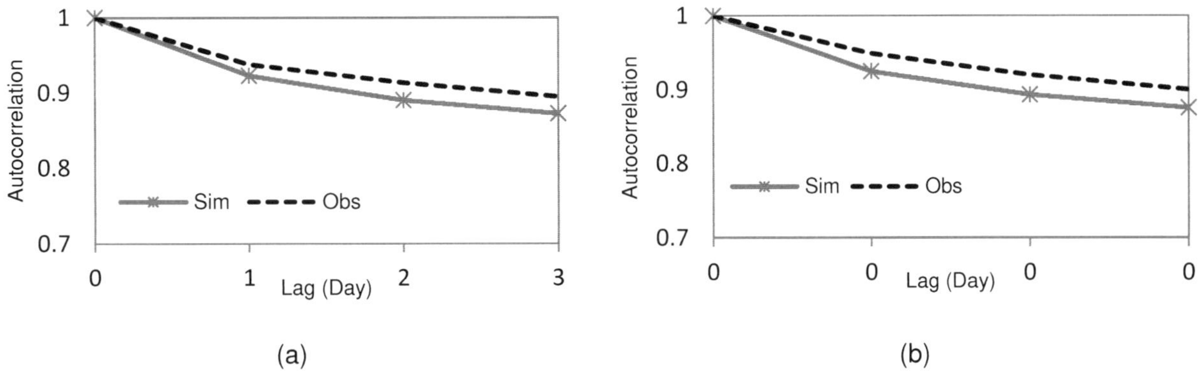

Figure 6: Same as Figure 3 but for Tmin at Dhaka station.

4 CONCLUSIONS

In this present study, an improved SD method has been proposed for downscaling daily Tmax and Tmin simultaneously at many sites concurrently. The proposed method was based on a combination of multiple linear regression models for describing the linkages between large scale atmospheric predictors and local scale daily maximum and minimum temperatures, and on the use of the SVD-MAR technique for capturing the observed statistical properties of the stochastic components of these regression models. The feasibility and accuracy of the proposed multivariate multisite SD approach have been assessed using two sets of daily temperature data available in two regions having different climatic conditions: the cold-climate southern Quebec-Ontario region in Canada and the tropical-climate region in Bangladesh. Results of these Illustrative applications have indicated the ability of the proposed SD to accurately reproduce the observed basic statistical properties of the historical temperature records, and the temporal and spatial dependences of daily temperature series at a given site as well as for different locations concurrently. The proposed multisite SD procedure could be used for improving the accuracy of climate change impact assessment studies.

Acknowledgements

The authors like to acknowledge the Data Access Integration (DAI) Team for providing the data. The DAI Portal (http://loki.qc.ec.gc.ca/DAI/) is made possible through collaboration among the Global Environmental and Climate Change Centre (GEC3), the Adaptation and Impacts Research Division (AIRD) of Environment Canada, and the Drought Research Initiative (DRI). The authors appreciate Institute of Water and Flood Management (IWFM) and Bangladesh Meteorological Department (BMD) for necessary information and data.

References

Chaleeraktrakoon, C. 1995. Stochastic Modelling and Simulation of Streamflow Processes, PhD. thesis, McGill University, QC,Canada.

Fowler, H.J., Blenkinsop, S. and Tebaldi, C. 2007. Linking Climate Change Modelling to Impacts Studies: Recent Advances in Downscaling Techniques for Hydrological Modelling. *International Journal of Climatology*, **27**(12): 1547-1578.

Green, P. E. (1978), Analyzing multivariate data, The Dryden Press, Hinsdale, Illinois, USA.

Islam, A.S. 2009. Analyzing Changes of Temperature Over Bangladesh Due to Global Warming Using Historic Data. *Young scientists of Asia conclave*, *Jawaharlal Nehru Centre for Advanced Scientific Research (JNCASR)*,15-17.

Kalnay, E., Kanamitsu, M., Kistler, R., Collins, W., Deaven, D., Gandin, L., Iredell, M., Saha, S., White, G. and Woollen, J. 1996. The NCEP/NCAR 40-Year Reanalysis Project, *Bulletin of the American Meteorological Society*, **77**: 437-471.

Khalili, M, and Nguyen, V.T.V. 2012. A Statistical Approach to Multi-Site Multivariate Downscaling of Daily Extreme Temperature Series for Climate-Related Impact Assessment Studies, *10th International Conference on Hydroinformatics - HIC 2012: Understanding Changing Climate and Environment and Finding Solutions*, Hamburg, Germany.

Khalili, M., Nguyen, V.T.V. and Gachon, P. 2013. A Statistical Approach to Multi-Site Multivariate Downscaling of Daily Extreme Temperature Series. *International Journal of Climatology,* **33**(1):15-32. Article first published online on 3 Nov 2011, DOI: 10.1002/joc.3402.

Khalili, M. and Nguyen V-T-V. 2016. An Efficient Statistical Approach to Multi-Site Downscaling of Daily Precipitation Series in the Context of Climate Change, *Climate Dynamics*, DOI 10.1007/s00382-016-3443-6, 18 pages.

Kistler, R., Kalnay, E., Collins, W., Saha, S., White, G., Woollen, J., Chelliah, M., Ebisuzaki, W., Kanamitsu, M., Kousky, V., Dool, Hvd., Jenne, R., Fiorino, M. 2001. The NCEP/NCAR 50-Year Reanalysis: Monthly Means CD-ROM and Documentation. *Bulletin of the American Meteorological Society*, **82**:247-267.

Nguyen, V-T-V., Nguyen, T-D., and Gachon, P. 2006. On the linkage of large-scale climate variability with local characteristics of daily precipitation and temperature extremes: an evaluation of statistical downscaling methods, *Advances in Geosciences, Vol. 4: Hydrological Science*, N. Park et al. (Ed.), World Scientific Publishing Company, pp. 1-9.

Nguyen, V-T-V., and Nguyen, T-D. 2008. *Statistical Downscaling of Daily Precipitation Process for Climate-Related Impact Studies*, Chapter 16 in "Hydrology and Hydraulics", V.P. Singh (Ed.), Water Resources Publications, pp. 587-604.

Rahman, M., Khalili, M. and Nguyen, V.T.V. 2013. A Multivariate Statistical Approach to Downscaling of Daily Extreme Temperature Processes for Different Locations. *3rd Climate Change Technology Conference*. Concordia University, Montreal, QC.

Srinivas, V.V., Basu, B., Kumar, D. N. and Jain, S. K. 2014. Multi-Site Downscaling of Maximum and Minimum Daily Temperature Using Support Vector Machine. *International Journal of Climatology*, **34**: 1538-1560.

Wilby, R.L., Dawson, C.W. and Barrow, E.M. 2002. SDSM – A Decision Support Tool for the Assessment of Regional Climate Change Impacts. *Environmental Modelling & Software*, **17**: 147-159.

Zareie, A, Rahman, M. and Nguyen, V.T.V. 2015. Comparison of Multisite and Single-site Temperature Downscaling Effects on Streamflow and Run-off Simulation. *22nd Canadian Hydrotechnical Conference, Water for Sustainable Development: Coping with Climate and Environmental Changes (L'eau pour le développement durable: adaptation aux changements du climat et de l'environnement)*, Montreal, Quebec.

Leadership in Sustainable Infrastructure
Leadership en Infrastructures Durables

Vancouver, Canada
May 31 – June 3, 2017/ *Mai 31 – Juin 3, 2017*

A STUDY OF FACTORS INFLUENCING THE SEDIMENT ACCUMULATION RATES IN STORMWATER MANAGEMENT PONDS

Darko Joksimovic [1], and Sidra Yasir Siddiqui [2]
[1] Department of Civil Engineering, Faculty of Engineering and Architectural Science, Ryerson University, Toronto, Canada
[2] Department of Civil Engineering, Faculty of Engineering and Architectural Science, Ryerson University, Toronto, Canada

ABSTRACT

Stormwater Management (SWM) ponds have been the most commonly used Best Management Practice (BMP) for controlling stormwater runoff in North America. The rate of accumulation of sediments and pollutants is uncertain, and highly dependant upon pond design, catchment characteristics and precipitation patterns. A wide range of sediment accumulation rates is reported in literature, ranging from 0.2 m^3/ha/year to 36.2 m^3/ha/year. Sediment removal from SWM ponds is costly and sometimes neglected by municipalities. Predicting the rate of accumulation, to aid municipalities in prioritizing their maintenance priorities and capital planning, relies on regular pond surveys. The long-term objective of the current project is to develop a data-driven modelling approach for forecasting of sediment accumulation rates in SWM ponds, based on data collected from a large number of surveyed ponds in Southern Ontario. This paper presents the methodology employed to collect the necessary information, consisting of 1) a literature review to determine the potential influencing factors, 2) a survey of relevant information availability in Ontario municipalities, 3) interviews with municipal staff, and 4) collection, organization and analyses of provided data. The results of analyses of collected information point to inconsistencies in availability and format of relevant data. A preliminary investigation of accumulation rates indicates lower values than those reported in similar studies in the past, although pointing to the same trend of decreasing performance of SWM ponds over time.

Keywords: Stormwater management ponds, sediment accumulation rates, detention ponds.

1. INTRODUCTION

Stormwater Management (SWM) in North America has evolved over the last several decades to include a number of Best Management Practices (BMPs) that reduce downstream impacts of runoff. The most commonly used BMPs in urban areas have been SWM ponds, designed to achieve one or more of objectives, including the minimization of downstream flooding and erosion, and minimization of pollutant discharges. Thousands of wet retention ponds, which feature a permanent pool in which the pollutants are removed through settling, have been constructed using a multitude of guidance documents. The remainder of this section presents a review of guidance recommendations for accumulated sediment removal, reported accumulation rates and factors influencing the accumulation rates reported in literature, and states the objectives of this research.

The performance of SWM ponds is time dependent and steadily decreases as the volume of accumulated sediment increases over time. This is generally recognized in guidance documents, albeit with different levels of detail and guidance on the frequency of accumulated sediment removal. Sediment accumulation depends on a number of factors including rain intensity, rainfall duration, construction activities, street-sweeping, storage volume, drainage area characteristics and internal hydraulics (Sivakumar and May, 2009). USEPA (2009) states that the removal of accumulated sediment will vary based on the pond design parameters, and suggests removal from forebays when 60% of original volume has been lost, or a 5-year cycle, and removal from main cells when 50% of original volume has been lost, or a 20-year cycle. The guidance provided in Ontario (MOE, 2003) recognised that limited data is available on sediment accumulation and provided a set of curves for determining the sediment removal frequency as a function of facility type, storage and level of catchment area imperviousness. These curves are based on continuous simulation and the guidance manual recommended that the maintenance frequency should be refined with operational and maintenance experience. In general, the removal of sediment was recommended to limit the reduction in the average annual Total Suspended Solids (TSS) removal of the facility to 5%. Finally, sediment accumulation rates with different level of imperviousness ranging from 0.6 m^3/ha/year (35% imperviousness) to 3.8 m^3/ha/year (85% imperviousness) were suggested in the absence of monitored data.

As stated earlier, sediment accumulation rates in SWM ponds vary over a wide range, depending on a number of factors related to the catchment area characteristics, precipitation patterns and the pond design and operation, and this is reflected in literature and guidance documents (Richard, 2010). Pond age, depth, layout, water levels, the number of measurement points, catchment, characteristics, size, shape, vegetation and flow monitoring are the factors that should be considered when performing sediment accumulation rates measurements. A study by Greenland International Consulting (1999) reported the sediment accumulation rates ranging from 0.2 to 5.9 m^3/ha/year for eight surveyed ponds in Ontario. However, Drake and Guo (2008) conducted a survey of Ontario municipalities, and found that the vast majority of ponds were not monitored for performance and had not yet been dredged. Subsequent studies in Ontario have reported the sediment accumulation rates ranging from 0.5 to 5.9 m^3/ha/year (TRCA and CH2M Hill, 2016), with recommendations to assess the sediment depth at least every 3 to 5 years. Another study that involved a survey of 98 ponds (LSRCA, 2011) found that the accumulation rates ranged widely from 0.26 to 36.17 m^3/ha/year, resulting in more than half of the surveyed ponds dropping in performance to levels below those that should have triggered corrective action (i.e. sediment removal).

There are also several reported efforts to develop relationships between catchment parameters and observed sediment accumulation rates. In USEPA (1986), a probabilistic model that determines the depth of accumulated sediment as a function of pond characteristics and long-term rainfall. The model was later compared against nine studied ponds in Florida by Yousef et. al. (1993), indicating a significantly higher accumulation rates for a range of pond surface area (A_{pond}) to contributing drainage area ($A_{drainage}$) ratios. The measured rates ranged from 1.08 cm/yr ($A_{pond}/A_{drainage}$ = 0.12) to 4.20 cm/yr ($A_{pond}/A_{drainage}$ = 0.014), with a sharp increase observed when the $A_{pond}/A_{drainage}$ ratio is below 2%. Heal et. al. (2006) reported sedimentation rates ranging from -0.2 to 1 cm/year for three monitored ponds in Scotland. For a sand/gravel inflow, bottom sediments comprised silt and clay, with high water content (up to 80%, by volume), and accumulated at a rate of 2 cm/yr (bulk sediment depth) (Marsalek and Marsalek, 1997). A number of other studies also demonstrated the uneven distribution of sediment deposits due to uneven and widely varied flow patterns in individual ponds (e.g. Montgomery et. al., 1983; Heal et. al., 2006; Dominic et. al., 2016).

Although the sediment accumulation rates can vary over a wide range, limited efforts were made until recently to collect the data that could inform the development of a more accurate approach. In the past few years, hundreds of SWM ponds in Southern Ontario have been surveyed. Thus, there is a potential to develop a generalized approach for forecasting the sediment accumulation rates that could be applicable to other areas as well. However, before this possibility could be explored, an investigation involving several steps needs to be undertaken. This objective of this paper is to present the results of such investigation that encompassed the identification of relevant information, investigation on data availability and processing needs, and a methodology to capture the information in a form usable for the development of a data driven forecasting model.

2. METHODOLOGY

To reach the stated objectives, the following activities were undertaken: 1) a questionnaire was developed based on the literature review that includes all the factors contribute to sediment accumulation in SWM ponds; 2) interviews were conducted with municipalities to follow-up of the survey for the collection of data; and 3) a database was developed considering the existing data provided by municipalities during the process of interview.

In this study, the factors contributing to sediment accumulation rates are divided into three parts: 1) Hydrologic Data, 2) Catchment Characteristics and 3) SWM pond characteristics. The issues for consideration are further described as follows:

1) Components that would be expected for gathering hydrologic data are historical precipitation records, snowfall, and snow melts measurement records, flow monitoring in ponds inlets and outlets, the level of pond monitoring records and records of water quality monitoring at pond inlet or outlet.
2) Catchment characteristics include the catchment areas to storm water management ponds, information regarding the road network and storm sewer collection network, information on impervious surfaces (roofs, roads/sidewalks, parking lots, driveways), land use breakdown (e.g. industrial, commercial, institutional, residential high/med/low density), information on catchment changes since SWM pond was constructed (e.g. construction activities), sediment control practices in the catchment (e.g. catch basin cleaning, street sweeping, road salting/sanding).
3) SWM pond characteristics include their design criteria (i.e. dry/wet pond, storage volume capacities, stage-storage curves, area, inlet/outlet structures, number of cells, sediment bay, length/width ratio, permanent pool depth, screening devices at inlet/outlet, etc.), sediment survey results (protocol followed in the survey, survey dates and results, sediment characterization, etc.), planting/ maintenance records and if ponds are used for irrigation purposes.

A questionnaire, shown in Table 1, was developed to clarify the information sought from municipalities that considered all the factors mentioned above. Almost fifty municipalities in Southern Ontario were contacted via email, with the questionnaire attached, of which five municipalities agreed to an interview and subsequent provision of data that they had available for the study. Once the data was collected the next step was to compile it in a usable format. The need for accurate and applicable data is the key factor for the development of the database. Master plans, reports, manuals, drawing digital images and spreadsheets were provided by municipalities. The database was populated with data provided by participating municipalities.

Table 1. Survey Questions

Group of factors	Survey questions
Hydrologic Data	1. Are historical precipitation records available? Yes - what duration (YEAR from - YEAR to, MONTHS of the year) and frequency (e.g. 15-min, hourly, daily) 2. Are there any records from snowfall and snowmelt measurements? 3. Are there any records of flow monitoring at pond inlet and/or outlet? 4. Are there any records of pond level monitoring? 5. Are there any records of water quality monitoring at pond inlet and/or outlet (either grab samples or concerted monitoring campaigns)?
Catchment Data	1. Are the catchment areas to SWM ponds delineated? Is this information available digitally? 2. Is information on the road network and storm sewer collection network available? Is this information available digitally? 3. Is information on impervious surfaces (roofs, roads/sidewalks, parking lots, driveways) available? Is this information available digitally? 4. Land use data available (industrial, commercial, institutional, residential high/med/low density)? 5. Is any information available on catchment changes since SWM pond was constructed (e.g. construction activities)? Are historic aerial images available? 6. Are there records available for catchment practices: catch basin cleaning, street sweeping, road salting/sanding practices, etc.?
SWM Pond Data	1. Are pond design or as-built drawings available? Have the ponds been characterized according to their design (i.e. dry/wet pond, storage volume capacities, stage-storage curves, area, inlet/outlet structures, the number of cells,

sediment bay, length/width ratio, permanent pool depth, screening devices at inlet/outlet, etc.?)

2. What information is available regarding the accumulated sediment (protocol followed in the survey, survey dates and results, sediment characterization, etc.)? Is this data available digitally?
3. Are planting/ maintenance records available?
4. Are any ponds used for irrigation?

3. RESULTS AND DISCUSSION

3.1 Survey Results

Information on 96 ponds was collected from the five municipalities. The ponds ranged in age from 3 to 33 years, with the average value of 15 years. Results from the survey and subsequent interviews indicated that all of five municipalities had implement some sort of monitoring programs for sediment accumulation in ponds. A sediment survey using the disk/rod technique is common practice by every municipality. Pond level monitoring and periodic water quality monitoring at inlet and outlet records were available for each pond. In addition, catchment area was calculated by all municipalities but only one of the five retained a record of catchment area delineation in digital format. It was generally acknowledged that some ponds within the municipality's jurisdiction will be approaching their designed capacity within the next three to five years, three municipalities reported planned clean-out projects and four are executing cleanouts at regular intervals. A wide variety of data shows the presence of accumulated sediments in each pond but many ponds have not been cleaned even though their capacities to remove solids at design levels are compromised. None of the municipalities had data readily available for catchment practices like street sweeping, road salting and catch basin cleaning. Similarly, no data was available for planting records and none of the ponds were apparently used for irrigation purpose.

3.2.1 Hydrologic Data

Precipitation records were generally not available from municipalities for the duration corresponding to age of the oldest ponds within their jurisdiction. In most cases, rainfall monitoring data were available for a few years, and in all cases the data covered non-winter months only. No snow measurements were available from any of the municipalities. Accounting for precipitation in a sediment accumulation forecast could be achieved using long-term records from Environment Canada, which operates gauges in or near each of the five municipalities, although these are typically not heated and removed in winter months. Only one of the municipalities surveyed maintains a network of level sensors across the ponds in their jurisdictions. Water quality sampling is carried out by the same municipality (grab samples at inlet and outlet) on an annual basis, while the same activity is performed by other municipalities less regularly and as a requirement of the facilities regulatory approvals. Sporadically collected water quality data are not suitable for inclusion in the eventual development of a sediment forecasting model.

3.2.2 Catchment Data

Catchment delineation for SWM ponds is generally available from all surveyed municipalities, typically from original design documents, but also in GIS format where delineation appears to have been updated. Similarly, road network and sewer network data is available, but again in a variety of formats. Impervious areas have been delineated in detail for one municipality, while the others provided the overall imperviousness of SWM pond drainage areas, in most cases from original design documents. Land use information is available for all municipalities. Although both vegetal cover and construction activities could have significant effect on sediment loading rates to SWM ponds, historic information on contributing catchments covering the period since their construction is not readily available. Information on catchment practices that also potentially influence sediment loadings to SWM ponds is very general and there is no historic information available in a usable format

3.2.3 SWM Pond Data

Therefore, the information on ponds and other BMPs was available, however, the information is not captured in a consistent and uniform format since different consulting companies carried out the inspections and document reviews. All municipalities are in the process of developing electronic databases of their SWM facilities, but again, there is no generally accepted format on how to record design information, inspections and maintenance records, sediment survey results, etc. Basic design information (e.g. type of facility, number of inlets/outlets, forebay presence, pool depth, volumes, etc.) is recorded in most instances, albeit not in a consistent format, requiring further inquiries on data interpretation. Although detailed calculations of sediment volumes were not available, it appears that several different methods are used, and include: cut/fill computations comparing pond designs and current elevations, calculations based on equations for different geometric shapes (e.g. truncated ellipse, truncated cone, etc.) and spatial interpolation calculations based on dense measurements across the pond. This point is particularly important, since values derived using different calculation approaches can yield substantially different results (e.g. Norton, 2007).

3.2.4 Sediment Survey Results

As indicated above, the data was provided by the five municipalities in a variety of formats, with some pieces of information missing. So far, measured accumulated sediment volumes for 39 SWM ponds have been assembled and analyzed. Figure 1 compares the distribution of observed sediment accumulation rates for these ponds, and compares them against the values reported by LSRCA (2011). Almost two thirds of ponds included in this study (23 out of 39) had accumulation rates less than 2 m^3/ha/yr, while the number of facilitates investigated by LSRCA that were in that range was substantially smaller (14 out of 49). Also, the high accumulation rates (> 11 m^3/ha/yr) observed by LSRCA were not observed for any of the ponds in this study.

Figure 1. Distribution of Measured Sediment Accumulation Rates

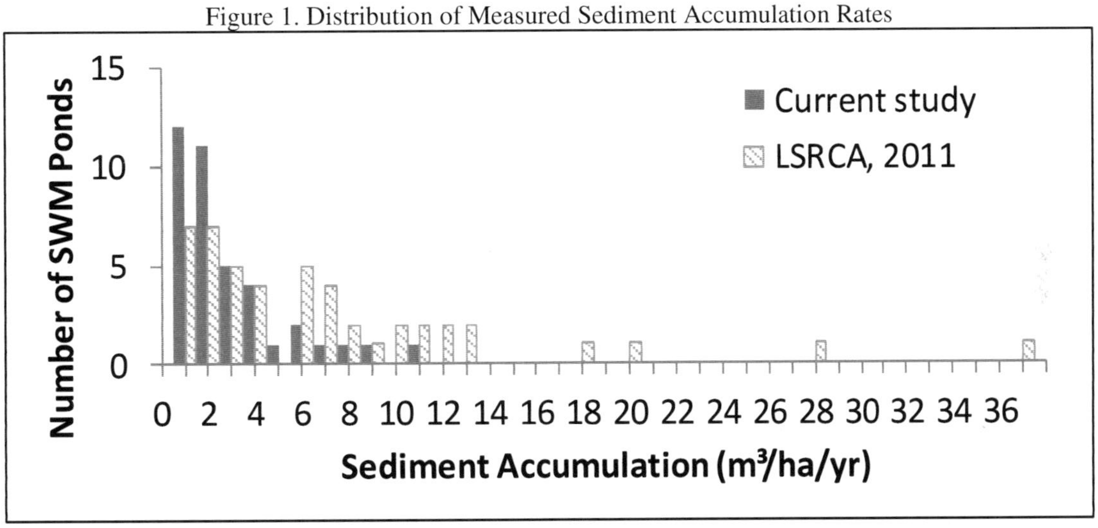

The lower overall sediment accumulation in ponds studied here, compared to LSRCA (2011) is also evident in Figure 2, which plots the unit accumulation rates against the pond age, along with the trend indicated with best-fit logarithmic lines. Although there is a high degree of data scatter, the lines point to the general trend of newer ponds being more efficient at capturing sediment. This could be attributed to a number of factors, including the improvement in pond design experience and practice over the last two decades, and reduced overall removal of older ponds due to accumulated sediment. It is not possible to determine other causes of reduced sediment accumulation rates over time without considering the multitude of other potentially influencing parameters discussed earlier. However, a general trend of decreased performance of ponds, in terms of their ability to provide their water quality function, is clearly observable. Although the contributing drainage area imperviousness has a major influence on the runoff volumes and peak discharge rates of SMW ponds, Figure 3 does not indicate any relationship between imperviousness and corresponding long-term average sediment accumulation rates.

Figure 2. Measured Sediment Accumulation Rates for Differently Aged Ponds

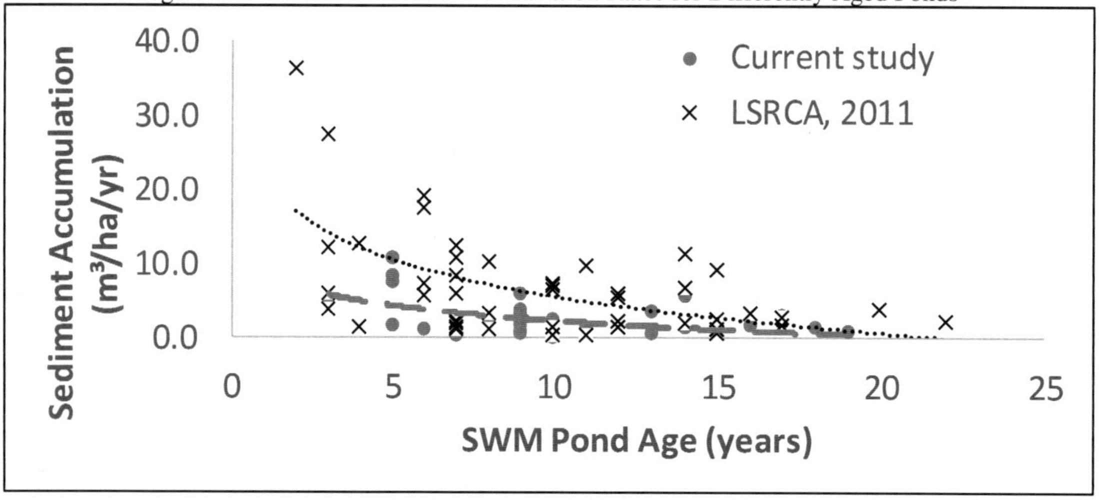

Figure 3. Catchment Imperviousness Influence on Sediment Accumulation Rates

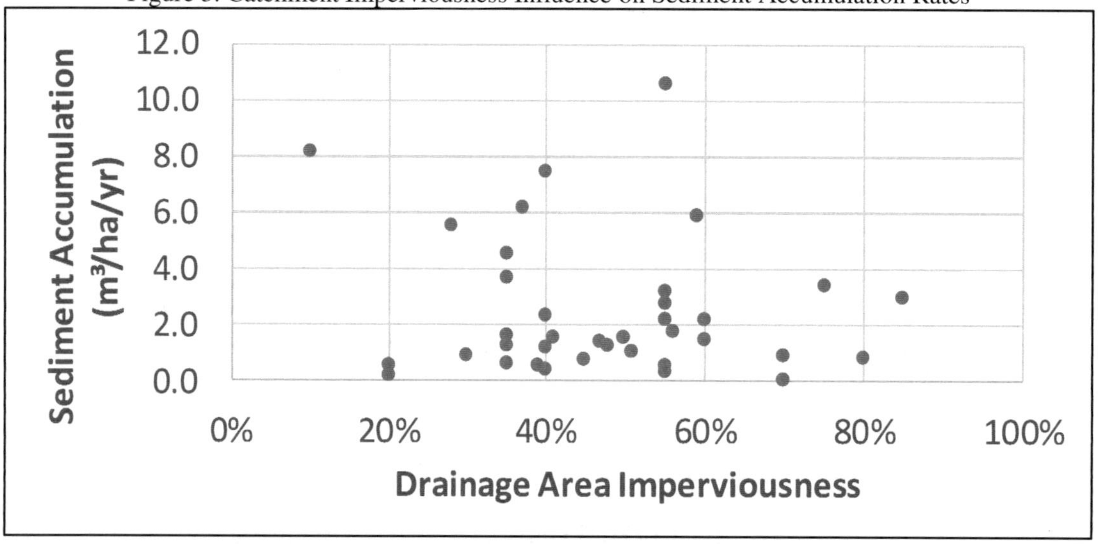

4. SUMMARY, CONCLUSIONS AND RECOMMENDATIONS

SWM ponds have been the most commonly used BMP over the past two decades. Clear and relatively simple design guidance, relatively low cost and end-of-pipe management of runoff from developed lands were all factors that led to their widespread adoption in Ontario and elsewhere, somewhat in in contrast to the current efforts to increase the adoption of Green Infrastructure and Low Impact Development approaches. Although early guidelines stated the need for regular monitoring to determine sediment accumulation rates and needs for pond cleanouts, these efforts have intensified only recently, with realization that many ponds need urgent removal of sediments to maintain their intended functions.

The available guidance on maintenance frequency is provided in several forms: 1) based on the loss of original volume alone, 2) based on facility type, provided storage and catchment imperviousness utilizing modelling results, and 3) based on the decrease in annual TSS removal percentages. With the presumed relationship between the loss of volume and the corresponding loss in performance (i.e. percent TSS removal), and impracticality and absence of continuous or even periodic monitoring to determine actual removal rates, the volume of accumulated sediment is clearly the most practical and adopted criterion applied to SWM pond cleanout decision making. With the sediment accumulation rates varying by an order of magnitude and near absence of studies that correlated accumulation rates with relevant hydrologic, catchment and pond characteristics, as the literature points out, capital planning of potentially very costly cleanout activities is difficult for stormwater managers.

From the limited analyses of sediment accumulation rates for ponds of different age conducted so far in this study, it is also evident that age of the pond has a major impact on the sediment accumulation rates. A single survey of the pond cannot be used to gain insight into either the temporal trend of its efficiency or the future sediment accumulation rate needed for capital planning. Another approach that municipalities are taking is to conduct two surveys of their ponds several years apart, in an effort to estimate the sediment loading rates. Although better than using a single survey, this approach also captures the performance (or accumulation rate) over a relatively short time window of pond operation, without considering the longer-term performance trend.

Prioritization of sediment removals has focused on most critical facilities, but a systematic approach for development of longer term capital programs to deal with years of accumulated sediment will require more frequent surveys or a numeric forecasting approach. In this respect, the quantity of data available from thousands of surveyed ponds offers interesting possibilities. However, as this study has illustrated, significantly more effort is needed to develop more standardized approaches, particularly for collecting information related to catchment hydrology and sediment surveys.

5. ACKNOWLEDGEMENTS

The authors gratefully acknowledge the contributions from stormwater personnel in five Ontario municipalities, who wished to remain unnamed, for providing insightful and forthright responses and data that made this study possible.

6. REFERENCES

Dominic, J.A., Aris, A.Z., Sulaiman, W.N.A., Tahir, W>Z.W.M. 2016 Discriminant analysis for the prediction of sand mass distribution in a holding pond using deposition thickness model of a single grain-sized particle. *Environmental Earth Sciences*, 75(9): 812.

Drake, J. and Guo, Y. 2008. Maintenance of Wet Stormwater Ponds in Ontario, *Canadian Water Resources Journal*, 33 (4): 351-368.

Greenland International Consulting Inc. 1999. *Storm Water Management Facility Sediment Maintenance Guide*. Concord, ON.

Heal K.V., Hepburn D.A., Lunn R.J. 2006. Sediment management in sustainable urban drainage system ponds, *Water Science & Technology*, 53(10): 219-27.

Lake Simcoe Region Conservation Authority (LSRCA). 2011. *Stormwater Pond Maintenance and Anoxic Conditions Investigation FINAL REPORT*. Lake Simcoe Region Conservation Authority, Newmarket, ON.

Marsalek, J. and Marsalek, P.M. 1997. Characteristics of Sediments from a Stormwater Management Pond. *Water Science and Technology*, 36(8-9): 117-122.

Ministry of Environment (MOE). 2003. *Stormwater Management Practices Planning and Design Manual*. Queen's Printer for Ontario, Toronto, ON.

Montgomery R.L., Thackston E.L., Parker F.L. 1983. Dredged material sedimentation basin design. *Journal of Environmental Engineering ASCE*, 109:466–484.

Norton, K. 2007. Spatial interpolation of patterns of sediment deposition in two urban stormwater detention ponds. Presentation, Environmental Science and Resources, Portland State University, accessed on February 8 2017 at http://web.pdx.edu/~jduh/courses/Archive/geog492s07/Projects/Norton_SpatialInterpolation.pdf.

Richard, R. 2010. *Investigation of Sediment Removal Frequency for Wet Detention Stormwater Management Ponds*. MASc Thesis, Ryerson University, Toronto, ON.

Sivakumar, M. and May, D. 2009. Optimum number of storms required to derive site mean concentrations at urban catchments. *Urban Water Journal*, 6 (2): 107-113.

Toronto and Region Conservation Authority (TRCA) and CH2M Hill Canada Ltd.. 2016. *Inspection and Maintenance Guide for Stormwater Management Ponds and Constructed Wetlands*. Toronto and Region Conservation Authority, Toronto, ON.

United States Environmental Protection Agency (USEPA). 1986. *Methodology for Analysis of Detention Basins for Control of Urban Runoff Quality, EPA 440/5-87-001* . USEPA, Washington, DC, USA.

United States Environmental Protection Agency (USEPA). 2009. *Stormwater Wet Pond and Wetland Management Guidebook, EPA 833-B-09-001*. USEPA, Washington, DC, USA.

Yousef, Y. A., Hvitved-Jacobsen, T., Sloat, J., and Lindeman, W. 1994. Sediment Accumulation in Detention or Retention Ponds. *Science of the Total Environment*, 146/147: 451-456.

Leadership in Sustainable Infrastructure
Leadership en Infrastructures Durables

Case Study

Vancouver, Canada
May 31 – June 3, 2017/ *Mai 31 – Juin 3, 2017*

ASSESSING THE BENEFITS OF VARIOUS LEVELS OF FLOOD MITIGATION IN A HIGHLY DEVELOPED SETTING

Lalji, Nadeer [1, 3], and Forsyth, Andrew [2]
[1,2] Water Resources, Associated Engineering, Canada
[3] laljin@ae.ca

As demonstrated during the 2013 Flood, The City of Calgary is at risk of overland flooding from the Bow and Elbow Rivers. The City of Calgary (The City) mitigates this risk through emergency response measures that include deployment of temporary barriers in anticipation of a flood event. However, The City has limited time to respond as flood events develop and propagate quickly within the steep, mountainous topography upstream. In addition, the variability of rainfall distribution in the Rocky Mountains makes it difficult to predict, with certainty, whether mountain rainfall events will impact Calgary. To reduce the necessary response efforts during an emergency, The City has considered implementing permanent flood barriers for communities at risk. As much of Calgary's floodplain is developed and densely populated, constructing permanent flood protection within The City is a costly and challenging task, with little room to construct barriers and significant flood depths to protect against. Additionally, the dynamics of groundwater flooding in Calgary are not well understood.

As such, The City retained Associated Engineering (AE) to develop a better understanding of overland and groundwater flooding, propose mitigation measures for vulnerable areas in Calgary, assess the benefits and costs of mitigation measures, and identify the optimal level of service for these locations. This analysis was performed for river flood events between the 1:20 year and 1:1000 year return periods, considering economic, social and environmental impacts as part of a Triple Bottom Line (TBL) analysis. This analysis included the following steps:

1. Groundwater Modelling & Inundation Mapping

AE completed groundwater modelling and mapping as a part of this project to estimate groundwater inundation depths and extents during high river events. AE modeled groundwater movement due to high river levels using two-dimensional modelling. A two-dimensional modelling approach was selected because of the limited amount of geological information available and the large area to be covered as part of the study. The geological information indicated the presence of a silt layer overlying a gravel layer overlying bedrock. The stratigraphy and related conductivity values applied to the model were based on a desktop review of available information.

Review of the groundwater modelling yielded three major findings:

1. Surface water intrusion was found to be predominantly horizontal from the river bed and banks. Vertical migration of water from flooded areas was limited by the presence of a low conductivity silt layer. As a result, two wetting fronts were observed and it was found that complete groundwater saturation did not occur beyond the overland flooding extents.

2. In certain scenarios, groundwater was predicted to rise to surface despite the presence of proposed overland flood protection. This finding indicated that, in certain locations mitigation of overland flooding may require accompanying groundwater flood mitigation to be effective.
3. For cutoff walls to be effective in mitigating groundwater flooding, they would need to be installed to bedrock. Cutoff walls not installed to bedrock were found to be ineffective.

AE extracted the two-dimensional groundwater profiles from the model and used them to develop groundwater inundation mapping across The City. These results were used to identify locations where flooding was expected to reach affect basements or the ground surface. AE prepared groundwater surfaces for multiple return periods for the existing condition with no overland flood protection in place and for a scenario with overland flood protection in place.

2. Flood Vulnerability Assessment

The City provided AE with overland flood inundation surfaces for the high river events in question, as well as a list of critical infrastructure to consider protecting from flooding. AE assessed this critical infrastructure for vulnerability to both overland and groundwater flooding and used this information to identify areas to analyze for flood protection.

3. Flood Mitigation Designs

AE generated conceptual level flood protection designs to mitigate overland and groundwater flooding for the sites identified in the vulnerability assessment. To mitigate overland flooding, AE considered earth fill dykes, flood walls and semi-permanent barriers. These barriers are proposed on City owned land where space is available; however, some barriers require land acquisition of private property.

To mitigate groundwater flooding, AE considered seepage collection trenches and cutoff walls to bedrock as modelling showed that partial cutoff walls were relatively ineffective. For either solution, accompanying pump stations would be required to manage groundwater flows but the pumping capacities would vary depending on the proposed solution.

4. Estimation of Flood Damage and Benefits of Flood Protection

AE estimated damages related to overland and groundwater flooding as part of this study. These flood damage estimates included economic, social and environmental impacts.

Depth-damage curves developed as a part of the "*Provincial Flood Damage Assessment Study*" (IBI Group, 2015) contain unit area damage information. These curves were developed for numerous building types and separated into structure and content curves. AE amalgamated the depth-damage curves into residential, commercial and industrial categories and applied them to all properties in the study area.

Overland and groundwater flooding damage estimates were developed for buildings and their contents by applying modelled flood depths to the aforementioned depth-damage curves.

Socio-environmental factors were quantified such that a TBL benefit-cost analysis of proposed flood protection could be conducted. The social and environmental factors included in the analysis were related to safety impacts, traffic impacts, loss of business, aesthetic impacts, and impacts to riparian areas.

5. Triple Bottom Line Benefit-Cost Analysis of Proposed Flood Mitigation

A TBL Benefit-Cost Analysis was conducted to assess the merits of proposed flood mitigation projects considering economic, social and environmental impacts of flooding and impacts averted by the proposed flood mitigation. The aforementioned damages averted by proposed flood protection projects were considered to be benefits of flood protection.

The results identified areas where providing flood protection is cost-beneficial (i.e. benefit-cost ratio of greater than 1) and identified optimal levels of service for flood protection. This enabled the City to make informed decisions with regards to city-wide flood protection, local flood protection, and prioritization of the proposed improvements.

Leadership in Sustainable Infrastructure

Leadership en Infrastructures Durables

Vancouver, Canada

May 31 – June 3, 2017/ *Mai 31 – Juin 3, 2017*

ASSESSMENT OF DEBRIS ISSUES IMPACTING DESIGN OF A FLOOD DIVERSION PROJECT IN A LARGE SCALE PHYSICAL MODEL

Knox, Paul[1,5], Cornett, Andrew[1], Provan, Mitchel[1], Menninger, John[2], Lux, Rick[3], and Abbas, Syed[4]

[1] National Research Council of Canada, Ottawa, ON, Canada
[2] Stantec, Cincinnati, OH, USA
[3] Stantec, Lexington, KY, USA
[4] Alberta Transportation, Edmonton, AB, Canada
[5] paul.knox@nrc.ca

Abstract: The Elbow River flows through southwest Calgary and is susceptible to flooding with catastrophic results. During the flooding event of June 2013, peak flow rates of the Elbow River reached approximately 1240 m^3/s while the natural capacity of the river is less than 200 m^3/s, and the losses experienced during the flood were valued in excess of $5B. The Springbank Off-stream Storage Project was conceived to divert and store the flows of the Elbow River during floods. Some of the main structural components of the project include a diversion and spillway structure that intersects the Elbow River, a diversion channel and an off-stream storage reservoir. A large scale (1:16) physical model study was subsequently commissioned to assist in assessing and improving the initial design for the diversion and spillway structures to ensure good performance under a range of flood conditions. The main objectives for the physical model study were to: determine the hydraulic performance of various key elements of the new structures for a range of operational and extreme flow conditions; assess the behaviour of sediments and woody debris within and around the structures; and help refine the proposed designs to improve conveyance, reduce the risk of erosion and sedimentation, reduce the risk of blockage by debris, improve constructability, and reduce costs where possible. This paper provides a full description of the physical model with a focus on the tests in which the impacts of woody debris were studied and assessed.

1 Introduction

The Elbow River flows through southwest Calgary and is susceptible to flooding with catastrophic results. During the flood of June 2013, the peak flow rate of the Elbow River entering the Glenmore Reservoir reached approximately 1240m^3/s. The capacity of the river without taking flooding precautions is 180m^3/s. Assessments made by the City of Calgary estimates that there is up to $942 million at risk in the event of a 1-in-200 year flood. As part of a flood mitigation study, the Springbank Off-stream Storage Project was conceived to divert part of the flows of the Elbow River during floods into an off-stream storage reservoir where the flood water would be stored and released gradually back into the Elbow River after the flood has passed. The project includes design, permitting and construction of a new diversion inlet structure that intersects the Elbow River west of Calgary, a diversion channel excavated through the adjacent land to transport the flow to the off-stream storage reservoir, a new earth dam to impound flood waters in the reservoir, and a new service spillway and sluiceway to control the flow through the Elbow

River at the diversion junction (see Figure 1). Figure 2 shows an artist's impression of the diversion inlet and sluiceway/spillway structures based on a preliminary conceptual design.

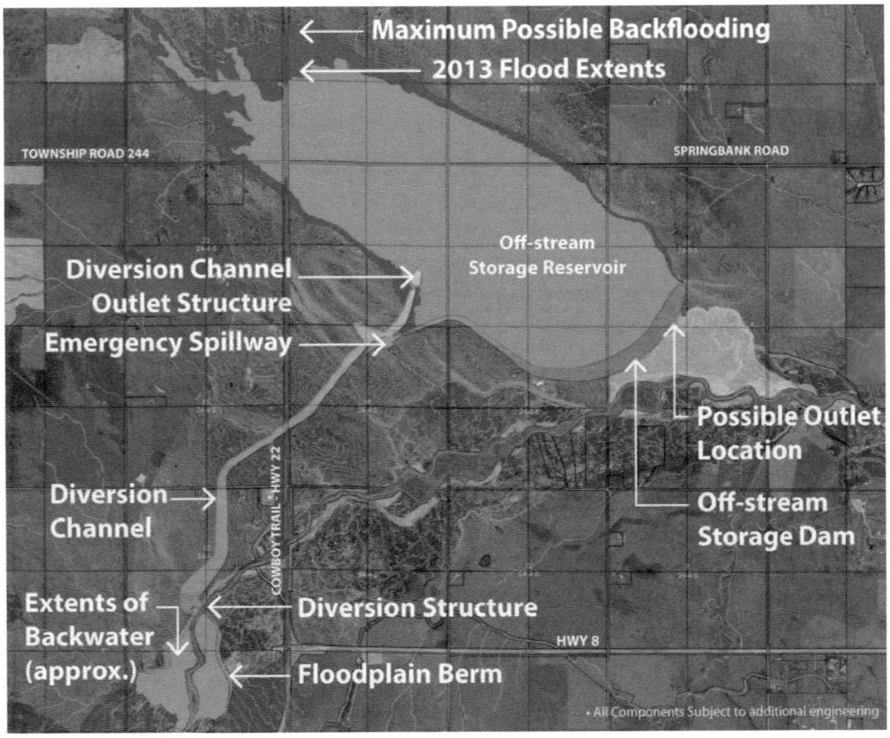

Figure 1. Location of the Springbank Off-stream Storage Project (Government of Alberta, 2016)

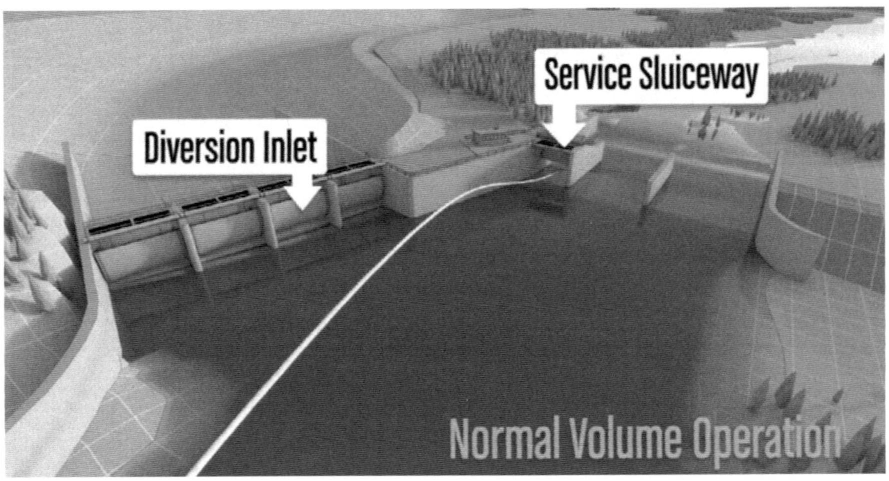

Figure 2. Artist's depiction of a preliminary concept design (Government of Alberta, 2016).

The Elbow River watershed extends into the Rocky Mountains west of Calgary, and large portions of the watershed are densely forested with mature conifers (Hemlock, Spruce, Lodgepole Pine and other similar species). The floodplain near the project site is littered with large clumps of old woody debris, see Figure 3, clear evidence of the many trees uprooted and transported downstream during previous flood events. Any future flooding will feature downstream transport of new and old woody debris, both as individual trees and clumps or rafts comprised of many pieces interlocked together. It is clear that any new flood control structures need to be designed to accommodate this debris load to the extent possible.

Field surveys were conducted (by others) to collect information on the sediments in the riverbed near the site and the type of woody debris to be expected during flood events. The field survey of the woody debris found that:

- 63% of logs had root wads;
- The mean log length was 16.5 m (standard deviation of 4 m);
- Mean diameter of the large end was 0.44 m (standard deviation of 0.09 m); and
- Mean diameter of the slender end was 0.14 m (standard deviation of 0.09 m).

Figure 3. Photos showing bottom sediments and woody debris on the Elbow River floodplain.

2 Diversion and Sluiceway/Spillway Structures

The diversion inlet and sluiceway/spillway structures were designed primarily to regulate the volume of water passing downstream through the diversion channel and the main (natural) river channel for various flood events, up to an including events with a peak discharge of 1,240 m^3/s. The desired flow split for several design flood discharges is summarized in Table 1. Flows in the Elbow River up to 160 m^3/s were directed through the main channel structures (the sluiceway and spillway). River flows above 160 m^3/s would have the excess flow directed through the diversion up to a maximum of total flow of 760 m^3/s (with 600 m^3/s into the diversion and 160 m^3/s into the sluiceway/spillway). The diversion structure is designed to accept 600 m^3/s, therefore flood flows above 760 m^3/s will have the remainder of the flow pass through the main channel structures as can be seen in the 1,240 m^3/s event where 600 m^3/s passes through the diversion inlet and 640 m^3/s passes through the sluiceway/spillway. The control structures were equipped with gates that could be raised and lowered to achieve the desired flow volumes.

Table 1. Target discharges for various design flood conditions (m^3/s).

Combined discharge	Diversion discharge	Sluiceway/Spillway discharge
160	0	160
320	160	160
760	600	160
1,240	600	640

The original (preliminary) structure designs are shown in Figure 4 to Figure 6. The diversion inlet structure featured four 10m wide bays separated by 2m wide piers, radial gates to control the flow, and energy dissipation chute blocks through the stilling basin. There were two structures to control the flow of the main river channel – a 10m wide sluiceway featuring a radial gate, a central pier, and a 31m wide service spillway with adjustable Obermeyer style crest gates. There were curved retaining walls on the west side of the diversion inlet and also on the east side of the service spillway (fronting the floodplain berm). A 31.63m straight-line retaining wall separated the diversion inlet and the sluiceway/spillway.

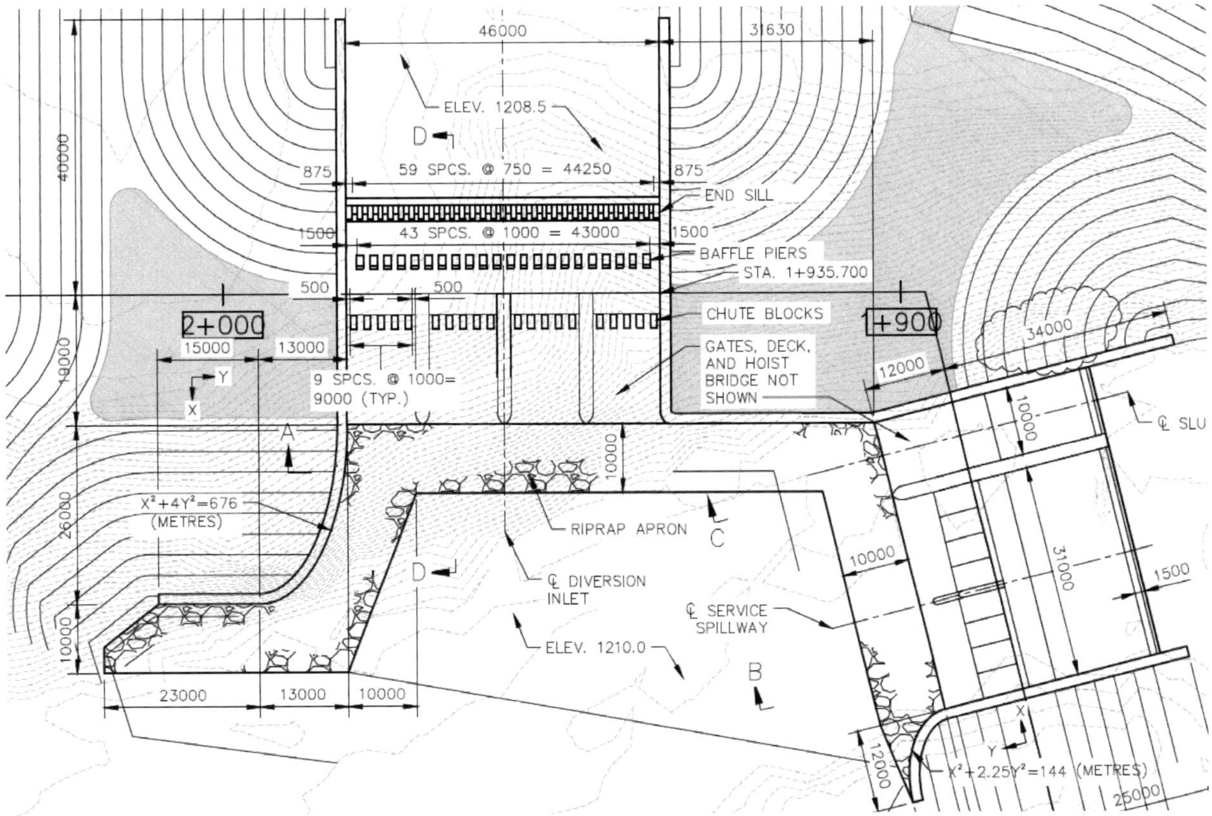

Figure 4. General arrangement of the initial designs for the diversion inlet and sluiceway/spillway structures (by Stantec).

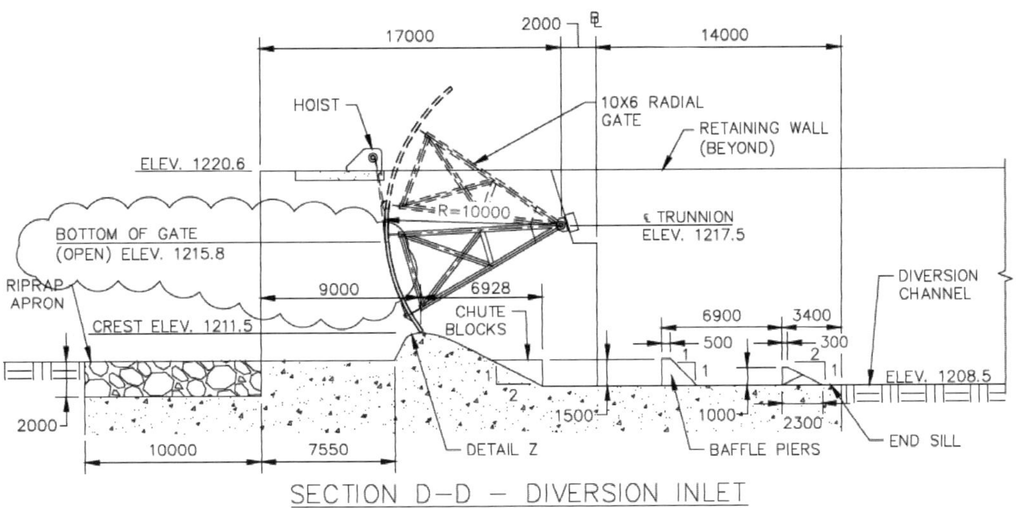

Figure 5. Cross-section views of the diversion inlet (by Stantec).

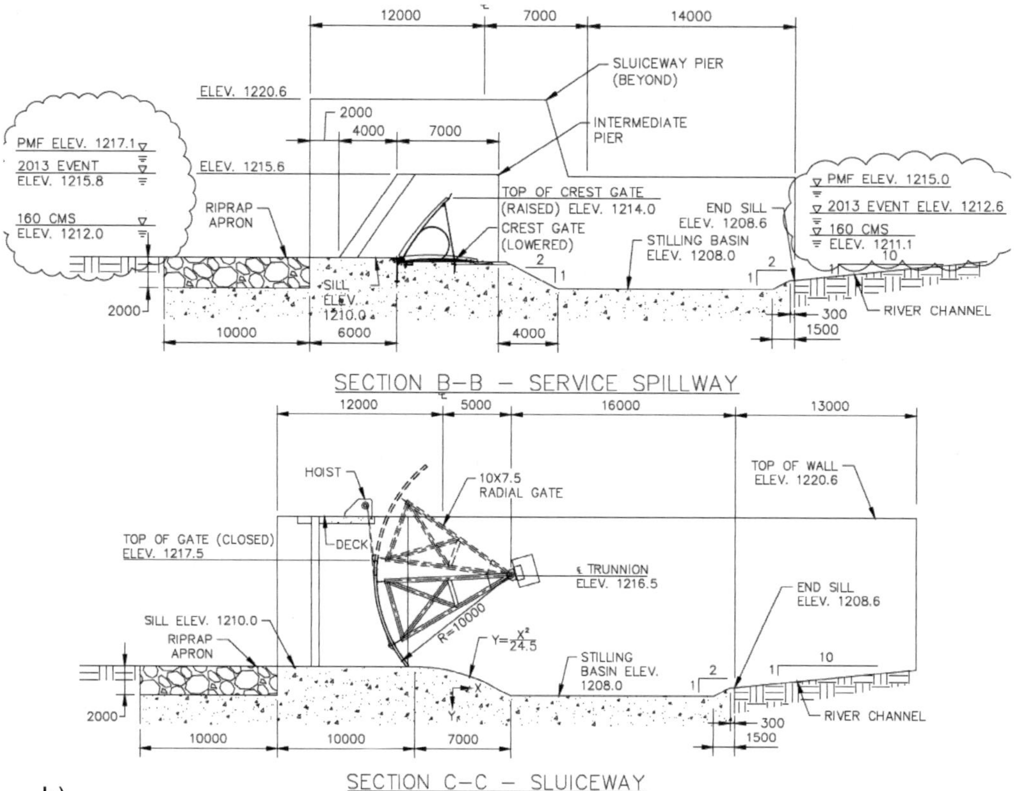

a)

b)

Figure 6. Cross-section views of: a) service spillway and b) sluiceway structures (by Stantec).

3 The Physical Model

A physical model study was subsequently commissioned to assist in assessing the performance of the initial designs for the diversion inlet, spillway and sluiceway structures, optimize them as necessary, and verify that these structures would perform well under flood conditions. Of particular concern was the behavior of sediments and the potential for sediment deposition, as well as the impact of woody debris transported downstream by the flood on the structures. Hence, the main objectives for the physical model study were to:

- Determine the hydraulic performance of various key elements of the new diversion inlet, service sluiceway/spillway and emergency spillway for a range of operational and extreme flow conditions;
- Assess the behaviour of sediments and the potential for sediment deposition within and around the diversion inlet, sluiceway, and service spillway;
- Assess the impact of debris on the performance of the diversion, sluiceway and service spillway;
- Help optimize the proposed designs to improve conveyance, reduce the risk of erosion and sedimentation, reduce the risk of blockage by debris, improve constructability, and reduce costs where possible; and
- Determine rating curves for the optimized hydraulic structures.

The current paper focuses on the role of the physical model in describing the impact of floating woody debris on the diversion inlet and sluiceway/spillway structures. The performance of the original designs as well as some of the design variants developed based on the results of the physical model are discussed.

An undistorted three-dimensional physical model of the proposed diversion inlet and sluiceway structures was designed and constructed at a length scale of 1:16 at the National Research Council of Canada laboratory in Ottawa. This scale represents a reasonable compromise between minimizing scaling and boundary effects, while maximizing the extent of the model domain and the accuracy of processes under investigation. This large scale physical model occupied a footprint of 50 m by 30 m and included realistic reproductions of the bathymetry and topography around the location of the diversion inlet, including the existing river channel and adjacent floodplain and proposed diversion channel as well as the diversion inlet and sluiceway/spillway structures. The physical model layout is sketched in Figure 7, while Figure 8 provides a general overview of the model in dry conditions (before a flood flow was generated). All aspects of the model, except for the mobile-bed sediment, were designed using Froude scaling criterion which assumes that gravitational and inertial forces are dominant in comparison to viscous forces.

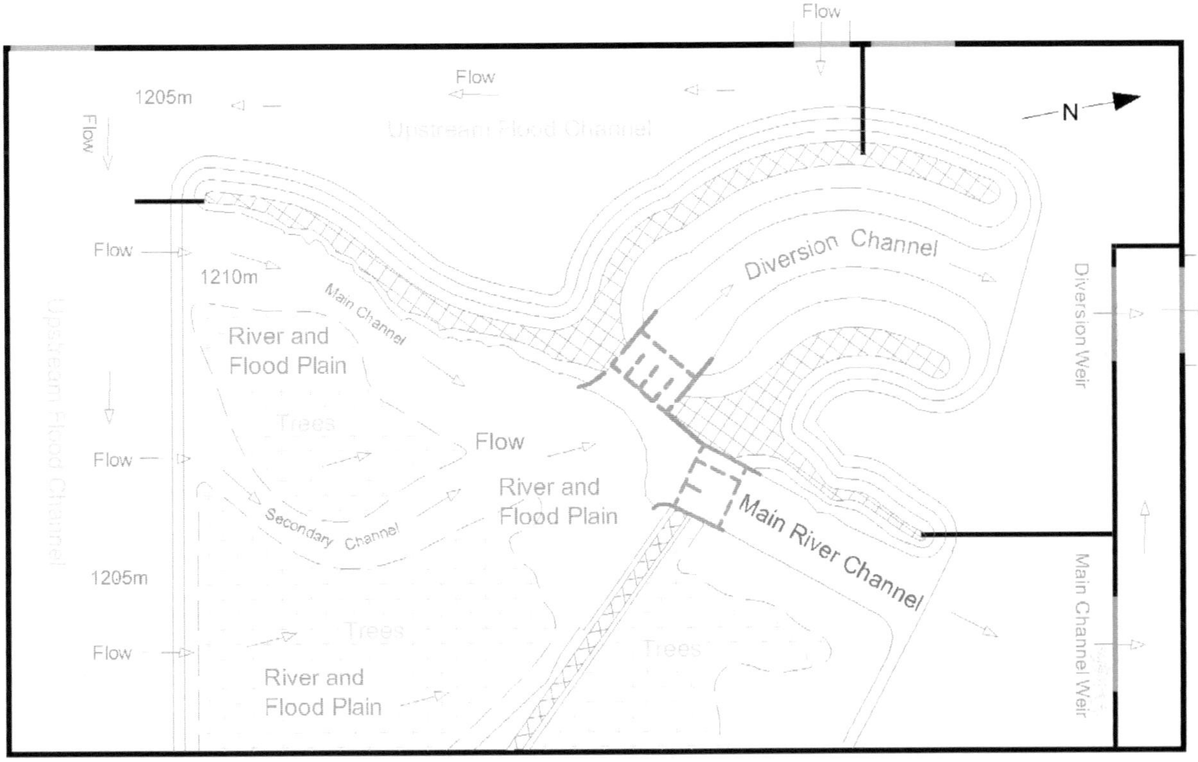

Figure 7. General layout of the 1:16 scale physical model in the 50m by 30m facility.

Figure 8. View of the physical model after construction (looking downstream).

The bathymetry and topography were modelled as a rigid surface formed in grout. Where appropriate, the concrete grout was given a stiff broom finish to roughen the bottom of the river channel, diversion channel and floodplain to approximate the hydraulic roughness of the prototype channels. Also, where the prototype floodplain was tree-covered, tree branches harvested from the forest were cast into the wet grout with a placement density similar to prototype conditions. A variable-pitch pump drawing from a large sump was used to supply a steady (adjustable) discharge into the upstream end of the model, and a pair of adjustable weirs were used to control the water level at the downstream end of both the river and diversion channels. Various flood flows up to the 1,240 m^3/s event were simulated by regulating the inflow discharge via the pump controller, while using the adjustable sharp-crested weirs to control the downstream tail-water levels. The model was outfitted with instruments for measuring water levels and flow velocities in multiple locations. The discharge passing through the diversion and downstream river channels was estimated by measuring the water levels above the two sharp crested weirs and applying a well-known relationship between head and discharge.

Faithful scaled reproductions of the diversion inlet structure, the sluiceway and the service spillway were designed and constructed, see Figure 9. The model sluiceway was equipped with an adjustable lift gate, while the model service spillway had adjustable bottom-mounted Obermeyer-type crest gates. These gates were adjusted as needed to control the water level in the head-pond area immediately upstream of the structures, as well as the portion of the incoming flow that was diverted into the diversion channel.

Figure 9. Physical model – views of the initial designs of a) the diversion inlet, b) the sluiceway and spillway, and c) both structures under the 760 m^3/s flow condition.

4 Woody Debris Interaction with the Structures

Over 300 pieces of woody debris were prepared for use in the physical model to simulate mature conifer trees that had been uprooted and transported downstream by previous floods and observed near the project site. The model debris was prepared from tree branches with mean diameters from 0.025 to 0.031 m (0.4 m to 0.5 m full scale) harvested in the forest and then cut to random lengths between 1.0

and 1.3 m (16 m to 21 m full scale). A small (~0.1 m) "X" shape made of timber was secured to one end of ~60% of the branches to simulate a root wad (or root ball).

The procedure for all test runs investigating debris behavior was similar – model tree debris was added at specific points in the upstream floodplain in combinations of single debris loads (single trees from one to four trees at a time), or rafts of 10, 40, and 100 trees. The fate of the debris was tracked thorought the model via overhead video camera, overhead time-lapse still photography, and isometric view video cameras. Observations were made on the debris-structure interactions, as well as the fate of the debris as it passed through the diversion inlet and service spillway structures as well as describing any debris jams as they formed. Also, streamlines were identified by tracking (with overhead cameras) the pathways followed by neutrally buoyant drogues introduced in several upstream locations. These streamlines were helpful in determining the position of the flow split line in the upstream floodplain (see Figure 10a). Many model trees were snagged on the upstream side of the diversion inlet pier noses and sluiceway/spillway structures and did not pass downstream (see Figure 10b). Individual trees oriented roughly parallel to the flow were more likely to pass through the structures, but any trees perpendicular to the flow, such as those randomly oriented in clumps or rafts, became more easily caught on the structures. Model debris with root wads was also more likely to become snagged. The narrow bays of the diversion and sluiceway (10m) were common catch points for the relatively longer (~20m) debris. The likelihood of debris passing over the Obermeyer crest gates in the spillway depended greatly on the freeboard of the flow over the gate. When the freeboard was low trees were more likely to catch on the gates, particularly those with root wads. In order to simulate conditions expected at the peak of the flood, the trees that became snagged were generally left in place while additional debris was added upstream. Once trees became snagged on the structures, other trees released upstream were more likely to snag and form a debris jam. Eventually, the size of the snagged debris raft became large enough to affect the discharge passing through the diversion inlet and the flow split between the diversion channel and the downstream river channel. Flow reductions on the order of 10% were observed for very large debris rafts comprised of ~150 model trees.

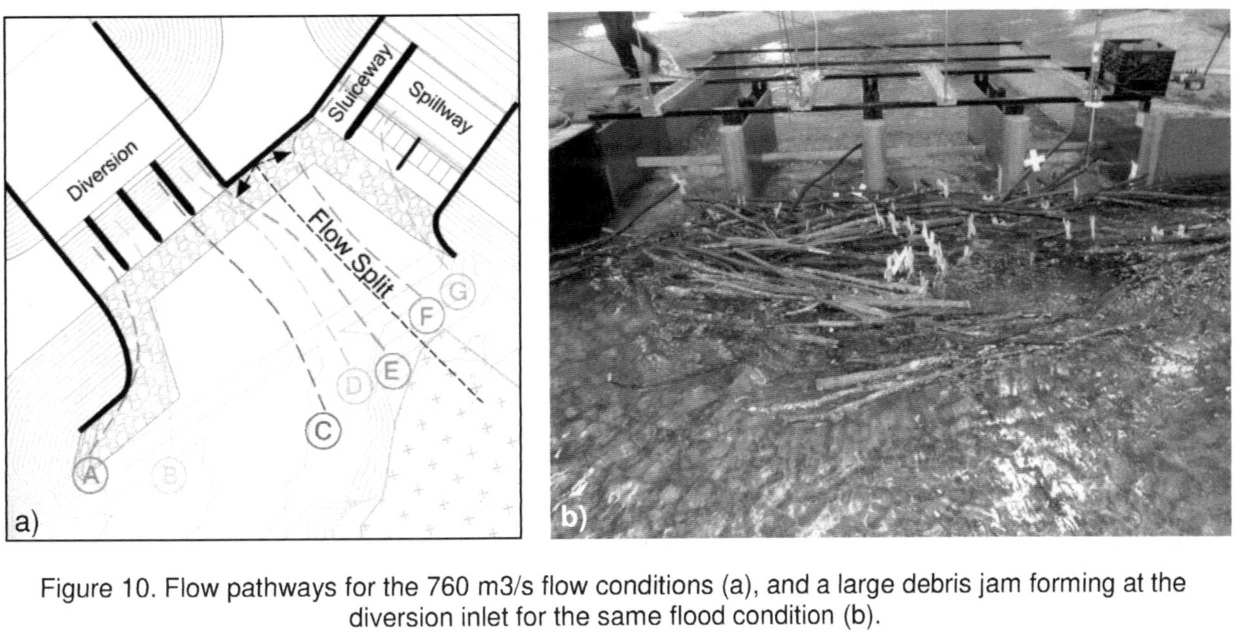

Figure 10. Flow pathways for the 760 m3/s flow conditions (a), and a large debris jam forming at the diversion inlet for the same flood condition (b).

Based on the observations and results of the early physical model test runs, the design of the diversion inlet and service spillway structures were revised with an aim to improve the hydrodynamics as well as the capability of the structures to handle tree debris. Some of the main alterations included:

- The four 10m wide bays of the diversion inlet were replaced with two 20m wide bays separated by a larger 4m wide pier. Note, the 4m central pier was outfitted with removable pier noses in the model to investigate the ability of different nose shapes and profiles to pass debris.
- The radial gates on the diversion inlet structure were replaced with vertical lift gates.
- The radial gate sluiceway section of the service spillway was removed.
- The main river channel was controlled by two independent Obermeyer crest gates (24m wide on the left hand side and 18m on the right hand side). This service spillway structure was also shifted 30m upstream of its original location (the flood plain berm to the east also shifted 30m).
- The retaining walls on the left hand side of the diversion inlet and right hand side of the service spillway were modified to improve the entrance flow characteristics.
- The breast walls in the diversion inlet bays were optimized to more easily pass debris.
- Six variations of debris barriers were installed in the optimized design model to assess their performance at controlling the movement of woody debris through the structures.

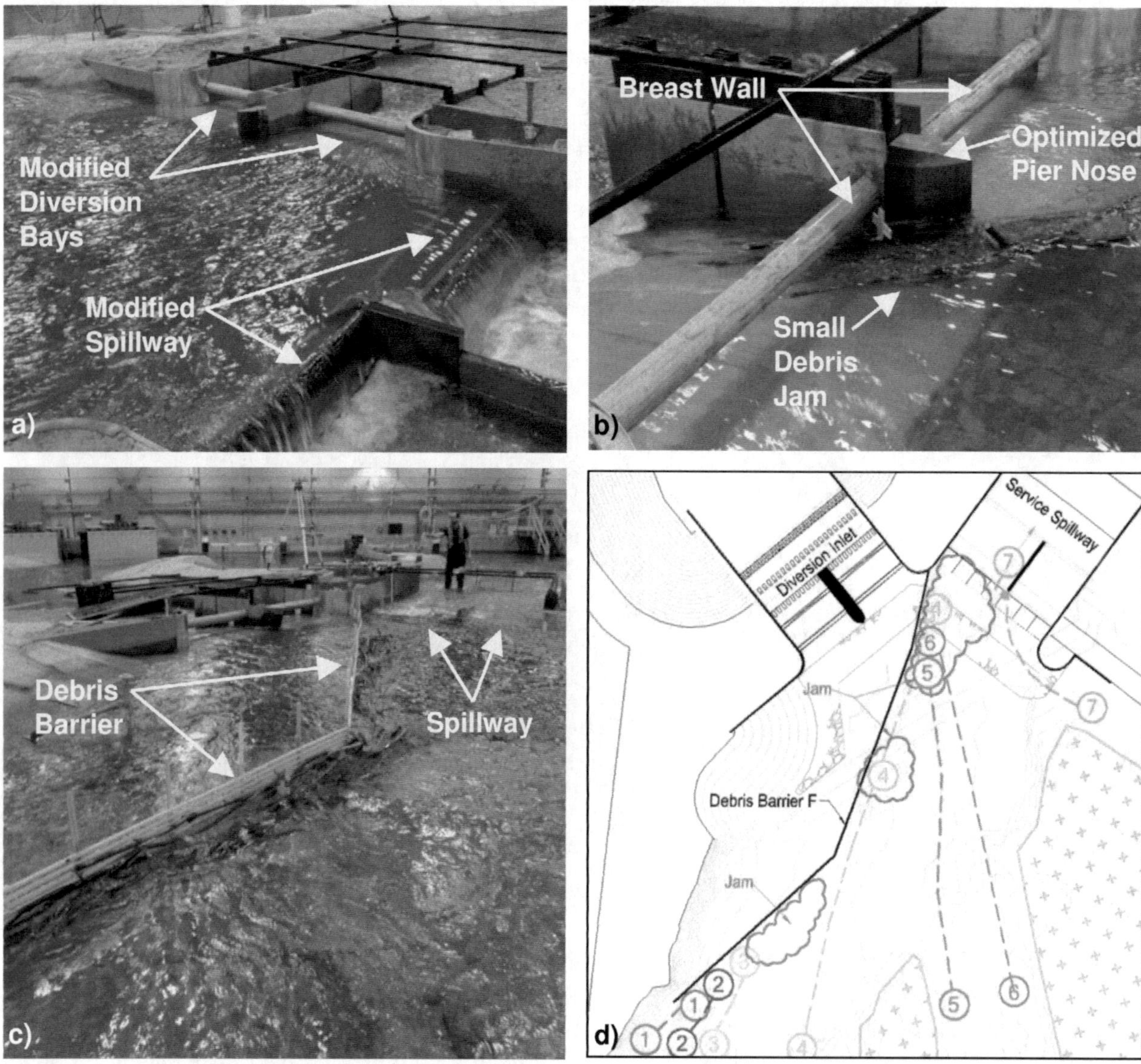

Figure 11. View of the modified structures in the physical model (a), a small debris jams forming during the pier nose optimization tests (b), a view of the model (c) and the flow pathways (d) for the 760 m^3/s flow conditions with a debris barrier installed in the upstream floodplain.

The revised designs were better adapted to accommodate the downstream transport of large woody debris and avoid the formation of problematic debris jams during flood events. Figure 11 shows photographs of some of the stages of the model testing where different design revisions were investigated. The diversion inlet's wider bays (Figure 11a), reduced number of central piers (b), and optimized breast wall (Figure 11b) proved effective at passing debris. Four different diversion inlet pier nose designs were tested in the model (Figure 11b), and the preferred designs passed 100% of the 40-element debris rafts through the diversion structure without a permanent jam. Debris occasionally became snagged on the Obermeyer gates and small central pier at the spillway, but likelihood and size of any debris jams were decreased with the new design. The physical modelling showed that lowering or raising the diversion and spillway gates were reasonably effective at removing any of the smaller debris jams caught on their structural elements. Six different debris barriers were installed in the upstream floodplain of the model (Figure 11c). These barriers were intended to screen debris from entering the diversion and potentially allow more cost effective designs of the inlet structures. The debris barriers ability to manage the flood debris and the effect of any jams on diversion and spillway hydraulics was investigated (Figure 11d). This data provided much insight into the construction costs associated with providing an effective debris barrier design.

Additional testing was undertaken where the behavior of the natural model woody debris was compared with the behavior of equivalent artificial debris more commonly used in previous physical model research studies (equivalent lengths of smooth wooden dowels with equal diameter). The results showed that the natural debris was more likely to become snagged in the model. Thus, the surface roughness and irregularity of the natural debris was found to be an important factor influencing the debris-structure interaction process. The use of model debris derived from natural sources having realistic surface roughness and irregularities is believed to be an improvement in the modelling process and provide a more realistic simulation of prototype conditions.

5 Conclusions

An undistorted physical model with a length scale of 1:16 was designed, constructed and operated to support the design of diversion inlet, sluiceway and service spillway structures proposed as part of the Springbank off-stream storage flood control project. The physical model generated valuable information on the hydraulic conditions around these structures for a range of flood flows, as well as providing crucial insight into the behaviour and structural interaction of sediments and large woody debris as it is transported downstream during flood events. The model used real tree branches with realistic surface roughness and irregularities to represent the standing trees in the model flood plain as well as the drifting tree debris transported by large floods; an approach that is believed to provide a realistic simulation of prototype conditions. Tests were conducted in which the behaviour of individual trees, a series of individual trees, and debris rafts comprised of 4, 10 and 40 overlapping pieces was investigated for several flood scenarios. This approach proved to be an efficient method for assessing debris-related issues, such as the tendency for large woody debris to accumulate on the upstream side of the structures. A large quantity of information was obtained on the performance of the diversion inlet and service spillway, the water levels and velocities encountered in the river and floodplain, the fate of the river sediments, and the interaction of tree debris with the proposed structures during floods. The physical model proved to be a very useful tool for assessing the performance of the proposed flood control structures, informing the development of refined structure designs, and confirming the performance of new structure designs, particularly with respect to the downstream conveyance of large water-borne woody debris during flood events.

6 References

Baraudrick, C. A., Grant, G. E., Ishikawa, Y., & Ikeda, H. (1997). Dynamics of wood transport in streams: a flume experiment. Earth Surface Processes and Landforms, 22(7), 669-683.

Baraudrick, C.A., Grant, G.E., (2001) Transport and deposition of large woody debris in streams: a flume experiment. J. Geomorphology, 41(4), 263-283.

Bocchiola, D., Ruili, M.C., Rosso, R. (2006) Transport of large woody debris in the presence of obstacles. J Geomorphology, 76(1-2), 166-178.

Government of Alberta (2016) Springbank Off-stream Reservoir Project. http://aep.alberta.ca/water/programs-and-services/flood-mitigation/flood-mitigation-projects/springbank-road.aspx. Website accessed December 20, 2016.

Gschnitzer, T., Gems, B., et al. (2013) Physical Scale Model Test on Bridge Clogging, Proc. 35th IAHR World Congress, Beijing: Tsinghua University Press.

Lyn, D.A., Coorper, T., et al. (2003) Debris Accumulation at Bridge Crossings: Laboratory and Field Studies, School of Civil Engineering, Purdue University, Joint Transportation Research Program, Report no. FHWA/IN/JTRP-2003/10.

Welber, M., Bertoldi, W. and Tubino, M. (2013), Wood dispersal in braided streams: Results from physical modeling, J. Water Resources Res., 49, 7388–7400.

Leadership in Sustainable Infrastructure
Leadership en Infrastructures Durables

Vancouver, Canada
May 31 – June 3, 2017/ *Mai 31 – Juin 3, 2017*

AUTOMATED FEATURE SELECTION FOR FUZZY NEURAL NETWORKS: AN APPLICATION FOR URBAN FLOOD PREDICTION

Khan, Usman T [1,2]
[1] York University, Canada
[2] usman.khan@lassonde.yorku.ca

Abstract: Urban floods are one of the most devastating natural disasters globally and improved flood prediction is essential for better flood management. Today, high-resolution real-time datasets for flood-related variables are widely available. This data can be used to create data-driven models for improved real-time flood prediction. However, data collected for extreme observations have large uncertainty, which must be accounted for in these models. In addition to this, typically the selection of input features and the architecture for data-driven models has been subjective. Addressing these concerns will improve flood prediction and will provide more accurate flood risk assessments. In this research, a new type of fuzzy neural network is proposed to predict peak flow in an urban river. The network uses fuzzy number to account for the uncertainty in the data and model parameters. An algorithm that uses possibility theory is used to train the network. An adaptation of the Automated Neural Pathway Strength Feature Selection (ANPSFS) method is used to select the input features. A search and optimisation algorithm is used to select the network architecture. A number of different inputs are considered including lagged precipitation and mean daily flow rate. The impact of this approach is that network training does not follow the typical ad hoc approach, but is based on objective criteria. Data for the Bow River in Calgary are used to train and test the network. Model performance using different features is used to compare the effectiveness of using the ANPSFS approach.

1 Introduction

In 2013, Calgary and southern Alberta experienced one of the worst natural disasters in Canada: floods caused approximately $6 billion in damage (Environment Canada, 2013), and were responsible for four deaths, and displaced more than 100,000 residents (Alberta Government, 2014; Khan & Valeo, 2016). The floods contributed to the transport of large amounts of sediment and the destruction of river banks, channels and aquatic ecosystems (Environment Canada, 2013). Better flood protection, and timely flood mitigation strategies may have prevented some of the destruction and casualties. Flood prediction models play an integral role in flood protection, and improved models are critically needed to help protect against more frequent and intense floods expected in the future.

The mechanisms behind extreme events in southern Alberta are generally understood and documented (Valeo et al., 2007) however, predicting floods remains a challenge because of the uncertainty in the numerical models. Flood prediction using physically-based and deterministic hydrologic models (such as a rainfall-runoff routing models) rely on simplified, conceptual representation of highly complex, correlated and spatially distributed processes that occur in a watershed (Wijiesekera et al., 2012; Vrugt et al., 2005). Uncertainty in the data used to calibrate model parameters and errors in model structure, compounds the complexity of flood models (Vrugt et al., 2005).

An alternative to physically-based models are data-driven models, which use generalized relationships between input and output datasets (Solomantine & Ostfeld, 2008). The models can characterize a system with limited assumptions and have similar (if not better) performance than physically-based models. A simpler model structure often means that the propagation of uncertainty from different sources is easier to quantify. These types of models (e.g. artificial neural networks, ANNs) have been widely used in hydrology, including for predicting flow rate (Adamowski & Sun, 2010; Duncan et al., 2011; Li et al., 2015). An advantage of using a data-driven approach is that data collected from real-time flow rate monitoring stations can be used to calibrate the model. This type of data is routinely collected by Environment Canada, which means that site specific surveys that are required for many physically-based methods are not necessary.

Data-driven models have intrinsic uncertainties associated with it that are not random or probabilistic in nature, thus, making it well suited for the use of fuzzy number theory (Ozbek & Pinder, 2006). Fuzzy numbers use fuzzy sets and possibility theory to describe uncertain or imprecise quantities, measurements or observations (Khan and Valeo, 2015). They are more suitable when data is missing, incomplete or vague, combined from multiple sites, and to represent uncertainty that is not random (Khan and Valeo, 2016). Using fuzzy numbers instead of probability-based methods means that some of the strict assumptions in many probability models can be relaxed (Kahraman et al., 2006). The literature demonstrates the utility and advantage of using fuzzy numbers in conjunction with data-drive methods. For example, Alvisi & Franchini (2011) proposed a method to create a fuzzy neural network (FNN), where the model coefficients and output of the ANN are fuzzy numbers rather than deterministic. These fuzzy numbers quantify the total uncertainty of the calibrated coefficients. This application is unique compared to the more typical use of fuzzy set based applications that involve ANN. Typical methods use fuzzy logic based methods where automated IF-THEN rules are used to create deterministic outputs (Abrahart et al., 2012; Alvisi & Franchini, 2011). These types of methods do not quantify the uncertainty of the predictions.

A major issue in implementing ANNs is that the choice of network architecture. This includes the selection of and the number inputs used, the number of hidden-layers, the number of neurons in each hidden-layer; the selection of the transfer functions, the amount of data in each subset, the training algorithm used; and the performance metric selected for optimisation. All of these important components of the network affect the values of the model coefficients, and the overall model performance. This means that each ANN model has uncertainty pertaining to the trained parameters. Recent reviews of ANN applications in hydrology have indicated that the lack of uncertainty quantification is a major reason for the limited appeal of ANN by water resource managers (Abrahart et al., 2012). Most ANN applications have a deterministic structure that generate point predictions without a quantification of the intervals corresponding to these predictions (Kasiviswanathan & Sudheer, 2013.). This means that end-users of these models may have excessive confidence in the forecasted values, and overestimate the applicability of the results (Alvisi & Franchini, 2011). However, the characterisation of uncertainty in a model is essential for both research and operational purposes. Without this, model results have limited value (Kasiviswanathan & Sudheer, 2013).

Thus, in this research a new method to account for the various uncertainties in ANN models is presented. First, an updated Automated Neural Pathway Strength Feature Selection (ANPSFS) method is used to select the optimum input data from a larger dataset. Second, a search algorithm is used to find the optimum network architecture (i.e. number of neurons in the hidden layer and fraction of data used for training). Lastly, a possibility-theory based FNN algorithm is used (based on Khan and Valeo, 2016b) to quantify the uncertainty of the model coefficients and outputs. Data from the Bow River in Calgary is used to train and validate the model, and results from the 2013 flood demonstrate the model.

2 Methods

2.1 Site description and data collection

The Bow River basin is located in southern Alberta and has an area of approximately 25,123 km^2. The basin provides significant economic activity for a major urban centre (the City of Calgary), as well as irrigation water for several irrigation districts in southern Alberta. The River originates in Bow Lake in the Rocky Mountains and flows south-easterly through Calgary (drainage area of 7870 km^2) and into Hudson Bay (Robinson et al., 2009). The River averages a 0.4% slope over its 645 km length. The Bow River is supplied by precipitation accumulated in the snowpack in the Rocky Mountains, precipitation and discharge from shallow groundwater. Runoff peaks in the spring (June) while low flows are seen in the winter (January). Just upstream of Calgary, the River is regulated by Bearspaw Dam. The Bow River flows through many residential communities and the commercial centre of Calgary. Thus, it is extremely important for accurate and timely peak flow prediction of this river for the safety of the residents, and for protection of assets in downtown Calgary. Therefore, for this research the "Bow River at Calgary" (Environment Canada WSC Station ID: 05BH004) flow measuring station was selected for peak flow rate prediction. This station is located centrally within Calgary.

Eleven years of hourly flow rate data for was obtained from Environment Canada for the period January 1, 2000 to December 31, 2010. The annual median flow rate and the annual peak flow rate varied between 47 – 85 m^3/s and 172 – 787 m^3/s respectively. The highest peak flow rate occurred in 2005 (787 m^3/s). The peak flow rates in 2005 are associated with a flood event in Calgary and southern Alberta, which estimated to have cost approximately $400 million in damage (Valeo et al., 2007). This flood event was the last major flood in southern Alberta prior to the floods in June 2013. The peak flow rate for the 2013 event was approximately 1600 m^3/s, significantly higher than annual peaks in the previous decade (Khan & Valeo, 2016). Data from 2013 was also collected and used for model testing. In addition to the hourly flow rate, the daily minimum and maximum temperature and daily precipitation for the same period was obtained from Environment Canada.

The hourly flow rate collected for the station was filtered by removing dates where shift corrections (usually due to ice conditions) were applied by Environment Canada. Any data points within the selected range that had corrections applied were removed from further analysis. In addition to this, a second filter was applied to remove the low flow (below media ice-free median) rate periods, since the primary objective of this research was to predict peak flow rates. This reduced the data from an original, unfiltered set of 4018 days of hourly flow rate data to 1860 for Calgary, i.e. representing about 40% of available data.

2.2 Fuzzy Neural Network

Alvisi & Franchini (2011) proposed a method to create a FNN where the coefficients and the output of the network are fuzzy rather than deterministic numbers. In this method, a feedforward Multi-Layer Perceptron neural network (see Section 2.2.1 for more details) is modified to predict an interval rather than a single value for the coefficients and outputs. Each interval corresponds to an α-cut interval (an interval corresponding to the lower and upper limits of a fuzzy set at a defined membership level α). This is repeated for several α-cut levels to build a discretized fuzzy number at a number of membership levels. A stepwise, constrained optimisation algorithm is used to define these intervals. The constraints are such that to find the upper and lower limits of each coefficient to minimize the width of the predicted interval. However, Alvisi & Franchini (2011) defined the amount of data to be included in each interval arbitrarily, setting a predetermined amount of data to be captured within each interval. Khan & Valeo (2016b) refined this method by using the relationship between possibility theory and probability theory (known as probability-possibility transformations) to define more objective constraints. Using the transformation principles proposed by Dubois et al (2004), this method selects (1 – α) fraction of data to be captured within each interval, where α corresponds to the α-cut interval of a fuzzy number. Theoretical and mathematical details of this method are available in Khan & Valeo (2016b).

This FNN method was implemented in MATLAB (version 2016a). The built-in MATLAB Neural Network Toolbox was used with a two-step optimisation approach: first, the Shuffled Complex Evolution algorithm (commonly known as SCE-UA, Duan et al., 1992) was used to find an initial solution to the minimisation problem. Then, further refinement of the solution was conducted using the built-in MATLAB minimisation function *fmincon*.

2.2.1 Network architecture

For this research a three layer, feedforward Multi-Layer Perceptron architecture was selected to model peak flow rate (Q_P) using several candidate inputs (see Section 2.2.2). This structure is one of the most common structures for ANNs and thus, was selected to be able to compare directly with previous studies

(He et al., 2011). For this architecture, two transfer functions are required, one between the input and hidden-layer which was selected as the hyperbolic tangent sigmoid function, and one between the hidden-layer and output-layer which was a pure linear function (following Alvisi & Franchini, 2011). Network training was completed using a backpropagation algorithm, the Levenberg-Marquardt method, minimising Mean Squared Error (MSE). The input and output data were pre-processed before training, validation and testing. The data was normalised so that input and output data fell within the interval [-1 1].

There is no consistent method for selecting two important components of ANN architecture: the number of neurons in the hidden-layer (n_H), and the amount of data used for training, validation and testing (known as data-division) (Abrahart et al., 2012). Typically, an ad hoc or trial-and-error method is used to select the number of neurons (Maier et al., 2010; Alvisi & Franchini, 2011; He et al., 2011). The n_H selected must balance the complexity and generalisation of the final model; too many neurons increase the complexity and hence the processing speed, while reducing the transparency of the model. Not enough neurons risk reducing model performance and forgoing the ability of modelling non-linear systems. Similarly, the issue of data-division, which can have significant impacts on final model structure, is also determined in an ad hoc or trial-and-error basis (Maier et al., 2010).

For this research, a coupled method to select the optimum n_H and data-division for the ANN model described is proposed. This method uses the independent test dataset for all inferences regarding the error statistic (MSE) to be minimized. The smallest number of neurons and the least amount of data for training is targeted. The first is to reduce computational effort. The second is to prevent the risk of over-fitting to the training data, and to a larger dataset for testing for robust statistical inference of that dataset.

First, the dataset is randomly split into a 50%-25%-25% ratio for training, validation and testing, respectively. For each subset, data is randomly sampled to group into each subset. Then, the network is trained using 1 to 20 neurons. This process is repeated 100 times to account for the different selection of randomly sampled data in each subset. This is because the random initialisation of the ANN can cause variability in overall model performance (Napolitano et al., 2011). The MSE for each test dataset is then calculated and compared, and the number of epochs for each model is measured. Epochs are the number of times each model coefficient is modified in the optimisation algorithm. The configuration that leads to the lowest MSE for the testing dataset and the lowest number of epochs was collected. This process is then repeated by sequentially increasing the amount of data used for training (and thus, reducing the amount of data equally partitioned for validation and testing) by increments of 0.5% from 50% to 75%, conducting each iteration of this change 100 times. Using this process, a number of different circumstances where the highest performance was calculated can be listed, and the best combination can be objectively selected. In doing so, both the processing time and the complexity of the system is accounted for in the final model architecture.

2.2.2 Automated Neural Pathway Strength Feature Selection

The ANPSFS method (Duncan, 2014) is a method designed to reduce the complexity of models by reducing the number of inputs (i.e., the features) from the total available amount to only those that are relevant to the model at hand. While most data-driven models are typically considered black-box models, input feature selection methods can provide additional insights into the system by identifying the features that are the most relevant. In the ANPSFS method, the strength of a particular pathway from a given input to the output (i.e. QP) is calculated using:

[1] $W_{IO} = W_{IH} \times W_{HO}$

where W_{IH} are the weights (the coefficients) between the input and hidden layer, and W_{HO} are the weights between the hidden layer and output later. The higher the value of the W_{IO}, the more significant the associated input. In the present research, W_{IO} is calculated using all the input data for each year individually. This array of W_{IO} (one for each year) is then used to create the ensemble interquartile range (EQR) for each of the inputs as follows:

[2] $EQR = \min(|Q_1|,|Q_3|)/\max(|Q_1|,|Q_3|) \times sign(Q_1).sign(Q_3)$

where $Q_1$and Q_3 are the first and third quartile of the W_{IO}. The EQR are ranked in descending order, and the top three ranked inputs are selected as the most significant features. Thus, the ANPSFS method uses W_{IO} and EQR to select the most important features for Q_P (output) prediction given a set of input data.

2.3 Model performance evaluation

The Nash-Sutcliffe efficiency (R^2) (Nash and Sutcliffe, 1970) and the root mean square error (RMSE) were used to asses the model performance for each year:

[3] $R^2 = 1 - \sum(y_m - y_o)^2/\sum(y_o - y)^2$

[4] $RMSE = sqrt[\sum(y_o - y_m)^2]$

where y_m is the model output, y_o is the observed value, and y is the mean of the observed data.

3 Results

3.1 Network architecture selection

Sample results of the proposed method to select the number of neurons in the hidden layer and the amount of data used for each component are shown in Figure 1. Figure 1a shows the mean MSE (solid black line) of the test dataset for the initial 50%-25%-25% data-division scenario, with n_H varying between 1 to 20 neurons. This simulation was repeated 100 times and the upper and lower limits of MSE for each of these simulations are shown in grey. This figure demonstrates that the number of neurons did not have a noticeable impact on the MSE for this configuration. The most significant outcome of this process is that the variability (the difference between the upper and lower limits) of the performance seems to decrease after $n_H = 6$ and increases again after $n_H = 12$, with the lowest MSE at $n_H = 10$. This result has two important implications: first, increasing the model complexity results in limited improvement of model performance, suggesting that a simpler model structure may be more suitable to describe the system. Second, the variability in performance indicates that the initial selection of data in each subset can highly influence the performance, especially at the lower (i.e., $n_H < 6$) and higher (i.e., $n_H > 12$) ends of the spectrum of n_H. This suggests that an optimum selection of hidden neurons lies within this range.

Figure 1b shows the change in the mean (solid black line) and the variability (in grey) of the number of epochs needed to train the network for the initial data-division scenario, as n_H increases from 1 to 20. The variability of the number of epochs drastically decreases as n_H increases from 1 to 5. This means that a simpler model structure may require more time to train, and the performance of these simpler architectures ($n_H = 1$ to 5) is more variable. This is likely because the initial dataset selection has a higher impact on the final model performance for less complex models. The lowest number of mean epochs for this analysis occurred at $n_H = 19$, with 26 epochs.

The impact of changing the amount of data used for training, validation and testing on the model performance (MSE) was generally inconclusive (see Figure 1c) as the amount of data used for training was increased from 50% to 75% at 0.5% intervals. For the sample shown, ($n_H = 10$) increasing the amount of data used for training has minimal impact on model performance, indicating that using the least amount of data for training (and thus having a higher fraction available for validating and testing) would be ideal. The number of epochs needed for training the network at different data-division scenarios was inconclusive (see Figure 1d for the $n_H =10$ sample case). The significance of this analysis is that the amount of computational effort (or time) does not necessarily decrease as a larger fraction of data is used for training.

Based on these results, $n_H = 10$ with a 50%-25%-25% data-division was selected as the optimum architecture for this research. The fact that the mean and the variability of MSE was the lowest at $n_H = 10$ makes it a preferred option over the $n_H = 19$ case, which as a lower number of mean epochs but had higher variability in MSE. Secondly no significant trend was seen as the amount of data used for training, validation and testing was altered, however lower MSE values were seen at $n_H = 10$ compared to other at n_H values. Thus, the option that guarantees the largest amount of data for validation and training is preferred.

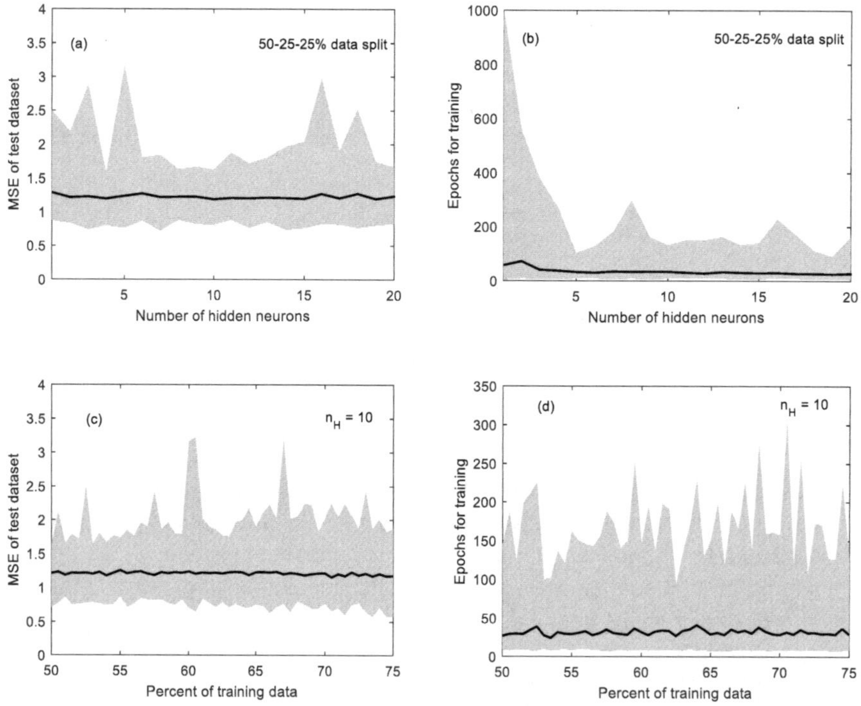

Figure 1: Sample results of the coupled method to determine the optimum number of neurons in the hidden-layer and percentage of data for training:

The overall outcome of this component of this research was that that instead of using the typical trial-and-error based approach to selecting neural network architecture parameters; the proposed method can provide objective results. Systematically exploring different numbers of n_H and division of data can help select a model with the best performance. Once these FNN parameters were identified, subsequent training of the FNN was completed and the results of this are presented in the next section.

3.2 Input feature selection using the ANPSFS method

The input data used for Q_P prediction in the Bow River at Calgary included mean daily flow rate (calculated using the collected hourly data), maximum, mean and minimum air temperature, and total daily precipitation. Each of these datasets were lagged by 1, 2 or 3 days, in order to be able to predict Q_P with 1, 2 or 3 days advance notice. Thus, the total amount of candidate input parameters were 18 (lagged versions of the above parameters, and lagged Q_P for each lag).

Results of the ANPSFS method are shown in Figures 2 – 4 below. Figure 2 shows the values of W_{IO} of each year for all 18 inputs. The figure shows that the importance (i.e. strength) of each candidate input parameter changes on an annual basis. The trend shows that there is no clear input parameter that is consistently ranked high, indicating the changing importance of each parameter in prediction Q_P. Figure 3 shows a box-plot of each candidate input parameter further highlighting this variability. When W_{IO} is below zero, this means that input parameter is not significant in prediction Q_P, i.e. it acts as an inhibitor in predicting QP. Several candidate variables have mean values of W_{IO} below 0 highlighting that their potential as an input is low.

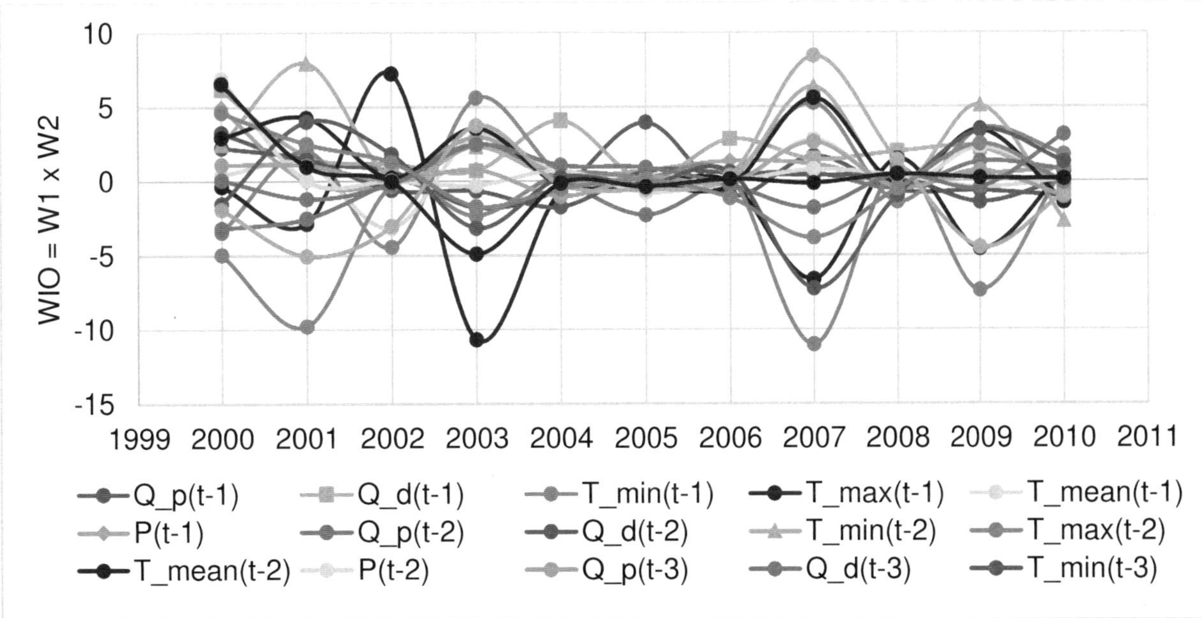

Figure 2: Values of WIO for the entire study period for each of the 18 input variables

Figure 4 shows the EQR values calculated using the W_{IO} for each feature from highest to lower. It shows that only three inputs have a positive (and thus non-inhibiting) effect: 1-day lagged mean flow, 1-day lagged precipitation and 2-day lagged minimum air temperature. Thus, for this research these three inputs are selected as the final inputs to the model to predict Q_P in the Bow River at Calgary. It is worth nothing that as additional data is available and added to the system it can be used to update the ANPSFS to select other input parameters to improve Q_P prediction. The significance of these results is that the method can be used to reduce the complexity of the model by reducing the number of inputs from 18 to 3, removing only those parameters that may not relevant.

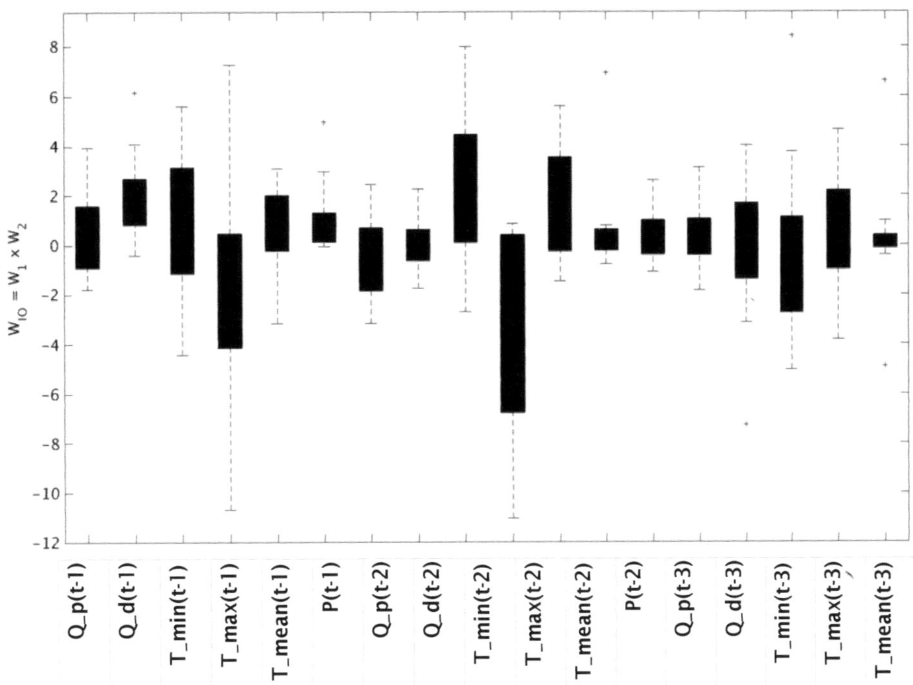

Figure 3: Box whisker plots for WIO for each of the input parameters

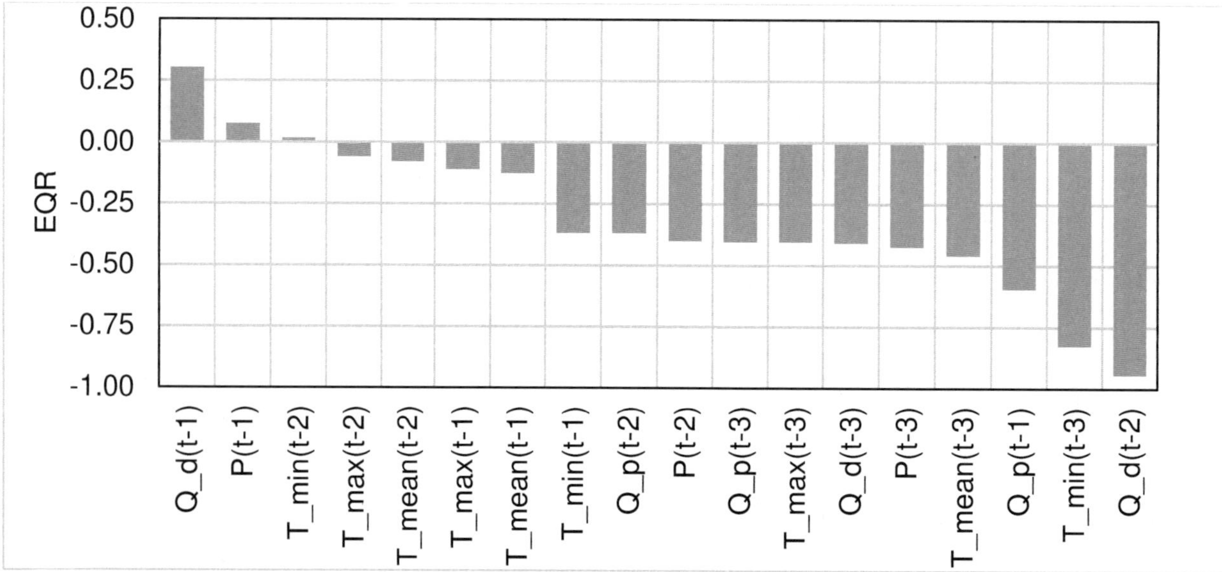

Figure 4: EQR values for each of the 18 candidate features to predict peak flow rate

3.3 Flood prediction

A FNN model was constructed using the method outlined in Section 2.2 using the three input parameters: 1-day lagged mean flow, 1-day lagged precipitation and 2-day lagged minimum air temperature. Figure 5 shows the R^2 and RMSE of predicting Q_P in the Bow River at Calgary for the eleven period. Network performance with these inputs was high: RMSE ranged between 7 and 47 m^3/s (about 10% of mean flow). The R^2 value was above 0.90 for all years except 2005 (which was also a flood year). This approach shows that using an objective rather than ad hoc method to design the architecture of a neural network (for input selection, n_H and data-division) can provide reliable results, which by design are ones with the highest performance. The trained FNN was used to predict the peak flow in the Bow River for the 2013 flood event, which is summarised in the next section.

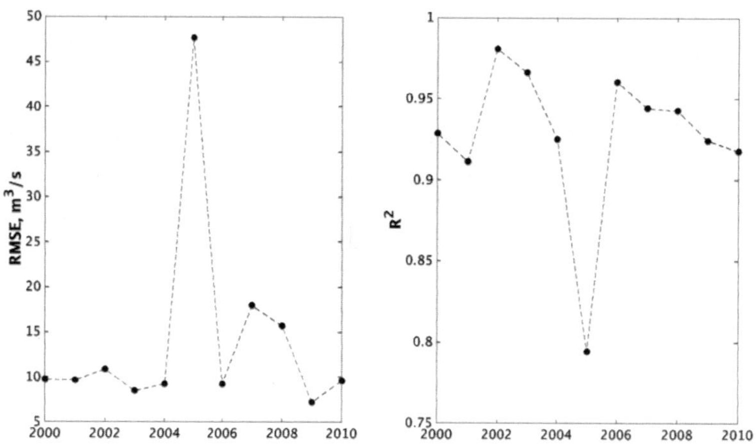

Figure 5: A summary of the RMSE and R2 values of the FNN mode

3.3.1 2013 Calgary floods

Figure 6 shows a comparison of predicted and observed daily peak flow in Calgary for 2013. Note that the predicted values for Q_P are fuzzy numbers rather than discrete numbers. So the Figure shows results for a membership level of 1 (the central value of the fuzzy number) and for membership level of 0 (the upper and lower bounds of the fuzzy number). The predicted intervals captured all daily peak flow rates, including the highest flows observed that year during the flood. This shows that using an FNN can predict Q_P in the Bow River up to a day in advance with high accuracy. Model outputs can be used for risk assessments to determine the probability of the predicted flow to exceed a given threshold. This data can be used by water resource managers to implement flood defence systems.

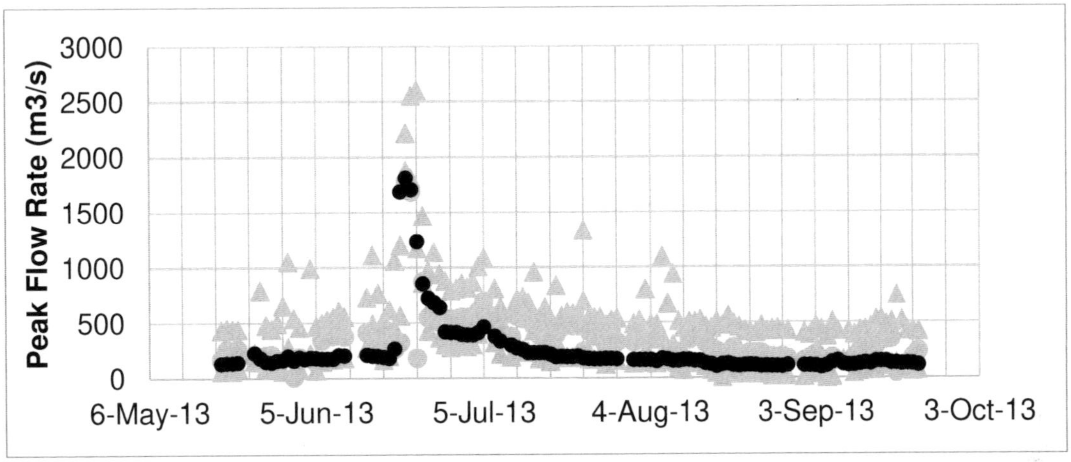

Figure 6: A comparison of observed (black circles) and predicted (grey markers) peak flow rate for Calgary in 2013

4 Conclusion

In this research, three different methods were used to reduce the overall uncertainty in an ANN model used to predict peak flow rate in the Bow River in Calgary. First, an existing fuzzy neural network was adapted using the principles of possibility theory. This allowed uncertainty in the model coefficients and output to be quantified by representing them as fuzzy numbers rather than discrete numbers. Second, an objective method to select the architecture of the FNN was developed. Lastly, the inputs used in the model were selected using a modified Automated Neural Pathway Strength Feature Selection (ANPSFS method). The impact of each of these methods was demonstrated by training a large dataset (11 years of data) for Calgary, and then tested on one extreme event (the 2013 floods in Calgary).

Acknowledgements

The author would like to acknowledge York University for funding, Dr S Alvisi (UNIFE) for the original FNN code, Dr A Duncan (Exeter) for discussions on ANNs; and Environment Canada for providing the data.

References

Alberta Government (2014). Respecting Our Rivers: Alberta's Approach to Flood Mitigation from https://pabappsuat.alberta.ca/albertacode/images/respecting-our-rivers.pdf (accessed July 1, 2014).

Adamowski, J., & Sun, K. (2010). Development of a coupled wavelet transform and neural network method for flow forecasting of non-perennial rivers in semi-arid watersheds. Journal of Hydrology, 390(1), 85-91.

Alvisi, S., & Franchini, M. (2011). Fuzzy neural networks for water level and discharge forecasting with uncertainty. Environmental Modelling & Software, 26(4), 523-537.

Abrahart, R. J., Anctil, F., Coulibaly, P., Dawson, C. W., Mount, N. J., See, L. M., Asaad Y. Shamseldin, A. Y., Solomatine, D. P., Toth, E., & Wilby, R. L.: Two decades of anarchy? Emerging themes and outstanding challenges for neural network river forecasting, Prog. Phys. Geog., 36, 480-513,

Duncan, A., Chen, A. S., Keedwell, E., Djordjevic, S., & Savic, D. (2011). Urban flood prediction in real-time from weather radar and rainfall data using artificial neural networks. In Weather Radar and Hydrology IAHS Red Book series no. 351. IAHS. Exeter, UK.

Duncan, A. P. 2014. The Analysis and Application of Artificial Neural Networks for Early Warning Systems in Hydrology and the Environment. PhD Thesis. University of Exeter, Exeter, UK.
Dubois, D., & Prade, H. (1997). Bayesian conditioning in possibility theory. Fuzzy Sets and Systems, 92(2), 223-240.
Dubois, D., Foulloy, L., Mauris, G., & Prade, H. (2004). Probability–possibility transformations, triangular fuzzy sets, and probabilistic inequalities. Reliable Computing, 10(4), 273-297.
Duan, Q., Sorooshian, S., & Gupta, V.: Effective and efficient global optimisation for conceptual rainfall-runoff models, Water Resour. Res, 28(4), 1015-1031,
Environment Canada (2013). Alberta's Flood of Floods on Canada's Top Ten Weather Stories for 2013, http://ec.gc.ca/meteo-weather/default.asp?lang=En&n=5BA5EAFC-1&offset=2&toc=show (accessed July 1, 2014).
He, J., Chu, A., Ryan, M. C., Valeo, C. & Zaitlin, B.: Abiotic influences on dissolved oxygen in a riverine environment. Ecological Engineering, 37, 1804–1814
Khan, U. T., & Valeo, C. (2016). Short-term peak flow rate prediction and flood risk assessment using fuzzy linear regression. Journal of Environmental Informatics. 28(2), 71-89.
Khan, U. T., & Valeo, C. (2014b). A new fuzzy linear regression approach for dissolved oxygen prediction. Hydrological Sciences Journal.
Kahraman, C., Beşkese, A. & Bozbura, F. T. (2006). Fuzzy regression approaches and applications. In C. Kahraman, (Ed.), Fuzzy Applications in Industrial Engineering, pp. 589-615. Springer Berlin Heidelberg. Berlin, Germany
Kasiviswanathan, K. S., & Sudheer, K. P.: Quantification of the predictive uncertainty of artificial neural network based river flow forecast models, Stoch. Env. Res. Risk A., 27, 137-146
Khan, U. T. and Valeo (2016b). Dissolved oxygen prediction using a possibility theory based fuzzy neural network. Hydrology and Earth System Sciences, 20(6), 2267
Li, Z., Huang, G., Han, J., Wang, X., Fan, Y., Cheng, G., Zhang, H., & Huang, W. (2015). Development of a Stepwise-Clustered Hydrological Inference Model. Journal of Hydrologic Engineering, (ahead-of-print).
Maier, H. R., Jain, A., Dandy, G. C., & Sudheer, K. P.: Methods used for the development of neural networks for the prediction of water resource variables in river systems: current status and future directions, Environ. Modell. Softw., 25, 891-909,
Napolitano, G., Serinaldi, F., & See, L.: Impact of EMD decomposition and random initialisation of weights in ANN hindcasting of daily stream flow series: an empirical examination, J. Hydrol., 406(3-4), 199-214,
Nash, J. E., & Sutcliffe, J. V.: River flow forecasting through conceptual models: Part I. A discussion of principles, J. Hydrol., 10(3), 282-290. 1970.
Ozbek, M. M., & Pinder, G. F. (2006). Non-probabilistic uncertainty in subsurface hydrology and its applications: an overview. Water, Air, & Soil Pollution: Focus, 6(1-2), 35-46.
Robinson, K. L., Valeo, C., Ryan, M. C., Chu, A., & Iwanyshyn, M. (2009). Modelling aquatic vegetation and dissolved oxygen after a flood event in the Bow River, Alberta, Canada. Canadian Journal of Civil Engineering, 36(3), 492-503.
Solomantine, D. P., & Ostfeld, A. (2008). Data-driven modelling: some past experiences and new approaches. Journal of Hydroinformatics, 10(1), 3-22.
Valeo, C., Xiang, Z., Bouchart, F. J.-C., Yeung, P. & Ryan, M. C. (2007). Climate Change Impacts in the Elbow River Watershed. Canadian Water Resources Journal. 32(4), 285-302.
Vrugt, J. A., Diks, C. G., Gupta, H. V., Bouten, W., & Verstraten, J. M. (2005). Improved treatment of uncertainty in hydrologic modeling: Combining the strengths of global optimization and data assimilation. Water Resources Research, 41, W01017.
Wijesekara, G. N., Gupta, A., Valeo, C., Hasbani, J. G., Qiao, Y., Delaney, P., & Marceau, D. J. (2012). Assessing the impact of future land-use changes on hydrological processes in the Elbow River watershed in southern Alberta, Canada. Journal of hydrology, 412, 220-232.

Leadership in Sustainable Infrastructure
Leadership en Infrastructures Durables

Vancouver, Canada
May 31 – June 3, 2017/ *Mai 31 – Juin 3, 2017*

AUTOMATIC DISCRETIZATION AND PARAMETERIZATION OF WATERSHEDS USING A DIGITAL ELEVATION MODEL

James, Robert W. [1], Finney, Karen A.[1,2], Perera, Nandana[1] and Xiao, Tiehong[1]
[1] Computational Hydraulics International (CHI)
[2] karen@chiwater.com

Abstract: Characterization of both overland and channelized flow is crucial for understanding the sources and fate of catchment runoff. It is essential that care is taken when delineating catchments to ensure that these two types of flow are distinguished from one another. Unfortunately, due to the level of complexity associated with large watersheds, obtaining detailed watershed data necessary to adequately characterize the hydrological process is not always possible. To represent stream response within in a watershed the sheet flow, channelized flow, flow path slope and contributing area must first be considered. This paper examines the Stormwater Management Model (SWMM) hydrology setup and parameterization for the Toronto and Region Conservation Authority's Don River flood forecasting model using an automatic watershed delineation, parameterization and discretization tool developed in PCSWMM. In addition, two scenarios, with varying parameterization and hydrologic and hydraulic resolutions, will be compared using observed rainfall and flow data.

1 INTRODUCTION

Modeling large watersheds requires the characterization of both overland and channelized flow. This can be achieved by creating a higher resolution model that better accounts for the watershed processes, particularly in headwater areas, by better representing additional processes occurring on watershed surfaces.

Due to common project time and budget constraints, model discretization is limited, both in the required data collection and model development time. Thus, watersheds are generally divided into a small number of coarse subcatchments with resolutions of 100 - 400 ha and parameters such as slope, imperviousness and infiltration averaged across each subcatchment.

Calibrating coarse watershed models to match observed flows at downstream locations can result in calibrating parameters beyond their reasonable bounds thus producing unrealistic head water hydrographs. For example, Figure 1 shows a section from a watershed model showing a subcatchment being routed two kilometers downstream of the outlet. It is likely that, in this instance, the transition from overland to channelized flow and the accompanying hydraulic lag did not do an adequate job representing the multitude of complex flow transitions within the subcatchment. In this case, the length of overland flow was artificially increased to match observed flows. And although this action improved the computed response time, it came at the expense of artificially inflating infiltration time.

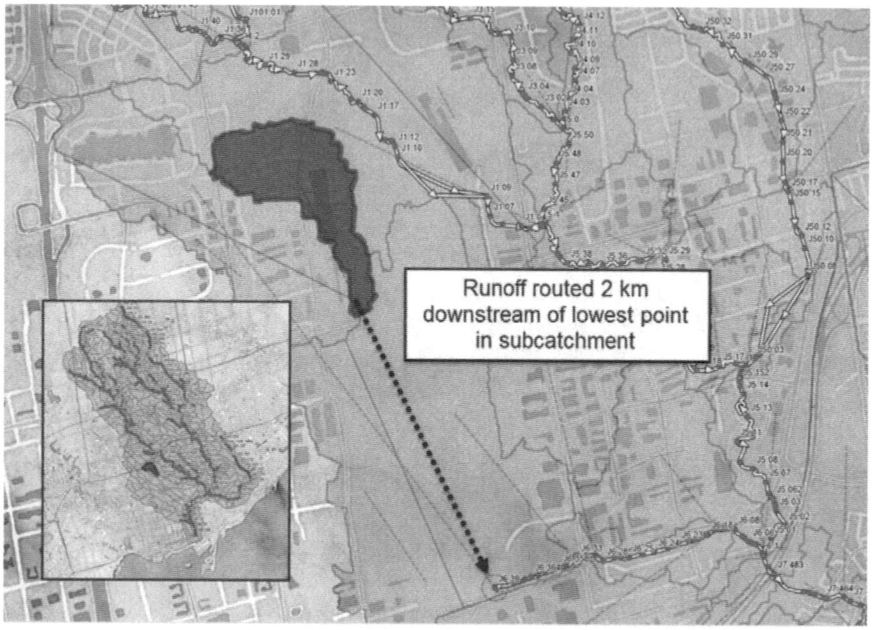

Figure 1: Example of subcatchment located significantly upstream of hydraulic network

While it is never possible to completely characterize watershed hydrological processes, separating sheet flow from concentrated flow and groundwater can better account for the overland processes without having to artificially change parameters to meet observed flows.

2 HYDROLOGICAL AND HYDRAULIC MODELING

The Stormwater Management Model (SWMM), originally developed by the United States Environmental Protection Agency (US EPA) in 1971, is a spatially distributed, rainfall-runoff simulation model used for single event or continuous simulation of runoff quantity and quality. SWMM has been updated many times since its first release, the most recent version being SWMM5.1. SWMM5 hydrology operates on a collection of catchments that receive precipitation and generate runoff and pollutant hydrographs, accounting for evapotranspiration, infiltration and groundwater percolation (Rossman 2008). SWMM5 models have been used for a wide variety of watershed applications (Bhaduri et. al. 2001; Davis et. al. 2007; Tränckner et. al. 2008).

SWMM5's non-linear reservoir formulation uses a kinematic wave approach to estimate overland flow. This approach requires that irregular subcatchments are converted into equivalent rectangles with a sloping plane, each with an associated width representing the width of the overland flow.

If the subcatchment width, W, is assumed to represent a true prototype width of overland flow, then the reservoir will behave as a rectangular catchment. In reality, subcatchments are delineated using topography/contour maps and land use layers and are unlikely to be perfect rectangles and the width (and the slope and roughness) may be considered calibration parameters.

Although SWMM5 uses a physically based approach to modeling surface hydrology, there is still only one flow equation representing runoff for each subcatchment, ostensibly representing sheet flow. Equation 1 shows the overland flow equation used in SWMM5.

[1] $Q = W * \frac{1.49}{n}\left(d - d_p\right)^{5/3} S^{1/2}$

Where:

W = Subcatchment width, ft.

n = Manning's roughness

d = water depth, ft.

d_p = depth of depression storage, ft.

S = Subcatchment slope, ft./ft.

Once runoff or overland flow reaches the subcatchment outlet, the flow becomes channelized and is transported through a system of pipes, channels, storage/treatment devices, pumps, and regulators.

Watershed delineation analysis tools, including the WDT, are available in PCSWMM and allow for seamless automated watershed and hydraulic network creation, parameterized for use in SWMM.

First released in 1984, PCSWMM is a spatial decision-support system for the US EPA SWMM program that was built upon a GIS engine, making it a powerful interface for developing models using GIS-based input data. The software provides all the hydrologic-hydraulic-water quality computational capabilities of US EPA SWMM5 while providing many additional tools for easier project management, model parameterization, calibration, data analysis, results inference and reporting.

3 Watershed Delineation Tool (WDT)

The Watershed Delineation Tool (WDT) provides a versatile way to delineate subcatchments based on a digital elevation model (DEM). It can be used to create models from scratch, or to increase the level of discretization for existing models.

In comparison to existing watershed delineation software, the WDT goes beyond watershed delineation, generating a dendritic SWMM network of subcatchments, junctions and conduits, partially parameterized from DEM features. If there are existing hydraulic entities (links and nodes) in the SWMM5 project, these entities are preserved and additional hydraulic components are generated as is necessary to connect the WDT-discretized subcatchments to the existing network.

In addition, while the watershed delineation tool may produce somewhat similar results as other watershed delineation tools, actual subcatchment delineations will differ as this approach uses the concept of a target subcatchment size, rather than a minimum area for channelization. The WDT also provides an option to burn-in streams (using a stream line layer) to create a hydrologically-corrected DEM, and can be set to discretize subcatchments to a set of predefined delineation (or "pour") points. At this time, elevation contours in a vector layer cannot be used to provide elevation data for the WDT.

The watershed delineation tool uses sequential computations of flow direction, flow accumulation, and stream definition based on a threshold and watershed delineation (Jenson and Domingue, 1988; Maidment, 2002). The results of each process are saved to a grid layer for review by the user. All created grid layers are saved to the same folder as the selected DEM layer in Arcinfo binary grid format (FLT).

If there are outfalls defined in the model, watersheds are delineated considering outfalls as drainage points. If there are no outfalls defined, watersheds will be delineated for the whole extent of the DEM. If there are existing drainage entities (links/nodes), the WDT created entities will be connected to the existing system where appropriate.

The remaining overland flow parameter, subcatchment slope, is also assigned by the WDT. The average subcatchment slope is estimated based on the flow direction for each cell. To ensure the WDT does not bias the slope because of micro-undulations in the DEM, a second coarser resolution DEM can be generated to reduce this effect and provide a more accurate hydraulically effective slope calculation.

To estimate flow length, the WDT uses the Guo and Urbonas (2009) method that applies a kinematic wave approach to convert irregular shaped subcatchments to equivalent rectangular planes. Using this method, subcatchment length values are estimated using a kinematic shape factor assigned based on the stream location in relation to the subcatchment and longitudinal elevation drop of the receiving concentrated flow channel. Thus, application of the WDT helps to avoid excessively long overland flow paths that violate the sheet flow approximation, and helps to overcome rectangular subcatchment shape limitations.

4 WDT processes

The following section goes through the WDT processes automated by the tool. In many cases, a resultant raster layer is created and used for the next process, ultimately delineating the hydrology and hydraulics networks.

4.1 Delineation points (optional)

The watershed delineation tool allows users to define an optional delineation points layer to generate individual subcatchments draining to each point in the layer. A delineation points layer (also called a pour points layer) is used to specify the outfall/outlet points, which will affect flow direction calculation. This option can be used for subcatchment or sewershed delineation in urban areas where surface runoff drains to manholes or catchbasins (Junctions), or to significant hydraulic structures such as ponds, bridges or culverts. Using a delineation points layer is not recommended for a model with no existing hydraulic network (i.e., minor drainage system).

4.2 Stream burn-in (optional)

The Stream burn-in option carves a trench in the original DEM along a specified line layer. The approach uses the AGREE DEM approach (Hellweger, 1997) for this purpose. It involves a sharp drop of 100 meters along the stream line and a smooth drop of 10 meters within a specified stream width. This allows for the removal of any dams created by captured DEM elevations of road crossings and actual dams.

4.3 Filling pits

Any local low spots in the burned DEM introduce difficulty for continuous flow over the DEM cells. Filling pits ensures a "hydrologically conditioned" DEM. The Planchon and Darboux (2001) approach is used to remove all the pits/depressions. During the pits removal operation, if outfalls exist, these outfalls will be used as a mask to preserve their elevations (this is to prevent the burnt-in stream from being filled up again). A new DEM layer, "WDT Fill Pits" will be saved.

4.4 Flow direction

Flow direction is defined for all cells by considering the direction of water flow out of the cells. This is based on the steepest slope to any of the eight adjacent cells. Thus, the methodology is called D8 (Jenson and Domingue, 1988; Tarboton, 1997).

Assigning flow direction is challenging within flat areas. Analyzing flow direction in flat areas is based on Barnes et al. (2014), and provides similar results to Garbrecht and Martz (1997), but is more efficient. The flow direction grid assigns each cell a number that indicates the direction of the flow, that is, the direction that water will flow from that particular cell. These grids are created and will find the lowest neighboring cell for every 3x3 cell neighbourhood. A new raster grid layer, "WDT Flow Direction" will be saved.

4.5 Slope

The Slope layer is used to generate average slope for delineated subcatchments. Subcatchment width is calculated using Guo and Urbonas (2009) method. Slope is calculated based on flow direction. For each cell, if the flow direction is given (there is elevation difference), slope can be calculated as elevDiff / distance. A new raster grid layer, "WDT Slope" will be saved.

4.6 Contributing area

Based on the flow direction grid, a "Contributing Area (or flow accumulation)" DEM layer (the number of grid cells draining through each grid cell) is developed. The contribution area calculates the flow into each cell by accumulating the cells that flow into each downslope cell. The contributing area for each grid cell is taken as its own contribution plus the contribution from upslope neighbours that drain to it according to the D8 flow model. A new raster grid layer, “WDT Contributing Area” will be saved.

4.7 Stream threshold

Each pixel is assigned with a value of 0 (shown in green) and 1 (shown in red). If a pixel is assigned with a value of 1 it means that the area draining into that pixel is significant enough to consider it part of the river network. A new raster grid layer, “WDT Stream Threshold” will be saved.

4.8 Flow Paths

The flow paths for runoff over the subcatchments based on the stream threshold layer used to determine subcatchment flow lengths. A new line layer, “WDT Flow Paths” will be saved.

4.9 Watershed creation

Watersheds are delineated using the stream grid and contributing area. Separate watersheds are delineated at any stream line confluence points. If there are outfalls, all outfalls will be relocated to nearest stream before watershed delineation. Subcatchment delineation is optimized to reduce small subcatchments by joining them and further discretizing larger subcatchments using a smaller stream threshold for the larger watershed areas (based on internal parameters). Subcatchments boundaries are smoothed to remove the jagged lines. A new raster grid layer, “WDT Watersheds” will be saved.

4.10 Creating a new model

Created watersheds will be used as subcatchments for the hydrologic model. Before watershed delineation, if there are existing subcatchments, they will be moved to a background layer called “Old Subcatchments”. It will be rendered as red lines with clear fill. The newly delineated subcatchments will be rendered as blue lines with clear fill.

After watershed delineation, the newly generated hydraulic network (junctions and conduits) will be merged into existing network if one exists. Existing conduits will be rendered as solid yellow lines, and newly generated DEM conduits will be rendered with dotted red lines. Some existing conduits could be split to allow connection to a new branch. The DEM layer is used to get invert elevation for new junctions.

4.11 Attributes for WDT entities

The following attributes are calculated by the WDT for all newly created entities. All other attributes values are populated based on the settings in the Defaults dialog. Existing SWMM5 entities are not edited unless a conduit is split to allow lateral inflow from the WDT network. If existing hydraulics are in place, the attribute values entered by the user remain the same.

5 Guo and Urbonas (2009) method

Guo and Urbonas conducted a study in 2009 to determine a method to better parameterize subcatchments for analysis using kinematic wave (KW) routing. In this study, the continuity and energy principles were interpreted as the preservation of watershed area and vertical fall over the receiving waterway's length. In the analysis, the watershed shape factor was used to correlate the waterway length in the natural watershed to the KW plane width. A parabolic one-to-one single valued function was derived to translate the watershed shape into its equivalent KW plane width, length, and slope. After numerous tests, it was confirmed that the watershed and KW shape factors provide a consistent and stable basis for watershed geometric conversion. (Guo and Urbonas, 2009)

The KW approach is to apply the one-to-one single valued rating curve relationship to the overland flow generation (Guo 2006). Similarly, the one-to-one relationship needs to be established between watershed and KW shape factors. The study had several conclusions which are summarized below:

The shape factor, X, of a natural watershed is defined by:

[2] $X = \frac{A}{L^2} \cong \frac{B}{L} (\leq K)$

Where:

A	=	tributary area [L^2]
B	=	watershed width [L]
L	=	length of collector channel [L]
K	=	limit for KW shape factor

The KW shape factor, Y, for the equivalent KW plane is defined by:

[3] $Y = \frac{L_W}{L}$

Where:

Lw	=	width of the KW plane [L]

The relationship between these two shape factors, X and Y, is described by Eq. 4 and 5 that was derived based on the channel alignment and the application limit on X. The area skewness coefficient, Z, varies from 0.5 for central channel to 1.0 for side channel. As recommended, the application limit, K, varies from 4 to 6.

[4] $Y = (1.5 - Z)\left(\frac{2}{1-2K}X^2 - \frac{4K}{1-2K}X\right)$

and

[5] $X = \left(\frac{Am}{A}\right)$

Where:

Z	=	the area skewness coefficient
A_m	=	dominating area [L2]

The energy potential along the water course is preserved by the Eq. 7 and 8 (shown below) between the natural watershed and KW cascading model.

[6] $\left(\frac{S_O}{S_W}\right) = \frac{X}{Y} + Y(X \leq K)$

and

[7] $Y = aX^2 + bX + c$

Where:

a	=	constant used in shape factor relationship
b	=	constant used in shape factor relationship
c	=	constant used in shape factor relationship
S_O	=	longitudinal slope along waterway through watershed
S_W	=	slope on KW plane [L/L]

Using this method, subcatchment length values are estimated using a kinematic shape factor assigned based on the stream location in relation to the subcatchment and longitudinal elevation drop of the receiving concentrated flow channel. Thus, application of the WDT helps to avoid excessively long overland flow paths that violate the sheet flow approximation, and helps to overcome rectangular subcatchment shape limitations.

6 Case Study

The following case study compares two similar models of the Don River watershed. Situated in Southern Ontario, the Don River watershed has an area of ~ 36000 ha and is home to 1.2 million residents. The first model, referred to heron in as the original model, has a relatively coarse resolution with 67 subcatchments, 2842 conduits and 2472 junctions. The second model was built using the WDT and had a significantly higher resolution, specifically in the hydrology portion of the model, with 475 subcatchments, 2938 conduits and 2581 junctions.

The following section reviews the model background, setup and comparison of the two models.

6.1 Model background and comparison

The hydraulic network in the original Don River SWMM5 model was assembled using 35 existing HEC-RAS basin models, with some modifications based on stormwater infrastructure updates. Stormwater management (SWM) ponds were included in the model and were based on a "lumped" SWM pond modelling procedure. The hydrology portion of the model was based on a subcatchment layer and the attribute values were estimated using area-weighting operations from a soils and land-use map. The remaining subcatchment attributes were estimated using a provided digital elevation model (DEM). The model was calibrated using observed rainfall and flow monitoring data to a level suitable for predicting the peak flows for large events.

The second model was built using the WDT and allowed every subcatchment to have an outlet node directly downstream. In this model, the original hydraulics were kept and additional branches were added to the upstream areas of the watershed. Lumped stormwater ponds from the original model were edited to match the resolution of the new WDT subcatchments. Hydrology parameters were assigned using the same soils and land-use layers used in the original model.

Both sets of model parameters were calibrated separately and optimized to best match observed flows, using the same level of effort.

A total of 10 events were selected for the comparison and ranged from 10.2 mm – 51.8 mm. Figures 2 and 3 show the resultant hydrographs for the two largest events compared. In both plots the peak flows in the detailed model better match the observed flows when compared to the coarse model. In addition, both the receding limbs and climbing limbs were improved in the detailed model.

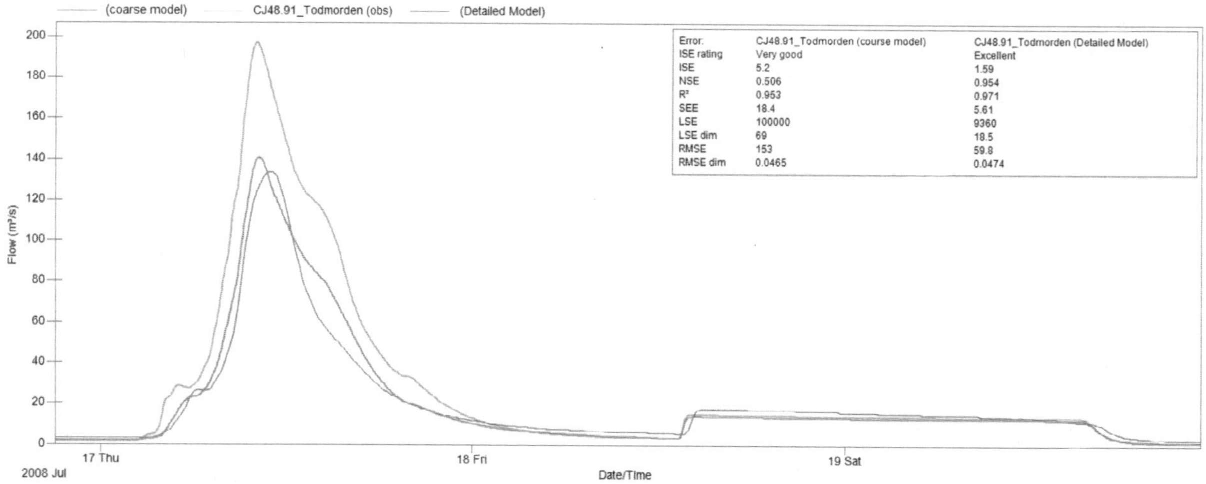

Figure 2 : Event 1 comparison showing hydrographs for both the coarse and detailed models

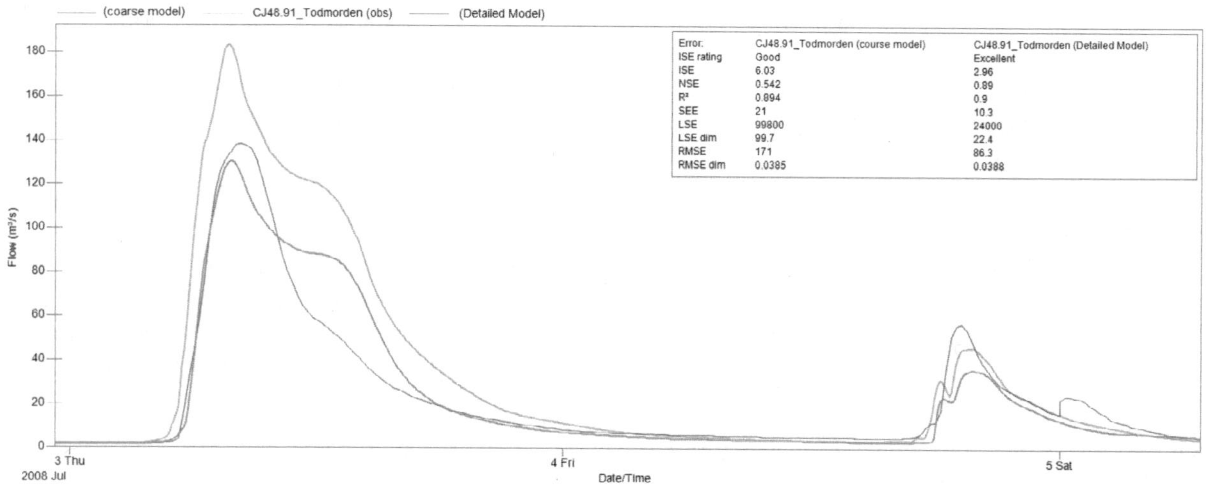

Figure 3 : Event 2 comparison showing hydrographs for both the coarse and detailed models

Figures 4 and 5 show the event comparison plots for the event peak and totals. In both comparisons, the detailed model better matched the observed flows when compared to the original model.

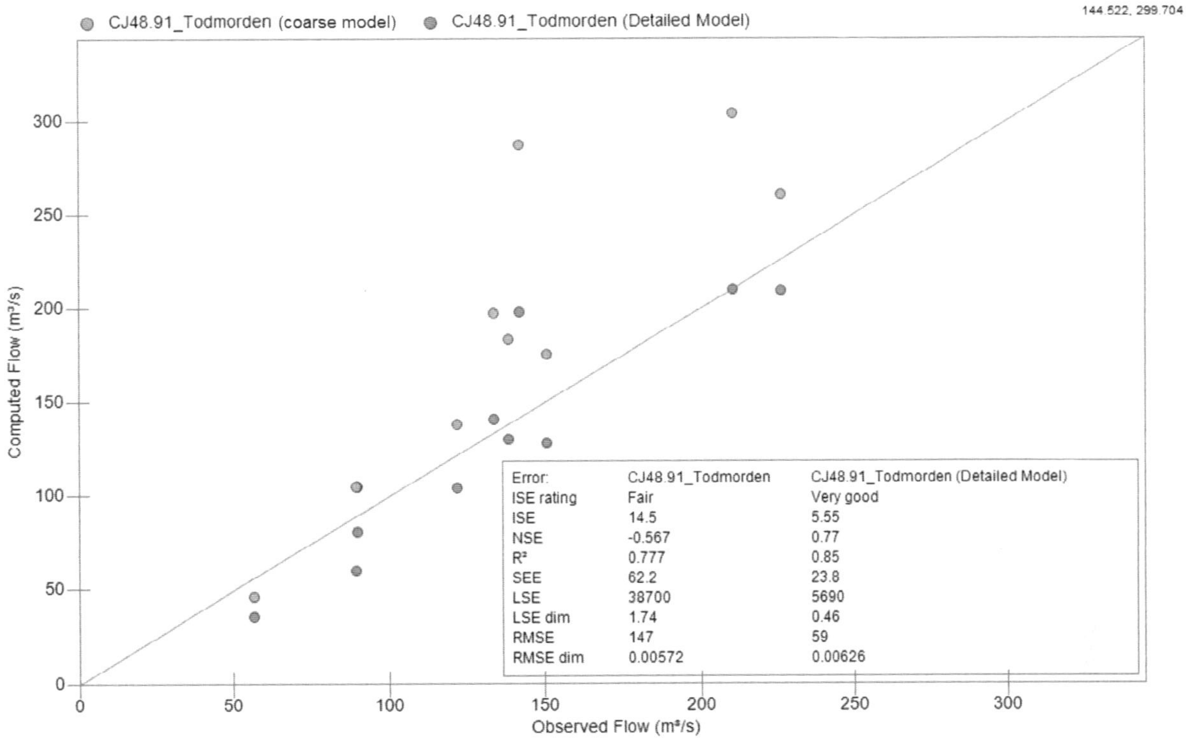

Figure 4 : Peak flow event comparison for coarse and detailed models

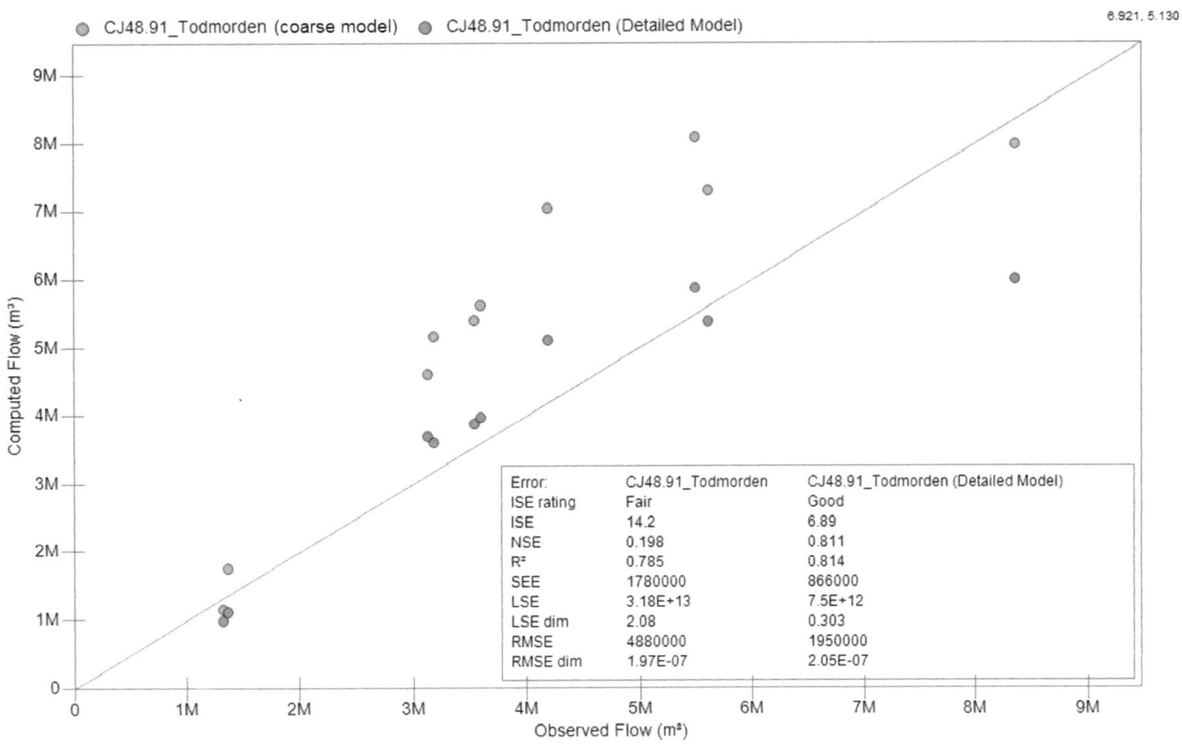

Figure 5 : Total flow event comparison for coarse and detailed models

7 Conclusion

The WDT automatically generates subcatchments and parameterizes elevation based attributes such as slope and overland flow length using a DEM layer and a user-defined target discretization value. By further discretizing the watershed, the processes involved with channelized flow and sheet flow can be better represented.

Although limited modeling budgets and short time lines likely are here to stay the technology and the data resources to improve model calibration and validation is available.

Creating a higher resolution model that better accounts for the watershed processes, particularly in headwater areas, result in a reliable tool that can be used for numerous future applications and automating this process can help reduce model error while staying within existing budget and time constraints.

Significant infrastructure expenditures are based on the application of stormwater models and it is therefore important to have a model capable of predicting flows for a variety of events. For the modeler, the benefits of a high-resolution calibrated / validated model include: better understanding of the modeled system; improved confidence and reliability in model results, model credibility and wider acceptance of engineering recommendations.

Acknowledgements

The authors would like to acknowledge that portions of the description of the WDT discussed in this paper is published on PCSWMM's support site (www.support.chiwater.com). Note the authors of paper were also the authors of the support material.

References

Barnes,R., Lehmanc, C., and D. Mulla. 2014. An efficient assignment of drainage direction over flat surfaces in raster digital elevation models. Computers and Geoscience, 62: 128-135

Garbrecht, J. and L.W. Martz. 1997. The assignment of drainage direction over flat surfaces in raster digital elevation models. Journal of Hydrology, 193: 204-213

Guo, J.C.Y. 2006. Storm water predictions by dimensionless unit hydrograph. Journal of Irrigation and Drainage Engineering, 132(4): 410–417

Guo, J.C.Y. and B. Urbonas. 2009. Conversion of natural watershed to kinematic wave cascading plane. Journal of Hydraulic Engineering. 14(8): 839-846

Hellweger, F. 1997. http://www.ce.utexas.edu/prof/maidment/gishydro/ferdi/research/agree/agree.html

Jenson, S. K., and J. O. Domingue. 1988. Extracting Topographic Structure from Digital Elevation Data for Geographic Information System Analysis. Photogrammetric Engineering and Remote Sensing, 54 (11): 1593–1600

Maidment, D.R. 2002. Arc Hydro: GIS for Water Resources. ESRI Press.

Planchon, O., and F. Darboux. 2001. A fast, simple and versatile algorithm to fill the depressions of digital elevation models. Catena, 46: 159-176

Rossman, Lewis A. 2008. Storm Water Management Model User's Manual Version 5.0. Pub. Water Supply and Water Resources Division, National Risk Management Research Laboratory, Cincinnati, OH.

Tarboton, D.G. 1997. A new method for determination of flow directions and upslope areas in grid digital elevation models. Water Resources Research. 33(2): 309-319

Leadership in Sustainable Infrastructure
Leadership en Infrastructures Durables

Vancouver, Canada
May 31 – June 3, 2017/ *Mai 31 – Juin 3, 2017*

BOUNDARY DAM TDG (TOTAL DISSOLVED GAS) ANALYSIS USING A CFD MODEL

Jalayeri, Nikou[1, 5] , Groeneveld, Joe [2], Bearlin, Andrew [3], and Gulliver, John [4]
[1] Engineer, Hatch, Vancouver, British Columbia, Canada
[2] Senior Engineer, Hatch, Calgary, Alberta, Canada
[3] Project Manager, Seattle City Light, Seattle, Washington, USA
[4] Consulting Engineer, University of Minnesota, Minneapolis, Minnesota, USA
[5] nikou.jalayeri@hatch.com

Abstract: The Boundary Dam is located on the Pend Oreille River in northeastern Washington. The project consists of a 340 ft. high concrete arch dam, seven low level sluiceway outlets, two high level overflow spillways, and a 660 MW powerhouse. The spillway and sluiceway discharge at the Boundary Hydroelectric Development have been shown to produce high total dissolved gas (TDG) concentrations in the tailwater of the spillway and the river reach downstream. Studies were commissioned to determine modifications to the project's spillway structures to help mitigate this gas production. Resolution of many of the hydraulic design issues for the study relied heavily on the results of numerical hydraulic models. These modifications were constructed and tested in the field. The CFD model that was developed in support of these studies was used to simulate flows through a number of the project's seven sluice gates and two overflow spillways. This model was also used to simulate the entry and movement of these flows through the project's downstream plunge pool and powerhouse area. The model was set up to track the pressure- and time-histories of representative air bubbles within the plunge pool and tailrace. These data were then used as input to a TDG predictive tool to help predict total dissolved gas production in the tailrace. The overall predictive performance was successfully calibrated and validated to actual prototype (field) TDG data.

1 Introduction

Boundary Dam is the third of five dams on the Pend Oreille River. The Project operates in a load-following mode, generating power during peak-load hours and curtailing generation during off-peak hours.

Figure 1 shows the key project features, consisting of an arch dam, reservoir, and underground power plant. Boundary Dam is a variable-radius concrete arch dam with a structural height of 340 feet, a crest length of 508 feet, and a total length of 740 feet. The arch dam varies in thickness from eight feet at the crest, which is at elevation 2,000 feet (NGVD 29) to 32 feet at the base at elevation 1,644 feet. The normal tailrace water surface elevation below the dam is 1,733 feet with approximately 261 feet of gross head for power purposes for a normal upstream pool level.

Boundary Reservoir extends 17.5 miles upstream to Box Canyon Dam. Total usable storage is approximately 41,000 acre-feet from the normal maximum water surface elevation of 1,994 feet to 1,954 feet as the 40-foot operating range authorized by the current license. Figure 2 shows a downstream view of the dam looking upstream. Spillway and sluiceway discharge at the Boundary project has been shown to produce high Total Dissolved Gas (TDG) concentrations in the river reach downstream. TDG is the amount of air held in saturation in the water and is quantified in the standards in terms of percent of saturation pressure relative to ambient barometric pressure. These high gas levels can be detrimental to the local fish species.

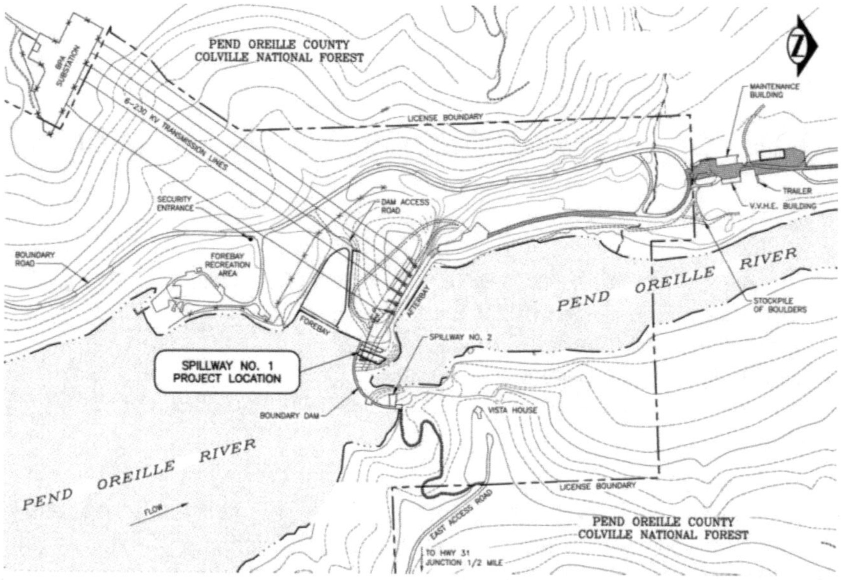

Figure 1 Boundary Dam Key Project Features

2 Planning of TDG Upgrades

The Seattle City Light (SCL) TDG team has convened a series of design workshops to brainstorm and develop alternatives, review Project progress, discuss technical aspects of the work, evaluate alternatives, and provide direction for ongoing development of operational and structural measures to reduce TDG downstream of the Project. The initial development of alternatives occurred during a two-day workshop in 2007, attended by SCL staff and an expert panel. During this initial workshop a large number of potential mitigation measures were proposed. SCL and the expert panel selected six alternatives that they considered to be feasible and likely to be effective, and presented these in relicensing study plan reports (SCL 2007 and SCL 2009a); these are the alternatives on which the SCL TDG team has focused their subsequent design development efforts. Since 2009, the TDG mitigation design team holds an annual design workshop, and provides input to or attends periodic agency meetings. The annual team meetings are a forum to review the years' progress and significant design advances, to brainstorm ideas and discuss challenges, and to establish the goals and prioritize the studies and design development for the following year.

The TDG abatement alternatives identified during the design workshops held in 2009 were investigated from the perspective of hydraulic performance and engineering feasibility. The three short-listed alternatives selected during this process (SCL 2009) were:

- Throttle Sluice Gates
- Roughen Sluice Flow
- Install Flow Splitters/Aerators on Spillways
    - Spillway deflectors, roughness elements, and/or dentated spillway

Figure 2 Downstream View of Boundary Dam Looking Upstream

Based on both physical and numerical hydraulic analysis, as well as engineering considerations, proposed modifications to Spillway No. 1 (SW1) and Spillway No. 2 (SW2) were selected for early implementation. SW2 modifications were scheduled first because SW2 had more favorable foundation conditions that would require a simpler analysis than that of SW1. Additionally, SW2 produced more TDG and was shorter, giving better access for construction of modifications.

The spillway modifications consisted of adding Roughness Elements (REs), similar to baffle blocks, to the spillway to induce turbulence, to accelerate break-up of the falling jet and to reduce jet penetration into the tailwater. The spillway REs were selected as the preferred first alternative to be implemented. Figure 3 shows the installed baffle blocks on the end of the SW2 chute.

Figure 3 View of REs Installed on SW2 Chute

3 Hydraulic Analysis and Design

Resolution of many of the hydraulic design issues for this concept relied heavily on the results of both physical and numerical hydraulic models. Both were used in complementary roles to maximize their particular strengths. The focus of this paper is on the numerical modeling portion of the work.

3.1 Computational Fluid Dynamics (CFD) Model

Full three-dimensional CFD models of SW1, SW2, a sluiceway, and the downstream plunge pool and river channel were developed in 2009. These models have been utilized as part of the suite of design tools used to develop an understanding of the governing hydraulic and hydrodynamic processes driving gas exchange in the tailrace of the existing project under spill conditions. In addition, these models been used to develop the designs of structural TDG mitigation alternatives, and in combination with the TDG predictive model described in subsequent sections of this paper, to predict the TDG performance of proposed TDG mitigation alternatives. The CFD models were used in conjunction with the physical model utilizing the relative strengths of the two analysis methods. The physical model was used to its greatest utility in iteratively developing physical geometry changes of the mitigation alternatives, while the CFD models were used to provide quantified information in terms of hydrodynamic loads for design purposes and flow fields for use in TDG prediction.

3.2 TDG Prediction model

TDG predictions were made for the project using a two step process: the CFD model was first applied to assess the plunge pool hydraulics and flow patterns, and then the hydraulic output of the CFD model was imported into the Plunge Pool Gas Transfer (PPGT) model on an Excel spreadsheet.

As a first step, the CFD model was run to simulate the hydraulics within the plunge pool area of the project for a particular configuration and flow scenario. The model used, FLOW-3D, is capable of simulating and tracking the movement of individual particles of air within a computational grid. This tracking feature provided a means to better understand how air, drawn in at the water surface, might circulate within the downstream channel and plunge pool. To do so, numerical models are initially run to achieve a quasi-steady flow regime downstream of the site where the mean velocities did not change appreciably with time. This is known as a "spin up" period. After this, the models are restarted and a number of discrete particles are released within the plunging jet. The particles are released in lines across the main flow stream at locations around the periphery of the jet or nappe to replicate the actual points at which air will become entrained.

The particles themselves are 0.02 ft in diameter, and are provided with a buoyancy that will propel them upwards at a velocity of 0.2 m/s (0.7 ft/s) in still water. This replicates the average still-water rise rate of entrapped bubbles. The model can then be used to track the movement of these particles. Where there are high downward vertical velocities, the particles tend to be drawn to depth. The buoyancy of the particles will eventually bring each back to the surface where they will be released into the atmosphere.

At the completion of each simulation, the particle data is extracted from the output database. The extracted data includes the position of the particle, the elapsed time, the particle velocity (x, y, and z directional velocities as well as total velocity magnitude) and the pressure at the particle location. Of these, the depth versus time and total velocity magnitude data are the most important for the subsequent calculation of TDG content. A large velocity generates greater gas transfer, and those particles that are drawn to depth, and remain at depth for periods of time, will contribute the most to TDG increases.

When the CFD simulations were completed, they were then imported into the PPGT model, where the transfer of air between the bubbles and the water is computed based on the overall depth/pressure time series that was tracked for the mass particles. When the overall mass transfer of the particles approaches zero, the steady state concentration of the tailwater is achieved in the computations. The PPTG model makes the following assumptions:

- There is sufficient air entrainment so that the rate of air entrainment is not a limiting factor. This means that there is sufficient air entrained to achieve a steady state concentration.
- The tailwater pool is well mixed.
- TDG concentration in the tailwater pool has reached steady state, in that any air added from the bubbles into the water at depth is balanced by an equal amount taken out of the water by the shallow bubbles, or that an overall mass balance has been achieved.
- The transfer of dissolved gas across the water surface is negligible. This assumption will cause the PPTG model to predict tailwater TDG that is above measurements.

Using this strategy, the transfer of air across a moving bubble-water interface is computed as:

$$[1]\ \frac{dM}{dt} = K_L A(C - C_E)$$

where;

dM = the amount of air transferred into the dissolved phase,

dt = the time interval,

K_L = transfer coefficient for nitrogen and oxygen,

A = the surface area of the bubble

C = the concentration of air within the water

C_E = the equilibrium water concentration, which is directly related to concentration in the bubble.

The problem is thus formulated so that each bubble is followed through the pool. The bubbles should be placed at the locations where the air would be entrained, and followed through the pool until they leave at the surface. Dividing Eq. (1) by the equilibrium concentration at atmospheric pressure, *Cs*, gives:

$$[2]\ \frac{1}{C_S}\frac{dM}{dt} = K_L A\left(\frac{C}{C_S} - \frac{C_E}{C_S}\right)$$

or

$$[3]\ V\frac{d\left({}^{C}/_{C_S}\right)}{d_t} = K_L A\left(\frac{C}{C_S} - \frac{C_E}{C_S}\right)$$

Where C/Cs and C_E/Cs are fractions of atmospheric saturation, and V is the volume in which C is determined ($M = C \cdot V$). The surface area of the bubble will decrease with the increase in pressure at depth:

$$[4]\ A \cong A_0\left(1 + \frac{depth(ft)}{34}\right)^{2/3}$$

where A_0 is the surface area of a given bubble at zero depth. An additional assumption in Eq. (4) is that the surface area of each bubble is independent of the amount of mass that has previously been transferred. This assumption may affect the surface area of each bubble, but would not violate the overall mass balance of all bubbles because the system is at steady state with regard to TDG concentration.

Equation (2) allows one to compute the concentration in the tailwater at steady state, when the sum of *dM/dt* for all bubbles is equal to zero. Although each bubble may have a non-zero value of *dM/dt,* the sum of *dM/dt* for all bubbles at all times must be equal to zero for the tailwater to be at steady-state TDG concentration. For example, with SW1 operating at 12,000 cfs, and SW2 operating at 10,000 cfs, the average residence time of the bubbles in the tailrace model domain is approximately 40 seconds. In the computations, bubble location and velocities were output every 0.5 seconds. Thus, with 50 bubbles, average residence time of 40 seconds, and mass transfer computations every 0.5 seconds, a total of 50 x 40 s / 0.5 s = 4,000 separate computations of mass transfer as the bubbles are followed through the flow. Some of these mass transfer calculations will be positive (adding TDG to the water) and some will be negative (stripping TDG from the water), depending upon the depth of the bubbles and the TDG concentration assumed for the tailwater. The value will also depend upon the velocity of the bubble and, to a lesser extent, upon water temperature. The tailwater TDG level is changed until the total sum of these 4,000 mass transfer calculations is zero, which means that steady state and an overall tailwater equilibration of mass transfer has been achieved.

For a bubble at depth, assuming limited change in the molar ratio of oxygen and nitrogen in the bubble,

[5] $$\frac{C_E}{C_s} \cong 1 + \frac{depth(ft)}{34}$$

Azbel (1981) developed an equation for the mass transfer coefficient from a bubble swarm, which Thompson and Gulliver (1997) altered to result in an equation that could describe the mass transfer from bubbles in a shear flow:

[6] $$K_L = (2\pi D)^{1/2} \frac{U^{\eta}}{L^{1-\eta} v^{\eta - 1/2}}$$

where;

D = diffusion coefficient of air in water ($\sim 2 \times 10^{-8}$ ft^2/s),

U = a representative turbulent velocity,

L = a representative length scale,

v = kinematic viscosity of the water ($\sim 10^{-5}$ ft^2/s)

η = a coefficient that generally varies between 0.5 and 1.

Azbel (1981) used η = 0.75. Thompson and Gulliver used η = 0.55. The difference in computed TDG with the two exponents is not great, so 0.75 is used in these computations.

The length scale, L, is assumed to be the same for both the sluice and spillway discharge, and is typically set at 60-75% of the smallest discharge dimension. L has a small effect on predicted TDG levels.

The turbulent velocity, U, is the velocity of the buoyant particle, calculated from FLOW-3D.

4 Validation of the TDG Tool

TDG measurements have been made at various locations and times of the year. These tests have been grouped into operational scenarios involving only sluice flow, and operational scenarios involving only discharge through on or more of the high-level spillways. The data has been further subdivided in to tests that showed some stripping of the TDG, tests in which downstream TDG levels remained neutral, and tests in which the downstream TDG was seen to rise significantly.

Upon reviewing the data, a number of test cases were selected to test the performance of the TDG predictive tool, i.e., the CFD and PPTG models. These cases involved operation of each spillway individually, as well as cases involving the combined operation of both.

One such test case involved the operation of SW1 in tandem with the project powerhouse facility. The first scenario represents an operating condition which occurred on June 9, 2007, in which 10,220 cfs was passed through SW1 and 45,110 cfs though the powerhouse. The incoming TDG level for this test was 121.5% and the outgoing TDG was measured as 127.5%. A second more recent test case was also simulated.

The model was setup to replicate both of these operational scenarios, and the results of these validation runs are shown in **Error! Reference source not found.** below. As shown, although the simulation tool appears to over-predict the TDG levels at the monitoring site in both cases, (by up to approximately 2.5 percent), the overall match was considered to be quite reasonable. It should be noted that the field data also show considerable variation in the measured TDG level at the USGS gauge, even for comparable flow conditions.

Table 1 Summary of Spillway 1 Validation Runs

Spillway 1 Q (cfs)	Powerhouse Q (cfs)	Total Q (cfs)	Upstream TDG	Tailrace TDG (Measured)	Tailrace TDG (TDG Tool)	Difference in TDG
10,000 (June 9, 2007)	45,110	55,110	121.5%	127.2%	129.7%	-2.5%
13,000 (June 20, 2016)	24,000	37,000	106.0%	124-125%	125.6%	-0.6%

4.1 Application of the TDG Tool

The validation tests have shown that the developed tool, based on the particle tracking modeling technique, does a reasonable job of predicting TDG levels downstream of the project. The next step in the assessment was to use the TDG tool to predict the TDG reduction that might be possible at the project's downstream pool with and without the proposed modifications in place. This allows a relative comparison to be made on the effectiveness of each proposed modification in reducing TDG levels.

For these tests, it was assumed that a flow of 48,000 cfs would be released through the powerhouse. The reservoir level for the tests was assumed to be at the project's Full Supply Level of elevation 1990 ft, and the incoming TDG level was assumed to be 106%. The following sections of this paper summarize test results for the proposed SW1 modifications, which involve the installation of a single row of roughness elements to the downstream chute.

4.1.1 Existing Case - No Modifications

The model was first run to simulate flow conditions for the existing or base case scenario with flows of 10,000, 13,000, and 20,000 cfs through SW1. The simulated hydraulic conditions for this test were analyzed. Bubble particles were then added to this model, the run was re-started, and the particles were tracked until they were able to reach the surface, and exhaust back into the atmosphere. Figure 4 shows the overall nature of the jet as it falls into the plunge pool for flows of 10,000 and 13,000 cfs. As shown, the spread of the jet is quite limited. The particle data was then extracted from the model, and input to the spreadsheet model.

The results for these tests are summarized in Table 2 below. Based on these results, it is expected that with SW1 operating, TDG levels downstream of the project will rise significantly. This is to be expected, since the maximum and average particle plunge depth increased with larger flow releases through SW1. For the smaller flow of 10,000cfs, the spillway jet is more easily broken up during its fall into the plunge pool, whereas for higher flows, while the edges of the jet may break up, a competent jet core persists through the jet's entry into the plunge pool. The larger, more competent core associated with these higher flows is able to penetrate to the pool bottom, bringing entrained air to a much greater depth.

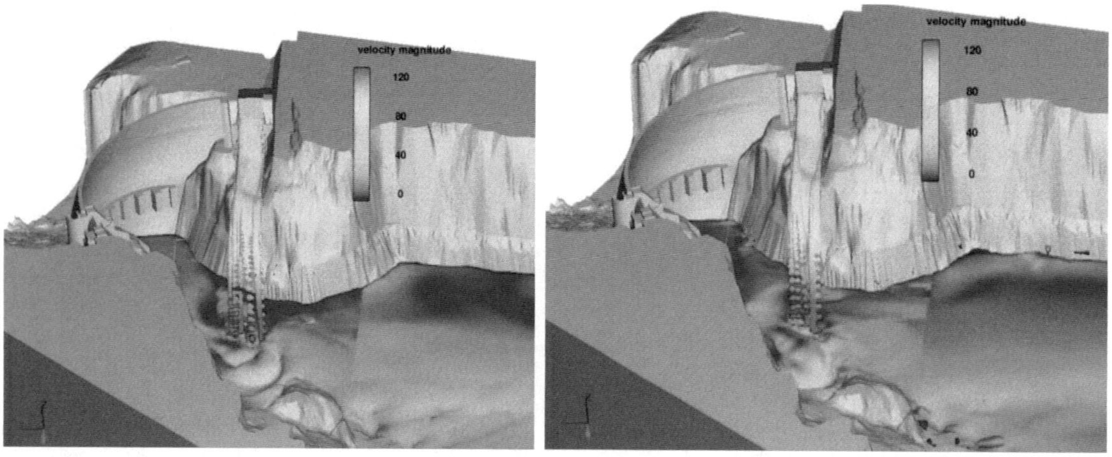

Figure 4 3-D View of the Unmodified Spillway 1 Jet : 10,000 cfs Flow (left), 13,000 cfs Flow (right)

4.1.2 SW1 Modification

Following the base case runs, various CFD simulations were conducted to assess the hydraulic conditions that would result from the introduction of REs on the downstream end of the SW1 chute. The introduction of these REs helps to break up the jet at the end of the chute more quickly and efficiently, accelerating boundary layer growth and resulting in the formation of small "packets" of water entering the plunge pool rather than coherent streams/jets.

Given concerns for potential cavitation damage on the spillway chute floor and on the REs themselves, additional runs were undertaken to test the effect on flow conditions at the REs if a ramp were to be installed immediately upstream of the roughness elements. The ramp would rise from the floor of the chute, to meet the upstream face of the first row of roughness elements, as shown in Figure 5 below.

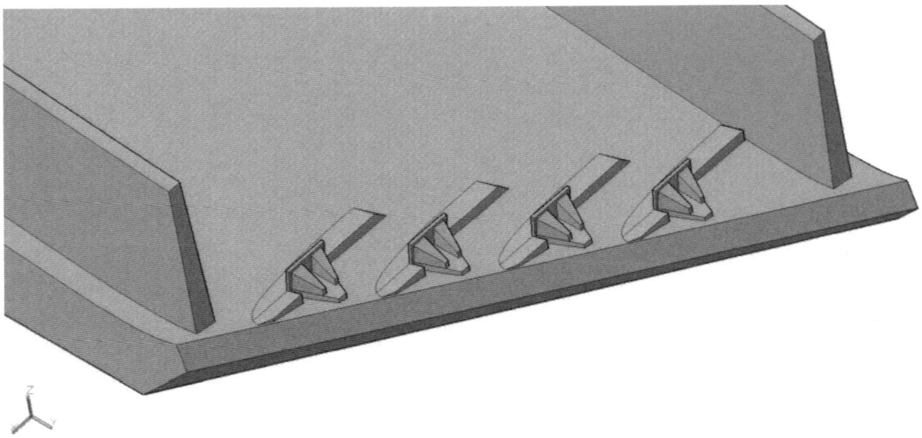

Figure 5 3-D View of Spillway No.1 Roughness Elements

The next step in the assessment of these schemes was to estimate the impact that the addition of these modifications would have on TDG levels downstream of the project under a range of flows. To do so, three predictive runs were made. For each run, it was assumed that 48,000 cfs would again be passed through the powerhouse. The reservoir level for each test was assumed to be at the project's Full Supply Level of elevation 1990 ft, and the incoming TDG level was again assumed to be 106 % of saturation. CFD runs were made with identical flow releases through SW1 of 10,000 cfs, 13,000 cfs, and 20,000 cfs, and bubble histories were extracted from the CFD results and input to the TDG predictive spreadsheet model.

The results of the CFD runs are presented in Figure 6 below. Comparing Figures 4 and 6 indicates that the plunging jet is more broken up and spread further with RE. **Error! Reference source not found.** also provides a final summary table for the RE runs, and compares key parameters associated with each flow condition, including the maximum plunge depth, the average plunge depth, the predicted TDG percentage at the USGS gauge downstream of the project, and the TDG contribution above incoming levels by the spill release.

As shown in Table 2 the proposed RE configuration appears to deliver the greatest TDG reduction when SW1 is operating at a flow of approximately 10,000 cfs. With the roughness elements in place, the reduction in TDG value for this flow is approximately 6.6 percent when compared to baseline conditions. For 13,000 cfs, the expected TDG reduction below the baseline condition is smaller, and estimated to be approximately 4.9 percent. For the 20,000 cfs, the expected reduction from the baseline condition is estimated to be approximately 2.4 percent. For the 20,000 cfs flow case, the ability of the roughness elements to break up the jet appears to be reduced, since the jet begins to override the roughness elements. This results in the formation of a more competent jet core that is able to penetrate the plunge pool to a greater depth.

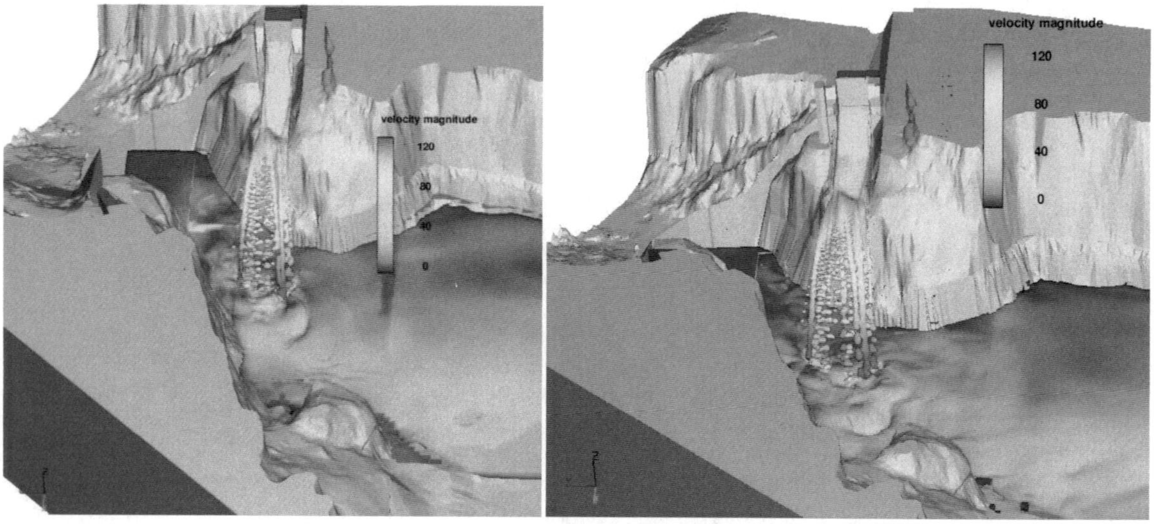

Figure 6 3-D View of the Modified Spillway 1 Jet : 10,000 cfs Flow (left), 13,000 cfs Flow (right)

Table 2 Summary of TDG Predictive Results: SPW1

Parameter	Values for Various Discharges					
	Baseline (cfs)			With Roughness Elements and Ramp (cfs)		
	10,000	13,000	20,000	10,000	13,000	20,000
Maximum Depth of Plunge (ft)	66	68	108	36	49	85
Average Depth of Plunge (ft)	40	42	92	18	22	43
TDG Computed at USGS Site (%)	117.6	117.9	127.4	111.0	113.0	125.0
TDG Contribution from Spill (%)	11.6	11.9	21.4	5	7	19
Change in Computed TDG Relative to Baseline	n.a.	n.a.	n.a.	-6.6	-4.9	-2.4

5 Summary

In summary, a tool has been developed to help support the design and evaluation of TDG mitigation projects for the Boundary project. This tool, which includes both a particle tracking CFD model and a spreadsheet (PPTG) analysis model has been validated through various field tests. Going forward, it will continue to be used to provide a means to assess the overall magnitude of gas reduction possible for various operational and structural options.

References

Azbel, D., 1981. *Two phase flows in chemical engineering.* Cambridge University Press, Cambridge, England, 27-50, 155-210

Flow Science Inc., 2005. FLOW-3D User's Manual. Flow Science Inc., Santa Fe, NM 87505, USA

SCL. 2016. Boundary Hydroelectric Project (FERC No. 2144) Total Dissolved Gas 2016 Annual Report. Nov 2016.

Thompson, E.J. and Gulliver, J.S.,1997. "Oxygen transfer similitude for a vented hydroturbine," J. Environmental Engineering, 123(6) 529-538.

Leadership in Sustainable Infrastructure
Leadership en Infrastructures Durables

Case Study

Vancouver, Canada
May 31 – June 3, 2017/ *Mai 31 – Juin 3, 2017*

CCDP: GIS-BASED DATA PORTAL FOR CLIMATE CHANGE IMPACT ASSESSMENT

Wang, Xiuquan[1,3] and Huang, Gordon[2]
[1,2] Institute for Energy, Environment and Sustainable Communities, University of Regina, Regina, Saskatchewan, Canada S4S 0A2
[3] XanderWang@uregina.ca

1 Background

While mitigating climate change would require substantial and sustained reductions in greenhouse gas emissions through worldwide consensus and collaborations, adapting to climate change has become a major focus of local policy makers and development practitioners. Sound decisions rely on impact-based modeling, but the coarse-resolution outputs of global climate models (GCMs) are unsuitable for driving impact models, which usually require finer resolution projections at both spatial and temporal scales. Effective downscaling of GCMs projections is thus required, but it is practically difficult due to the lack of computational resources and/or long-term reference data. Such difficulty has become a major barrier preventing informed climate change adaptation planning at regional scales. To address this challenge, a GIS- and web-based data portal will be established to allow intuitive and open access to high-resolution climate scenarios, thus to support decision making of effective climate policies at both regional and local scales.

2 Framework of CCDP

The developed data portal is named Climate Change Data Portal (CCDP) and is available to the public at http://ccdp.network. CCDP now contains three entries to allow free access to high-resolution climate data for Canada (i.e., Canada CCDP, available at http://canadaccdp.ca), China (i.e., China CCDP, available at http://chinaccdp.org), and the Province of Ontario (i.e., Ontario CCDP, available at http://ontarioccdp.ca), respectively. CCDP offers functions of visual representation through geospatial maps and data downloading for a variety of climate variables (e.g., temperature, precipitation, relative humidity, solar radiation, and wind) at multiple temporal resolutions (i.e., annual, seasonal, monthly, daily, and hourly). The vast amount of information this portal encompasses can provide a crucial basis for assessing impacts of climate change on local communities and ecosystems and generating relevant adaptation strategies. As rainfall intensity-duration-frequency (IDF) curves are widely used to summarize the characteristics of extreme rainfall events characteristics, which are critical for designing flood protection structures and many other civil engineering structures (e.g., stormwater sewers, dams, and bridges). Therefore, in addition to the high-resolution projections for typical climate variables, CCDP also includes projected IDF curves at grid point scales under future climatic conditions. The projected IDF curves allows development practitioners to convert climate projections into water flow and flood risks for supporting engineering design under a changing climate.

3 Summary

Since its initial launch in January 2014, CCDP has received more than 22,000 downloading requests from over 310 registered users, including academia, municipal and provincial agencies, non-government agencies, and private sectors. The climate data provided by CCDP has been widely used for different research purposes, such as agricultural impact and risk assessments, water quality and quantity forecasting, infrastructure design and operations, wind power applications, and many others. The increasing visits to the CCDP demonstrate that easy and open access to high-resolution regional climate data is essential for promoting further impact studies.

Acknowledgements

This research was supported by the National Key Research and Development Plan (2016YFA0601502 and 2016YFC0502800), Natural Sciences Foundation (51190095, 51225904), the Program for Innovative Research Team in University (IRT1127), the 111 Project (B14008), the National Basic Research Program (2013CB430401), Ontario Ministry of the Environment and Climate Change, and the Natural Science and Engineering Research Council of Canada.

References

Wang, X., G. Huang, J. Liu, Z. Li, and S. Zhao 2016. GIS-based Data Portal for Climate Change Impact Assessment. *Bulletin of the American Meteorological Society*, 97: 17-18.

Wang, X., G. Huang, J. Liu, Z, Li, and S. Zhao 2015. Ensemble Projections of Regional Climatic Changes over Ontario, Canada. *Journal of Climate*, 28: 7327-7346.

Leadership in Sustainable Infrastructure
Leadership en Infrastructures Durables

Vancouver, Canada
May 31 – June 3, 2017/ *Mai 31 – Juin 3, 2017*

CFD STUDY OF THE AIR-ENTRAINMENT CHARACTERISTICS OF A SUBMERGED HYDRAULIC JUMP

Jesudhas, V [1], Roussinova, V [2], Balachandar, R [3], Barron, R [4] and Williams, P [5,6].
[1,2,3,4,5] University of Windsor, Canada
[6] williamq@uwindsor.ca

Abstract: Hydraulic jumps are associated with strong turbulence, free-surface fluctuations, air entrainment and energy dissipation. Over the last century, several researchers have attempted to study these characteristics experimentally. However, due to limitations of conventional measuring techniques in the air-water flow of the hydraulic jump, they were only partially successful. This paper presents the results of a three-dimensional, unsteady, detached eddy simulation of a submerged hydraulic jump with an inlet Froude number 8.2 and a submergence factor of 0.24. Volume of fluid (VOF) multiphase model is used for capturing the free-surface and the air-entrainment characteristics. The results of the simulations are validated with experimental results. The flow features of the submerged hydraulic jump captured by the simulation is presented with pertinent discussions. Since, the submergence factor of the submerged hydraulic jump is low, considerable amount of air was entrained into the flow. The mechanism of air entrainment in a submerged hydraulic jump is also analysed.

1 INTRODUCTION

When the flow transitions from a supercritical to subcritical flow in an open channel, hydraulic jumps are formed. Hydraulic jumps occurring in rectangular channels are called as classical hydraulic jumps (CHJ). They are characterized by a sharp increase in the free-surface height accompanied by strong turbulence, air entrainment and energy dissipation. The super-critical depth d_1^* and the sub-critical depth d_2^* are called sequent depths and are a function of the inlet Froude number F_1 (Chow, 1959), defined as:

$$\frac{d_2^*}{d_1^*} = 0.5 \left(\sqrt{1 + 8F_1^2} - 1\right) \quad (1)$$

where F_1 is the inlet Froude number. The point where the free surface increases abruptly is called the jump toe. Starting at the jump toe, there is a region of recirculation near the free surface of CHJ called as the roller. When the tail-water depth d_2 is higher than the sub-critical sequent depth d_2^*, the toe of the hydraulic jump moves toward the sluice gate. Both the toe and the roller become submerged. This is called as the submerged hydraulic jump (SHJ). A dimensionless submergence factor $S = (d_2 - d_2^*)/d_2$ was defined by Rajaratnam (1967). A schematic of the complex flow field of SHJ is presented in Fig. 1. There is a region of forward flow near the bed resembling a wall-jet. Above the wall-jet region is the recirculation region or the roller. A shear layer is developed between the wall-jet flow region and the roller, which expands in the vertical direction as we move downstream. The distance from the sluice gate to the end of the roller is called as the length of the roller L_r, which is an important parameter in the design of stilling basins for energy dissipation in hydraulic structures. After the end of the roller region the flow transitions into a regular open channel flow.

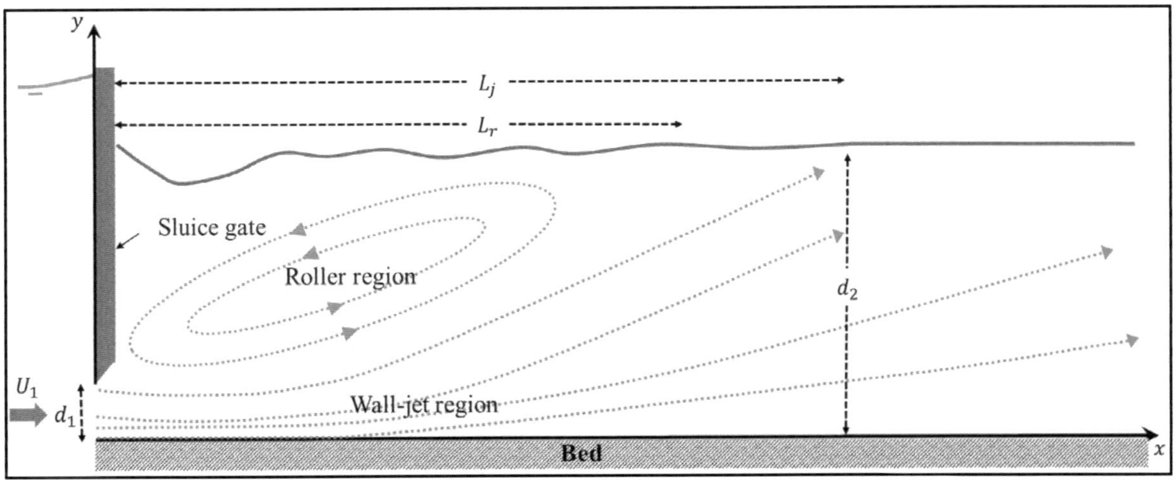

Figure 1: Schematic representation of the submerged hydraulic jump (Jesudhas et al, 2016)

To study the flow characteristics, Long et al. (1990) carried out an experimental study on SHJ using laser Doppler velocimetry (LDV). They reported the velocity field not only in the streamwise direction but also in the lateral direction. Their studies concluded that the interaction between the roller and wall-jet causes the SHJ flow field to become three-dimensional in nature. Several numerical studies were subsequently conducted on SHJ (Long et al., 1991, Ma et al., 2001 and Javan and Eghbalzadeh, 2013). However, these were two-dimensional computations and were not able to fully capture the complex three-dimensional flow field of the SHJ. To address this, a three-dimensional, unsteady, detached eddy simulation (DES) of SHJ with an inlet Froude number 8.2 and a submergence factor of 0.24 was carried out at the Department of Civil and Environmental Engineering, University of Windsor, Canada. Volume of fluid (VOF) along with high resolution interface capturing technique (HRIC) was used for capturing the free surface. A detailed analysis of the flow field along with turbulence characteristics and coherent structures was presented in Jesudhas et al. 2016.

Air entrainment in hydraulic jumps have received considerable attention by researchers. Air entrainment in SHJ not only depends on inlet Froude number F_1 but also on the submergence factor S. Since the submergence factor of the SHJ presented is low, considerable amount of air was entrained in the flow. The present paper discusses the air-entrainment characteristics of SHJ simulated in Jesudhas et al. 2016 after briefly discussing the flow features of SHJ. The mechanism of air entrainment and the instigating factor for entraining air is also analyzed.

2 THE MODEL

The DES method uses Reynolds-averaged Navier-Stokes (RANS) models close to the wall and large eddy simulation (LES) in the outer regions of the flow. Improved delayed detached eddy simulation (IDDES) available in the commercial solver STAR-CCM+ was used in the present study. The k-ω detached eddy model was selected as it is known to perform well in an adverse pressure gradient flow like the hydraulic jump. The complete formulation of this model is presented in Jesudhas (2016) and STAR-CCM+ v8.06 User Guide and hence not repeated here for brevity. The schematic of the domain used in the simulations is shown in Fig. 2. The size of the domain was selected based on the experiments of Long et al. (1990) to enable comparison with the results.

The size of the domain was 2.5 m × 0.467 m × 0.5 m. The Cartesian coordinates x, y and z are adopted as streamwise, vertical (wall-normal) and transverse directions, respectively. The origin of the coordinate system is positioned at the beginning of the inner wall of the flume. The boundary conditions used in the simulation are also presented in Fig. 2. The mesh used in the present simulation consists of about 5.6 million hexahedral cells. Refinements were made in the regions of interest and also prism layers were used near all the walls to capture the wall effects. The transient simulations were run with a time step δt = *0.001 s*. The solution was considered to be converged when the residuals of continuity and momentum fell below 10^{-6}. The velocity statistics of the present simulation have been calculated by averaging the data for 10 s following convergence.

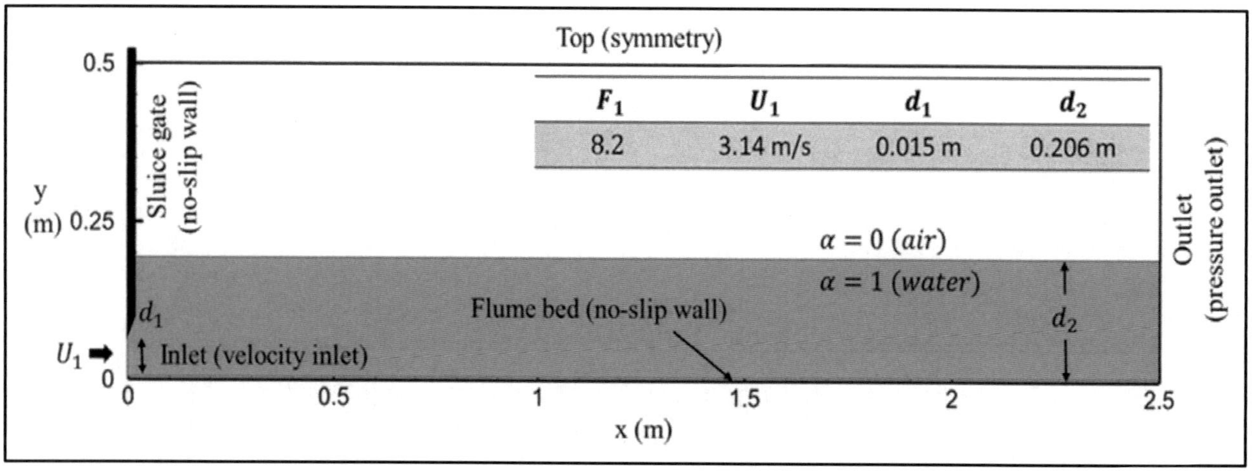

F_1	U_1	d_1	d_2
8.2	3.14 m/s	0.015 m	0.206 m

Figure 2: Schematic representation of the simulation domain at *t = 0* s (Jesudhas et al, 2016)

3 VALIDATION

An exhaustive validation of the flow and turbulence characteristics is presented in Jesudhas et al. (2016). A sample of the validation of results is presented in Fig. 3. Fig. 3(a) shows the comparison of the mean free surface profile predicted by the present simulation to the experimental results of Long et al. (1990). The free surface decreases in height initially due to the high-speed wall jet emerging from the sluicegate, after which it increases to the tail-water depth d_2. The location of the minimum depth in the free surface is predicted accurately by the simulations.

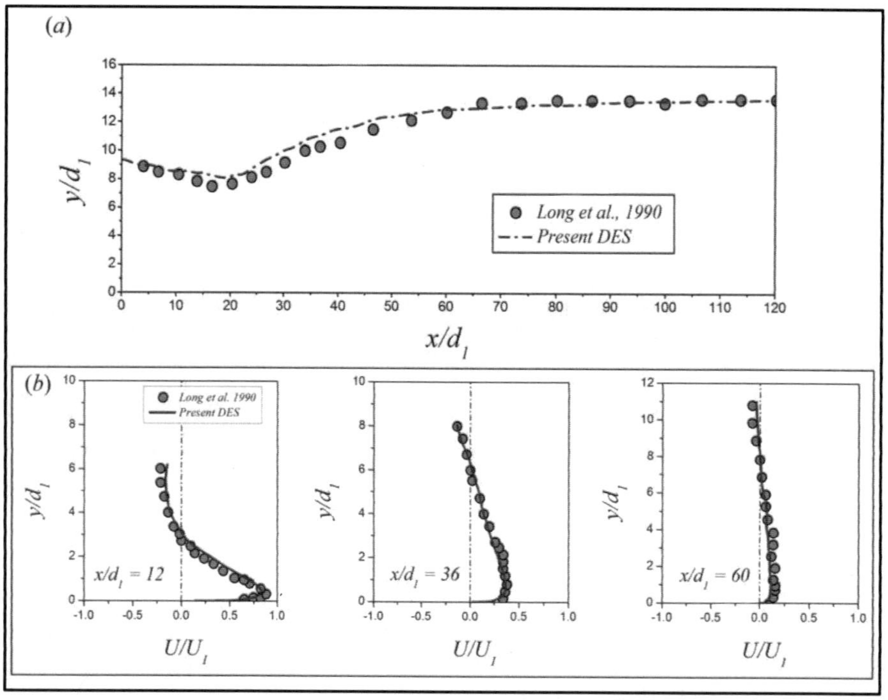

Figure 3: Validation of simulations results with the experiments of Long et al. (1990).

Immediately downstream of the minimum depth, there is a difference of 5% between the experimental and simulation results. As, mentioned earlier, considerable amount of air is entrained in the flow due to the low submergence factor of SHJ. A difference of up to 10% is considered to be satisfactory in such flow fields (Chachereau and Chanson, 2011). Fig. 3(b) presents the comparison of mean streamwise profiles between experimental and simulation results at three streamwise locations x/d_1 =12, 36 and 60. It can be seen that the simulation results agree well with the experimental results of Long et al. (1990). The mean streamwise velocity increases from zero at the bed to a positive maximum velocity U_{max} in the wall-jet region before it decreases and becomes negative in the roller region. From Fig. 3(b) it is also apparent that the value of U_{max} decreases as we move downstream i.e., from x/d_1 =12 to x/d_1 = 60. This is caused by the decay and expansion of the wall-jet flow emanating from the sluice gate. From Fig. 3 it is apparent that the results of the present simulations agree well with the experimental results of Long et al. (1990).

4 RESULTS AND DISCUSSIONS

Fig. 4 shows the three-dimensional instantaneous free surface of the SHJ. The free surface is colored using the contours of instantaneous vorticity magnitude. Initially, there is a reduction in free surface height. Beyond the region of minimum depth the free surface is undulated intensely as observed in Fig. 4. This is the location were the submerged roller reaches the free surface. These undulations are caused by vortices from the shear layer impacting the free surface (Sarpkaya, 1996), which is discussed in the later part of this paper. This is evidenced by the maximum vorticity intensity in this region. Downstream of the roller, the free surface perturbations are reduced and the free surface becomes almost flat farther downstream (Fig. (4)).

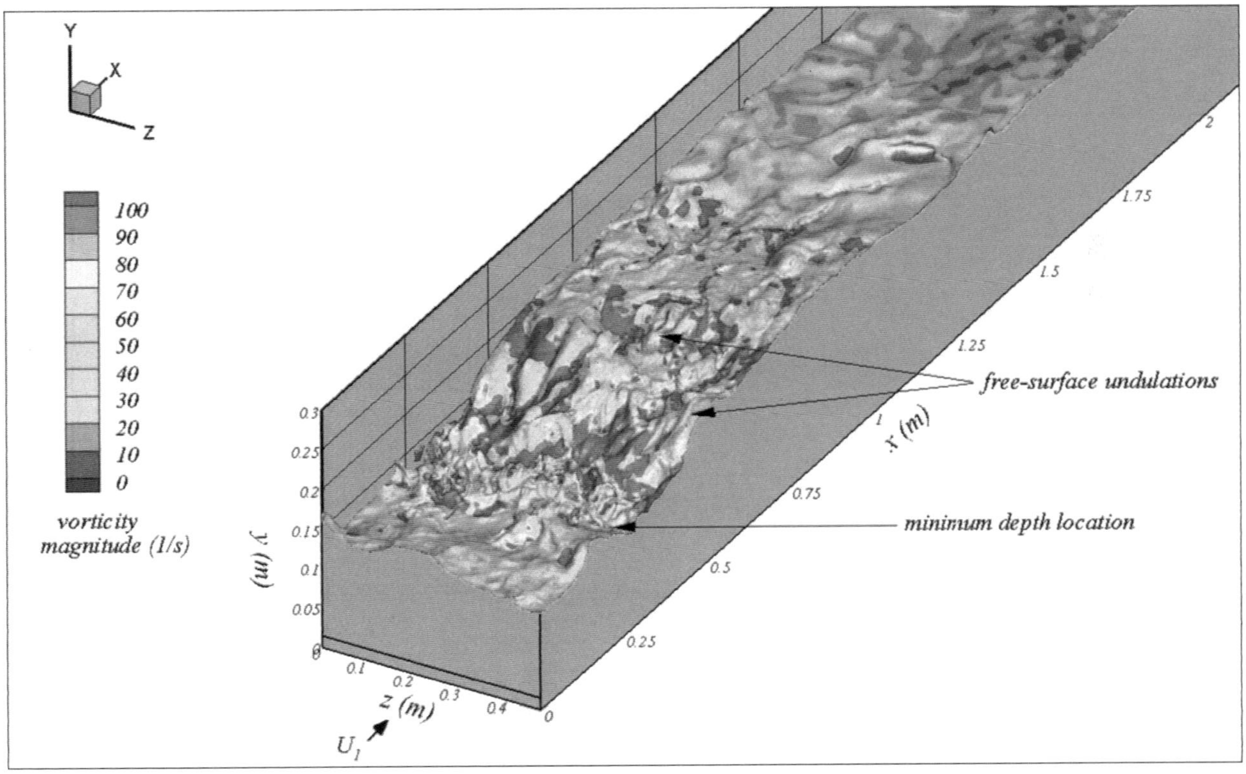

Figure 4: Three-dimensional free surface of the SHJ.

Fig.5(a) shows the contours of mean streamwise velocity in the central plane of the SHJ. The large negative region near the free-surface is part of the roller region. It extends from x/d_1 =0 to x/d_1 = 60 (Fig 5(a)). Beyond this region, the flow slowly transitions to open channel flow. However, the present simulation domain is not long enough for the flow to fully recover to a fully developed open channel flow.

The wall-jet flow emanates from the sluice gate and the region up to $x/d_1 = 10$ resembles the potential core of a wall-jet. The wall-jet decays and expands beyond $x/d_1 = 10$ as observed from Fig. 5(a). The shear layer is formed between the roller region and wall-jet region and expands in the vertical direction as we move in the streamwise direction. Fig. 5(b) shows the mean vector plots in the central plane of SHJ. The recirculation in the roller is clearly visible in Fig. 5(b). Also, plotted in Fig. 5(b) are dashed lines denoting $U = 0$ and $U = U_{max}$. The represent the upper and lower extremities of the shear layer (Jesudhas et al., 2016).

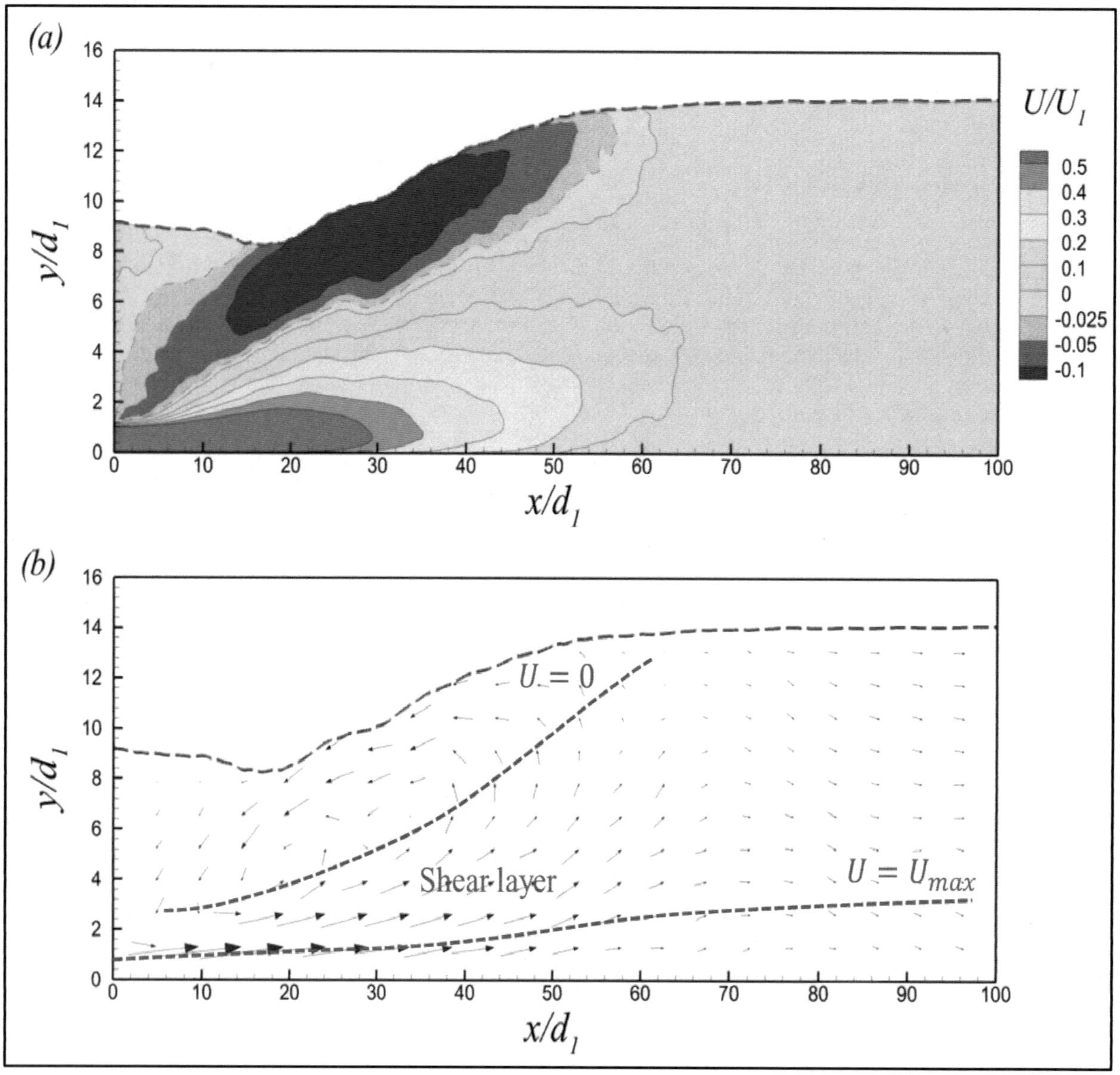

Figure 5: (a) Contours of mean streamwise velocity in the central plane of SHJ (b) mean vector plot in the central plane of SHJ.

Fig. 6(a) shows the contours of the mean Reynolds stress in the central plane of the SHJ. It can be clearly seen that the maximum values are generated in the shear layer due to the interaction of the roller and the wall-jet. Also, the turbulence generated by the bed is not as significant as the turbulence generated in the shear layer. Figs. 6(b) and (c) shows the contours of mean stream wise and vertical fluxes of the turbulent kinetic energy given by $t_{ku} = 0.5(u^3 + uv^2 + uw^2)$ and $t_{kv} = 0.5(v^3 + u^2v + vw^2)$, respectively. As indicated by the arrows in Figs. 6(b) and (c) it is evident the turbulent kinetic energy generated in the shear layer by the interaction of roller and wall jet is transported in the streamwise and vertical directions. This is significant because for air entrainment to occur the surface tension of the free surface must be overcome by the turbulence that is generated below the free surface (Chanson, 1995).

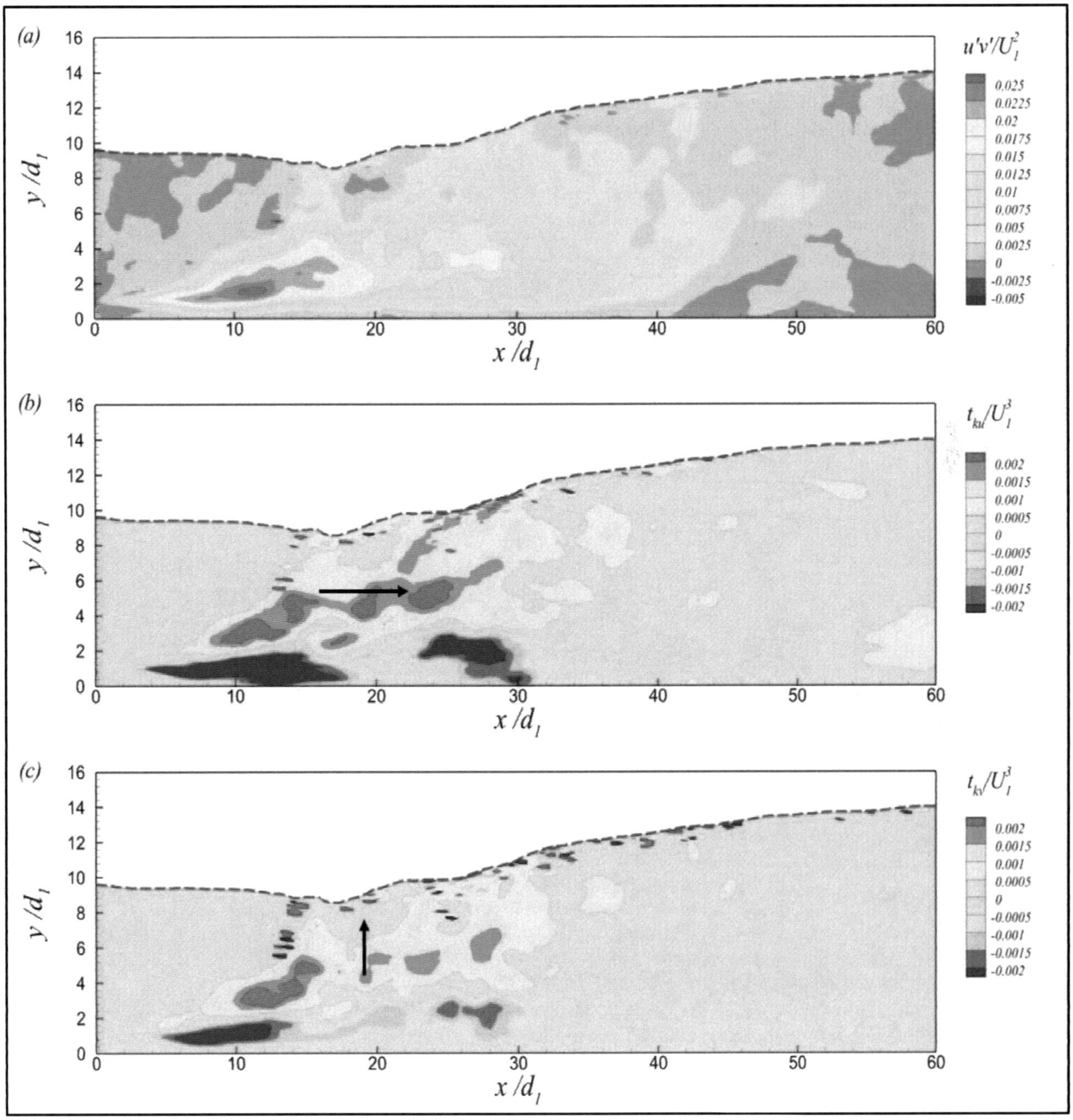

Figure 6: (a) mean Reynolds stress distribution in the central plane (b) Streamwise flux of turbulent kinetic energy (c) Vertical flux of turbulent kinetic energy.

Fig. 7(a) shows the contours of mean air concentration C in the central plane of the SHJ. It can be seen that significant amount of air is entrained into the flow. The maximum air concentration happens near the sluice gate within $x/d_1 = 15$. Beyond this there is negligible air entrainment into the flow. The variation of average air concentration C_{ave} at any streamwise location normalized by the maximum value C_{max} along the central plane is presented in Fig. 7(b). The field measurements of Valle and Pasternack (2006) for a SHJ is also plotted in Fig. 7(b) for qualitative comparison. The streamwise distance is normalized using the aeration length L_a. The peak air concentration for the SHJ occurs near the sluice gate. Also, plotted in Fig. 7(b) is a typical C_{ave} plot of a CHJ from Hager (1995). In contrast to the submerged jump, the toe of the classical hydraulic jump occurs downstream of the sluice gate consequently the maximum air concentration occurs downstream of the toe. Overall the amount of the air entrained by the CHJ is higher than the SHJ, as observed from Fig. 7(b).

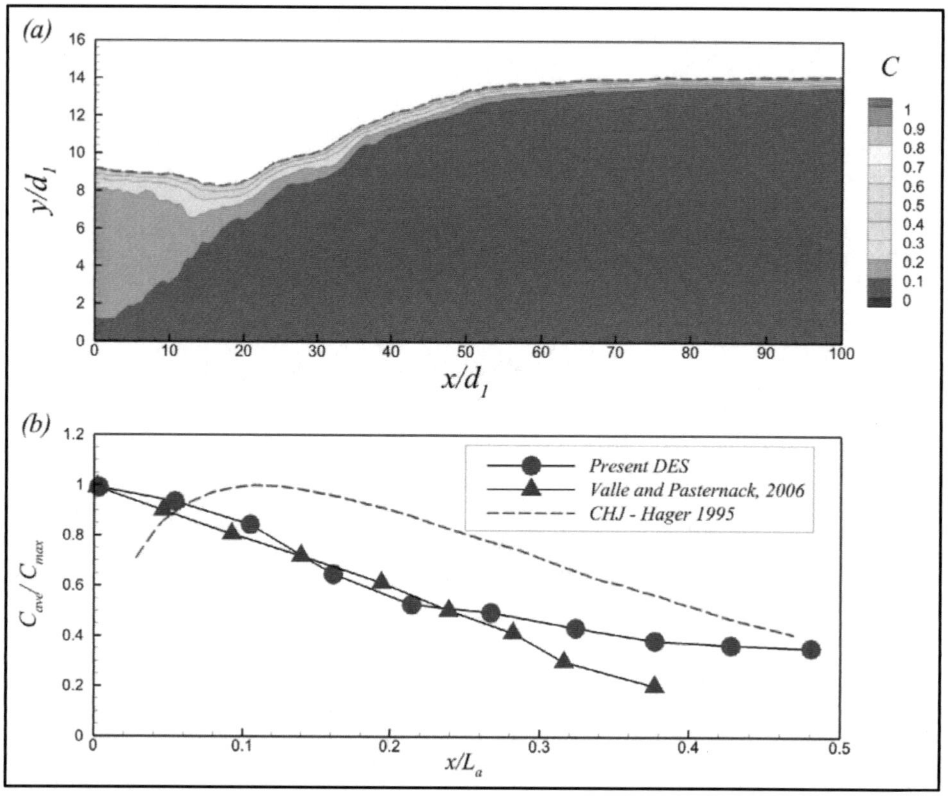

Figure 7: (a) Distribution of air concentration in the central plane of SHJ (b) Variation of average air concentration in the streamwise direction

In order to better understand the mechanism of air entrained in SHJ, numerical pressure probes, to collect time-dependent pressure data, were positioned at a height of d_1 at different streamwise locations along the channel. FFT analysis was carried out on the data collected for 10 s and the frequency of vortex shedding from the gate was found to be $f=13\ Hz$. The time period $T= 1/f$ of the vortex shedding was split into eight time intervals. The contours of z-vorticity at the end of each time interval are presented in Fig. 8, to highlight the vortex shedding from the gate and to understand the interaction of these vortices with the free surface. At time $t = T/8$, it can be seen that a vortex has just been shed from the gate. The vortex shed from the gate is broken down into multiple smaller vortices 'A' and 'B' at $t = 2T/8$, due to the impact of the roller. The smaller vortex 'A' is pushed up towards the free surface by the accelerating influence of the roller, as observed at times $t = 3T/8$ and $t = 4T/8$. The flux of turbulent kinetic energy by the vortex is sufficient to cause the free-surface cusp to collapse and break the free-surface ($t = 5T/8$). This causes air entrainment into the SHJ. If the submergence factor of SHJ is large, the smaller vortices are completely broken down before they reach the surface and hence do not have enough intensity to cause the collapse and break-up of the free surface, leading to a reduction in the amount of air entrained. At $t =$

3T/8 it can also be seen that the smaller vortex 'B' is pushed towards the bed by the downwash caused by the roller and interact with the counter-rotating structures near the bed. This vortex is further broken down by the interaction with the bed and moves downstream without disturbing the free surface. Hence, it is evident Fig. 8 that the turbulent kinetic energy carried by the vortices are responsible for the breaking up of free surface and entraining air in the SHJ.

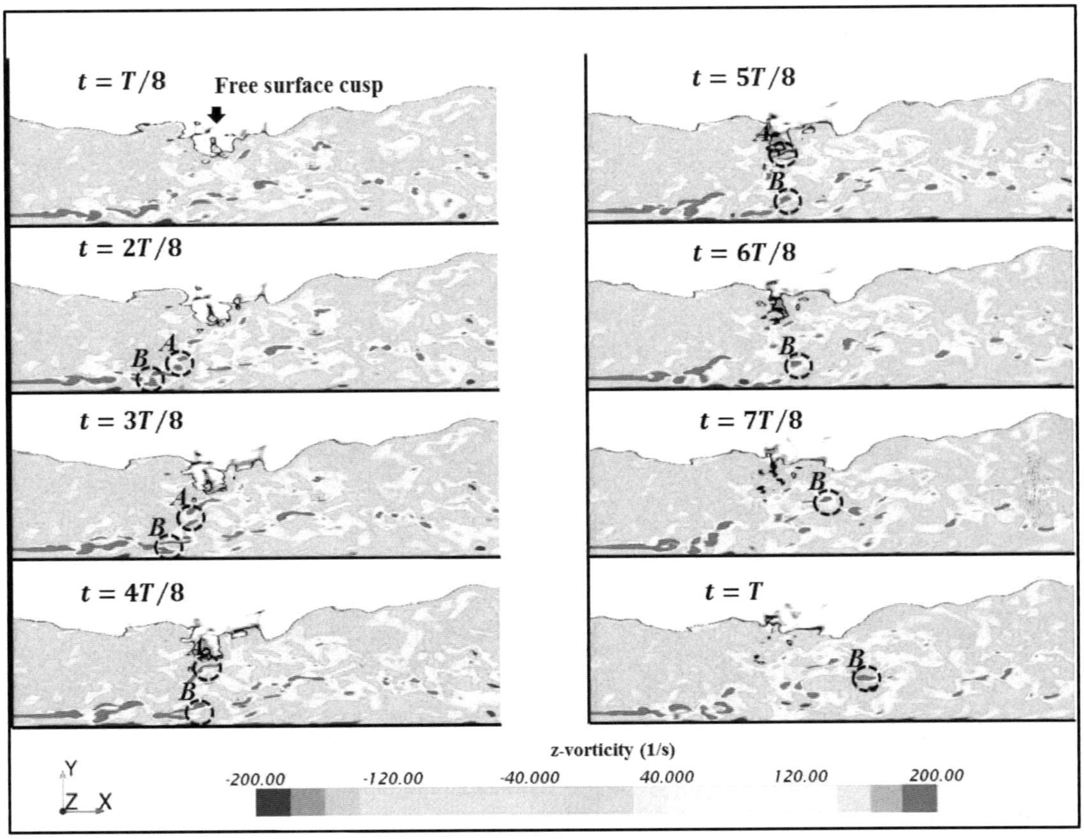

Figure 8: Mechanism of air entrainment in a submerged hydraulic jump

5 CONCLUSIONS

This study presents the results of a three-dimensional, unsteady, detached eddy simulation of a submerged hydraulic jump (SHJ) with an inlet Froude number 8.2 and a submergence factor of 0.24. The results showed that the numerical simulations were accurate in capturing the flow features of the SHJ. Since, the submergence factor of the present simulations is low, considerable amount of air is entrained into the flow. The simulations were accurate in predicting the air concentration in the flow. Large vortical structures shed from the sluice gate are broken down in the shear layer and the smaller vortices are pushed towards the free surface by the action of the roller. The turbulent kinetic energy carried by these vortices are sufficient to over come the surface tension, resulting in air entrainment.

REFERENCES

Chachereau, Y., and Chanson, H. 2011. Free-Surface Fluctuations and Turbulence in Hydraulic Jumps. *Experimental Thermal and Fluid Science* 35 (6): 896–909.

Chanson, H. 1995. Air Bubble Entrainment in Free-Surface Turbulent Flows: Experimental Investigations. https://espace.library.uq.edu.au/view.php?pid=UQ:9030.

Javan, M., and Eghbalzadeh, A. 2013. 2D Numerical Simulation of Submerged Hydraulic Jumps. *Applied Mathematical Modelling* 37 (10): 6661–6669.

Jesudhas, V. 2016. *Modeling of free-surface flows with air entrainment.* University of Windsor, Electronic Theses and Dissertations. Paper 5738

Jesudhas, V., Vesselina, R., Balachandar, R. and Barron, R. 2016. Submerged Hydraulic Jump Study Using DES. *Journal of Hydraulic Engineering*, 04016091.

Long, D., Steffler, P. M. and Rajaratnam, N. 1990. LDA Study of Flow Structure in Submerged Hydraulic Jump. *Journal of Hydraulic Research* 28 (4): 437–460.

Long, D., Steffler, P. M. and Rajaratnam, N. 1991. A Numerical Study of Submerged Hydraulic Jumps. *Journal of Hydraulic Research* 29 (3): 293–308.

Ma, F., Y. Hou, and P. Prinos. 2001. Numerical Calculation of Submerged Hydraulic Jumps. *Journal of Hydraulic Research* 39 (5): 493–503.

Rajaratnam, N. 1967. Hydraulic Jumps. *Advances in Hydroscience* 4 (1): 197–280.

Sarpkaya, T. 1996. Vorticity, Free Surface, and Surfactants. *Annual Review of Fluid Mechanics* 28 (1): 83–128.

STAR-CCM+, *User Guide v8.06*, 2013. CD-adapco.

Te Chow, Ven. 1959. *Open Channel Hydraulics*. McGraw-Hill Book Company, Inc; New York.

Vallé, Brett L., and Gregory B. Pasternack. 2006. Submerged and Unsubmerged Natural Hydraulic Jumps in a Bedrock Step-Pool Mountain Channel. *Geomorphology* 82 (1): 146–159.

Vischer, D. L., Willi H. Hager, and others. 1995. *Energy Dissipators*. Balkema Rotterdam.

Leadership in Sustainable Infrastructure
Leadership en Infrastructures Durables

Vancouver, Canada
May 31 – June 3, 2017/ *Mai 31 – Juin 3, 2017*

DEBRIS ENTRAINMENT DYNAMICS DURING EXTREME FLOODING EVENTS

Stolle, Jacob[1,4], Goseberg, Nils[2], Nistor, Ioan[1], and Petriu, Emil[3]
[1] Department of Civil Engineering, University of Ottawa, Canada
[2] Ludwig-Franzius-Institute for Hydraulic, Estuarine and Coastal Engineering, Leibniz Universität Hannover, Germany
[3] School of Electrical Engineering and Computer Science, University of Ottawa, Canada
[4] jstol065@uottawa.ca

ABSTRACT: Extreme coastal flooding events, such as the 2005 Hurricane Katrina and 2011 Tohoku Tsunami, have led to an increased emphasis placed on the design of resilient infrastructure. While the primary emphasis has focused on hydraulic load, less consideration has been given to debris loads, particularly in the assessment of the risk of debris loads occurring. The following study examines debris motion in a dam-break wave, physically modelling a tsunami-like event, evaluating the influence of debris configuration and hydrodynamic conditions on debris (lateral) spreading. The debris in this study was modelled as scaled-down shipping containers. The study found that deeper, faster flow conditions limited the influence of the bed surface, resulting in less debris spreading. The initial orientation of the debris also influenced the debris spreading, as the debris rotated to reach an equilibrium orientation. A probabilistic analysis showed that the debris spreading was normally distributed around the mean trajectory of the debris. The objective of this study was to provide a framework for debris load assessment for future design considerations.

Keywords: Debris Transport, Natural Disasters, Tsunami, Dam-Break, Flooding, Hydraulic Engineering

1 INTRODUCTION

With rising sea levels and the prevalence of extreme weather events, increased flooding and extreme hydrodynamic events have been observed worldwide (IPCC 2014). As a result, areas that had previously been not considered a high-risk for flooding have become vulnerable. Meanwhile, the demand for these high-risk areas for tourism, economic, or residential uses has continued unabated (Cooper 2013), therefore regulatory and government agencies have needed to adapt to protect these areas while balancing the other societal needs. One of the primary considerations when assessing the protective needs of a community is to determine the vulnerability of these communities to the increased loads associated with the flooding events.

To this point, the focus of research has been on evaluating the risk of hydraulic loads through flood hazard mapping or tsunami inundation maps (Sorensen 2000). A recent example of developing highly detailed tsunami hazard maps and evacuation modeling for the Indonesian city of Padang was presented by Taubenböck et al. (2013) and (2009). While the hydraulic loads are certainly among the most prevalent loads, loads from debris have been observed in many different flooding events. Several examples of river flooding in Switzerland have shown debris accumulation on structures, such as bridges, resulting in the failure of certain bridges (Schmocker and Hager 2013). Saatcioglu et al. (2005) noted the failure of several structures as a result of debris impact during the 2004 Indian Ocean Tsunami. Robertson et al. (2007) found examples of damaged building piles and columns as a result of debris impact and accumulation during a field survey of the 2005 Hurricane Katrina.

Debris loads are generally addressed as two different loads: a dynamic load associated with debris impact (F_{di}) and a static load associated with debris accumulation (F_{da}). Aghl et al. (2015) developed an equation for maximum debris impact using the conservation of energy principle:

[1] $F_{di} = u\sqrt{m_d k}$

where u is the impact velocity of the debris (m/s), m_d is the mass of the debris (kg), and k is the contact-stiffness between the debris and the structure (N/m). Equation 1 has been used in building standards in the United States for the design of buildings for flooding and tsunami loads (FEMA 2012, Chock 2016). Debris accumulation has been an overlooked topic in relation to flooding events, however highway design standards have often addressed this issue as a result of debris accumulation at bridge piers. In the Canadian Highway Bridge Design Code (2006), the debris accumulation loads are calculated by adjusting the drag coefficient to represent a square pier ($C_D = 1.4$).

To this point, debris loads are designed for throughout a hazard area (FEMA 2012, ASCE 2016). However, in extreme flooding events, large debris, such as vehicles, shipping containers, and shipping vessels, can be entrained within the inundating flow (Robertson et al. 2007, Naito et al. 2014). Designing for these loads throughout a flood hazard area would be expensive and oftentimes unnecessary.

As larger debris tend to have clear source areas (ports, shipping yards, parking lots, etc.) assessing the area at risk for debris impact is viable with an understanding of debris spreading dynamics in extreme flooding conditions. Debris studies in steady-state conditions have often mentioned that debris tended to propagate within the deepest, fastest section (thalweg) of the flow (Bocchiola et al. 2006, Matsutomi 2009, Schmocker and Hager 2013). Naito et al. (2014) presented a method of assessing at-risk areas for debris impact based on a field study from the 2011 Tohoku Tsunami. By determining the source of shipping vessels and surveying the final position, Naito et al. (2014) proposed a ±22.5° envelope from the debris source in the inundation direction. The envelope contained all the shipping vessels identified by the limited field survey.

Nistor et al. (2016) assessed the proposed method by Naito et al. (2014) over flat, horizontal surface with scaled-down shipping containers. Nistor et al. (2016) found that the ±22.5° spreading envelope was a conservative estimation of debris spreading and the spreading angle (θ) was dependent on the number of debris (N) present at the debris source.

[2] $\mp\theta = \mp 3.69 \mp 0.80N$

Goseberg et al. (2016b), using the same scaled-down shipping containers, evaluating cases where obstacles (scaled-down buildings) were placed on the flat, horizontal surface. The study found that the obstacles had limited influence on the spreading area as the debris tended to propagate within the deeper, faster sections between the obstacles, resulting in little interaction between the debris and obstacles.

Building guidelines have tended to go with a more conservative estimation of debris motion and have continued to always consider potential debris impacts, however this could be an expensive approach when applied at a wider scale. With the eventual goal to provide an engineering tool to assess the risk associated with large debris impacts during an extreme flooding event, the objectives of this paper are:

- Examine the effect of the hydrodynamic boundary condition on debris spreading.
- Examine the influence of the initial debris configuration on debris spreading.
- Compare the debris spreading to current techniques for analyzing debris spreading.

The study presented here focuses on one type of hydrodynamic boundary condition (a dam-break wave) and examines the influence of the hydrodynamic boundary condition by changing the impoundment depth in the reservoir. The study also only examines one type of debris (shipping containers). The experimental facilities and the experimental protocol are further discussed in Section 2. The results of the experiment are described in Section 3. The results of the experiments are placed within context of the wider literature and scale effect are discussed in Section 4 and 5.

2 EXPERIMENTAL SETUP

2.1 Experimental Facility

The experiments were performed at the Water Resources Laboratory in the Department of Civil Engineering, University of Ottawa, Canada. The experiments were performed in a dam-break flume (DBF), 30 m long, 1.5 m wide, and 0.70 m deep (Figure 1). The reservoir was 21.55 m long and the water was impounded behind a swinging gate, placed on a 0.20 m false floor. The false floor was covered with a fixed layer of 0.001 m sand grains, which resulted in an experimentally-determined Darcy-Weisbach friction factor (f) of 0.014. The swinging gate was manually opened to generate a dam-break wave; a counter weight was placed on the top of the gate to aid in the opening of the gate.

The spatial origin of the experiment (0,0) was considered to be the center of the flume at the upstream edge of the swinging gate. The y-axis was chosen positive in the flow direction, using a right-hand coordinate system, the positive x-direction was to the right and the positive z-direction was up.

The debris was placed on the false floor with the centroid of furthest downstream debris placed center-flume (x = 0 m) at y = 3.20 m. The structure was placed 7.03 m downstream from the gate and was used to measure impact forces for concurrent studies. This aspect of the experiment will not be discussed in the context of this study.

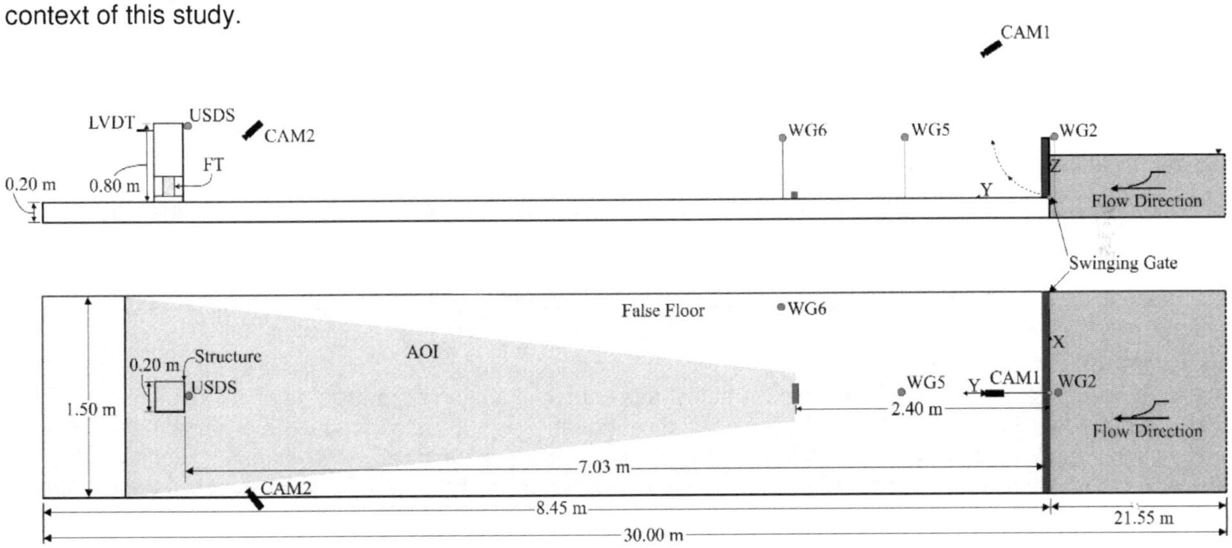

Figure 1. Dam-break Flume (DBF) at the University of Ottawa. The debris source is indicated with a green rectangle. The wave gauge (WG) positions are indicated with a red circle.

2.2 Instrumentation

The time-histories of the water depth were measured using capacitance-type wave gauges (WG) (Akamina Technologies WG-50) with a sampling rate of 1200 Hz. Before installation of the WG, the WG were calibrated with a calibration coefficient greater than 0.99 for each probe. WG2 was placed within the reservoir at y = -0.01 m. The reference time for each experiment was determined as the time that the water depth at WG2 began to decrease, signaling the opening of the gate. An ultra-sonic wave gauge (USDS) (MassaSonic M-5000) was placed at the front face of the structure at a distance of 0.80 m from the false floor.

Two HD cameras (CAM) (Basler AG pi1900-32 gc) were located to analyze the debris trajectories, impact conditions, and bore front profile. The CAM were externally triggered with a 25 Hz sampling rate using an output block signal from a DAQ (National Instruments USB-6009). The area of interest (AOI) of CAM1 is indicated in Figure 1 as the light grey area. Six reference points were placed within the AOI to geo-reference the images from CAM1 to the flume coordinate system.

2.3 Model Debris

The debris consisted of idealized 20-foot shipping containers (ISO668/688) down-scaled based on Froude scaling at a length scale of 1:40. The down-scaled shipping containers were manufactured from positively buoyant polyethylene (PE-HMW, 920 kg/m^3). Each down-scaled shipping container had overall dimensions of 0.06 x 0.06 x 0.15 m with an approximate draft of 0.025 m. The weight of the container, from three weightings, was determined to be 0.234 kg. The debris uniformly manufactured with a hollow inner shell. A full description of the model debris can be found in Goseberg et al. (2016a).

2.4 Experimental Protocol

The experiments performed in the DBF at the University of Ottawa examined three different experimental configurations, each configuration was repeated at least 10 times (Table 1). The experiments varied the impoundment depth in the reservoir and the orientation of the long-axis of the debris. The impoundment depth was varied between 0.20 m and 0.40 m. The orientation of the long-axis of the debris was placed perpendicular (0°) and parallel (90°) to the flow direction.

For each experiment, the center of the debris was placed 2.40 m from the inner edge of the dam-break gate. The placement of the debris was facilitated by position markers to ensure repeatability between the experimental repetitions. The debris was sealed with petroleum jelly to prevent water from entering the inner volume of the debris. The reservoir was filled to the impoundment depth and allowed to rest to ensure a repeatable hydrodynamic boundary condition. The dam-break gate was then released to form the dam-break wave. Before the beginning of the following experiment, the water was removed from the surface of the false floor.

Table 1. Experimental Protocol

Experimental Category	Impoundment Depth [m]	Debris Orientation [°]	Experimental ID [#]	Repetitions [#]
E01	0.40	0	254-263, 274-283	20
E02	0.20	0	294-303	10
E03	0.40	90	264-273, 284-293	20

2.5 Debris Tracking

The method for tracking the debris used an automated object tracking algorithm developed by Stolle et al. (2016). The algorithm uses a camera-based color thresholding algorithm to identify the green-colored debris within each image, implemented in MATLAB. Due to overtopping of the debris as the surge initially impacted the debris, a Kalman-filter based tracking algorithm was used to maintain the tracking identifier of the debris even when the debris was obscured by the surge. Previous evaluation of the algorithm was found to track the debris with ±0.03 m accuracy (Stolle et al. 2016).

3 RESULTS

3.1 Hydrodynamics

For the experiments, the hydrodynamic boundary condition was a dam-break wave. Contrary to dam-break waves propagating freely throughout the downstream section, in this case, the bore front interacted with the downstream-mounted vertical structure. Although not addressed herein, the flow-structure interaction did affect upstream water levels for conditions where Froude numbers fell short of unity. Figure 2 shows the normalized water elevation profiles from the four WG present in each of the experiments (WG2, WG5, WG6, and USDS). WG2 was placed within the reservoir as close a possible to the face of the gate. The initial decrease in the water surface elevation was used as the opening time of the gate and considered to be the zero-time instance. WG5 and WG6 were placed at 2.00 m and 3.20 m from the gate, respectively. WG6 was placed at the upstream face of the debris at x = 0.61 m (0.14 m from the flume wall).

As can be observed in Figure 2, the hydrodynamic boundary was repeatable between the various experiments. For the different impoundment depths, the time-averaged standard deviation for WG6 was 0.0055 m and 0.0019 m for the 0.40 and 0.20 m impoundment depths, respectively. As the debris propagated predominantly in the bore front, the two impoundment depths showed similar profiles at that position. However, the further from the gate (WG6), the smaller impoundment depths lagged behind the larger impoundment depths. This was potentially due to viscous effects within the bore, as noted by Lauber and Hager (1998) for impoundment depths less than 0.25 m.

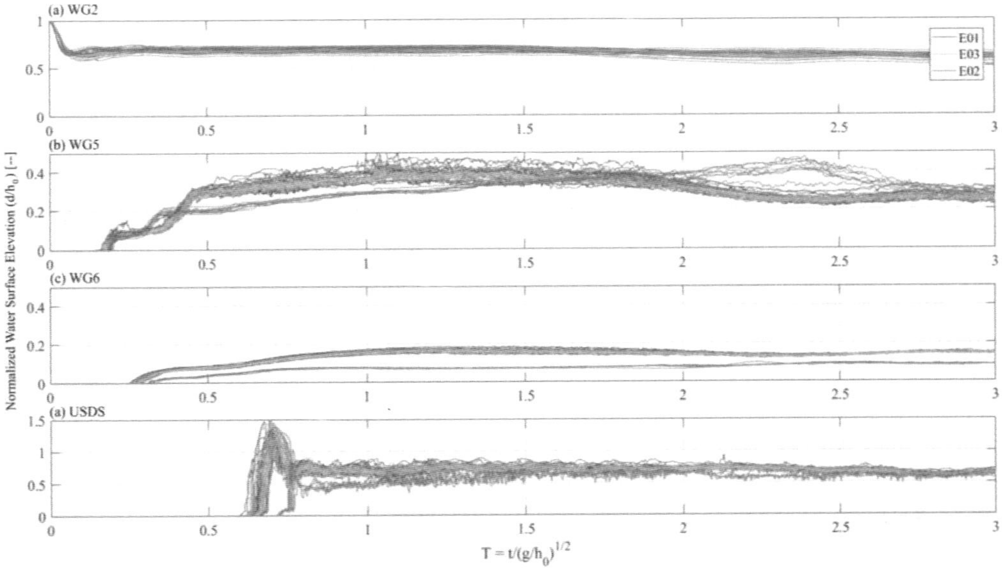

Figure 2. Recorded water surface elevations (D) as a function of time (T) for WG. (a) WG2; (b) WG5; (c) WG6; and (d) USDS.

3.2 Debris Trajectory

Using the debris tracking algorithm discussed in Section 2.5, the trajectory and orientation of the debris were tracked throughout the AOI. As the debris always propagated the length of the flume, the debris spreading (in the lateral direction) was the focus of this study. The mean trajectory of the debris was plotted as a solid line with the 95% confidence interval in Figure 3. The trajectories were compared to the methods proposed by Naito et al. (2014) and Nistor et al. (2016). In all cases, the trajectory of the debris were contained within the methods proposed by Naito et al. (2014) and Nistor et al. (2016).

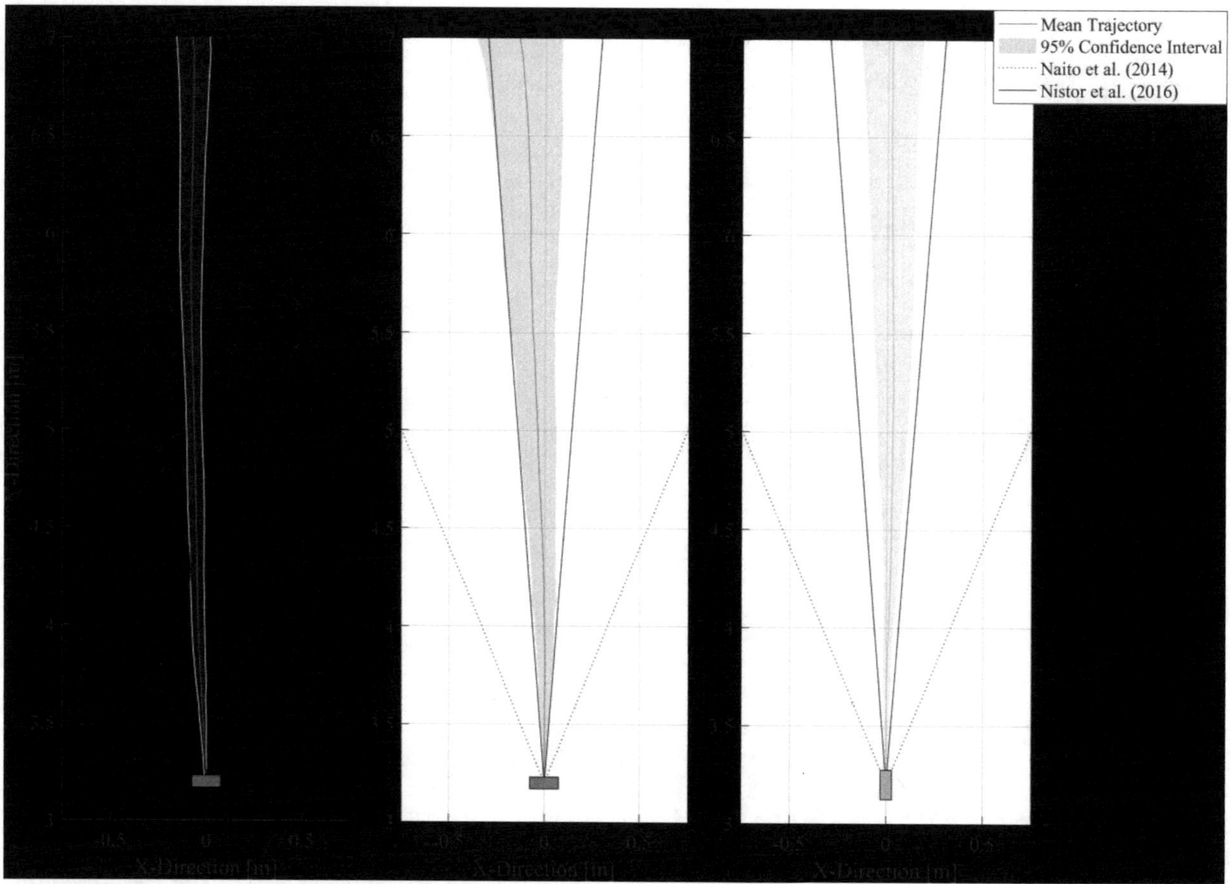

Figure 3. The debris trajectory with the mean trajectory plotted as a solid line, the 95% confidence interval plotted as a faded area. The trajectories are compared to previous literature: Naito et al. (2014) as a dotted line, Nistor et al. (2016) as a solid black line. (a) Impoundment depth of 0.4 m, initial orientation of debris 0°; (b) Impoundment depth of 0.20 m, initial orientation of debris 0°; and (c) Impoundment depth of 0.40 m, initial orientation of debris 90°.

Using an ANOVA test (Freund and Littell 1981), there was significant differences between the three cases (p-value << 0.05). Qualitatively comparing the three cases, less spreading was observed with the larger impoundment depth and the long axis of the debris perpendicular to the flow direction (Figure 3a). The equilibrium propagation orientation of the debris was with the long axis of the debris perpendicular to the flow direction. As a result, there was limited rotation of the debris. In the case with the smaller impoundment depths, the smaller water depths and slower flow velocities resulted in the debris interacting with the bed of the flume. As the debris interacted with the bed, the debris' forward momentum was impeded, forcing the debris to rotate around the point of the debris-bed contact. Similarly, in the case with the long axis of the debris parallel to the flow, the debris rotated towards an equilibrium orientation. The rotation of the debris resulted in increased lateral movement of the debris. Due to the relatively small number of cases examined in this paper, a revised estimate of the spreading angle was not provided as both Naito et al. (2014) and Nistor et al. (2016) equations provide a sufficient conservative estimate for the set of cases presented here.

Figure 4 compares the mean velocity profiles and acceleration distances of the debris for each case, normalized using the theoretical bore front velocity (Chanson 2006). The equilibrium debris velocities closely matched the theoretical flow velocity of a dam-break wave ($\sqrt{gh_0}$ or one-half of the theoretical bore front velocity). In the initial stages of the debris motion, significant overtopping occurred obscuring the debris. Therefore, resulting in larger error associated with the initial stages of debris motion.

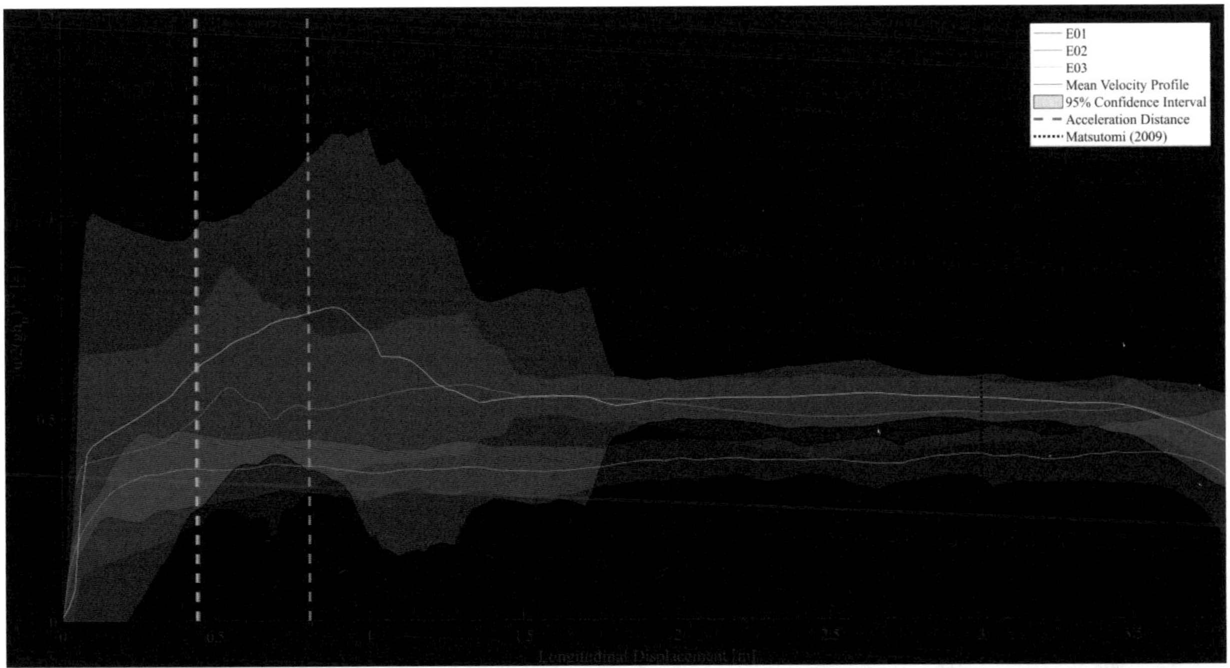

Figure 4. Debris velocity (u) as the debris propagated through the AOI. The solid line represents the mean velocity of the case, the shaded area indicates the 95% confidence interval. The dashed lines show the acceleration distance compared to Matsutomi (2009) (black dotted line).

A comparison between the velocity profiles show an insignificant difference in the acceleration distance and equilibrium velocities of the two cases with 0.40 m impoundment depth (E01 and E03) (Two-sample t-test, p-value >> 0.05). As the debris rapidly rotated to have the long axis perpendicular to the flow direction, the driving (drag) force was similar between the two cases. For the case with the 0.20 m impoundment depth, the acceleration distance and equilibrium velocity were significantly lower. As was discussed in Section 3.1, viscous effect may have resulted in the slower velocities. The acceleration distances for the dam-break case in this experiment were significantly less than the steady-state acceleration distances found by Matsutomi (2009), 20 times the length of the debris (black dotted line).

3.3 Probabilistic Analysis of Debris Trajectory

Both the methods for assessing debris spreading in Section 1 assigns the same risk to all structures within the spreading area. In a similar problem, the Eurocode 1 – Action on Structures (EU 2006) addressed accidental impacts, where an object has a typical trajectory (e.g. car travelling along a road, a ship passing under a bridge), and assessing the risk of the object impacting critical structure members with a Gaussian probability function. Based on the three cases presented, a Shapiro-Wilk test (Shapiro and Wilk 1965) was performed to determine that the spreading was normally distributed. The results of the test around the central y-axis found that the spreading was normally distributed (p-value >> 0.05). In this experiment, the mean of the spreading had a slight bias in the negative x-direction, likely due to inconsistencies in the bed profile of the flume. Figure 5 presents the normal probability density functions of the debris spreading, assuming that the debris was not biased by the bed profile.

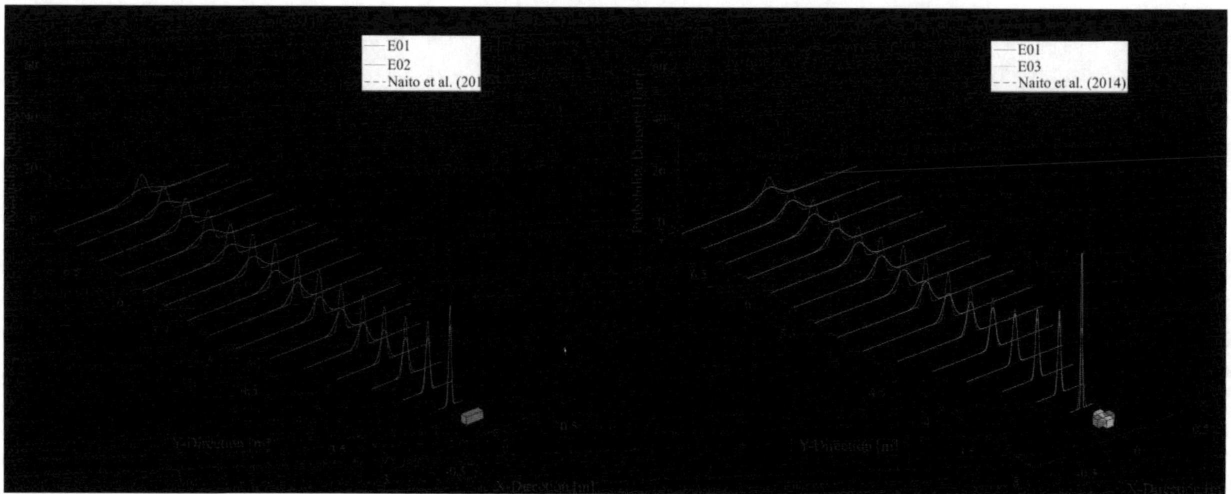

Figure 5. The probability density function for the propagation paths of the debris. (a) compares cases E01 and E02; (b) compares cases E02 and E03.

4 DISCUSSION

The use of dam-break waves has become common to tsunami engineering for the physical modelling of a tsunami wave, particularly in the analysis of loading conditions on structures (Nouri et al. 2010, Al-Faesly et al. 2012). Chanson (2006) revealed, using examples from the 2004 Indian Ocean Tsunami, that a dam-break wave well represented the bore profile of an inundating tsunami wave on a flat coastal plain. Additionally, the dam-break wave can have significantly longer flow duration, and therefore can be more representative of a tsunami wave, compared to a solitary wave, which historically been used for near-shore tsunami wave impingements (Madsen et al. 2008). A dam-break wave is limited by the length of the reservoir, as the negative wave formed at the release of the gate propagates through the reservoir, reflects off the back wall, then interferes with the wave profile in the AOI. As the debris generally propagated through the AOI in less than 5 s, the negative wave, which reached the AOI in > 20 s (Lauber and Hager 1998) should have had no impact on the presented results.

One aspect of the dam-break wave that potentially influences the presented results was the viscous effects in the bore front. A comprehensive dam-break experiment, performed by Lauber and Hager (1998), noted that at impoundment depths of less than 0.25 m, a comparison of the normalized water surface elevation profiles showed significant deviations. Lauber and Hager (1998) indicated that the deviations were likely caused by the viscous forces in the bore front. However, the study presented by Lauber and Hager (1998) focused on the bore front profiles at one normalized distance from the dam-break gate. Further research is needed to understand the influence of the viscous forces in the near- and far-field conditions. As the viscous forces pertain to this study, the reduction in flow velocity could result in the decreased normalized flow velocity observed in Figure 4. The debris velocity was expected to be around the celerity of the disturbance caused by the dam opening ($\sqrt{gh_0}$), which was observed for the 0.40 m impoundment depth. The lower flow velocity for 0.20 m impoundment depth indicate that there are potential viscous effects influencing the flow propagation.

The analysis of debris spreading showed that the spreading of the debris in an ideal flat bottom could be approximated using a normal distribution function, with the spreading angle increasing with distance from the debris source. The determination of the change in spreading angle should be performed at a larger scale to eliminate the potential scale effects related to viscous and friction forces. Further research is required and facilities to conduct those tests need to become available where larger scales could be realized. The combination of the probability density function for debris spreading and the debris acceleration profiles would allow for the development of risk assessment maps for large-scale debris impacts (Figure 6). The risk assessment maps could reduce the cost of designing debris impact resistant structures, as fewer structures would need to be designed for such extreme loads. Figure 6 combines the probability density functions with the mean velocity profile of case E01 by multiplying the force (calculated using Eq. [1]) by the probability density function. In this case, risk is considered to be the probability of a large impact force occurring at a given position.

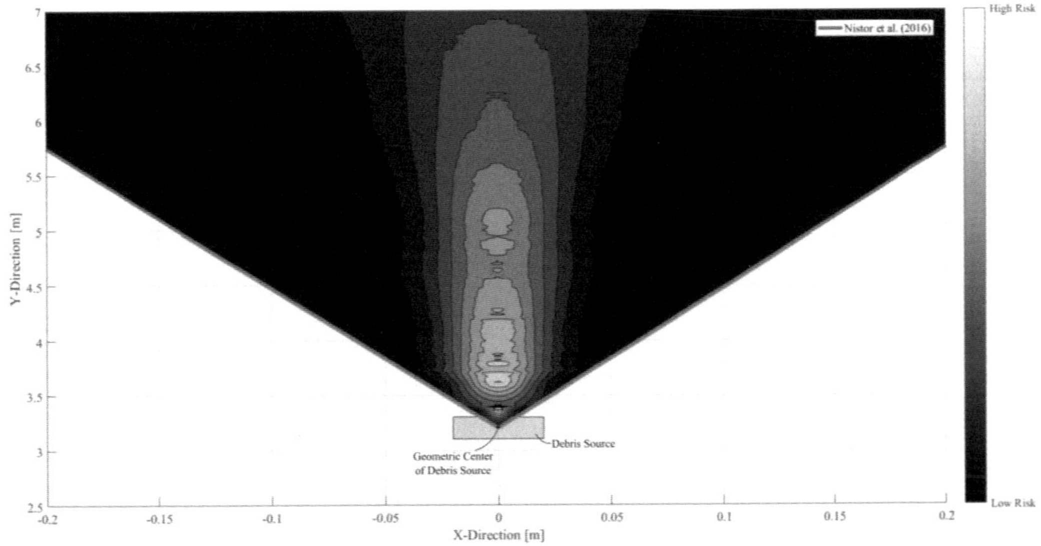

Figure 6. Risk analysis plot for debris loads in extreme flooding events. E01 was used as an example case for these experiments. The risk analysis was constrained by the maximum spreading angle determined by Nistor et al. (2016).

As was observed in Figure 3, as the bed surface was not perfectly flat, the inconsistencies in the bed profile influenced the mean trajectory of the debris. In ideal conditions, the mean trajectory of the debris would be expected to be straight. As the bed profile did have a significant effect, the consideration to the mean trajectory used in determining the central point of the probability density function would be needed. Detailed studies analyzing the influence of bed profile and the influence of built environments should be used to better evaluate the mean trajectory of the debris.

5 CONCLUSIONS

The study presented here analyzes the physical modelling of debris spreading during coastal flooding events, such as storms surges and tsunamis. The experiments were performed at the dam-break facility at the University of Ottawa. The debris was considered as uniform scaled-down shipping container (1:40 length scale), and the motion of the debris was tracked using a camera-based optical tracking algorithm (Stolle et al. 2016). Based on the results, presented in Section 3, the authors concluded:

- Debris spreading increased as a result of the rotation of the debris towards the equilibrium propagation orientation of the debris.
- Deeper, faster flow conditions resulted in less debris spreading due to less interaction with the bed surface.
- The debris spreading area was well within the methods proposed by Naito et al. (2014) and Nistor et al. (2016).
- The spreading of the debris can be considered to be normally distributed around the mean trajectory of the debris.

Debris motion studies, such as the experiment presented here, can be used for better assessment of the risk of large-scale debris impact. Additionally, the work could also be applied to the recovery of valuable equipment and objects in the aftermath of coastal flooding events to aid in narrowing the scope of search operations. However, as discussed in Section 4, there are scale effects related to the relatively small scale of such debris studies. Therefore, further work at larger scales will be needed to address scale effects for these results to be used in real-world applications.

6 ACKNOWLEDGEMENTS

J. Stolle would like to acknowledge the NSERC Canadian Graduate Scholarship (CGS-D) for supporting ongoing research projects at the University of Ottawa. N. Goseberg acknowledges that this research was supported by a Marie Curie International Outgoing Fellowship within the 7th European Community Framework Programme.

7 REFERENCES

Aghl, P. P., C. Naito, and H. Riggs. 2015. Estimation of demands resulting from inelastic axial impact of steel debris. Engineering Structures 82:11–21.

ASCE. 2016. Minimum design loads for buildings and other structures. . American Society of Civil Engineers.

Association, C. S., and others. 2006. Canadian highway bridge design code. . Canadian Standards Association.

Bocchiola, D., M. Rulli, and R. Rosso. 2006. Transport of large woody debris in the presence of obstacles. Geomorphology 76:166–178.

Chanson, H. 2006. Tsunami surges on dry coastal plains: Application of dam break wave equations. Coastal Engineering Journal 48:355–370.

Chock, G. Y. 2016. Design for tsunami loads and effects in the ASCE 7-16 standard. Journal of Structural Engineering:04016093.

Cooper, N. J. 2013. Innovative Coastal Zone Management: Sustainable Engineering for a Dynamic Coast. Pages 146–147 Proceedings of the Institution of Civil Engineers-Maritime Engineering. . Thomas Telford Ltd.

EU. 2006. Eurocode 1: Actions on structures—.

Al-Faesly, T., D. Palermo, I. Nistor, and A. Cornett. 2012. Experimental modeling of extreme hydrodynamic forces on structural models. International Journal of Protective Structures 3:477–506.

FEMA. 2012. P646 Guidelines for Design of Structure for Vertical Evacuation from Tsunamis. Federal Emergency Management Agency.

Freund, R. J., and R. C. Littell. 1981. SAS for linear models: a guide to the ANOVA and GLM procedures. . Sas Institute.

Goseberg, N., I. Nistor, T. Mikami, T. Shibayama, and J. Stolle. 2016a. Nonintrusive Spatiotemporal Smart Debris Tracking in Turbulent Flows with Application to Debris-Laden Tsunami Inundation. Journal of Hydraulic Engineering:04016058.

Goseberg, N., J. Stolle, I. Nistor, and T. Shibayama. 2016b. Experimental analysis of debris motion due the obstruction from fixed obstacles in tsunami-like flow conditions. Coastal Engineering 118:35–49.

IPCC. 2014. Climate Change 2014–Impacts, Adaptation and Vulnerability: Regional Aspects. . Cambridge University Press.

Lauber, G., and W. H. Hager. 1998. Experiments to dambreak wave: Horizontal channel. Journal of Hydraulic research 36:291–307.

Madsen, P. A., D. R. Fuhrman, and H. A. Schaffer. 2008. On the solitary wave paradigm for tsunamis. Journal of Geophysical Research: Oceans 113.

Matsutomi, H. 2009. Method for estimating collision force of driftwood accompanying tsunami inundation flow. Journal of Disaster Research 4:435–440.

Naito, C., C. Cercone, H. R. Riggs, and D. Cox. 2014. Procedure for site assessment of the potential for tsunami debris impact. Journal of Waterway, Port, Coastal and Ocean Engineering 140:223–232.

Nistor, I., N. Goseberg, T. Mikami, T. Shibayama, J. Stolle, R. Nakamura, and S. Matsuba. 2016. Hydraulic Experiments on Debris Dynamics over a Horizontal Plane. Journal of Waterway, Port, Coastal and Ocean Engineering:04016022.

Nouri, Y., I. Nistor, D. Palermo, and A. Cornett. 2010. Experimental investigation of tsunami impact on free standing structures. Coastal Engineering Journal 52:43–70.

Robertson, I., H. R. Riggs, S. C. Yim, and Y. L. Young. 2007. Lessons from Hurricane Katrina storm surge on bridges and buildings. Journal of Waterway, Port, Coastal, and Ocean Engineering 133:463–483.

Saatcioglu, M., A. Ghobarah, and I. Nistor. 2005. Effects of the December 26, 2004 Sumatra earthquake and tsunami on physical infrastructure. ISET Journal of earthquake technology 42:79–94.

Schmocker, L., and W. H. Hager. 2013. Scale modeling of wooden debris accumulation at a debris rack. Journal of Hydraulic Engineering 139:827–836.

Shapiro, S. S., and M. B. Wilk. 1965. An analysis of variance test for normality (complete samples). Biometrika 52:591–611.

Sorensen, J. H. 2000. Hazard warning systems: Review of 20 years of progress. Natural Hazards Review 1:119–125.

Stolle, J., I. Nistor, and N. Goseberg. 2016. Optical Tracking of Floating Shipping Containers in a High-Velocity Flow. Coastal Engineering Journal:1650005.

Taubenböck, H., N. Goseberg, G. Lämmel, N. Setiadi, T. Schlurmann, K. Nagel, F. Siegert, J. Birkmann, K.-P. Traub, S. Dech, and others. 2013. Risk reduction at the Last-Mile: an attempt to turn science into action by the example of Padang, Indonesia. Natural hazards 65:915–945.

Taubenböck, H., N. Goseberg, N. Setiadi, G. Lämmel, F. Moder, M. Oczipka, H. Klüpfel, R. Wahl, T. Schlurmann, G. Strunz, and others. 2009. Last-Mile" preparation for a potential disaster–Interdisciplinary approach towards tsunami early warning and an evacuation information system for the coastal city of Padang, Indonesia. Natural Hazards and Earth System Science 9:1509–1528.

Leadership in Sustainable Infrastructure
Leadership en Infrastructures Durables

Vancouver, Canada
May 31 – June 3, 2017/ *Mai 31 – Juin 3, 2017*

DETERMINISTIC LAKE AND STREAM TEMPERATURE MODELLING: MAINTAINING OPTIMAL WATER TEMPERATURES FOR KOKANEE AND RAINBOW TROUT THROUGH INFORMED DESIGN

Shewan, Alana[1], Akkerman, Anna[1] and Martin, Violeta[1,2]
[1] Knight Piésold, Ltd. Vancouver, Canada
[2] vmartin@knightpiesold.com

Abstract: Stream temperature is an important indicator of habitat quality for fish of all life stages. However, stream temperature is often overlooked when designing mitigation systems, with the focus being on the effects of development on stream flow, depth and velocity. A number of simple deterministic water temperature models exist for lakes, channels, and pipelines that can be utilized to optimize key design parameters for mitigation systems. Knight Piésold Ltd. (KP) used water temperature modelling to support the preliminary design of a water supply system for a proposed mine in BC. The water supply system will augment flows to an existing stream to support kokanee and rainbow trout populations. The proposed water supply system includes an intake in a nearby lake, a pump and pipeline conveyance system, and a small reservoir with low level and surface outlets. KP modelled the water temperatures throughout the supply system and in the receiving stream using three models: a one-dimensional Fresh Water Lake model (FLake), a heat transfer pipeline model, and the Stream Segment Temperature model (SSTEMP). Baseline models of the existing lake and stream were completed to assess model suitability and to calibrate model parameters. Operation models were then developed using long-term site specific climate inputs for the project. The models were run for various design considerations to investigate if stream temperatures could be maintained within optimal ranges to support various life stages of kokanee and rainbow trout populations. Modelling identified periods in which conditions similar to baseline could not be fully achieved and additional mitigation measures needed to be considered.

1 INTRODUCTION

Flow mitigation is required to minimize the impacts of a proposed mining project in British Columbia on kokanee and rainbow trout habitat. A Water Supply System (WSS) will be used to augment stream flow reductions in a fish bearing stream (the Creek). The WSS will consist of an intake in a nearby lake (the Lake), a pipeline with a booster pump station, a Water Reservoir (WR) located on the Creek, and outlet works from the WR. Water from the WSS will provide the minimum seasonal Instream Flow Needs (IFN) for both fish species. The purpose of this study was to use water temperature modelling to support the preliminary design of the WSS by determining if the seasonal variation in water temperatures could be maintained within baseline conditions to support various life stages of the kokanee and rainbow trout populations. This paper describes the temperature modelling approach, the results of the modelling, and the recommendations that resulted from the study.

2 WATER TEMPERATURE MODELLING APPROACH

There are three major components that contribute to the overall thermal regime of the WSS: the Lake, the pumped pipeline and the WR. The combined thermal regimes of these components ultimately contribute to the water temperatures in the Creek supporting kokanee and rainbow trout. Water temperatures for each of the elements shown on Figure 1 were modelled separately and in sequence following the downstream direction.

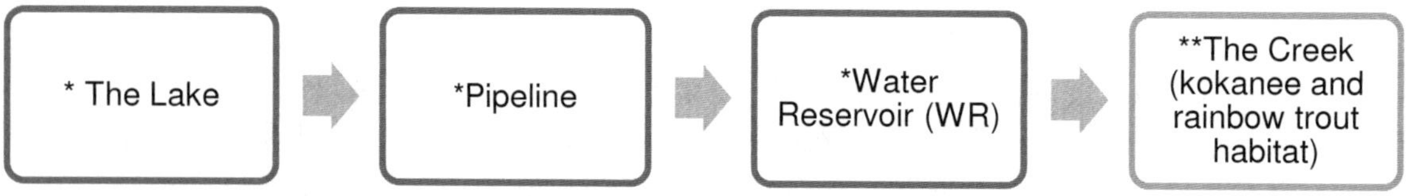

* Denotes elements within the WSS

** Denotes the open channel stream downstream of the WSS

Figure 1: Elements modelled separately and in sequence following the downstream direction.

Temperature assessments were conducted to address the possible range of water temperature conditions that could occur within each of these elements by considering the long-term mean monthly, and minimum and maximum daily values for various key parameters. The input climate parameters for each model were based on long-term climate data (about 40 years) for the site that was developed through regression analysis with regional data based on the short-term data collected on site. Using the minimum and maximum daily temperatures derived from the long-term natural climate record provided a thermal range over which the WSS is predicted to operate.

2.1 Temperature Modelling for the Fresh Water Source and Water Reservoir

Modelling of the daily thermal changes in the fresh water source, the Lake, and the WR was conducted using the FLake fresh water lake model. FLake, developed in Germany (Mironov 2008), is capable of estimating the vertical temperature structure and mixing conditions in lakes over time. The model is based on a two-layer parametric representation of the evolving temperature profile and on integral budgets of heat and kinetic energy for the layers. The model was developed as a lake parameterization scheme for use in numerical weather predictions and climate modelling, but it has also been used as a stand-alone lake model for many studies in Europe and North America (Mironov et al. 2005, Kirillin 2010, Kirillin et al. 2011, Bernhardt et al. 2012, Semmler et al. 2012). It is incorporated into the Canadian Regional Climate Model version 5 (CRCM5) developed by Le Centre ESCER in Montreal (Martynov et al. 2012).

The FLake model for the Lake was calibrated using two years of measured temperature data in the Lake and concurrent climate data as shown on Figure 2. The calibrated parameters from the Lake model were also used for the WR model. The calibrated models were then run with the long-term synthetic climate data series developed for the project. Daily results were used to calculate the mean monthly, and daily minimum and maximum epilimnion and hypolimnion temperatures, and epilimnion thicknesses for each water body.

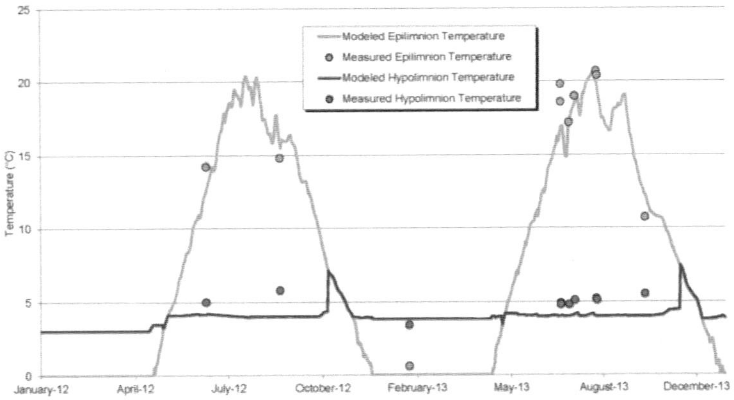

Figure 2: Temperature calibration of the Lake model.

2.2 Temperature Modelling for the Pipeline

Outputs from the Lake temperature model were used as inputs to the pipeline temperature model. Assessments of water temperature changes along the pipeline were conducted using a heat transfer model developed by KP after Kwon (1998) that calculates steady-state heat transfer to or from the water in a pipeline, including temperature increases resulting from pumping inefficiencies. The model considers heat transfer effects from conduction, convection and radiation, and takes into account the pipeline material, flow magnitude as well as the water, air and soil temperatures. A sensitivity analysis was conducted to evaluate the impact of different pipeline materials, soil temperatures, and soil thermal conductivity on the modelled pipeline water temperatures.

2.3 Temperature Modelling for the Creek with Kokanee and Rainbow Trout Habitat

Temperature variations in the Creek with kokanee and rainbow trout habitat were simulated using the Stream Segment Temperature Model (SSTEMP) Version 2.0 developed by the U.S. Geological Survey (USGS). SSTEMP was developed to aid biologists and engineers in predicting the consequences of stream manipulation on water temperatures. SSTEMP is a mechanistic, one-dimensional heat transport model that predicts water temperatures as a function of stream distance and environmental heat flux (Bartholow 2004). The SSTEMP model was calibrated using concurrent water temperature and stream flow data at two locations along the Creek. An example of the calibration is presented on Figure 3. The calibrated model was then run to estimate the baseline and project impacted temperatures for various flow conditions and at two locations, one for rainbow trout habitat and a second for kokanee habitat located farther downstream.

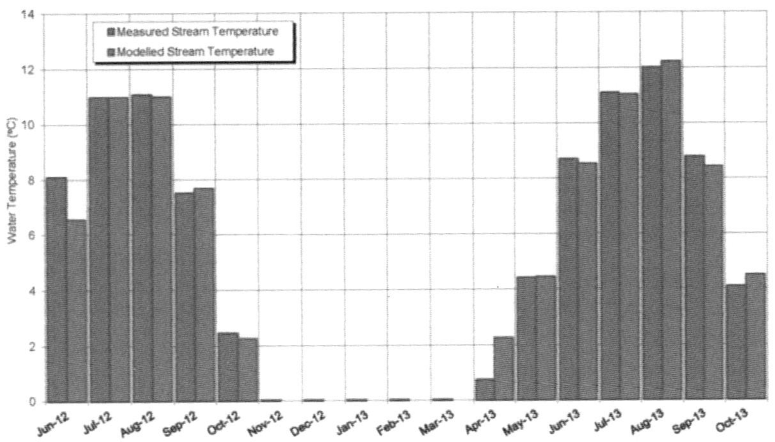

Figure 3: Temperature calibration of the Creek model.

Both the baseline and project conditions were modelled using the long-term mean monthly, and minimum and maximum daily air temperatures in each month derived from the synthetic climate data set developed for the project. These air temperatures were used in combination with various outflow options from the WR (low level and surface outlet), or direct discharge from the pipeline. The resulting water temperature ranges were then used to evaluate changes between the baseline and augmented flow conditions in the Creek to assess the various WR outlet and pipeline discharge options.

3 Temperature Modelling Results

3.1 Temperature in the Water Supply System

The temperature modelling for the Lake confirms in-situ observations of the Lake being dimictic with full overturn in the spring and fall, and with the thermocline (epilimnion thickness) deepening throughout the fall as the Lake cools and overturns. The epilimnion temperatures increase substantially in the summer, while the epilimnion thickness exceeds 5 m from August to November. This result indicates that if the intake was located shallower than 10 m, it is likely that the warm epilimnion water would be drawn into the WSS. The modelled long-term epilimnion and hypolimnion temperatures are illustrated on Figure 4. The maximum and minimum daily values represent the highest and lowest daily values for each month within the entire modelling record, respectively, and show the full range of temperatures that may potentially occur in the Lake.

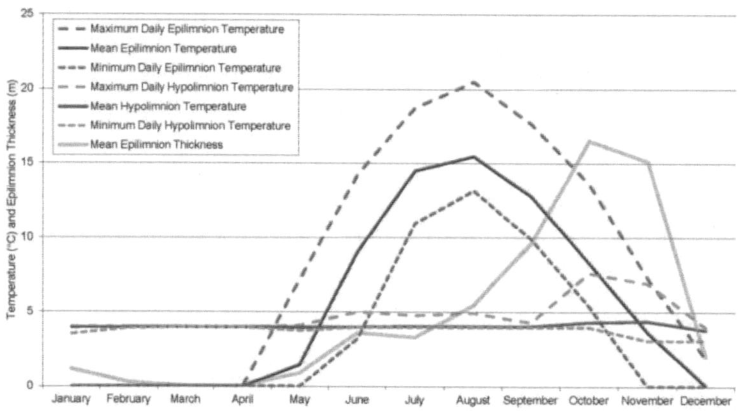

Figure 4: Range of long-term epilimnion and hypolimnion temperatures in the Lake

The water temperature change in the pipeline is predicted to be within ±1°C based on mean monthly intake water and air/soil temperatures. The general trend for the pipeline is cooling in the winter months and heating in the summer months. The analysis demonstrated that under extreme conditions the winter heat loss within the pipeline could be such that freezing may occur if the intake water temperature was between 0°C and 3°C. For this reason, an intake in the Lake epilimnion is not recommended.

The WR is predicted to be stratified throughout the winter and early summer, and overturn in the spring and late summer/fall. Results of the WR temperature modelling are illustrated on Figure 5. The WR will be smaller and shallower than the Lake, and as a result, it is predicted to undergo a weaker stratification in the winter and summer, and remain fully mixed for longer periods of time in the spring and late summer/fall. Temperatures throughout the depth of the WR are predicted to equalize faster than in the Lake, which is consistent with observed water temperatures in an existing smaller lake close to the Project area. As expected, the modelling confirmed that the climate has a larger impact on the smaller water body compared to the much larger Lake.

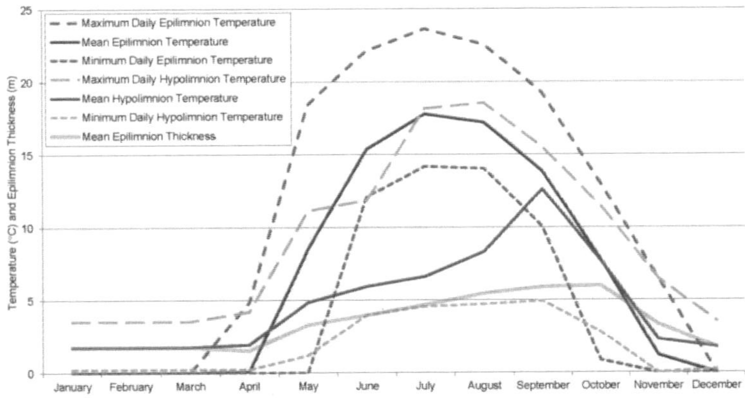

Figure 5: Range of long-term epilimnion and hypolimnion temperature in the Water Reservoir

3.2 Temperatures in the Creek with Kokanee and Rainbow Trout Habitat

Three potential sources of water from the WSS may be used to provide flow to the Creek to meet the IFN. These three sources were assessed separately to evaluate the temperature conditions in the Creek under various air temperature conditions as follows:

- Flows from the WR low level outlet
- Flows from the WR surface outlet, and
- Flows from the pipeline discharging directly to the Creek at the same location as the WR.

The results of the water temperature modelling for each discharge option are illustrated on Figure 6 for the rainbow trout habitat location and Figure 7 for the kokanee habitat location farther downstream. The mean monthly water temperatures are shown as bars, while the minimum and maximum daily water temperatures are indicated as whisker plots. Guidelines for the optimum water temperatures for various life stages of rainbow trout (Figure 6) and kokanee (Figure 7) are also shown for reference (BCMOE 2001). Comparison of baseline and optimum water temperatures indicates that this Creek is cooler than optimal for rainbow trout and kokanee; however, the species living in the Creek are adapted to these cooler temperatures.

The results indicate that the augmented flows in the Creek would be within 1°C of baseline water temperatures in most months, and that this could be achieved by using a combination of various proposed discharge options (e.g. the WR low level outlet, the WR surface water outlet, or direct discharge from the pipeline). An exception would occur in October, when both the Lake and the WR undergo fall turnover and cool slower than the shallow waters in the Creek. Without further mitigation, releases from the WSS in October would result in temperatures 3°C to 5°C above baseline, depending on the distance downstream from the release point. This finding led to additional design considerations to incorporate diversions of undisturbed catchment areas upstream of the project. These diversions would help augment the fall water temperatures, considering they have temperatures similar to baseline temperatures in the Creek.

The effect of water temperatures released from the WSS on water temperatures in the Creek decreases with distance downstream due to the increased direct influence of air temperatures and other climate parameters. The predicted differences in water temperatures between project and baseline conditions are smaller at the downstream kokanee habitat location than at the rainbow trout habitat location.

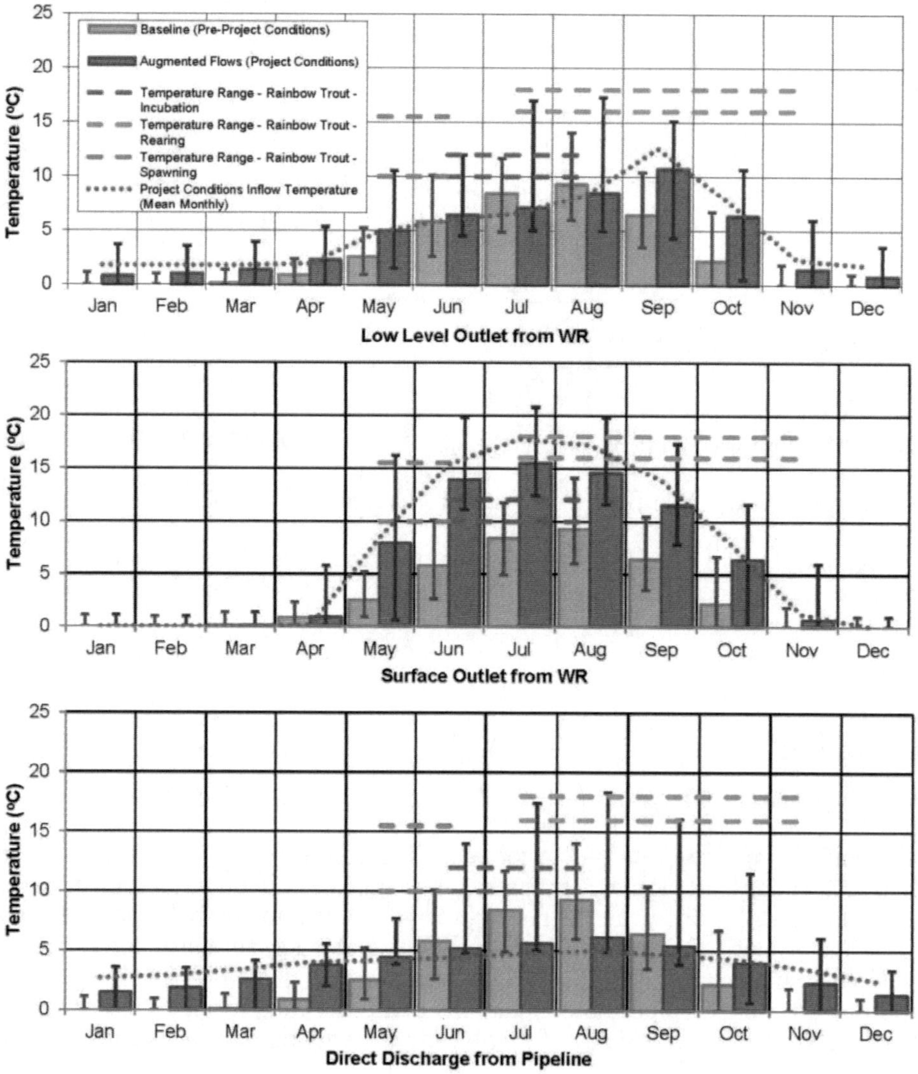

Figure 6: Simulated Creek water temperatures at rainbow trout habitat location

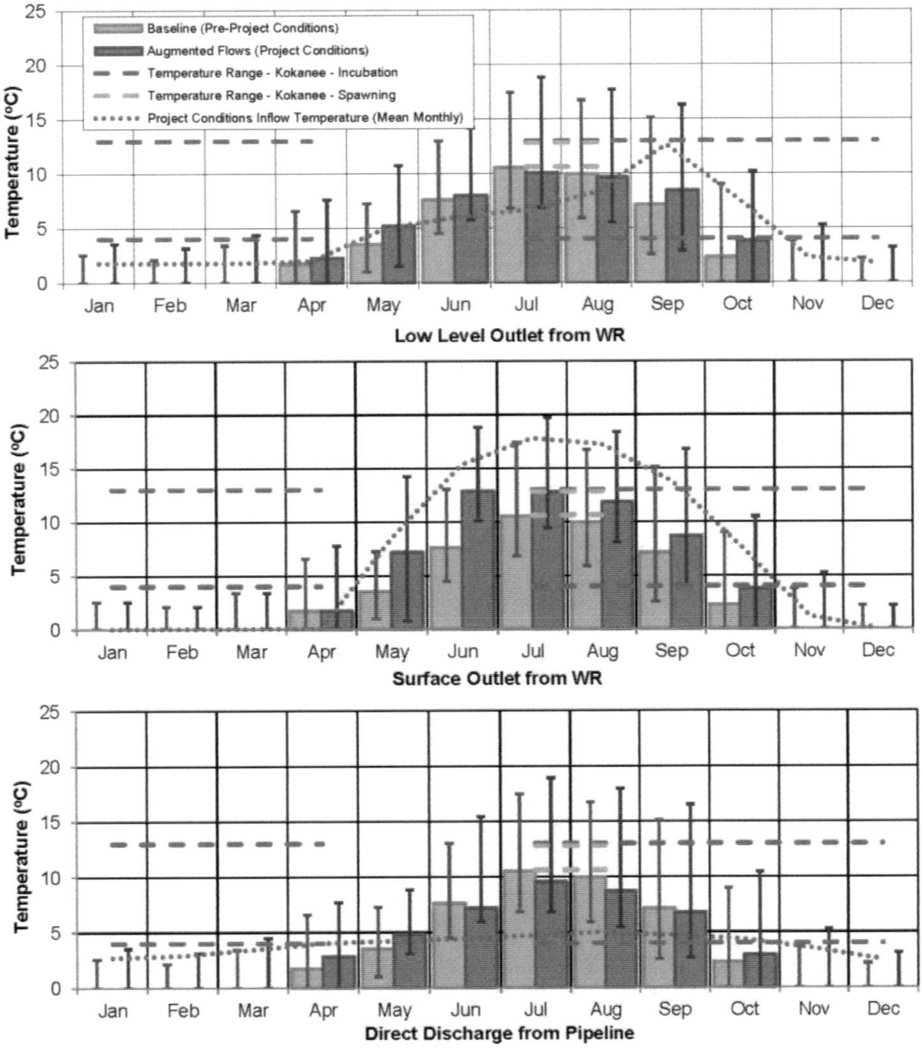

Figure 7: Simulated Creek water temperatures at kokanee habitat location

4 Discussion and Conclusions

The water temperature modelling conducted for the WSS indicates that flows and water temperatures could be augmented within 1°C during project conditions using a combination of the three discharge options from WSS for most of the year, except in the fall. The Creek temperatures would overall be consistent with baseline conditions, or would be within optimum species temperatures for various life stages (BCMOE 2001). The natural baseline temperatures are sub-optimal for certain life stages of rainbow trout and kokanee; however, the local fish populations are adapted to these suboptimal conditions. The effects of water temperatures released from the WSS on water temperatures in the Creek decrease with distance downstream due to the increased direct influence of air temperatures and other climate parameters.

The findings of the initial modelling led to additional design considerations in terms of the placement of the intake in the Lake and incorporation of upstream diversions to augment the fall water temperatures. The water temperature modelling of the Lake and pipeline indicated that the intake in the Lake should be located in the hypolimnion to help minimize the potential for freezing in the pipeline in the winter, while providing cooler temperatures in the summer. Furthermore, incorporating diversions of undisturbed catchment areas upstream of the project that have water temperatures similar to baseline temperatures in the Creek, provides additional mitigation options and aids in managing the fall temperatures bringing them in line with the observed baseline temperatures.

This study shows the importance of considering not only flows, but also water temperatures in early stages of the design of mitigation systems, so that additional adaptive management opportunities could be considered as necessary.

5 Acknowledgments

The authors gratefully acknowledge the opportunity to conduct these temperature assessments and learn from this experience. They wish to thank their clients, regulators, and colleagues for their valuable discussions, input and support during the study and preparation of this manuscript.

6 References

Bartholow, J.M, 2004. SSTEMP for Windows: The Stream Segment Temperature Model (Version 2). US Geological Survey computer and documentation. Available on the internet at https://www.fort.usgs.gov/

Bernhardt, J., C. Engelhardt, G. Kirillin, and J. Matschullat, 2012. Lake ice phenology in Berlin-Brandenburg from 1947–2007: observations and model hindcasts. *Clim. Change*, **112**, 791-817. **doi:10.1007/s10584-011-0248-9**

British Columbia Ministry of Environment (BCMOE), 2001. *Water Quality Guidelines for Temperature.* Available on the internet at http://www.env.gov.bc.ca/wat/wq/BCguidelines/temptech/temperature.html

Kirillin, G., 2010. Modeling the impact of global warming on water temperature and seasonal mixing regimes in small temperate lakes. *Boreal Env. Res.*, **15**, 279-293.

Kirillin, G., J. Hochschild, D. Mironov, A. Terzhevik, S. Golosov and G. Nützmann 2011. FLake-Global: Online lake model with worldwide coverage. *Environ. Modell. Softw.*, **26**, 683-684. **doi:10.1016/j.envsoft.2010.12.004**

Kwon, K.C., 1998. Heat Transfer Model of Above and Underground Insulated Piping Systems. A document prepared for ASME Conference – Heat Exchanger Committee Meeting and International Joint Power Generation Conference. Baltimore, MA, USA, Aug. 23-26, 1998.

Mironov, D., S. Golosov, E. Heise, E. Kourzeneva, B. Ritter, N. Scheider, and A. Terzhevik, 2005. FLake - a lake model for environmental applications. *Proc. of the 9th Workshop on Physical Processes in Natural Waters*, A. Folkard and I. Jones, Eds., Lancaster University, UK, 73.

Mironov, D.V., 2008. Parameterization of lakes in numerical weather prediction. Description of a lake model. COSMO Technical Report No. 11. Deutscher Wetterdienst, Offenbach am Main, Germany, 44 pp.

Martynov, A., L. Sushama, R. Laprise, K. Winger, and B. Dugas, 2012. Interactive lakes in the Canadian Regional Climate Model, version 5: the role of lakes in the regional climate of North America. *Tellus A.* 64 16226, doi: 10.3402/tellusa.v64i0. 16226.

Semmler, T., B. Cheng, Y. Yang, and L. Rontu, 2012. Snow and ice on Bear Lake (Alaska) – sensitivity experiments with two lake ice models. *Tellus A*, **64**, 17339. doi:10.3402/tellusa.v64i0.17339

Leadership in Sustainable Infrastructure
Leadership en Infrastructures Durables

Vancouver, Canada
May 31 – June 3, 2017/ *Mai 31 – Juin 3, 2017*

DEVELOPING DEBRIS-FLOOD MITIGATION CONCEPTS TO ADDRESS COMPLEX FLOODING IN MOUNTAIN COMMUNITIES LOCATED ON ALLUVIAL FANS

Tang, Gaven[1,2], and Kitagawa, Ron[1,3]
[1] Golder Associates Ltd., Calgary, Alberta, Canada
[2] Gaven_Tang@golder.com
[3] Ronald_Kitagawa@golder.com

Abstract: The June 2013 flood in the Hamlet of Exshaw occurred over the course of a three-day rainfall event. The recurrence interval of the rainfall event was estimated to be approximately 300 years. This resulted in flooding of both Exshaw Creek and Jura Creek with recurrence intervals of 500 years and 350 years, respectively. The Hamlet of Exshaw was affected by various sources of flooding including: debris-floods on Exshaw Creek and Jura Creek; local catchment runoff; and flooding of the Bow River. In November 2015, the MD of Bighorn No.8 retained Golder Associates to provide debris-flood mitigation solutions for Exshaw Creek and Jura Creek for the MD of Bighorn No. 8. Work completed in 2016 and continuing into 2017 includes: concept development and validation (Phase 1), preliminary design (Phase 2), and detailed design (Phase 3). This paper focuses on the work completed during the concept development and validation stage, Phase 1. This initial phase included the development of flood risk reduction concepts for both Exshaw Creek and Jura Creek based on design criteria developed by the previous consultant, who performed debris-flood hazard assessments for both creeks. At the conclusion of Phase 1, four design concepts were developed for Exshaw Creek and five design concepts were developed for Jura Creek. These design concepts were presented to all stakeholders to garner feedback on design, feasibility, cost, environmental impacts, and regulatory constraints, prior to making recommendations to the MD of Bighorn No. 8 in proceeding to the next phase of work.

1 INTRODUCTION AND BACKGROUND

1.1 Project Setting

The Hamlet of Exshaw is located about 10 km east of Canmore on an alluvial fan between the Bow River to the south and steep mountain slopes to the north. The alluvial fan is subject to slow growth as evidenced by mature subalpine fir trees in the non-active parts of the fan area. The Hamlet is affected by two active areas, Exshaw Creek and its floodway immediately west of the Hamlet, and Jura Creek and its floodway located on an adjacent alluvial fan about one kilometre east of the Hamlet. The active area of Exshaw Creek is relatively narrow (about 50 m), being constrained by the Hamlet on the east side and Lafarge Cement Plant on the west side. The active area of Jura Creek is much wider (about 100 m) as it is historically unconstrained on both sides. Figure 1 shows the project setting for the Hamlet of Exshaw.

Figure 1: Hamlet of Exshaw – Project Setting

1.2 Flood of 2013

The June 2013 flood in the Hamlet of Exshaw occurred over the course of a three-day rainfall event. The Hamlet was affected by debris-floods on Jura Creek and Exshaw Creek, local catchment runoff and flooding of the Bow River as discussed below.

Jura Creek

Jura Creek has a drainage area of 16.0 km^2. It began flowing at flood stage in the evening of June 18, 2013 shortly after the commencement of intense rainfall. The creek began overtopping its banks at 2:00 AM on June 19th. This spillage flowed along the north embankment of Highway 1A towards the lowland area on the eastside of the Hamlet. A second source of spillage occurred about 200 m north of Highway 1A resulting in a flow path directly to the east side (Heart Mountain Drive) of the Hamlet. Jura Creek was the first source of flooding to reach the Hamlet. To stem spillage from Jura Creek, Alberta Transportation breached Highway 1A at the Jura Creek culvert crossing to allow more flow to pass to the south side of the highway in an effort to reduce flooding of the Hamlet. Flooding of Jura Creek lasted for approximately 12 hours. The 2013 Jura Creek flood was estimated to have a recurrence interval of 350 years (BGC 2015). Photo 1 was taken during the crest of the flood on Jura Creek.

Photo 1: Jura Creek Flooding

Exshaw Creek

Exshaw Creek has a drainage area of 33.3 km^2. It began flowing at flood stage at 8:00 AM on June 19, 2013 resulting in extensive deposition of bed material that raised the creek bed and partially filled the flow path beneath the three bridges (Highway 1A, Canadian Pacific Railway (CP) and Diamond Drive) near the confluence of Exshaw Creek and the Bow River. Exshaw Creek continued to flow at flood stage for approximately 36 hours. During this period, creek banks were eroded and creek bed levels were raised to the extent that flow levels exceeded the banks in several places. The upstream spillage on the west bank into the Lafarge Cement Plant occurred near the apex (head) of the alluvial fan. Some spillage over the east bank in the vicinity of Mt Kidd Crescent resulted in flooding through the Hamlet along Windridge Road and Heart Mountain Drive. As the floodwaters of Exshaw Creek crested, debris blocked the passage ways of the three bridges. This resulted in some spillage eastward just upstream of Highway 1A bridge flowing down Heart Mountain Drive. More significantly, blockage of the CP bridge was so severe that CP diverted a portion of the Exshaw Creek flows immediately upstream of CP bridge eastward to prevent washout of the abutments. This action diverted the water eastwards towards a lowland area between Highway 1A and the CP Railway embankment, resulting in flow overtopping Highway 1A at the low point, and contributing to further flooding of residences along Heart Mountain Drive. The 2013 Exshaw Creek flood was estimated to have a recurrence interval of 500 years (BGC 2015). Photos 2 and 3 were taken during the crest of the flood on Exshaw Creek.

Photo 2: Exshaw Creek Flooding

Photo 3: Exshaw Creek Pedestrian Bridge

Drainage from Exshaw Mountain

The local drainage area of the lowlands on the eastside of the Hamlet is approximately 1.0 km^2 including the residential area of the Hamlet and steep mountain sides of Exshaw Mountain on the north side of the Hamlet. The contributing drainage area of the mountain causes significant runoff during precipitation events because of considerable areas of shallow bedrock and exposed rock faces. During the June 2013 flood event, the entire catchment was saturated causing drainage from the mountain to flow through the Hamlet and along Heart Mountain Drive. This contributed to flooding in the Hamlet.

Bow River

The Bow River has a drainage area of 4,120 km^2 at the mouth of Exshaw Creek. The 2013 flood on the Bow River was considered a 100-year flood event. According to the existing flood hazard study for the Bow River at Exshaw (Acres 1996) and the post 2013 flood LiDAR surface (provided by Alberta Environment and Parks), the elevation of the lowlands immediately east of the residential area is lower than that of the 100-year flood level. The elevated levels of the Bow River during the 2013 flood (beginning the evening of June 20) following peak flows on Exshaw Creek, impeded drainage of the floodwaters along Heart Mountain Drive and contributed to prolonged inundation of the residential area.

These sources of flooding mentioned above led to elevated groundwater levels that may have contributed to flood damages of the residential areas along Heart Mountain Drive that were inundated.

1.3 Emergency Response

Various flood mitigation measures were enacted in response to the flooding from Exshaw Creek and Jura Creek as follows:

- Excavators removed debris from the Highway 1A and CP bridges that cross Exshaw Creek (Photo 4).
- Rockfill and large riprap were placed on the banks of Exshaw Creek to contain the flow in the creek and to prevent bank failure.
- Temporary dykes were constructed to redirect floodwaters back into Exshaw Creek.
- Excavators breached the Jura Creek culvert crossing at Highway 1A to maintain flow across Highway 1A and to prevent flow westward towards the residential area along Heart Mountain Drive (Photo 5).

Photo 4: Exshaw Creek Emergency Response

Photo 5: Jura Creek Emergency Response

1.4 Previous Studies

Following the June 2013 flood, Matrix Solutions (Matrix) was engaged to design a dyke to protect the north part of the Hamlet immediately downslope of the apex of the alluvial fan. BGC Engineering Inc. (BGC) was engaged to perform debris-flood hazard and risk assessments of debris-flooding caused by Exshaw Creek and Jura Creek. All previous work was thoroughly reviewed by the Golder project team prior to developing flood mitigation concepts for Exshaw Creek and Jura creek.

2 DESIGN CRITERIA

A comprehensive analysis of appropriate design criteria was undertaken by BGC following the 2013 flood to determine the criteria for flood mitigation for the Hamlet of Exshaw. The results of the analysis indicate that the debris-flood mitigation works at Exshaw Creek and Jura Creek should be based on a debris-flood event associated with the 300-year to 1000-year recurrence interval, considering the vulnerability of the residential area of the Hamlet of Exshaw, critical infrastructure such as the CP bridge, Highway 1A bridge and Diamond Drive bridge crossings, and the Lafarge Cement Plant. This criterion exceeds the level of flood protection that is typically adopted for flooding on rivers and creeks in Alberta, because debris-floods are far more likely to cause fatalities due to their rapid response to rainfall with minimal flood warning and the large quantities of flowing bed material. The resulting recommended design flood flows and sediment yield estimates follow in Table 1.

Table 1: Exshaw Creek and Jura Creek Discharges and Sediment Yield (BGC 2015)

Creek	2013 Peak Discharge (m^3/s)	2013 Sediment Yield (m^3)	Recommended Design Discharge (m^3/s)	Recommended Sediment Yield (m^3)
Exshaw Creek	70	58,000	130	79,000
Jura Creek	40	26,000	70	48,000

The design criteria recommended by BGC was adopted by Golder for the conceptual design of the debris-flood risk reduction measures. This design criteria was reconsidered following completion of the conceptual design and determination of associated costs.

3 EXSHAW CREEK DEBRIS-FLOOD MITIGATION CONCEPTS

Figures 2 to 5 present the Exshaw Creek debris-flood mitigation concepts developed for this project. Costs of each concept are provided in Table 2. The following provides a description of each concept.

Option 1: Excavate the reservoir behind the existing Lafarge Dam to impound the required 79,000 m^3 of debris for the design event. The reservoir behind the existing Lafarge Dam has been completely filled with sediment to the dam crest level from past flood events. Excavation of the material will generate the required storage capacity. The downstream channel will need to be enlarged and armored to convey floodwaters and to prevent entrainment of sediment. A small sediment trap will be constructed immediately upstream of Highway 1A to capture any entrained sediment and prevent blockage of the three bridges. The Highway 1A bridge will need to be upgraded to allow for conveyance of the design discharge, and the Diamond Drive bridge will need to be removed to eliminate a constriction.

Option 2: Raise the crest of the existing Lafarge Dam to store the required 79,000 m^3 of debris for the design event above the level of the existing filled-in reservoir. This option is similar to option 1, except the small sediment trap is provided by the capacity created by raising the dam.

Option 3: Construct a new dam at the apex of the alluvial fan to provide 79,000 m^3 of sediment storage between the Lafarge Dam and the new dam. This option is similar to option 1, except a new dam is constructed at the apex of the alluvial fan, and the sediment storage is located between the existing Lafarge Dam and the new dam.

Option 4: Excavate a large sediment trap to capture the 79,000 m^3 of sediment at the downstream end of the alluvial fan, immediately upstream of the three bridges. The channel upstream of the sediment trap will need to be enlarged, but not armored as the flows coming down the creek will be saturated with sediment and will not entrain additional sediment. The channel banks will be armored to prevent lateral migration of the channel. The creek will need to be lowered under the three bridges to allow for conveyance of the design discharge and to eliminate any constrictions. The channel downstream of the sediment trap will be armored to enhance capacity and prevent erosion.

Table 2: Exshaw Creek Debris-Flood Mitigation Concept Costs (Golder 2016)

Debris-Flood Mitigation Option	Capital Construction ($ Millions)	Mobilization & Contingency ($ Millions)	Total Estimated Cost ($ Millions)	Annual Operations and Maintenance ($ Thousands)
1	9.2	4.3	13.5	86
2	11.5	5.4	16.9	102
3	13.5	6.4	19.9	112
4	7.0	3.4	10.4	53

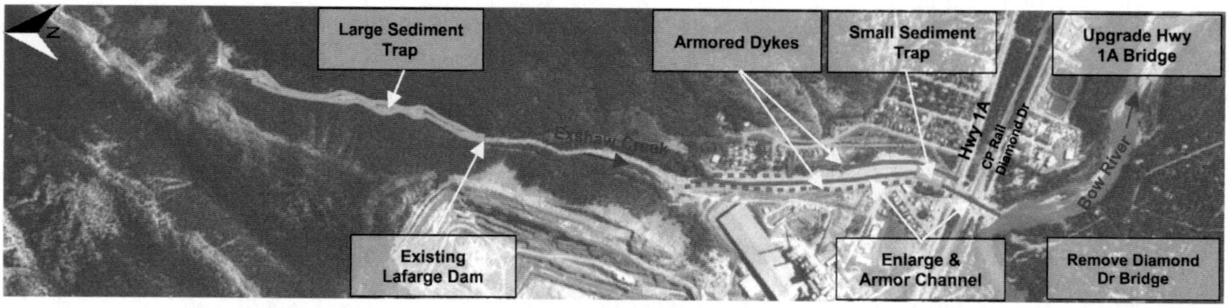

Figure 2: Exshaw Creek Debris-Flood Mitigation Concepts – Option 1

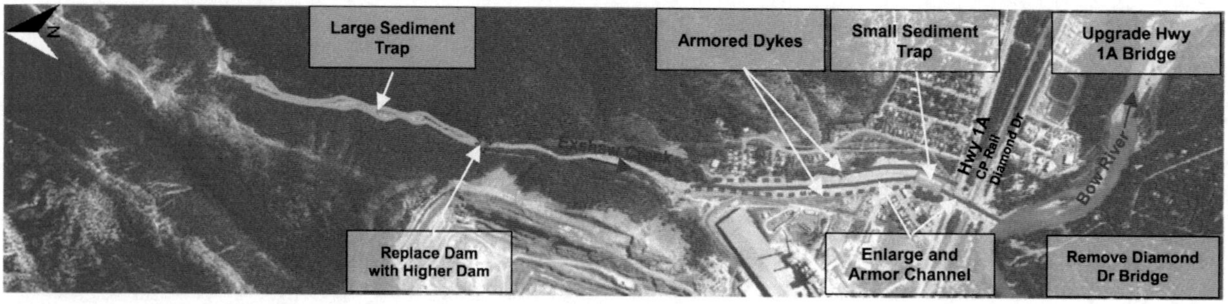

Figure 3: Exshaw Creek Debris-Flood Mitigation Concepts – Option 2

Figure 4: Exshaw Creek Debris-Flood Mitigation Concepts – Option 3

Figure 5: Exshaw Creek Debris-Flood Mitigation Concepts – Option 4

4 JURA CREEK DEBRIS-FLOOD MITIGATION CONCEPTS

Figures 6 to 10 present the Jura Creek debris-flood mitigation concepts that were developed for the project. Estimated costs of each concept are included in Table 3. The following provides a description of each concept.

Option 1: Use dyking strategies to protect the residential area along Heart Mountain. The Highway 1A culvert crossings will be upgraded and a new channel will be excavated downstream of the Highway 1A culvert crossing to allow water to reach the Bow River. When the Highway 1A culvert infills with sediment, the east dyke will force avulsion of the Jura Creek channel to occur northward along the west bank. The three small deflector dykes will keep the flood flows north of the highway within the trees such that flow velocities decrease and all of the sediment is deposited. The west L-shaped dyke will then direct any overland spillage over Highway 1A towards the existing culvert crossings (that will be upgraded) along the CP embankment and Diamond Drive. A sump and automated pumping facility will evacuate any overland flow that collects to the west of the L-shaped dyke.

Option 2: Use dyking strategies to protect the residential area north of Highway 1A. This is similar to option 1, except the south side of the L-shape dyke is removed and the existing Highway 1A culvert crossing is upgraded. A flap-gate will be installed to prevent backwater from the Bow River from flooding the residential area. The sump and automated pumping facility is replaced by the excavation of a new channel to the Bow River.

Option 3: Use dyking strategies to protect the residential area along Heart Mountain Drive. This is similar to option 2, except the west dyke is shifted eastward such that water is redirected towards the Bow River and away from the residential area. This option requires a new culvert crossing on Highway 1A and a new channel to direct water that collects in the residential area.

Option 4: Install a large sediment trap at the apex of the Jura Creek alluvial fan. This option requires a sediment trap sized for the design sediment yield to be constructed immediately downstream of the Jura Creek slot canyon. The channel downstream of the sediment trap will be enlarged and armored to convey the design flood and prevent any sediment from being entrained. The new armored channel will be constructed downstream of Highway 1A and the Bow River. The crossings at Highway 1A and the CP embankment will be upgraded. This option does not require dyking measures because no culvert infilling or avulsion of Jura Creek is anticipated.

Option 5: Install a large sediment trap immediately upstream of Highway 1A and a small sediment trap immediately upstream of the CP embankment. This option involves enlargement of the channel between the Jura Creek alluvial fan apex and the large sediment trap. Significant enlargement of the channel will ensure that no avulsion of Jura Creek occurs. The large sediment trap is expected to collect all of the design sediment yield and clean water will pass through the Highway 1A culvert crossing. The clean water will entrain some sediment between Highway 1A and the CP embankment, which will be collected by the small sediment trap upstream of the CP culverts. No culvert infilling is anticipated.

Table 3: Jura Creek Debris-Flood Mitigation Concept Costs (Golder 2016)

Debris-Flood Mitigation Option	Capital Construction ($ Millions)	Mobilization & Contingency ($ Millions)	Total Estimated Cost ($ Millions)	Annual Operations and Maintenance ($ Thousands)
1	8.4	3.9	12.3	83
2	9.1	4.3	13.4	92
3	8.6	4.1	12.7	87
4	10.9	5.2	16.1	87
5	3.2	1.4	4.6	58

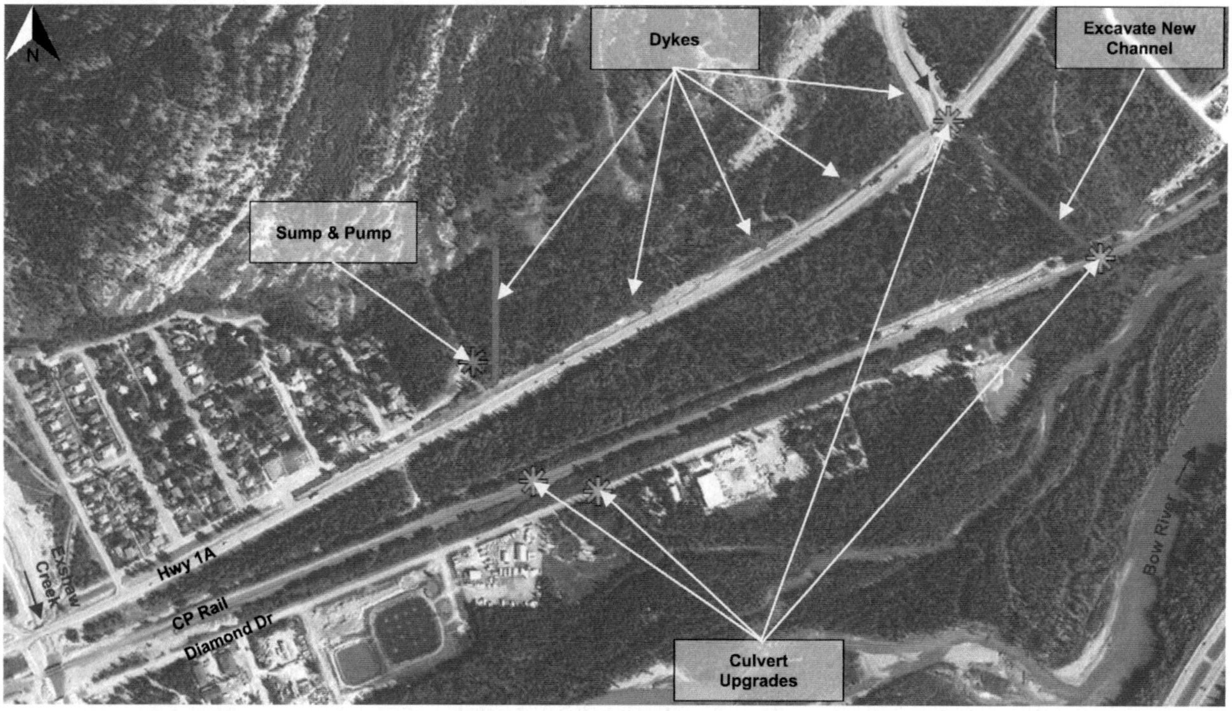

Figure 6: Jura Creek Debris-Flood Mitigation Concepts – Option 1

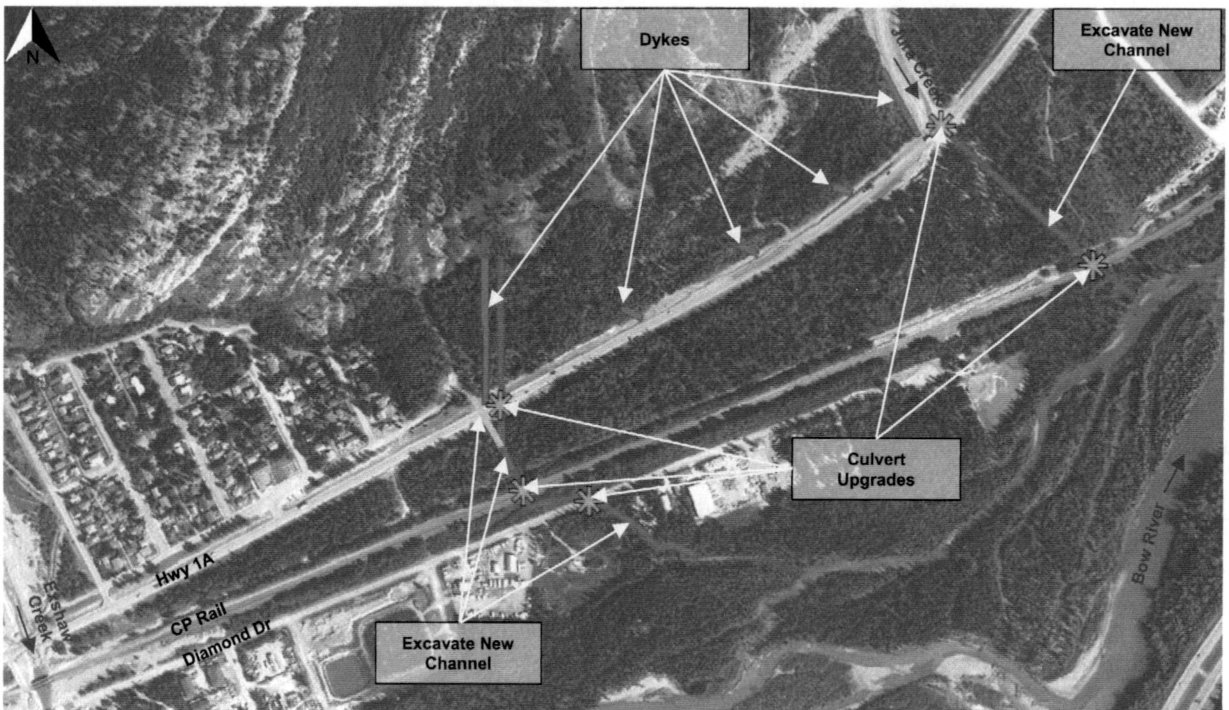

Figure 7: Jura Creek Debris-Flood Mitigation Concepts – Option 2

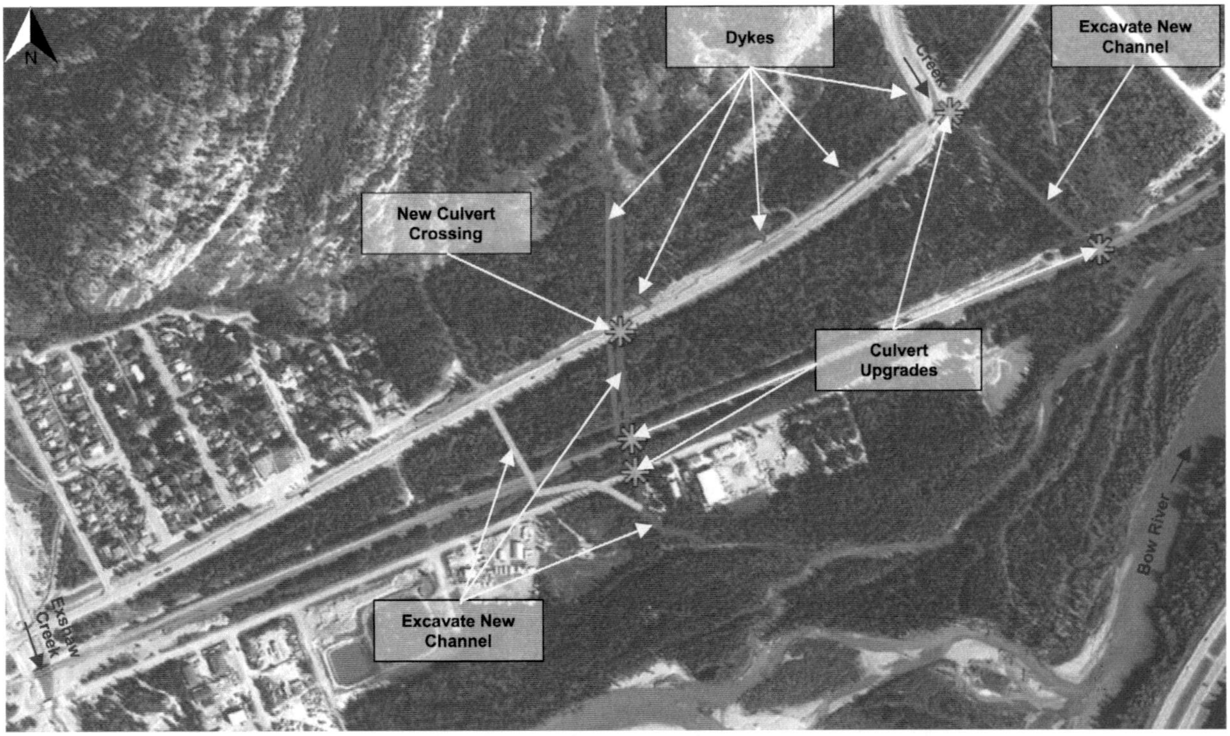

Figure 8: Jura Creek Debris-Flood Mitigation Concepts – Option 3

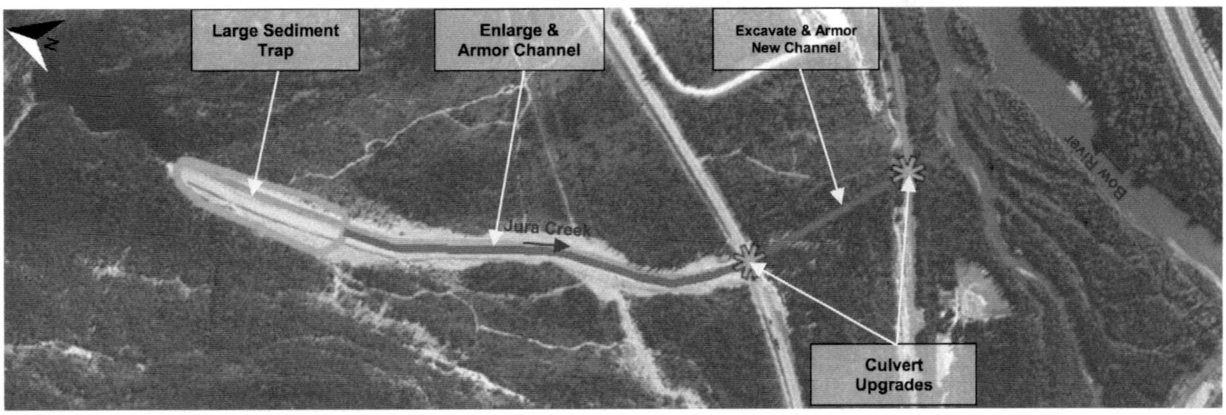

Figure 9: Jura Creek Debris-Flood Mitigation Concepts – Option 4

Figure 10: Jura Creek Debris-Flood Mitigation Concepts – Option 5

5 STAKEHOLDER ENGAGEMENT

MD of Bighorn No. 8: The MD of Bighorn No. 8 has a small tax base, totalling approximately $4 million per year. The current funding provided by the Alberta Community Resilience Program (ACRP) requires partial contribution from the MD of Bighorn No. 8 and other stakeholders for any selected flood mitigation concept. Therefore the selected design concept is required to be practical and cost effective.

LafargeHolcium: LafargeHolcium has indicated a willingness to contribute to the Exshaw Creek debris-flood mitigation with a discounted price of riprap for erosion protection and has used their relationship with a local contractor to obtain favorable quotes for hauling excavated sediment and debris. LafargeHolcium will also pay the cost to complete construction of the dyke on the west side of Exshaw Creek.

Alberta Transportation: Alberta Transportation is a significant stakeholder because of the new and upgraded creek crossings and replacement of the Highway 1A bridge that may be required at Exshaw Creek. They also own the property located between Highway 1A and the CP alignment along Jura Creek.

CP: CP prefers options for Jura Creek that have the least change in risk to the railway from natural flood conditions. CP is open to upgrading of existing crossings or to construction of new crossings across their railway embankment.

Private Landowners for Properties Directly Impacted: All relevant stakeholders have been engaged, and these concepts were developed in consultation with each of the respective property owner.

Residents of the Hamlet of Exshaw: The MD of Bighorn No. 8 has held several open houses to engage the residents of the Hamlet for the development of the options and the preferred selected option. The Golder design team was present at each of the open houses to explain the flood mitigation concepts to the residents and address any questions or concerns.

6 SUMMARY AND RECOMMENDATIONS

The work performed during the Phase 1 (Concept Development and Validation) supports the overall goal to reduce the risk of Exshaw Creek and Jura Creek debris-floods. The project involved identifying and evaluating four debris-flood mitigation options for Exshaw Creek and five debris-flood mitigation options for Jura Creek. All of the options were considered feasible, but vary greatly in cost, environmental impacts, and requirements for regulatory review.

From the cost summaries, the lowest cost options are Exshaw Creek Option 4 and Jura Creek Option 5. The MD of Bighorn No. 8 was advised to proceed with these options for Phase 2 (Preliminary Design). It was also noted and recommended that reduction of design criteria could result in significant cost savings, which was valuable to the MD of Bighorn No. 8 to reduce the burden to their relatively small tax base due to the ACRP cost-sharing framework.

ACKNOWLEDGEMENT

The authors would like to acknowledge and thank Martin Buckley, Chief Administrative Officer, at the Municipal District of Bighorn No. 8 (Exshaw, Alberta) for allowing this work to be showcased at the 2017 CSCE Conference in Vancouver, BC.

REFERENCES

Acres International Ltd. (Acres). 1996. Bow River – MD of Bighorn: Flood Risk Mapping Study. *A report prepared for the Alberta Environmental Protection River Engineering Branch.*

BGC Engineering Inc. (BGC). 2015. Exshaw Creek and Jura Creek Preliminary Debris-Flood Mitigation Design. *A report prepared for the Municipal District of Bighorn No. 8.* Retrieved from http://mdbighorn.ca/flood-hazard-management/ on August 18, 2015.

Golder Associates Ltd. (Golder). 2016. Conceptual Design for Exshaw Creek and Jura Creek Debris-Flood Risk Reduction. *A report prepared for the Municipal District of Bighorn No. 8.*

Leadership in Sustainable Infrastructure
Leadership en Infrastructures Durables

Case Study

Vancouver, Canada
May 31 – June 3, 2017/ *Mai 31 – Juin 3, 2017*

DEVELOPMENT AND FIELD APPLICATION OF THE RCC JACK JETTY FOR RIVER TRAINING

Nayak, Anupama[1], Sharma, Nayan[2] and Mazurek, Kerry[3,4]
[1] Water Resources Department, Government of Odisha, Bhubaneswar, Odisha, India
[2] Professor, Department of Civil Engineering, IIT Roorkee, Roorkee, India
[3] Associate Professor, Civil Engineering, University of Saskatchewan, Saskatoon, Saskatchewan
[4] Kerry.Mazurek@usask.ca

1 INTRODUCTION

Jack Jetties were river training structures used for bank protection and creating sediment deposition in the United States in the 1920's. A Jack was a structure that looked like the jacks that children had played with earlier in that century. The Jacks were made of steel and were strung together to form lines much like the groynes that are used today. A Jetty is a line of Jacks and Jetty Field was a series of Jetties. The structures were seen to be reasonably effective at producing sedimentation but design procedures were ad hoc. Ultimately, their use in North America did not continue because of their potential for disrupting navigation. In India, however, problems with river engineering are significant; rivers can be large. As such, any river engineering measure developed should be inexpensive due to the scale of the work that can be required. Jack Jetties are of interest for renewed investigation for application in Indian rivers as an inexpensive river engineering structures. Because of the costs of steel, it was thought instead the jacks could be made of cheaper reinforced concrete (RCC). Figure 1 gives a sketch of a Jack and photo of a RCC Jack Jetty field.

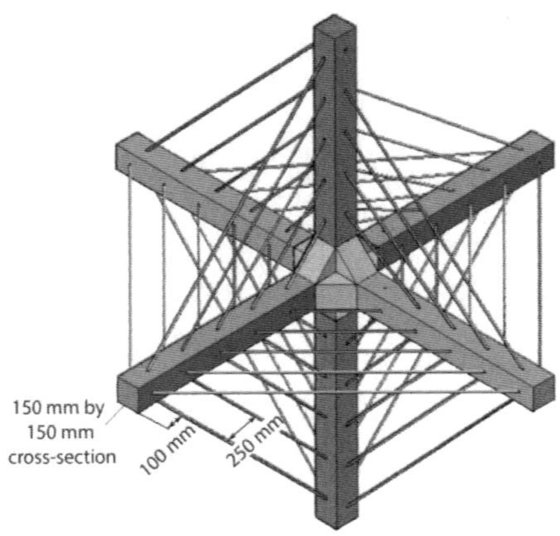

(a)

(b)

Figure 1: (a) RCC Jack Jetty of 3 m length and 0.15 m by 0.15 m cross-section (adapted from Nayak *et al.* 2017) and (b) RCC Jack Jetty field in the River Ganga at Nakhwa

A study of the efficacy of the Jack Jetties was therefore conducted to determine their potential to create sediment deposition in large rivers in India. First, measurements were conducted in the laboratory to examine the flow field behind single and multiple Jacks set parallel with, then perpendicular to, the flow. A further laboratory study was conducted to study the optimum configuration of Jacks for sediment deposition to develop guidelines for how much sediment deposition might be expected for different flow, sediment concentrations, and submergence conditions of the Jacks. Finally, a field test was conducted in the River Ganga at Nakhwa to test the effectiveness at producing sedimentation in a Jack Jetty field configuration expected to produce a significant sedimentation from the laboratory experiments.

2 LABORATORY EXPERIMENTS AND FIELD TRIAL

The laboratory experiments to examine the flow field behind a Jack or multiple Jacks either inline or as Jetty were conducted in two parts. The first set of experiments was conducted at the Indian Institute of Technology (IIT) Roorkee, in Roorkee, India, within a 22.5 m long, 1.2 m wide, 0.6 m high flume with a sand bed of mean grain size of 0.25 mm. Jacks of 0.08 m or 0.1 m height were placed individually, with 2-4 placed in a single or double row parallel to the flow, or 2-4 placed in a single to quadruple row across the flume, with a depth of flow of 100 or 200 mm. The flow rate was 0.013 to 0.039 m^3/s and the Froude number was kept constant at 0.11. A second set of experiments were conducted within a 25 m long, 1.25 m wide, 0.6 m deep flume in the Hydraulics Laboratory at the University of Saskatchewan in Saskatoon. A fine sand of 200 mm depth served as the flume bed with an average grain size of 0.59 mm. Single or multiple Jacks of 0.2 m or 0.3 m height were placed in the flume in a line of 1 to 3 Jacks parallel to the flow. The flow rate range from 0.071 to 0.107 m^3/s, with a flow depth of 300 or 400 mm, and a constant Froude number of 0.1. Velocity measurements in both sets of experiments were carried out in clear water and taken just upstream and behind the Jacks using an Sontek 16 MHz MicroADV. The objective was to observe how much the velocities behind the structures were reduced by installation of the Jacks. Results indicated up to a 60 % velocity decrease with the Jack in the flow and this occurred immediately behind the obstacle.

Further experiments were conducted in the Hydraulics Laboratory at IIT Roorkee to examine how much sediment would deposit for different configurations of Jack Jetty fields for varied sediment concentration in the flow (250 to 1000 ppm). The Jacks were 0.1 m high and were set in a 0.5 m wide flume on a sand bed of 0.15 m thickness. Two sands of average grain size of 0.248 mm and 0.59 mm were used, the flow rate ranged from 0.013 to 0.0225 L/s, and the Froude number ranged from 0.11-0.15. Tests were first run for 30 minutes with clear water, after which the bed profile was measured. Then, flow in the flume was reinitiated and sediment was injected for 4 hours, after which the flow was again stopped and the bed profile measured to judge the amount of sediment deposition on the bed. From the work, guidelines were developed for the design of a Jack Jetty field configuration to provide different (qualitative) amounts of sediment deposition (Nayak *et al.* 2017) for varied suspended sediment concentrations in the flow. This is done in terms of a bed deposit factor, the ratio of the bed deposition to the total depth of water, and the Jetty Field Density Index, which is the ratio of the length of a Jack Jetty to the Jetty centre-to-centre spacing.

Next, a field trial was conducted to test the efficacy of a RCC Jack Jetty field in the River Ganga at Nakhwa, where the channel is split into two. The left channel was to be reduced to force more flow through the right channel to improve navigability. After installation of a Jack Jetty field at the site, due to deposition, there was a reduction in width from 246 m to 50 m over two years in the left channel of the river.

3 CONCLUSION

The Jack Jetty appear to be an effective, inexpensive means of forcing sediment deposition river and reducing the width of the river to provide sufficient draught for navigation.

REFERENCES

Nayak, A., Sharma, N., Mazurek, K.A., and Kumar, A. 2017. Design Development and Field Application of RCC Jack Jetty and Trail Dykes for River Training, in *River System Analysis and Management*, N. Sharma, ed., Springer, Singapore.

Leadership in Sustainable Infrastructure
Leadership en Infrastructures Durables

Vancouver, Canada
May 31 – June 3, 2017/ *Mai 31 – Juin 3, 2017*

EFFECT OF LONGTERM NAVIGATION CHANNEL LOWERING ON SCOUR AND DEGRADATION PROCESSES ON LOWER FRASER RIVER

Andrew D. Nelson [1,5], Ilana Klinghoffer [2], Matt Gellis [3], and David McLean [4]
[1] Northwest Hydraulic Consultants, USA
[2,3,4] Northwest Hydraulic Consultants, Canada
[5] anelson@nhcweb.com

ABSTRACT

The Fraser River, which is the largest river on the west coast of Canada, terminates in a large, sand-bedded and tidally-influenced delta that extends from New Westminster to the Strait of Georgia. This lower reach of the river has been extensively modified over the last century to provide flood protection for surrounding areas and adequate draft for navigation of ocean-going vessels. Sediment removal by dredging the navigation channel below New Westminster, in combination with river training, bank hardening, and scour protection, has significantly altered the channel hydraulics, sediment transport characteristics, and morphology of the river. This paper uses the extensive historical record of bathymetric surveys that date back to the early 20th century to characterize the complex channel response that has occurred over time. The resulting bed lowering and degradation extend well beyond the limits of the localized navigation channel improvements. Importantly, bed lowering has led to increasing exposure of non-alluvial channel boundary material including riprap structures that were installed previously as scour protection measures, as well as cohesive delta foreset beds and less erodible Pleistocene deposits such as till and outwash. These hard points generate additional turbulence and plunging flow that have induced additional local scour, effectively amplifying the effect of the initial degradation. Consequently, deep local scour (over 20 m) has developed at some sections of the river.

Scour, Large river, Dredging, River training, River morphology, Channel bed facies

1. INTRODUCTION

River management activities, such as dredging, diking, and construction of river training structures and scour protection counter-measures may alter long-term channel stability and morphology. Furthermore, the presence of non-alluvial bed material —both natural and constructed engineering works— are increasingly recognized as having a significant influence on processes such as scour in lowland alluvial rivers (e.g. Nittrouer et al., 2011). Many previous studies have used historic and recent bathymetric charts to examine changes in estuary morphology (Van Der Wal and Pye, 2003) and some of these have explicitly explored estuary response to human modifications (e.g. Nichols and Howard-Strobel, 1991; Lane, 2004; Wang et al., 2013), but relatively few studies have used historical bathymetry data to evaluate long-term changes in channelized areas of the lowest reaches of large alluvial rivers.

This paper examines historical changes in channel morphology of the lower Fraser River, to characterize the complex channel response to anthropic activities over the last century. From a local perspective, this case study highlights the importance of harmonizing scour protection and navigation channel improvement activities. In the general case, it highlights the importance of understanding the local geomorphic controls and their potential interaction with intensive channel management activities on large lowland alluvial rivers.

2. THE LOWER FRASER RIVER

Geologic context

The Fraser River is the largest river on the west coast of Canada, with a basin draining about 232,000 km^2 of British Columbia. Following Pleistocene Glaciation, when glaciers scoured the Lower Mainland and left deposits of till and outwash, the river constructed an alluvial delta, which initially prograded through subaerial hills of Pleistocene glacial sediment and subsequently extended across buried glacial sediment out into the Strait of Georgia (Figure 1) to form the modern delta that extends from the channel trifurcation at New Westminster into the Strait.

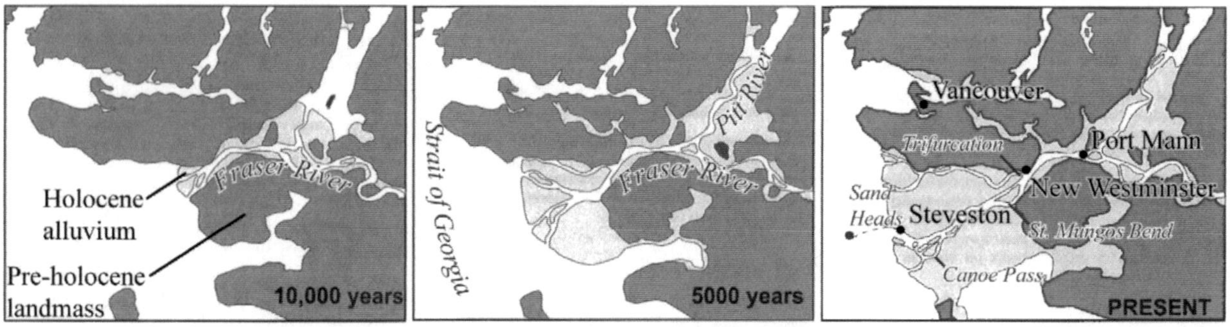

Figure 1: Holocene progradation of Fraser River into the Strait of Georgia (adapted from Clague et al., 1991) with present conditions showing key points referenced in the text.

The study area extends from the edge of confined flow at Steveston, upstream to the confluence of the Fraser and Pitt Rivers (Figure 1). From the Trifurcation (RK 34) east to the gravel-to-sand transition at the toe of the modern gravel fan near Sumas Mountain (RK 100), the river has a meandering, sand bedded channel with a few local flow splits around large, stable islands formed by chute cutoffs. The channel locally interacts with banks and bed substrate composed of Pleistocene glacial sediment, including overconsolidated advance outwash and pre-Vashon sediment (Quadra marine and advance outwash deposits), till (Vashon drift), glaciomarine drift (Ft. Langley formation), and gravelly recessional outwash (Capilano sediments) (Armstrong and Hicock, 1980; Armstrong and Kovachic, 1980).

The river splits into several distributary branches downstream of New Westminster as it flows through the modern delta, which extends 15 to 23 km in a broad plain. The top surface of Pleistocene sediment is typically 150 m and deeper below the delta (Figure 1) but it is irregular—ranging in elevation from approximately -300 m to -15 m (geodetic) (Clague et al., 1998). The Holocene delta is comprised of four primary facies: fine-grained bottomset beds deposited in deep water, foreset beds, medium to coarse sand river channel deposits, and floodplain and bog deposits at the surface (Clague et al., 1983, 1991). The river channel deposits form a nearly continuous 8-20 m thick sheet across the delta surface with a surface contact typically near mean sea level (Monahan et al., 1993); this sediment has very similar properties to active lower Fraser bed material. Sediment below the sand sheet was deposited on the delta foreslope and consists of fine sand and mud (Clague et al., 1983); this material has a much finer grainsize distribution than active lower Fraser bed material and will interact differently with scouring processes due to its compaction history, cohesive properties, and tendency to move into the river's wash load once eroded.

An early bathymetric chart (Richards, G.H. and Officers of the H.M.S. Plumper, 1860) provides a view of the Lower Fraser bathymetry prior to major human modification. The river ranged in width from 600 m to 1.2 km and had a general morphology like that of the present-day river, with alternating deep pools (-10 to -18 m) and shallow crossings (-2 to -6 m). The deepest observed sites occurred in the narrow channel at New Westminster (-19 m) and in a sharp bend in the area that is now Canoe Pass, with scour down to approximately -23 m.

Hydrology

Due to its large, mostly interior and high-elevation basin, the Fraser River has a snowmelt-dominated flow regime. Discharge typically rises in April, peaks between May and July (the freshet period) then recedes during the autumn and winter months. At Hope, 180 km upstream from the mouth (RK 180), the river has a mean annual discharge of about 2,800 m^3/s. Winter low flows are typically around 1,000 m^3/s, and annual freshet flood peaks typically range from about 7,000 to 15,000 m^3/s (average peak flow is about 8,700 m^3/s occurring in June). There is no secular change in peak flow intensity over the period of record for the Fraser, which extends back to 1912. Decadal-scale periodicity in flood peaks occurs on the river (NHC, 2015), which are related to Pacific Decadal Oscillation climate variability (Whitefield et al. 2010). These have been slightly higher than average for the past 10 years.

Sediment supply

Sediment flux in the lower Fraser River is wash load dominated. Most sediment flushes out to the delta front in the Strait of Georgia, but the sand component of the flux interacts with the bed. Observations of sediment loads on the Lower Fraser River were collected by Water Survey of Canada during the period 1965 to 1986, and the total suspended load averaged 17.3 million tonnes/year (range 12.3 million to 31.0 million), with the load consisting of 35% sand, 50% silt and 15% clay (McLean and Tassone, 1990; McLean et al., 1999). The suspended bed material load of sediment larger than 0.18 mm averaged 2.8 million tonnes/year (range from 1.2 to 9 million). Detailed measurements collected during 2010 are consistent with (but on the low side of) these historical observations, introducing the possibility that the load has decreased since the cessation of systematic transport measurement (Attard et al., 2014).

An estimate of the annual bed material load between Mission and the river mouth at Sand Heads was made by McLean and Tassone (1990) using a sediment budget approach based on measurements at Mission, observed channel topographic changes, and dredging removal records. That analysis indicated the annual bed material load delivered to the mouth of the river at Sand Heads was roughly 30% of the load at Mission during the period 1972 to 1997, indicating a major bed material sediment budget deficit.

Hydraulics

The river is tidally affected throughout the year as far as the gravel to sand transition at RK 100. The relative influence of the tide is governed by the fresh water inflow, magnitude of the tide, and distance from the river mouth. The lower river is subject to reversing flow during rising (flood) tides, particularly during periods of low or moderate freshwater inflows. At Port Mann (44 km upsteam from the river mouth), reversing flow occurs up until freshwater inflows exceed approximately 7,000 m^3/s. A salt water wedge can extend upstream to about RK 30 during winter when freshwater inflows are low and the tidal range is large (Ward, 1976). High freshwater discharges during the June freshet prevent the salt wedge from extending upstream of Steveston (approximately 10 km from the river mouth). The salt wedge can strongly influence near-bed hydraulics and sedimentation patterns in the lower 10 km (e.g. Kostaschuk, 2002; Dashtgard et al., 2012).

Dredging, river training history, and consequent general channel changes

The Lower Fraser is an important commercial corridor, with the river accounting for approximately one-quarter of the total cargo tonnage handled by Port Metro Vancouver—the largest port in Canada and fourth busiest port by tonnage in North America (Richmond Chamber of Commerce, 2014). Major flood protection and river training structures control the channel, which has been dredged since the early part of the 20th Century to provide deep water navigation from New Westminster downstream. The combined effect of sediment removal from dredging and constriction of the channel has significantly altered river processes extending a considerable distance beyond the limits of the localized dredging operations (e.g. McLean et al., 2006; NHC, 2008).

Flood control dikes, constructed initially between 1913 and 1919 (Dashtgard et al., 2012) now isolate the entire historical floodplain downstream of New Westminster (except a few small mid-channel islands) and isolate 90% of the river downstream of Hope. Training structures have been built to control the flow split into distributaries, concentrate scour and increase the channel draft, and protect infrastructure along and crossing the channel as detailed by Hales (2000). Most training structures were installed prior to 1936; notable exceptions include the trifurcation project (1967-70) and Sapperton Wing Dams (1955). Scour and bank protection features have been installed more gradually through time, these include bank revetments, riprap collars around bridge piers, and riprap aprons over pipelines and tunnels.

Channel dredging has been conducted to increase the depth of a navigation channel for deep draft vessels along the river up to New Westminster. It has also included industrial borrow dredging, mostly in the period between 1976 and 1988. Dredging began in 1885 when the main channel was only 2.7 m deep at the river mouth (Dashtgard et al., 2012), but the main systematic effort began around 1945. Figure 2 illustrates the quantity of material removed from the river through time and historic change in mean bed elevation. The large removal volumes (4-6×10^6 m^3/year) between 1975 and 1992 reflect the borrow dredging activities on the river. Dredging decreased to approximately 2×10^6 m^3/year between 1997 and 2003, then increased to approximately 3×10^6 m^3/year in recent years. The total dredged volume over the last 50 years amounts to 1.5×10^8 m^3. Channel response to the dredging and river training, which have imposed a persistent bed material budget deficit, has been persistent long-term channel degradation beginning in the early 20th century and continuing to the present day (e.g. McLean and Tassone, 1990; McLean et al., 2006; NHC, 2015). Between 1965 and 1988, when sediment removals were at their peak, the bed along the navigation channel centerline lowered at an average rate of about 8 cm/year. Comparison of change in bed elevation between 1988 and 2014 shows that that the pattern of channel degradation has continued through the past two and a half decades, but notably slowed to an average rate of about 2 cm/year since the late 1980s or early 1990s.

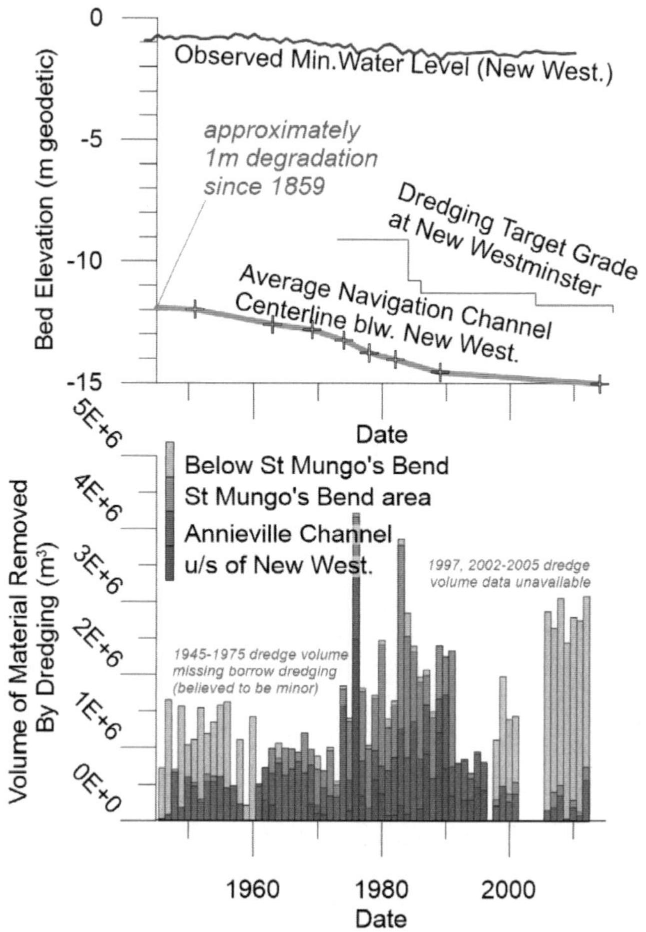

Figure 2: Historic changes in general bed elevation (top) and pattern of channel dredging activity (bottom).

Figure 2 also shows how the river bed lowering at and downstream of the project site has caused the minimum observed water level to systematically drop at New Westminster, as well as how the average channel grade has typically been 3 to 4 m below the target grade through time. This offset between the target grade and average channel centerline grade reflects the fact that the target grade is the maximum allowed elevation at shallow current crossover locations and that deeper pool areas are common along the channel.

3. METHODS

Bathymetric changes in time have been evaluated by compiling, digitizing, registering, and differencing historic sounding information at two distinct scales along the river. General channel changes have been evaluated at the reach-scale from Steveston (RK 9) to the Pitt River confluence (RK 44), as summarized in Table 1. In order to better understand the processes responsible for the observed reach-scale changes in channel morphology, bathymetric data were analyzed in detailed annual or semi-annual temporal resolution at two sites; St Mungo's Bend (RK 28-32) and in the area of New Westminster from the channel Trifurcation upstream to the Pattulo Bridge (RK -34-38). Local data from 1898 to 1955 were digitized from bathymetric charts included in the UBC Library Fraser River Model Project fonds (Jarlan, 1956); from printed navigation charts published by Public Works Canada from 1962 through 1985; and digital data provided by Public Works and General Services Canada (PWGSC) from 1989 through the present. Prior to 1999, the digital dataset consisted of surveyed cross-sections—typically with a spacing of 75-100 m— of about one fifth of the navigation channel width. From 1999 to 2011, multi-track surveys were conducted by producing grid coverage with an average spacing of around 1 m. Since 2011, extremely detailed multi-beam bathymetric data are available. These data were transformed from point data into a grid format with a common vertical reference.

Bathymetric changes at the reach-scale were analyzed by dividing the channel into 100 m segments and evaluating the pattern of elevation change—with emphasis on the relation between change in mean channel depth and maximum scour depth. In addition, the comparison between late 1980s and 2014/15 mosaics was stratified by the initial depth of the channel and overall changes in each 100 m segment. Results of this comparison at the reach-scale were plotted both spatially (to illustrate patterns along the channel) and as cumulative distributions in order to show the magnitude and relationships between bathymetric parameters along the channel.

Areas of non-alluvial channel boundary material exposed at the present (2014/15) time were mapped based on the surface texture visible in multi-beam bathymetry data (see for example Nittrouer et al., 2011). This is possible because recently acquired sub-metre topography can reveal bedforms indicating some depth of alluvial material coverage, erosional features such as flutes or kettles, slopes higher than the angle-of repose indicating exposed cohesive material, and high local roughness and a blocky texture which suggests riprap cover. Because only recent data show these features, the timing of initial non-alluvial material exposure is generally unknown. Therefore, analysis of the relationship between scour amplification and non-alluvial material focuses on the period between the late 1980s and 2014/15.

Table 1: Summary of data sources used in river-scale bathymetric comparison.

Period	Year	Extent	Mosaic Extent	Mosaic Comparison
1953	1953	RK 34 to RK 44	RK 34 to RK 44	NA
Early 1970s	1971 1972 1974	RK 9.7 to RK 20.3 RK31.6 to RK 44.6 RK 20.2 to RK 31.4	RK 9 to RK 44	Change between 1953 and 1974 from RK 34 to RK 44
Late 1980s	1988 1989 1992	RK9 to RK 44 with gaps fill gaps in 1988 dataset	RK 9 to RK 44	Change between 1974 and 1988 from RK 9 to RK 44
2014/15	2014 2015	RK 9 to RK 35 RK 35 to RK 46	RK 9 to RK 44	Change between 1988 and 2015 from RK 9 to RK 44

Data sources: 1953 through 1974 PWGSC Charts; 1988-1992 PWGSC single beam soundings; 2014-2015 PWGSC multi-beam soundings. The source data are referenced to a local low water datum, which is a stepped datum along the lower river that has changed through time as hydraulic conditions have responded to the general bed lowering. For the purposes of this study, data were reduced to geodetic datum.

4. RESULTS

Reach scale changes

The spatial and temporal patterns of bathymetric change along the Lower Fraser River support the hypothesis that human-forced channel bed lowering is increasingly exposing non-alluvial channel boundary material with the consequence of increasing heterogeneity in channel bed elevation through time, which is dominantly expressed through scour depths increasing faster than average channel depth (Figure 3, Figure 4, and Table 2).

Comparison of bathymetry between 1953 (approximately the beginning of major channel lowering) and the early 1970s is restricted to RK 34 to 44 due to the spatially limited extent of the 1953 dataset. During this period, the average bed elevation in individual segments lowered by 1.1 m and the median maximum local depth increased by 1.9 m in this area. Over the whole study reach between 1859 and the early 1970s, the average bed elevation lowered by about 3 m, and median maximum local depth increased by about 2 m, probably mostly between 1945 and 1974 given the known history of navigation centerline channel lowering (Figure 2). Between the early 1970s and late 1980s, a tremendous volume of sediment was evacuated from the Lower Fraser, with a cumulative bed material deficit of around 30 million m^3 between RK 44 and RK 9. The average bed elevation dropped by 2.5 m and the median maximum local depth increased by 3.5 m.

The mean bed elevation change has been about -0.37 m between the late 1980s and 2014/15, resulting in a cumulative bed material deficit of about 4.5 million m^3 from RK 44 to RK 9. The average bed elevation changes were relatively modest; however, local maximum depths continued to rapidly lower. The maximum depths (observed during non-flood conditions) increased by 1 to 3 m along most of the lower river, with some areas experiencing up to an additional 13 m of bed lowering (Figure 3).

Figure 4 shows how scour holes have increasingly become more sensitive to channel lowering than the general bed level. Scour hole depth and morphology has become exaggerated in areas where the deepest portion of the channel is reaching below -20 m—the approximate bottom of the Fraser Delta sand sheet and corresponding historic maximum scour depth. Figure 5 shows the characteristic relation of these areas of bed lowering in deep channel areas to the presence of non-alluvial channel boundary materials.

Cumulative distribution plots of changes in the maximum, mean, and minimum bed elevation calculated for each 100 m segment are a way to summarize the magnitude of bed changes that have occurred along the representative reach of the lower river (Figure 6). These illustrate the magnitude of possible bed changes to be expected with moderate lowering of the average channel grade and show that the scour amplification ratio has increased through time. Most recently, between the late 1980s and 2014/15, minimum bed elevations have been very sensitive to the modest general bed lowering: 50% of sites experienced more than a 1.4 m increase in maximum depth (about triple the median change in mean bed elevation), and 16% of sites experienced more than 2.8 m increase in maximum depth. In comparison, between 1974 and 1988, 50% of sites experienced more than 3.5 m increase in maximum depth (about 1.2 times the median change in mean bed elevation), and 16% of sites experienced more than 5.2 m increase in maximum depth. Between 1953 and 1974, 50% of sites experienced more than 1.9 m increase in maximum depth (about 1.3 times the median change in mean bed elevation), and 16% of sites experienced more than 4.1 m increase in maximum depth.

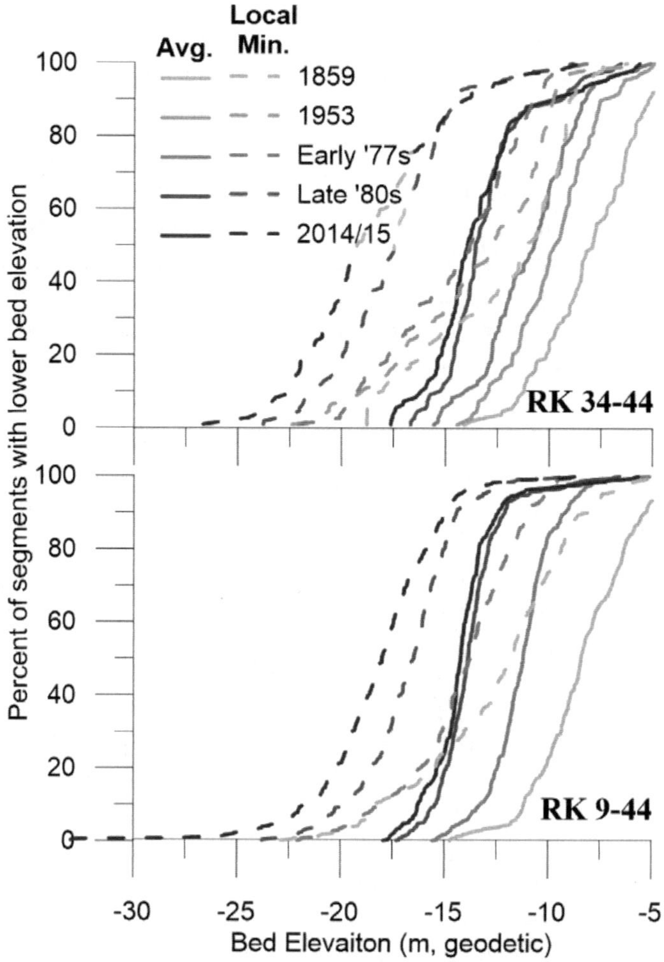

Figure 3: Cumulative bed elevation distribution plots for portions of the lower Fraser River showing much faster lowering of minimum bed elevations than mean bed elevations.

Table 2: Average bed elevation change and cumulative bed material deficit from RK 9 to RK 45.

Period of Comparison	Average Bed Elevation Change (m)	Cumulative Volumetric Bed Change (m^3), by RK 44 to 34	44 to 9	Change in Maximum Local Depth (m), by percentile 16th	50th	84th	Scour Amplification Ratio**
1953 to 1970*	-1.06*	-4,376,000	n/a*	*-0.45	-1.9*	-4.1*	2.7*
1974 to 1988	-2.5	-8,404,000	-29,637,700	-2.2	-3.5	-5.2	1.4
1988 to 2015	-0.37	-1,394,000	-4,495,700	-1.0	-2.3	-3.6	6.2

* Available data to evaluate changes for the period 1953 to 1974 restricted to RK 34-44.

** Sour amplification ratio defined as the ratio of 50th percentile change in local maximum depth to average bed elevation change.

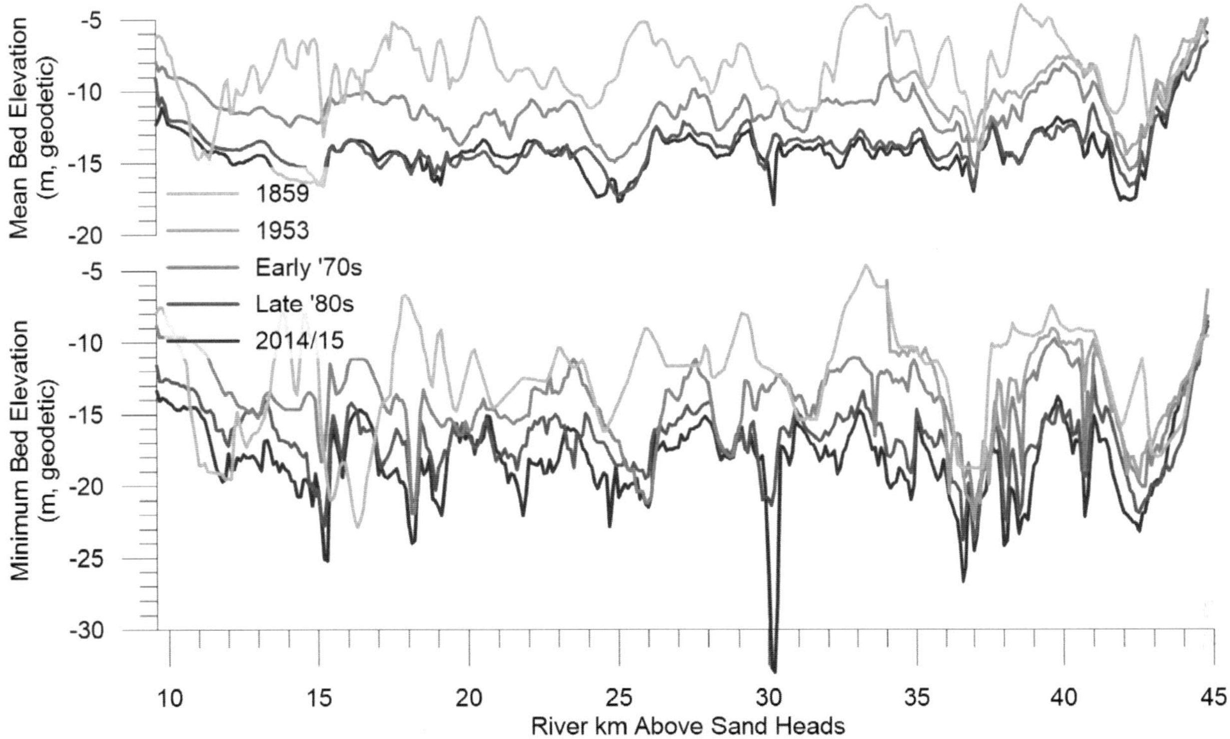

Figure 4: Profiles showing historical change of bed elevation summarized by 100 m segments along the study reach. Downstream of RK 19.3, the 1859 alignment diverges from the present alignment; bed elevation shown correlated to the nearest segments of the present alignment.

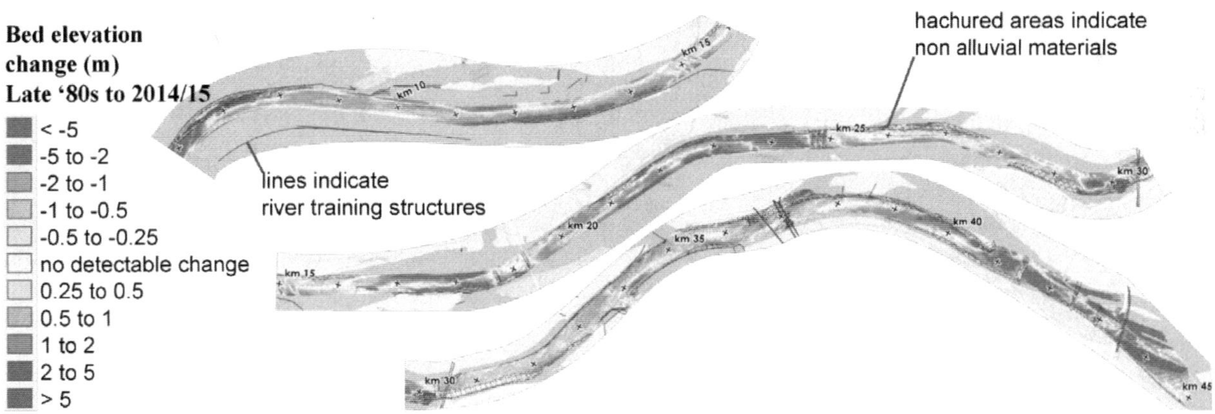

Figure 5: Map view of recent (late 1980s to 2014/15) bathymetric changes showing relationship to areas with non alluvial materials mapped from the 2014/15 multi-beam bathymetry.

Case studies

Two local case studies (Figure 7) that motivated the preceding reach-scale analysis provide helpful insight regarding some of the processes governing scour depth amplification along the lower Fraser River. The first of these is St. Mungo's Hole (RK 30 Figure 4), which has scoured to a depth of -33 m (geodetic), about 10 m below the historic maximum scour depth for the reach.

The Annieville Channel upstream of the hydraulic influence of Alex Fraser Bridge has been persistently degrading as it tracks general bed lowering. From the 1972 survey to the present, the channel has degraded by approximately 5 m. The reach-wide pattern of bed lowering upstream of Alex Fraser Bridge must be kept in mind when interpreting bed-level changes at and downstream of the bridge, as it may compound or otherwise interact with local hydraulic perturbations in those areas. One particularly important change through the whole project reach has been increasing exposure of non-alluvial bed material along the left bank toe area, which has increased the hydraulic complexity of the channel through this reach.

The dominant change in the channel configuration downstream of Alex Fraser Bridge has been the initial development and subsequent metamorphosis of the St. Mungo's scour hole near the centre of the channel downstream of the bridge. Interaction of shear layer flow shedding from the Alex Fraser Bridge's left abutment in the horizontal plane and diving flows downstream of the submerged apron protection of South Surrey Interceptor in the vertical plane are believed to have created the hole. The initial development tracked the predicted morphology (Western Canada Hydraulic Laboratories Ltd, 1982) for the feature following installation of the bridge abutments. Around 1994, the prominence of riprap apron protection over a sewer line upstream increased as the channel degraded (and likely, additional riprap placed), forming a 1.5 m high sill on the bed. Immediate deepening of the hole followed this perturbation. Continued channel degradation upstream has increased the effective height of the sill and generated continued scour hole deepening.

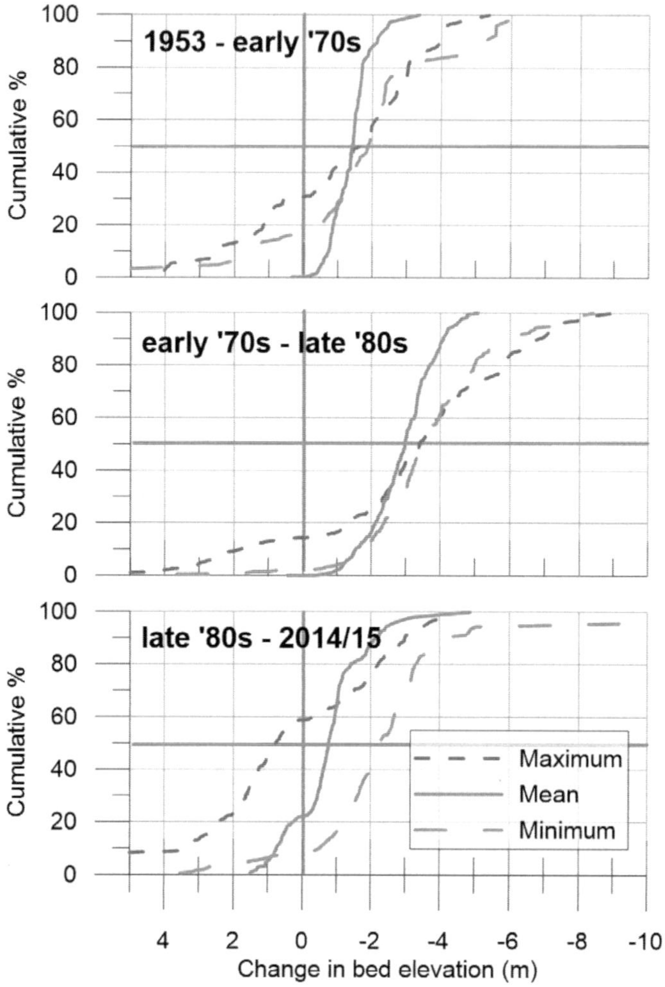

Figure 6: Cumulative distribution plots illustrating the changes in maximum, mean, and minimum bed elevation (max. scour depth) that have occurred along the Lower Fraser. 1953-early 70s comparison restricted to RK 34 to RK 44.

While changes at St. Mungo's Bend illustrate an extreme response to increased relative prominence of non-alluvial features (glacial sediment and riprap), changes near New Westminster (RK 34-38), illustrate a more typical channel degradation pattern. Initial channel-narrowing efforts (1900-1970) caused typical regime-type response of general channel lowering in response to constriction. Intensive dredging beginning in the mid-1970s has been focused on the left bank, where persistent deposition of material on a point bar causes shoaling; maintenance is required to facilitate shipping access to the Fraser Surrey Docks. In contrast, deeper portions of the channel have persistently and monotonically degraded through time, which is interpreted to be a result of the reach-scale sediment

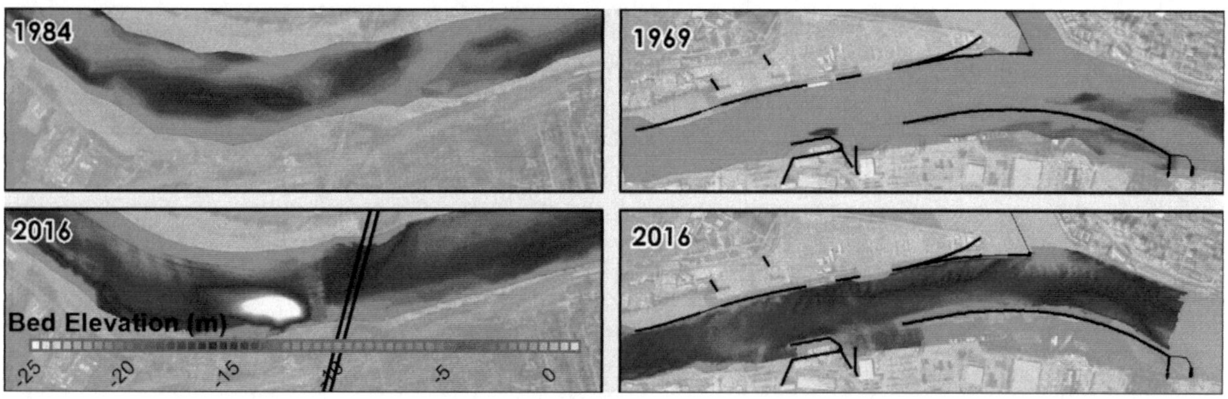

Figure 7: Overview of bed-level changes at Mungo's Bend (left) and the Trifurcation (right).

budget deficit. The historic persistent deposition of material on a point bar causes shoaling and maintenance is required to facilitate shipping access to the Fraser Surrey Docks. In contrast, deeper portions of the channel have persistently and monotonically degraded through time, which is interpreted to be a result of the reach-scale sediment budget deficit. The historic bathymetry also shows persistent scour associated with the natural constriction formed by Pleistocene sediment at Pattullo Bridge, and local scour associated with bridge piers (including scour formation following construction of the Skytrain Bridge).

5. DISCUSSION

Combining observations from the detailed case studies and morphodynamic history with understanding of the geologic context and river management history suggests that human activities have caused a fundamental change in the factors governing channel topography and scour along the Lower Fraser River. Over the last several decades, maximum depths along the channel have been increasing at a faster rate than general (average) bed lowering. Today, general channel levels approach the historic maximum "natural" scour depth, which corresponds to the typical stratigraphic boundary between alluvial sand with similar characteristics to the active bed material in the river, and Pleistocene glacial sediment or Holocene delta foreset beds with contrasting properties. Thus, the channel boundary has shifted from one underlain by deep alluvial cover, to one with intermittent alluvial cover over much finer grained, partially cohesive delta foreslope deposits over the delta, and intermittent alluvial cover over Pleistocene glacial sediments from New Westminster upstream. Furthermore, scour protection features that were once buried or only exposed during high-flow now form prominent sills above the channel bed. These sills, in conjunction with other river training features and increasingly exposed Pleistocene deposits, have caused strong secondary currents to locally develop. Such localized currents are responsible for the deepest observed scour—which presently exceed comparable historic scour depths by 10 to 15 m.

This interpretation raises important considerations regarding the strategy for long-term management of the river. Continued aggressive navigation channel dredging to maintain or increase the available draft in the channel may be expected to trigger disproportionate increases in scour depths along the river. If past trends continue—which is not guaranteed considering the inference of a fundamental change in the typical channel boundary materials—increasing the navigation channel draft by 1-2 m may trigger a disproportionate increase in scour depths of 2 to 6 m. At many sites, scour protection measures are, ironically, one of the principal factors driving intense local channel deepening. Future construction of such features should be designed to minimize hydraulic obstruction, assuming that channel degradation will continue into the future—otherwise feedback between the structures and channel degradation may result in a self-defeating dynamic.

In the general case, these observations from the Fraser River highlight the importance of understanding the local geomorphic context and predicting long-term geomorphic processes interaction with intensive channel management of large lowland alluvial rivers. Such studies need to consider potential perturbations to the sediment budget balance through altered hydraulics, sediment supply, or storage (e.g. dredging); the geologic history and stratigraphic context of the channel boundary; and —crucially— the cumulative impact of diverse management activities over a reasonable management timeframe of decades to centuries.

6. REFERENCES

Armstrong, J. E., and Hicock, S. R. 1980. *Surficial Geology, New Westminster, West of Sixth Meridian, British Columbia*. 1:50,000. [online] Available from: http://geogratis.gc.ca/api/en/nrcan-rncan/ess-sst/ec9bff7c-bd6f-5707-94ab-7130a25cc334.html (Accessed 3 January 2017).

Armstrong, J. E., and Kovachic, H. 1980. *Map 1485A Surficial Geology, Mission, British Columbia*. 1:50,000. Geological Survey of Canada. [online] Available from: http://geogratis.gc.ca/api/en/nrcan-rncan/ess-sst/3f5525ec-88f3-5045-b965-ecf2e0c983c9.html (Accessed 2 December 2014).

Attard, M. E., Venditti, J. G., and Church, M. 2014. Suspended sediment transport in Fraser River at Mission, British Columbia: New observations and comparison to historical records. *Canadian Water Resources Journal/Revue canadienne des ressources hydriques*, 39(3): 356–371.

Clague, J. J., Luternauer, J. ., Monahan, P. A., Edwardson, K. A., Dallimore, S. R., and Hunter, J. A. 1998. Quaternary stratigraphy and evolution of the Fraser delta. *Geology and natural hazards of the Fraser River delta, British Columbia*. Geological Survey of Canada, Ottawa.

Clague, J. J., Luternauer, J. L., and Hebda, R. J. 1983. Sedimentary environments and postglacial history of the Fraser Delta and lower Fraser Valley, British Columbia. *Canadian Journal of Earth Sciences*, 20(8), 1314–1326.

Clague, J. J., Luternauer, J. L., Pullan, S. E., and Hunter, J. A. 1991. Postglacial deltaic sediments, southern Fraser River delta, British Columbia. *Canadian Journal of Earth Sciences*, 28(9), 1386–1393.

Dashtgard, S. E., Venditti, J. G., Hill, P. R., Sisulak, C. F., Johnson, S. M., and La Croix, A. 2012. Sedimentation Across the Tidal-Fluvial Transition in the Lower Fraser River, Canada. *The Sedimentary Record*, 10(4). doi:10.2110/sedred.2012.4.

Hales, W. J. 2000. *Sedimentation in the Fraser River and its estuary, southwestern British Columbia (Canada)*. University of British Columbia, Vancouver.

Jarlan, G. E. 1956. *Influence of proposed structures for the improvement of the shipping channel at New Westminster on the local water stages.*

Kostaschuk, R. 2002. Flow and sediment dynamics in migrating salinity intrusions: Fraser River Estuary, Canada. *Estuaries*, 25(2), 197–203.

Lane, A. 2004. Bathymetric evolution of the Mersey Estuary, UK, 1906–1997: causes and effects. *Estuarine, Coastal and Shelf Science*, 59(2), 249–263.

McLean, D. G., Church, M., and Tassone, B. 1999. Sediment transport along lower Fraser River: 1. Measurements and hydraulic computations. *Water Resources Research*, 35(8), 2533–2548.

McLean, D. G., and Tassone, B. L. 1990. A sediment budget of the lower Fraser River. *Unpublished Environment Canada Report, 8p*. [online] Available from: https://pubs.usgs.gov/misc/FISC_1947-2006/pdf/1st-7thFISCs-CD/5thFISC/5Fisc-V1/5Fisc1-2.PDF#page=34 (Accessed 24 June 2016.

McLean, D., Mannerstrom, D. G., and Lyle, T. 2006. Impacts of human interventions on the lower Fraser River. Vancouver, B.C., pp. 10.

Monahan, P. A., Luternauer, J. L., and Barrie, J. V. 1993. A delta plain sheet sand in the Fraser River delta, British Columbia, Canada. *Quaternary International*, 20, 27–38.

NHC 2008. *Fraser River Model Update Final Report*. Report prepared by Northwest Hydraulic Consultants for the BC Ministry of Environment.

NHC 2015. *Annacis Island WWTP Transient Mitigation and Outfall Project Overview of Fluvial Geomorphology at Project Site -Progress Report* (3000386). Memorandum prepared by Northwest Hydraulic Consultants for CDM Smith on Behalf of Metro Vancouver.

Nichols, M. M., and Howard-Strobel, M. M. 1991. Evolution of an urban estuarine harbor: Norfolk, Virginia. *Journal of Coastal Research*, 745–757.

Nittrouer, J. A., Mohrig, D., Allison, M. A., and Peyret, A.-P. B. 2011. The lowermost Mississippi River: a mixed bedrock-alluvial channel: Mississippi River - mixed bedrock-alluvial channel. *Sedimentology*, 58(7), 1914–1934. doi:10.1111/j.1365-3091.2011.01245.x.

Richards, G.H., and Officers of the H.M.S. Plumper 1860. *Fraser River and Burrard Inlet*. 1:72960. Hydrographic Office.

Richmond Chamber of Commerce 2014. *The Economic Importance of the Lower Fraser River*. [online] Available from: http://www.richmondchamber.ca/external/wcpages/wcwebcontent/webcontentpage.aspx?contentid=10641 (Accessed 4 January 2017).

Van Der Wal, D., and Pye, K. 2003. The use of historical bathymetric charts in a GIS to assess morphological change in estuaries. *The Geographical Journal*, 169(1), 21–31.

Wang, Y., Dong, P., Oguchi, T., Chen, S., and Shen, H. 2013. Long-term (1842–2006) morphological change and equilibrium state of the Changjiang (Yangtze) Estuary, China. *Continental Shelf Research*, 56, 71–81.

Ward, P., R. B. 1976. Seasonal salinity changes in the Fraser River Estuary. *Canadian Journal of Civil Engineering*, 3, 342–348.

Western Canada Hydraulic Laboratories Ltd. 1982. Hydraulic Model Studies of Annacis Island Bridge Crossing Over Annieville Channel. Report prepared by Western Canada Hydraulic Laboratories LTD for Ministry of Transportation and Highways.

Whitfield, P. H., Moore, R. D., Fleming, S. W., and Zawadzki, A. 2010. Pacific Decadal Oscillation and the hydroclimatology of western Canada—review and prospects. *Canadian Water Resources Journal*, 35(1), 1–28.

Leadership in Sustainable Infrastructure
Leadership en Infrastructures Durables

Vancouver, Canada
May 31 – June 3, 2017/ *Mai 31 – Juin 3, 2017*

EFFECTS OF CLIMATE CHANGE ON STREAM EROSION IN A SMALL WATERSHED

Brennan, Colin P.[1,5], Parsapour-moghaddam, Parna[2], Rennie, Colin D.[3] and Seidou, Ousmane[4]
[1] University of Ottawa, Canada
[2,3,4] University of Ottawa, Canada
[5] cbren101@uottawa.ca

Abstract: The effects of climate change on stream erosion are evaluated in Watts Creek basin, located in Ottawa, Ontario, Canada. The subwatershed has a drainage area of 21 km^2 with land use split between urban development (68%), forest (12%) and agricultural lands (20%). The SWMHYMO platform was used to develop a lumped hydrologic model of the area. The model was calibrated using field data collected between May and August 2015 and May and October 2016. Precipitation time series simulated by the Canadian Regional Climate Model 4 (CanRCM4) regional climate model ran under Representative Concentration Pathway (RCP) 85 for the 2041-2080 period at the MacDonald Cartier International Airport were downscaled using quantile matching and then used as input to the hydrologic model. A cumulative effective work index in response to reach-averaged shear stress was calculated in a reach of Watts Creek for both the historic (1967 – 2007) and projected future (2041-2080) flows. Results suggest an increase of 240% in the work index compared to historic conditions for the average measured bed material critical shear stress for entrainment of 3.7 Pa. The increased work is shown to occur in fewer, relatively more intense events, suggesting a significant change to the flow and erosion regime in Watts Creek.

1 INTRODUCTION

River engineers and geoscientists have engaged in river analysis and restoration design for several decades. These endeavors have focused largely on mitigating or reversing the negative effects of human activity on watercourses due to land-use changes associated with urbanisation and its many related industries (Hammer 1972; FISRWG 1998). An increasing number of studies have identified potential changes in both precipitation frequency and intensity in the future. These changes may lead to further impacts on our watercourses (IPCC 2012). Therefore, there is a need to consider how potential climatic change can be incorporated into river analysis and restoration design by current practitioners.

The purpose of this project is to assess the impacts of future rainfall patterns on the erosion potential within Watts Creek using continuous simulations from a lumped hydrologic model and a shear stress exceedance routine to assess erosion potential. This study builds on previous work assessing the existing state of the Creek resulting from ongoing effects of past land-use changes (agricultural and urban) on channel stability and fish habitat utilisation (Maarschalk-Bliss 2014; Parsapour-moghaddam et al. 2015; Rennie 2014; Salem et al. 2014).

Watts Creek is located in Ottawa, Ontario, Canada, and has a drainage area of approximately 21 km^2 with land use split between urban development (68%), forest (12%) and agricultural lands (20%). The system has been found to support important fish habitat, but is undergoing active erosion that can potentially lead to degraded water quality (Rennie 2014). The Creek also has meander bends adjacent to a rail bed, raising the question of risk to infrastructure due to erosion. This study aims to assess how the erosive regime may change in the future as a result of climate change. Figure 1 shows the Watts Creek subwatershed general location. The river drains in a generally north-easterly direction to the Ottawa River. The studied reaches are on lands owned by the National Capital Commission (NCC) that form part of the Capital Greenbelt.

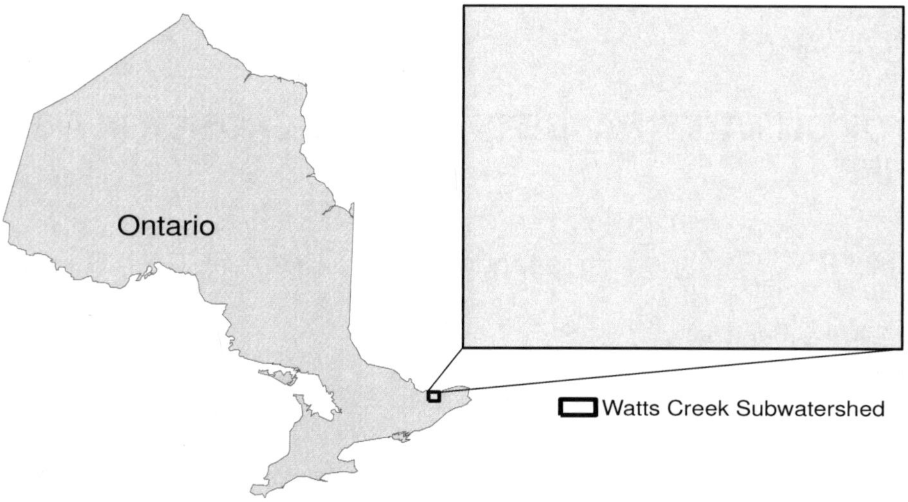

Figure 1: Watts Creek general location and subwatershed

2 METHODS

The study was designed as a joint field monitoring and numerical exercise. Since Watts Creek is an ungauged watershed, continuous flow and precipitation data were collected during the 2015 and 2016 summer seasons to serve for the calibration of the hydrologic model. Given the short length of the observation time series, the model was not validated.

2.1 Field Measurements

Watts Creek, like many small subwatersheds in Ontario, is ungauged and, as such, historic streamflow data or rating curves are unavailable. Water levels were measured at the upstream (-75°52'37.75"W, 45°20'25.62"N) and downstream (-75°52'34.29"W, 45°20'25.72"N) extents of the study reach (referred to as reach M3 following the convention from previous studies by Maarschalk-Bliss (2014) and Rennie (2014)). Pressure transducers (Onset® Hobo water level loggers) were installed within the Creek at both locations (upstream site named WC3 and downstream site named WC4). The total pressure readings from the underwater transducers were compensated for air pressure based on concurrent readings from a third transducer which was installed nearby above the water line. The underwater loggers were fastened to rebar posts that were hammered into the creek substrate. The compensated water depth readings were converted to water level values based on periodic in-field depth measurements at the rebar posts. The geodetic elevation of the top of rebar was surveyed using a Hemisphere® Real Time Kinematic (RTK) GPS on Sept 18, 2015. It was assumed that the top of rebar elevation did not change throughout the study period.

The pressure transducers were installed from May 31 to Aug 21, 2015 and May 30 to October 23, 2016. The loggers were set to record every five (5) minutes. Subsequently in this paper, when referring to continuous field data or hydrologic model output, the data have a measurement period of five (5) minutes.

Periodic flow measurements were collected at WC4 in 2014 and 2015 as part of a previous study on sediment erodibility in Watts Creek (Parsapour-moghaddam et al. 2015; Rennie 2014). Those five (5) flow measurements [range 40 L/s to 1,368 L/s] were augmented with a summer low flow measurement [12 L/s] on Aug 11, 2016 (using a propeller meter and the velocity-area method) and a mid-range flow measurement [391 L/s] on Oct 23 2016 using a SonTek RiverSurveyor® M9 acoustic Doppler current profiler (aDcp). The previous flow measurements in 2014 and 2015 were also recorded with the SonTek M9 aDcp. Note that some flow measurements were taken within a few hundred metres upstream of WC4, though an assumption of constant discharge between the measurement location(s) and WC4 is reasonable based on field observations.

The continuous water level records at the downstream station (WC4) were converted to flow series for the 2015 and 2016 seasons using a rating curve derived by fitting a power function to the paired depth-flow data described above.

2.2 Hydrologic Modelling

The SWMHYMO platform (JFSA 2000) was used to develop a hydrologic model of the Watts Creek watershed to simulate historic and future flow series at the study reach (M3). SWMHYMO is a lumped hydrologic model with routines for simulating runoff from watersheds with mixed land use (urban, agricultural and natural), and has the ability to perform multi-year simulations with low computational cost. The model uses the unit hydrograph method to convolute runoff volume into a runoff hydrograph at the sub-catchment level. Sub-catchment runoff hydrographs can be routed through: reservoirs to simulate stormwater detention facilities; pipes to simulate travel through storm sewer systems and culverts; and channel sections to simulate watercourses. Conceptually, the model consists of a series of hydrographs (flow-time series) that can be added to one another and routed along the various hydraulic features within the subwatershed. At all nodes for hydrograph generation, addition and routing, the model returns the total drainage area, runoff volume, maximum peak flow and time to peak flow. At hydrograph generation nodes the runoff coefficient (ratio of runoff volume to rainfall volume) is also returned. The user has the option to save the hydrograph time series at any or all nodes. The natural catchment hydrograph routines include a baseflow subroutine, which stores infiltrated volume in a groundwater reservoir. This groundwater reservoir can be depleted by baseflow discharge to the watercourse (volume returned to runoff hydrograph) or recharge to the deep aquifer (volume not returned as runoff). The groundwater reservoir behavior is user-defined for each catchment. The platform reads in rainfall data files and can accept design storm or continuous data. The continuous option has been used for this work. SWMHYMO also has an erosion routine to calculate erosion potential, which was modified as part of the present research program to be based on a shear stress exceedance method, see section 2.2.2 below for a description of the modifications.

Input data for the hydrologic model were collected from a variety of sources. Detailed information on storm sewer catchment delineation, underground and surface storage (detention) facilities, drainage direction and catchment imperviousness for the urban areas were based on a stormwater management (SWM) report prepared for the City of Ottawa (AECOM 2015). The hydrologic model prepared in that 2015 study was provided for use in this analysis by the City of Ottawa. Drainage boundaries for the agricultural and natural catchments were delineated based on LiDAR information provided by the NCC, which was converted into a digital elevation model (DEM) using ArcGIS©. The SCS runoff method option, see Eq. 1, has been used to calculate runoff excess (and infiltration) in response to precipitation in the model.

[1] $$Q = \frac{(P-I_a)^2}{P-I_a+S}$$

In Eq. 1 above, Q is the runoff volume, P is the total precipitation, I_a is the initial abstraction and S is soil storage; all values have units of volume, expressed in mm (depth over the catchment area). Soil storage is defined as a function of the non-dimensional curve number (CN), see Eq. 2.

[2] $$CN = \frac{25\,400}{254+S}$$

In Eq. 2 above, CN is a non-dimensional value that can range from 1 to 99 and all values on the right-hand-side have units of mm.

Initial abstraction is defined as a fixed depth by the user to represent interception and depression storage losses. This is a modified form of the original SCS method, in which I_a was defined as 20% of S. The modification allowed the method to be expanded to simulate runoff volumes from frequent (small volume) storms as well as infrequent flood events, as expressed in various USDA, Natural Resources Conservation Service (NRCS) documents (USDA 1986; Woodward et al. n.d.). With the modification of I_a from the original method, a modification of CN and or S is also required to maintain the runoff volume response for a given precipitation volume. Hawkins et al. (2001) propose a straightforward equation to modify (reduce) literature CN values for this purpose; equation 9 from their paper is reproduced below as Eq. 3.

$$[3] \qquad CN_{0.05} = \frac{100}{1.879\left[\frac{100}{CN_{0.20}}-1\right]^{1.15}+1}$$

In Eq. 3 above, $CN_{0.05}$ is the modified CN value to account for a reduced I_a and $CN_{0.20}$ is the typical CN that one would extract from literature tables prepared for the SCS method.

The CN values for the non-urban catchments were selected from literature tables included in the product's user's manual (JFSA 2000) based on an area weighting of hydrologic soil group (HSG) from the Soil Survey Complex map data (OMAFRA 2015) and land-use based on Provincial Landcover map data (OMAFRA 1999) for each catchment. The corresponding modified CN value was calculated using Eq. 3 and input to the model. Topographical characteristics like catchment length and slope were extracted from the DEM and used to calculate the time to peak (t_p) response for each catchment using several time of concentration (t_c) methods and the 85/10 method suggested by the Ministry of Transportation Ontario (MTO 1997, Ch 8). Time to peak is equal to 2/3 the time of concentration value. The model response to various t_c methods was assessed as a calibration exercise, and the method that resulted in the closest timing match with the measured data, Bransby-Williams, was selected (Eq. 4).

$$[4] \qquad t_c = \frac{0.605*L}{A^{0.2}S^{0.1}}$$

The time of concentration (t_c) is in minutes, catchment length (L) is in km, drainage area (A) is in km^2 and average catchment slope (S) is in percent.

The Watts Creek hydrologic model has 50 catchments for the urban area, 32 of which (1197.5 ha) have significant impervious cover (residential, commercial or industrial land use) with an average total imperviousness of 39%. The remaining 18 urban catchments (234.9 ha) contain mostly pervious areas (golf courses, parks, etc.) with an average CN of 62. Some of the urban catchments drain to SWM facilities that provide some runoff control, as such, there are 17 reservoir commands in the model. The balance of the watershed is comprised of agricultural and forested areas (678.9 ha), which are simulated with 9 sub-catchment routines and have an average CN of 63.

2.2.1 Model calibration

Model calibration was conducted to improve the match with the 2015 and 2016 field flow data. Total runoff volume was assessed first, as the model was initially overestimating volume for both years. The directly connected impervious areas (originally 87% of total impervious areas) were reduced to 64%. This reduction was applied for areas with relatively deep lots and large yards (rooftops assumed to be non-directly connected (NDC) areas) where the roads had no curb or sidewalk (some driveways considered NDC areas). This configuration applies to at least 60% of the urban area.

The model timing was adjusted by selection of a t_c method as described above and by increasing the Manning's n values for the impervious surfaces in catchments where ditches or creeks comprised part of the drainage network. The catchment length parameter (LGI) was also increased for some large urban catchments to consider the length of the drainage network. Watts Creek, and its largest tributary the Kizell Drain, were simulated using channel routing commands. The cross-section shape, channel lengths and roughness values were originally based on the model prepared previously by others. The channel lengths were investigated in detail and modified where appropriate based on aerial imagery. The channel and overbank roughness (*n*) values were increased in some areas (from 0.04 to 0.085) to reflect the field conditions observed in 2016 (overall range of *n* 0.03 to 0.085). Those routing modifications slowed the runoff response, which improved the match to the field flows. Table 1 summarises the calibration results for the 2015 and 2016 data. The Nash-Sutcliffe efficiency for 2015 is 54% while 2016 is 46%. The calibration was stopped since the model was providing a slight underestimate for the June and July 2015 data but an overestimate for August 2015 and all months in 2016.

Table 1: Hydrologic Model Flow and Runoff Volume Results Compared to Field Values

a) 2015

	FIELD FLOW				MODEL FLOW		
MONTH	Rainfall Volume (mm)	Runoff Volume (mm)	Average Flow (L/s)	Maximum Peak Flow (m^3/s)	Runoff Volume (mm)	Average Flow (L/s)	Maximum Peak Flow (m^3/s)
JUNE	88.6	16.9	137.4	5.2	11.4	92.8	1.7
JULY	65.0	10.4	81.9	2.3	8.2	64.6	2.5
AUG	71.4	3.3	39.9	3.2	8.9	108.9	3.6
TOTAL	225.0	30.5	92.1	5.2	28.5	86.0	3.6

b) 2016

	FIELD FLOW				MODEL FLOW		
MONTH	Rainfall Volume (mm)	Runoff Volume (mm)	Average Flow (L/s)	Maximum Peak Flow (m^3/s)	Runoff Volume (mm)	Average Flow (L/s)	Maximum Peak Flow (m^3/s)
JUNE	63.0	4.4	35.9	5.2	13.9	112.9	7.6
JULY	91.2	4.9	38.6	3.3	16.8	132.3	4.4
AUG	119.1	21.4	260.9	17.4	35.4	432.1	22.0
TOTAL	273.3	30.7	92.5	17.4	66.0	199.1	22.0

The less than 1% increase in calculated field flow volume from 2015 to 2016 shown in Tables 1a and 1b is less than expected based on the measured rainfall. Work to improve this calculation is ongoing, the issue seems to be based in part on the differing hydraulic conditions during 2015 and 2016. Figure 2a and 2b show the measured rainfall, field and model flows for July of 2015 and 2016 respectively.

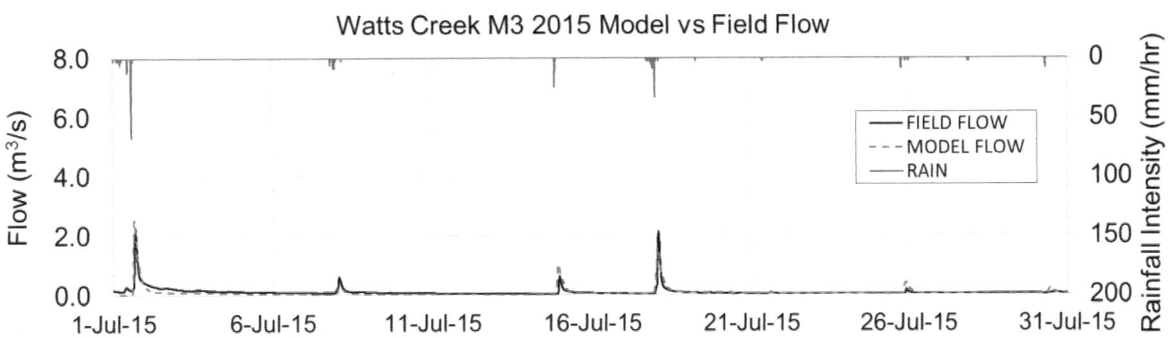

Figure 2a: Hydrologic model calibration results field flow versus model flow at reach M3 2015

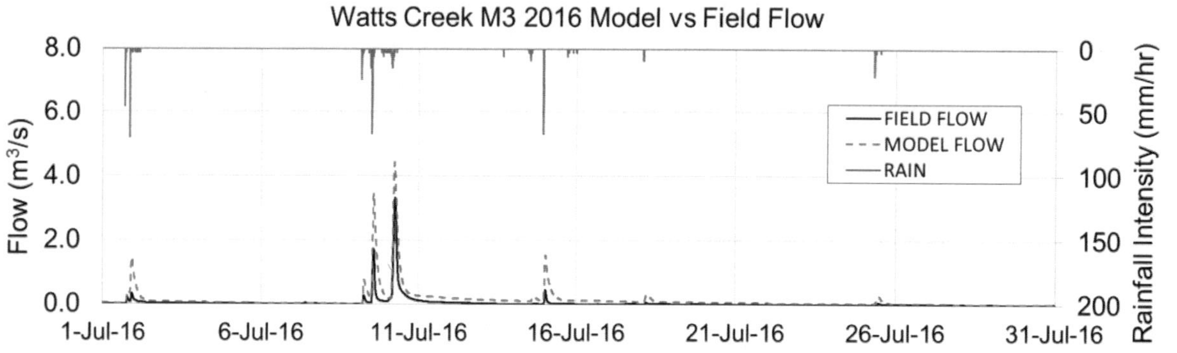

Figure 2b: Hydrologic model calibration results field flow versus model flow at reach M3 2016

The model was used to generate continuous hydrographs at the study reach for the periods when historic hourly rainfall data are available, 1968 to 2007 inclusive (excluding 2001 and 2005 due to missing data). The input rain data used for the historic runs were the hourly rainfall data (HLY03, element 123) collected by Environment Canada at the Ottawa International Airport Gauge. The data do not include frozen precipitation, so the model runs were constrained from April 1 to Oct 31 for each year.

The same model was also used to generate future hydrographs for 2041 to 2080. Derivation of the future hourly rainfall dataset is discussed in section 2.3.

2.2.2 Modified Erosion Index Routine

To assess the effects of climate change on erosion at the study site, both the historic and future hydrographs were analysed with the erosion index routine.

The erosion index routine has been modified to include a shear stress exceedance subroutine. The routine requires a 1-dimensional cross-section definition: station-elevation points, Manning's *n* roughness values for the segments of interest, bed slope and floodplain slope, an inflow hydrograph and the critical shear stress value for bed material entrainment.

The depth-flow relationship is derived from the Manning's equation, Eq. 5:

[5] $$Q = \frac{1}{n} \mathrm{R_h^{\frac{2}{3}} S^{\frac{1}{2}} A}$$

In Eq. 5 above, Q is flow (m^3/s), n is the Manning's roughness parameter, R_h is hydraulic radius (m), S is the channel slope (m/m) and A is the flow area (m^2). Shear stress is based on the 1-dimensional reach-averaged shear stress equation, Eq. 6:

[6] $$\tau = \gamma \mathrm{R_h S}$$

In Eq. 6 above, τ is the bed shear stress (Pa) and γ is the specific weight of water (N/m^3). The cumulative effective work index method, Eq. 7, was presented by the Toronto and Regional Conservation Authority (TRCA) in their SWM Criteria document (TRCA 2012) as a suggested method to estimate erosion potential using a stream power approach.

[7] $$W_i = \sum(\tau - \tau_c) v \Delta t$$

W_i is the cumulative effective work index, τ_c is the user-entered critical shear stress (Pa), τ is the bed shear stress (Pa), v is the main channel mean velocity (m/s) and Δt is the model time step. The erosion index routine returns the work index, the total hours and total flow volume when shear stress exceeds the critical, and the number of events that cross the threshold. These erosion indicators have been used to quantify potential changes to erosion in Watts Creek under climate change induced future rainfall patterns.

2.3 Statistical Downscaling (Quantile Quantile)

The future rainfall data were generated by downscaling precipitation time series simulated by the Canadian Regional Climate Model (CanRCM4) under the RCP85 radiative forcing. RCP85 refers to a scenario where the radiative forcing on the earth surface reaches or exceeds 8.5 W/m^2 by 2100, and continues to rise. The simulation was conducted by the Canadian Centre for Climate Modelling and Analysis (CCCma) in the CORDEX (Coordinated Downscaling Experiment:(Scinocca et al. 2015)); the experiment output data were accessed from their website (Environment and Climate Change Canada 2014).

The quantile-quantile (QQ) downscaling method (or quantile matching) was selected in this study because of its ability to correct the whole distribution of each relevant hydrologic variable (Maraun et al. 2010). It adjusts the regional model results to match the local distribution for the historical period (1961 – 2011) and applies that same bias correction to the predicted future results (2012-2100). The numerical code to perform the downscaling was previously developed in Shirkani et al. (2015) and Seidou et al. (2011) and modified for this study to read in the CanRCM4 RCP format precipitation output. The downscaling approach calculated a QQ transform to fit the model output to the station data for half of the historic period (even years). The odd years in that period were used as a validation dataset. The QQ downscaling was applied to the CanRCM4 outputs at the Ottawa International Airport station to generate precipitation time series that represents the climate in the Watts Creek subwatershed for the 2041-2080 period.

The effect of the bias correction on precipitation occurrence for the historic and future period is shown in Figure 3.

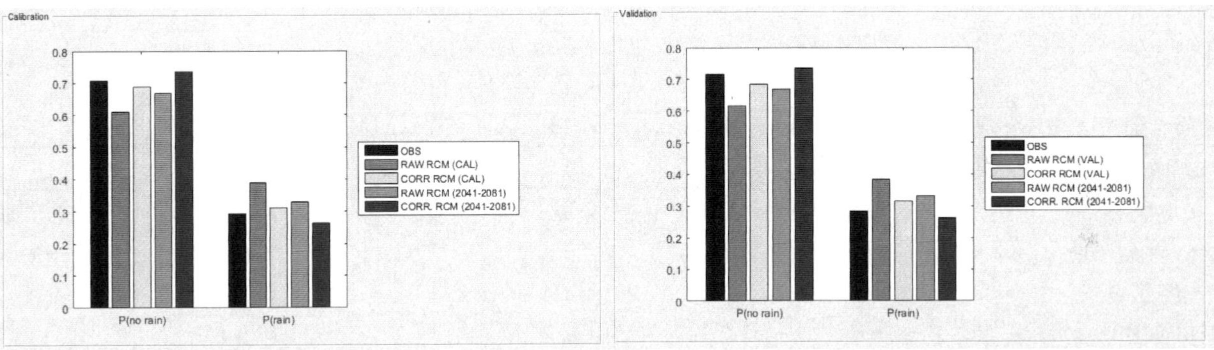

Figure 3: Precipitation occurrence in July, observed and modelled (raw and downscaled) during historical period for calibration (odd years) and validation (even years), and modelled (raw and downscaled) during future period. P(rain) indicates likelihood of precipitation in the daily data set.

Since the subwatershed is approximately 20 km^2, daily data are too infrequent to capture peak flows if used in a hydrologic model. To overcome this obstacle, the daily volume for each wet future day was compared to the historic rainfall data from the airport station. When a historic match was found, the recorded hourly distribution was returned to define the future distribution. The comparison was conducted monthly. This process resulted in a set of hourly future rainfall data in response to the climate change impacts predicted by CanRCM4 RCP 85. A forty year future window from 2041 to 2080 was selected for the erosion analysis.

Precipitation indices from the Expert Team on Climate Change Detection and Indices (ETCCDI) provide consistent metrics for assessing the potential impact of future increase in extreme weather events across climate studies (Zhang et al. 2011). The future results have average increases in both total annual precipitation (PRCPTOT) and number of days with rainfall exceeding 20 mm (R20). The cumulative dry days (CDD) are expected to increase slightly while the consecutive wet days (CWD) decrease very slightly, see Table 2.

Table 2: Average Precipitation Indices for Historic Data and Future Predictions

PERIOD	PRCPTOT (mm)	CDD (days)	CWD (days)	R20 (days)
HISTORIC (1968-2007)	664.1	16.6	4.4	7.8
FUTURE (2041-2080)	805.7	17.8	4.3	11.5
PERCENT DIFFERENCE	21%	7%	-2%	47%

3 RESULTS

The continuous hydrographs at reach M3 for the historic (1968-2007) and future (2041-2080) periods were analyzed with the erosion index method to quantify the expected changes in erosion potential due to the climate change induced future rainfall patterns. Previously measured critical shear stress values for Watts Creek bed material samples ranged from 0.9 Pa to 5.1 Pa (Rennie 2014; Salem et al. 2014) showing significant spatial variability. The average of the two measurements at reach M3 was 3.7 Pa. Table 3 below compares the erosion potential at reach M3 compared to a critical shear stress value of 3.7 Pa.

Table 3 Comparison of Future and Historic Erosion Potential at Reach M3, Critical Shear Stress of 3.7 Pa

PERIOD*	Average Flow (L/s)	Runoff Duration (hours)	Erosion Hours (hours)	Exceedance Percent	No. of erosion events	Total Exceedance Volume (10^3 m^3)	Cumulative Work Index (kPa.m or kJ/m^2)
1968 to 1977	119	5436	63	1%	12	327	292
1978 to 1987	147	5554	84	2%	16	503	442
1988 to 1997	117	5522	65	1%	12	363	322
1998 to 2007**	139	5497	82	1%	15	532	452
All historic	130	5503	73	1%	14	426	373
2041 to 2050	183	5743	110	2%	18	742	682
2051 to 2060	246	5768	143	2%	21	1,290	1,205
2061 to 2070	207	5757	123	2%	21	883	816
2071 to 2080	198	5723	120	2%	17	920	862
All Future	208	5748	124	2%	19	959	891
Percent Difference	60%	4%	70%	63%	36%	125%	139%

*The rainfall input to the hydrologic model is from April 1 to Oct 31 inclusive.
**The historic record is missing data for 2001 and 2005.

As expected, given the increase in average annual precipitation volume and particularly in number of days with greater than 20 mm for the future period shown in Table 2, the future rainfall results in increased erosion potential at reach M3 in Watts Creek.

It is worth noting that the results presented in Table 3 depend upon the estimate of critical shear stress for Watts Creek bed material. In addition to the assessment at a critical shear stress of 3.7 Pa, responses for τ_c ranging from 1 Pa to 10 Pa were assessed. This sensitivity test was conducted due to the difficulties in calculating a precise critical shear stress for clay channels, and the high spatial variability in bed material strength along Watts Creek. Still, the average measured value of τ_c = 3.7 Pa is reasonable.

Results from the sensitivity analysis show that the percent increase for future conditions rise to a maximum at 7 and 8 Pa before falling. This indicates that the highest relative increase in work occurs at flows corresponding to critical shear stress values that are approximately double the measured bed strength at reach M3 of Watts Creek.

Bankfull depth has been estimated as 1.2 m at reach M3 based on the fairly abrupt change in vegetative cover and transverse slope on both banks. The highest discharge measured (1,368 L/s) by Parsapour-moghaddam et al. (2015) corresponded to a depth of approximately 1.1 m at reach M3. Therefore, the bankfull discharge is approximately 1,400 L/s. The erosion index routine returns the critical flow value corresponding to the specified critical shear stress. For the 3.7 Pa estimate of bed strength, the model calculates a flow of 1,230 L/s. This result corresponds quite well with the visual estimate of bankfull discharge, suggesting that 3.7 Pa is likely a good estimate for the strength of the controlling bed layer in this reach and that the model produces a reasonable relationship between flow and potential erosion.

4 DISCUSSION AND CONCLUSION

A lumped hydrologic model developed for the 21 km^2 mixed-use Watts Creek subwatershed was calibrated based on field data collected in 2015 and 2016. While the calibration was reasonably successful, work is ongoing to improve the field flow estimate for 2016. The hydrologic model was setup to perform continuous simulations and has been used to produce flow hydrographs at the study reach for a 40-year historic period (1968 to 2007) and a 40-year future period (2041 to 2080). The future and historic hydrographs were assessed with the model's modified erosion routine, which calculates the cumulative effective work index (CWI) above a defined critical shear stress threshold.

Future rainfall time series were generated at the Ottawa Airport Station by downscaling the output of the regional CanRCM4 model simulated under the RCP85 representative concentration pathway. This produced an hourly future rainfall dataset. The model was used to simulate flow and erosion in the historical period and for a 40-year future period spanning from 2041 to 2080.

Potential changes to the erosion regime at the study reach were assessed for long-term annual average results, accounting for changes to wet and dry periods as well as volume, with low computational cost. This annual average approach has the benefit of overcoming, smoothing out, some of the uncertainties inherent in predicting future stream flows and provides a broader assessment than using a single representative (bankfull) flow approach.

The erosion routine results show an expected increase in the average CWI by approximately 240% with a critical bed shear stress of 3.7 Pa. The total flow volume in exceedance is 225% higher than the historic results while the number of above-critical events per year is only expected to increase by 136%. This suggests that not only will the net erosion more than double, the erosive work will be done by fewer relatively more intense events.

This predicted increase in erosive potential could change the classification of reach M3 from relatively stable to unstable. This, in turn, could have negative consequences for fish habitat and potentially increase risk to nearby infrastructure. These results suggest the need for further study to understand in more detail what the potential effects could be and how to either mitigate them or adapt. Particularly, future work is being considered to incorporate continuous flow hydrograph results for current and future conditions into a 3-D morphodynamic and habitat suitability model.

ACKNOWLEDGEMENTS

Several University of Ottawa colleagues provided valuable support: Sam (Hicham) Salem and Carlos Custodio provided bed material shear strength data, Sean Ferguson, Zhina Mohammed, Philipe April LeQuéré and Mark Lapointe provided field support and Abdullah Alodah provided technical guidance for downscaling. We are grateful to the CCCma for preparing the CanRCM4 and making the results available, and similarly to the World Climate Research Programme's Working Group on Coupled Modelling for CMIP phase 5. The City of Ottawa provided information on urban drainage, and the NCC provided detailed geographic information. The project was supported by funding through an NSERC engage grant involving J.F. Sabourin and Associates Inc. and Dr. Colin Rennie. I would like to thank my co-authors for their encouragment, support and technical guidance throughout this project.

REFERENCES

AECOM. 2015. *Shirley's Brook & Watt's Creek Phase 2 Stormwater Management Study*. Report prepared by AECOM for the City of Ottawa, Ottawa, Ontario.

Environment and Climate Change Canada. 2014. *Climate Modelling and Analysis Canadian Centre for Climate Modelling and Analysis. Government of Canada website.* http://ec.gc.ca/ccmac-cccma/default.asp?lang=en&n=82DD0FCC-1. Accessed 1 January 2017.

Hammer, T. R. 1972. Stream Channel Enlargment Due to Urbanization. *Water Resources Research*, 8(6), 1530–1540.

Hawkins, R. H., Jiang, R., Woodward, D. E., Hjelmfelt Jr., A. T., & VanMullem, J. E. 2001. *Runoff Curve Number Method: Examination of the Initial Abstraction Ratio.*Retrieved from NRCS website: https://www.wcc.nrcs.usda.gov/ftpref/wntsc/H&H/CNarchive/CNinitAbstr.doc

IPCC. 2012. *Managing the Risks of Extreme Events and Disasters to Advance Climate Change Adaptation. A Special Report of Working Groups I and II of the Intergovernmental Panel on Climate Change.* (C. B. Field, V. Barros, T.F. Stocker, D. Qin, D.J. Dokken, K.L. Ebi, et al., Eds.). Cambridge: Cambridge University Press.

J.F. Sabourin and Associates Inc. (JFSA) 2000. *SWMHYMO User's Manual.* Self-published, Ottawa.

Maarschalk-Bliss, S. 2014. *Seasonal Variation in Assemblage Structure and Movement of Small Stream Fish in an Urban Environment.* M.S. Thesis, Biology, Carleton University.

Maraun, D., Wetterhall, F., Chandler, R. E., Kendon, E. J., Widmann, M., Brienen, S., et al. 2010. Precipitation downscaling under climate change: Recent developements to bridge the gap between dynamical models and the end user. *Reviews of Geophysics*, 48(RG3003) 1-38.

MTO. 1997. *Drainage Management Manual. Ministry of Transportation Ontario.* Self-published, Ontario.

Ontario Ministry of Agriculture Food and Rural Affairs. 1999. *Provincial Landcover 2000 - 27 Classes. Land Information Ontario - metadata tool.* https://www.ontario.ca/page/land-information-ontario. Accessed 3 June 2015.

Ontario Ministry of Agriculture Food and Rural Affairs. 2015. *Soil Survey Complex. Land Information Ontario - metadata tool.* https://www.ontario.ca/page/land-information-ontario. Accessed 1 December 2015.

Parsapour-Moghaddam, P., Rennie, C. D., Midwood, J., Cvetkovic, M., & Cooke, S. J. 2015. Influence of Channel Erosion on Fish Habitat Utilization. *36th IAHR World Congress, Hague, Netherlands.*

Rennie, C. D. 2014. *Linking sediment erodibility and channel stability to utilization of available habitats.* Report written by the University of Ottawa for the National Capital Commission, Ottawa.

Salem, H. (S)., Rennie, C. D., & Custodio, C. Z. 2014. Influence of Pore Pressure on Clay Erosion. *River Flow 2014, Lausanne, Switzerland.*

Scinocca, J. F., Kharin, V. V., Jiao, Y., Qian, M. W., Lazare, M., Solheim, L., et al. 2015. Coordinated Global and Regional Climate Modeling*. *Journal of Climate*, 29(1), 17–35.

Seidou, O., Ramsay, A., & Nistor, I. 2011. Climate change impacts on extreme floods II: improving flood future peaks simulation using non-stationary frequency analysis. *Nat Hazards,* 60(2): 715-726.

Shirkhani, H., Seidou, O., Mohammadian, A., & Qiblawey, H. 2015. Projection of Significant Wave Height in a Coastal Area under RCPs Climate Change Scenarios. *Natural Hazards Review*, 17(1) 1-15.

The Federal Interagency Stream Restoration Working Group (FISRWG) (15 Federal agencies of the US gov't). 1998. *Stream Corridor Restoration, Principles, Processes, and Practices.*

Toronto and Region Conservation Authority. 2012. *Stormwater Management Criteria.*

USDA, N. R. C. S. 1986. *Urban Hydrology for Small Watersheds.* Self-published, Technical Release 55.

Woodward, D. E., Hawkins, R. H., Hfelmfelt Jr., A. T., Van Mullem, J. A., & Quan, Q. D. (n.d.). *Curve Number Method: Origins, Applications and Limitations.* Retrieved from NRCS website: https://www.wcc.nrcs.usda.gov/ftpref/wntsc/H&H/CNarchive/CNorigins.doc

Zhang, X., Alexander, L., Hegerl, G. C., Jones, P., Tank, A. K., Peterson, T. C., et al. 2011. Indices for monitoring changes in extremes based on daily temperature and precipitation data. *Wiley Interdisciplinary Reviews: Climate Change*, *2*(6), 851–870.

Leadership in Sustainable Infrastructure
Leadership en Infrastructures Durables

Vancouver, Canada
May 31 – June 3, 2017/ *Mai 31 – Juin 3, 2017*

EXAMINATION OF POOLED FLOOD FREQUENCY ANALYSIS FOR CANADIAN CATCHMENTS

Mostofi Zadeh, Shabnam[1,2] and Burn, Donald H.[1]
[1] University of Waterloo, Canada
[2] smostofizadeh@uwaterloo.ca

Abstract: Floods are known as one of the most damaging forms of natural hazards with devastating influence on people and the environment. Accurately estimating flood frequencies is essential for effective design of flood mitigation systems, reservoir management, and pollution control. Estimation of these frequencies is difficult since extreme events are by definition rare and the length of recorded data is often short. In such situations, extreme flow information from a number of similar sites is combined (pooled) to augment the available at-site information. Pooled flood frequency analysis is a well-known approach used to improve the estimation of extreme flow quantiles at sites with short data records. Identification of pooling groups that will effectively transfer extreme flow information is thus important. This paper aims to explore approaches for obtaining improved flood quantile estimates based on pooled frequency analysis for extreme flow events. This study focuses on the region of influence (ROI) approach among different regionalization techniques to identify homogeneous pooling groups. The ROI is a site focused pooling approach that identifies a potential set of stations for each catchment that constitutes the region for that catchment. Instead of using either catchment characteristics or flood statistics, this study will explore several statistics representing the timing and variability of peak flow events. These flood seasonality measures will be employed in the definition of similarity/dissimilarity between sites. The dissimilarities are defined as a single numeric that represents the separation of two catchments in seasonality space. Different distance metrics can be considered to describe the closeness of each catchment to every other catchment. The discussed pooling techniques were employed for a collection of catchments in the Canadian prairies. The effectiveness of these techniques in identifying extreme flow quantiles was explored for the catchments under study.

1 Introduction

While floods are inevitable natural events, their impact on people and the environment can be reduced by putting mitigation measures in place. Effective mitigation measures require a solid understanding of the frequency of floods. It is crucial to accurately estimate the relationship between extreme flow quantiles and associated recurrence interval to design appropriate infrastructure and plan river engineering works. For these purposes, a sufficiently long streamflow record is required at the site of interest; however, at many hydrometric gauges, the observation period is shorter than is desired. To compensate for the short data record, the trade-off between the spatial and temporal characterization of extreme flows can be effected through the use of regional (pooled) flood frequency analysis (Zrinji and Burn, 1994). In such situations, extreme event information from a collection of sites is combined for the estimation of extreme event quantiles for a site of interest.

Pooled frequency analysis is carried out in a series of steps as follow: data screening; formation of pooling groups; assessment of pooled group homogeneity; revisions to regions; choice of frequency distribution; estimation of pooled growth curve and pooled quantiles.

An important requirement for regional (pooled) flood frequency analysis is the identification of the region or pool of catchments, not necessarily geographically contiguous, that can be considered similar in terms of catchment hydrologic response (Burn, 1997). The defined regions therefore, should be representing a homogenous region with respect to extreme flow characteristics. Research has explored various ways of delineating homogenous regions. Burn (1990) developed the region-of-influence (ROI) approach which is a site focused pooling approach and that avoids the use of fixed regions. This approach allows each catchment to have a potentially unique set of stations that constitute the region for the target site. This approach, and its modifications, has been extensively applied as a regionalization approach in flood frequency analysis (e.g., Tasker et al., 1996; Castellarin et al. 2001; Shu and Ouarda, 2008; Micevski et al. 2015).

Selection of variables to define pairwise similarity (or dissimilarity) among catchments is a necessary step in the ROI approach (Burn et al., 1997). Physiographic characteristics of catchments (such as drainage area, slope, etc.) and flood statistics (such as statistical measures estimated from flow series) are known as two general types of variables (Burn and Goel, 2000). Burn (1997) discussed the difficulties that can occur using these two types of variables and suggested using seasonality statistics of the catchment flood as a similarity measure between sites. In addition, the ROI approach requires the choice of a threshold value that functions as a cut-off point for a dissimilarity measure. Sites with dissimilarity measure greater than the threshold value are excluded from the region of influence for that particular site.

This study is focused on pooled flood frequency analysis of a selection of catchments in Alberta, Canada, and employs the ROI approach for delineation of homogeneous regions. The next section of this paper describes different seasonality measures that have been used as similarity measures between catchments. This is followed by a description of the regionalization strategy in this study. The results of application of these approaches are summarized and discussed next.

2 Seasonality analysis

The timing and regularity of flood events have been introduced as a measure of similarity in the catchment hydrologic response (Bayliss and Jones, 1993; Burn 1997; Cunderlik et al., 2004). Catchments with similarities in the timing and regularity of flood response might be expected to also have similarities in important physiographic and hydrologic response characteristics. Thus, such catchments can be considered as potential members of the same pooling group for pooled flood frequency analysis (Ouarda et al. 2006). Seasonality measures describe the timing and regularity of flood events and can be defined using directional statistics (Mardia, 1972). Circular statistics are the simplest case of directional data.

As per Burn (1997), the date of occurrence of the peak flow for a flood event is defined as a directional statistics by converting the Julian date, where January 1 is the day 1 and December 31 is day 365 (366), of the flood occurrence of the event i to an angular value using:

$$[1]\ \theta_i = (\text{Julian Date})_i \frac{2\pi}{\text{lenyr}}$$

where θ_i is the angular value (radians) for the date for event i and lenyr is the number of days in a year. From a sample of n events, the x- and y-coordinates of the mean date can be determined as:

$$[2]\ \bar{x} = \frac{1}{n}\sum_{i=1}^{n}\cos(\theta_i);\ \bar{y} = \frac{1}{n}\sum_{i=1}^{n}\sin(\theta_i)$$

where $\bar{x}$ and $\bar{y}$ represent the x- and y-coordinates of the mean event date. The mean event date is then can be defined from:

$$[3]\ \text{MD} = \tan^{-1}\left(\frac{\bar{y}}{\bar{x}}\right)\left(\frac{\text{lenyr}}{2\pi}\right)$$

where MD represents the average date of occurrence of the flood event. A measure of the regularity of the n extreme event occurrence, can be determined through:

[4] $\bar{r} = \sqrt{\bar{x}^2 + \bar{y}^2}$

where $\bar{r}$ characterizes the dimensionless spread of the data in a given catchment and range from 0 (low regularity) to 1 (high regularity).

Chen et al. (2013) discussed the importance of information of the flood magnitude in the identification of flood seasonality. They suggest using flood magnitudes (q_i) as weights to take into account their effect in defining timing and regularity of flood events as follows:

[5] $\bar{x}' = \frac{\sum_{i=1}^{n} q_i \cos(\theta_i)}{\sum_{i=1}^{n} q_i}$; $\bar{y}' = \frac{\sum_{i=1}^{n} q_i \sin(\theta_i)}{\sum_{i=1}^{n} q_i}$

Values of MD and $\bar{r}$ can then be estimated using the new defined weighted seasonality measures, $\bar{x}'$and $\bar{y}'$.

3 Similarity measure

The seasonality statistics discussed above can now be employed in the definition of the dissimilarity between catchments. The starting point for regionalization is the selection of a distance metric defining the closeness of each station to every other station (Burn, 1990). This yields a single numerical value that will be used to define the separation of two catchments in seasonality space. Different distance metrics have been introduced in the literature. In this study the Euclidean distance between catchments in the seasonality space was used to define separation between two catchments as follow:

[6] $D_{ij} = \left[\sum_{m=1}^{M}\left(x_m^i - x_m^j\right)\right]^{1/2}$

where D_{ij} is the distance between station i and j and x_m^i is the value of attribute m for station i. Combination of attributes that were used in this study are $\bar{x}$ and $\bar{y}$; MD and $\bar{r}$; and also the weighted modifications of these seasonality measures. The seasonality measures discussed in the previous section were used to define the pairwise closeness of all catchments under study.

4 Pooling scheme

In this analysis, the region of influence (ROI) approach is employed as the basis for the regionalization of catchments. This approach allows each catchment to have a potentially unique set of stations that constitute the region for that catchment (Zrinji and Burn, 1994). The effective identification of a pooling group is governed by two fundamental principles, the homogeneity of the group and its target size (Casterllarin et al., 2001). The collection of catchments should be hydrologically homogeneous so that the extreme flow information can be effectively transferred from sites within the region to the site of interest (Burn, 1997). Hosking and Wallis (1997) demonstrated the homogeneity of a group generally decreases as the group size increases.

The approach taken herein is to use the first 25 stations with minimum pairwise dissimilarities with the target site, defined in the previous section, as a cut-off point for including the station in ROI for the target site. For each site under study four different initial ROIs were constructed using the four types of seasonality measure discussed previously. The resulting collection of catchments is next evaluated using a homogeneity test.

4.1 Assessment of regional homogeneity

The pooling groups resulting from application of the described regionalization approach are subsequently evaluated for hydrologic homogeneity. The aim of regional analysis is to form groups of sites that approximately satisfy the homogeneity condition (Hosking and Wallis, 1997). The homogeneity test proposed by Hosking and Wallis (1993) was used in this analysis to evaluate the homogeneity of each constructed pooling group. For definition and further details refer to Hosking and Wallis (1997).

In this homogeneity test, a statistic (H) based on the weighted variance of the L-coefficient of variation $(\mathrm{L}-\mathrm{CV})$ is derived. A region can be considered homogeneous if $\mathrm{H} < 1$, possibly heterogeneous if $1 \leq \mathrm{H} < 2$, and definitely heterogeneous if $\mathrm{H} \geq 2$. Hosking and Wallis (1997) stated that the H-value criteria are useful guidelines and approximate homogeneity is sufficient to ensure that regional frequency analysis is much more accurate than at-site analysis. The goal in this study is to successfully delineate homogeneous ROIs for the catchments under study using different seasonality measures.

4.2 Revisions of regions

The discussed regionalization technique is useful for initial pooling group formation, but it may be determined that these groups are heterogeneous and require revisions. Where the ROI cannot be considered as a homogeneous group, screening of the pooling group and revisions to the group are required.

For the purpose of screening, Hosking and Wallis (1997) proposed a discordancy measure for identifying unusual sites in a region. Discordancy is a measure of how dissimilar a site is from the group as a whole in terms of the sites' L-moments. Critical values for the discordancy statistic and deciding whether a site is discordant are dependent on the number of sites in a region, and can be found in Hosking and Wallis (1997).

In the case of a heterogeneous pooling group, the discordancy measure is used to identify catchments with unusual flood series in comparison with the rest of the pooling group. Catchments with large discordancy values could sequentially leave the pooling group to enhance the homogeneity of the group of sites. Next, the homogeneity of the revised ROI should be re-evaluated. The region revision process continues until the pooling group can be considered homogeneous.

5 Selecting the regional frequency distribution

After identifying a region, the next stage in the specification of a statistical model is the choice of an appropriate regional frequency distribution. There are many families of distribution that might be candidates for fitting to a regional data set. Their suitability as candidates can be evaluated by applying a goodness-of-fit test. In this analysis, the statistical test described by Hosking and Wallis (1997) was used to select the frequency distribution with best fit to the regional data. The selected distribution can be used to estimate the flood quantiles for different return periods for a target site. For further details please refer to Hosking and Wallis (1997).

6 Description of data set and study area

The seasonality measures discussed previously to define region of influence for pooled flood frequency analysis were employed on a collection of hydrometric gauges in Alberta, Canada. Stations with continuous record series, unregulated flow and minimum of 20 years of flow record were initially selected. The annual series of maximum daily flows were extracted for each site.

For the selected sites in the region, trends in the maximum flow magnitude were evaluated using the Mann-Kendall (Kendall, 1975; Mann, 1945) non-parametric test for trend. In this work, the block bootstrap (BBS) (Önöz and Bayazit, 2012) approach was used in conjunction with the trend test. The BBS approach involves resampling data in blocks to estimate the significance of the test statistic from the data sample while reflecting the serial correlation present in the data set. Sites were evaluated for trends in the annual maximum series, and sites exhibiting significant increasing or decreasing trends were removed from the collection of catchments under study. Nonstationary frequency analysis should be considered for these sites.

A total of 60 hydrometric stations passed the data screening and were selected for further analysis. Figure 1 shows the location of these catchments. The catchments in this study experience similar climate conditions. The range of flow record length is from 20 to 92 years with a mean record length of 45years. The catchment drainage areas range from 3 to 132,588 km^2 with an average of 7034 km^2.

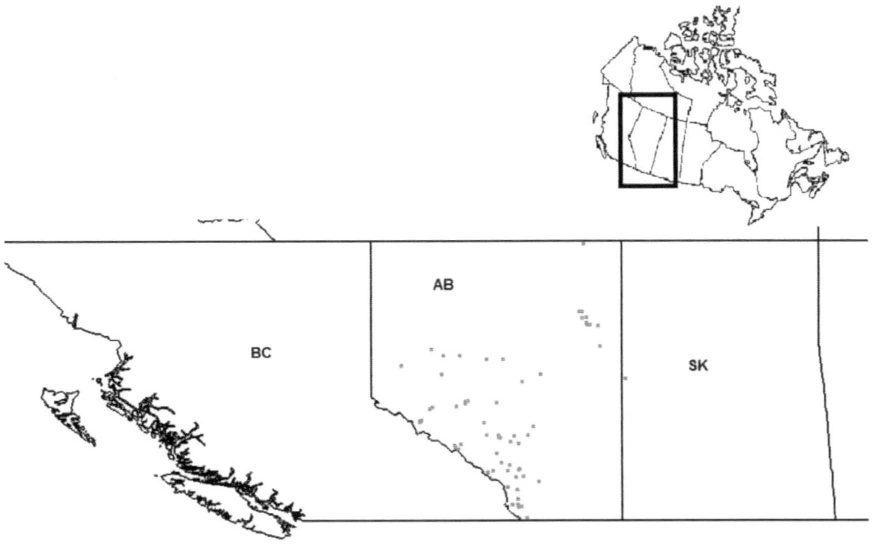

Figure 1: Location of selected gauging stations

7 Results and discussion

The region of influence approach was employed here to identify a potential set of stations for each catchment that constitutes the region for that catchment. Seasonality variables defined in Section 2 were used to define similarity (or dissimilarity) among catchments. The aim in this study is to employ and compare the performance of these different seasonality measures in successfully constructing ROIs for the hydrometric sites under study.

Figure 2 displays the results from calculating the seasonality statistics for the catchments under study. Each catchment is plotted as a point in the space defined by $\bar{x}$ and $\bar{y}$. Figure 2 reveals that the flooding regime for the collection of catchments exhibit a high degree of seasonality as there is a relatively small range in the values for the mean date of occurrence of floods. These stations experience similar climatic conditions and flood events of these catchments mostly arise from spring snowmelt. Due to these natural conditions, the seasonality measures should be an effective means for identifying similarity between catchments.

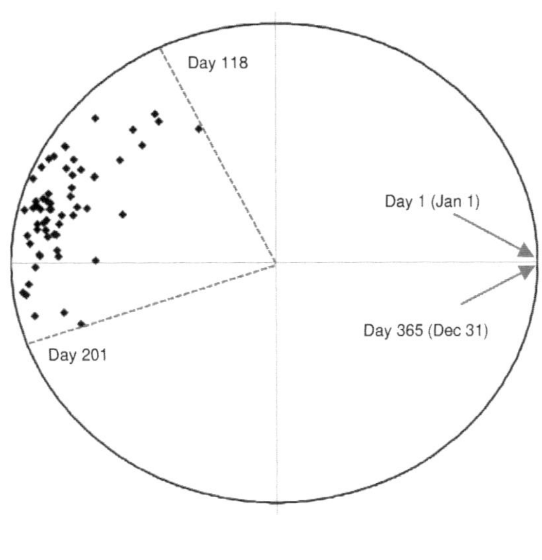

Figure 2: Seasonality space of the gauging stations

7.1 Results of pooling group formation

The seasonality statistics, $\bar{x}$ and $\bar{y}$; MD and $\bar{r}$; and also the weighted modifications of these seasonality measures were employed respectively in the definition of dissimilarity between each pair of catchments using Euclidean distance in the seasonality space. As discussed before, the first 25 stations identified with minimum dissimilarity in each defined seasonality space, were included in the initial ROI for a target site. The resulting collection of catchments was then evaluated for their homogeneity. Revisions were performed where required.

Table 1 summarizes results of the regionalization approach employed in this study for different seasonality measures. This table reveals the number of stations (from total 60 stations) for which the identified ROI was determined as homogeneous, possibly homogeneous, and definitely heterogeneous based on Hosking and Wallis (1997) definition.

Table 1: Summary of homogeneity test for ROIs

Seasonality Statistics	$H < 1$	$1 \leq H < 2$	$2 \leq H < 3$	$H > 3$
$\bar{x}$ & $\bar{y}$	38	10	2	10
MD & $\bar{r}$	42	7	1	10
weighted $\bar{x}$ & weighted $\bar{y}$	42	8	1	9
weighted MD & weighted $\bar{r}$	44	6	1	9

Comparison of the results presented in Table 1 shows that both the weighted seasonality measures, $\bar{x}$ and $\bar{y}$, and also MD and $\bar{r}$ perform better than non-weighted seasonality measure as they were able to successfully construct greater number of homogeneous ROIs. Weighted MD and $\bar{r}$ display slightly better performance, resulting in identifying 44 homogenous ROIs, 6 possibly heterogeneous ROIs, and total of 10 heterogeneous ROIs. It has been noted by Hosking and Wallis (1997) that moderately heterogeneous regions may still offer valuable information concerning quantiles for return periods of rare events.

7.2 Quantile estimation

In the next step, for different combinations of seasonality measures, the identified ROIs and their respective best fitted distribution, were used to the estimate the quantiles of flow for different return periods for sites within the regions. Generalized normal, generalized extreme value, and generalized logistic distribution were regional distributions identified more frequently among others.

At-site flow quantiles were also determined for different return periods, however the at-site analysis should be limited according to the length of flow records at the site, to obtain reliable results. Flood Estimation Handbook (1999) suggested that at-site quantiles can be determined only for return periods where record length is larger than $2T$, where T is the return period of interest. If the record length is less, quantile estimates obtained from regional analysis are more reliable.

7.3 Uncertainty analysis

A measure of uncertainty associated with quantile estimation is essential in this analysis for both at-site and pooled quantiles. This can be estimated through the calculation of confidence intervals (CI) for estimated quantiles. Several approaches have been identified to quantify the uncertainty in either pooled or single site quantile estimates. One approach is the use of resampling or bootstrap to construct confidence intervals.

In this study, the balanced resampling technique introduced by Burn (2003) was used to estimate CIs for the associated at-site quantile estimates. Hosking and Wallis (1997) provided a parametric resampling approach to quantify uncertainty in the estimated pooled quantiles. This parametric approach can reflect the heterogeneity of the pooling and has been reported to provide more realistic estimate of confidence intervals. This approach was employed in this study. Interested reader is referred to Hosking and Wallis (1997) for further information.

The pooled approach to estimate flood quantiles based on using different seasonality measures, was compared with the results of applying single site estimates. The primary basis of comparison was the width of 95% confidence interval. A narrower confidence interval corresponds to more precise estimate and is preferred to an estimate with wider confidence interval (Burn, 2014). Therefore, the ratio of confidence interval width for two estimates were investigated for the case of different ROIs.

Table 2 provides the ratio of width of at-site CI estimated for 6 long record sites over the pooled quantiles. Long recorded sites were used to demonstrate the merits of the pooled approach versus at-site quantile estimates. The results from this table display the successful use of these seasonality measures in constructing effective ROIs for different return periods. The ratios of regional CI to at-site CI are less than 1 for most of the return periods and types of constructed ROIs. Averaging these ratios over different return periods reveals that ROIs constructed using weighted $\bar{x}$ and weighted $\bar{y}$ are most effective among others.

Table 2: Ratios of confidence interval width for the pooled and at-site quantiles

Site	Seasonality Statistics	Return Period								average
		2	5	10	20	30	50	100	500	
05BJ004	$\bar{x}$ & $\bar{y}$	1.06	0.95	0.72	0.57	0.49	0.44	0.39	0.31	0.62
	MD & $\bar{r}$	1.01	0.88	0.65	0.52	0.44	0.40	0.35	0.27	0.57
	weighted $\bar{x}$ & weighted $\bar{y}$	1.01	0.89	0.67	0.52	0.46	0.41	0.36	0.27	0.57
	weighted MD & weighted $\bar{r}$	1.04	0.89	0.67	0.53	0.46	0.41	0.35	0.26	0.58
05CC001	$\bar{x}$ & $\bar{y}$	0.88	0.78	0.78	0.72	0.66	0.62	0.55	0.42	0.68
	MD & $\bar{r}$	0.85	0.72	0.75	0.68	0.62	0.57	0.51	0.41	0.64
	weighted $\bar{x}$ & weighted $\bar{y}$	0.86	0.73	0.74	0.68	0.62	0.58	0.52	0.41	0.64
	weighted MD & weighted $\bar{r}$	0.90	0.88	0.85	0.76	0.68	0.61	0.51	0.36	0.69
05DB002	$\bar{x}$ & $\bar{y}$	0.96	0.90	0.97	0.95	0.93	0.88	0.84	0.78	0.90
	MD & $\bar{r}$	0.96	0.95	1.00	0.94	0.92	0.89	0.85	0.83	0.92
	weighted $\bar{x}$ & weighted $\bar{y}$	0.95	0.86	0.87	0.82	0.79	0.73	0.68	0.61	0.79
	weighted MD & weighted $\bar{r}$	0.86	0.82	0.87	0.92	0.95	0.96	0.94	0.92	0.91
05FA001	$\bar{x}$ & $\bar{y}$	0.79	0.87	0.83	0.74	0.68	0.60	0.54	0.45	0.69
	MD & $\bar{r}$	0.70	0.80	0.77	0.71	0.65	0.59	0.53	0.47	0.65
	weighted $\bar{x}$ & weighted $\bar{y}$	0.66	0.86	0.62	0.77	0.60	0.65	0.56	0.44	0.64
	weighted MD & weighted $\bar{r}$	0.70	0.87	0.79	0.71	0.64	0.55	0.47	0.38	0.64
07AG003	$\bar{x}$ & $\bar{y}$	1.07	0.89	0.69	0.55	0.49	0.43	0.37	0.29	0.60
	MD & $\bar{r}$	1.14	0.99	0.78	0.64	0.56	0.48	0.42	0.33	0.67
	weighted $\bar{x}$ & weighted $\bar{y}$	1.04	0.90	0.71	0.58	0.52	0.45	0.39	0.31	0.61
	weighted MD & weighted $\bar{r}$	1.08	0.95	0.78	0.62	0.56	0.49	0.43	0.35	0.66
07BF002	$\bar{x}$ & $\bar{y}$	0.89	0.91	0.90	0.82	0.75	0.70	0.64	0.62	0.78
	MD & $\bar{r}$	0.87	0.91	0.88	0.80	0.73	0.68	0.63	0.59	0.76
	weighted $\bar{x}$ & weighted $\bar{y}$	0.85	0.87	0.86	0.80	0.73	0.68	0.65	0.63	0.76
	weighted MD & weighted $\bar{r}$	0.85	0.89	0.88	0.82	0.77	0.74	0.70	0.66	0.79

8 Conclusion

This study explored approaches for obtaining improved flood quantile estimates based on pooled frequency analysis for extreme flow events. The region of influence (ROI) approach was employed as a regionalization techniques to identify homogeneous pooling groups. ROI approach requires selection of variables that are used to define the pairwise similarity (or dissimilarity) for the catchments. Geographic catchment characteristics and flood statistics are two common types of attributes used to define inter-site similarities. Following recent research, flood seasonality measures were employed in this analysis in the definition of similarity (or dissimilarity) between sites. Four combinations of seasonality based measures representing timing and regularity of peak flows were used in this study. The seasonality based regionalization was performed on a selection of catchments in the province of Alberta. The performance of these different seasonality based regionalization techniques were evaluated regarding the hydrologic homogeneity of their produced ROIs. It was revealed that ROIs constructed using weighted seasonality measures had greater number of homogeneous pooling groups among all the sites under study. In addition, the pooled approach to estimate flood quantiles based on using different seasonality measures was compared with the results of applying single site estimates. The ratio of width of 95% confidence interval for regional and at-site estimate of flood quantile was the basis of this comparison. The results of these analyses showed that all the seasonality based measures were successful in constructing effective pooling groups. On average for different return periods, the ROIs constructed using weighted coordinates of mean flood date were the most effective regionalization technique among others with the narrowest confidence intervals.

References

Bayliss, A.C. and Jones, R.C. 1993. *Peaks – Over – Threshold Database: Summary Statistics and Seasonality.* IH Report no. 121, Institute of Hydrology, Wallingford, UK.

Burn, D.H. 1990. Evaluation of regional flood frequency analysis with a region of influence approach. *Water Resources Research*, **26**(10), 2257-2265.

Burn, D.H. 1997. Catchment similarity for regional flood frequency analysis using seasonality measures. *Journal of Hydrology*, **202**(1–4), 212-230.

Burn, D.H. 2003. The use of resampling for estimating confidence intervals for single site and pooled frequency analysis. *Hydrological Sciences Journal*, **48**(1), 25-38.

Burn, D.H. 2014. A framework for regional estimation of intensity-duration-frequency (IDF) curves. *Hydrological Processes*, **28**: 4209-4218.

Burn, D.H. and Goel, N.K. 2000. The formation of groups for regional flood frequency analysis. *Hydrological Sciences Journal*, **45**(1), 97-112.

Burn, D.H. Zrinji, Z. and Kowalchuk, M. 1997. Regionalization of Catchments for Regional Flood Frequency Analysis. *Journal of Hydrologic Engineering*, **2**(2), 76-82.

Castellarin, A., Burn, D.H. and Brath, A. 2001. Assessing the effectiveness of hydrological similarity measures for flood frequency analysis. *Journal of Hydrology*, **241**(3–4), 270-285.

Chen, L., Singh, V.P., Guo, S., Fang, B. and Liu, P. 2013. A new method for identification of flood seasons using directional statistics. *Hydrological Sciences Journal*, **58**(1), 28-40.

Cunderlik, J., Ouarda, T.B.M.J. and Bobée, B. 2004. Determination of flood seasonality from hydrological records. *Hydrological Sciences Journal*, **49** (3), 511–526.

Hosking, J.R.M. and Wallis, J.R. 1993. Some statistics useful in regional frequency analysis. *Water Resources Research*, **29**(2), 271-281.

Hosking, J.R.M. and Wallis, J.R. 1997. *Regional Frequency Analysis: An Approach Based on L-moments*, Cambridge University Press, Cambridge UK.

Kendall, M.G. 1975. *Rank Correlation Methods.* London: Griffin.

Mann, H.B. 1945. Nonparametric Tests against Trend. *Econometrica*, **13**(3), 245-259.

Mardia K.V. 1972. *Statistics of directional data.* Academic Press: New York, NY.

Micevski, T. Hackelbusch, A. Haddad, K. Kuczera, G. and Rahman, A. 2015. Regionalisation of the parameters of the log-Pearson 3 distribution: a case study for New South Wales, Australia. *Hydrological Processes*, **29**(2), 250-260.

Önöz, B. and Bayazit, M. 2012. Block Bootstrap for Mann-Kendall trend test of serially dependent data. *Hydrological Processes*, *26*, 3552-3560.

Ouarda, T.B.M.J., Cunderlik, J.M., St-Hilaire, A., Barbet, M., Bruneau, P., and Bobée, B. 2006. Data-based comparison of seasonality-based regional flood frequency methods. *Journal of Hydrology*, **330**(1–2), 329-339.

Shu, C. and Ouarda, T.B.M.J. 2008. Regional flood frequency analysis at ungauged sites using the adaptive neuro-fuzzy inference system. *Journal of Hydrology*, **349**(1–2), 31-43.

Tasker, G.D., Hodge, S.A., and Barks, C.S. 1996. Region of influence regression for estimating the 50-year flood at ungauged sites. *Journal of the American Water Resources Association*, **32**(1), 163-170.

Zrinji, Z., and Burn, D.H. 1994. Flood frequency analysis for ungauged sites using a region of influence approach. *Journal of Hydrology*, **153**(1), 1-21.

Leadership in Sustainable Infrastructure
Leadership en Infrastructures Durables

Vancouver, Canada
May 31 – June 3, 2017/ *Mai 31 – Juin 3, 2017*

EXPERIENCES IN JET TESTING OF COHESIVE SOILS

Cossette, Daniel[1], Amin, Mahmud[2] and Mazurek, Kerry[3,4]
[1] British Columbia Ministry of Transportation and Infrastructure, Coquitlam, British Columbia, Canada
[2] University of Alberta, Edmonton, Alberta, Canada
[3] University of Saskatchewan, Saskatoon, Saskatchewan, Canada
[4] Kerry.Mazurek@usask.ca

Abstract: This paper discusses the issues encountered over many years of soil erodibility testing using the jet erodibility test, which has been used in Canada and internationally. The test uses a submerged circular turbulent impinging jet, set at a large impingement height, to create scour in a soil sample and the time development of the depth of scour along the jet centerline is measured. The scouring rate and estimation of the equilibrium depth of scour then can be related respectively to the erodibility coefficient of the soil, which is a coefficient used in the relationship between erosion rate and excess shear stress on the soil, and the soil's critical shear stress. There have been issues with the diameter and depth of the test sample, fissures and rocks in the samples, vegetation, setting the appropriate jet velocity, and ultimately the underlying theory of the test. For the test theory, it has been seen that original analysis of the test significantly underestimates the equilibrium depth of scour and therefore under predicts the critical shear stress of the soil. Further, the analysis procedures assumed a linear relationship between the erosion rates of the soil and the excess shear stress on the soil, which is not the optimum model for this relationship. It also appears that the shape of the scour hole can impact results; however, the standard method of analysis of the test does not consider effects of the scour hole shape.

1 INTRODUCTION

The estimation of potential soil erosion by water flows is important to a number of problems: (i) design of hydraulic structures such as bridges, dams, weirs, drops, and culverts; (ii) determination of the river bank erosion in assessment of needs for bank protection; (iii) estimation of soil losses from agricultural fields and the associated pollutant transport into rivers; (iv) assessment of the stability of mine waste covers; (v) determination of potential shoreline erosion in reservoirs at dams; (vi) assessment of the potential for sediment mobilization in streams; and (vii) in the assessment of the stability of canals and irrigation channels. For "cohesionless soils" such as clean, coarse sands, there are reasonably reliable procedures for determining a soil's erosion resistance based on its characteristics, such as its mean particle size, gradation, density, shape, packing (Raudkivi 1998). However, there is currently no widely accepted method for determining the critical shear stress and erosion rates for clay-rich or "cohesive soils". Equations for estimating the erosion resistance of cohesive soils have been difficult to develop. This is partly because of the wide number of parameters that can affect the erosion resistance of a cohesive soil: soil density; clay content; clay mineralogy; pore and eroding fluid chemistry; temperature; fabric; water content; organic content; and matric suction (Paaswell 1973; Hanson and Cook 1998). In natural soils, there are also the additional complications of biological activity and vegetation (Black *et al.* 2002). As such, many have chosen to instead measure the erosion resistance of the material of interest in some way.

There exists a wide variety of apparatuses and testing methods for determining a cohesive soil's erodibility, yet none of them are currently accepted as a standard. Testing has been conducted in open flumes of various sizes (Enger 1963; Hanson 1990; Haralampides and Rodriguez 2006); in closed flumes such as the Erosion Function Apparatus (Briaud *et al.* 2001); within concentric rotating cylinders (Lim and Khalili 2009); by drilling a hole in the sample and treating the flow as a pipe flow as in the Hole Erosion and Slot Erosion tests (Wan and Fell 2002; Luthi *et al.* 2011); and by using a submerged circular turbulent impinging jet (Hollick 1976; Hanson 1991; Hanson and Cook 2004; Clark and Wynn 2007; Thoman and Niezgoda 2008; Mazurek 2010; Simon *et al.* 2010). The Jet Erodibility Test methodology is defined in ASTM Standard D5852 (ASTM Standard D5852, 2007) and Hanson and Cook (2004). The advantages of this test over other techniques has been that it can be used both in-situ and in the laboratory, with different scales of apparatuses, using both large and relatively smaller soil samples, and for a broad range of soil erosional strengths.

The Jet Erodibility Test was originally developed by Greg Hanson and his colleagues at the United States Agricultural Research Station in Stillwater, Oklahoma, based on early work with jet testing discussed in Hanson (1991). The earlier analysis procedures of the test given in Hanson (1991) were later modified to adapt the work of Stein *et al.* (1993), who studied the time development of scour in soils by obliquely impinging plane jets, to scour by circular impinging jets (Hanson *et al.* 1998; Hanson and Cook 2004). The resulting analysis then produced two important parameters for describing soil erodibility: the critical shear stress of the soil and its erodibility coefficient. The critical shear stress is the shear stress created by flow on the soil's surface at which the soil is first observed to erode. The erodibility coefficient is the ratio of the erosion rate of the soil to the excess shear stress on the bed.

Over many years of experience with Jet Erodibility Test (JET) and in analyzing the scour produced by vertical, circular turbulent jets, the authors have encountered issues with both conducting the test and with its underlying theory. These issues are discussed herein.

2 CONDUCTING A JET ERODIBILITY TEST

2.1 Standard Jet Erodibility Test Procedures

In the Jet Erodibility Test, the soil sample and jet are submerged in water. The jet is created by flow through a plenum and then through a nozzle. The jet of diameter, d, is set in such a way that it impinges at 90° to the soil surface at a height, H, above the sample. Figure 1 gives a definition sketch. In the original standard, which is summarized in ASTM Standard D5852(2007), a 13 mm diameter nozzle is suspended 0.22 m above the sample surface. Water is pumped into a constant head tank to feed the nozzle, which produces the jet. The recommended head on the jet is 0.91 m though other heads could be used depending on the soil conditions. A test timing sequence of 10, 20, 30, and 60 minutes is used, with a total test duration of 2 h. At each time, the jet flow is stopped and the profile of the scoured soil bed is taken using a pin profiler. Alternatively, a point gauge may be lowered through the nozzle to facilitate measuring the centreline scour depth, ε_{cl}, instead of measuring the profile of the bed.

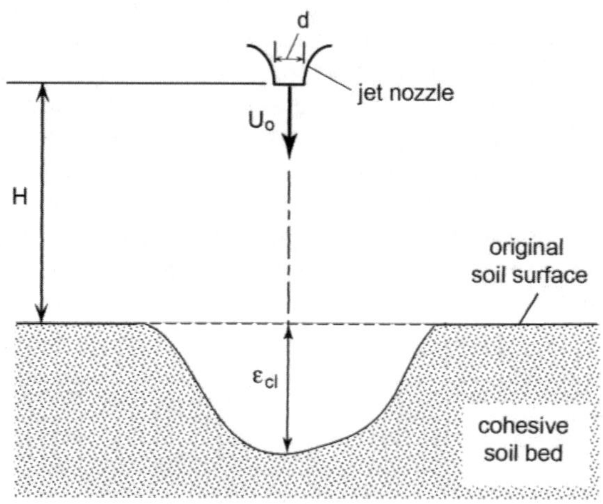

Figure 1: Definition of jet test variables

In further developing the jet erodibility test procedures, Hanson and Cook (2004) noted that the jet could be set at a height above the sample from 6 to 35 nozzle diameters, with a recommended initial height equal to 12 diameters, and a jet diameter of 6.4 mm. The head of water above the nozzle could be adjusted depending on the soil conditions. Measurements of the scour depth at the jet centreline should be taken at intervals of 5 to 10 minutes, with a recommendation of taking at least 10 to 12 readings.

2.2 Experimental Setup and Experiments

The results from two sets of experiments in the scour of cohesive soils by submerged vertical, circular, turbulent impinging jets are discussed herein. In the first set of tests, the maximum and centreline scour depths were measured with time until scour hole was deemed to have reached equilibrium. The entire scour hole was then measured. In the second set of experiments, the entire scour hole was additionally measured at intermediate times of scouring until equilibrium was reached. For all these tests, the samples were set on a table within a 1.2 m deep, 1.1 m wide octagonal tank that was filled with water. A circular jet was created by flow through a 0.95 m long plenum, with internal flow straighteners, then a nozzle, which was designed so that that the flow contraction occurred smoothly without any significant head loss and the velocity profile across the nozzle was uniform; both the plenum and nozzle were designed following the standards laid out in American Society of Mechanical Engineers (1990). To create the jet, City of Saskatoon tap water was pumped from a constant tank through a magnetic flow meter and then finally to the top of the jet plenum. The plenum was suspended from a steel frame on hinges so that it could be moved to the side of the jet tank during scour measurements. The water in the tank overspilled the top of the tank around all its edges so that a constant submergence of the jet was maintained. The apparatus is shown in Figure 2.

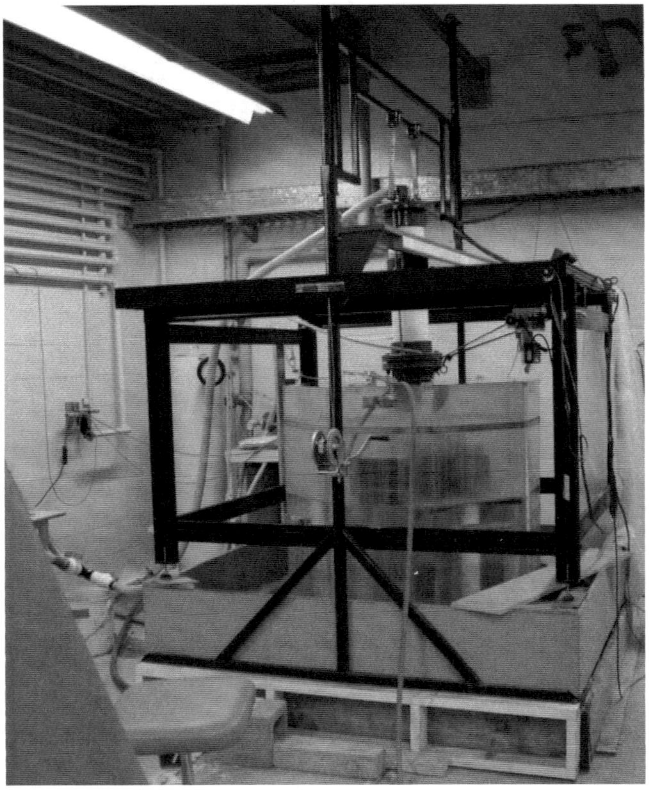

Figure 2: Experimental apparatus

For measurements of the entire scour profile, an optoNCDT 1700-750 Laser Triangulation Displacement Sensor was mounted above the jet tank on a two-dimensional computer-controlled motorized traverse system that could move up to 1.2 m in each direction of the horizontal plane. The scour profile measurements could be taken automatically by setting the limits of the measurements. The system was controlled using LabView. Measurements were taken on a 2 mm by 2 mm grid and the exact position of the laser was tracked using encoders. Details of the measurements are respectively given in Cossette (2016) and Amin (2016).

In the first set of experiments, each soil sample was contained within a 152 mm diameter stainless steel cylinder. Jet testing was conducted on 14 natural samples collected from rivers in Ontario, 4 samples from a field in Saskatoon, Saskatchewan, and 10 samples of manufactured pottery clays. Details of the samples properties are given in Cossette (2016). A nozzle diameter of 8 mm was used, with the impingement height ranging from 59-123 mm and a velocity at the nozzle, U_o, in the range of 1.1-14.8 m/s. In the natural samples, measurements of the centerline scour depth and maximum scour depth were taken at times of 5 min, 10 min, 20 min, 30 min, 60 min, 2 h, 4 h, and 8 h from the beginning of testing. After this, measurements were taken at about 24-48 h intervals until the scour hole was deemed to have reached an equilibrium state. For the pottery clays, the testing sequence was modified to include more measurements early in a test, with measurements at 5 min, 10 min, 15 min, 20 min, 30 min, 40 min, 60 min, 1.5 h, 2.5 h, 4.5 h, and 8.5 h from the beginning of testing. After this, measurements were taken at about 24-48 h intervals until the scour holes were deemed to have reached equilibrium. For the decision about whether a scour hole had reached equilibrium, the scour depth data was plotted with time and graphical assessment of whether the scour hole was continuing to grow was made (see the discussion on this type of decision in Amin (2016)). The time to equilibrium ranged from 42 to 759 h, with most tests taking about 240 h. After equilibrium was reached and the test was stopped, the entire profile of the scour hole was measured using the laser displacement meter.

In the second set of experiments, using the laser displacement meter, the entire scour hole profile was measured at times of 5 min, 10 min, 15 min, 20 min, 30 min, 40 min, 50 min, 1 h, 1.5 h, 2 h, 4 h, 8 h, 16 h, 24 h, and then after every 24 h thereafter until the equilibrium condition was achieved. During the test, the eroded volume was plotted against the logarithm of time to assess whether the resulting curve started to become parallel to the time axis, thus indicating that a state of equilibrium had been reached (the graphical approach used by Amin (2016)). For this testing, 10 samples of pottery clays were used. The jet impingement height was 85 mm, the jet diameter was 7.76 mm, and the jet velocity at the nozzle ranged from 4.4-11.4 m/s. Further details of these experiments are given in Amin (2016).

It should be noted that the underlying analysis of the jet test does not restrict the conditions to those used in the ASTM standard or Hanson and Cook (2004).

3 JET ERODIBILITY TEST ANALYSIS TO OBTAIN THE SOIL ERODIBILITY PARAMETERS

3.1 Determining the Critical Shear Stress of the Soil

Following Hanson and Cook (2004), to estimate the critical shear stress of a soil from JET data, first, the shear stress created by the flow on the soil surface must first be estimated. For this work, the variation of the velocity with distance from the nozzle is considered. At the nozzle, it is assumed that the velocity profile is uniform and equal to U_o. A shear layer forms at interface between the edge of the jet and the surrounding fluid as the jet exits the nozzle. At a certain distance from the nozzle, x_p, the shear layer penetrates through the entire diameter of the jet. The distance x_p is the length of the potential core of the jet. If the distance from the nozzle along the jet centreline, x, is greater than the length of the potential core (*i.e.* $x > x_p$), the maximum velocity of the jet varies with distance from the nozzle as (Rajaratnam 1976):

[1] $$\frac{u_m}{U_o} = C_d \left(\frac{d}{x} \right)$$

where u_m is the maximum velocity at a distance x from the nozzle and C_d is a diffusion coefficient, which has a value of $C_d \approx 6.3$ (Rajaratnam 1976). This therefore assumes that the velocity of the jet is decreasing with distance from the nozzle in a similar way to a free circular jet (*i.e.,* one that is not impinging on a boundary or is affected by a boundary). It is known that this relationship (Eq. 1) remains valid to about 86

% of the distance from the nozzle to the flat, rigid boundary for an impinging jet (Beltaos and Rajaratnam 1974).

What one would estimate as the velocity of the jet at the boundary using Eq. 1 is converted to an estimate of the maximum shear stress on the bed, τ_{om}, using the local skin friction coefficient, c_f:

[2] $$_{om} = c_f \; u_m^2$$

where ρ is the eroding fluid density. The value of c_f = 0.00416, which was determined by Hanson *et al.* (1990) in a study of the shear stresses created by a circular impinging jet along a smooth, flat boundary. This value of c_f compared favourably with c_f = 0.00403 found by Beltaos and Rajaratnam (1974), who did not assume free jet behavior in developing their estimate but did assume that the jet is at a large impingement height (H/d>8.3) (Beltaos and Rajaratnam 1977). Use of c_f = 0.00416 assumes that the soil bed is smooth, non-porous, and rigid. It is known, however, that c_f for an impinging jet depends on the roughness of the bed (it is larger for a rough bed) (Rajaratnam and Mazurek 2005).

For a bed that has scoured to a depth of ε_{cl}, which is the depth of scour at the centreline relative to the original, unscoured soil surface, with an original height of the jet above the soil bed of H, the total distance from the jet to the scoured-out bed is H+ε_{cl} (see Figure 1). The maximum shear stress on the bed at this depth is then:

[3] $$_{om} = c_f \left[C_d U_o \frac{d}{H + \;_{cl}} \right]^2$$

When the scour depth reaches equilibrium, or $\varepsilon_{cl} = \varepsilon_{cl\infty}$, it is assumed that the maximum shear stress on the bed, τ_{om}, is equal to the critical shear of the soil, τ_c, or:

[4] $$_c = \;_{om} = c_f \left[C_d U_o \frac{d}{H + \;_{cl\infty}} \right]^2$$

Hanson and Cook (2004) did not run their tests until the scour hole reached an equilibrium depth, since doing so can take several hundred hours and this is impractical. Instead, the equilibrium scour depth is estimated by the curve-fitting approached developed by Blaisdell *et al.* (1981). A hyperbolic equation is fitted to the scour data from the jet test to estimate the scour depth at very long times, which is taken to be the equilibrium depth. Blaisdell *et al.* (1981) provide a step-by-step methodology for this technique, which is also described in Hanson and Cook (2004) and Cossette (2016).

3.2 Determining the Erodibility Coefficient of the Soil

The erodibility coefficient of the soil comes from the relationship of the soil erosion rate of the soil, $\dot{E}$, to the excess shear stress on the bed:

[5] $$\dot{E} = K(\tau_{om} - \tau_c)^n$$

where K is the erodibility coefficient and n is a power. To determine K, the data for time development of the centerline scour depth are used. The method of analysis was developed by Hanson and Cook (1997). It adapts the work of Stein *et al.* (1993) (see also Stein (1990)) for the time development of scour by obliquely impinging plane wall for the vertical circular jets used in the jet erodibility test.

First, the erosion rate of the soil is taken as the rate of change of the scour depth along the jet centerline, or:

$$\dot{E} = \frac{d(H+\varepsilon_{cl})}{dt} = \frac{dH_t}{dt} \qquad [6]$$

where t is the time from the start of scouring and $H_t = H+\varepsilon_{cl}$. Next, for cohesive soils, it has been assumed that n=1 for the jet erodibility test analysis, although Stein *et al.* (1993) did not require this assumption. They only noted that for the cohesive soil tested n=1 gave the best fit to the time development of scour data. Hanson and Cook (1997) showed that the relationship between the time of scouring measured from the start of a test and the scour depth can be expressed as

$$t = T_r\left(0.5\ln\left(\frac{1+H_*}{1\ \ H_*}\right)\ \ H_*\ \ 0.5\ln\left(\frac{1+H_{i*}}{1\ \ H_{i*}}\right)\ \ H_{i*}\right) \qquad [7]$$

where T_r is a reference time given as $T_r = H_{t\infty}/(K\tau_c)$; $H_{t\infty}$ is value of H_t when scour has reached equilibrium (or $H_{t\infty} = H+\varepsilon_{cl\infty}$); H_* is the dimensionless scour depth, defined as $H_* = H_t/H_{t\infty}$; and H_{i*} is the initial height of the soil surface expressed in dimensionless form, where $H_{i*} = H/H_{t\infty}$.

To determine K, the critical shear stress was first determined by the method described above. Then, measurements for the time from the start of the test and centerline scour depth measured at that time, expressed as H_*, were plotted. Equation 7 is then fit to the data by minimizing the least square errors to find T_r using the Solver function in Excel. The K value is then found from T_r.

4 ISSUES WITH CONDUCTING A JET ERODIBILITY TEST

One of the main issues with conducting a jet erodibility test in its standard form is that it can be difficult to know what velocity to set the jet *a priori*. The test is conducted to determine the erodibility of the sediment so the erodibility of the material is therefore unknown. When using a sample taken from the field, there will be a limited diameter of the sample. It is very easy then to accidentally "blow out" the sample by setting the jet velocity too high. One of the ways the authors have avoided this is to make a visual assessment of the velocity to initiate erosion at the start of the test. At the beginning of the test, the flow rate is slowly stepped up every 5 minutes in small increments to visually assess the point of erosion. Cohesive soils tend to erode in chunks so that the observer is looking for the first chunk removal. From the flow conditions at this critical flow, either one can estimate a critical shear stress from Eq. 4 then use the formulas for estimating the equilibrium size of the scour hole presented in Mazurek *et al.* (2001) to assess whether the scour hole would fit within the sample container or simply multiply the flow by 1.2-1.3 times. Samples that scour to the edge of the container tend to be asymmetric and this means that the assumptions for the jet decay of velocity given in Eq. 1 are questionable in their application.

Cohesive soils tend to erode in chunks and the size of the chunks are small (on the order of a few millimeters in dimension) to several centimeters in dimension. When a large chunk is eroded, the scour depth tends to remain constant with time for several measurements. This can impact the quality of curve fitting for the time development of scour data. In that case, it may be necessary to cut the sample in some way and repeat the test if the enough of the sample is available for testing. Similar issues often occur with the time development of scour of a sample when the scouring ceases due to the presence of a rock or roots in natural samples. Although measurements of jet erodibility have been made for samples with roots used previously (Clark and Wynn 2007), there is some question of whether the scour test is representative for the entire sample, as the roots introduce heterogeneity in the sample.

5 ISSUES WITH ANALYZING A JET ERODIBILITY TEST

5.1 Estimates of the Equilibrium Depth of Scour

The most important quantity to determine from the jet erodibility test is the critical shear stress of the soil, which, as noted, is done by determining the equilibrium depth of scour and calculating the critical shear stress using Eq. 4. Determining the equilibrium scour depth is normally done using the curve fitting approach of Blaisdell *et al.* (1981), however, for the experiments described herein the scour experiments were conducted until the sample reached equilibrium. For the first set of experiments described herein, the equilibrium scour depth was estimated using Blaisdell *et al.* (1981) and then compared to the measured values.

In estimating the equilibrium scour depth using the Blaisdell et al. (1981) approach, it was found that the number of data points used in the curve-fitting procedure highly impacted results. For the soil sample called BS(1) (see Cossette (2016) for details of its characteristics), using all data points from the first 120 minutes of testing (8 points) gave a value for the equilibrium depth of scour 274 mm. However, using the first 10 data points from testing gives an equilibrium scour depth of 300 mm. Using all the depth readings collected during testing (20 points) gives a value for the equilibrium scour depth of 261 mm. Although the equilibrium scour depth varied by up to 39 mm depending on how many data points were used, Blaisdell *et al.*'s (1981) method still produced values over 123 mm above what was measured as equilibrium depth for BS(1), as seen in Figure 3. Figure 4 shows a similar finding for sample LC(1), where the addition of subsequent data points early on had a large impact on the equilibrium scour depth value. Using the first 120 minutes of testing (6 points) produced a value for the equilibrium scour depth that is about 1.3 times the magnitude of the measured depth, whereas using all the available data produced a value about 7 times the measured depth. Table 1 shows a comparison of the critical shear stress evaluated from the measured equilibrium scour depth, τ_{ce}, to the critical shear stress estimate from the Blaisdell *et al.* (1981) using the first 120 minutes of test data, τ_{cB}, as would normally be done in the standard jet test analysis, for 10 of the tested soil samples. Thus, it necessary to find an alternate model for estimating equilibrium scour from early scour measurements for the jet test.

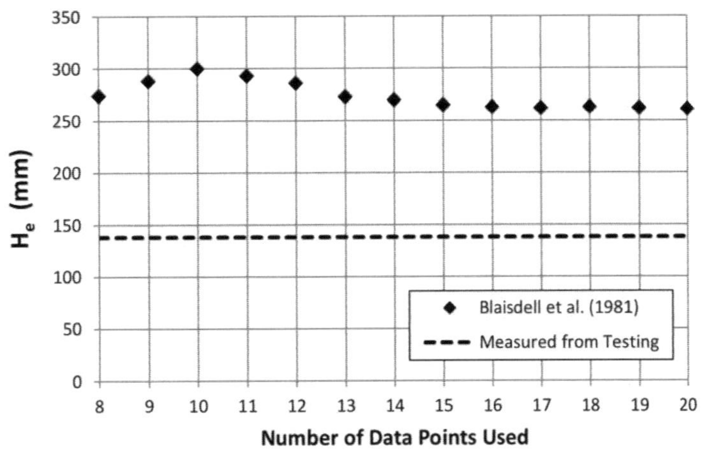

Figure 3: Effect of the number of data points on the equilibrium scour depth estimate for sample BS(1)

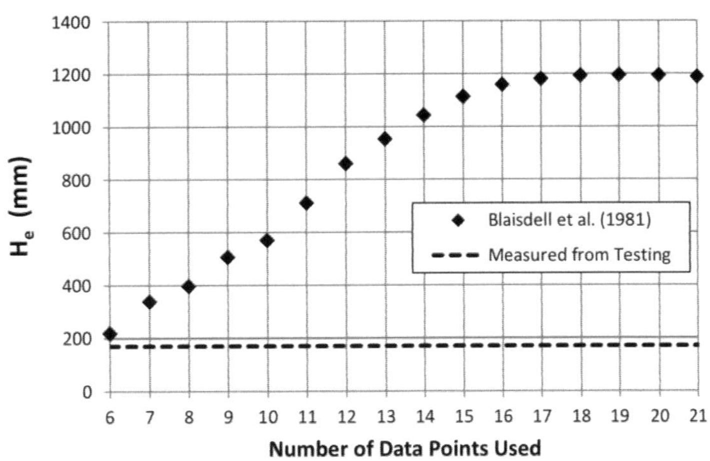

Figure 4. Effect of the number of data points on the equilibrium scour depth estimate for sample LC(1).

5.2 Determination of the Equilibrium Depth of Scour

In the second series of experiments, where the entire scour hole profile was measured with time, it was observed that the scour hole did not obtain an equilibrium state even though it appeared that the centerline depth had reached that value (Amin 2016). The decision about equilibrium is better to result from measurement of scour depths at several locations instead of simply relying on measurements made along the jet centreline.

5.3 Assumption of Linear Relationship between Erosion Rate and Excess Shear Stress

One of the assumptions made for the jet test is that there is a linear relationship between erosion rate and excess shear stress. The basis for this original assumption was reviewed by Mazurek *et al.* (2016); there was data that suggested that the linear relationship might be true, but it was not rigorously verified. Recently, it has come into question *(e.g.* Knapen *et al.* 2007; Walder 2015).

The linear assumption was tested for the first set of experiments described herein. The assumption of linearity (n=1 in Eq. 5) was loosened and the equation for the time development of scour given in Eq. 7 was redeveloped for powers of n=0.5, 1.5, and 2.0. Details are given in Cossette (2016). It is seen that although n=1 was a reasonable fit, it was not the optimum value, which was found by minimizing the mean absolute relative error of the model from the data. The optimized n values are given for 10 of the samples tested in Table 1. In Cossette (2016), the best-fit n value was typically found to be n=2. Any erosion model considered in future for analyzing jet test data in future, therefore, needs to avoid the assumption of linearity between erosion rate and excess shear stress.

Table 1: Comparison of the values for the critical shear stress estimated from the measured equilibrium scour depth and that determined from the estimate using Blaisdell *et al.* (1981) and best-fit n value

Soil Sample	Critical Shear Stress from Equilibrium Depth, τ_{ce}, (Pa)	Critical Shear Stress from Blaisdell Method Estimates, τ_{cB} (Pa)	Optimized n value
SN(1)	8.8	2.8	1.0
WC(1)	3.8	2.0	2.0
LC(1)	1.3	0.8	1.0
SC(2)R	10.1	3.3	1.5
BB(1)	1.2	0.9	0.5
JR(1)	5.8	2.4	1.5
RR(1)	13.9	7.4	1.5
M390(2)	99.3	23.5	2.0
M370(1)	77.0	14.5	2.0
LW(1)	10.4	9.2	2.0

5.4 Estimation of Shear Stress on the Bed

The jet test analysis relies on estimation of shear stresses through estimation of jet velocities. The estimation of velocities relies on jet theory for a jet unaffected by any boundaries. It is known that the boundaries do affect jet shear stresses significantly (Rajaratnam *et al.* 1993; Camino *et al.* 2012; Weidner

2012; Weidner *et al.* 2012). When the jet is redirected upon itself because of the concave shape of a scour hole, the shear stresses that would otherwise be observed will be reduced.

Even if the estimated jet velocities were correct, an assumption in estimating shear stresses is that the bed is smooth. However, although the soil surface may be cut so that it is initially reasonably smooth at the start of testing, this does not remain so once erosion begins. The roughness of boundary impacts the bed shear stress (Rajaratnam and Mazurek 2005). The bed roughness will change as the sample erodes since this is usually in the form of chunks or lumps of soil for cohesive soils. As such, it appears necessary to incorporate some measurement of shear stresses into the jet test as the test is conducted..

6 CONCLUSIONS

There are many problems with the existing jet erodibility test analysis: (i) the assumption of linearity between erosion rates and excess shear stress on the bed; (ii) the estimation of the equilibrium scour depth; and (iii) the estimation of the stresses on the bed. Although the jet test has utility in erosion testing because of its ease of setup and operation and its potential for use for a broad range of sediment erosional strengths, it appears there are a number of ways that it could be improved to more accurately estimate sediment erodibility.

7 ACKNOWLEDGMENTS

The authors gratefully acknowledge funding support from the Natural Sciences and Research Council of Canada for this work. The authors also acknowledge the contributions of Colin Rennie of the University of Ottawa in collecting the Ontario soil samples used in the first set of experiments described herein.

REFERENCES

American Society of Mechanical Engineers (ASME). 1990. *Measurement of Fluid Flow in Pipes Using Orifice, Nozzle, and Venturi.* (ASME MFC-3M-1990), ASME, New York, NY, USA.

Amin, M.R. (2016). *Development of Scour in Cohesive Soils by Normal Impingement of a Submerged Circular Turbulent Jet.* M.Sc. Thesis, Dept. Civ., Geol., & Env. Eng., Univ. Saskatchewan, Saskatoon.

ASTM Standard D5852-00 (2007). *Standard Test Method for Erodibility Determination of Soil in the Field or in the Laboratory by the Jet Index Method.* ASTM International, West Conshohocken, Pennsylvania, USA.

Beltaos, S. and Rajaratnam, N. 1977. Impingement of Axisymmetric Developing Jets, *J. Hydraul. Res.*, **15**(4), 311-326

Beltaos, S., and Rajaratnam, N. 1974. Impinging Circular Turbulent Jets. *J. Hydraul. Div, ASCE*, **100** (HY10): 1313-1328

Black, K.S., Tolhurst, T.J., Paterson, D.M., Hagerthey, S.E. 2002. Working with Natural Cohesive Sediments. *J. Hydraul. Eng.*, **128**(1): 2-8.

Blaisdell, F.W., Anderson, C.L., Hebaus, G.G. 1981. Ultimate dimensions of local scour. *J. Hydraul. Div., ASCE,* **107**(HY3): 327-337.

Briaud, J.L., Ting, F.C.K., Chen, H.C., Cao, Y., Han, S.W., Kwak, K.W. (2001). "Erosion function apparatus for scour rate predictions." *J. Geotech. Geoenviron. Eng.*, 127(2): 105-113.

Camino, G.A., Zhu, D., and Rajarantam, N. 2012. Jet Diffusion inside a Confined Chamber, *J. Hydraul. Research*, **50**(1): 121-128.

Clark, L.A. Wynn, T.M. 2007. Methods for Determining Streambank Critical Shear Stress and Erodibility: Implications for Erosion Rate Predictions. *Trans. ASABE*, **50**(1): 95-106.

Cossette, D. (2016). *Erodibility and Scour by a Vertical, Submerged, Circular, Turbulent Impinging Jet in Cohesive Soils.* M.Sc. Thesis, Dept. of Civil, Geol., & Env. Eng., Univ. Saskatchewan, Saskatoon.

Enger, P.F. 1963. Canal Erosion and Tractive Force Study - Analysis of Data from a Boundary Shear Flume. *Report No. Hyd-532*, United States Department of the Interior Bureau of Reclamation, Denver, Colorado, December, 1963.

Hanson, G.J. 1991. Development of a Jet Index to Characterize Erosion Resistance of Soils in Earthen Spillways. *Trans. ASAE*, **34**(5): 2015-2020.

Hanson, G.J. 1990. Surface Erodibility of Earthen Channels at High Stresses: Part I - Open Channel Testing. *Trans. ASAE*, **33**(1): 127-131.
Hanson, G.J., Cook, K.R. 1998. Relationship of Soil Suction and Erodibility of a Compacted soil. *1998 ASAE International Meeting*, July 12-16, 1998, Orlando, Florida, ASAE (Paper. No. 982065).
Hanson, G. J., Cook, K. R. 1997. Development of Excess Shear Stress Parameters for Circular Jet Testing. *1997 ASAE Annual Int. Meeting*, Minneapolis, Minnesota, ASAE (Paper 972227).
Hanson, G.J., Cook, K.R. 2004. Apparatus, test procedures, and analytical methods to measure soil erodibility in situ. *Applied Engineering in Agriculture*, **20**(4): 455-462
Haralampides, K., Rodriguez, A. 2006. Erosional properties of the sediments in the Petitcodiac River estuary at Moncton, New Brunswick. *Can. J. Civ. Eng.*, **33**(9): 1209-1216.
Hollick, M. (1976). Towards a routine assessment of the critical tractive forces of cohesive soils. *Trans. ASAE*, 19(6): 1076-1081.
Knapen, A., Poesen, J., Govers, G., Gyssels, G., and Nachtergaele, J. 2007. Resistance of Soils to Concentrated Flow Erosion: A Review. *Earth-Science Reviews*, **80**: 75-109.
Lim, S.S., Khalili, N. 2009. An improved rotating cylinder test design for laboratory measurement of erosion in clayey soils. *Geotech. Test. J.*, **32**(3): 1-7.
Luthi, M. 2011. *A Modified Hole Erosion Test (HET-P) to Study Erosion Characteristics of the Soil.* M.Sc. Thesis, Univ. of British Columbia, Vancouver, British Columbia.
Mazurek, K.A. 2010. Repeatability in Determining Erodibility Coefficients for Cohesive Soils using a Submerged Circular Turbulent Impinging Jet. *Proceedings of the 9th Federal Interagency Sedimentation Conference,* Las Vegas, Nevada, USA.
Mazurek, K.A., Amin, M.R., Cossette, D. 2016. *Discussion of* "Dimensionless Erosion Laws for Cohesive Sediment" by Joseph S. Walder, *Journal of Hydraulic Engineering* (DOI: 10.1061/(ASCE)HY.1943-7900.0001068, 04015047).
Mazurek, K.A., Rajaratnam, N., Sego, D.C. 2001. Scour of Cohesive Soil by Submerged Circular Turbulent Impinging Jets. *J. Hydraul. Eng.,* **127**(7): 598-606.
Paaswell, R.E. 1973. Causes and Mechanisms of Cohesive Soil Erosion: State of the Art. *Soil Erosion: Causes and Mechanisms, Prevention and Control*, Highway Res. Board, Washington, D.C., 52-74.
Rajaratnam, N. 1976. *Turbulent Jets.* Elsevier, New York, NY, USA.
Rajaratnam, N., Johnston, G.A., Barber, M.A. 1993. Energy Dissipation by Jet Diffusion in Stormwater Dropshafts. *Can. J. Civ. Eng.*, **20**: 374-379.
Rajaratnam, N., and Mazurek, K.A. 2005. Impingement of circular turbulent jets on rough boundaries, *Journal of Hydraulic Research*, **43**(6): 688-694.
Raudkivi, A.J. 1998. *Loose-Boundary Hydraulics*, A.A. Balkema, Rotterdam, The Netherlands.
Simon, A., Thomas, R.E., and Klimetz, L. 2010. Comparison and Experiences with Field Techniques to Measure Critical Shear Stress and Erodibility of Cohesive Deposits. *Proceedings of the 9th Federal Interagency Sedimentation Conference,* Las Vegas, Nevada, USA.
Stein, O. R. 1990. *Mechanics of Headcut Migration in Rills.* Ph.D. Thesis, Colorado State University, Fort Collins, Colorado, USA.
Stein, O.R., Julien, P.Y., and Alonso, C.V. 1993. Mechanics of Jet Scour Downstream of a Headcut. *J. Hydraul. Res.*, **31**(6): 723-738.
Thoman, R.W., and Niezgoda, S.L. 2008. Determining Erodibility, Critical Shear Stress, and Allowable Discharge Estimates for Cohesive Channels: Case Study in the Powder River Basin of Wyoming. *J. Hydraul. Eng.,* **134**(12): 1677-1687.
Walder, J. 2015. Dimensionless Erosion Laws for Cohesive Sediment. *J. Hydraul. Eng.*, DOI: 10.1061/(ASCE)HY.1943-7900.0001068.
Wan, C.F., Fell, R. 2002. Investigation of Internal Erosion and Piping of Soils in Embankment Dams by the Slot Erosion Test and the Hole Erosion Test. *UNICIV Rep. R-412*, UNSW, Sydney, Australia.
Weidner, K. 2012. *Evaluation of the Jet Test Method for Determining the Erosional Properties of Cohesive Soils; A Numerical Approach.* M.Sc. Thesis, Department of Civil Engineering, Virginia Polytechnic Institute and State University, Blacksburg, Virginia, USA.
Weidner, K., Petrie, J., Diplas, S., Nam, S., Gutierrez, M., Ellenberg, M. 2012. Numerical Simulation of Jet Test and Associated Soil Erosion. *Proc. ICSE'6,* Paris, France, Aug. 27-31, 2012.

Leadership in Sustainable Infrastructure
Leadership en Infrastructures Durables

Vancouver, Canada
May 31 – June 3, 2017/ *Mai 31 – Juin 3, 2017*

EXPERIMENTAL AND NUMERICAL MODELING OF HYDRODYNAMIC LOADING ON PIPELINES DUE TO EXTREME HYDRODYNAMIC CONDITIONS

Ghodoosipour, Behnaz[1,4], Nistor, Ioan[3] and Mohammadian, Abdolmajid[3]
[1] PhD student, University of Ottawa, Canada,
[2] Professor, University of Ottawa, Canada
[3] Associate professor, University of Ottawa, Canada
[4] bghod068@uottawa.ca

Abstract

Pipelines in coastal areas are used for gas and oil transportation, as well as for disposal of wastewater to water bodies. Installation of pipelines in coastal areas is of great importance, and requires consideration of different engineering design criteria. A new topic in pipeline design has emerged lately due to the recent extreme hydrodynamic events such as tsunami and storm surges. Therefore, the primary objective of this study was to investigate the hydrodynamic forces induced on pipelines during such extreme waves. A comprehensive program of physical model experiments were conducted in the Hydraulic Flume at the University of Ottawa. The tests attempted to measure the hydrodynamic forces exerted on pipe due to tsunami-like bores replicated using dam break flow. Different pipe placement configurations using different gap ratios (e/D), (e being distance to bottom and D pipe's Diameter), were tested under various flow conditions in wet and dry bed. For the case of dry bed conditions, experimental results show that only for e/D ≤ 0.3, two peaks are noticeable in force time-history, whereas, for other relative gap ratios, no distinct peak was noticed. The main reason for such behavior when using small e/D ratios is the suppression of the vortex shedding which forms around the pipe. For the case of wet bed conditions, the time-history of the drag force shows no change even when drastically changing the relative gap ratio. The 1-D shallow water equations were also used to study the dam break flow over the flat flume bed and its impact with the circular cylinder downstream of the flow. The drag coefficient was then derived using experimental and numerical results and showed relatively good agreement between the two. Results from this study will be used for the assessment of the current design recommendations with the ultimate goal of improving them.

1. Introduction

The design of pipelines to withstand external loading exerted on them requires a proper understanding of the external hydrodynamic forces. Existing literature on this important subject goes back to Morison et al. [1950], who proposed a widely-accepted formulation of the forces on water surface-piercing piles due to ocean waves. Morison equation could be applied to derive the forces exerted on the center of a pipe as:

$$[1.1]\quad F = \underbrace{\rho C_m V \dot{u}}_{F_I} + \underbrace{\frac{1}{2}\rho C_d A u|u|}_{F_D}$$

where F_I is inertia force due to the accelerating flow, C_m is the inertia (mass) coefficient, $\dot{u}$ is the flow acceleration, V is the volume of the pipe, F_D is the drag force caused by fluid viscosity and the velocity gradient within the thin boundary layer next to the cylinder's surface, C_d is the drag coefficient, *A* is the area of the pipe projection perpendicular to the direction of the flow and u is the horizontal flow velocity. Variations of the above formula to calculate the hydrodynamic forces in steady and oscillatory flows exerted on horizontal cylinders were reported in other articles. Flow around circular cylinders is complex: vortex shedding suppression occurs at Reynolds numbers above critical values, subsequently changing the pressure distribution around the cylinder and causing extra fluctuations of the drag and lift forces to act on the cylinder (Lei et al [1999]). Drag and lift coefficients are non-dimensional and dependent on both the flow and the cylinder geometry and location. For cylinders placed near the bottom boundary, bed roughness, turbulent behaviour and velocity gradient of the boundary layer play an important role on the force exerted on the cylinder. Several experimental studies focused on issues related to cylinders located near a plane boundary. Most studies revealed a critical distance to bottom ratio (e/D=0.3) defined as the distance for which vortex shedding suppression occurs. Bearman et al. [1978], Angrilli et al. [1982] and Zdravkovich, [1985] proposed this hypothesis by looking into the power spectra of hot wire signals occurring in the wake region of a cylinder. However, Buresti & Lanciotti's [1992] results show e/D=0.4 to be the critical relative gap ratio. The existence of vortex shedding causes the occurrence of a peak in the power spectra which disappears due to the vortex shedding suppression in a specific gap ratio called vortex shedding suppression gap ratio.

In recent years, a new criteria has emerged in pipeline design due to the recent catastrophic events such as tsunamis and storm surges which caused massive damaging consequences of such infrastructure. The 2011 Tohoku Tsunami in Japan and the 2012 Typhoon Haiyan in the Philippines, raised an increased interest of research groups to re-evaluate the existing design and safety codes such that they would consider effects of such extreme events. The American Society of Civil Engineers (ASCE), through its ASCE7 Tsunami Loads and Effects Subcommittee, has developed a new standard for tsunami impacts and loading (Chock [2016]) and has emphasized the need to investigate tsunami loads on pipelines located in tsunami-prone areas.

Different studies have focused on investigating forces induced by a hydraulic bore on infrastructure: among them Stoker [1957], and Cross [1967]. Stoker [1957] have used nonlinear shallow water equations to study the behaviour of bores for a moving hydraulic jump surging over still water layer. Guidelines for the Design of Structures for Vertical Evacuation from Tsunamis FEMA P-646 [2012] introduces forces exerted on a body during a tsunami as drag and impulse force. The lateral component of hydrodynamic force is called drag force and is a combination of the pressure force and the friction force due to the fluid. The drag force can be calculated using

$$[1.2] \quad F_d = \frac{1}{2}\rho_s C_d B(hu_{max}^2)$$

Where ρ_s is the fluid density, Cd is the drag coefficient, B is the width of the structure in the plane normal to the direction of flow, h is flow depth, and u is the flow velocity at the location of the structure. For forces on structural components, B is taken as the width of the component. The drag coefficient depends mainly on structure geometry. hu_{max}^2 is the maximum momentum flux per unit volume where u_{max} is the maximum velocity during tsunami and h is the corresponding water depth at the instant when velocity is maximum.

The impulse force is caused by bore front impacting the structure. Experiments conducted on measuring impact force show that in dry bed conditions, there is no considerable peak in the force time-history at the time of impact. However, experimental work recorded a considerable force peak in wet bed condition (Ramsden [1993]). Arnason [2005], estimated the maximum impact force to be approximately 1.5 times larger than the corresponding drag force.

Due to the need for increased protection of pipelines against extreme hydrodynamic events, the main objective of this study is to investigate the induced hydrodynamic forces on pipelines during extreme waves and flood events. The drag coefficient plays an important role in the magnitude of the drag force. Suggested values in existing guidelines for drag coefficient are mostly derived for steady or oscillatory flows. Because of the unsteady behavior of tsunami-induced inland inundation, the drag coefficient changes with time in contrast to the case of steady flow conditions in which the drag coefficient is constant (Arnason [2005]). This study proposes values for drag coefficient in the case of a transient dam break flow for different relative gap ratios.

2. Experimental Methodology

The objective of this experimental study was to characterise hydrodynamic forces exerted on pipelines due to extreme events, modelled using dam break waves. A comprehensive experimental approach was conducted for this purpose. Different parameters were varied during the experiments namely, reservoir depth (h), tailewater depth (d), pipe distance to bottom (e) to diameter (D) ratio (e/D). More than 500 tests were conducted at the hydraulic flume in university of Ottawa.

2.1. Experimental setup

2.1.1. Flume and impoundment reservoir

The experiments were conducted in the Hydraulic Flume in the Hydraulic Laboratory at the Department of Civil Engineering at the University of Ottawa. The flume is 30.1m in length, 1.5 in width, 0.5m in height.The Flume is constructed of concrete. Water is pumped into the flume and a swing gate was installed 21.5m downstream from flume inlet separating the reservoir from the actual flume test section. Dam break waves with different characteristics were generated by the rapid opening of the gate (in a very short time (t = 0.06–0.08 s). A movable steel gate installed at the downstream end of the flume ensures that a layer of water can be maintained for various wet bed conditions (different water levels). Figures 1 and Figure 2(b) illustrate the flume in plan and side view respectively. The pipe used in the tests is a steel pipe with a 10 cm diameter, which is located at x=6 m downstream of the swing gate. Experiments were conducted with different relative gap ratios, e/D=0.3, 0.6, 0.8, *e* being distance to the flume bed and *D* (pipe diameter), different tailwater depths for wet bed condition (d=0-22cm) and with three different reservoir (impoundment) depths (h=30cm, 40cm & 50 cm) and different components of the force exerted on the pipe were measured.

2.2. Instrumentation

Figure 1, shows a schematic of the location of the instrumentation used in the tests, while Figure 2(a) & 2(b) show images of the instrumentation and its location.

2.2.1. Wave gauges

To measure the time-history of the water level at different locations, level three wave gauges (Akamina technology) were used. The wave gauge 1 was installed right upstream from the swing gate and is used to record the water level variation upstream of the gate. The recorded data can be used for the purpose of helping synchronize all othere recordings.
The other two wave gauges (2 and 3) are located at x=3.0m and x=5.6m downstream of the gate.

2.2.2. Acoustic Doppler Velocity meter (ADV)

A high-resolution acoustic Doppler velocity meter (Vectrino, SE8660) was used for velocity measurements. The ADV is able to measure water velocities along three axis using Doppler processing technology. The ADV's sampling rate was set to 200 Hz and it was located at x=5.6 m downstream of the swinging gate and placed 10 cm upstream from outer edge of the pipe. Each experimental test with a given configuration was repeated 3 times and the ADV head was vertically placed at different elevations corresponding to the level of the center of the pipe, its lower side and 5 cm away from the bottom of the flume.

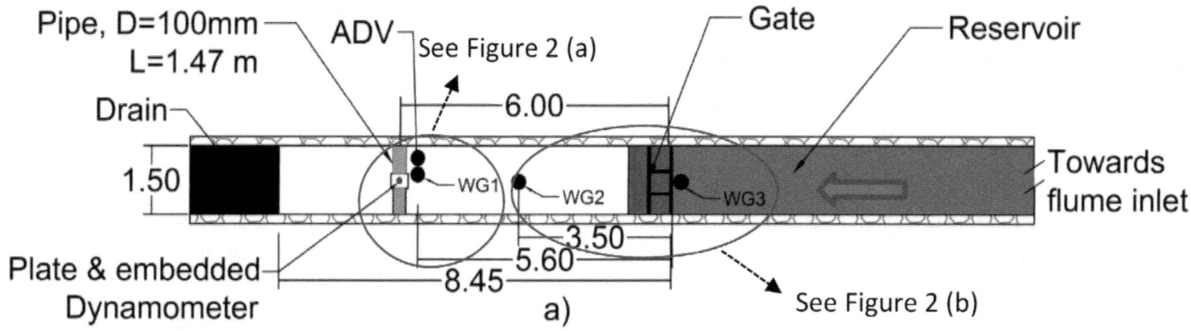

Figure 1: Flume and instrumentation. a) Plan view (unless otherwise specified, all dimensions are in m).

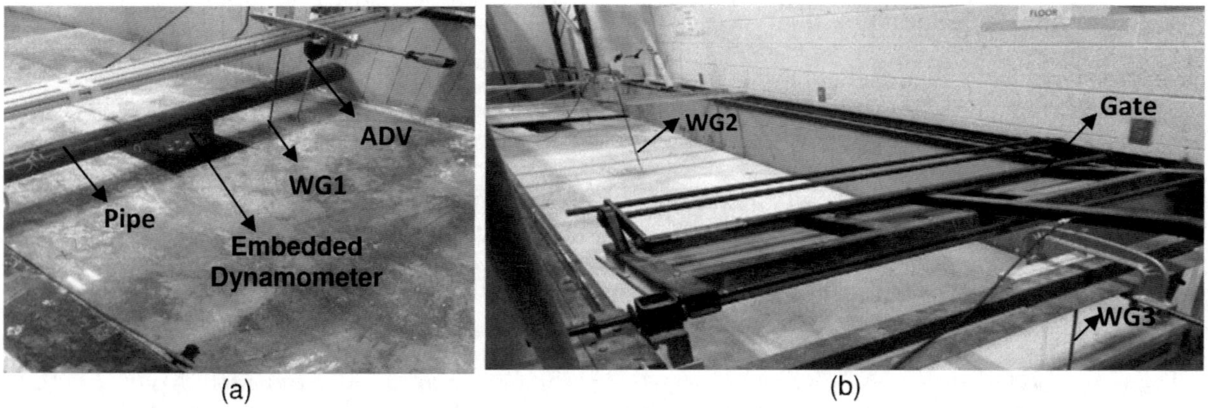

Figure 2: a) Side view, pipe, ADV and wave gauge, b) side view - flume and gate.

2.2.3. Dynamometer

In order to record time-history of the forces exerted on the pipe due to the dam break wave, a six degrees of freedom (6DOF) dynamometer (Interface- 6A68E) was used. This dynamometer is able to simultaneously measure forces moments along the 3 axis of coordinates (*X-Y-Z*). Six full bridges provide mV/V output on 6 independent channels. The dynamometer was installed beneath the concrete flume floor by cutting the concrete, placing the device and embedding it again with concrete around as shown in Figure 3. Steel plate was placed on top in a way that the plate was leveled with the flume (Figure 2.a). The pipe is attached to the plate using two thin plates. Extra drag forces exerted because of the plates were tested and proved to be negligible. The loading exerted on the pipe was transmitted to the dynamometer through four bolts which connected the plate on which he pipe rested with the dynamometer.

2.2.4. Data acquisition system

The analog voltage signals from the instruments used in the experiments were converted to digital format and saved into data files using a QuantumX data acquisition system (MX840B, 8-channel universal amplifier & MX1601B with 16 individually configurable channels with sample rates up to 19.2 kS per channel and a bandwidth of up to 3 kHz.). All data were time-synchronized.

Figure 3: Dynamometer embedded in the flume floor

3. Experimental results and discussion

Results for drag force from testing different reservoir and tailwater depth values for different e/D values in wet and dry bed conditions, are presented below. As mentioned in Introduction section, the ratio e/D=0.3 has been shown to be the critical ratio for vortex shedding suppresion in previous studies deadling with the hydrodynamic loading on circular cylinders in constant and oscilatory flows. Results from this study demonstrated the validity of the critical e/D ratio for dam break flows over a dry bed.

3.1. Validation of the model

Multiple tests with identical initial conditions were carried out to verify the repeatability of the tests. Figure 4, represents the wave height in dry and wet bed conditions at x=3.5 m. Figures for both dry and wet bed conditions show good agreement in a way that oscillations remain the same in different tests with same release and bed conditions.

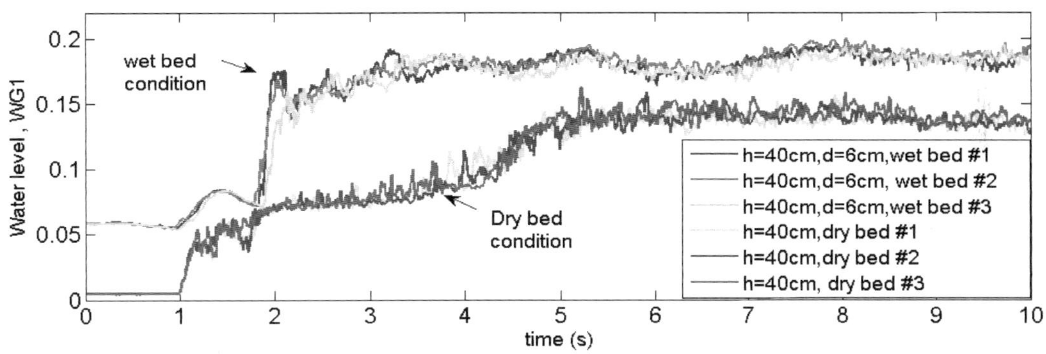

Figure 4: Repeatability of tests in dry and wet bed conditions

3.2. Evaluation of drag force for dry bed conditions and different (e/D) ratio

Analyzing the force time history in a dry case and different distance to bottom ratios, in Figure 5(b), one can observe that for e/D=0.3, the drag force behaviour is considerably different when compared to data obtained using other e/D ratios. For this particular value of e/D=0.3,two force peaks are noticed and they are defined as impulse and run up force while for values of e/D of 0.6 and 0.8, no such distinct impulse peak is observed. This phenomenon could be explained by vortex shedding suppression for such small e/D values. The vortex shedding deflects part of the fluid from upstream of the plane boundary over the top of the circular cylinder and thus reduces the flow passing through the gap (Lin et al. 2009). For values of e/D=0.6 and 0.8, the flow first passes through the gap between the pipe and the flume bed and and then reaches vertically the pipe in a very short time: hence, the presence of a rising time observed for the force to reach its maximum value.

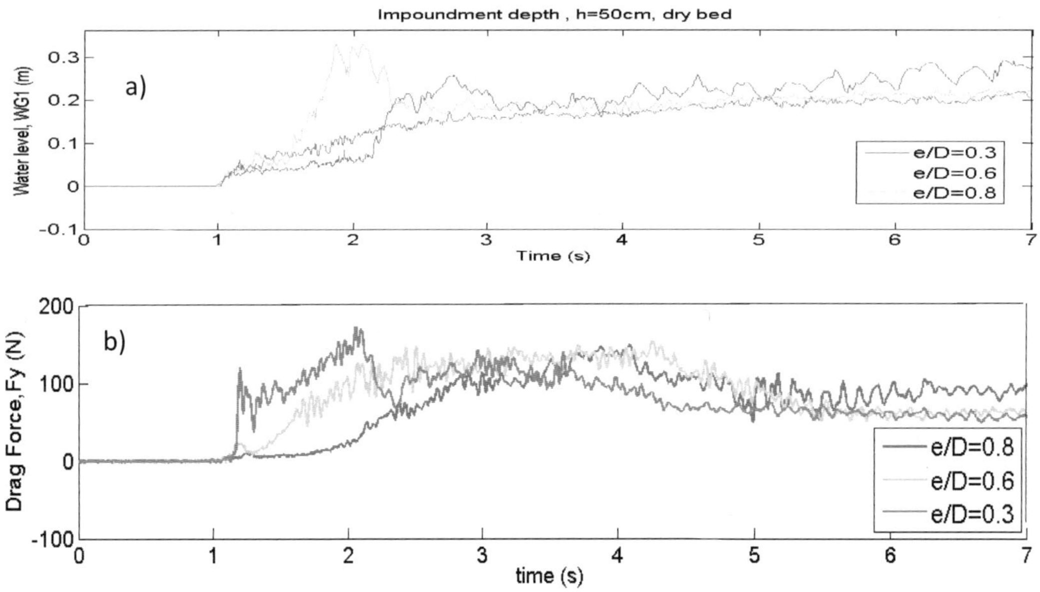

Figure 5: Time-history of the a) water level, WG1 b) drag force for dry bed condition, impoundment depth h=50cm, e/D=0.3,0.6,0.8.

3.3. Evaluation of drag force for wet bed conditions and different (e/D) ratio

Figures 6.a and 6.b illustrate hydrodynamic force components in the case of wet bed condition with *h/d=5*.0. The hydrodaynamic force time-history in wet bed condition shows a different pattern comapring to the case when using dry bed condition at the time of bore impact: the impulse force is significant comparing to the case of the force magnitude at the same instant when usign a wet bed. According to Yeh, [2007] impulse force increases due to the impact of a steeper bore front onto the object: hence the absence of a clear impulse force for the case of the dry bed condition can be attributed to the comparatively milder slope of the bore front for the dry bed condition. From Figure 5, one can reasonably conclude that drag force pattern does not change drasticaly when changing the gap ratio (e/D). Only the horizontal impulse force (Fy) decreases as the pipe submerges into the water and by placing it closer to the flume floor (decreasing the gap). The time-history of the force under similar conditions but with wet bed and with different h/d ratios, i.e, h/d=6.25, 4.16 etc, shows that impulse force is the maximum force in all these configarations. This is different from the results presented by Palermo et al. [2012] for a vertical square model in which the water reflecting from the obstacle after the first impact rejoins the incoming bore, surges up the model and causes larger forces to be exerted onto the object, a comonent that the authors identified as *run up force*. The reason behind the runup force being larger than the impulse force for vertical obstacles, is the relatively small surge depth to obstacle width ratio which causes a transient amount of added ponded water to surge against the front wall of the obstacle until it finally passes around side walls. In this study, the pipewidth is equal to flume width, therefore there is no opportunity for the water to flow past the obstacle, thus decreasing the force. Hence, the runup peak force is not recorded in this case.

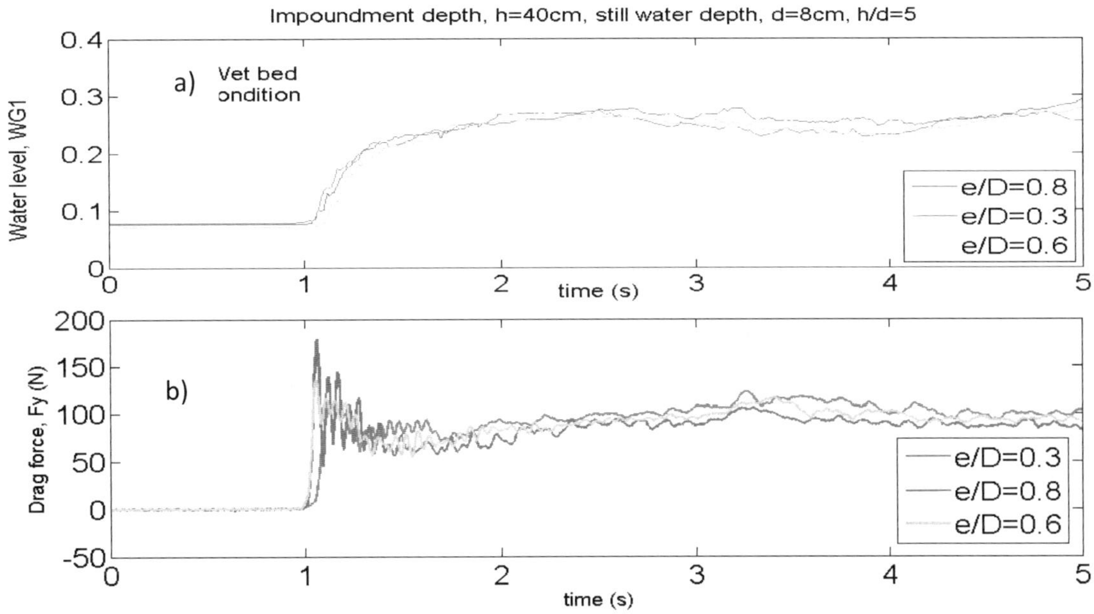

Figure 6: Time-history of the a) water level, WG1 b) drag force, for wet bed condition, h/d=5, e/D=0.3,0.6,0.8.

3.4. Drag coefficient

Results for horizontal velocity and the calculated drag coefficient for the circular cylinder impacted by the dam break wave are shown in Figure 7.a and b, respectively. In Figure 7.a colored lines show weighted average measured velocity at different elevations (z=5cm, z=8cm, z=12cm), for different relative gap ratios, i.e. e/D=0.8, e/D=0.6 and e/D=0.3.The black line shows the velocity measured in the absence of pipe. The ADV results for the first two seconds following bore impact exhibited extensive noise and were not considered in this analysis. Bore front velocity which is important for calculating the impulse or slam coefficient will be derived using image processing and analysis technics in the next step of this comprehensive experimental program.

In figure 7.b, one can be observe that the magnitude of the velocity in the absence of pipe is larger than for all cases with the pipe installed with any relative gap ratio. The average velocity decreases as the gap ratio increases from e/D=0.3 to e/D=0.8, which indicates that larger gap ratios increases momentum loss. Figure 7.b shows the time-history of the drag coefficient, *Cd*, which was calculated using the experimentally-measured drag force and the weighted depth averaged velocity U in the vicinity of pipe as in Equation [5.1]

$$[5.1] \quad C_D = \frac{1}{2}\rho U^2 / F_D$$

Where F_D is the horizontal component of the force exerted on the pipe, U is horizontal velocity and ρ is water density considered as 1000 kg/m^3.

The highest and lowest Cd values occur for the case of the pipes with e/D=0.8 and e/D=0.3, respectively, which once again, indicated more momentum loss within the case of larger gap ratios. The values of Cd derived using the velocity in the absence of pipe are lowest of all other configurations which indicates that using this velocity to drive the drag coefficient for pipeline design may result in underestimating the drag force and may hence lead to possible damage to the pipeline in the case of an extreme hydrodynamic event. Safety factors should be considered for this reason in pipe design guidelines.

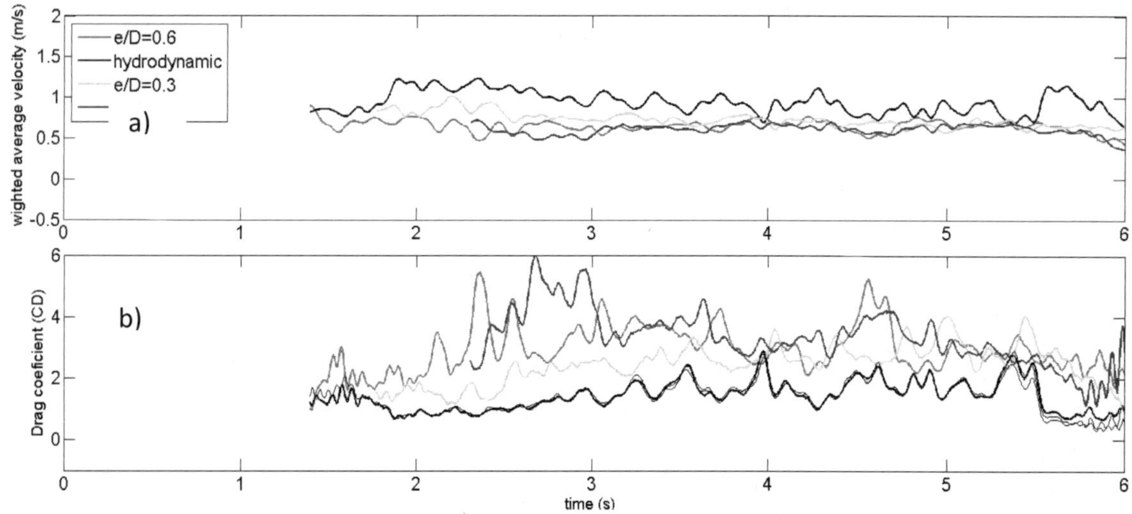

Figure 7: Time-history of: a) Horizontal velocity measured using ADV, b) Calculated drag coefficient using measured drag force and the horizontal velocity. Colored: different e/D ratios, black: pure hydrodynamic.

4. Numerical model

The 1-D shallow water equations were used to simulate the dam break flow over a flat flume and its flow pasta circular cylinder located downstream of the swinging gate. The 1-D Navier-Stockes equations consist of the conservation of mass.

$$[4.1] \quad \frac{\partial h}{\partial t} + \frac{\partial uh}{\partial x} = 0$$

$$[4.2] \quad \frac{\partial uh}{\partial t} + \frac{\partial (u^2 h + 0.5gh^2)}{\partial x} = gh(S_0 - S_f)$$

Where, *u(x,t)* is the flow velocity, *h(x,t)* is the instantaneous water depth, *S*(*x*) is the bed slope and $S_f(x,t)$ is the friction slope.

An isolated obstacle impacted by the flow, will add a new momentum source term to Eq. 4.2 as the drag increases in the vicinity of the obstacle. This extra momentum was incorporated in the momentum equation as the second term to the right side of Equation 4.3, using drag coefficient C_D.

$$[4.3] \quad \frac{\partial uh}{\partial t} + \frac{\partial (u^2 h + 0.5gh^2)}{\partial x} = gh(S_0 - S_f) + C_D u^2$$

The dam break flow in the presence of a half pipe with its major axis=0.16 and minor axis=0.1 was numerically simulated by solving the modified 1-D shallow water equations [4.1] & [4.3] using a MATLAB code developed by the authors. The results of the water surface time history obtained with the proposed numerical model were compared with the experimentally-recorded data for the case with dam break flow over a horizontal circular cylinder with diameter=0.1 and gap ratio e/D=0.6d. Reservoir depth was set to $d_0 = 40cm$.

5. Numerical results

Figure 8.a & b show results for water level from numerical model using 1D shallow water equations, at t=0 and t=10s, while, Figure 8. shows the comparison between results for water surface elevation time history derived from numerical simulation with different Cd values used in equation [4.3] and experimental results in e/D=0.6.

As can be seen in the Figure 9. results for drag coefficient from numerical and experimental study with same impoundment depth do not match perfectly. As the Cd value is increased from 0.8 to 4 the peak in water surface time history is shifted to the right and results get closer. The reason behind this difference is mainly the fact that, 1D equations are unable to simulate vertical velocities accurately. Drag coefficient is mostly dependent on the shape of the obstacle and this caused the results from simulations to be inaccurate. However, such simulations could be used for obtaining a rough estimate for a constant drag coefficient for design purposes.

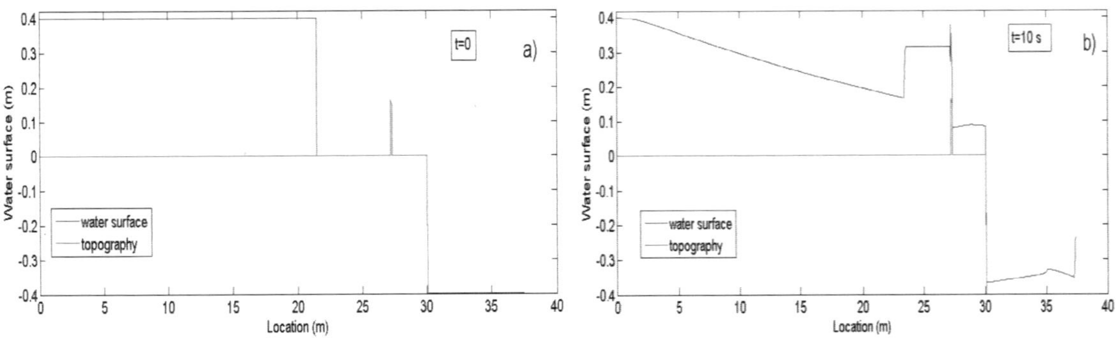

Figure 8: Results of the numerical models using the 1-D shallow water equation a) t=0, b) t=10s.

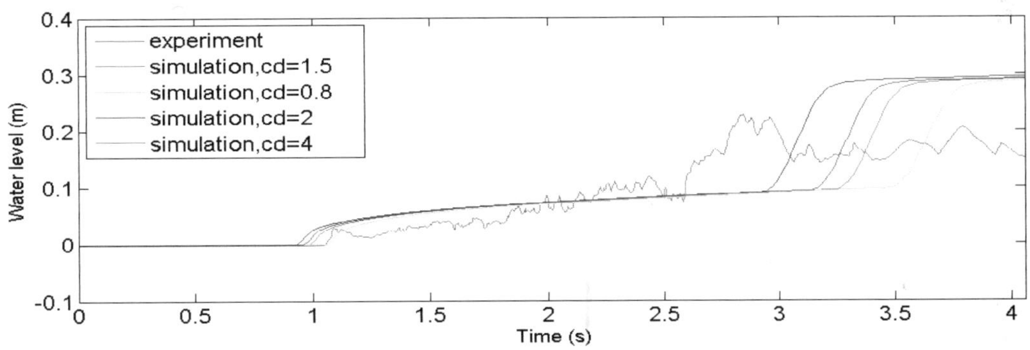

Figure 9: Comparison between experimental data and numerical time –history of the water surface of a dam break wave generated by an impoundment depth of 40 cm.

6. Conclusions

An experimental program of physical model testing was conducted to study the evolution of the hydrodynamic forces exerted on a pipe due to dam break flow which could be representative for an extreme hydrodynamic event such as tsunami-induced inundation. Force time-history was measured and analyzed for different experimental conditions with respect to changing both the tested pipe distance to bottom using gap ratio (e/D) and flow characteristics using different reservoir and both dry and wet bed conditions. Most important remarks from this study are:

- The impulse force at the time of the initial bore impact is more pronounced for the case of wet bed condition. For the case of the dry bed condition, the hydrodynamic force increases gradually until it reaches its maximum value.
- For the dry bed condition, time history for different relative gap ratios (e/D=0.3, 0.6, 0.8) was studied. Only for $e/D \leq 0.3$, a peak force, termed impulse force at the beginning of the bore impact and later a larger peak as run up force were observed. Vortex shedding suppression at such small gap ratios causes this results to be observed. This result confirms previous finding of other

researchers for steady and oscillatory flows which confirm that e/D=0.3 is indeed a critical gap ratio for vortex shedding suppression.

- Drag force time-history for wet bed condition shows no variation, irrespective of the gap ratio.
- The drag coefficient calculated using the time-history of the velocity obtained from ADV measurements in the absence of the pipe exhibits lower magnitudes compared to the measurements when the pipe is installed. This indicates that calculations based on the latter, as suggested in many studies, may lead to underestimating the drag force for pipe design purposes.

- The 1-D shallow water equations are able to reproduce the dam break wave formation and propagation and can be used to derive the drag coefficient.

Measuring the lift force exerted onto pipelines due to dam break flows as well as calculating lift coefficient will be the focus of future studies on this topic. Image analysis technics will be used for deriving the slam coefficient at the instance of bore impact. The 3-D shallow water equations will be also employed for a more accurate numerical simulation of the problem.

References

Angrilli, F., Bergamaschi, S. and Cossalter, V., 1982. Investigation of wall induced modifications to vortex shedding from a circular cylinder. *Journal of Fluids Engineering*, *104*(4): 518-522.

Arnason, H., 2005. *Interactions Between an Incident Bore and a Free- Standing Coastal Structure, Ph.D. dissertation, University of Washington, Seattle, Washington.*

Bearman, P.W. and Zdravkovich, M.M., 1978. Flow around a circular cylinder near a plane boundary. *Journal of Fluid Mechanics*, *89*(01): 33-47.

Buresti, G. and Lanciotti, A., 1992. Mean and fluctuating forces on a circular cylinder in cross-flow near a plane surface. *Journal of Wind Engineering and Industrial Aerodynamics*, *41*(1-3): 639-650.

Chock, G. Y. 2016. Design for tsunami loads and effects in the ASCE 7-16 standard. Journal of Structural Engineering, ASCE:04016093.

Cross, R. H., 1967. Tsunami surge forces. *Journal of Waterways and Harbor Division, ASCE, 93(4): 201-231.*

FEMA. 2012. P646 Guidelines for Design of Structure for Vertical Evacuation from Tsunamis. Federal Emergency Management Agency.

Lei, C., Cheng, L. and Kavanagh, K., 1999. Re-examination of the effect of a plane boundary on force and vortex shedding of a circular cylinder. *Journal of Wind Engineering and Industrial Aerodynamics*, *80*(3): 263-286.

Lin, W.J., Lin, C., Hsieh, S.C. and Dey, S., 2009. Flow characteristics around a circular cylinder placed horizontally above a plane boundary. *Journal of engineering mechanics*, *135*(7):697-716.

Morison, J.R., Johnson, J.W. and Schaaf, S.A., 1950. The force exerted by surface waves on piles. *Journal of Petroleum Technology*, *2*(05): 149-154.

Palermo, D., Nistor, I., Al-Faesly, T. and Cornett, A., 2012. Impact of tsunami forces on structures: the university of Ottawa experience. *Fifth international tsunami symposium,Ispra, Italy,*: 3-5.

Ramsden, J.D., 1993, Tsunamis: Forces on a Vertical Wall Caused by Long Waves, Bores, and Surges on a Dry Bed, Report No. KH-R-54, W.M. Keck Laboratory, California Institute of Technology.

Stoker, J. J., 1957. Water Waves: The Mathematical Theory with Applications. *Wiley-Interscience, New York.*

Yeh, H. 2007, Design Tsunami Forces for Onshore Structures. *Journal of Disaster Research*, 2(6): 531-536

Zdravkovich, M.M., 1985. Forces on a circular cylinder near a plane wall. *Applied Ocean Research*, *7*(4): 197-201.

Leadership in Sustainable Infrastructure
Leadership en Infrastructures Durables

Vancouver, Canada
May 31 – June 3, 2017/ *Mai 31 – Juin 3, 2017*

FIELD OBSERVATIONS AND ANALYSIS OF THE GENERATION OF TOTAL DISSOLVED GASES DOWNSTREAM OF THE HUGH L. KEENLEYSIDE DAM

Billay, Graeme[1], Zhu, David[2], Bruce, James[3], Leake, Alf[4], and Crossman, James[5]
[1, 2] Department of Civil and Environmental Engineering, University of Alberta, Canada
[3] Creekside Aquatic Sciences, Canada
[4,5] BC Hydro, Canada
[1] gbillay@ualberta.ca

Abstract: Supersaturation of total dissolved gases (TDG) can occur in rivers downstream of hydropower facilities as a result of dam operations. The subsequent release of dissolved gases in downstream waters can result in fish mortality and other impacts to the aquatic environment. In order to determine the effect of different dam operations on TDG generation detailed field measurements were conducted both at and downstream of the Hugh L. Keenleyside dam on the Columbia River. TDG generation was evaluated as a result of six different operating scenarios that included releases of water through different dam structures including low-level outlet gates, the spillways, and a combination of the two types of structures. All scenarios utilized the low-level outlets and one utilized the spillways. Results indicate that the type of spill mode has significant effect on TDG generation, with southern low-level outlet gates increasing TDG 12% more than the northern low-level outlet gates for a smaller discharge. Accordingly, it is possible to optimize the spill operations in an effort to reduce TDG levels by focusing on the northern low-level outlet gates for spill discharge. In addition, the mixing of the spill water in the tailrace is important to the TDG level in the downstream river region as seemingly complete mixing occurs, decreasing the supersaturation when both southern and northern low-level outlets are operating.

1 Introduction

The act of releasing water through dam structures and the subsequent plunging and turbulent mixing of the reservoir water with the downstream river entrains a significant amount of atmospheric gases. This entrained air can create a supersaturated condition whereby there is a higher concentration of gases dissolved in the water relative to the concentration of gases in the atmosphere. If the levels of dissolved gas increase above certain thresholds, fish that inhabit the downstream river are at risk of gas bubble trauma (GBT) which can result in mortality if severe enough (Weitkamp and Katz, 1980). The zone of supersaturated gases can extend kilometers downstream of the facility possibly impacting the aquatic environment depending how gas concentration changes as a result of river hydraulics. Understanding the concentration of gases that are generated as a result of different dam operations, and how these concentrations persist downstream, is important to determine possible environmental impacts.

When air is entrained into the stilling basin at the base of a spillway, the bubbles that are formed are forced downward by the momentum of the plunging jet. Typically, in the stilling basin region, the surface area for gas transfer is along the free-surface of the river, with the introduction of many small bubbles, the surface area available for gas transfer greatly increases (Hibbs and Gulliver, 1997). As the bubble descends, the hydrostatic pressure of the water column and the dynamic pressure of the water jet acts on it to force the air from within the bubble into solution. With an increase in hydrostatic pressure, also comes an increase in saturation concentration of the gas within the bubbles and therefore, the gases ability to be dissolved in the water (Gulliver et al., 1997). Different dam operations can influence the levels of total dissolved gas in a number of ways. If water velocity at the base of a spillway is large, bubbles will be forced further down into the stilling basin increasing the hydrostatic pressure acting on the bubble. This effect is larger if the tailwater depth is also large. Different types of spill structures can create different flow patterns and levels of surface turbulence as well in the stilling basin. Low-level outlets will produce different flow patterns that impact how quickly bubbles rise to the surface. If air is entrained via low-level outlets the bubble sizes may also be different from those produced by the spillways. Bubble size distribution is an important parameter affecting TDG generation (Qu et al., 2011b) as it affects the rate of gas transfer. Modeling bubble characteristics is an important part of a physically-based predictive TDG model, as it is known that most of the gas transfer in a stilling basin will occur through the bubble-water interface. Therefore, bubble residence time and coalescence are other important parameters. Computational models have been increasingly used to model gas transfer and estimate complex hydraulic flows in the stilling basin and tailraces of dams (Politano et al., 2005; Urban et al., 2008).

Since there are many factors specific to each structure or facility that influence the generation of TDG, determining a predictive equation that applies to all dams has been challenging. However, there have been models developed to predict downstream TDG for several lower Columbia River dams in the United States (University of Washington, 2000). These models rely on coefficients determined for each of the dams the model is to be applied to, and in many cases, this requires a huge amount of effort in the way of field observations. Studies conducted by the United States Army Corps of Engineers have looked into various solutions to the problem of increased TDG levels with some success via the construction of spillway flow deflectors (USACE, 1996; Orlins and Gulliver, 2000).

As water moves downstream from the dam, the TDG concentrations may change as a result of mixing with facility generation flow and gas transfer across the surface of the water. What may also be important is if the dam operational scenario has an impact on the gradient of TDG across the width of the river or if the spill flows mix completely within the tailrace before moving downstream. This may affect measurements downstream of the tailrace where the measured values may be less than the actual TDG generated in the stilling basin.

This study will focus on the TDG issues that arise at the Hugh L. Keenleyside (HLK) Dam on the Columbia River (Figure 1). It is approximately 7 km upstream of Castlegar, BC and 60 km upstream of the Canada – United States border. It is the third and final dam on the Columbia River before flows enter the United States. The reservoir held behind HLK is the Arrow Lakes Reservoir (ALR). The main objective of this study is to quantify the TDG levels as a result of different operational scenarios (combinations of spillway and low-level outlet discharge). This study will benefit hydropower facilities in optimizing the operations of spillways and low-level outlets. Since the problem of supersaturated TDG has been difficult to correct in existing dams (USACE, 2001), predictive tools and clearer relationships between dam spill events and TDG generation will benefit designers of new facilities to meet environmental regulations. A better understanding of the operational conditions that lead to TDG increases at HLK will assist the dam operators in reducing the potential risk to fish. Another important factor in determining TDG increase across operational scenarios that utilize more than one type of spill gate is the mixing in the tailrace. Qualitative conclusions regarding the level of mixing within the tailrace may be possible as a result of the field observations. However, a computational model may be the only way to determine specific flow patterns within this region.

2 Field Program

TDG measurements were taken across several transects near the dam. The first 2 km of the river downstream of HLK are of particular interest. Additional data collected at downstream transects are available in a companion paper on TDG mixing and dissipation (Kamal et al., 2017). Measurements were also collected approximately 1.6 km upstream of the dam in the forebay.

Data collected at each transect consisted of local barometric pressure, air temperature, water temperature, total gas pressure, discharge, and velocity. Two methods were used to collect the TDG data (which included stationary continuous monitoring and spot measurements taken along transects from a boat). The probes that were used were the Pentair Point Four Lumi4 DO/TGP probe (for spot measurements) and the Pentair Point Four TGP probes (for continuous monitoring). These two types of probes have a measurement accuracy of +/- 4% saturation. Continuous monitoring stations were set up near the banks of the river at 1 km (left bank), and 4 km (right bank). Spot measurements were taken along those transects as well as at transects 0.5 km, 2 km, and for some scenarios (SC) at the end of the tailrace (around 300 m downstream of the spill gates). In all cases, the TDG probes were located approximately 1 m below the water surface for data collection. These probes were placed in a water tight container situated on a custom-built PVC frame that was anchored in the river. The monitoring stations were limited to being located near the banks of the river where the current was less strong and where it did not require extremely long lengths of chain and rope to anchor them in place.

In an effort to collect meaningful data about the TDG generation as a result of spilling operations at HLK dam, different operational scenarios were conducted and the resulting TDG measured. These scenarios consisted of different numbers of gates and patterns and were tested over a one-week period with as many measurements taken as possible at different transects. Between each scenario, the ratio of spill rate to generation flow was held nearly constant. The scenarios were chosen to give a range of specific discharges that came along with different numbers of gates operating at once. An effort was made to also ensure that all types of structures were utilized to test a large range of possible dam operations. The timing between scenarios was important because it was desirable to have the flow of a particular scenario completely replace the flow of the previous scenario, ensuring the measurements were reflective of only one scenario at a time. This also limited the number of scenarios that could be tested during the field work. A summary of the times of each scenario along with its operational combination and gate discharges are listed in table 1.

Figure 1: HLK dam site on the Columbia River and downstream river transects, tailrace (TR) measurement location, and Bruce (2016a) meter 2 (M2)

HLK dam conveys water through two different types of gates. Four radial spillway outlet gates (labeled SPOG 1 – 4) are between two sets of four low-level outlet gates (labeled LLOG 1 – 8). LLOGs labeled 1 – 4 are commonly referred to as the northern LLOGs (or NLLOGs) and those labeled 5 – 8 are the southern LLOGs (or SLLOGs), referring to their position relative to the SPOGs (see figure 1). In total, all spill gates are capable of discharging 10,500 m^3/s. Several gates may be in operation at once, with the exception of LLOGs 1 and 8. These two gates are reserved for emergency flows only. The Arrow Lakes Generation Station (ALGS) was commissioned in 2002 and is currently the main source by which flows pass through the dam. It has a discharge capacity of 1,150 m^3/s.

The continuous monitoring stations were set up over the course of two days and recorded TDG data in 2-minute intervals. Meter 2 (M2) is a continuous monitoring station tied to a berm on the south end of the tailrace and collected data at 5-minute intervals (Bruce, 2016a). The spot measurements were conducted at various transects over the course of the six different spill scenarios, but not every scenario was able to have transect TDG data collected because of the tight spill schedule and the amount of time it took to travel up and down the 20 km river reach by boat. For spot measurements taken from the boat, the TDG probe was kept in the flow for approximately 3 – 7 minutes to allow the reading to stabilize. The stabilized reading was recorded for each point. A typical data series for one spot measurement is illustrated in figure 2.

Table 1: HLK Operational Scenarios

SC	Start Time	End Time	Operational Pattern	Gate Number (m³/s)	Spill Rate (m³/s)	ALGS (m³/s)
1	July 26, 12:00	July 27, 14:00	3 NLLOGs	2(186.38), 3(188.41), 4(650.81)	1025.6	1085
2	July 27, 14:00	July 27, 16:00	2 SPOGs	4,3(490.52)	981.04	1085
3	July 27, 16:00	July 28, 14:00	3 SLLOGs	5,6,7(311.58)	934.74	1085
4	July, 28, 14:00	July 28, 16:00	2 SLLOGs	5,6(644.92)	1289.84	1100
5	July 28, 16:00	July 29, 16:00	2 SLLOGs + 2 NLLOGs	5,6(277.4), 3,4(277.85)	1110.5	1100
6	July 29, 16:00	July 30, 16:00	1 SLLOG + 2 NLLOGs	5(191.71), 3(186.12), 4(645.48)	1023.31	1081

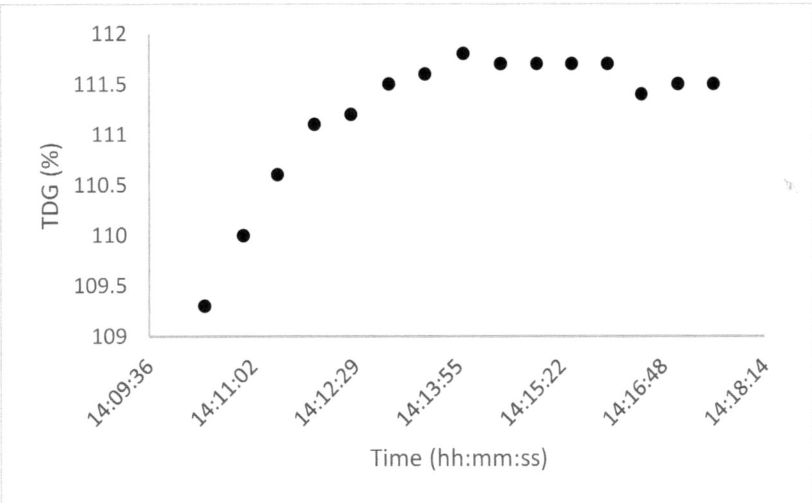

Figure 2: TDG data series for single spot measurement

There is a stabilization period between scenarios where the spill gates are in a transition between operating scenarios. With the exception of scenario 2, all scenarios had been in operation for at least 16.5 hours before the first transect measurement was taken, most scenarios began in the mid-late afternoon and the measurements were carried out the following morning. Figure 3 shows the continuous TDG data recorded at the 4 km transect (right bank) for the duration of the field work. TDG levels below approximately 109% indicate the period of time that the instrument was recording data while being in the open air before placing it into the water for the first time and taking it out at the end of the week. The vertical lines delineate the scenarios, however, at 4 km downstream these lines are slightly behind the location on the graph that would indicate the TDG level at the time the scenario's flow would have passed. It is because of this that scenario 2 appears to have the same TDG measurements as scenario 1.

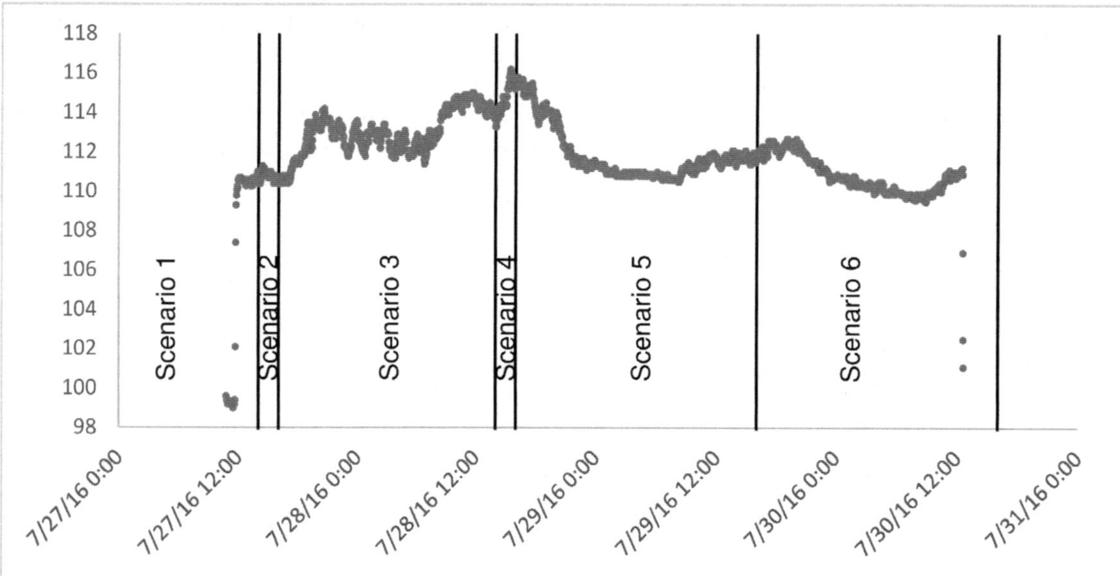

Figure 3: TDG data series for continuous monitoring station at 4 km (right bank)

3 Results and Discussion

Figure 4 illustrates the cross-sectional spot measurements recorded for each scenario for the first three transects. Table 2 is a summary of measured TDG values. The dashed lines represent straight-line interpolations between measured points. In reality, it is expected to vary with the mixing functions suggested by Kamal et al. (2017). For instance, figure 4(a) suggests that SC 3 showed an increase in TDG much closer to the left bank than other scenarios, this is simply because the second data point was not collected at the same location as other scenarios. Collecting data at consistent locations with appropriate density was a challenge due to the fact that the boat could not always select the same point of measurement between each scenario and the time constraints of the work. The right-most data point for SC 2 in figure 4(c) is likely in error, this deviates dramatically from the trend seen in other measurements.

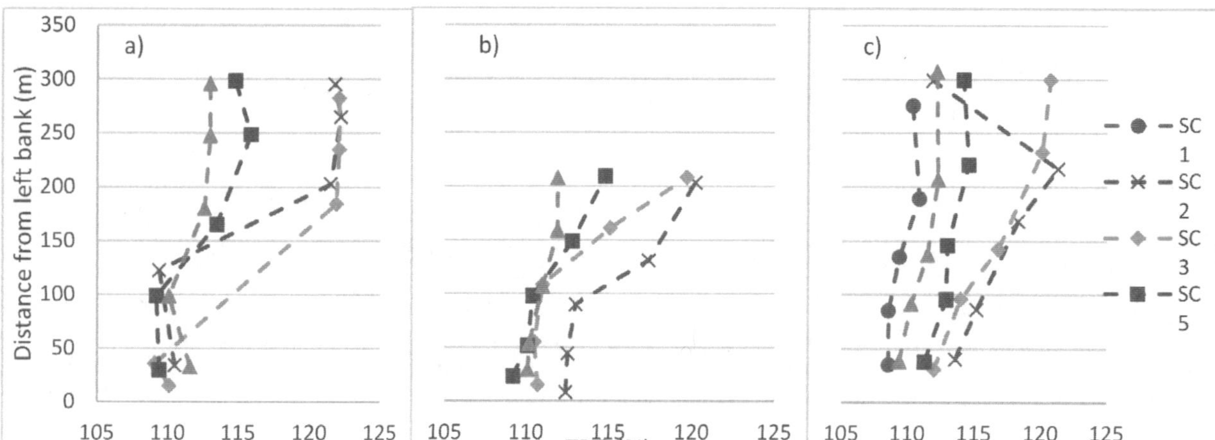

Figure 4: TDG measurements at a) 0.5 km, b) 1.0 km, and c) 2.0 km

Table 2: TDG measurements (%) recorded from the Columbia River at transects downstream of HLK Dam for six different operational scenarios

Transect	SC 1		SC 2		SC 3*		SC 4		SC 5		SC 6	
	LB	Max	LB	Max	LB	Max	LB	Max	LB	Max	LB	Max
M2	-	112	-	119	-	117	-	115	-	114	-	111
TR**	-	-	-	-	-	124	-	-	-	116	-	113
0.5 km	-	-	111	122	110	122	-	-	109	116	110	113
1.0 km	-	-	112	120	111	120	-	-	109	115	110	112
2.0 km	108	111	114	121	112	121	-	-	111	115	110	112

*Note that no measurements were taken for scenario (SC) 4, this was a short scenario in which not enough time was available to gather complete measurements from SC 3 in addition to measurements from SC 4. The operating patterns were also similar to SC 3 and therefore not considered to be critical.

** Tailrace measurement (TR) from boat

From figure 4, it is apparent that there is a strong TDG gradient between the left bank and the right bank of the downstream river. This is explained by the two different types of discharge in the river at these locations, the spill discharge (staying mostly near the right bank) and the ALGS discharge (staying mostly near the left bank). The two separate flows were visually distinguishable from each other during some of the measurements in the field. Another result of the measurements that can be seen in table 2 is the difference in maximum TDG levels between the scenarios. The maximum TDG (typically near the right bank are attributed primarily to spill flow) during scenarios 1, 5, and 6 are markedly less than scenarios 2 and 3. This shows a strong increase in TDG is attributed to the spillways and southern low-level outlets in particular and TDG generation by northern low-level outlets is apparently much less. Figure 5 below shows a starker comparison between the maximum TDG generated for each scenario at 0.5 km and at the tailrace (TR). The data point for SC 1 (112%) is measured from station M2 as a comparison since no measurements were taken for this scenario at 0.5 km or in the tailrace.

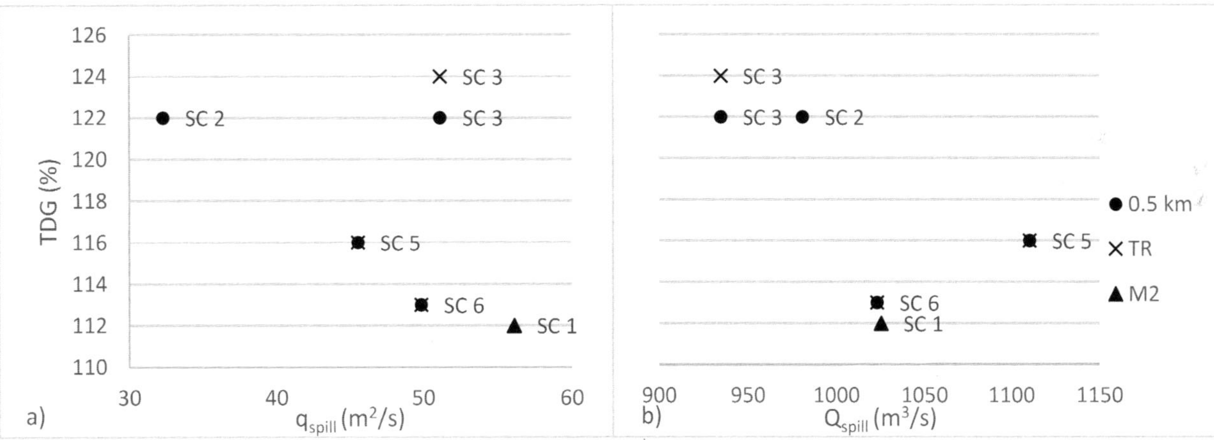

Figure 5: Maximum measured TDG vs a) specific discharge (q_{spill}) and b) total spill (Q_{spill}) for each scenario

Figure 5 takes into account the specific discharge of each scenario. Where different gate openings were used between each gate, an average specific discharge was calculated. This demonstrates the difference particularly well between the southern LLOGs and northern LLOGs. Scenarios 1 and 3 both utilize three LLOGs to spill water, yet SC 1 (3 NLLOGs) generates significantly less TDG than SC 3 (3 SLLOGs) for a greater specific discharge. SC 6 included one less northern LLOG and one southern LLOG. The specific discharge decreased and the TDG generated increased. Further, in SC 5, which consisted of 2 SLLOGs, and 2 NLLOGs, a decrease in specific discharge occurred followed by an increase in TDG. This, along with the measurements pertaining to SC 3, seems to indicate that the southern LLOGs generate much more TDG than the northern LLOGs, a difference of 10% TDG. Considering the differences between figure 5(a) and figure 5(b), another conclusion can be made that not only is TDG generation dependent on the type of structure, but the number of gates in operation as well. This has consequences for the way dam operators would prioritize gate use for spilling a given flow rate. If the same flow rate can be spilled with the use of more northern LLOGs as opposed to using southern LLOGs, this may significantly reduce TDG generation.

Another interesting note is that the SLLOGs (SC 3) were observed to produce a similar level of TDG supersaturation for a slightly less total discharge than the SPOGs (SC 2). Although the SLLOGs have a higher specific discharge than the SPOGs for these scenarios (32.271 m^3/s and 52.133 m^3/s respectively). This is contrary to conventional wisdom due to the fact that it is easily observed at the water surface much more turbulence and therefore air entrainment caused by the plunging jets of the spillways, relative to the observed entrainment due to the LLOGs. Perhaps an unseen mechanism of air entrainment is occurring between the low-level outlet gate and the stilling basin that leads to higher levels of TDG than one might expect. Anecdotal evidence of similar findings is mentioned in Johnson and King (1975) and Qu et al (2011a).

During scenarios 3, 5, and 6, tailrace measurements approximately 300 m downstream of the spill gates, were taken. These measurements are sure to contain spill water only. From figure 5 it would appear that these tailrace measurements are equivalent to the maximum measured values for the 0.5 km transect. With the exception of SC 3, which showed a slightly higher level of TDG in the tailrace. It would appear that the right bank measurements at 0.5 km give good approximations for the TDG leaving the tailrace region before entering the main flow of the river. Figure 4(a) also suggests that regardless of the spill mode, at about 100 – 110 m from the left bank, the spilled water becomes apparent. Therefore, whatever combination of spill gates are open, the individual spills mix together before 0.5 km. The fact that SC 3 seems to show an earlier increase in TDG is most likely due to the lack of measurement around 100 m. For figure 4(b and c) it can be shown that it does follow the same path as other scenarios. It can be assumed, then, that the spill flow leaving the tailrace is fully mixed before encountering the ALGS flow.

Forebay measurements upstream of the dam indicated an average of 109% TDG. This value is taken as the background TDG that remained very nearly constant and is representative of the TDG levels in the ALGS flow as indicated by left bank transect measurements and the 1 km continuous monitoring station (situated near the left bank). With this in mind, over the range of tested scenarios, the northern LLOGs generate the least amount of TDG, a 2-3% increase. The southern LLOGs increased TDG levels by a maximum of 17% and the SPOGs increased TDG by a maximum of 15%. It is possible that due to the short spill time associated with SC 2 that the maximum TDG measured may have been taken slightly ahead of the true maximum TDG flow. In the cases of SC 5 and 6, where there are two types of spill gates operating in tandem (SLLOGs and NLLOGs) an estimation of the TDG increase over 109% can be made. If the measurement from M2 (115%, or a 6% increase over background levels) is used for SC 4 (2 SLLOGs) and a 3% increase in TDG is assumed from the NLLOGs, a mass balance can be used (assuming complete tailrace mixing of the two flows) to estimate a TDG increase of 4.5% for SC 5. The total TDG generated is 113.5%, less than the maximum 116% measured at the edge of the tailrace region. Using the M2 value may underestimate the TDG in the tailrace, this was observed in all scenarios and may be a product of the instrument being different than the one used for spot measurements. Similarly, for SC 6, assuming a 6% increase from the SLLOGs and a 3% increase from the NLLOGs, the total increase is 3.5%. Yielding a total generation of 112.5%. This is closer to the maximum value measured of 113%.

Between 1992 and 1994, TDG measurements were taken downstream of HLK. The database collected has a number of useful parameters that were measured, including the operational scenarios and conditions that resulted in the generated TDG downstream, along with initial forebay measurements. The main issue with the measured TDG data is that it was collected approximately 5.7 km downstream of HLK (Bruce and Plate, 2013). Therefore, any fitted parameter would not truly capture generation mechanisms occurring in the tailrace. Furthermore, at the time of the data collection, the Arrow Lakes Generation Station had not been constructed, this means that mixing mechanisms would also play a role in predicting the TDG levels of the most recent field observations, making the verification of the predictions difficult. Nonetheless, Bruce (2016b) has utilized the existing database to determine a predictive model for TDG generation that incorporates some physical parameters such as the Froude number of the spillway jets. This model is the result of a regression analysis of pertinent parameters that has the added benefit of specifying equations for the spillway outlet gates, southern low-level outlet gates, and the northern low-level outlet gates. Then, to take into account the ALGS flow (and assuming complete mixing across the river), a mass balance is used to determine a final TDG value. The main equations used in the model are shown below.

Northern LLOG: [1] $dP_{TR} = 1.1511 \cdot dP_{FB} + 11.051 \cdot Fr_G - 10.129$

Southern LLOG: [2] $dP_{TR} = 1.1511 \cdot dP_{FB} + 11.051 \cdot Fr_G + 54.825 \cdot TWD_S - 986.67$

Where dP_{TR} is the difference in TDG pressure between water and atmosphere at the tailrace (mmHg), dP_{FB} is the difference in TDG pressure between water and atmosphere at the forebay (mmHg), Fr_G is the Froude number downstream of LLOG, and TWD_S is the tailrace depth at southern LLOG end sill (m).

Spillway: [5] $H = FB_{EL} - 424.9 - 0.5 \cdot d_G$

If $H \geq 6.30$,

[6] $TDG_{TW} = 10.84 \cdot TWD + 896.3$

If $H < 6.30$,

[7] $TDG_{TW} = -37.395 \cdot Fr_{TW} + 14.956 \cdot H + 936.28$

Where TDG_{TW} is the total dissolved gas in the tailrace below the spillway (mmHg), H is the hydraulic head upstream of SPOG (m), FB_{EL} is the forebay elevation (m), TWD is the tailrace depth at SPOG end sill (m), and Fr_{TR} is the Froude number at the tailrace surface. Table 3 lists the results of this model and gives a comparison of measured values between 4 and 7 km transects.

Table 3: TDG Predictions

Scenario	**SLLOG Generation (% TDG)**	**SPOG Generation (% TDG)**	**NLLOG Generation (% TDG)**	**ALGS Generation (% TDG)***	**TGP Mixed (% TDG)**	**Δ TDG (%)****
1	-	-	112	108	111	1
2	-	124	-	109	117	-
3	142	-	-	110	125	9
5	144	-	114	109	119	6
6	143	-	112	108	114	2

*This value was taken as the mean TDG measurement from the 1 km continuous monitoring station over the period of time the field measurements were taken for a scenario near the tailrace and is reflective of forebay TDG levels.

**This is the difference between the predicted value (TGP mixed) and the average measured value from 4 km and 7 km.

It can be seen that there is some deviation between predicted and measured values particularly for scenarios 3 and 5. This is primarily caused by the inclusion of the SLLOG in the scenario. Bruce (2016b), indicates that there was the most uncertainty in the predictive equations for the SLLOG as a result of having much fewer data points (and a narrower range of gate openings) from which to base the regression equation on. Compared to the predictions made with the NLLOG, which make for a much more reasonable agreement. Unfortunately, measurements beyond 2 km were not taken for SC 2 due to time constraints Although the left bank-right bank variability in TDG was found to be 114 – 121% at 2 km for this scenario, these numbers are unlikely to change very much by the time the flow reaches 5.7 km, putting the predicted value of 118% within reasonable proximity.

A Klohn-Crippen Integ (1994) TDG reduction study reflects many of the results observed during this field work campaign. It was found in their report that the spillways generated TDG levels up to 140% and routinely generated levels around 125%. The report also recognized that the NLLOGs generated nearly no TDG when compared to the forebay levels, which were measured to be around 110% and fluctuated mostly with time of year. The SLLOGs produced supersaturated waters up to 125%. The observed results of this past study seem to agree with the conditions found in the current work.

4 Conclusion

Field work was conducted on the lower Columbia River near Castlegar, BC to determine if operational procedures and patterns at HLK dam impact the level of total dissolved gas generated in the tailrace. Overall, it does appear that specific dam discharge structures and patterns contribute differently in generating TDG. It was found that scenarios with a significant portion of the flow attributed to SLLOG operation yielded high levels of TDG (SC 3 and SC 5). Scenarios with a significant portion of flow attributed to NLLOG operation yielded relatively low TDG levels (SC 1 and SC 6). It is suggested that from measurements taken near the tailrace that there may be a significant diluting effect caused by the operation of the NLLOGs on the high TDG levels created by the SLLOGs, this would theoretically extend to a scenario where the NLLOGs are operated in tandem with the SPOGs. If significant dilution is occurring between the SLLOGs and the NLLOGs, then it is reasonable to assume that much of the two flows are mixed before leaving the stilling basin and enter the downstream river along with the ALGS flow. Only a single scenario was tested with spillways in operation and a characteristically high level of TDG was recorded. Although different operational scenarios determine the magnitude of TDG leaving the tailrace, it has been shown that regardless of which gates are in operation, the spill flow mixes before leaving the tailrace. Because of the highly turbulent nature of the stilling basin and the narrowing geometry of the tailrace, this is perhaps to be expected.

The regression equations developed for this site have some limitations that make conclusions about generation specific results difficult. The reason for this is primarily the location of which the measurements were taken, at 5.7 km downstream of the dam on the left bank. This location and therefore the equations based on the data collected there are not true predictions of generation, however, because at the times these data were collected, ALGS was not in operation. Therefore, these values are unlikely to have changed very much and no ALGS dilution would have taken place. Instead, a mass balance equation is used to estimate this dilution. This model still seems to agree quite closely with the measured results between 4 and 7 km of the current study, taking into consideration the uncertainty in the SLLOG predictions and instrument uncertainty.

It has been determined with relative confidence based upon field measurements and initial analysis that the pattern or mode of spill discharge is an important consideration in predicting TDG generation. For future field work, more effort should be put towards ensuring hydraulic variability, not only in gate combinations but in the amount of water that is spilled (i.e. gate opening).

Acknowledgements

The field work described in this paper was the result of the effort made by BC Hydro to accommodate the work carried out by the University of Alberta research team and Creekside Aquatic Sciences. We would also like to acknowledge the significant contribution of Golder Associates Ltd. for providing a boat and operator that were invaluable to the research effort.

References

Bruce, J. A., and E. Plate. 2013. Total Dissolved Gas Supersaturation Downstream of Hugh Keenleyside Dam: A Model-Based TGP Risk Assessment of Facility Operations that Reduce Reliance on Low Level Ports for Water Conveyance. Report prepared for BC Hydro, Burnaby BC. Creekside Aquatic Sciences *Report No. CAQ-001.*

Bruce, J.A. 2016a. TDG flow studies at HLK/ALGS in 2016 – Interim Technical Note. Prepared by Creekside Aquatic Sciences for A. Leake, BC Hydro, Environmental Risk Management. Burnaby B.C. November 21, 2016.

Bruce, J.A. 2016b. (DRAFT). Fate Modeling of Air Supersaturated Waters in the Columbia River: Phase 1 – Model Development and Information Gap Analysis. Report Prepared for BC Hydro, Burnaby, BC, by Creekside Aquatic Sciences and LGL Limited. *Report No. CAQ-017.*

Geldert, D.A., et al. 1998. Modeling dissolved gas supersaturation below spillway plunge pools. *Journal of Hydraulic Engineering*, **124**(5), 513-521.

Hibbs, D.E., and Gulliver, J.S. 1997. Prediction of effective saturation concentration at spillway plunge pools. *Journal of Hydraulic Engineering,* **123**(11), 940-949.

Kamal et al., 2017. Field study of supersaturated total dissolved gases in the lower Columbia river downstream of Hugh L. Keenleyside dam. Paper submitted for the CSCE conference 2017, Vancouver, Canada.

Klohn-Crippen Integ. 1994. [Draft Report] Hugh Keenleyside Dam Total Gas Pressure (TGP) Reduction Study: 1994 Status Report. Prepared for BC Hydro. Klohn-Crippen Integ Report No. KCI-173.

Orlins, J.J., and Gulliver, J.S. 2000. Dissolved gas supersaturation downstream of a spillway II: Computational model. *Journal of Hydraulic Research*, **38**(2), 151-159.

Politano, M., et al. 2005. Prediction of total dissolved gas downstream of spillways using a multidimensional two-phase flow model. *Mecanica Computacional*, **24**, 285-295.

Qu, L., et al. 2011a. Field observation of total dissolved gas supersaturation of high-dams. *Sci China Tech Sci*, **54**, 156-162.

Qu, L., et al. 2011b. Experimental study on total dissolved gas supersaturation in water. *Water Science and Engineering*, **4**(4), 396-404.

Roesner, L.A., and Norton, W. R. 1971. A nitrogen gas (N_2) model for the lower Columbia River. *Report No. 1-350,* Water Resources Engineers, Inc., Walnut Creek, California.

University of Washington. 2000. Columbia River Salmon Passage Model CRiSP.1.6 Theory and Calibration.

Urban, A.L., et al. 2008. Modeling total dissolved gas concentration downstream of spillways. *Journal of Hydraulic Engineering*, **134**(5), 550-561.

U.S. Army Corps of Engineers, Portland District, Walla Walla District. 1996. Dissolved Gas Abatement Study Phase I, *Technical Report.*

U.S. Army Corps of Engineers, Portland District, Walla Walla District. 2001. Dissolved Gas Abatement Study Phase II, *Technical Report.*

Weitkamp, D.E., and Katz, M. 1980. A review of dissolved gas supersaturation literature. *Transactions of the American Fisheries Society*, **109**(6), 659-702.

Leadership in Sustainable Infrastructure

Leadership en Infrastructures Durables

Vancouver, Canada

May 31 – June 3, 2017/ *Mai 31 – Juin 3, 2017*

FIELD STUDY ON THE DECAY OF SUPERSATURATED TOTAL DISSOLVED GASES IN THE LOWER COLUMBIA RIVER DOWNSTREAM OF HUGH L. KEENLEYSIDE DAM

Kamal, Rajib[1,5], Zhu, David Z.[2], Leake, Alf[3] and Crossman, James[4]
[1,2] University of Alberta, Canada
[3,4] BC Hydro, Canada
[5] rajib@ualberta.ca

Abstract: Reliable prediction of the decay of supersaturated total dissolved gases (TDGs) produced by hydropower operations is crucial to evaluating potential hydro-environmental-ecological impacts downstream of facilities. For quantitative prediction of TDG decay under different gate operation scenarios, detailed measurements of TDG and river hydraulics were collected in the Lower Columbia River downstream of Hugh L. Keenleyside Dam near Castlegar, British Columbia. The measured data in conjunction with an analytical approach was utilized to estimate the mixing and decay rates for four different operational conditions of low level outlet gates. For dimensionless mixing coefficients of 0.6 and 0.2 for the two sub-reaches of the river, the average decay rate was 0.017 hr^{-1}. Comparison with available decay equations and some widely used reaeration models indicated that further investigation would be useful to assess their applicability in the case of supersaturated TDGs and develop a predictive tool to estimate decay.

1 INTRODUCTION

Supersaturation of total dissolved gases (TDGs) often occurs in hydropower facilities as a result of voluntary (e.g. facilitating non-turbine fish passage) or involuntary (releasing excess water) spills, and may pose adverse environmental and ecological effects (Weitkamp, 2008). It is generally caused by the entrainment of air in the form of bubbles during spillway releases and the subsequent transfer of atmospheric gasses (primarily nitrogen and oxygen) into water. As the flows move out of the tailrace, the net mass transfer reverses resulting in TDG dissipation. Supersaturated dissolved gases dissipate through decay at the free surface and mixing with adjacent waters which are dependent on factors like velocity, turbulence, available surface area, and river depth (Geldert et al., 1998). The excess TDG decays through gas exchange in the air-water interface and requires a long time to diffuse out of the river. Therefore, elevated or even supersaturated gas levels may persist far downstream from the source of supersaturation and can result in fish mortality (e.g. gas bubble disease) and other impacts to the aquatic environment.

The decay of TDG comprises of various physical, biological, and chemical processes where the physical component is driven by gas exchange at the air-water interface. This transfer process (i.e. decay) across the gas-liquid interface is of fundamental importance to understanding the distribution of TDG in the natural environment. The process becomes more complicated when spilled water with high TDG concentrations mixes with generation discharge downstream, and with any tributary inflows. A limited number of studies have been reported on the direct quantification of TDG decay. Most of the previous studies considered the decay rate by incorporating surface gas transfer rate at the air-water interface by replacing it with semi-empirical reaeration equations based on the surface renewal or the energy dissipation model. Given the complexity of the generation and dissipation process, the decay rate should be evaluated directly in order to assess the impact of TDGs in fish, fish habitat, and the downstream environment. Therefore, the objective of the present study is to assess the decay rate of TDG for different operational conditions in hydropower facilities. This study is based on the extensive field work carried out on the Lower Columbia River downstream of Hugh L. Keenleyside Dam (HLK) at Castlegar, BC. Detailed monitoring of TDG as well as hydraulic measurements were carried out for different combinations of low level gate (LLOG) operations of the dam to assess the TDG decay rate for these conditions.

2 Theoretical Background

In general for a non-conservative substance like TDG, the depth averaged concentration (C) can be described by the advection-dispersion equation with a first-order decay term (Fischer et al., 1979):

$$[1]\ \frac{\partial C}{\partial t}+u\frac{\partial C}{\partial x}+v\frac{\partial C}{\partial y} = E_l\frac{\partial^2 C}{\partial x^2}+E_t\frac{\partial^2 C}{\partial y^2} - kC$$

where u and v are velocities in x and y-directions respectively; E_l and E_t are the corresponding dispersion coefficients and k is the decay rate. For accurate prediction of TDG distribution in the river, both mixing and decay rate should be evaluated based on field measurements. When the flow is not completely mixed, i.e. at the near and intermediate mixing zone of a river, the transverse mixing coefficient (E_t) usually governs the concentration distribution across the channel. It is usually difficult to control the quality of field data due to the non-conservative nature of the substance, inadequate number of samplings, or measurement errors. Therefore, the field data should be used in conjunction with analytical methods to estimate E_t (Zhang and Zhu, 2011). Using the concept of cumulative discharge, the simplified, steady-state distribution of a non-conservative substance can be written as (Gowda, 1984):

$$[2]\ \frac{\partial C}{\partial x} = D\frac{\partial^2 C}{\partial q^2} - \frac{kC}{U}$$

where q is the cumulative discharge; U is the mean velocity in x-direction and D is the factor of diffusion. Downstream of hydropower facilities, TDG concentrations can vary laterally and the two-dimensional distribution can be described by Eq. (2). Assuming a line source of high TDG stretching from $q = q_1$ to $q = Q$ (as in the case of spill), the analytical solution of Eq. (2) for a conservative case can be written as:

$$[3]\ C_i(x, q) = C_b+\left(\frac{C_o-C_b}{2}\right)\left[\mathrm{erf}\left(\frac{q-q_1}{\sqrt{4Dx}}\right)-\mathrm{erf}\left(\frac{q-Q}{\sqrt{4Dx}}\right)\right]+\text{Image sources}$$

where Q is the total discharge; C_o is the maximum TDG concentration at the initial section and C_b is the background concentration corresponding to generation flow or tributary inflow. Image sources were included to account for the effect of banks on concentration distribution. In order to evaluate transverse mixing coefficient (E_t), the factor of diffusion can be expressed as follows (Rutherford, 1994):

$$[4]\ D = \psi E_t U H^2$$

where H is the average depth and ψ is the dimensionless shape factor that has been reported to be in the range from 1.0 – 3.6 (Rutherford, 1994). When supersaturation occurs, the exchange of mass at the air-water interface (i.e. decay) during transport through the river system drives TDG levels towards equilibrium conditions with the atmosphere. Since the high concentrations could dissipate by both mixing and decay, Eq. (3) can be utilized to estimate the transverse variation in TDG concentration. Then the final concentration of TDG incorporating decay can be obtained as follows:

$$[5]\ C = C_s+(C_i-C_s)\exp(-kt)$$

where t is the travel time of water ($t = x/U$) and C_s is the saturation concentration and is usually 100% at atmospheric pressure. For gases of low solubility (such as nitrogen and oxygen), the decay of TDG across the air-water interface depends on the molecular and turbulent transport processes in the water layer similar to the process of reaeration. For such cases, two widely used semi-empirical treatments of gas transfer are the surface renewal model and the energy dissipation model. For isotropic turbulence, the gas transfer rate according to the simplified surface renewal model of O'Connor and Dobbins (1958) can be expressed as follows:

[6] $k \propto U^{1/2} H^{-3/2}$

In the energy dissipation model, the dependence of gas transfer on turbulent transport and diffusivity can be modeled in the form (Moog and Jirka, 1999) as follows:

[7] $k \propto Sc^{-1/2} R_*^{-n}$

where S_c is the Schmidt number and R_* is the turbulent Reynolds number. The Reynolds number exponent (n) is ½ for the large eddy model and ¼ for the small eddy model.

3 FIELD WORK

3.1 Description of the Study Site

The Hugh Keenleyside Dam (HLK) and Arrow Lakes Generating Station (ALH) form the lower-most of three Hydroelectric Projects (along with Mica Dam and Revelstoke Dam) on the Canadian portion of the Columbia River. The dam is located 7 km upstream of the City of Castlegar, BC and approximately 57 km upstream from the BC-Washington border. The 52 m high dam impounds the Arrow Lakes Reservoir and was constructed for the Columbia River Treaty in 1968, with generating capacity at ALH added in 2002. There are two styles of outlet gates at HLK - four radial spillway outlet gates (SPOGs) and eight low level outlet gates (LLOGs). Relative to the position of the spillway, the first four low level gates are referred to as northern LLOGs while the other four are denoted as southern LLOGs. These units are collectively capable of discharging up to 10,500 m³/s (Bruce and Plate, 2013). The ALH Generating Station is located directly north of the HLK facility (Figure 1). It receives water through a power canal inlet, which was located approximately 900 m upstream of HLK. Maximum turbine discharge through ALH is 1,150 m³/s. The confluence of the Lower Columbia River and the Kootenay River is located approximately 10 km downstream of HLK. The Brilliant Dam is the furthest downstream dam on the Kootenay River, located about 2.5 km upstream from the confluence. The region for the current field work comprised of the lower reach of the Columbia River from HLK to about 20 km downstream including the Kootenay River confluence.

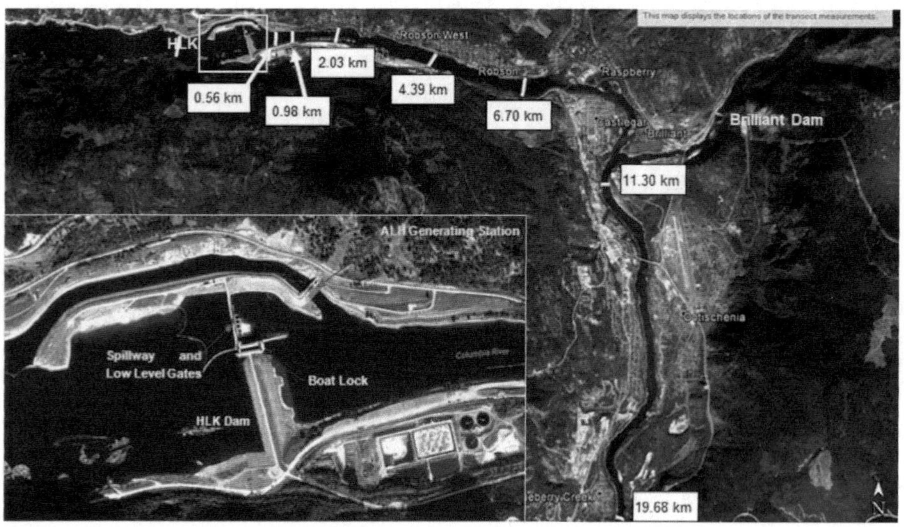

Figure 1: Study site at the Lower Columbia River with measurement locations

3.2 TDG Monitoring and Hydraulic Measurement

The field work in the Lower Columbia River was conducted from July 26 to July 30, 2016 for different combinations of gate settings at the HLK. A total of four scenarios of different LLOGs operations were implemented for the present study which were operated for longer duration. Detailed measurements were carried out for these scenarios. Table 1 shows the operational conditions of different gates for these scenarios along with the flow conditions. The spill rates (Q_s) mentioned in the table were the total flow through the gates. Flow through individual units and height of gate opening were different for these scenarios. Detailed operational modes and their effect on the generation of TDG will be found in a companion paper (Billay et al., 2017).

Table 1: Various gate operational scenarios for the field work

Scenario	Gate operation	**Spill rate** Q_s (m^3/s)	**Generation flow** Q_g (m^3/s)	**Kootenay R. inflow** Q_K (m^3/s)
1	3 Northern LLOGs	1025.6	1085	585.21
2	3 Southern LLOGs	934.7	1100	573.33
3	2 Southern + 2 Northern LLOGs	1110.5	1100	530.43
4	1 Southern + 2 Northern LLOGs	1023.3	1081	528.50

Measurements of TDG for the operational scenarios were conducted at various locations through combination of spot measurements from the boat as well as continuous monitoring from floating platforms designed to capture temporal variation of TDG. The total gas pressure (TGP), barometric pressure (BP) and water temperature was collected by two types of probes – the Lumi4 DO-TGP probe and PT4 Smart TGP probe (manufactured by Point Four Systems Inc.). These probes are capable of recording TGP and BP readings with an accuracy of ±2 mmHg resulting in an overall accuracy of 4% for TDG measurement (Pentair Aquatic Eco-Systems, 2014).

Using floating platforms, TDG was monitored continuously at the left bank at 0.98 km; at the right banks of 4.39, 6.70, 11.30 and 19.68 km, and upstream on the Kootenay River with PT4 Smart TGP probes. In addition, two continuous monitoring stations were installed on the right bank of the HLK tailrace (courtesy: James Bruce). Before installation and deployment, each of the probes was calibrated and set to record data continuously at 2 minute intervals at about 1 m depth. The spot measurements were taken along seven transects downstream of HLK (as shown in Figure 1). In addition, measurements were also taken along a transect in the forebay and at a point close to the spillway gates in the tailrace. The spot measurements were taken by the Lumi4 DO-TGP probe at a depth of about 1 m below the water surface. For each measurement, sufficient time (5 – 15 minutes) was allowed so that the probes could acquire stable TGP readings. These measurements were carried out in all seven transects for scenarios 2 and 3. Due to time limitations and other constraints it was not possible to take measurements at 0.56 and 0.98 km for scenario 1 and at 11.30 and 19.68 km for scenario 4. In all cases, the locations of TDG measurement were recorded using a handheld GPS. Uncertainty in TDG measurements could arise due to the inaccuracy of the instruments, of the measurement, or operator error. From the recorded TDG at the continuous monitoring stations, the fluctuation in percent saturation over time for the same condition (related to precision error) was typically ≤ 0.2%. The bias error, associated TGP and BP measurements, was about 0.4%. The overall uncertainty was well within the accuracy of TDG measurement.

In order to understand TDG dissipation mechanisms and their relation to various hydraulic features, a RiverRay 600 kHz Acoustic Doppler Current Profiler (ADCP) was used to measure velocity and flow at the HLK tailrace and downstream Lower Columbia River. The ADCP was equipped with a universal mount, a trimaran and an external GPS (DGPS) to record the locations. The ADCP measurements were carried out in each of the tailrace and downstream river transects as shown in Figure 1. The measurements were completed by driving the boat slowly across the transects allowing for simultaneous measurement of velocity and discharge. Each of these transects were repeated 3 to 4 times for accuracy and comparison purposes. These measurements obtained transverse variations of velocity which were utilized to calculate cumulative flow at each transect.

During the period of field work, the HLK releases varied from 934.7 m^3/s (scenario 2) to 1110.5 m^3/s (scenario 3), while the ALH flow remained consistent at about 1081 to 1100 m^3/s resulting in a spill-to-generation flow ratio of about 1.0 for all the scenarios. The discharge measured by ADCP at the first five transects showed similar flow rates (Table 2), except for the transect at 0.98 km. In this location it was not possible to cover the whole cross section due to the presence of log booms along the right bank. As a result, about 50 m length was not measured by the ADCP and hence the discharge was found to be lower compared to other transects. At transect 11.30 and 19.68 km, the discharges were about 2630 and 2445 m^3/s, indicating the additional inflow coming from the Kootenay River. Table 2 outlines some of the basic hydraulic parameters at different transects obtained from the ADCP measurements.

Table 2: Basic hydraulic parameters collected in the transects

Reach	Trans. (km)	*Q* (m^3/s)	*U* (m/s)	*B* (m)	*H* (m)	S_o (×10-4)	U^* (m/s)	*Ψ*	*Fr*
1	0.56	2069.9	0.43	309.6	15.89	0.492	0.086	2.34	0.03
	0.98	1736.0	0.61	259.5	13.18		0.078	1.63	0.05
	2.03	2111.4	0.48	338.8	13.62		0.080	2.59	0.04
	4.39	2089.9	0.48	328.9	13.42		0.080	1.89	0.04
	6.70	2086.5	0.49	409.8	10.31		0.070	1.96	0.05
2	11.30	2629.5	1.82	199.6	7.42	4.218	0.174	1.83	0.21
	19.68	2444.4	0.75	200.0	17.29		0.075	3.25	0.06

Note: Q = total discharge; U = mean velocity; B = channel width; H = mean depth; S_0 = slope of the reach; U^* = shear velocity = $(gRS_o)^{0.5}$; Ψ = shape factor; Fr = Froude number = $U/(gH)^{0.5}$

4 RESULTS AND DISCUSSION

4.1 TDG Distribution and Flow Hydraulics

In the Lower Columbia River, flows exiting from the HLK showed marked distinction in TDG concentrations between the spill and generation discharges, with spill releases containing high level of TDG compared to the powerhouse releases (Figure 2). The discharge coming from the HLK (through LLOGs) was similar to the flow rate of ALH, with similar spill-to-generation flow ratios for all the scenarios. For these scenarios, TDG concentrations near the left bank (corresponding to ALH generation flow) were comparatively low with initial TDGs (at 0.56 and 0.98 km) in the range of 108-110% which were similar to the measured TDG in the forebay. From continuous monitoring station at the tailrace, it was found that the maximum TDG for scenario 1 was 112%. Depending on the gate operation, the maximum TDG at 0.56 km, i.e. generated TDG, was 122, 116 and 113% for scenarios 2, 3 and 4 respectively. The TDG was mixed between the two flows and simultaneously decayed as the water moved downstream. As a result, the TDGs along the left bank were found to increase with distance. For most of the downstream transects, the maximum TDG was found near the right bank. At 7 km downstream, the maximum TDG was 110% for scenario 1 and 117% for scenario 2. ADCP measurements showed the cross-sectional mean velocities at the transects from 0.56 to 6.70 km varied from 0.43 to 0.61 m/s (Table 2). The Froude number in this reach ranged from 0.03 to 0.05.

The flow condition changed after the Kootenay River confluence with inflow coming from the release of Brilliant Dam. The velocity was very high at 11.30 and 19.68 km (Table 2) with this additional flow passing through comparatively narrower and shallower cross sections. At these transects, the Froude numbers were 0.2 and 0.06 respectively. Downstream of the confluence, the TDG distribution was also found to be affected by the two converging rivers with different TDG levels. Continuous monitoring and some spot measurements showed that the Kootenay inflow had TDG concentrations in the range of 106 to 109%. Spot measurements at 11.30 km showed that the maximum TDG was 114% for scenario 2 and 111% for scenario 3, while the left bank had TDGs of 109% for both cases. At 19.68 km, the maximum concentrations varied from 109-114% depending on the scenarios.

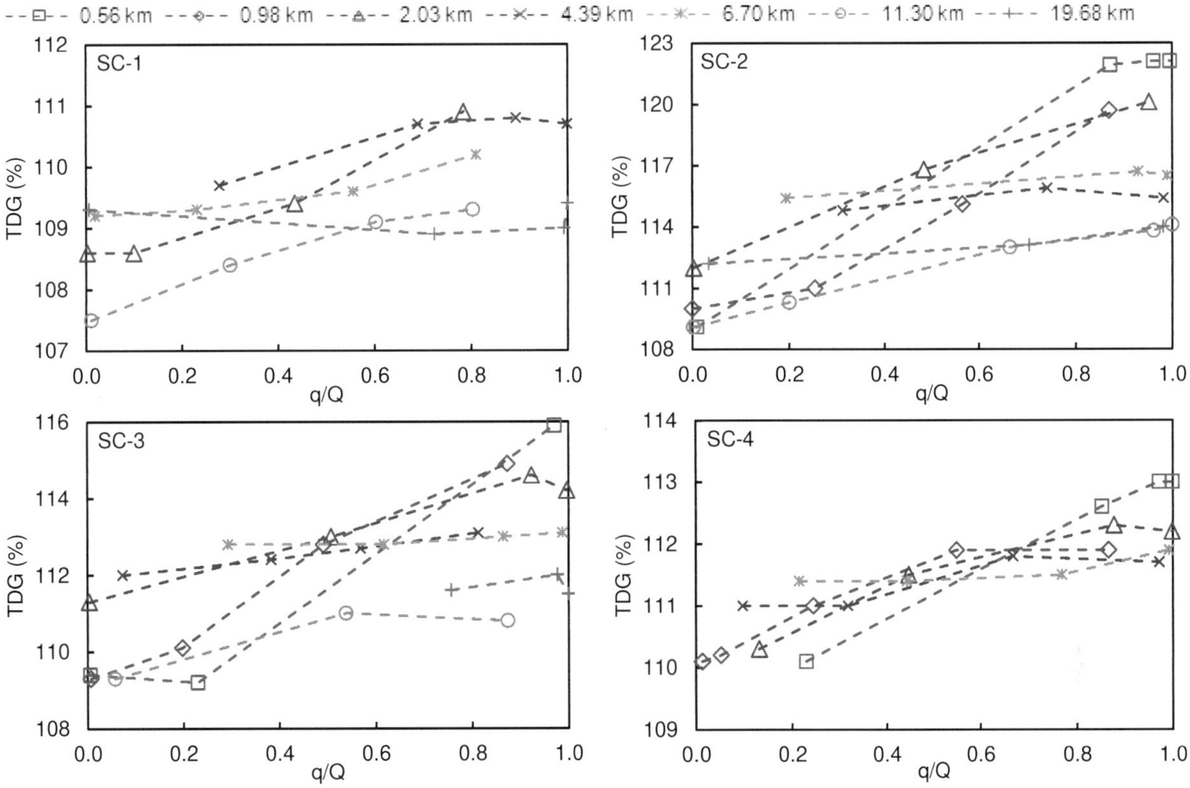

Figure 2: TDG at the downstream transects for different scenarios

4.2 Analysis on Transverse Mixing

As seen from Figure 2, the two flows of different TDG concentration changed downstream, as TDG was dissipated through mixing within the water and decay at the air-water interface. Since the release from HLK (spills) mixed with ALH flow initially and then the flow received inflow from Kootenay River, the two-dimensional distribution of TDG can be described by Eq. (2). Then using Eq. (3) and (4), the transverse mixing coefficient could be obtained from the lateral distribution of the measured TDG across the individual cross-sections. Most of the existing methods to evaluate mixing coefficient place emphasis on the use of raw field data. Because of the sparsely measured data with limited number of samplings (Figure 2) and non-conservative nature of TDG, such methods might not be useful for the current case.

In the present study, various dimensionless transverse mixing coefficients (E_t/HU^*) were assumed and compared with the measured data to obtain the mixing coefficient. In natural rivers, this dimensionless coefficient can vary from 0.3-0.6 for regular channels, 0.6-0.9 for gently meandering channels and 1-3 for sharp curved channels (Fischer et al. 1979; Rutherford 1994). Figure 3 shows theoretical concentration profiles calculated for dimensionless mixing coefficients ranging from 0.1 to 0.9 in different transects and their comparison with the measured data of scenario 2. Since the confluence of the Kootenay River adds additional flow, the study reach was divided into two sub-reaches. For these reaches, the factors of diffusion (D) were evaluated from E_t assuming the coefficients would remain same with the slightly varying discharge conditions for different scenarios. A comparison of Figure 2 and Figure 3 showed that the value of E_t/HU^* was about 0.6 and 0.2 for the two above mentioned reaches. In reach 1 (upstream of confluence), the corresponding mixing coefficient was 0.631 m^2/s, while E_t was found to be 0.313 m^2/s in reach 2 (downstream of confluence). The higher value of the mixing coefficients at the upstream reach was not surprising, considering the highly turbulent conditions in the downstream of the dam.

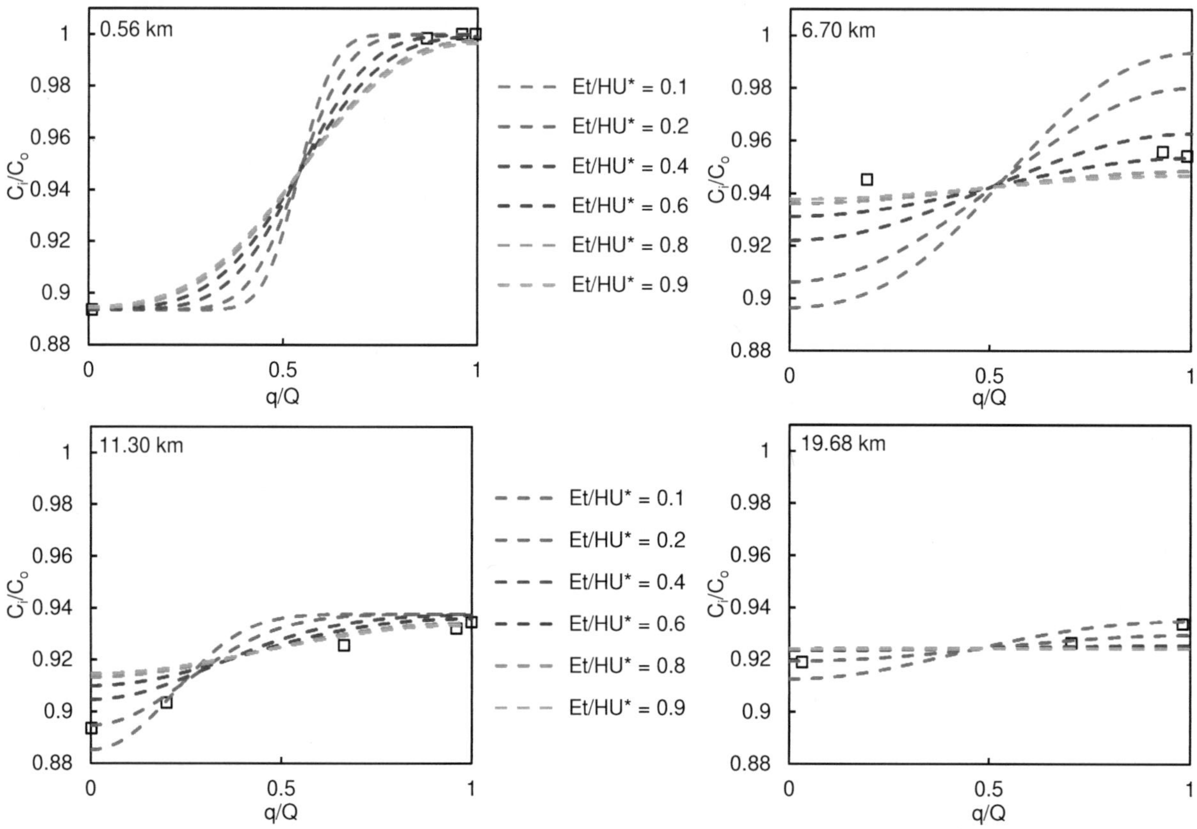

Figure 3: Analysis of transverse mixing coefficient for the two sub-reaches (square symbol represents the measured data for scenario 2)

4.3 Estimation of Decay Rate

Measurements showed that lateral gradients in TDG existed in the Lower Columbia River due to differential concentrations of spillway and powerhouse releases. However as seen from Figure 2, the change in concentrations along the river was not similar across the transects. As a result of mixing and decay, the TDGs along the left banks increased with downstream distance up to the confluence of the Kootenay River. On the other hand, TDG concentrations along the right banks (associated with high TDGs from spills) tended to decrease. Therefore in the present study, the TDG decay was estimated by following the flow of high TDG water released from the low level gates. Since the high concentrations could dissipate by both mixing and decay, Eq. (3) was utilized to estimate the transverse variation in TDG concentration. Then the final concentration of TDG incorporating decay was obtained using Eq. (5).

Since background concentration in the river was 108 to 110%, the saturation concentration was taken as 108% for the present analysis. Eq. (5) was employed to estimate the decay rate of TDG for various scenarios. TDG concentrations measured at 0.56 km was applied as the initial or reference concentration, while the mean velocities (Table 2) obtained from the ADCP measurements were used in the calculation. Then the decay rates were obtained by fitting the calculated concentrations with measured data based on the least square method. Using this method, the rates of decay were 0.012, 0.013, 0.02 and 0.019 hr^{-1} respectively for scenarios 1-4 with an average value of 0.016 hr^{-1}. Figure 4 shows the calculated TDG concentrations at various transects for scenarios 2 and 3 along the spill flows (from $q = Q_g$ to $q = Q$, where $Q_g = Q - Q_s$). These calculated profiles incorporated both mixing and decay.

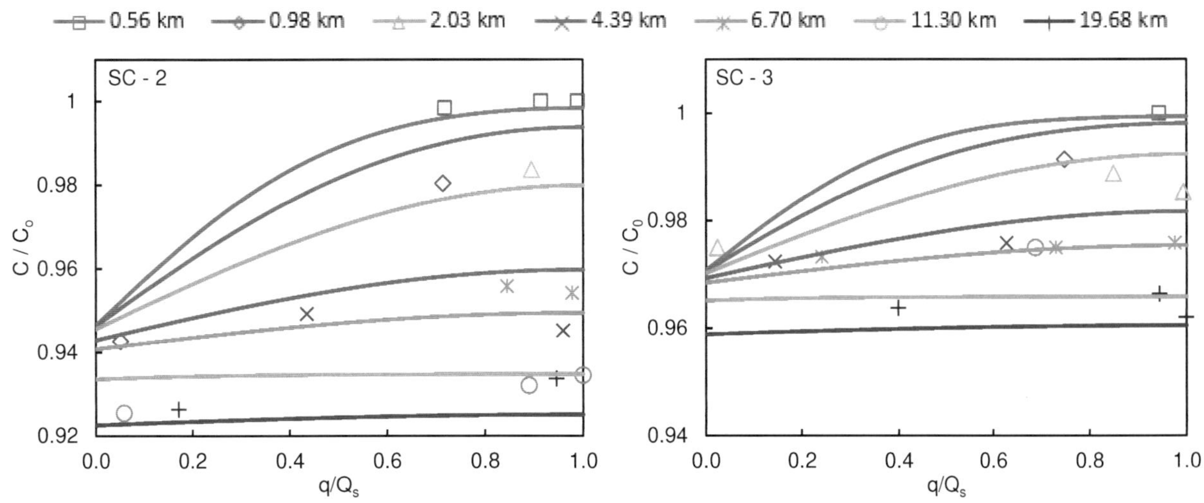

Figure 4: Calculated TDG incorporating mixing and decay for scenarios 2 and 3 with measured data

The variation of temperature could affect the rate of decay and usually it increases with rising temperature (Shen et al., 2014). Therefore the usual practice is to present decay rates corresponding to a standard temperature of 20 ^{0}C. The values of k could be temperature corrected relative to 20 ^{0}C with the simplified equation as follows –

[8] $k = k_{20}(1.0241)^{(T-20)}$

where *T* is the measured river water temperature. The temperature corrected decay rates are shown in Figure 5. The corresponding average decay rate was 0.017 hr^{-1}.

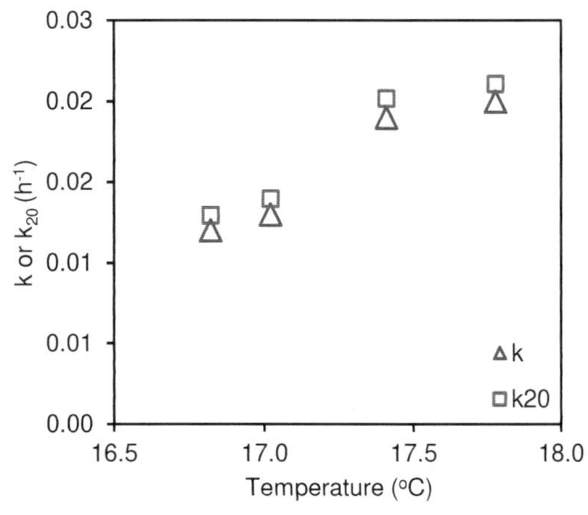

Figure 5: Decay rates for different scenarios in relation with temperature

The US Army Corps of Engineers (USACE 2001) proposed an equation to estimate decay based on surface renewal theory. Recently Feng et al. (2014) developed a semi-empirical relationship by including the effect of water depth, shear velocity, hydraulic radius, and Froude number. The decay rate obtained in the present study was found to be much higher compared to the calculated rates using these methods (Table 3). The USACE equation simply considered the velocity and depth as governing parameters which might not always be applicable for turbulent flow. The equation developed by Feng et al (2014) underestimated the decay rate as well. It might be due to the reason that the equation was formulated for relatively shallow rivers with Froude Numbers ranging from 0.11 to 0.54, which might not be applicable to deep rivers like the Lower Columbia River. However, the form of the model could be useful to relate decay with the variation of flow.

Table 3: Decay rate of TDG using different methods

Method	Equation	Decay rate, k (hr^{-1})
Present study	-	0.017
USACE (2001)	$k = \left(\frac{D_m U}{H^3}\right)^{1/2}$	0.0033
Feng et al. (2014)	$k = 0.0037 \frac{U^*}{R} \left(\frac{H}{R}\right)^{2.02} Fr^{1.73}$	0.0026

4.4 Comparison with Gas Transfer Theories

Gas transfer across the air-water interface is difficult to measure in the field and therefore, empirical methods based on the fundamental gas transfer theories are generally used. Numerous equations based on mean velocity, shear velocity, channel slope, depth, and Froude number have been developed to estimate the stream reaeration coefficient, K_a. A number of studies have also included the effect of wind. In the present study, some of the most widely used formulae have been used as outlined in Table 4. These predictive equations referred to a standard temperature of 20 °C. The field data obtained from the hydraulic measurements in the Columbia River was utilized in these equations in order to determine the values of K_a and compare them with the estimated decay rate.

Analysis of the reaeration equations presented an overall poor performance, with most of the equations under or over predicting the decay rate except the equation given by Thackston and Krenkel (1969). Such variation likely resulted due to different hydraulic conditions from which the study data were collected and the fact that the principal gas transfer mechanisms could vary with hydraulic conditions. The reaeration rates calculated by the methods given by O'Connor and Dobbins (1958), Churchill et al. (1962) and Bennett and Rathbun (1972) was found to be much lower compared to the estimated decay in the current study. These equations primarily used velocity and depth as key variables which might not be always representative for turbulent flow conditions as in the case of downstream regions of hydropower facilities. The methods that incorporated channel slope (Tsivoglou and Wallace, 1972 and Moog and Jirka, 1998) over predicted the decay rate indicating that the slope might not be a governing variable for our case. According to the equation of Thackston and Krenkel (1969), the value of K_a was 0.011 hr^{-1} which was close to the estimated decay rate. This implied that addition of parameters that relate turbulence such as the shear velocity and Froude number, could be useful as predictor variables to estimate decay.

Table 4: Estimation of reaeration coefficient (K_a) using predictive equations

Reference	**Equation** (1/day at 20 °C)	***K_a* (hr^{-1})**		
		Reach 1	Reach 2	Average
O'Connor and Dobbins (1958)	$K_a = 3.9\, U^{0.5} H^{-1.5}$	0.002	0.004	0.003
Churchill et al. (1962)	$K_a = 5.01\, U^{0.969} H^{-1.673}$	0.001	0.004	0.003
Thackston and Krenkel (1969)	$K_a = 24.9\,(1 + Fr^{0.5}) U^* H^{-1.0}$	0.007	0.014	0.011
Bennett and Rathbun (1972)	$K_a = 32.5\, U^{0.413} S_0^{0.273} H^{-1.408}$	0.002	0.005	0.004
Tsivoglou and Wallace (1972)	$K_a = 22500\, S_o U$	0.023	0.506	0.265
Moog and Jirka (1998)	$K_a = 1740\, U^{0.46} S_0^{0.79} H^{0.74}$	0.141	1.130	0.636

5 CONCLUSIONS

The current study presents a methodology to estimate the decay rate of supersaturated TDG in a large regulated river through a combination of field measurements and analytical modelling. Detailed field monitoring of TDG in the Lower Columbia River downstream of Hugh L. Keenleyside Dam indicated differential TDG concentrations between the spill releases and the generation flows. Variations in concentrations were also observed between the Columbia River flow and Kootenay River inflow downstream of the confluence. For four different spill scenarios of low level outlet gates with similar hydraulic condition, the maximum TDG were 112, 122, 116 and 113% while the initial concentrations corresponding to the generation flows were in the range of 108-110%. Downstream of the Kootenay River confluence, the maximum TDG varied from 111 to 114% depending on the gate operations. Analyses of the measured TDG for the operational scenarios showed that the decay rate varied from 0.013 to 0.02 hr^{-1}. For dimensionless mixing coefficients of 0.6 and 0.2 for the two sub-reaches of the river, the average decay rate was 0.017 hr^{-1} at standard temperature. Due to greater depth and low Froude Number in the river, the estimated decay rate was found to be much higher compared to the calculated rates based on other available methods. Comparison with some widely used reaeration equations indicated that shear velocity and Froude number could be considered in the predictive equations to estimate decay. However, all of these analyses were carried out for similar hydraulic conditions. TDG prediction and decay rate estimation can be improved by incorporating data for different hydraulic conditions and employing more detailed analytical and numerical approaches.

References

Bennett, J. P. and Rathbun, R. E. 1972. Reaeration in open-channel flow. *USGS Prof. Pap. 737*, U.S. Government Printing Office, Washington, DC.

Billay, G., Zhu, D., Bruce, J., Leake, A. and Crossman, J. 2017. Field Observations and Analysis of Total Dissolved Gas Generation on the Lower Columbia River Downstream of the Hugh L. Keenleyside Dam. Paper submitted to the 23rd Hydrotechnical Conference, Vancouver, Canada.

Bruce, J.A. and Plate, E. 2013. Total Dissolved Gas Supersaturation Downstream of Hugh Keenleyside Dam: A Model-based TGP Risk Assessment of Facility Operations that Reduce Reliance on Low Level Ports for Water Conveyance. *Report Prepared for BC Hydro, Burnaby BC.* Creekside Aquatic Sciences Report No. CAQ-001: 40 p + app

Churchill, M. A., Elmore, H. L. and Buckingham, R. A. 1962. The prediction of stream reaeration rates. *J. Sanit. Eng. Div.*, 88(4), 1–46.

Feng, J., Ran, L., Gian, M. and Le-Le, W. 2014. Experimental and field study on dissipation coefficient of supersaturated total dissolved gas. *J. Cent. South Univ* 21:1995–2003

Fischer, H. B., List, E. G., Koh, R. C. Y., Imberger, J. and Brooks, N. H. 1979. *Mixing in Inland and Coastal Waters.* Academic Press, New York, NY.

Geldert, D.A., Gulliver, J.S. and Wilhelms, S.C. 1998. Modeling dissolved gas supersaturation below spillway plunge pools. *J. Hydrol. Eng.* 124(5): 513-521.

Gowda, T. P. H. 1984. Water quality prediction in mixing zones of rivers. *J. Environ. Eng.*, 110(4), 751–769.

Moog D. and Jirka, G. 1998. Analysis of reaeration equations using mean multiplicative error. *Journal of Environmental Engineering*, 112(2): 104–110.

Moog, D. and Jirka, G. 1999. Air-Water Gas Transfer in Uniform Channel Flow. *J. Hydraul. Eng.*, 125:1(3), 3-10.

O'Connor, D. J. and Dobbins, W. E. 1958. Mechanism of reaeration in natural streams. *Trans. Am. Soc. Civ. Eng.*, 123, 641–684.

Pentair Aquatic Eco-Systems. 2014. PT4 TGP Probe Technical Manual.

Rutherford, J.C. 1994. *River mixing.* John Wiley & Sons, West Sussex, England.

Shen, X., Liu, S.Y., Li, R. and Ou, Y.M. 2014. Experimental study on the impact of temperature on the dissipation process of supersaturated total dissolved gas. *Journal of Environmental Science*, 26(9), 1874-1878.

Thackston, E. L. and Krenkel, P. A. 1969. Reaeration prediction in natural streams. *J. Sanit. Eng. Div.*, 95(1), 65–94.

Tsivoglou E.C. and Wallace J.R. 1972. Characterization of Stream Reaeration Capacity. *Report No. EPA-R3-72-012*, US Environmental Protection Agency: Washington, DC.

USACE. 2001. Dissolved Gas Abatement Study: Phase II. *Technical Report*, US Army Corps of Engineers, North Pacific Division, Portland District and Walla Walla District.

Weitkamp, D.E. 2008. Total Dissolved Gas Literature 1980-2007, An Annotated Bibliography Parametrix, Bellevue, Wa.

Zhang, W. and Zhu, D. 2011. Transverse Mixing in an Unregulated Northern River. *J. Hydraul. Eng.*, 137(11), 1426-1440.

Leadership in Sustainable Infrastructure
Leadership en Infrastructures Durables

Vancouver, Canada
May 31 – June 3, 2017/ *Mai 31 – Juin 3, 2017*

FREQUENCY ANALYSIS AND PLOTTING POSITIONS

Bertrand Massé [1, 2]
[1] SNC-Lavalin, Canada
[2] bertrand.masse@snclavalin.com

ABSTRACT

Various formulas have been proposed for estimating the probabilities of occurrence or return periods of the various elements comprised in a sample of historic hydrological data. Most of these formulas are based on the equation $P = (m-\alpha)/(N+1-2\alpha)$, in which *m* is the rank of the element; *N* is the number of elements in the ranked sample; and α is a parameter whose value has been a subject of dispute among hydrologists for several decades. Proposed values of α vary between 0 and 0.5 and, presumably, depend on the probability distribution considered. A visual representation of the return period of the most extreme value in a ranked series is proposed, which shows that the most appropriate value of the parameter α is $\alpha = 0.31$ and that it's independent of the probability distribution.

RÉSUMÉ

Différentes formules ont été proposées pour estimer les probabilités empiriques ou périodes de récurrence pour des échantillons de variables hydrologiques. La plupart de ces formules prennent la forme $P = (m-\alpha)/(N+1-2\alpha)$, dans laquelle m est le rang d'une observation donnée, N est le nombre d'éléments dans l'échantillon et α est un paramètre dont la valeur a suscité beaucoup de discussions au cours des années. Les valeurs proposées de α varient entre 0 et 0.5 et dépendent présumément de la distribution statistique considérée. Une représentation visuelle de la période de récurrence de l'événement le plus grand dans un échantillon ordonné est présentée, laquelle montre que la valeur la plus appropriée du paramètre α est $\alpha = 0.31$ et que cette valeur est indépendante de la distribution considérée.

Keywords: - Frequency analysis, plotting positions, probability, return period

1. INTRODUCTION

Frequency analysis is commonly used to estimate return periods of meteorological or hydrological variables such as precipitation or river discharge. Nowadays, the analysis is carried out using specialized software such as HYFRAN-PLUS (Bobée and El Adlouni 2015), PeakFQ (Veilleux *et al.* 2014) or HEC-SSP (USACE 2016). The software often proposes several statistical distributions which can be tested and the user has to select the distribution that offers the best fit to the actual data. The selection is made by comparing the various distributions using a special probability paper. Results of an analysis using the normal probability distribution will plot as a straight line on a normal probability paper. A straight line will also be obtained with the lognormal distribution when the variables are logarithmically transformed. Similarly, the results of the application of the Gumbel distribution will also plot as a straight line on a Gumbel probability paper. All other distributions will plot as curved lines, whether concave upward or concave downward.

Two-parameter distributions like those mentioned above most of the time do not fit well hydrological data and this explains why three-parameter distributions (generalized extreme value (GEV), log-Pearson III, etc.) were introduced. When plotted on normal or Gumbel probability paper these distributions most of the time show curved lines.

2. PLOTTING DATA ON A FREQUENCY PLOT

Let a sample of N values of a variable X be ranked in order of magnitude from the largest to the smallest:

$x_1 \geq \ldots \geq x_m \geq \ldots \geq x_N$

If $y_m = F(x_m)$ is the cumulative distribution function of X then the probability of exceedence of variable x_m will be decreasing:

$y_1 \geq \ldots \geq y_m \geq \ldots \geq y_N$

Fitting data to a frequency plot requires that a probability be assigned to each element of the sample. Gumbel (1958) has proposed the following conditions which should be satisfied for any plotting formula:

1. The plotting positions must be such that all members can be plotted;
2. The return period (plotting position) of a value equal to or larger than the largest observation should converge towards N, the number of observations;
3. The observations should be equally spaced on the frequency scale;
4. The plotting position formula ought to be simple and have an intuitive meaning; and
5. The plotting positions of observation i should lie between the observed frequencies (m-1)/N and m/N and should be universally applicable, i.e. it should be distribution-free.

Gumbel has not offered any proofs as to the necessity of the conditions. Cunnane (1978) has discussed them and expressed disagreements about conditions (2), (3) and (4). For example, he states that condition (2) is not in keeping with statistical fact and is misleading. He also questioned the necessity and desirability of conditions (3) and (4) but did not reject them.

A general plotting formula for ranked observations is the following:

$$[1] \quad P = \frac{m - \alpha}{N + \beta}$$

in which P is the probability of occurrence of observation m; m is the rank of the observation; N is the number of observations in the sample; and α and β are parameters to be determined. In order to estimate the value of parameter β one assumes that the probabilities will be symmetrically distributed around probability 0.5. If the observations are ranked from the largest (m = 1) to the smallest (m = N), this means that the probability of exceedence of observation m will be equal to the probability of non-exceedence of observation N - (m - 1), or

$$[2] \quad \frac{m - \alpha}{N + \beta} = 1 - \frac{N - (m - 1) - \alpha}{N + \beta}$$

Resolution of the equation yields $\beta = 1 - 2\alpha$ and the plotting formula becomes:

$$[3] \quad P = \frac{m - \alpha}{N + 1 - 2\alpha}$$

Different values of parameter α have been proposed and Table 1 presents some of the commonly used formulae.

Hazen's formula, Eq. 4, is more than a century old and is not based on a theoretical development but was rather proposed to circumvent the ambiguity posed by the then familiar equations m/n and (m-1)/N. Hazen avoided this problem by using $\alpha = 0.5$.

The Weibull formula, Eq. 5, has long been used because it was said to be unbiased for any statistical distribution but Cunnane (1978) showed that it gives a biased overestimation at the upper end of a skewed distribution. In fact, the Weibull formula is unbiased only for the uniform distribution.

Blom (1958) showed that for the normal distribution, an optimal unbiased estimate of the standard deviation σ may be obtained by using $\alpha = 0.375$ (Eq. 6). The same formula would also apply to the two-parameter lognormal distribution when the variables are logarithmically transformed. Using a similar method, Gringorten (1953) proposed to use $\alpha = 0.44$ (Eq. 8) for the exponential (or extreme value) distribution. Cunnane (1978) suggested using $\alpha = 0.40$ (Eq. 10) as a compromise if the formula is to be used with all distributions.

Tukey (1962) thought that the differences between the various equations is probably not important and proposed Eq. 7 because it is simple and represents an adequate approximation to what is claimed to be an optimum.

A computation method equivalent to Eq. 9 was initially proposed by Beard (1943), based on the idea that a natural estimate of the unknown variable $y_m = F(x_m)$ is the median of its distribution, the value with equal probabilities of being exceeded or not being exceeded. Later on, Chegodaev (1953) and Benard and Bos-Levenbach (1953) both used a variant of Eq. 9 where α has been rounded up to 0.30 instead of 0.31. Eq. 9 has been adopted by Jenkinson (1977) based on the same principle.

Table 1. Plotting position formulae

Equation	Plotting position	Formula	Value of α
[4]	Hazen (1914)	$P = \frac{m-0.5}{N}$	0.5
[5]	Weibull (1939)	$P = \frac{m}{N+1}$	0
[6]	Blom (1958)	$P = \frac{m-0.375}{N+0.25}$	0.375
[7]	Tukey (1962)	$P = \frac{m-0.33}{N+0.33}$	0.33
[8]	Gringorten (1963)	$P = \frac{m-0.44}{N+0.12}$	0.44
[9]	Jenkinson (1977)	$P = \frac{m-0.31}{N+0.38}$	0.31
[10]	Cunnane (1978)	$P = \frac{m-0.4}{N+0.2}$	0.4

More recently, In-Na and Nguyen (1989) have modified the general plotting formula to include the coefficient of skewness while investigating three-parameter distributions. For the generalized extreme value (GEV) distribution they established the following approximately unbiased plotting formula:

$$[11] \quad P = \frac{m-0.13\tau-0.27}{N-0.08\tau+0.38}$$

in which τ is the coefficient of skewness.

At the same time, Nguyen *et al.* (1989) developed a similar unbiased formula for the Pearson III distribution:

$$[12] \quad P = \frac{m - 0.42}{N + 0.3\tau + 0.05}$$

Eq. 12 would also apply to the logPearson III distribution when the variables are logarithmically transformed.

3. VISUAL INTERPRETATION OF THE PLOTTING FORMULAE

Only the plotting formulae conforming to the general formula expressed by Eq. 3 are considered in the following sections.

The different plotting formulae can be visualized as follows:

Let us assume a hydrological event, say peak instantaneous flow, which has a return period equal to T_r. Then the probability that this event will be exceeded on any year is $1/T_r$ and the probability it will not be exceeded is $1 - 1/T_r$. Now if we have a sample of N annual observations of that event, the probability that all these events will have a return period less than T_r will be $(1 - 1/T_r)^N$. Therefore, the probability that at least one observation in that sample will have a return period exceeding T_r will be:

$$[13] \quad P = 1 - (1 - 1/T_r)^N$$

The latter expression has been computed for different values of the return period, T_r, and of the sample size, N. The results are shown in Table 2 and illustrated on Figure 1.

Table 2. Probability that a flood of return period T_r years will be exceeded at least once over a period of N years

Return period (years)	Sample size, N					
	10	20	50	100	1000	10000
2	0.999	1.000	1.000	1.000	1.000	1.000
5	0.893	0.988	1.000	1.000	1.000	1.000
10	0.651	0.878	0.995	1.000	1.000	1.000
20	0.401	0.642	0.923	0.994	1.000	1.000
50	0.183	0.332	0.636	0.867	1.000	1.000
100	0.096	0.182	0.395	0.634	1.000	1.000
200	0.049	0.095	0.222	0.394	0.993	1.000
500	0.020	0.039	0.095	0.181	0.865	1.000
1000	0.010	0.020	0.049	0.095	0.632	1.000
2000	0.005	0.010	0.025	0.049	0.394	0.993
5000	0.002	0.004	0.010	0.020	0.181	0.865
10000	0.001	0.002	0.005	0.010	0.095	0.632

3.1 Relationship between the return period of the largest element and the sample size

The relationship between the return period of the largest event (m = 1) in a sample of N elements for the different plotting formulae is obtained as follows:

$$[14]\quad \frac{1-\alpha}{N+1-2\alpha}=\frac{1}{T_r}$$

which can be rearranged as:

$$[15]\quad \frac{T_r}{N}=\frac{1-1/N-2\alpha/N}{1-\alpha}$$

For large samples, Equation 15 yields

$$[16]\quad \frac{T_r}{N}=\frac{1}{1-\alpha}$$

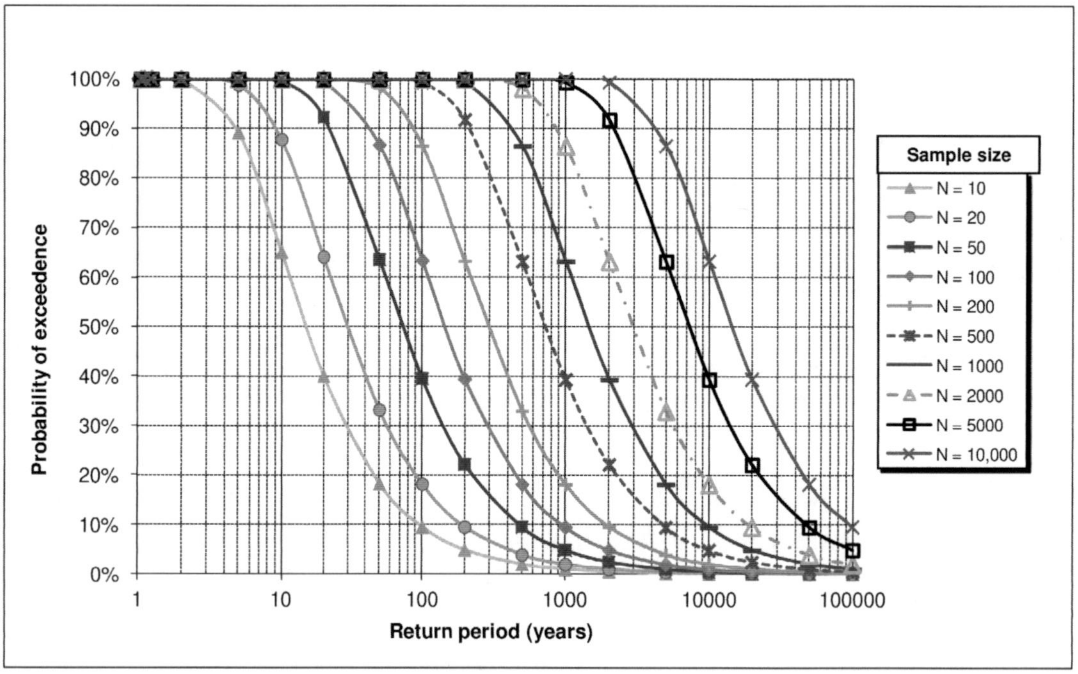

Figure 1. Probability that the T_r return period event will be exceeded at least once over a period of N years

The ratios T_r/N for the largest element in the sample as a function of the plotting formulae are shown in Table 3.

Table 3. Ratios Tr/N and probabilities of exceedence of the largest element in large samples

Formula	Tr/N	Probability of exceedence of largest element
Weibull	1.00	0.632
Jenkinson	1.44	0.500

Tukey	1.50	0.487
Blom	1.60	0.465
Cunnane	1.67	0.451
Gringorten	1.79	0.428
Hazen	2.00	0.393

3.2 Probability of exceedence of the largest element in a sample

The probability of exceedence of the largest element in the sample is given by Eq. 13. When combined with Eq. 15 this probability can be estimated for different sample sizes and the results for the various plotting formulae are illustrated in Figure 2.

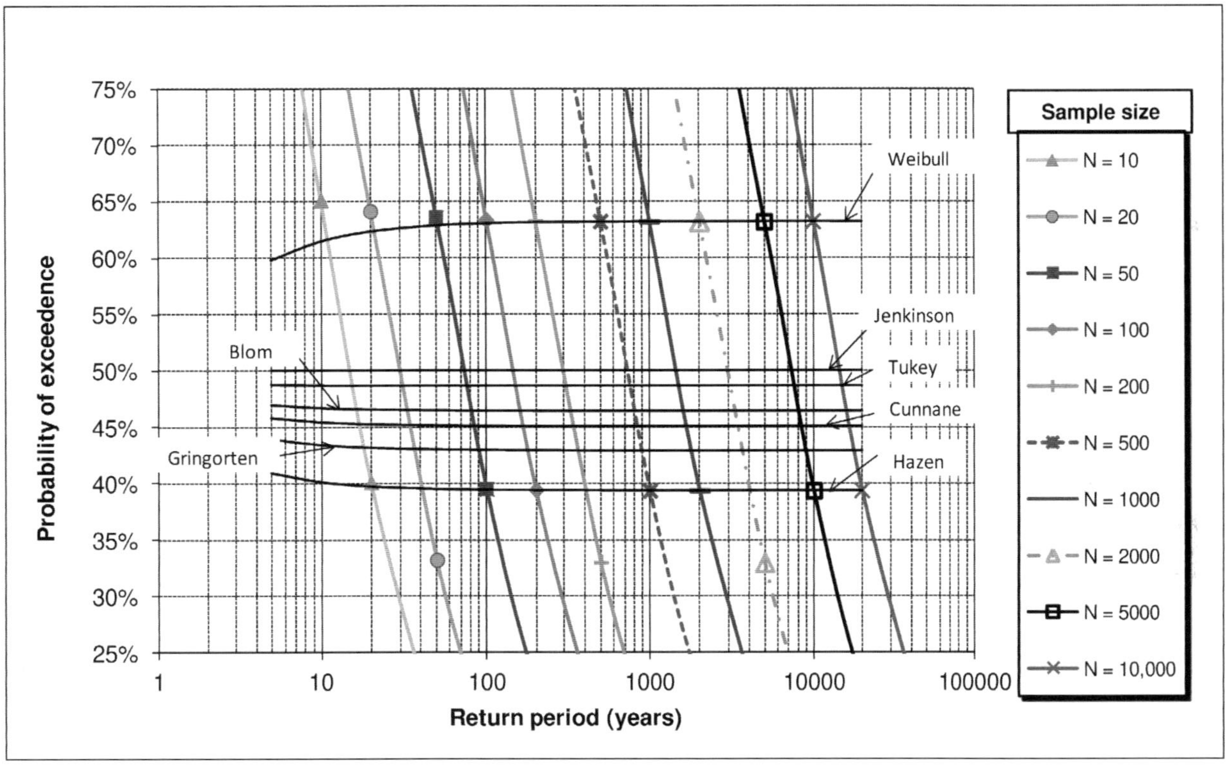

Figure 2. Probability of exceedence of the largest element for the various plotting formulae

Figure 2 shows clearly that the probability of exceedence of the largest event using Weibull's formula is over 60% which means that this formula tends to overestimate the magnitude of the event for a given return period. On the contrary, all other formulae, but Jenkinson's, tend to more or less underestimate the event. Hazen's formula is the most extreme in that regard. Only Jenkinson's formula appears neutral.

Even though Blom's or Gringorten's formula have been declared optimal for the normal or extreme value distributions, respectively, a hydrologist never knows which distribution will provide the best fit for the data and, therefore, there is no statistical reason to adopt these formulae a priori. Jenkinson's formula, by definition, can be applied whatever the statistical distribution of the sample.

It is possible that a formula other than Jenkinson's may provide a better fit for a given sample but, when used with a large number of samples, Jenkinson's formula is the one that will provide, most of the time, values for a given return period which are neither overestimated nor underestimated.

Combining Equation 16 with Equation 13 for large samples is equivalent to

[17] $P = 1 - e^{-N/T_r} = 1 - e^{\alpha - 1}$

Probabilities of exceedence of the largest element in the sample and for the various plotting formulae are given in Table 3.

Figure 3 shows a comparison of Jenkinson's formula with other ranking formulae for a sample of size N = 50 using return periods for $46 \leq m \leq 50$. It is clear that a significant difference for return periods is only apparent for the two or three highest ranks. Weibull's formula gives clearly a lower return period for the largest event. All other formulae give relatively similar results except Hazen's which gives the highest estimate.

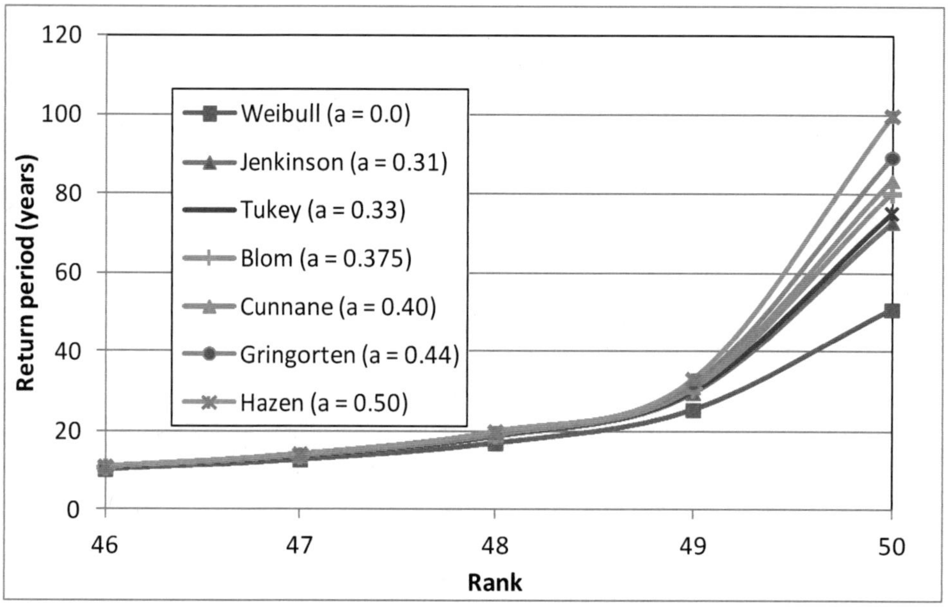

Figure 3. Comparison of Jenkinson's formula with other ranking formulae

4. THE JENKINSON FORMULA

The Jenkinson formula is different from the other plotting formulae because it is the only one that is defined based on a specific probability which corresponds to the median. In other words, the probability of an element obtained with the Jenkinson has 50% chance of being exceeded and 50% of not being exceeded. It is, therefore, independent of the statistical distribution.

Combining Eq. 13 and Eq. 14 and setting P = 0.5, one gets

[18] $$\alpha = \frac{N\left(2^{1/N} - 1\right) - 1}{2^{1/N} - 2}$$

Table 4 shows the value of α for different sample sizes.

Table 4. Values of α for different sample sizes

Sample size	α
10	0.3041
50	0.3063
100	0.3066
1000	0.3068

It can be seen that for small samples ($N < 20$) α is about equal to 0.30, the value proposed by Chegodaev (1953) and Benard and Bos-Levenbach (1953). For $N \geq 20$ the value of α is closer to 0.31.

Using Equation 17 with $P = 0.5$ it can be shown that for large N, $T_r/N = 1 / \ln 2$. Similarly, Eq. 17 can be used to get an estimate of parameter α for large sample sizes: $\alpha = 1 - \ln 2$, a result already obtained by Folland and Anderson (2002).

5. CONCLUSION

Jenkinson's formula is the only one having a statistical meaning and being applicable with any statistical distribution. All other formulae generally either overestimate or underestimate the magnitude of the largest element in a sample. It is still possible that a formula other than Jenkinson's may provide a better fit for a given sample but when performing frequency analyses for a large number of samples the Jenkinson formula is likely to provide, most of the time, values for a given return period which are neither overestimated nor underestimated.

6. REFERENCES

Beard, L.R., 1943. Statistical analysis in hydrology. *Trans. Am. Soc. Civ. Eng.*, 108 : 1110-1160.

Benard, A. and Bos-Levenbach, E.C., 1953. The plotting of observations on probability paper (in Dutch). *Statistica Neerlandica*, 7: 163-173.

Blom, G., 1958. *Statistical Estimates and Transformed Beta Variables.* Wiley, New York, N.Y., pp.68-75 and 143-146.

Bobée, B. And El Adlouni, S., 2015. *Éléments théoriques d'analyse fréquentielle - Utilisation du logiciel HYFRAN-PLUS.* INRS-ÉTÉ, Québec.

Chegodaev, N.N., 1953. *Computation of surface runoff on small catchments* (in Russian). All-Union Scientific Research Institute for Railway Construction and Planning. Rep. 37, State Rail Transport Publishing House, 76 p.

Cunnane, C., 1978. Unbiased plotting positions – a review. *Journal of Hydrology*, 37: 205-222.

Folland, C. And Anderson, C., 2002. Estimating Changing Extremes Using Empirical Ranking Methods. *Journal of Climate.* (15): 2954-2960.

Gringorten, I.I., 1963. A plotting rule for extreme probability paper. *J. Geophys. Union*, 68(3): 813-814.

Gumbel, E.J., 1958. *Statistics of Extremes.* Columbia University Press, New York, N.Y.

Hazen, A., 1914. Storage to be provided in impounding reservoirs for municipal water supply. *Trans. Am. Soc. Civ. Eng.*, Paper 1308, 77: 1547-1550.

In-Na, N. And Nguyen, V.T.V., 1989. An Unbiased plotting position formula for the general Extreme Value distribution. Journal of Hydrology, 106 : 193-209.

Jenkinson, A.F., 1977. The analysis of meteorological and other geophysical extremes. *Met. Office Synoptic Climatological Branch Memo*. 58, 41 pp.

Nguyen, V.T.V., In-Na, N. and Bobée, B., 1989. New plotting position formula for Pearson type-III distribution, J. Hydraul. Eng., 115 : 709-730.

Tukey, J.W., 1962. The future of data analysis. *Ann. Math. Stat.*, 33(1):21-24.

US Army Corps of Engineers, 2016. *HEC-SSP Statistical Software Package V. 2.1. User's manual*. Institute for Water Resources, Hydrologic Engineering Center, CPD-86, Davis, CA.

Veilleux, A.G., Cohn, T.A., Flynn, K.M., Mason, R.R., Jr., and Hummel, P.R., 2014, *Estimating magnitude and frequency of floods using the PeakFQ 7.0 program*. U.S. Geological Survey Fact Sheet 2013-3108, 2 p., http://pubs.usgs.gov/fs/2013/3108/

Weibull, W., 1939. A statistical theory of strength of materials. *Ing. Vetenskaps Akad.*, Handlingar NR 51, Stockolm.

Leadership in Sustainable Infrastructure

Leadership en Infrastructures Durables

Case Study

Vancouver, Canada

May 31 – June 3, 2017/ *Mai 31 – Juin 3, 2017*

HARVIE PASSAGE ON BOW RIVER: USING A PHYSICAL MODEL TO EVALUATE POST-FLOOD REHABILITATION OPTIONS

Habibzadeh, Ali[1]; Shepherd, Darren[2]; Slack, Chuck[3]; Hughes, Brian[4]; and Wirzba, Carl[5]

[1] Project Engineer, Northwest Hydraulic Consultants Ltd., 30 Gostick Place, North Vancouver, BC, Canada V7M 3G3, AHabibzadeh@nhcweb.com

[2] President, SG1 Water Consulting Ltd., 7303 118A St NW, Edmonton, AB, Canada T6G 1V3, Darren@SG1water.ca

[3] Principal, Klohn Crippen Berger Ltd., 500-2618 Hopewell Place NE, Calgary, AB, Canada T1Y 7J7, CSlack@klohn.com

[4] Principal, Northwest Hydraulic Consultants Ltd., 30 Gostick Place, North Vancouver, BC, Canada V7M 3G3, BHughes@nhcweb.com

[5] Team Lead, Alberta Transportation, 3rd Floor, 909 3rd Avenue North, Lethbridge, AB, Canada T1H 0H5, Carl.Wirzba@gov.ab.ca

The "Calgary Weir" (Figure 1, left image) on the Bow River in Calgary, Alberta has been a major component of the Western Headworks and is bounded by Pearce Estate Park on the right (south) bank and Deerfoot Trail on the left bank. Transforming the Calgary Weir into an attractive and navigable component of the City's parks system involved redeveloping the weir to: (i) eliminate the extreme drowning hazard that it created; and (ii) enable passage for fish and non-motorized boats while maintaining the weir's primary function of facilitating diversion from the Bow River to the Western Irrigation District (Figure 1, middle image). This transformation allowed safer passage for river users in non-motorized boats, improved fish passage, and created a more natural-looking river. Harvie Passage is comprised of a High Water Channel (HWC) located in the centre and left side of the river and a Low Water Channel (LWC) on the right side.

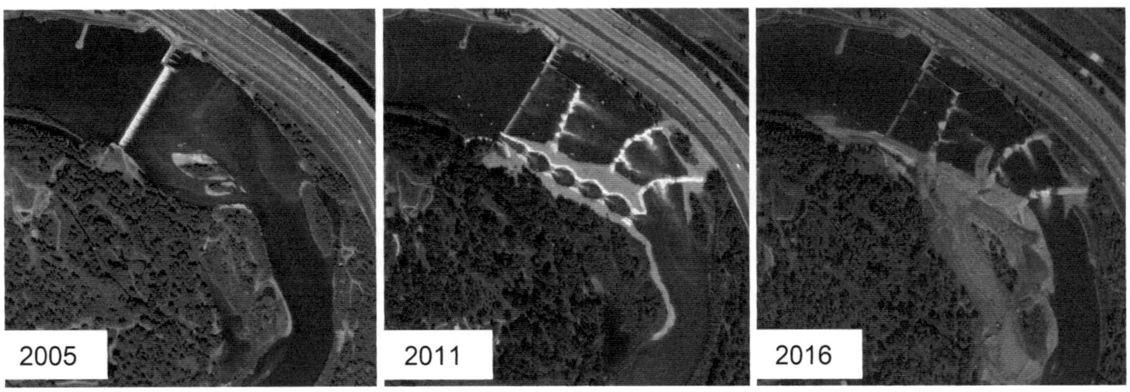

Figure 1: Aerial photo of the Calgary weir (left) and Harvie Passage before the 2013 flood (middle) and after the 2013 flood (right). *Imagery source: Google Earth.*

The unprecedented flood of June 2013 resulted in the near 1:100-year flood discharge being conveyed through Harvie Passage, which caused major infilling of the LWC, minor damage to the HWC drop structures, and reduced tailwater levels below HWC Drop #4 (Figure 1, right image). Rehabilitation measures currently underway include repairing damaged areas, eliminating unsafe hydraulic conditions, re-establishing fish passage, and construction of a new LWC.

A 1:14 scale physical model was utilized to evaluate the existing design of HWC Drop #4 and to develop modifications to its geometry that will: (i) eliminate the safety hazard that resulted from tailwater reductions within the Bow River; (ii) re-establish upstream fish passage; and (iii) not be adversely impacted by future changes in tailwater levels. The study objectives were expanded during the model testing phase to include the development of potential design modifications to HWC Drops #3L and #3R, all of which were intended to improve their recreational hydraulic performance.

Figure 2: Side view looking at the recommended design of the rehabilitation of HWC Drop #4 (left) and the wave forming on this drop at a river discharge of 150 m^3/s (right).

Different geometric modifications to HWC Drop #4 were evaluated in the physical model. The recommended final design modifications included: (i) adding a 16H:1V sub-plate below the existing drop exit followed by a horizontal sub-plate; and (ii) extending the high-flow channels downstream to the Bow River (Figure 2). The recommended final design of HWC Drop #4 performed satisfactorily over the range of flow conditions examined in the model. The safety hazard below HWC Drop #4 was eliminated and replaced by non-retentive hydraulic features that are expected to flush boaters and other types of river users downstream. Furthermore, the proposed extensions to the high-flow channels of HWC Drop #4 are expected to re-establish upstream fish passage at this drop structure.

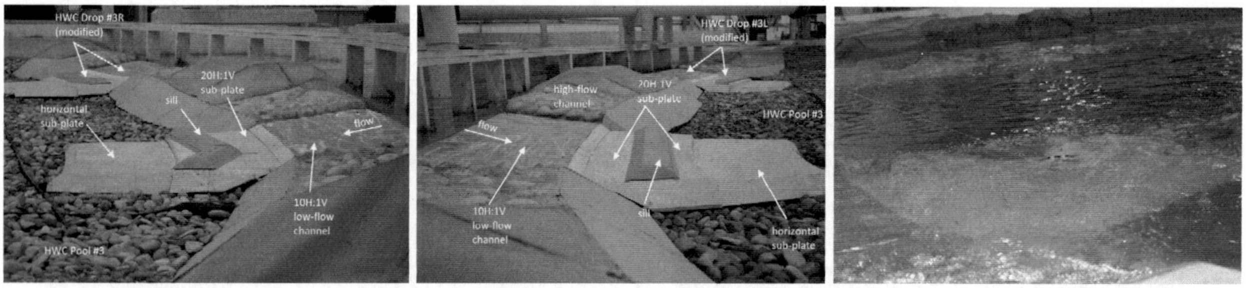

Figure 3: View looking downstream at the recommended design of HWC Drop #3L (left) and HWC Drop #3R (middle); and the playwave forming on HWC Drop #3R at a river discharge of 150 m^3/s (right).

Different design modifications were also evaluated on HWC Drops #3L and #3R. The recommended final design modifications to these drops consisted of: (i) reducing the longitudinal slope of the existing low-flow channel from 8H:1V to 10H:1V; (ii) installing two sub-plates in series below the low-flow channel, the first one comprised of a 20H:1V longitudinal slope and the downstream one set horizontal; and, (iii) installing a sill (V-shaped in planform on HWC Drop #3L and linear on HWC Drop #3R) across the width of the 20H:1V sub-plate near its downstream end (Figure 3). Model testing confirmed that the recommended design modifications successfully improved the hydraulic conditions at the drops by making them less retentive, thus more appealing to a wider range of recreational users. An enhancement to the existing design of these two drop structures included the formation of a secondary hydraulic that may be suitable for recreational usage under certain flows.

Leadership in Sustainable Infrastructure
Leadership en Infrastructures Durables

Case Study

Vancouver, Canada
May 31 – June 3, 2017/ *Mai 31 – Juin 3, 2017*

HYBRID ABSORBABLE LANDSCAPES: TREATMENT TRAINS FOR URBAN STORMWATER MANAGEMENT – MONITORING, MODELLING, DESIGN AND IMPLEMENTATION

Valeo, Caterina[1,2] and Gupta, Rishi[1]
[1] University of Victoria, Canada
[2] valeo@uvic.ca

Low-Impact Development (LID) is a form of urban infrastructure that provides sustainable management of urban stormwater. LIDs treat stormwater, both from a water quality and quantity perspective, at the source by imitating the absorptive capacity of the natural landscape in pre-developed form. Examples of popular LIDs include bioretention cells (including bioswales and rain gardens), green roofs, and permeable pavements (including porous asphalt, porous concrete, pavers, grass and concrete grids and innovative, highly permeable materials). Practitioners often place various forms of LIDs in the same development with parking lots consisting of permeable asphalt surfaces, for example, interrupted with bioswales between adjacent parking stalls. Interestingly, most research on the efficacy, design and maintenance of these LIDs is conducted on only one type of LID at a time. Treatment trains is a term that refers to two or more LIDs (of the same type or of multiple types) used in combination to treat the stormwater flow over its total flow path (either to a receiving body, as evapotranspired water or as recharge). Not surprisingly, little research has been conducted on the optimum technical configuration for complete stormwater treatment (Brown et al. 2012) for LIDs implemented in combination.

Permeable pavements and bioretention technology (LIDs that use vegetation whether in a rain garden or a green roof) can be considered two distinct categories of LID with the obvious distinguishing feature of living vegetation as an integral part of the LID's function. Each type of LID has different advantages and disadvantages in terms of water quality treatment, initial and developing absorptive capacity over the intended life of the LID, and maintenance and cost. Both types can be designed to provide large amounts of water quantity attenuation at very small sizes in comparison to the drainage area, and as well, each has the capacity varying degrees of physical, chemical and biological water treatment processes. While a wealth of knowledge exists for each category of LID, currently the critical questions to wide scale implementation for both types continues to be the reduction in absorptive capacity over time and maintenance requirements; and for vegetated LIDs, additional questions involve vegetation (both type and extent) requirements given local conditions as well as soil mix designs. While critical gaps remain, treatment trains are a more sustainable method of stormwater management and current knowledge can provide insight into how to best configure a treatment train for any Canadian climate. In addition, given the continued development of highly absorptive materials for permeable pavements, soil mix designs and admixtures, research is necessary to test the optimum treatment train given local requirements.

To this end, the authors have created the Hybrid Absorbable Landscape (HAL) Project at the University of Victoria. HAL is a field scale development to test LIDs alone or in treatment trains as suitable options for stormwater mitigation and sustainable urban landscape design. The user (municipalities, etc.) defined objectives steering development of the site involve providing maintenance schedules, recommendation of vegetation, optimum configuration and suitability of new innovative materials coming onto the market. The research objectives include developing scaling tools, determining clogging and bio-clogging rates, and changes in hydraulic conductivity for all LID types. The site is supported by both the Biofilm Research Laboratory (BRL) and the Facility for Innovative Materials and Infrastructure Monitoring (FIMM), both at the University of Victoria. The facility consists of four bio-retention cells and five types of permeable pavement as seen in Figure 1. Three of the bio-retention cells, A, B, and D, are individually connected to the porous asphalt, Porous Pave®, and grass grid pavers, respectively, via a pipe system (shown as short, thick black lines in Figure 1). The retention cells are individually lined with an impermeable rubber liner. Cell C was maintained as a control with no pavement system connection. All four bioretention cells can be tested individually or in tandem, with or without the permeable surfaces. Surfaces, vegetation and soil media can be replaced at any time and the site is monitored by meteorological, hydrological and ecological equipment and an autosampler that can sample at any sampling point including the sediment trap (ST) which is the final drainage point. FIMM and BRL also contain lab equipment for testing water quality and material parameters. Testing at the site thus far has spawned numerous issue related to scale and test methods for determining site performance (Jernslet et al, 2016).

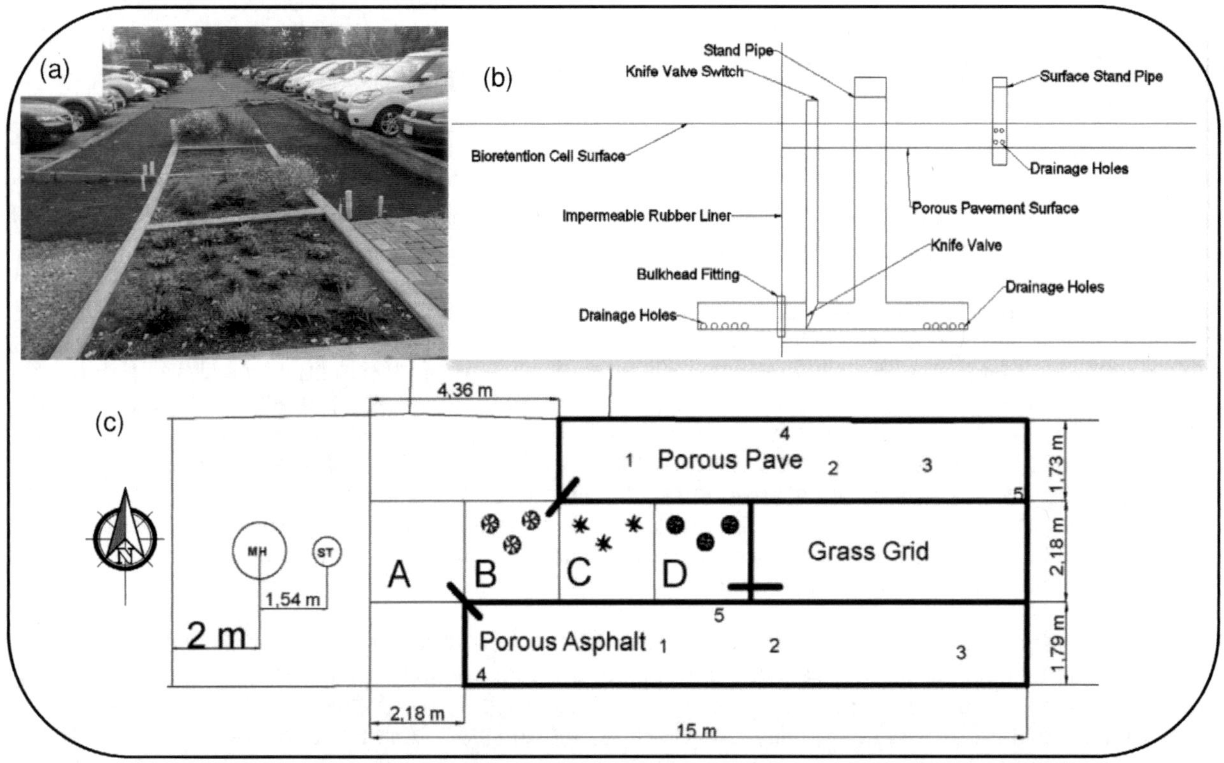

Figure 1: (a) View of site from manhole looking east; (b) schematic of sampling points placed at solid hatch marks of each permeable surface; (c) bird's eye view of the site.

Acknowledgements

The authors would like to thank the Canadian Foundation for Innovation and NSERC for funding the equipment and materials necessary to develop HAL. The authors would also like to thank Butler Brothers, Porous Pave, Island Paving and Don Mann Construction of Vancouver Island for their generous donation of materials and services; and Facilities Management, Rhonda Rose, Armando Tura and many of the students who worked to construct HAL at the University of Victoria.

References

Brown, RA, Line DE, Hunt WF. 2012. LID treatment train: pervious concrete with subsurface storage in series with bioretention and care with seasonal high water tables. J. of Env. Eng. 138(6):689-697.

Jernslet, M, Valeo, C and Gupta R. 2016. Methodology and scale issues in the ASTM infiltration test for permeable pavements. Submitted to Urban Water, December, 2016.

Leadership in Sustainable Infrastructure
Leadership en Infrastructures Durables

Case Study

Vancouver, Canada
May 31 – June 3, 2017/ *Mai 31 – Juin 3, 2017*

IMPLEMENTATION OF BIORETENTION AS A LOW IMPACT DEVELOPMENT OPTION FOR STORMWATER MANAGEMENT IN COMMUNITIES: A BARRIER ANALYSIS

Valeo, Caterina[1,4], He, Jianxun[2], Amell, Bernie[3] and Chu, Angus[2]
[1] University of Victoria, Canada
[2] University of Calgary, Canada
[3] Source 2 Source Inc., Canada
[4] valeo@uvic.ca

Conventional pipe systems that constitute the bulk of Canada's "stormwater management" have been shown to be unsustainable. The construction of these management systems is costly and their rehabilitation and replacement can be cost-prohibitive. Managing or controlling stormwater pollution at, or in proximity to its source likely enables more efficient and sustainable management. Consequently, many regions have turned to Best Management Practices (BMPs) or Low Impact Development technologies (LIDs), which work with the natural landscape to decentralize source control. LIDs manage urban stormwater runoff in both quantity and quality by targeting pre-developed levels. These technologies can be both environmentally and economically favourable in contrast to conventional stormwater infrastructure. Among a variety of such technologies, bioretention systems, also known as rain gardens, have been demonstrated to counteract the increase in volume and degradation in stormwater runoff quality posed by urbanization. A bioretention system generally consists of a soil bed of mixed sand/soil/organic media, a surface mulch layer below the vegetation layer and occasionally, an underground drainage system. Bioretention functions by temporarily ponding stormwater, allowing it to infiltrate into the sub-soil where it is stored and eventually it is either evapotranspired between storm events or allowed to drain out through a sub-drain to surrounding soil. The storage attenuates the stormwater (both in terms of runoff peak and volume) and the combination of soil media and vegetation promotes pollutant removal by various processes within the system, therefore, making bioretention systems highly beneficial to stormwater quantity and quality management. Numerous research studies conducted over the last two decades confirm the dual benefits of bioretention systems but they have also revealed that this technology is still immature for wide-scale application for a variety of reasons, some of which are technological and some are sociological. There is a lack of attention on how to more efficiently evaluate bioretention performance under different scenarios and how to optimize bioretention design. This is a key barrier hindering municipalities from implement these solutions as close to the stormwater source as possible – a residence.

Bioretention, which was introduced for stormwater management by Prince George's County as early as the 1990's, is a technology that integrates knowledge from multiple disciplines including engineering, hydrology, hydrogeology, ecology, soil science and landscape architecture. While the technology has been around for over two decades, we are still some distance away from implementing this technology in a distributed, pervasive way. To this end, a barrier analysis was conducted recently to further explore how to make bioretention technology a pervasive standard for stormwater infrastructure in our communities. The barrier analysis here refers to the method for determining critical paths to achieving an objective. The process begins with a desired objective, which is phrased into the form of a question: How might we achieve the objective? This question is posed to the invitees who brainstorm to identify the first hurdle or barrier that must be overcome to achieve that objective. Further hurdles are then identified by defining the action/hurdle/barrier to achieving the hurdles previously identified. For example, a group of stakeholders may have the objective of maintaining a healthy watershed and healthy supply of water for their community. The goal is framed as a question: *1. How might we achieve a healthy watershed and healthy supply of water?* The first hurdles that must be overcome right before achieving that objective might be: *1.1 Develop a watershed management plan; 1.2 Protect watershed and its supply;* and *1.3 Achieve political support, compromise and good will.* The questions are then asked as to what are the first steps to achieving each of these three barriers. Thus, the first hurdle to step *1.2* may be identified as, *1.2.1 Identify watershed stakeholders* and so on. The barrier analysis is continued until the session ends and in this way, the bottom of the tree, which may be the last few items identified in the session, is actually *the first steps* that must be taken to achieve the objective that began the session. A session that is only 30 minutes long may produce 50 separate hurdles, but inevitably in such an exercise, many hurdles require the same first steps to be overcome. The facilitator then analyses the hurdles to produce a route map initiated by a group of common first steps that reduces redundancies (effectively trimming the tree).

On September 24, 2015, researchers at the University of Calgary and the University of Victoria together with Source 2 Source Inc. engaged members of the community in a one day workshop to determine the barriers to widespread implementation of bioretention technologies in local communities. Over 20 people from industry, local government, and NGOs attended the workshop ranging in expertise from landscape architecture to ecology to municipal engineering. A barrier analysis was then conducted on the question:

> *1. How might we make bioretention technology a pervasive standard for stormwater infrastructure in our communities?*

This spawned two hurdles:

> *1.1 Make it easier to implement bioretention technology administratively and politically;*
>
> *1.2 Provide regulatory support for stormwater infrastructure with "enabling policy."*

Over an approximately 90 minute period, discussion on *1.1* produced 17 separate barriers and *1.2* produced 21 separate barriers that would have to be conducted prior to achieving *1.1* or *1.2*. Analysis of the hurdles and their relationships led to the following first steps at the bottom of the tree:

1.1.#.#.#.1 GET APPROVAL ENGINEERS TO BE SUPPORTED BY POLICY AND TECHNICAL INFORMATION 1.1....# GET MORE DATA AND RESEARCH FINDINGS INTO APPLICATION 1.1...#.# Increase needed research in the field 1.1...#.# Develop research outcomes that build on existing tools ......#.#.# *Align stakeholder and researcher interests* ...#.#.#.# *Increase communication between researchers and practitioners*	*1.2.#.....# Increase understanding of scale dependence of bioretention technology* *1.2.#.....# Increase information and data on bioretention performance* *1.2.#.....# Test sites for longer than five years*

Figure 1: First steps to achieve 1.1 (left) with last step at the bottom to be achieved firstly and 1.2 (right) where all three steps must be achieved at the same time.

This barrier analysis shows the common first steps often encountered in such analysis related to water resources problems: (a) develop a greater integration and working relationship between researchers progressing the field and practitioners governed by policy; and (b) increase funding directed towards research of critical gaps. Barrier (a) suggests engineers and engineering researchers play a more pivotal role in policy formation. The barriers specific to this analysis not shown in Figure 1 provided good insight into the relationship between the home owner as a user of bioretention technology and local municipalities.

Acknowledgements

The authors would like to thank all the participants in the workshop, which included members from the Town of Okotoks, the City of Calgary, the Bow River Basin Council, Urban Systems Limited, and the Alberta Low Impact Development Group.

Leadership in Sustainable Infrastructure
Leadership en Infrastructures Durables

Vancouver, Canada
May 31 – June 3, 2017/ *Mai 31 – Juin 3, 2017*

INVESTIGATION OF SPATIAL AND TEMPORAL SEASONAL RAINFALL PATTERNS OVER SINAI PENINSULA

Elsanabary, Mohamed H.[1,4], Khafagy, Hadeer E.[2] and Abdellah, Sherif[3]
[1] Assistant Professor, Civil Engineering Department, Port Said University, Egypt
[2] Graduate Student, Civil Engineering Department, Port Said University, Egypt
[3] Associate Professor, Civil Engineering Department, Port Said University, Egypt
[4] mohamed.elsanabary@eng.psu.edu.eg

Abstract: Heavy rainfall events are one of the essential tough climate events especially in steep and mountainous regions with small drainage area such as Sinai Peninsula, Egypt, which can lead to devastating flash water causing massive damages to life and infrastructure. The main goal of this paper is to investigate rainfall spatial and temporal patterns over Sinai Peninsula for 1981 – 2015 and its teleconnection to El Niño region. This study employed the wavelet analysis and wavelet principal component analysis (WPCA) to analyze the seasonal rainfall gridded data over Sinai Peninsula. The paper focused on two seasons: from January to March (JFM) and the other from October to December (OND). Results showed that Sinai rainfall can be delineated into three zones: northern zone of Sinai bounded by the Mediterranean Sea and Suez Canal, the central plateau and the southern triangular zone between Suez and Aqaba gulfs. The study revealed that the dominant frequency of seasonal rainfall ranged between 2-4 years for both seasons. For (JFM) season, northern Sinai suffered from a serious decrease in rainfall from 1991 to 2005, whereas, the peaks of rainfall occurred during the periods 1981-1991 and 2005-2012. For (OND) season, southern Sinai witnessed rainfall decrease within the period 1981-1992, rainfall increase within the period 1993-2015. Moreover, the whole Peninsula was exposed to an abundance of rainfall in the period 1981-1992 and rainfall decrease from 1993-2015. Finally, the study found a strong teleconnection between El Niño region and the northern zone of Sinai, represented by El Arish city rainfall.

1 INTRODUCTION

Sinai Peninsula is the only part of Egypt located in Asia acting as a land bridge between Africa and Asia. In Sinai, the main activities of the Bedouin tribes are herding and rain cultivating (Elewa and Qaddah, 2011). Sinai Peninsula is characterized by steep and mountainous regions with small drainage area. For example, El Arish city is subjected to heavy rainfall events which can lead to devastating flash water causing massive damages to life and infrastructure. Therefore, studying rainfall spatial and temporal variability was a demand to reduce the impact of such events. Furthermore, rainfall Forecasting, using oceanic anomalies with a lead time, is very necessary to provide early warning before flash water events which in turn can minimize the resulting disasters. The mountainous nature of Sinai Peninsula and the war that hit the region for a long time making rainfall gauge data unavailable enough for rainfall forecasting. Previous studies, related to Sinai Peninsula rainfall, used hydrological and hydraulic models such as rainfall-runoff models to study flash flood risk in Sinai Peninsula considering the rainfall has a stationary characteristic such as (Cools et al. 2012; El-Sammany, 2010; Elsayyad et al., 2011; Masoud, 2009; Youssef et al., 2010; Abuzied et al., 2016). Cools et al. (2012) have developed an early warning system (EWS) for flash floods for Wadi Watir, a part of Sinai Peninsula, based on field measurements, simulations, and remote sensing images. The results for 2010 flash flood show that 90% of the total rainfall volume was lost due to infiltration and transmission losses. Using the Weather Research and Forecasting Model (WRF), El-Sammany (2010) forecasted a total rainfall of 11.6 mm that cause a flash flood which took place over Wadi Watir, Sinai Peninsula, on October 24th, 2008 while the real measured one was 10.8 mm. El-Sammany (2010) showed significant consistency between WRF model and real measurements results. Some studies mapped flash flood risk using topographic, hydrological, and geological factors based on Rainfall-Runoff models. (Abuzied et al. 2016) assessed flash floods risk in the Nuweiba area located in Sinai Peninsula using soil Conservation Service (SCS) Rainfall-Runoff model then spatially integrated all the risk factors. Youssef et al. (2010) used GIS-based morphometry and satellite imagery to estimate the flash flood risk levels within the Wadi Feiran basin in Sinai Peninsula. Masoud (2009) analyzed the SRTM3 digital elevation model (DEM) and based on hydrologic characteristics and soil type during on-site visits, to derive morphometric parameters and the channel networks, in order to estimate flash flood of wadi catchments in Southern Sinai. Among previous studies, Elsayyad et al. (2011) investigated risk assessment of flash flood over Sinai Peninsula using similar techniques based on hydrological, metrological and hydraulic parameters.

This study takes into consideration the nonstationary variation of Sinai rainfall. Wavelet Principal Component Analysis (WPCA), a nonstationary technique, is applied to analyze the rainfall variability in space and time by applying the technique on rainfall gridded data of Sinai Peninsula. Elsanabary et al. (2014) applied wavelet Principal Component Analysis (WPCA), the non-stationary technique, to examine the spatial and temporal patterns of Ethiopian Highlands's rainfall. This paper performed wavelet principal component analysis (WPCA) for analyzing the nonstationary variations of Sinai Peninsula rainfall during the period (1981-2015) and define the spatial and temporal patterns. The main goal of this paper is to improve our understanding of the spatio-temporal rainfall variability over Sinai Peninsula, and find the teleconnection between such rainfall and El Niño region. The paper consists of six main sections. In section 2 the research objectives are shown, a brief description of the study area and rainfall data in section 3, the results and discussions in Section 5, and the summary and conclusions are given in Section 6.

2 RESEARCH OBJECTIVES

To meet the above goals, the objectives of this study are to employ the non-stationary technique of wavelet principal component analysis (WPCA) in order to investigate the non-stationary variation of Sinai seasonal rainfall for the two seasons (JFM) and (OND) from the period (1981-2015), Using the relations resulted to divide Sinai Peninsula into homogenous spatial patterns according to the nonstationary rainfall variability and determine the spatial localization of Sinai rainfall and investigate the periods of droughts, floods and their frequencies. Moreover, investigate the teleconnections between El Arish rainfall and climate anomalies such as El Niño.

3 STUDY AREA AND DATA

Sinai Peninsula is located in the northeastern part of Egypt and is bounded by longitudes 32°20′–34°52′E and latitudes 27°45′–31°10′N. It occupies an area of about 61,000 km^2 or about 6% of Egypt's total area with a population of about 400,000, which is mainly Bedouin (60%) and the rest are located in small cities such as El Arish and Sharm El Sheikh. The Peninsula has a triangular shape; its apex is to the south at Ras Mohammed (south of latitude 28°), whereas its base is to the north extending along the Mediterranean coast between Port Said and Rafah for about 210 km (Elewa et al., 2011). Sinai Peninsula is one of the coldest regions in Egypt because of its high altitudes and mountainous topography. Sinai Peninsula is characterized by two main rainy seasons. First one is from January to March (JFM) and the second is from October to December (OND). The rest of the year is dry. Precipitation data were collected for every year from October to March during the study period from 1981 to 2015. The data are monthly means of daily forecast accumulations. In order to have the whole Peninsula rainfall, another source of data was collected from ECWMF ERA-Interim dataset website at (http://apps.ecmwf.int/datasets/data/interim-mdfa/levtype=sfc/?). These data covers Sinai Peninsula over

(27.5°N-31.5°N) and (31.5°E-35.5°E) with grid resolution of (0.25° x 0.25°) from 1981 to 2015. Data contains 289 grid points. For verification of ERA- Interim dataset, gauge data were also collected from the Egyptian Meteorological Authority (EMA) for the period (2006 – 2015) for four gauge stations which are; Beer Elabd station at (30.59°N, 32.41°E), Al Melez station at (30.26°N, 33.10°E), El Arish station at (31.07°N, 33.46°E) and Sharm El_Sheikh station at (27.55°N, 34.19°E) as shown in Figure 1. Wars and Political conditions of Sinai Peninsula made gauge data of Sinai Peninsula rainfall isn't available enough to make a Long-term study and difficult to forecast the rainfall occurrence. The gauge data from the EMA was compared with ECMWF ERA-Interim gridded data, used in the study, in order to evaluate its reliability. Data comparison between the gridded data from ECWMF dataset and gauge data from EMA showed high correlation of 60% between El Arish station and gridded data at point (31°N, 33.50°E) with location difference of (-0.07°N, +0.04°E), high correlation (82%) between Sharm El_Sheikh station and gridded data at point (27.50°N, 34.25°E) with location difference of (-0.05°N, +0.06°E), medium correlation (50%) between Al Melez station and gridded data at point (30.25°N, 33°E) with location difference of (-0.01°N, -0.1°E) and low correlation (16%) between the last station Beer Elabd and gridded data at (30.50°N, 32.50°E) with location difference of (-0.09°N, +0.09°E), see Figure 2. The purpose of making this correlation is to investigate conformity of gridded data that will be used in the study with real gauge data. In order to confirm the teleconnection between Sinai rainfall and El Niño anomaly, El Niño SSTs data were downloaded from climate prediction center NOAA dataset at (http://www.cpc.ncep.noaa.gov/data/indices/oni.ascii.txt).

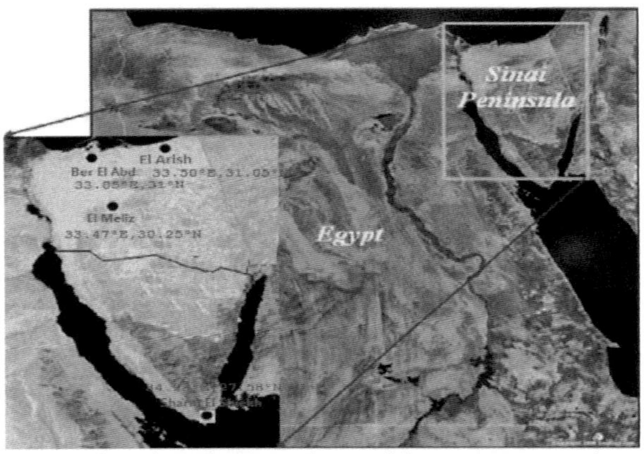

Figure 1: Sinai Peninsula's map showing the locations of four gauge stations of monthly rainfall data from EMA used in this study

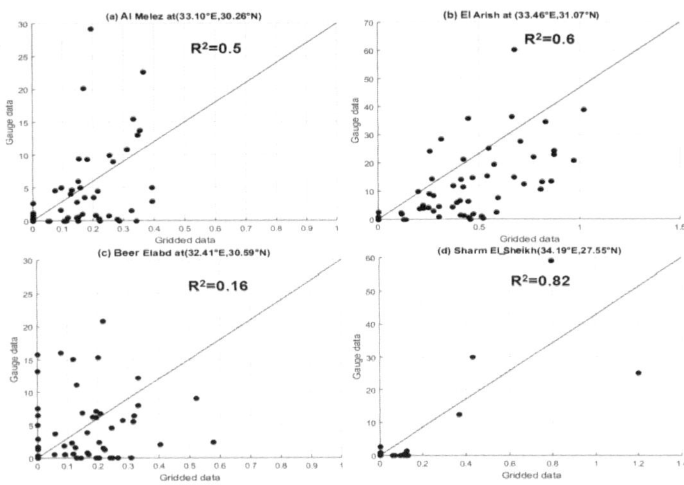

Figure 2: correlation between gridded data from ECWMF dataset and gauge data from EMA at four stations (a) Al Melez station, (b) El Arish station, (c) Beer Elabd station and (d) Sharm El_Sheikh station

4 METHODOLOGY

It is useful to have an idea about the possible techniques that can be used to identify spatial and temporal rainfall characteristics over the Sinai Peninsula. These possible techniques include Fourier Spectral Analysis, Hilbert-Huang transformation and Wavelet Analysis (WA). Fourier analysis technique considers the time series as a stationary time series while Hilbert-Huang transformation is suitable for analysing the nonlinear-nonstationary characteristics of rainfall. Considering the nonstationary variation of the rainfall over Sinai Peninsula, the nonstationary technique of Wavelet Analysis (WA) was used (Torrence and Compo 1998). WA is a mathematical means for performing signal analysis when signal frequency varies over time (Torrence and Compo 1998). Wavelet is convenient for studying nonstationary characteristics of any patterns such as rainfall (Elsanabary et al. 2014). The Morlet wavelet is a good choice because it is commonly used, simple and looks like the rainfall signals. Principal Components Analysis (PCA) is a way of identifying patterns in data and expressing the data in such a way to highlight their similarities and differences. The main advantage of PCA is compressing the data by reducing the number of dimensions without much loss of information. However, PCA method cannot eliminate out noise well enough. Therefore, a better solution is to combine wavelet analysis with PCA, called Wavelet PCA, which can improve the results. Wavelet power spectra and global wavelet spectra were used to define the prevalent frequency period. After that SAWP was calculated for each of the 289 rainfall grid points. Finally, Wavelet coherence was applied to El Arish gridded rainfall data and El Niño region to investigate the teleconnection between the two variables.

5 RESULTS

Wavelet and global spectra for all 289 grid points of the two seasons were calculated but couldn't be shown all. Figure 3(a) and Figure 3(b) show examples of wavelet and global spectra of Sinai rainfall at (33.5°E, 30.25°N) and (33.50°E, 30.75°N) respectively during the studying period 1981–2015 at OND season. The SAWP for rainfall was computed using the 2–4, 5-7 and 2-8 years frequency range. According to Figure 3 and other figures not shown, statistically significant power in JFM and OND Sinai rainfall was found to have occurred within the 2-8 years frequency band during the study period 1981–2015.

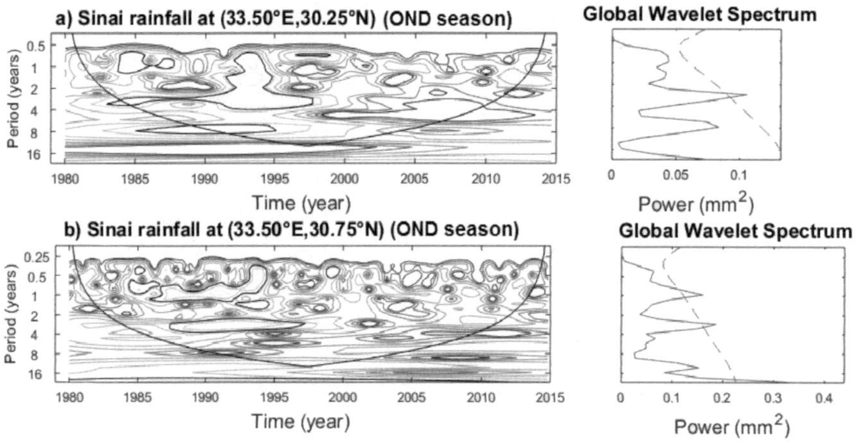

Figure 3: Examples of wavelet power spectra and global wavelet spectra plot at (a) (33.50°E, 30.25 °N) and (b) (33.50°E, 30.75°N)

To determine the dominant spatial and temporal characteristics of Sinai rainfall, wavelet principal component analysis (WPCA) was applied to the SAWP of rainfall (Elsanabary et al. 2014). Before applying PCA, all rainfall time series were standardized by subtracting the mean and then dividing by the standard deviation. Eigenvectors in principle component analysis provide us with information about the patterns in the data by extracting axes that characterize the data. In general, once eigenvectors are found from the covariance matrix, the next step is to order them by eigenvalues, highest to lowest. The eigenvector with the highest eigenvalue is the principle component of the data set. When using a filtered time series, such as the SAWP, a useful way to determine the number of signals presented in a time series can be done by observing spatial patterns of the eigenvectors. The number of eigenvectors determines the number of signals which represent the time series and the discarded eigenvectors considered as noise. Figure 4 shows that Sinai rainfall seasons JFM and OND may have up to two distinct signals represented by two essential eigenvectors (i.e. the first and the second principal component) representing more than 97% of the total variance for January to March (JFM) season and 95% of the total variance for October to December (OND) season for 2-8 years frequency band. For 2-4 and 5-7 years frequency bands the first and the second principal component representing more than 93% and 98% respectively of the total variance for JFM season and more than 97% and 98% respectively of the total variance for OND season.

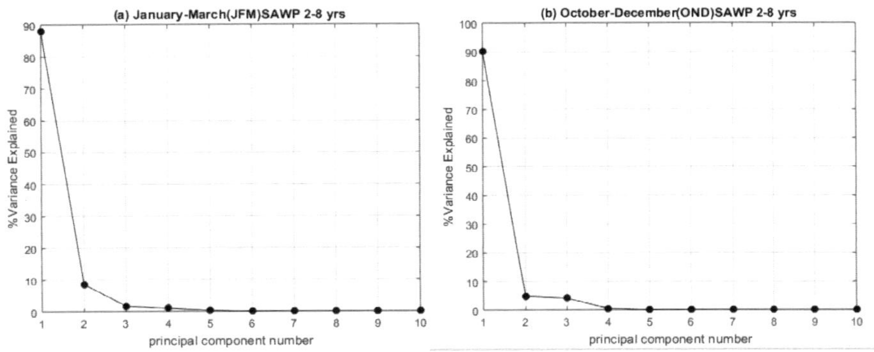

Figure 4: Scree plot of Sinai rainfall data of the first ten principal components which show that the first two principal components generally explain more than 97% of the total variance (a) for January to March and more than 95% of the total variance (b) for October to December for 2-8 years frequency band

5.1 Spatial patterns of Sinai rainfall

5.1.1 January to March (JFM) Rainfall

Figure 4 shows WPC1 and WPC2 accounted for, approximately, 97% of the rainfall variability where WPC1 and WPC2 explaining about 88% and 9% respectively for 2-8 years frequency band. For the analysis, 2-8 years frequency band was chosen as it is the most convenient for our data and is the same frequency of El Niño region. Figure 5 shows a topographic map of Sinai Peninsula that will be used to compare the rainfall patterns with the topography of the whole map. Figure 6 (a and b) display the correlation between each of the 289 grid point SAWP with WPC1 and WPC2 respectively for JFM season within the frequency band 2-8 years. According to figure 6 (a), WPC1 is positively correlated with SAWP for all the area of Sinai (all grid points) except the zero correlation found in the south western between (33.25°E-34°E) and (28°N-29°N). The strongest positive correlation was found in the eastern north overlooking the Mediterranean Sea between (33°E-34.5°E) and (30.25°N-31.25°N). The correlation decreases southwards towards the triangular part in the south. Figure 6 (b) describes the WPC2 correlation with SAWP where negative correlation was found in the majority of Sinai Peninsula except for the positive correlation in the western north area between (32.5°E-34°E) and (30°N-31°N) including one gauge station under study Al Melez at (30.26°N, 33.1°E). Figure 6 (a) appears more similar to figure 6 (e) than figure 6 (c) this clarifies that spatially 5–7 years frequency period is more dominant within the 2–8 years frequency band than the 2–4 years frequencies. Whereas, figure 6 (d) appears more similar to figure 6 (b) than figure 6 (f) showing that 2–4 years frequency period is more dominant within the 2–8 years frequency band than the 5–7 years frequencies.

The spatial correlation between SAWP with WPC1 and WPC2 for JFM season produced zones that are similar to that of the topographic contour map of the Sinai Peninsula that is shown in Figure 5. The rainfall over the southern region with elevated mountains of Sinai seems to have low correlation with WPC1. The hilly region in northern Sinai seems to have high positive correlation with WPC1. The central plateau seems to range in positive correlation values descending southwards. Generally, rainfall seems to decrease southwards while the elevation of the terrain increases and vice versa.

According to the spatial patterns of the three frequency bands 2-8, 2-4 and 5-7 years Sinai's January to March (JFM) rainfall season rainfall can be delineated into three zones: northern zone of Sinai bounded by the Mediterranean Sea and Suez Canal, the central plateau and the southern triangular zone between Suez and Aqaba gulfs.

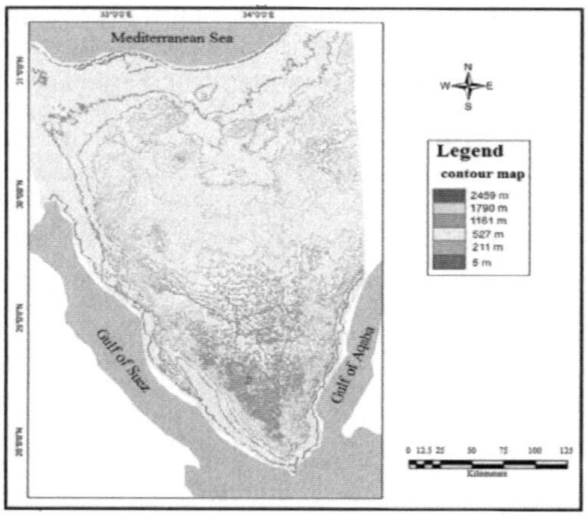

Figure 5: Topographic contour map showing land elevations in Sinai Peninsula from (Elsayyad et al. 2011)

5.1.2 October to December (OND) Rainfall

Figure 7 shows the correlation, between each of the 289 grid point SAWP with WPC1 and WPC2 for OND season. WPC1 (representing 90% of the gridded rainfall data)is positively correlated to SAWP for the total area of Sinai Peninsula within 2-8 years frequency band as shown in Figure 7 (a). While Figure 7 (b) shows that WPC2 (representing 5% of the gridded rainfall data) is negatively correlated to SAWP at almost all parts of the Sinai Peninsula, whereas it seems to be positively correlated to a small part of the northern coast of Sinai at 31°N-31.25°N and 34°E-34.5°E. Figure 7 (a) appears more similar to figure 7 (c) than figure 7 (e) and also figure 7 (d) appears more similar to figure 7 (b) than figure 7 (f). The results indicate that spatially 2–4 years frequency band is more dominant within the 2–8 years frequency band than the 5–7 years frequency band. Similarly, for the JFM rainfall season, the energy within the 2–4 years for OND rainfall season is dominating the rainfall over Sinai Peninsula.

There is similarity between results shown in figure 7 and the topographic map in Figure 5. Generally, the positive correlation between SAWP and WPCs, of OND season, decreases southward as the altitude increases and vice Versa. The 2-4 years frequency band WPC1 and WPC2 represents 91% and 6% of SAWP data respectively. Moreover, the 2-8 years frequency band WPC1 and WPC2 represents 90% and 5% of the data respectively. Generally, the spatial patterns of JFM and OND rainfall seasons divide Sinai Peninsula into three zones: northern zone of Sinai bounded by the Mediterranean Sea and Suez Canal, the central plateau and the southern triangular zone between Suez and Aqaba gulfs.

5.2 Temporal Patterns

5.2.1 JFM Rainfall

Figure 8 shows the temporal variations of the leading eigenvectors (i.e. WPC1 and WPC2) of Sinai Peninsula for JFM rainfall season based on 2–4, 2-8 and 5-7 years frequency bands. For JFM rainfall season, Figure 8 (a) shows a positive gradient in WPC1 that dominated northern Sinai Peninsula for the period 1981-1991 and 2005-2012. Sinai rainfall was in stagnation for the period 1991 to 2005 which appears as a negative gradient in WPC1. Generally, WPC1 varies with high energy from 1981 to 1991 and from 2005 to 2012. The highest rainfall values were in 1987, 1991 and 2012 whereas the least rainfall values were in 2000 and 2005. WPC1 is high positively correlated to northern Sinai within the JFM rainfall, the variations in energy show that northern Sinai suffered from a serious decrease of rainfall for 1992-2005. On the other hand, there is an abundance of rainfall in the periods (1981 – 1991) and (2005 – 2012). Figure 8 (b) shows the temporal variations of WPC2 where rainfall increased slightly in (1981 – 1983), (1992 – 2003) and (2009 – 2013), while rainfall decreased in (1983 – 1991) and (2003 – 2009). Moreover, the rainfall peaks values were in 1983, 1998, 2003 and 2013 while the lowest rainfall values were in 1992 and 2009. WPC2 is highly negatively correlated with the southern triangular part of Sinai Peninsula. As a result, the southern triangular part of Sinai Peninsula was exposed to abundance periods of rainfall in (1983 – 1991) and (2003 – 2009). Also, the southern triangular part was exposed to rainfall decrease in (1981 – 1983), (1992 – 2003) and (2009 – 2013). For 2-8 years frequency band, WPC1 and WPC2 accounted for 97% of total rainfall variability where WPC1 and WPC2 explaining about 88% and 9% respectively. In Figure 8 (a), there is similar pattern of WPC1 for the 2 – 4 and 2-8 frequency bands during JFM season. WPC1 was largely dominated by the 2–4 years cycle rather than 5–7 years cycle, as well as WPC2 was dominated by 2 – 4 year cycle, see Figure 8 (b).

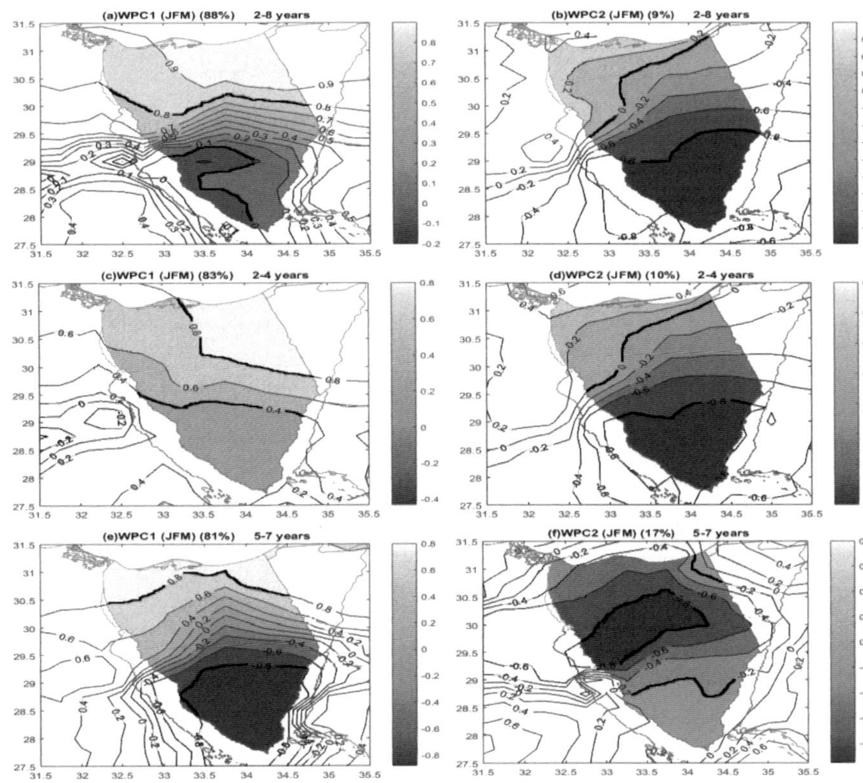

Figure 6: Spatial correlation of SAWP with WPC1 (a) 88% (c) 83% (e) 81% of gridded rainfall data within the frequency bands 2-8, 2-4 and 5-7 year respectively and with WPC2 (b) 9% (d) 10% (f) 17% of gridded Sinai rainfall data within the frequency bands 2-8, 2-4 and 5-7 year for JFM season respectively

5.2.2 OND Rainfall

According to Figure 9 when using 2–8, 2-4 and 5-7 year frequency bands to compute WPC1 and WPC2 at OND rainfall season. WPC1 shows a positive gradient in the period (1981-1992); see Figure 9 (a). Also, a negative gradient was observed in the rest of the study period (1992 – 2015). The rainfall peak was observed in in 1992. WPC1 is highly positive correlated with the majority of Sinai within the OND rainfall season. The variations in energy show that approximately, the total area of Sinai has an abundance of rainfall in the period (1981-1992) and rainfall decrease in (1993 – 2015) while WPC2 is highly negative correlated with the southern triangular area of Sinai Peninsula. Therefore, it suffered from rainfall decrease within the period of (1981 – 1992). Rainfall increases within the period of (1993 – 2015) and the decreases during 1992. For 2-4 year frequency band, WPC1 and WPC2 represent 91% and 6% respectively of SAWP data while for 2-8 year frequency band WPC1 and WPC2 represent 90% and 5%

of the data respectively. Comparing Figure 9 (a) and (b) we can notice that the frequency band 2-4 year within 2-8 years frequency band have a convenient representation of the data better than the 5-7 year frequency band for both WPC1 and WPC2.

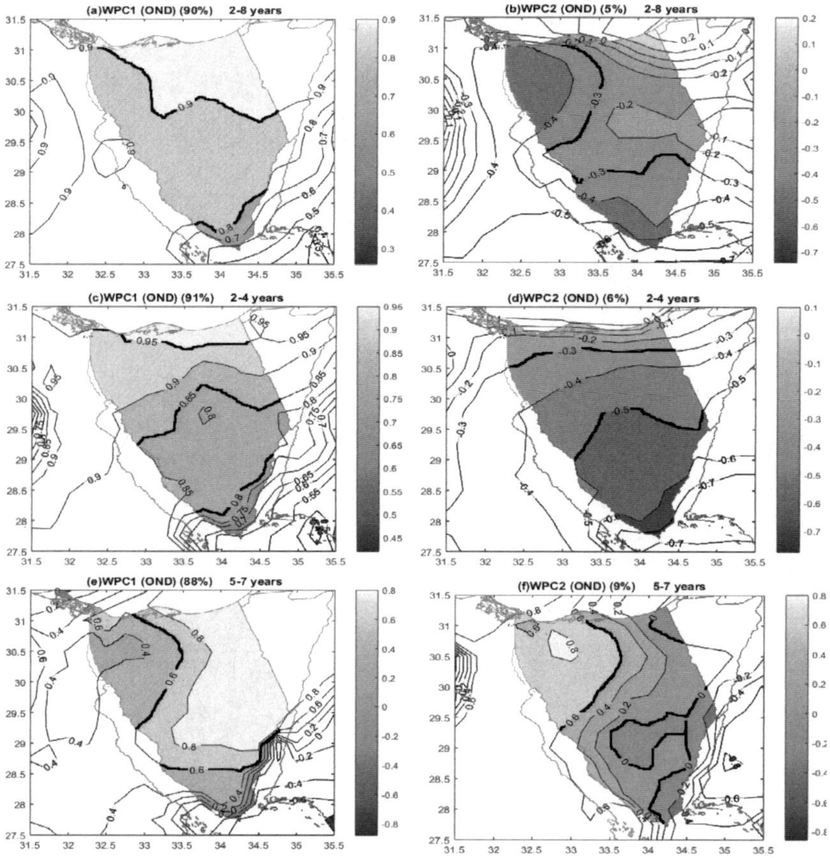

Figure 7: Spatial correlation between SAWP and WPC1 (a) 90% (c) 91% (e) 88% of gridded rainfall data within the frequency bands 2-8, 2-4 and 5-7 year respectively and between SAWP and WPC2 (b) 5% (d) 6% (f) 9% of gridded rainfall data within the frequency bands 2-8, 2-4 and 5-7 year for OND season respectively

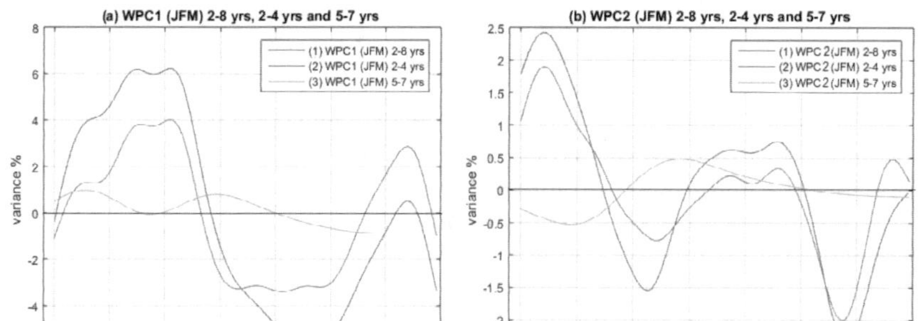

Figure 8: Time series of (a) WPC1 and (b) WPC2 of Sinai January to March (JFM) rainfall season computed using 2–8, 2-4 and 5-7 year frequency bands.

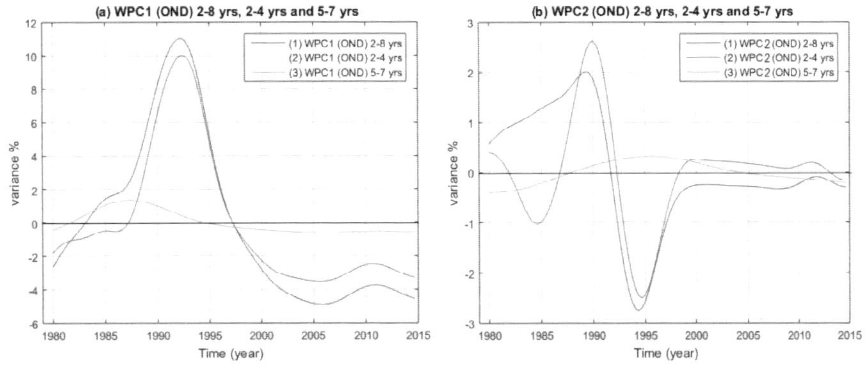

Figure 9: Time series of (a) WPC1 and (b) WPC2 of Sinai October to December (OND) rainfall season computed using 2–8, 2-4 and 5-7 year frequency bands.

5.3 Coherence between El Arish rainfall and El Niño

To understand the correlation between El Niño region and El Arish rainfall, the 3-month lead time, the July-September (JAS) and OND El Niño anomalies were used to help with forecasting of OND and JFM El Arish rainfall respectively. The association was assumed to exist if the temporal variation of the El Niño and rainfall were well correlated (Mwale et al. 2007). Figure 10(a) shows the wavelet coherence between El Arish (JFM) rainfall and El Niño (OND). The coherence between El Arish (JFM) rainfall and El Niño (OND) remains stable for the period 1990 to 1998. The rainfall over El Arish (JFM) shows an inter-annual coherence of over 0.8 with El Niño (OND) for the period 1998-2010. It seems that El Niño (OND) SST is leading the significant coherence. Figure 10(b) shows the wavelet coherence between El Arish (OND) rainfall and El Niño (JAS) SST. There is in phase relationship between El Niño (JAS) and El Arish (OND) rainfall for time periods of significant coherence from 1995 to 2010.

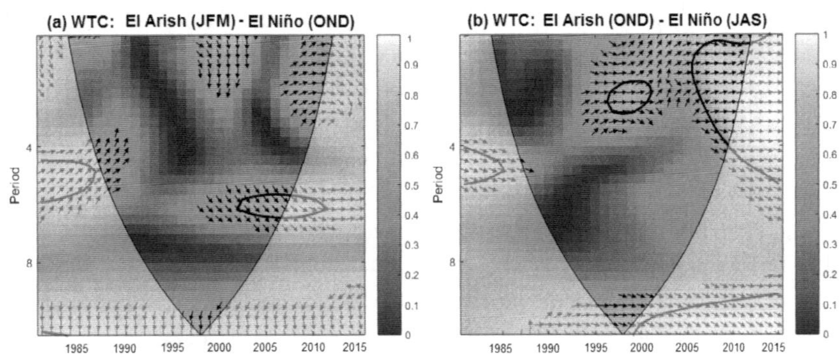

Figure 10: Wavelet Coherence between (a) El Arish (JFM) rainfall and El Niño (OND) SST (b) El Arish (OND) rainfall and El Niño (JAS) SST

6 SUMMARY AND CONCLUSIONS

This paper discusses the spatial and temporal seasonal rainfall characteristics over the Sinai Peninsula during two rainy seasons including January to March (JFM) season and October to December (OND) season. Non-stationary analysis technique was employed by applying wavelet analysis and WPCA on gridded rainfall data of the Sinai Peninsula in order to analyze the variability of rainfall in space and time due to the non-stationary variation of rainfall. These techniques were also applied to El Niño anomaly to teleconnect Sinai rainfall and El Niño region. According to the study, Sinai Peninsula was delineated into three zones: northern zone of Sinai bounded by the Mediterranean Sea and Suez Canal, the central plateau and the southern triangular zone between Suez and Aqaba gulfs. Temporal characteristics of the leading WPC signals showed that northern Sinai within the (JFM) rainfall season had been in stagnation in the period from 1991 to 2005, the peaks of rainfall occurred during the periods 1981-1991 and 2005-2012 while southern Sinai (OND) rainfall season witnesses rainfall decrease in 1981-1992. The dominant frequency in Sinai rainfall ranged between 2 and 8 years. By examining both (5-7years) and (2- 4years) frequency regimes to understand rainfall nonstationary, dominated behavior seems to be (2- 4years) frequency band. The teleconnection between El Niño region and El Arish rainfall was found by applying Wavelet Coherence between El Arish (JFM) and (OND) rainfall and El Niño (OND) and (JAS) respectively. This will help in forecasting floods and droughts over El Arish city.

7 ACKNOWLEDGEMENTS

The authors would like to thank the Egyptian Meteorological Authority for providing the gauge data of the four rainfall gauges used in the study.

8 REFERENCES

Abuzied, Sara, May Yuan, Samia Ibrahim, Mona Kaiser, and Tarek Saleem. "Geospatial risk assessment of flash floods in Nuweiba area, Egypt." Journal of Arid Environments, 2016.

"A cross wavelet and wavelet coherence toolbox for MATLAB." Aslak Grinsted. Accessed April 20, 2017. http://www.glaciology.net/wavelet-coherence.

Climate Prediction Center. Accessed February 18, 2017. http://www.cpc.ncep.noaa.gov/data/indices/oni.ascii.txt.

Cools, J., P. Vanderkimpen, G. El Afandi, A. Abdelkhalek, S. Fockedey, M. El Sammany, G. Abdallah, M. El Bihery, W. Bauwens, and M. Huygens. "An early warning system for flash floods in hyper-arid Egypt." Natural Hazards and Earth System Science , 2012.

Elewa, Hossam H., and Atef A. Qaddah. "Groundwater potentiality mapping in the Sinai Peninsula, Egypt, using remote sensing and GIS-watershed-based modeling." *Hydrogeology Journal*, 2011.

ERA-Interim, Monthly Means of Daily Forecast Accumulations. Accessed February 09, 2017. http://apps.ecmwf.int/datasets/data/interim-mdfa/levtype=sfc/.

El-Sammany, Moustafa. " Forecasting of Flash Floods over Wadi Watier – Sinai Peninsula Using the Weather Research and Forecasting (WRF) Model". World Academy of Science, Engineering and *Technology International Journal of Environmental, Chemical, Ecological, Geological and Geophysical Engineering*, 2010.

Elsanabary, Mohamed Helmy, Thian Yew Gan, and Davison Mwale. "Application of wavelet empirical orthogonal function analysis to investigate the nonstationary character of Ethiopian rainfall and its teleconnection to nonstationary global sea surface temperature variations for 1900-1998." *International Journal of Climatology*, 2014.

Elsanabary, Mohamed Helmy, and Thian Yew Gan. "Investigation of Seasonal Rainfall Variability over the Ethiopian Highlands: Teleconnection between the Upper Blue Nile Basin Rainfall and the Oceanic Anomalies." *Annual Conference of the Canadian Society for Civil Engineering*, 2012.

Elsayyad, Mohamed Ashraf, Gamal Ibrahim Kotb, Abd El-Moniem Sanad. "Risk assessment of flash flood in Sinai ". *Arab Academy For Science, technology and Maritime Transport,* 2011.

Masoud, Alaa A. "Runoff modeling of the wadi systems for estimating flash flood and groundwater recharge potential in Southern Sinai, Egypt." *Arabian Journal of Geosciences*, 2009.

Mwale, Davison, Thian Yew Gan, Samuel S. P. Shen, Ting Ting Shu, and Kyu-Myong Kim. "Wavelet Empirical Orthogonal Functions of Space-Time-Frequency Regimes and Predictability of Southern Africa Summer Rainfall." *Journal of Hydrologic Engineering*, 2007.

Torrence C, Compo GP. " A practical guide to wavelet analysis". *Bulletin of the American Meteorological Society*, 1998.

Youssef, Ahmed M., Biswajeet Pradhan, and Abdallah Mohamed Hassan. "Flash flood risk estimation along the St. Katherine road, southern Sinai, Egypt using GIS based morphometry and satellite imagery." *Environmental Earth Sciences*, 2010.

Leadership in Sustainable Infrastructure
Leadership en Infrastructures Durables

Vancouver, Canada
May 31 – June 3, 2017/ *Mai 31 – Juin 3, 2017*

LABORATORY AND FIELD MEASUREMENTS OF TURBULENCE USING AN ACOUSTIC DOPPLER CURRENT PROFILER

Kimiaghalam, Navid[1,2] and Clark, Shawn[1]
[1] University of Manitoba, Canada
[2] Navid.Kimiaghalam@umanitoba.ca

Abstract: Acoustic Doppler current profilers (ADCPs) have been used widely in river engineering applications, in particular for measuring discharge and bathymetry along open channels. Reduced measurement time and simple deployment techniques are among the advantages of such devices compared to Acoustic Doppler Velocimeters (ADVs). While the measurement accuracy of ADCP discharge measurements has been proven in several studies, there are few studies that have focused on using ADCP data to calculate turbulent flow characteristics within open channels, which is the focus of this paper. A four-beam M9 Sontek ADCP was used to measure and calculate stationary velocity profiles, turbulent kinetic energy (TKE), and Reynolds stress (RS) in a flume in the Hydraulics Research & Testing Facility at the University of Manitoba and within the Assiniboine River in the City of Winnipeg. Vertical profiles of velocity, TKE, and RS as well as depth-averaged values of these parameters were calculated under different flow conditions using measured ADCP data. Laboratory and field results were evaluated and compared with corresponding measurements using a Sontek MicroADV under the same flow conditions. Results showed that a four-beam ADCP can be used to calculate turbulent flow characteristics with a reasonable accuracy in a very short period of time compared to ADVs. It is anticipated that these field-based measurements of river turbulence will be helpful in predicting the type of river ice that may occur during freeze-up on rivers in cold regions.

1 Introduction

Turbulent flow characteristics significantly affect open channel flow hydraulic processes. Measuring or calculating turbulence characteristics are essential to understand river morphodynamics, mixing, and energy dissipation. Turbulence is also a primary variable that influences the type of ice that will form on a river during freeze-up. Field measurements of turbulence can be a costly and time-consuming task depending on measurement techniques and expected resolution. Acoustic Doppler Velocimeters (ADVs) are a common measurement device used in hydraulics laboratories to quantify turbulence; however, using such devices in the field is time consuming and costly due to the single point velocity measurement of ADVs and special deployment arrangements such as preparing an appropriate traversing mechanism. Moreover, offshore measurement of turbulent flow characteristics requires the use of a boat. Acoustic Doppler Current Profilers (ADCPs) can measure velocity profiles over depth and along a cross section easily and quickly with a simple deployment technique and using a hydro-board. A few recent studies focused on evaluation of turbulence measurements using different ADCPs; however, more studies and experiments are required to evaluate the ability of these devices to measure turbulence under different flow conditions.

During discharge measurements using an ADCP, several other flow parameters are measured such as signal-to-noise ratio and beam velocity components. Studies have been shown that these parameters can be used to calculate several flow characteristics such as sediment concentration and turbulence characteristics (Kimiaghalam et al. 2016; Kimiaghalam et al. 2015; Stacy 2000). Three and four beam

ADCPs have recently become the most common devices that are used for discharge measurements of large rivers. Nystrom et al. (2007) conducted an experimental study to evaluate turbulence measurements using the 1.5 MHz three beam Nortek ADCP and the 600 kHz four beam RD Instrument Rio Grande ADCP. They used beam coordinates velocity data to calculate the time-averaged streamwise velocity component ($\bar{u}$), turbulent kinetic energy (TKE), and Reynolds stress ($\overline{u'w'}$). They concluded that the calculated velocity and Reynolds stress profile using the four-beam ADCP had a good agreement with ADV measurements; however, corrections are required to calculate TKE. The mean velocity varied between 20-30 cm/s in their experiments and they suggested to perform further field measurements to evaluate turbulence measurements and calculations.

Several error sources were identified while using an ADCP for turbulence measurements (Nystrom et al. 2002). Sidelobe interference can affect near-bed velocity measurements and cause errors in measured velocity (Appell et al., 1991). Moreover, flow disturbance and ringing can cause errors in measurements near ADCP transducers (Gartner and Ganju 2002). In addition, extra assumptions were considered to calculate turbulence parameters using ADCPs since all calculations were conducted based on three or four measured velocity components along beams.

Recently a group of researchers have started to conduct a study on river ice processes in collaboration with Manitoba Hydro and the Natural Sciences and Engineering Research Council of Canada (NSERC). There was a need to understand the effects of turbulent flow characteristics on ice formation, freeze-up, and break-up processes. A Sontek River Surveyor M9 that is a four-beam ADCP was used to collect flow data within the Assiniboine and Red Rivers. It was essential to evaluate turbulence measurements and calculations using this ADCP since there was no study regarding the ability of this ADCP to measure turbulence, in particular in the field. This study focused on field and experimental evaluation of turbulence measurements using an M9 ADCP. Results from this study will be helpful for the ongoing research in the field of river ice engineering.

2 Measurements

Velocity profiles were measured using the 16 MHz Sontek MicroADV and the M9 Sontek ADCP in a flume in the Hydraulic Research & Testing Facility (HRTF) of the University of Manitoba and at two locations within the Assiniboine River in the City of Winnipeg. The flume is 13 m long, 0.95 m wide, and 0.7 m deep. The flume has glass sidewalls and a smooth bed with approximate slope of 0.0053. Water level can be regulated using a gate at the downstream end of the flume. Preliminary measurements demonstrated that at a location 8 m downstream of the flume inlet the flow was fully developed; therefore, measurements were performed at this location at the center of the flume. Velocity profiles were collected at the centerline when flow rates were 42, 120, 200, and 295 L/s and the water depth ranged between 0.45-0.48 m. Field measurements were performed within the Assiniboine River at the Hugo Dock and the Forks where performing measurements were possible without having a boat (Figure 1). Measurements were conducted on November 15, 2016 when the discharge was around 155 m^3/s and water depth varied between 0.8-1 m at the measurements locations. The first set of measurements were performed at a sampling rate of 50 Hz using the 16 MHz Sontek MicroADV. Three minutes of velocity data (9000 samples) were collected for each measurement points over depth and the ADV was moved in 3-5 cm intervals to cover the water depth. The sampling distance for this ADV was 5 cm from the probes. For field measurement, the ADV was mounted to a metal pole (4 cm by 4 cm in cross-section) that was long enough to reach the river bed. The second set of measurements were performed at a sampling rate of 1 Hz using the River Surveyor M9 ADCP. Ten minutes stationary velocity profile data (600 samples) were collected for each measurement cell over depth. The ADCP was mounted to a hydro-board during the measurements.

3 Data processing and calculations

Measured ADCP velocity data from the experimental and field measurements were exported to MATLAB files in beam coordinates and XYZ coordinates. A modified version of the Goring and Nikora (2002) method was used to remove noise and spikes from the raw measured ADCP and ADV velocity data. $\bar{u}$, TKE, and ($\overline{u'w'}$) were the turbulent flow characteristics that were measured and calculated using both ADV and ADCP data. The ADCP has four beams with symmetrical orientation in the Janus configuration and each beam

Figure 1: Field measurements location within the Assiniboine River in Winnipeg (Google Earth Pro, 2017)

has an angle of θ (25° for the M9) with the vertical plane (Figure 2). With the assumption of beams 3 and 4 being aligned with the stream-wise axis and homogeneity in the mean and variance of the velocity signals (Stacy et al. 1999):

[1] $u_3 = usin\theta - wcos\theta$

[2] $u_4 = -usin\theta - wcos\theta$

where u_3 and u_4 are the measured velocity in beam 3 and 4 respectively, u is the stream-wise velocity component, and w is the vertical velocity component. For both ADV and ADCP field data in XYZ coordinates, the turbulence parameters were calculated. Using Equation 3 and 4, $\bar{u}$ and $\overline{u'w'}$ can be calculated as follows using data in beam coordinates:

[3] $\bar{u} = \frac{\overline{u_4} - \overline{u_3}}{2\ sin\theta}$

[4] $\overline{u'w'} = \frac{\overline{u_4{'}^2} - \overline{u_3{'}^2}}{4\ sin\theta\ cos\theta}$

where $\overline{u_3}$ and $\overline{u_4}$ are the measured time-averaged stream-wise velocity components in beam 3 and 4, respectively, u_3' and u_4' are the turbulent intensities in beam 3 and 4, respectively. For data in beam coordinates, TKE was calculated using the following equation with the assumption of instantaneous homogeneity (Nystrom et al. 2007):

[5] $TKE = (TKE)_{error} + 0.5\left(1 + \frac{1}{4}tan^2\theta\right)S^2 + \frac{1}{16cos^2\theta}[\overline{u_1'u_3'} + \overline{u_1'u_4'} + \overline{u_2'u_3'} + \overline{u_2'u_4'}) + (1 + 4cot^2\theta)(\overline{u_1'u_2'} + \overline{u_3'u_4'})]$

where $S^2 = \frac{1}{4sin^2\theta}\sum_{i=1}^{4}\overline{u_i{'}^2}$.

A sensitivity analysis was performed to calculate the effects of having the ADCP beams misaligned from the stream-wise and span-wise axis based on Equations 1 and 2 (Table 1). This analysis is important, in particular for field measurements and the results showed that the accuracy of the measurements are reasonable if the misalignment is less than 10°.

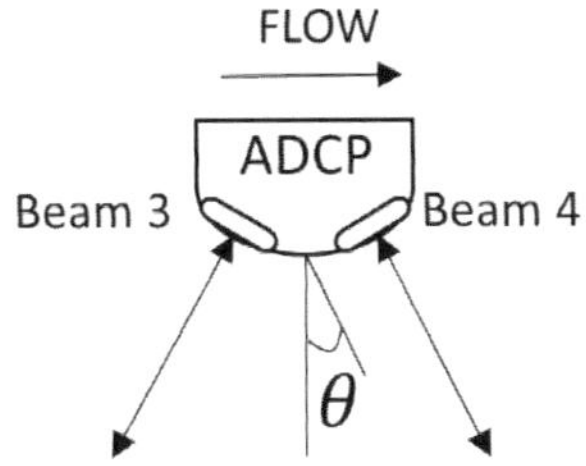

Figure 2: ADCP configuration

Table 1: Uncertainty in measured flow parameters with respect to the ADCP misalignment from the stream-wise axis

Δα (°)	Δu (%)	Δw (%)	Δu'w' (%)	ΔTKE (%)
5	0.4	0.4	0.1	0.1
10	1.5	1.5	2.4	2.4
15	3.5	3.5	12.4	12.4
20	6.4	6.4	41.2	41.2

4 Results

Figures 3a, 3b, 3c, and 3d show the turbulence characteristics measured and calculated using both the ADCP and the ADV in the laboratory. Table 2 shows depth-averaged turbulent flow characteristics with an error percentage. There was a good agreement between measured time-averaged stream-wise velocity profile using the ADCP and the ADV with 3% average error . Depth-averaged Reynolds stress error varied between 12-69% for different flow rates and always the average calculated Reynolds stress using the ADCP was less than the ADV. The depth-averaged TKE error varied between 9-47% and with increasing the flow rate, error decreased (Table 2).

Figures 4a and 4b show the turbulent flow characteristics measured and calculated using both the ADCP and the ADV in the field within the Assiniboine River in the City of Winnipeg. There was a good agreement between measured time-averaged stream-wise velocity with average error around 3%. There was a good agreement between calculated TKE using the ADCP and the ADV and also error in the depth-averaged TKE varied between 3-24%. ADCP's and ADV's calculated Reynolds stress were different with 56-61% error in depth-averaged Reynolds stress. Also, similar to the experimental results, the ADCP's calculated Reynolds stress were less than the ADV's calculated Reynolds stress.

While results showed that the calculated Reynolds stress based on the ADCP had a high error compared to the ADV measurements; the calculated Reynolds stress based on the ADCP measurements still showed the same trend; ie. that there is a range of Reynolds stresses that were a function of flow rate. For example from Figure 3, clearly one can understand that when the discharge were 42, 120, 200, and 295 L/s, $\overline{u'w'}$ varied between 0-0.06, 0.1-0.6, 0.2-2, and 1-12 cm^2/s^2, respectively. Also, the same trend can be found between calculated TKE using the ADCP and the ADV.

The experimental study by Nystrom et al. (2007) using the four-beam 600 kHz RD Instruments Rio Grande ADCP showed that the four-beam ADCP can be used to calculate Reynolds stress profile with a high accuracy. However, the present study suggests that the Sontek M9 ADCP can be used to calculate TKE profile with a reasonable accuracy but calculated Reynolds stress accuracy varied with different flow rates, in particular, field measurements showed that error in calculating Reynolds stress is higher than error in calculating TKE in the field. Since there was a good agreement between the measured ADCP and ADV stream-wise velocity profile, it can be concluded that the source of error in $\overline{u'w'}$ is the measured vertical velocity component. Moreover, since the stream-wise velocity component has the most significant effect on TKE compared to the vertical velocity component, the error in TKE was decreased with increasing flow rate and stream-wise velocity. For steady, uniform, and turbulent flow friction velocity (u*) can be calculate as follows:

[6] $-\overline{u'w'} = u^{*2}(1 - \frac{z}{h})$

where z is the distance from the measurement location from the bed and h is the water depth. Results from this study suggests that using Equation 6 to calculate u* in the field using the ADCP results in a high inaccuracy in the calculated u* and bed shear stress. Therefore, the log-law (Equation 7) based on the measured stream-wise velocity is the best way to calculate u* since there was a high accuracy in the measured stream-wise velocity component:

[7] $\frac{u}{u^*} = \frac{1}{k}(\frac{30z}{k_s})$

Where k is the Von Karman's constant and k_s is the bed roughness.

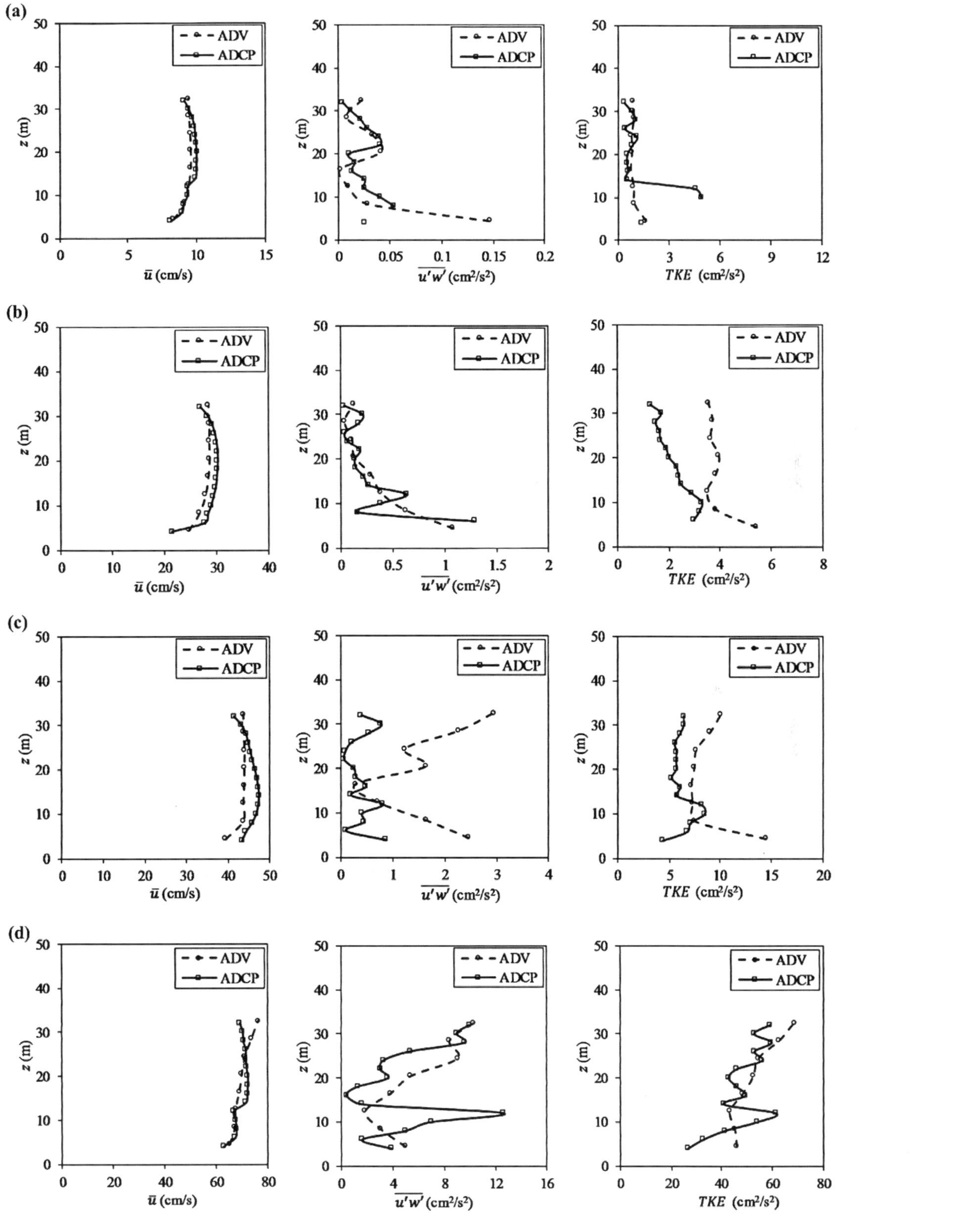

Figure 3: measured and calculated stream-wise velocity, Reynolds stress, and TKE using both the ADCP and the ADV in the HRTF: (a) Q=0.042 m^3/s; (b) Q=0.12 m^3/s; (c) Q=0.2 m^3/s; (d) Q=0.295 m^3/s

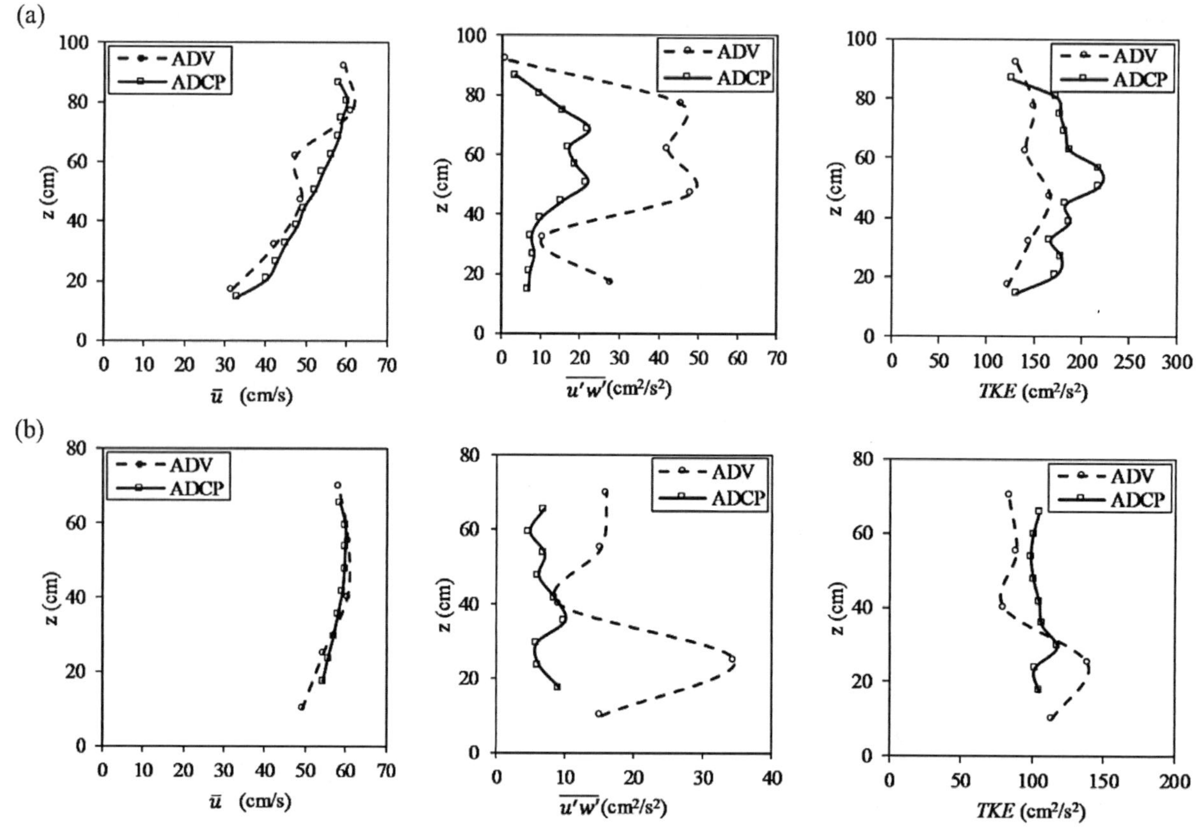

Figure 4: measured and calculated stream-wise velocity, Reynolds stress, and TKE using both the ADCP and the ADV in the field (Q=155 m^3/s): (a) Hugo Dock; (b) Forks

Table 2: Comparison of the depth-averaged stream-wise velocity (U), Reynolds stress (U'W'), and TKE

		Laboratory Testing				Field Testing		Average Error
	Q (m^3/s)	0.042	0.12	0.2	0.295	150	150	
U	ADV (cm/s)	9.3	27.8	42.4	70.1	56.7	48.4	3.0
	ADCP (cm/s)	9.5	28.6	45.4	69.7	58.2	50.3	
	Error (%)	2.2	2.9	7.1	-0.6	2.6	4.1	
U'W'	ADV (cm^2/s^2)	0.0	0.4	1.7	5.9	1.8	29.0	-41.6
	ADCP (cm^2/s^2)	0.0	0.3	0.5	5.2	0.7	12.7	
	Error (%)	-30.6	-20.0	-69.7	-11.9	-61.1	-56.2	
TKE	ADV (cm^2/s^2)	0.9	4.0	8.9	52.8	101.2	142.5	26.0
	ADCP (cm^2/s^2)	1.3	2.3	6.3	48.2	104.9	176.7	
	Error (%)	47.3	-41.9	-29.2	-8.7	3.7	24.0	

5 Summary and conclusions

Experimental and field measurements were conducted to evaluate turbulence measurements using the Sontek River Surveyor M9 ADCP. The purpose of this study was to understand possible uncertainty and error in the measured and calculated turbulence parameters using the ADCP including stream-wise velocity component, Reynolds stress, and TKE. Calculated Reynolds Stress using the ADCP was always underestimated compared to the ADV; however, the ADCP can be used successfully to find a range for Reynolds stress as a function of discharge. TKE was successfully calculated using the ADCP with a reasonable error, in particular, in the field. It can be expected that the uncertainty of the turbulence measurements would decrease if the measurements were performed off shore due to the distance from riverbanks and deeper water. Riverbanks can cause the similar signal interference in the results as a flume sidewalls. Moreover, deeper water will reduce the effect of sidelobes near the bed and ringing near the transducer; therefore, the agreement between the turbulence parameters would improve.

Generally, the Sontek River Surveyor M9 ADCP can be used to measure and calculate turbulence parameters to save time and cost of ADV measurements for practical purposes; however, considerations from this study may be applicable in terms of accuracy of the reported results.

Acknowledgements

This research was supported by Manitoba Hydro and the Natural Sciences and Engineering Research Council of Canada.

References

Appell, G. F., Bass, P. D., and Metcalf, M. A. 1991. Acousic Doppler Current Profiler Performance in Near-Surface and Bottom Boundaries. IEEE Journal of Ocean Engineering, **16** (4): 390-396.

Gartner, J. W., and Ganju, N. K. 2002. A Preliminary Evaluation of Near-Transducer Velocities Collected with Low-Bank Acoustic Doppler Current Profiler. Hydraulic Measurements and Experimental Methods Specialty Conference (HMEM), American Society for Civil Engineering, Estes Park, Colorado, United States, 13 p. paper in CD-ROM proceedings.

Goring, D. G., and Nikora, V. I. 2002. Despiking Acoustic Doppler Velocimeter Data. Journal of Hydraulic Engineering, **128** (1): 117-126.

Kimiaghalam, N., Goharrokhi, M., and Clark, S. P. 2015. Estimating Cohesive Sediment Erosion and Deposition Rates in Wide Rivers. Canadian Journal of Civil Engineering, **43** (2): 164-172.

Kimiaghalam, N., Goharrokhi, M., and Clark, S. P. 2016. Assessment of Wide River Characteristics Using an Acoustic Doppler Current Profiler. Journal of Hydrologic Engineering, **21** (12): 06016012.

Nystrom, E. A., Oberg, K. A., and Rehmann, C. R. 2002. Measurement of Turbulence with Acoustic Doppler Current Profilers - Sources of Error and Laboratory Results. Hydraulic Measurements and Experimental Methods Specialty Conference (HMEM), American Society for Civil Engineering, Estes Park, Colorado, United States, 10 p. paper in CD-ROM proceedings.

Nystrom, E. A., Rehmann, C. R., and Oberg, K. A. 2007. Evaluation of Mean Velocity and Turbulence Measurements with ADCPs. Journal of Hydraulic Engineering, **133** (12): 1310-1318.

Stacey, M. T., Monismith, S. G., and Burau, J. R. 1999. Measurements of Reynolds Stress Profile in Unstratified Tidal Flow. Journal of Geophysical Research, **104** (C5): 10933-10949.

Stacey, M. T. 2000. Estimation of Turbulence Parameters Using an ADCP. Joint Conference on Water Resource Engineering and Water Resources Planning and Management, American Society for Civil Engineering, Minneapolis, Minnesota, United States, 14 p. paper in CD-ROM proceedings.

Leadership in Sustainable Infrastructure

Leadership en Infrastructures Durables

Vancouver, Canada

May 31 – June 3, 2017/ *Mai 31 – Juin 3, 2017*

LOCAL INFLOW CALCULATION FOR A CASCADE RESERVOIR

Sadeque, Faheem[1,3] and Taylor, John[2]
[1] Hydrotechnical Engineering, BC Hydro, Canada
[2] Allium Hydrotech Ltd., Canada
[3] Muhammad.Sadeque@bchydro.com

Abstract: Local inflows from ungauged tributaries of a cascade reservoir are often calculated as the residual from the hydraulic balance for the reservoir due to discharges from the upstream and downstream facilities. Level pool reservoir routing is commonly used for its simplicity of application. However, the accuracy of level pool reservoir routing principle has been found questionable in many cases, especially during peak facility discharges. The slope of the water surface increases as the discharge through the reservoir increases. Therefore, the storage volume estimated based on the measured reservoir elevation at the downstream end is no longer accurate. The change of reservoir volume in each time-step is under-estimated during the spill ramp-up period. This generates significant negative local inflows as output of the hydraulic balance equation. The additional storage volume under the sloping reservoir surface does not change if the spill discharges from the upstream and downstream facilities remain steady. As a result, the hydraulic balance equation provides reasonable output for local inflows during steady spill periods. Again, as the spill discharges ramp-down the slope of the reservoir surface diminishes releasing the additional volume of water. This unaccounted change of storage volume generates significant positive local inflows as output of the hydraulic balance equation. Hydrodynamic modeling was carried out for a cascade hydropower reservoir to demonstrate the inaccuracies of level pool reservoir routing calculations. The model results showed that the inaccuracies in storage volume estimates appear to be insignificant (<<1%), but the calculated local inflows could be significantly erroneous. Some simple solutions were proposed to eliminate significant errors in local inflow calculations using hydraulic balance equation.

1 INTRODUCTION

Two or more reservoirs are often impounded by dams along the same course of a river to form a cascade facility. Flows from an upstream reservoir are released to a downstream reservoir by various discharge facilities; e.g. spillways and hydropower plants. Generally, the upstream reservoir receives water from a large drainage basin and it serves as the storage reservoir. The downstream reservoirs are generally smaller than the upstream reservoir. They do not provide any long-term storage. The downstream hydropower plants are considered run of river plants, typically operating in daily hydraulic balance with upstream facilities. The average daily discharges from the cascade facilities are approximately equal. The drainage basins for the downstream reservoirs are generally small and often ungauged. However, it is essential to estimate the local inflow for the downstream reservoirs for optimum operation of the hydropower plants and above all for the safety of the downstream facilities.

The objective of this study was to identify the sources of error in local inflow calculation for a cascade facility. BC Hydro's Peace River Hydropower Project was considered as a test case for the purpose of investigation.

2 BACKGROUND

2.1 Peace River Project

The Peace River Project is located in the northeastern side of British Columbia. It includes Williston Reservoir, W.A.C. Bennett Dam, G.M. Shrum Generating Station (GMS), Dinosaur Reservoir and Peace Canyon Dam and Generating Station (PCN). Williston Reservoir is impounded by W.A.C. Bennett Dam and has a drainage area of about 70,000 km^2. Peace Canyon Dam impounds Dinosaur Reservoir, about 20.5 km downstream of Bennett Dam. The drainage area between Bennett Dam and Peace Canyon Dam is very small (860 km^2). Flow regulation is provided by Williston Reservoir storage and GMS releases with limited short-term storage provided by Dinosaur Reservoir for load factoring purposes only. PCN is considered a run of river plant, typically operating in daily hydraulic balance with GMS. Therefore, average daily discharges from GMS and PCN are approximately equal.

2.2 Dinosaur Reservoir Local Inflow

BC Hydro uses an internal tool called CRO/Flocal to record plant operations and compute inflows and outflows from all its plants. Some operational data comes directly from measurements (e.g. reservoir forebay level, tailwater level, gate position etc.); some are estimated based on measured data and standard relationships (e.g. turbine discharge, spillway discharge etc.); and some data are back-calculated based on other measurements and estimations (e.g. natural inflows to the reservoirs).

Dinosaur Reservoir local inflows are calculated as the residual from the hydraulic balance for the reservoir due to discharges from the GMS and PCN facilities. An analysis of the operational records showed that Dinosaur Reservoir daily average local inflow values are frequently (~25% of the time) negative. These negative values are generally small (0 to -40 m^3/s for 75% of the occurrences) during normal operating conditions (i.e. without spill) at GMS and PCN. Even larger negative values (as low as -160 m^3/s) can appear during spill events. Negative inflows are more common and significant when calculated at an hourly interval. While small negative values were considered to arise from the imprecision of the measurement of reservoir elevations, large negative inflows were suspected to occur due to error in spillway discharges from Bennett Dam as high correlations were observed during GMS spilling. Similar negative local inflows were observed at other BC Hydro reservoirs (e.g. Revelstoke, Seven Mile) when calculated in hourly and daily intervals. This paper describes the detailed engineering study undertaken to identify the precise source of error, and proposes potential solutions.

3 REVIEW OF THE AVAILABLE INFORMATION

3.1 Method of Calculation

The forward calculation of the propagation of a flood wave in an open channel, known as flood routing, is a problem of applied hydrology that has been studied extensively. The relevant methods of solution, all within the framework of one-dimensional free-surface flow, span the spectrum from numerical solutions of the hydrodynamic equations of Saint-Venant to storage routing models of the diffusion-wave. Local inflows from ungauged tributaries of a cascade reservoir are often calculated as the residual from the hydraulic balance for the reservoir due to discharges from the upstream and downstream facilities; i.e. reverse flood routing. Computation of diffusive wave propagation in reverse time is ill-posed; i.e. the solution exists but it is unstable and may not be unique (Koussis et al. 2012). In such cases, small data perturbations cause large spurious solution. Therefore, the reverse solution must be smoothness-constrained towards stability and uniqueness (regularized).

Level pool reservoir routing is commonly used for its simplicity of application. It is based on the continuity equation which simply states that; the inflow minus the outflow equals the rate of change in storage.

[1] $I(t)-Q(t)=dS/dt$

where $I(t)$ is the inflow, $Q(t)$ is the outflow and $S(t)$ the storage at time t.

The first order Euler's scheme produces the following simple expression for level pool routing:

[2] $I(t)-Q(t)= [S(t+\Delta t)-S(t)]/\Delta t$

This explicit scheme is conditionally stable. However, the solution is unstable when there is a rapid change of storage. Therefore, level pool reservoir routing has been found questionable for local inflow calculation for cascade reservoirs, especially during peak facility discharges (e.g. during spill periods).

CRO/Flocal uses a hydraulic balance equation, similar to Eqn. [2], to calculate Dinosaur Reservoir local inflow:

[3] NI_DNR_CRO = Delta_Str_ DNR*(10^6/3600) – Rel_ WSR + Rel_DNR

where NI_DNR_CRO = Natural inflow to Dinosaur Reservoir in the CRO/Flocal records in m^3/s;

Delta_Str_ DNR = Change in storage volume over the measurement interval in Dinosaur Reservoir in million m^3 units calculated from the forebay elevation change. [(10^6/3600) is to convert the hourly million m^3 values to the m^3/s units];

Rel_ WSR = sum of all releases from Williston including releases from all the 10 turbines and the spills from all of the 3 radial gates if any, 9 sliding gates if any and 4 low level outlets if in service, in m^3/s; and

Rel_DNR = sum of all releases from Dinosaur Reservoir including releases from all the 4 turbines and the spills from all of the 6 radial gates if any, m^3/s.

The principal assumption of this reservoir routing approach is that the reservoir level remains horizontal from the upstream to the downstream boundary for each time step of the flow routing period. The accuracy of this calculation depends on the validity of this principle (i.e. diffusive wave propagation) and the accuracy of each pertinent input data and assumptions. Dinosaur Reservoir local inflow (i.e. *NI_DNR_CRO*) can be negative if either or a combination of the following inaccuracy occurs:

a) change (increase) in storage volume (i.e. *Delta_Str_ DNR*) is underestimated,

b) sum of releases from Williston Reservoir (i.e. *Rel_ WSR*) are overestimated; and

c) sum of releases from Dinosaur Reservoir (i.e. *Rel_ DNR*) are underestimated

3.2 Input Data for Hydraulic Balance Equation

- **Water Levels**

Forebay and tailwater levels are measured continuously. For CRO/Flocal calculations data is taken at the top of each hour as instantaneous values. Use of instantaneous values could introduce randomness in the Flocal results.

- **Turbine Discharge Calculations**

A powerhouse with multiple units can have highly variable turbine operations within an hour. Flow through a turbine is a function of generator output (MWh for the hour), generator state information (e.g. generating or offline), forebay and tailwater elevations, and unit characteristics (i.e. flow as a function of head and power). The hourly average flow for a single unit is calculated on the basis of hourly average gross head (forebay minus tailwater) on the unit, the number of minutes in the hour that the unit is in a generating state and the average MW generated in that period. If a unit operates for a short period within an hour, the estimated hourly average head obtained from the instantaneous values of forebay and tailwater levels at the top of the hour may not be representative due to the variation of discharges from the adjacent units. Therefore, turbine discharge calculations may produce some random errors in Flocal calculated reservoir inflow.

- **Non-Power Release Calculations**

Non-power release facilities at GMS and PCN consist of spillways with radial gates. Calculation of spillway discharges in CRO/Flocal is based on forebay elevation, spillway gate position (recorded to a minute time resolution) and the spillway gate rating curves. Average forebay elevations are determined for each gate position-time interval within an hour. This is done via linear interpolation of the elevations between the beginning and end of the hour. Using the elevation-position-flow relationship, the flow through the gate for each interval within the hour is determined. Then an hourly average spillway discharge is calculated as, average flow = ∑ (interval duration in minutes) * (calculated interval flow) / 60, where the sum is over all position-intervals in the hour.

Due to its large storage volume Williston Reservoir level at the forebay does not change noticeably within an hour even during spillway operation. Moreover, the spillway approach channel is about 1.5 km away from the power intake tower where the forebay gauge is located. Therefore, no cross-flows are expected in front of the spillway.

PCN powerhouse is located adjacent to the spillway gates. There is a 280 m long anti-vortex dyke across the reservoir at ~400 m upstream of the dam. The purpose of this dyke is to create cross-flows near the dam to sweep away air-entraining vortices from the power intakes. Therefore, the flow travels across the forebay from the right side to the powerhouse located on the left side of the concrete dam. This cross-flow may affect the calculation of PCN spillway discharge.

- **Review of the Spillway Rating Curves**

The spillway rating curves at PCN and GMS were developed on the basis of physical hydraulic model studies undertaken during the design of the projects. Hydraulic models were built at small scales (1:84 for GMS and 1:60 for PCN) that did not include all the details of the spillways. In addition, design modifications are common during the construction period and this was the case for both GMS and PCN spillways.

A review of the prototype modifications for GMS spillways indicates that there are only minor unaccounted alterations in the spillway discharge rating curves. Due to these minor alterations GMS spillway discharge could be slightly underestimated. However, a similar review for PCN spillways shows that there are many unaccounted alterations (e.g. the anti-vortex dyke, gate radius and trunnion position, shape of the piers and side walls). The magnitude of the effects of all changes on the rating curves is difficult to estimate. The unaccounted issues appear to be insignificant when considered separately, but when taken together they act in the same direction and suggest that actual discharge may be higher than the rating curve. Because of these inaccuracies PCN spillway discharge may be underestimated.

- **Review of the Storage Volume Calculation**

Dinosaur Reservoir storage volume is calculated hourly in CRO/Flocal based on instantaneous measurement of forebay level at the top of each hour and the storage curve (volume vs. elevation relationship). The reservoir surface is assumed to be flat for each time step.

The Dinosaur Reservoir storage curve provides the relationship to estimate volume of water for reservoir elevations from El. 491.3 m to El. 504.5 m. The accuracy of storage curve is unknown. It was developed on the basis of pre-dam survey data. The accuracy of live storage volume near the operating level (El. 500 m to El. 503 m) would be critical for Dinosaur Reservoir local inflow calculation.

- **Review of the Backwater Effects**

An engineering study was carried out in 1970 to calculate the backwater effects of PCN Dam to GMS powerhouse. The standard step method was used to calculate steady water surface profiles from PCN forebay to the GMS manifold for various flow conditions and assumed forebay levels. The calculation showed that the water surface elevations just downstream of GMS tailrace could be 0.4 m, 1.8 m, 4.8 m above the PCN headpond level for 1274 m^3/s, 1982 m^3/s and 6230 m^3/s discharges, respectively (assuming PCN forebay level = 502.92 m). The slope of the water surface is relatively steep at the upstream reach of the reservoir. Therefore, for average powerhouse flows of 1200 m^3/s the water surface in the main body of the reservoir is essentially horizontal and normal reservoir routing would be adequate. But the difference in water level could be significant for higher powerhouse discharges and during spillway operation, and hence the assumption of a flat reservoir surface is not valid during these high flow events.

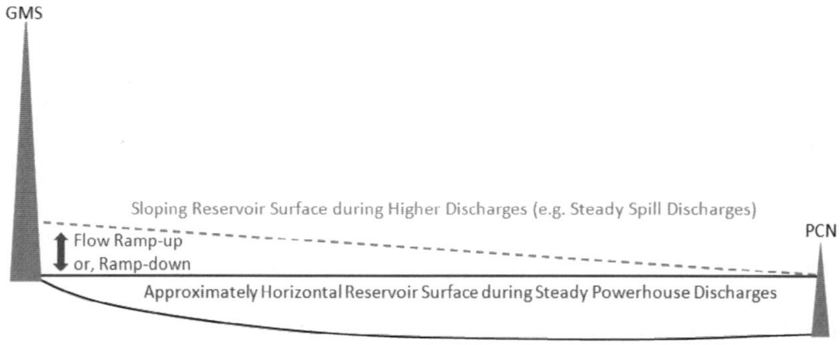

Figure 1: Schematic diagram of the water surface above PCN forebay

Figure 1 shows a schematic diagram of Dinosaur Reservoir surface for various flow conditions. During flow ramp-up periods (both inflows and outflows), upstream water levels in the reservoir will be elevated higher than the downstream water levels. This will provide additional storage in the reservoir. When inflows (i.e. GMS flow releases and local inflow) and outflows (i.e. PCN outflows) are in balance the difference between upstream and downstream water levels will remain constant. Therefore, storage volume will remain constant during steady flows. Again, when the flows ramp-down the reservoir level differences will be diminished. Thus the reservoir will flush out the additional storage developed during flow ramp-up. Therefore, the assumption of a flat reservoir level would impart errors in storage volume estimation, especially at the beginning and end of a balanced spill event.

An underestimation of the hourly change of Dinosaur Reservoir storage volume by 0.36 Mm^3 may lead to 100 m^3/s negative local inflow. A water level difference of about 0.1 m between upstream and downstream of the main body of Dinosaur Reservoir can hold 0.36 Mm^3 water (assuming an average width of reservoir = 350 m). This estimate shows the sensitivity of the input parameters on Dinosaur Reservoir local inflow calculation.

3.3 Review of the Historical Data

Hourly operational data for GMS and PCN facilities obtained from CRO/Flocal for three spill events in 1996, 2002 and 2012 were reviewed. Figure 2 shows the historical records during the spill period of 1996.

The following observations were made from the historical data review:

- Large negative local inflows to Dinosaur Reservoir were observed during balanced spill ramp-up conditions at GMS and PCN. GMS tailwater levels were similar to PCN forebay levels prior to spill initiation. During spill ramp-up the differences between GMS tailwater level and PCN forebay level increased. This indicates additional storage development during spill ramp-up period and it was coincident with the observation of large negative local inflow.

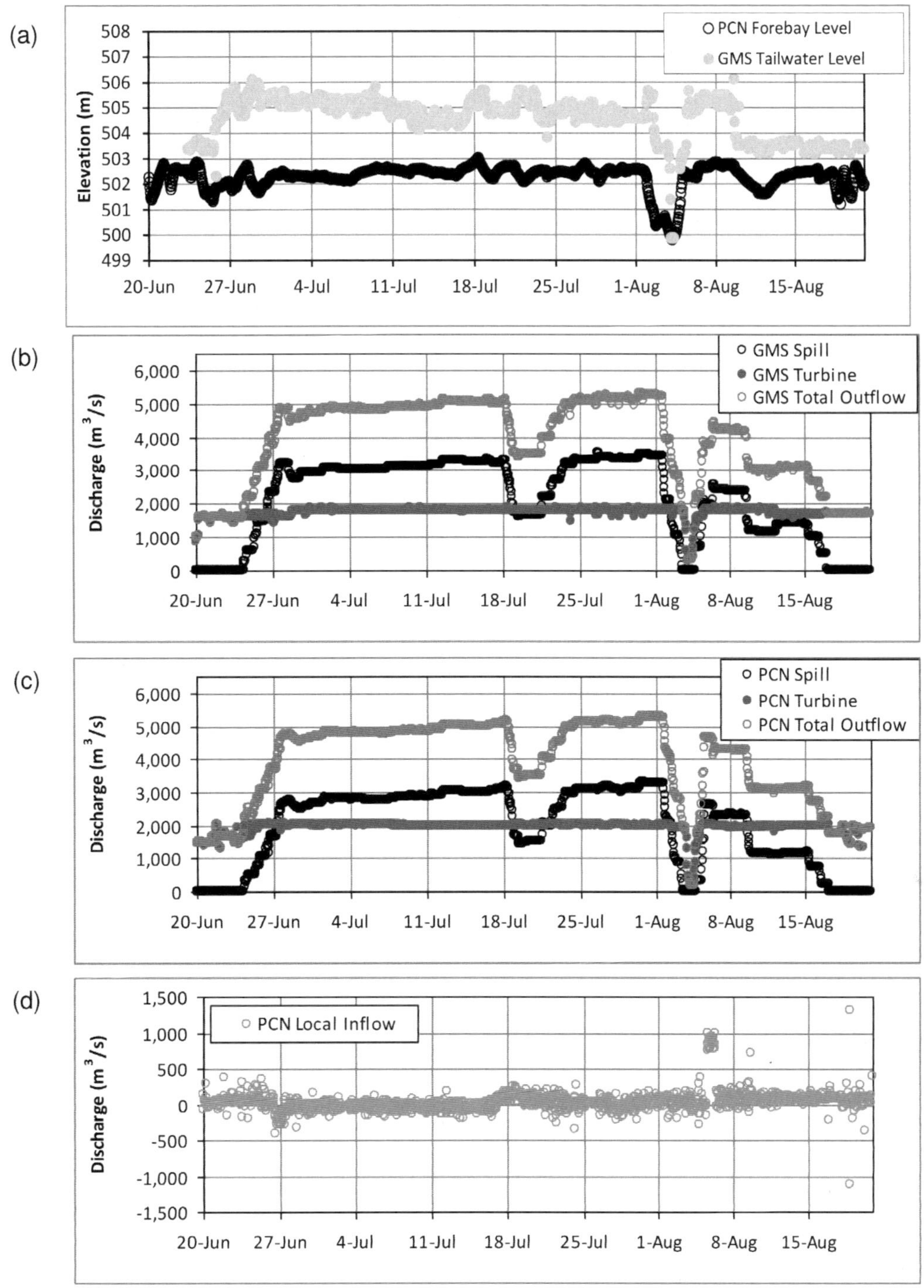

Figure 2: Review of historical data for 1996 spill event

- Large positive local inflows were calculated during spill ramp-down conditions when the difference between GMS tailwater level and PCN forebay level diminished. This indicates flushing of additional storage developed upstream of the reservoir.

- Considerably large negative and positive local inflows were also calculated for short durations during powerhouse flow ramp-up and ramp-down conditions, respectively.

- Consistently large negative inflows were observed during steady spill discharge conditions at GMS and PCN. Storage volume did not change much during steady spill conditions. Therefore, these negative inflows were caused by the inaccuracy of GMS and/or PCN discharges. Underestimation of PCN outflows and/or overestimation of GMS release could likely result in negative local inflows to Dinosaur Reservoir. Review of spillway rating curves in Section 3.2 shows that PCN spillway discharge could be underestimated, especially during partial gate opening.

4 HYDRODYNAMIC FLOW ROUTING FOR SPILL EVENETS

4.1 Hydrodynamic Model

The MIKE 11 model, a robust one-dimensional hydrodynamic model, was used in this study for hydrodynamic flow routing of Dinosaur Reservoir to confirm that this type of modelling can mitigate the negative inflows. GMS tailrace channel and Dinosaur Reservoir were represented in the MIKE 11 model with 34 cross-sections. Bathymetry for GMS tailrace channel (about 2 km reach) were obtained from BC Hydro survey data collected in October 2002. Bathymetry for the remaining 18.5 km reach of Dinosaur Reservoir was based 1984 B.C. Ministry of Environment maps. Topographic data were collected from B.C. Provincial TRIM (Terrain Resource Inventory Mapping) DEM (Digital Elevation Model).

The upstream boundary condition for the MIKE 11 model was the inflow to Dinosaur Reservoir (i.e. GMS releases from the powerhouse and spillway). The downstream boundary was set at PCN forebay with measured water levels during the period of simulation. No lateral inflows (i.e. local inflows) were assumed for model simulations.

Manning's coefficients for channel roughness was set in the model as 0.035 to represent deep channel through natural rocks with limited vegetation and obstructions (Chow 1959).

4.2 Model Accuracy

The MIKE 11 model representation of the Dinosaur Reservoir storage was checked by comparing the modelled volume with the CRO/Flocal storage curve between Maximum Normal Reservoir Level (El. 502.92 m) and El. 500 m. The modelled volumes were within 0.1% of the CRO storage curve volume, and therefore acceptable.

4.3 Modelling Spill Events

Hydrodynamic routing of Dinosaur Reservoir was performed for three spill events in 1996, 2002 and 2012. These results are compared with CRO/Flocal recorded quantities. Figure 3 show the time series plots of various parameters for the spill event in 1996.

Observations are provided below:

- Estimations of storage volume in Dinosaur Reservoir for normal operating conditions (i.e. before and after the spill period) are found to be generally similar for both reservoir routing (Flocal approach) and hydrodynamic routing (MIKE 11 model) methods. However, during spill discharges the hydrodynamic model estimated storage volumes are higher than the reservoir routing method.

- The difference in estimated storage volume between reservoir routing and hydrodynamic routing remains essentially unchanged for steady spill discharge conditions. The difference is found to grow during spill ramp-up and diminish during spill ramp-down.

- Modelled hourly outflows at PCN appear to be very similar to CRO/Flocal records of PCN outflow for the entire simulation period (i.e. with and without spill) for three events. But these apparently small differences between model outflow and CRO/Flocal records of PCN outflow are in very good agreement with CRO/Flocal estimated Dinosaur Reservoir local inflow. Local inflows to Dinosaur Reservoir were assumed to be zero for model simulations. Since local inflows are practically very small (in the order of 10 m^3/s), these differences are actually the errors in CRO/Flocal calculation.

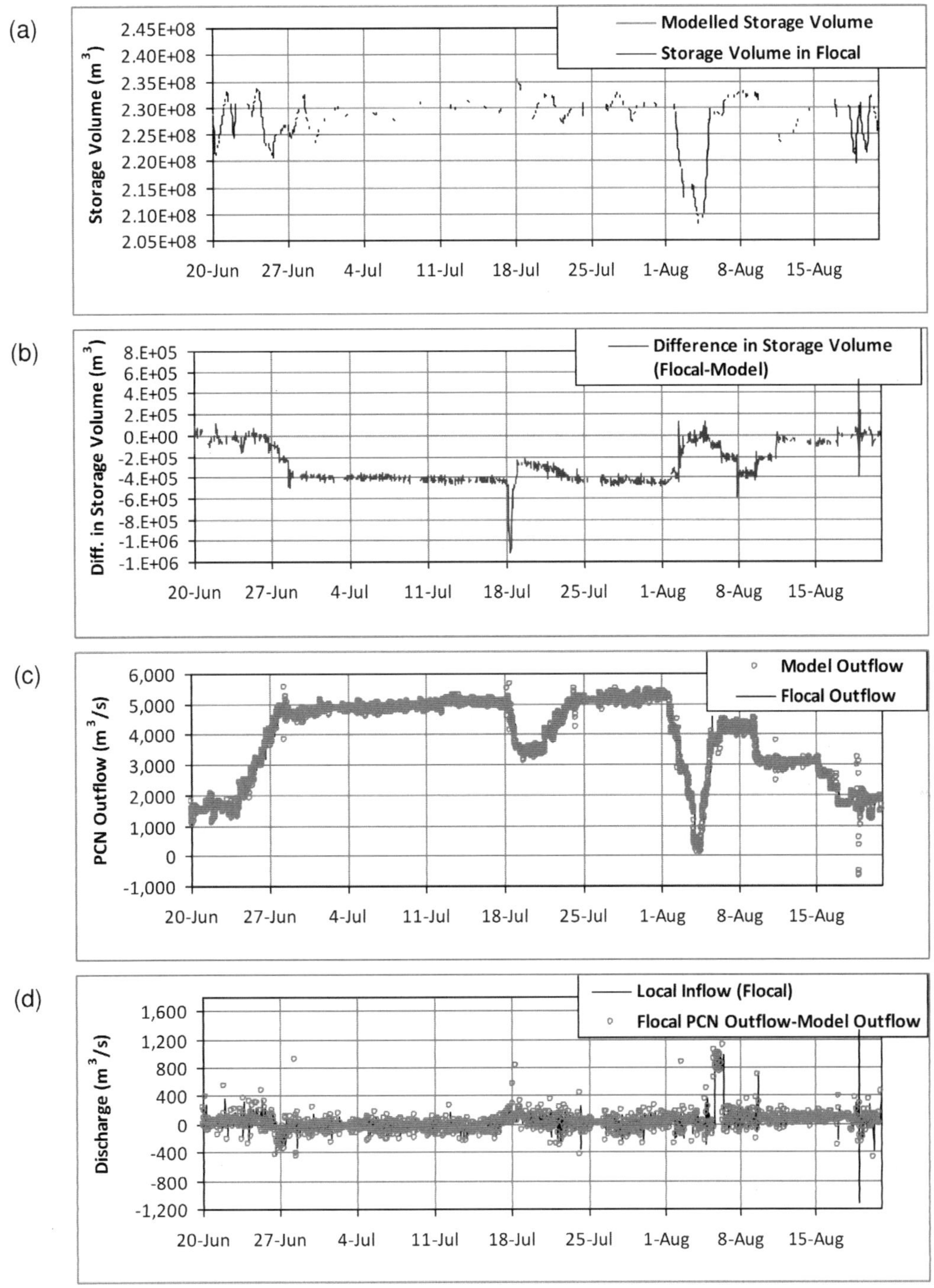

Figure 3: Comparison of Flocal calculations and model results for 1996 spill event

5 METHODS TO REDUCE ERRORS IN LOCAL INFLOW CALCULATION

Reverse routing can be performed using the shallow water wave equations (Chow 1959) in an iterative technique. At each time step the inflow is approximated and routed downstream. The simulated and measured downstream discharges are compared and the inflow is adjusted until the desired accuracy is obtained. It is recognized that the solution of the shallow water wave equations should provide the most accurate results, but this method is computationally inefficient and not suitable for operational record keeping. Therefore, a literature review was conducted to prescribe a suitable reverse routing method to reduce errors in local inflow calculation for a cascade reservoir.

The main features of flood wave motion are translation and attenuation, the latter due to (hydraulic) wave diffusion; both are nonlinear features resulting in nonlinear deformation. The main flood body travels as a kinematic wave. A channel routing method (e.g. the Muskingum scheme) approximates the propagation of the diffusion wave efficiently when the kinematic wave mode dominates (Koussis et al. 2012). Therefore, it can reduce large errors associated with rapid change of storage in a reservoir. The principle of this method is similar to reservoir routing, but it considers both prism storage (for horizontal water surface) and wedge storage (storage above a horizontal water surface) (Chaudhry 2008). In this method, the storage in a reach is expressed as a linear function of the weighted mean of inflow and outflow:

[4] $S=K[xI+(1-x)Q]$

where K and x are two numerical parameters.

An essential requirement to ensure accuracy in such finite difference calculations is that the finite time interval (Δt) must be small relative to K. An exact method of solution under the storage assumption was developed by Nash (1959). He derived modified equations for the Muskingum parameters. These equations are true even when Δt is not small relative to K.

Koussis et al. (2012) showed that the accuracy of a reverse Muskingum routing method could be as good as the reverse solution of the St. Venant equations of flood wave motion at a fraction of the computing effort compared to a hydrodynamic solution. However, they recommended low-pass filtering (symmetric, second order, five-point Savitzky-Golay filter) of the imperfect outflow hydrographs to control noise amplification and assemble true data from spurious information. For superior solution the results can be optimized by using objective function, e.g. absolute error.

The literature also shows that a non-linear implicit scheme was traditionally used by hydrologists for its unconditional stability, but its solution was derived by an iterative procedure (Zoppou 1999). This approach is computationally inefficient and the solution of $I(t)$ is dependent on $I(t-\Delta t)$, estimated at the previous time step and its associated errors.

There are some simplified approaches of reverse reservoir routing that can reduce errors in local inflow calculation for reservoirs between two dams. For example, the reverse level pool routing equation can be expressed in central difference scheme as,

[5] $I(t)=Q(t)+[S(t+\Delta t)-S(t-\Delta t)]/(2\Delta t)$

The advantage of this scheme over the implicit scheme is that the inflow $I(t)$ is not dependant on the inflow $I(t-\Delta t)$. Zoppou (1999) showed that this simple explicit scheme can eliminate random errors caused by small inaccuracies in the input data, and it is not necessary to employ computational effort in filtering the imperfect flow records obtained from the field.

Therefore, reverse reservoir routing using the Muskingum method can be adopted for local inflow calculation for a cascade reservoir. This will provide a practical balance between simplicity and accuracy for operational calculations. In addition, a central difference scheme can be used to eliminate random errors caused by small inaccuracies in the input data.

6 CONCLUSIONS

- This study has shown that there are multiple sources of errors that caused negative local inflows into Dinosaur Reservoir. The large errors are correlated to flow ramping conditions due to inaccuracies in storage volume estimation on the basis of flat reservoir level assumption. In addition, there is an indication that the PCN spillway rating curves are underestimating discharge. Some random errors in local inflow calculation are also caused by inaccuracy of hourly turbine discharge.

- The principle of level-pool reservoir routing is acceptable to calculate local inflows for reservoirs between two dams if the hydrodynamic effects are negligible (i.e. for relatively small flows through the reservoir). For higher flows between the dams, the reservoir water surface develops a gradient like a river. In such cases, reservoir routing will generate errors and hydrodynamic flow routing would be most appropriate.

- There are some simplified approaches for reverse flood routing that can reduce errors in local inflow calculation for reservoirs between two dams. A central difference scheme can eliminate random errors caused by small inaccuracies in the input data. Channel routing using Muskingum method can reduce large errors associated with the wedge storage for reservoir inflow calculation.

7 RECOMMENDATIONS

- This study recommends a simple channel routing approach (Muskingum method) to eliminate large errors (both positive and negative) in the estimated local inflows for a cascade reservoir between two dams. The use of an explicit central difference scheme can be considered to avoid the need for low pass filtering of the imperfect input data.

- Time-averaged values of forebay and tailwater levels should be used as input for operating discharge calculations.

- Turbine discharge calculations can be improved by using average forebay and tailwater levels measured for each generating state. When there are multiple units and the generating conditions change frequently, turbine discharges should be calculated at shorter intervals.

- A Computational Fluid Dynamics (CFD) model study is recommended to investigate the effects of the unaccounted issues on PCN spillway discharge. The spillway rating curves can be revised based on the CFD model results for future applications.

References

Chaudhry, M.H. 2008. *Open-Channel Flow.* 2nd ed., Springer, New York, NY, USA.

Chow, V.T. 1959. *Open-Channel Hydraulics.* 1st ed., McGraw Hill, New York, NY, USA.

Koussis, A.D., Mazi, K., Lykoudis, S. and Argiriou, A.A. 2012. Reverse Flood Routing with the Inverted Muskingum Storage Routing Scheme. *Natural Hazards and Earth System Sciences*, **12**: 217-227.

Nash, J.E. 1959. A Note on the Muskingum Flood-Routing Method. *Journal of Geophysical Research*, **64**(8): 1053-1056.

Zoppou, C. 1999. Reverse Routing of Flood Hydrographs using Level Pool Routing. *Journal of Hydrologic Engineering*, ASCE, **4**(2): 184-188.

Leadership in Sustainable Infrastructure

Leadership en Infrastructures Durables

Vancouver, Canada

May 31 – June 3, 2017/ *Mai 31 – Juin 3, 2017*

MESH-FREE TWO-PHASE MODELLING OF HIGHLY-DYNAMIC SEDIMENT TRANSPORT

Shakibaeinia, Ahmad[1,3], Ehsan Jafari-Nodoushan[2]
[1] Polytechnique Montreal, Canada
[2] Semnan University, Iran
[3] ahmad.shakibaeinia@polymtl.ca

Abstract: Conventional numerical methods for the simulation of sediment transport solve single-phase hydrodynamics and advection-diffusion equations in combination with the empirical relations. These simplified methods are not capable of dealing with the highly-dynamic sediment movements which often induced by highly-erosive flows (e.g., floods) or powerful outer forces (e.g., landslides). Two-phase methods have recently proven their potential for such sediment flow problems. Among them the multiphase mesh-free particle methods provide a unique opportunity for the simulation of large interfacial deformations and fragmentations involved in two-phase sediment-water system. This article briefly overviews example applications of a multiphase mesh-free particle model for two-phase flow of water and sediments in different configurations. The model treats the sediment material as a continuum (a visco-plastic fluid) whose behaviour is predicted using a stress-dependent rheological model. The model is validated and applied for highly-dynamic movement of sediments in cases such as submarine landslide and sediment erosion. The results of this study, evaluate the capabilities of mesh-free particle methods for complex sediment transport problems and provide a more thorough understanding of their complex mechanism and processes.

1 INTRODUCTION

Sediment transport has a crucial role in many natural and engineering processes such as fluvial/coastal morpho-dynamics and ecosystem. The foundations of contemporary sediment research can be found in many experimental and field studies, while with the advancement in computing powers, the numerical simulations have recently gained much attention. Traditional approaches for simulation of sediment transport, have mostly been based on the single-phase modelling, where the flow equations are coupled with advection-diffusion equation and (semi-) empirical relations (such as Meyer-Peter and Müller, 1948; Van Rijn, 1984). These methods, which are the base of most of the existing models, have proved their effectiveness in some of past researches. However, they are not suitable for highly-dynamic movement of sediment material as they are not able to describe the underlying physics in detail. Such conditions are commonly initiated by highly-erosive flows or under powerful outer forces (e.g., gravity). Sediment transport during flood, dam-break, ice-break, river bank failure, and landslide are few examples of the highly-dynamic sediment movements.

Two-phase models, where the sediment and water act as separate phases, have proven to be capable of these sediment transport simulations. Two-phase models are either use continuum-based mesh-based Eulerian methods such as Finite Element or Finite Volume, or discrete-based Lagrangian methods such as Discrete Element Method, DEM. The first class of methods, has problems in dealing with flows with large interfacial deformations (Shakibaeinia and Jin, 2011). The second class of methods are computationally expensive for real-case large-scale problems as they simulate each individual sediment grain. The mesh-free particle (Lagrangian) methods developed for continuum mechanics such as Smoothed Particle Hydrodynamics, SPH (Gingold and Monaghan, 1977) and Moving Particle Semi-implicit method, MPS (Koshizuka and Oka, 1996) have the advantage of both classes of methods. Therefore, they provide an opportunity for dealing the complexities involved in highly-dynamic sediment transport problems.

The present paper will overview the development and implementing multiphase mesh-free model for the simulation sediment transport problems in authors' recent works. The method treats the multiphase system as a multi-viscosity and multi-density continuum. An exponentially-regularized visco-plastic rheological model with the Mohr-Coulomb yield criteria is applied to predict the non-Newtonian behavior of the sediment phase. The model is applied and evaluated for the simulation of one flow-driven and one gravity-driven sediment dynamic test cases.

2 NUMERICAL MODEL

1.1 Flow Governing equations

Considering the multiphase system of water and sediments as a multi-density multi-viscosity system, a single set of flow governing equations will describe both phases. The flow governing equations for a weakly-compressible flow system, including mass and momentum conservation and the motion in Lagrangian coordinates are as follows:

$$[1]\quad \begin{cases} \dfrac{D\rho}{Dt} + \rho\nabla\cdot\mathbf{u} = 0 & \text{Mass conservation} \\ \dfrac{D\mathbf{u}}{Dt} = -\dfrac{1}{\rho}\nabla p + \dfrac{1}{\rho}\nabla\cdot\boldsymbol{\tau} + \mathbf{f} = 0 & \text{Momentum conservation} \\ \dfrac{D\mathbf{x}_p}{Dt} = \mathbf{u} & \text{Lagrangian motion} \end{cases}$$

where, $\mathbf{u}:(u,v)$ is velocity vector, t is time, ρ is density, p is pressure, $\mathbf{f}$ represents body forces, $\mathbf{x}:(x,y)$ is position vector and $\boldsymbol{\tau}$ is shear stress tensor.

1.2 Sediment rheology

The sediment martial is treated as a visco-plastic continuum (Figure 1). Here the $\mu(I)$ rheological model proposed by GdR MiDi (2004) and Jop et al. (2006) is used to describe the non-Newtonian behaviour of sediment continuum. In this model the shear stress tensor is related to the inertial number, I, and the normal stress (pressure) between sediment grains as:

$$[2]\quad \boldsymbol{\tau} = \frac{\mu(I)p}{\|\mathbf{E}\|}\mathbf{E} \;\; where\; \mu(I) = \mu_s + \frac{\mu_2 - \mu_s}{I_0/I + 1}; \quad I = \frac{\|\mathbf{E}\| d_s}{\sqrt{p/\rho_s}}; \quad \mathbf{E} = \frac{1}{2}\left(\nabla\mathbf{u} + (\nabla\mathbf{u})^T\right) \; ; \; \|\mathbf{E}\| = \sqrt{II_E} = \sqrt{\frac{1}{2}\mathbf{E}:\mathbf{E}}$$

where, $\mathbf{E}$ is strain rate tensor, $||\mathbf{E}||$ and II_E are its magnitude and second invariant, respectively, and μ_s and μ_2 are first and second friction coefficients (i.e. tanϕ, ϕ being friction angle).

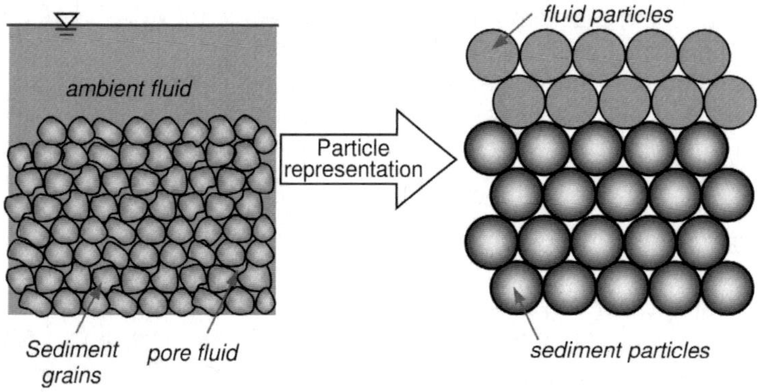

Figure 1:Particle representation of multiphase sediment continuum

1.3 MPS method

Here a weakly-compressible MPS method (Shakibaeinia and Jin 2010) is used for solving the governing equations. In MPS method, the continuum is represented by a set of mobile particles. Each particle possesses a set of field variables (e.g., velocity and pressure). MPS is based on a local weighted averaging of quantities, vectors and derivatives. The target particle i with position vector $\mathbf{x}_i$, interacts with particle j in its vicinity using a weight function, $W(x_{ij}, r_e)$, with $x_{ij}=|\mathbf{x}_j-\mathbf{x}_i|$ being the distance between particle i and j. Particles located within distance r_e from the target particle interact with that particle. A dimensionless parameter, the particle number density, n, represents the density of particles around a certain particle as:

$$[3] \quad \langle n \rangle_i = \sum_{j\neq i} W\left(x_{ij}, r_e\right)$$

The pressure gradient term (Shakibaeinia and Jin 2010) and the divergence-free viscous term for multiphase flow (Shakibaeinia and Jin, 2011a, 2012) are given by:

$$[4] \quad \langle \nabla p \rangle_i = \frac{d}{n_0} \sum_{j\neq i} \left(\frac{p_j - \hat{p}_i}{x_{ij}} \mathbf{e}_{ij} W\left(x_{ij}, r_e\right) \right); \quad \hat{p}_i = \min_{j\in J}(p_i, p_j)\ ;\ J = \left\{ j : W\left(x_{ij}, r_e\right) > 0 \right\}$$

$$[5] \quad \langle \nabla(\mu \nabla \cdot \mathbf{u}) \rangle_i = \frac{4d}{\lambda n_0} \sum_{j\neq i} \left(\frac{\mu_i \mu_j}{\mu_i + \mu_j} (\mathbf{u}_j - \mathbf{u}_i) W\left(r_{ij}, r_e\right) \right)$$

The MPS approximation of the strain rate tensor, **E**, can be written as:

$$[6] \quad \mathbf{E}_i = \begin{bmatrix} \dot{e}_{xx} & \dot{e}_{xy} \\ \dot{e}_{yx} & \dot{e}_{yy} \end{bmatrix}_i = \frac{1}{n_0} \begin{bmatrix} 2\sum_{j\neq i} \left(\frac{u_{ij} x_{ij}}{r_{ij}^2} W_{ij} \right) & \sum_{j\neq i} \left(\frac{u_{ij} y_{ij} + v_{ij} x_{ij}}{r_{ij}^2} W_{ij} \right) \\ \sum_{j\neq i} \left(\frac{u_{ij} y_{ij} + v_{ij} x_{ij}}{r_{ij}^2} W_{ij} \right) & 2\sum_{j\neq i} \left(\frac{v_{ij} y_{ij}}{r_{ij}^2} W_{ij} \right) \end{bmatrix}$$

where n_0 is the initial value of particle number density, d is the number of space dimensions and $\mathbf{e}_{ij}=\mathbf{x}_{ij}/x_{ij}$ is the unit direction vector. The viscous term, considers the multiphase force due to the discontinuity in the viscosity field. A fractional-step time integration methods is employed, where the velocity in a new time step is the sum of the predicted velocity and corrected velocity.

$$[7] \quad \mathbf{u}_i^{k+1} = \mathbf{u}_i^{*} + \mathbf{u}_i' \quad \textit{where} \quad \mathbf{u}_i^{*} = \mathbf{u}_i^{k} + \frac{\Delta t}{\rho_i}\left(\mathbf{f}_i + \mathbf{f}_{si} + \mu_{ij} \nabla^2 \mathbf{u}_i^{k} - (1-\alpha) \nabla p_i^{k} \right)\ ;\ \mathbf{u}' = -\frac{\alpha \Delta t}{\rho_i} \nabla p_i^{k+1}$$

where α is a relaxation factor varying between 0 and 1. Particles are moved based on their predicted velocity and therefore a new particle number density, n^*, is calculated. The n^* used to calculate the pressure value. The pressure gradient term is implicitly deployed to calculate the velocity correction. As in the WC-MPS (Shakibaeinia and Jin, 2010) the fluids are assumed to be weakly compressible and the Tait's equation of state is used for prediction of the pressure, as:

$$[8] \quad p_i^{n+1} = \frac{\rho_i c_0^2}{\gamma}\left(\left(\langle n^* \rangle_i / n_0\right)^\gamma - 1\right)$$

where c_0 is the numerical sound speed. The typical value of γ=7 is used in this relation. Since an explicit algorithm is used, the time intervals in each time step are selected in the way to satisfy the stability condition. To avoid the sudden pressure, jump across the interfaces created by the large density discontinuity, Shakibaeinia and Jin (2012) used the smoothed value of density for the pressure calculation. To deal with the large Reynolds number flow, the model uses the Sub-Particle-Scale (SPS) LES model. The inflow/outflow and solid boundaries are treated by the method proposed by Shakibaeinia and Jin (2010, 2011b).

3 TEST CASES

Model of this study can be applied to any multiphase sediment transport problem. Here we consider two benchmarks. The first benchmark mimics the submarine landslide or river bank failure. The second benchmark is study of sediment erosion resulted from a water jet. Figure 2 shows the schematic of the first benchmark. It is based on the experiments done by Rzadkiewicz et al (1997) on slide of a wedge of sand on a 45 degrees inclined surface under water. The computational domain is represented by particle with size of 0.015 m. Figure 3 show the snapshots of the numerical results in different times. In this figure two color sediment particles have been used to study the internal deformation of sediments. At t=0 s, the sediments that are bounded by a gate are released and start to slide down the surface. This creates a water wave that propagates downstream. Going down the slope, the wage of sediments stretches along the slope under friction with the solid surface. Eventually the sediments form a pile down the slope. The numerical results show the ability of the model in representing sediment and water surface deformations. Figure 4 compares (for an example time) the surface profile and sediment profile resulted from the model of this paper with those from experimental and numerical work of Rzadkiewicz et al (1997). It shows better compatibility of present model results with the experiments (bed profile Err.=0.041m) comparing the model of Rzadkiewicz et al (1997) (Err.=0.053m).

In the second benchmark the sediment erosion resulted from an un-submerged jet is studied. The experimental data are taken from Pagliara et al. (2006). Figure 5 shows the geometry of the problem and snapshots of the numerical results. The jets' momentum (and the resulting circulations) induces the erosion of the sediments and deposits them further downstream. This creats an scour hole, which developes over time. Figure 6 compares the numerical and exerimental bed profile and the evolution of the scour hole depth. The comparosons show a good compatibility.

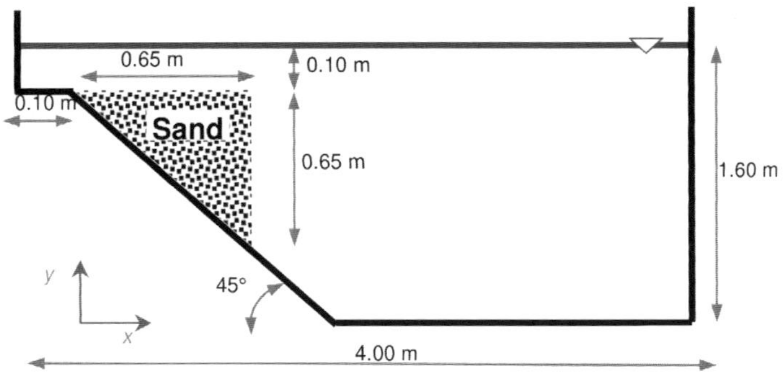

Figure 2: schematic of submarine landslide test case

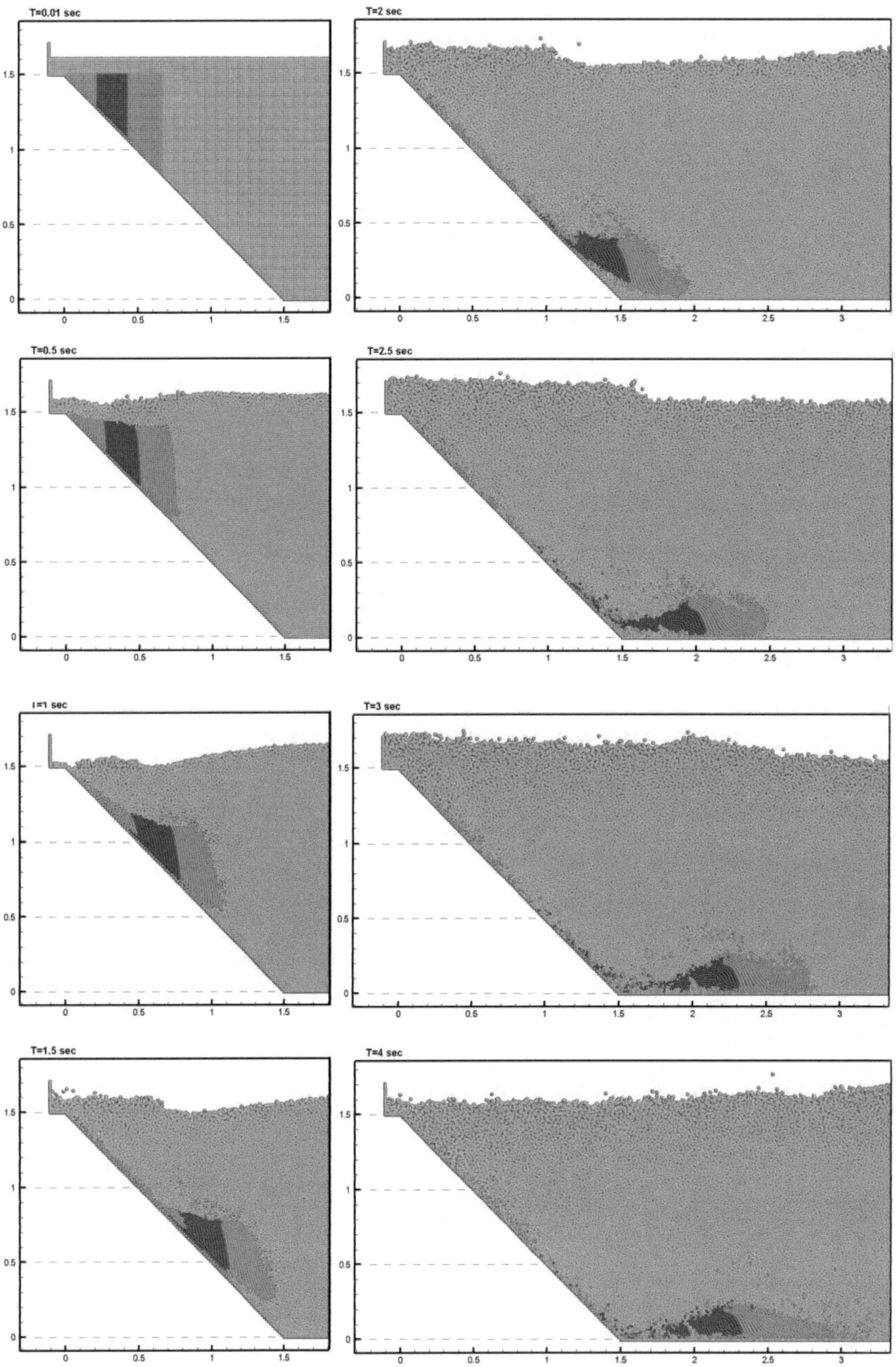

Figure 3:Time evolution of numerical submarine landslide

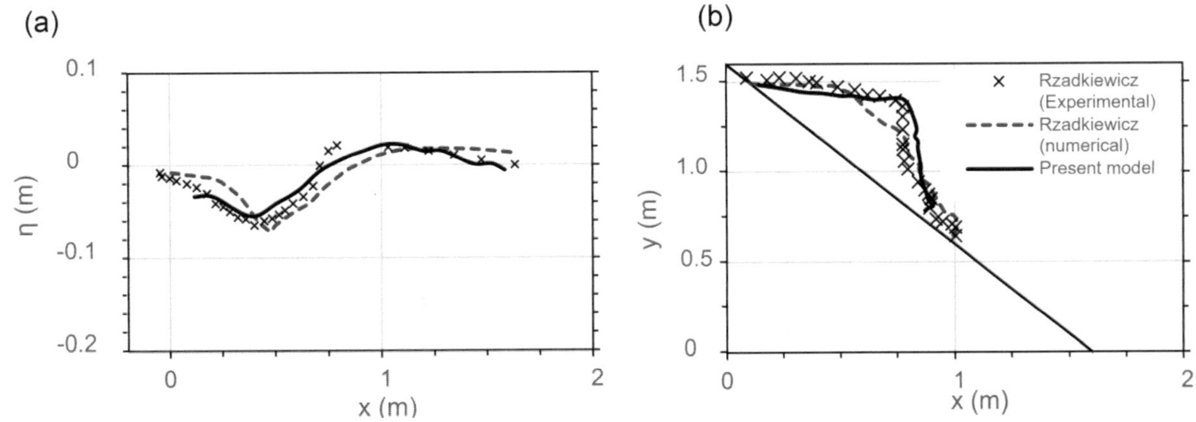

Figure 4: comparison of experimental and numerical (a) surface wave profile and (b) sediment profile at t=0.4 s.

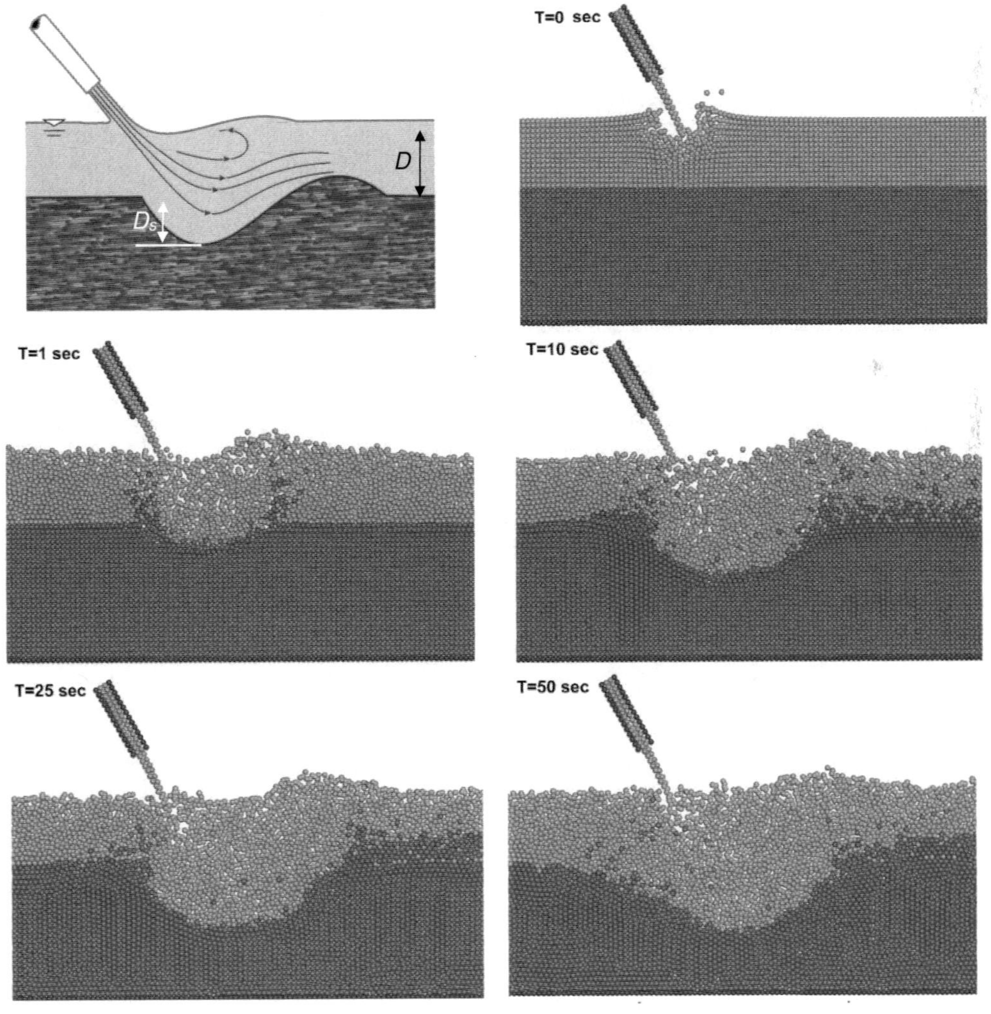

Figure 5:Geometry and time sequence of numerical results for the jet-erosion test case.

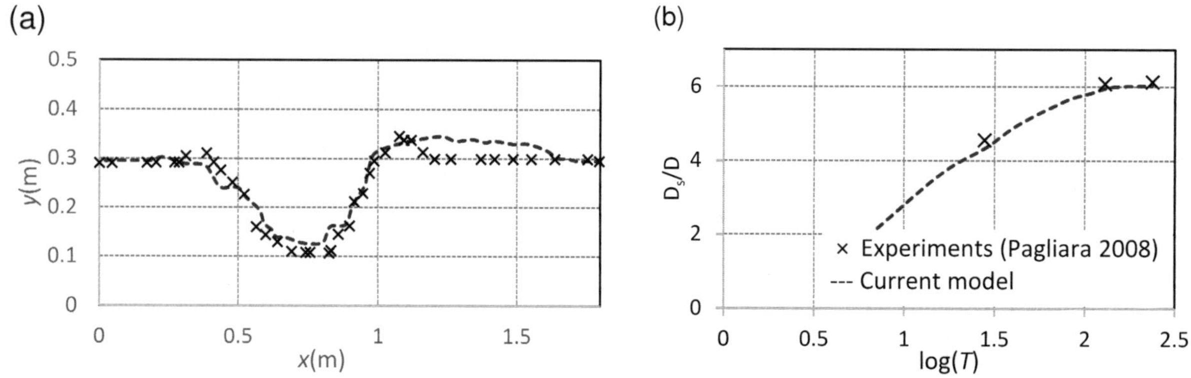

Figure 6: (a) final bed profile, and (b) time evolution of the maximum scour depth (dimensionless)

4 CONCLUSION

A multiphase mesh-free particle model base on the MPS mesh-free particle method is evaluated for two-phase flow of water and sediments in two example test cases. The sediment material is treated as a continuum (a visco-plastic fluid) whose behaviour is predicted using $\mu(I)$ rheological model. The model is validated and applied for highly dynamic movement submerged sediments in various regimes in cases such under-water landslide and erosion (resulted from a water jet). The results of this study, proved the capabilities of mesh-free particle methods for accurate simulation of the complex sediment transport problems for the test cases of this study. The model of this study can be applied to the other complex sediment dynamics problems such as sediment slump, mobile-bed dam-break, and sediment scouring.

References

Gingold, R. A., & Monaghan, J. J. (1977). Smoothed particle hydrodynamics: theory and application to non-spherical stars. *Monthly notices of the royal astronomical society*, *181*(3): 375-389.

Jop, P., Forterre, Y., & Pouliquen, O. (2006). A constitutive law for dense granular flows. *Nature*, *441*(7094): 727-730.

Koshizuka, S., & Oka, Y. (1996). Moving-particle semi-implicit method for fragmentation of incompressible fluid. *Nuclear science and engineering*, *123*(3): 421-434.

Meyer-Peter, E., & Müller, R. (1948, June). Formulas for bed-load transport. IAHR.

MiDi, G. D. R. (2004). On dense granular flows. *The European Physical Journal E*, *14*(4): 341-365.

Pagliara, S., Hager, W. H., & Minor, H. E. (2006). Hydraulics of plane plunge pool scour. *Journal of Hydraulic Engineering*, *132*(5): 450-461.

Rzadkiewicz, S. A., Mariotti, C., & Heinrich, P. (1997). Numerical simulation of submarine landslides and their hydraulic effects. *Journal of Waterway, Port, Coastal, and Ocean Engineering*, *123*(4): 149-157.

Shakibaeinia, A., & Jin, Y. C. (2011). A mesh-free particle model for simulation of mobile-bed dam break. *Advances in Water Resources*, *34*(6): 794-807.

Shakibaeinia, A., & Jin, Y. C. (2010). A weakly compressible MPS method for modeling of open-boundary free-surface flow. *International journal for numerical methods in fluids*, *63*(10): 1208-1232.

Shakibaeinia, A., & Jin, Y. C. (2012). Lagrangian multiphase modeling of sand discharge into still water. *Advances in Water Resources*: *48*, 55-67.

Van Rijn, L. C. (1984). Sediment transport, part I: bed load transport. *Journal of hydraulic engineering*, *110*(10): 1431-1456.

Leadership in Sustainable Infrastructure
Leadership en Infrastructures Durables

Vancouver, Canada
May 31 – June 3, 2017/ *Mai 31 – Juin 3, 2017*

MODELING OF HYDRAULIC AND WATER QUALITY PERFORMANCE OF BIORETENTIION CELLS

Huang, Jian[1], He, Jianxun[1,3], Chu, Angus[1] and Valeo, Caterina[2]
[1] University of Calgary, Canada
[2] University of Victoria, Canada
[3] jianhe@ucalgary.ca

Abstract: The use of Low Impact Development (LID) technologies for managing urban stormwater runoff has shown promise in reducing or eliminating the need of traditional stormwater infrastructure. Bioretention cells, also called rain gardens, are one type of LID that collects, stores and treats stormwater runoff on site. To date, few models are available for simulating the behavior of bioretention cells. Therefore, this paper proposed an event-based model for predicting hydraulic and water quality performance of bioretention cells. The model was developed using the data collected from a field-scale bioretention cell located in the City of Calgary, Alberta. A total of eight storm events were used in model calibration and validation. The performance of the model was assessed in terms of both hydraulic parameters including time to peak, peak flow, and volume reduction, and water quality parameter - the removal efficiency of total suspended solids (TSS). The results demonstrate that the proposed model is capable of capturing the temporal variation of hydraulic performance of bioretention and simulating outflow hydrographs with the coefficient of determination (R^2) above 0.77 and the normalized root-mean-square deviation (*NRMSD*) ranging from 8.5% to 15.6%. The percent errors between measured and modeled hydraulic and water quality parameters have a maximum value of 10.0%. As a result, the proposed model has great potential as a practical modeling tool for assessing the performance of bioretention cells and designing.

1 INTRODUCTION

Urbanization has led to an increase of stormwater runoff as natural land surfaces are being replaced by impervious surfaces such as roads, buildings, parking lots, and many other urban infrastructures. The continuous increase of stormwater runoff potentially challenges the capacity of existing stormwater infrastructures and jeopardizes urban water environment (Lee and Bang 2000, Muthanna et al. 2008). Bioretention cells, also called rain gardens, are one of Low Impact Development (LID) technology and offers a practical solution to manage increasing stormwater runoff. When surface runoff is routed to bioretention cells, they provide temporary detention and promote infiltration so that the stormwater runoff volume and peak can be potentially attenuated (Davis et al. 2009, Trowsdale and Simcock 2011). The existing body of knowledge demonstrates the effectiveness of bioretention in delaying the time to peak and reducing the runoff peak and volume (Davis 2008, Hunt et al. 2008, Khan et al. 2012a). Previous studies also demonstrate that bioretention cells can substantially reduce the concentrations of total suspended solids (TSS), nutrients and heavy metals. Their removal efficiency was found to be associated with several variables including inflow pollutant concentration, magnitude of the storm event, as well as environment temperature (Braga et al. 2007, Hatt et al. 2009, Hsieh and Davis 2005, Khan et al. 2012b).

In the literature, many previous studies have primarily focused on assessing bioretention performance through analyzing and interpreting field and laboratory observations. There are also a few studies that have utilized several modeling techniques (e.g., data-driven and physically-based models and commercial software) to simulate the hydraulic and/or water quality performance of bioretention cells. For instance, Brown et al. (2013) employed an agricultural drainage model to simulate the water storage of bioretention cells. Palhegyi (2009) developed a water balance model to estimate water storage and assist in sizing bioretention cells. Khan et al. (2013) developed a regression model to predict the outflow of bioretention. He and Davis (2010) applied Richard's equation to simulate flow through bioretention cell using a finite element model built on the platform of COMSOL Multiphysics. In addition, Akan (2013) adopted the Green Ampt equation to model water movement within a bioretention cell. To date, very few studies have been conducted to model water quality performance of bioretention cells. The water quality modeling has been particularly focused on TSS removal by a soil or gravel layer. Yao et al. (1971) proposed a physically-based model for a gravel media, which is actually a rapid sand filter, to simulate the major governing removal mechanisms including sedimentation, interception and diffusion. Wu (1994) developed a model to simulate TSS removal by a gravel media given a constant water pressure head. Wong et al. (2006) introduced a first order kinetic decay model to estimate the overall TSS removal by a gravel media. Most recently, Li and Davis (2008) utilized the classical macroscopic depth filtration model to simulate TSS removal by individual layers of a bioretention cell. No study has been found to apply the developed methodologies from filter units to model TSS removal by bioretention, regardless of the recognition of their similarity in TSS removal.

The majority of previous studies on hydraulic modeling for bioretention primarily addressed a particular variable such as runoff storage, flow through layers, or water balance; while the water quality models have mostly been developed for gravel or soil media individually instead for an entire bioretention cell, which often consists of both soil and gravel layers. Due to the popularity of bioretention, several municipalities in Canada have formulated design standards for bioretention cell, which use reduction of peak flow and runoff volume, delay in time to peak, and TSS removal as design criteria (e.g., Greater Vancouver Sewerage & Drainage District 2012, The City of Calgary 2011, Toronto and Region Conservation Authority 2010). Thus, there is a need to develop a modeling tool that can provide comprehensive information to aid in engineering design of bioretention. To fulfill the need, the objectives of this paper are to: 1) develop a modeling tool that can simulate both hydraulic and water quality performance of bioretention for assisting in its design, and 2) to validate the model by applying it to a bioretention cell, which was tested in a field study conducted in the City of Calgary, Canada.

2 MODEL DEVELOPMENT

A bioretention cell usually consists of multiple soil layers, which have various particle sizes for rooting of vegetation, and a gravel layer often with an under-drain lying underneath of the soil layers. The soil layers are designed with high porosity to promote infiltration; while the gravel layer is added not only to provide temporary storage for infiltrated water but also to support the upper soil layers and the under-drain (Roy-Poirier et al. 2010). Flow movement within, and sediment removal by the soil and gravel layers can be modeled based on the same theory considering their individual particle sizes and porosities. In this paper, the model developed for permeable pavements by Huang et al. (2016) was adopted and further revised to model the performance of bioretention cells, considering their similarity in structure and mechanisms governing flow and TSS removal. The conceptual scheme of the model is shown in Figure 1, while the flow chart of the model is illustrated in Figure 2. The model was briefly described below and detailed information about the model can be found in Huang et al. (2016). The model was developed for receiving both direct precipitation and runoff from surrounding area.

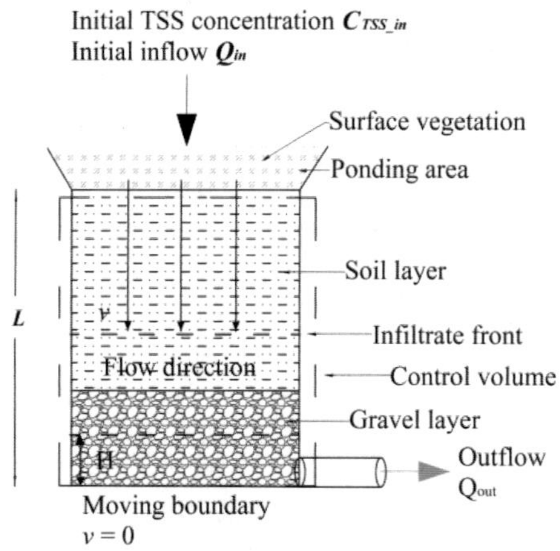

Figure 1. Conceptual scheme of the hydraulic model

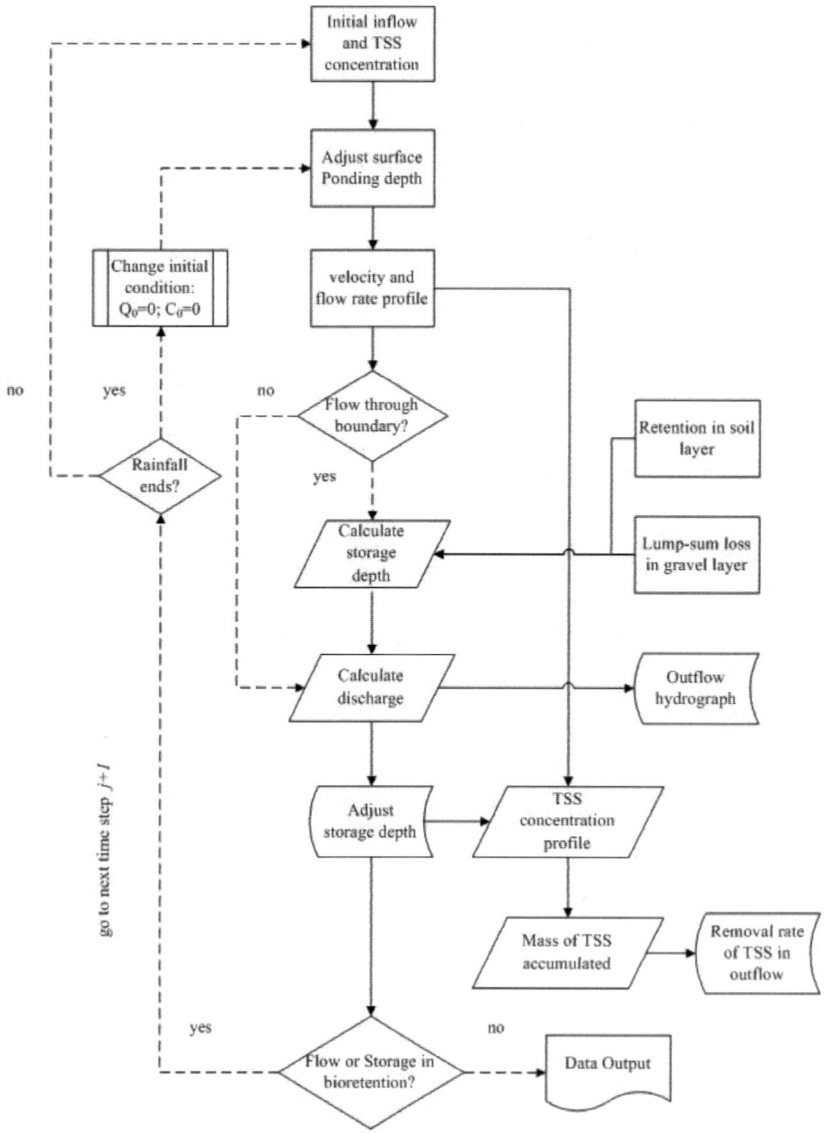

Figure 2. Flow chart of the model

2.1 Flow and sediment removal equations

The modified Kozeny-Carman equation (Eg. 1) is employed to simulate the infiltrated flow through a bioretention cell. The infiltrated flow, which moves downward, is assumed to be vertical and outflow is generated when the downward moving water front reaches the under-drain (Figure 1).

[1] $$\frac{\partial v}{\partial L} = \frac{1}{180}\frac{m^3}{(1-m)^2} \cdot (\varphi d)^2 \cdot \frac{g}{\mu L}$$

where v is the vertical flow velocity through the soil and gravel layers [m/s]; m is the porosity of soil or gravel layers; φ is the sphericity of particle, which ranges from 0 to 1 depending on the shape of particle; d is the particle diameter [m]; g is the gravitational acceleration [m/s^2]; μ is the kinematic viscosity of infiltrated water [m^2/s]; and L is the depth of bioretention cell [m].

Yao et al. (1971) developed a sediment removal model for sand filter media in water and wastewater treatment processes. This model is utilized to simulate the processes of sediment removal by the soil and gravel layers of bioretention cell.

[2] $$\frac{\partial C}{\partial L} = -\frac{3}{2}\frac{(1-m)}{d}\alpha\left[0.9\left(\frac{kT}{\mu d_p d v}\right)^{2/3} + \frac{3}{2}\left(\frac{d_p}{d}\right)^2 + \frac{(\rho_p-\rho)g {d_p}^2}{18\mu v}\right]C$$

where C is the concentration of TSS [mg/L]; and α represents contact efficiency, which is the number of contacts that succeed in producing adhesion divided by the number of collisions that occur between suspended particles and soil or gravel layer. The coefficient α ranges from 0 to 1; while ideally α equals to 1 if the suspended particles are completely mixed in water. k is Boltzmann constant (1.3805×10-23 J/K); T is absolute temperature of stormwater [K]; d_p and ρ_p are the diameter [m] and the density [kg/m^3] of particle; and ρ is the density of infiltrated water [kg/m^3].

2.2 Numerical approximation

The finite difference approximation is applied to solve Eqs. 1 and 2. The numerical calculation of v is given in details below, while the calculation of TSS removal can be found in Huang et al. (2016).

The velocity, v [m/s], at the time step j is expressed by:

[3] $$v_{i+1}^j = v_i^j + \frac{1}{180} \cdot \frac{{m_i}^3}{(1-m_i)^2} \cdot (\varphi d_i)^2 \cdot \frac{g}{\mu L}\Delta L$$

where the subscript i is the section number after discretization of the depth of bioretention cell L; and ΔL represents the distance between sections i and $i+1$.

During a storm event, a bioretention cell undergoes a "filling process" as stormwater runoff continuously enters the ponding area for temporary detention. At the same time, water in the ponding area starts to infiltrate the soil and/or gravel layers. When infiltrated water exceeds the storage limit of the cell, namely reaching the under-drain, excess water is discharged out of the cell. The storage depth, from the top of the cell to the boundary, where $v = 0$, moves either upward or downward (Figure 1). The surface ponding depth at time step j can be estimated based on the ponding depth at the previous time step $j-1$ and the change of ponding depth under the current time step j:

[4] $${D_{pond}}^j = {D_{pond}}^{j-1} + a \cdot D_{in}^j - D_{out}^j$$

where D_{pond} represents the ponding depth [m]; a is the water loss coefficient due to interception by vegetation on the top of the bioretention cell; D_{in} and D_{out}, which are the inflow and infiltrated water depths [m], respectively. D_{in} and D_{out} at time step j are calculated by:

[5] $${D_{in}}^j = {Q_{in}}^j \Delta t / A$$

[6] $${D_{out}}^j = {f_0}^j \Delta t$$

where A represents the surface area of the bioretention cell [m^2]; Δt represents the time interval between time step j and $j+1$ [s]; f_0 is the surface infiltration rate of the soil layer [m/s]; and Q_{in} is the inflow rate [m/s].

When infiltrated water flows through the cell, a portion of infiltrated water is retained by the soil layer. The amount of water retained is dependent on the field capacity and antecedent moisture level of the layer. The depth of water retained within the layer (D_{soil}) is expressed by:

[7] $$D_{soil}{}^{j} = \begin{cases} \frac{v_b^j \cdot \Delta t \cdot L_{soil}}{m_{soil} \cdot L}, & \sum_1^j D_{soil} < \text{cap} \cdot L_{soil} \cdot \theta_{initial} \\ 0, & \sum_1^j D_{soil} \geq \text{cap} \cdot L_{soil} \cdot \theta_{initial} \end{cases}$$

where v_b represents flow velocity near the bottom [m/s]; L_{soil} is the depth of the soil layer [m]; m_{soil}, cap and $\theta_{initial}$ denote the porosity [m^3/m^3], the filled capacity [m^3/m^3] and the antecedent moisture level of the soil layer, respectively. Eq. 7 implies that infiltrated water is retained in the soil layer before its field capacity is reached.

Since gravel merely retains water due to its physical nature, a water loss coefficient b is used to take account of the lump-sum loss when water moves through the gravel layer (D_{gravel}).

[8] $$D_{gravel}{}^{j} = \frac{b \cdot v_b^j \cdot \Delta t \cdot L_{gravel}}{m_{gravel} \cdot L}$$

where L_{gravel} and m_{gravel} are the depth [m] and the porosity [m^3/m^3] of the gravel layer.

Then the storage depth (H [m]) in the bioretention cell at time step j can be estimated based on the storage depth at the previous time step $j-1$ and the change of water storage (in depth) under the current time step j:

[9] $$H^j = H^{j-1} + h_{in}^j - h_{out}^j - D_{soil}^j - D_{gravel}^j$$

[10] $$h_{in}^j = {v_b^j \Delta t}/{m_{gravel}}$$

[11] $$h_{out}^j = {Q_{out}^j \Delta t}/{A \cdot m_{gravel}}$$

[12] $$Q_{out}^j = \delta A_{pipe} \sqrt{2gH^{j-1}}$$

where h_{in} and h_{out} are the infiltrated water depth [m] that is ready for discharge and the water depth [m] discharged from the under-drain pipe, respectively; Q_{out} represents the outflow rate from the sub-drain pipe [m^3/s]; δ and A_{pipe} are the transient loss coefficient and the cross-sectional area of the under-drain pipe (0.008 m^2 in the paper), respectively for calculating Q_{out}.

3 SITE DESCRIPTION AND DATA COLLECTION

The pilot-scale bioretention cell is located in Currie Barracks, a residential area in the southwest of Calgary. As the cell was constructed for testing purposes only, it does not receive any stormwater runoff from its adjacent areas. The cell is 32 m^2 in surface area (8.0 m long and 4.0 m wide). Vegetation sits on the mulched surface layer which is 75 mm deep. Major types of vegetation include *Salix bebbiana*, *prunus pensylvanica* and *potentilla fruticosa*. This layer along with the surface ponding area can provide a temporary detention up to 200 mm. The cell consists of a 300 mm upper rooting zone, a 600 mm deep rooting zone, and a 300 mm gravel layer from the top to the bottom. The upper and deep rooting zones have similar soil types (70% of sand and 30 % of silt and clay) and particle characteristics. Thus, these two zones were treated together in the model and referred to as the "soil layer". The gravel layer consists of 40 mm drainage rocks and a 100 mm diameter under-drain leading to an adjacent monitoring manhole. The underlying soil is fairly impermeable as it has a measured infiltration rate of 0.58 mm/hr. In addition, woven-geotextile is placed between the soil and gravel layers to separate and stabilize the two layers. A layer of non-woven geotextile was installed to prevent pollutant migration to the underlying soil. Thus, water loss due to infiltration into the underlying soil was considered negligible.

Bioretention cells are considered as a part of major drainage system, which should provide a service level for 100-year storm event (The City of Calgary 2011). The hydraulic and water quality data collected in eight 100-year storm events, which were simulated considering the impervious to pervious ratio of 4. The simulations were conducted from August 2008 to July 2009 and were used to calibrate and validate the developed model. In each simulated event, street sediments were added to stormwater withdrawn from a nearby stormwater pond to simulate the stormwater runoff from an urban catchment with various TSS levels. Outflow from the under-drain was continuously monitored and soil moisture was measured at several depths to capture soil moisture levels prior to each event. Table 1 summarizes inflow characteristics including intensity (i) and duration (t) of each storm event, event mean TSS concentration ($C_{TSS\text{-}in}$), antecedent soil moisture of the soil layer ($\theta_{initial}$), and water temperature (T). Table 1 also shows the outflow characteristics including peak flow (Q_p), time to peak (t_p), volume (V_{out}), volume reduction (VR), modeled event mean TSS concentration ($C_{TSS\text{-}out}$), and TSS removal efficiency (R_{TSS}).

Table 1. Inflow and outflow characteristics observed in each simulated storm event

Test No.	Test date	Inflow characteristics					Outflow characteristics					
		i (mm/hr)	t (min)	$C_{TSS\text{-}in}$ (mg/L)	$\theta_{initial}$	T (°C)	Q_p (L/s)	t_p (min)	V_{out} (L)	VR (%)	$C_{TSS\text{-}out}$ (mg/L)	R_{TSS} (%)
1	06/08/2008	134.29	21	33	0.307	26.4	0.039	96	52	99.31	13	99.73
2	15/08/2008	80.58	36	290	0.328	25.4	0.105	70	239	96.91	15	99.84
3	27/08/2008	96.08	31	292	0.308	24.1	0.270	116	575	92.76	15	99.63
4	07/10/2008	90.77	34	295	0.410	11.6	0.166	42	396	95.19	20	99.67
5	24/10/2008	95.16	32	111	0.430	14.3	0.232	50	728	91.03	8	99.35
6	09/06/2009	119.46	27	206	0.315	18.3	0.290	92	897	89.57	20	98.99
7	08/07/2009	92.59	34	135	0.385	21.1	0.188	54	403	95.20	7	99.75
8	22/07/2009	81.69	36	128	0.334	19.5	0.180	74	477	93.92	19	99.10

4 MODEL CALIBRATION AND VALIDATION

4.1 Model calibration

In the model, several parameters require estimation. Some of these parameters can be obtained from field measurements, whereas others are determined through model calibration due to a lack of information and/or difficulty in measuring in the field. These parameters are selected or determined as following:

- Δt and ΔL are selected to ensure the stability of numerical results.
- The kinematic water viscosity is determined according to measured T of inflow.
- φ is set as 0.25 and 0.1 for soil particle and gravel, respectively, according to their shape.
- Based on inflow particle size distribution (PSD) (Khan et al. 2012b), d_p of 120 μm is used.
- From lab measurements, ρ_p is approximately 1,090 kg/m^3.
- δ is in the range of [0 1] and is determined through model calibration.
- a and b are initially assumed to be 0.05 and 0.08, respectively, and their further calibration may be needed in model calibration.
- cap is estimated to be 0.275 m^3/m^3 (Khan et al. 2012a).
- m_{soil}, which is not a constant, is determined as a function of the operation time of the cell in the model calibration as it is expected to vary with time due to the deposition of solids in the layer; whereas m_{gravel} is treated as a constant (0.4) due to the absence of obvious external loading (e.g., traffic), which could lead to the degradation of m_{gravel} (Huang et al. 2016).

The eight simulated storm events (Table 1) were divided into two groups: Events 1, 3, 5, 6 and 8 for the model calibration and Events 2, 4 and 7 for the model validation. Before the model calibration, sensitivity analyses were performed to identify the acceptable ranges of the calibration parameters. The model was then calibrated manually using the trial-and-error approach. In the model calibration, the parameters were determined to yield acceptable percent errors (e.g., less than 10%) in a variety of modeled hydraulic and water quality variables. The model performance was also evaluated using two statistical metrics: determination coefficient (R^2) and normalized root-mean-square deviation (*NRMSD*).

The simulated and measured outflow hydrographs are presented in Figure 3 for Events 3 and 6 as examples. In the model calibration, R^2 ranges from 0.84 to 0.93; while *NRMSD* is in the range between 9.65% and 14.08%. These results indicate a good agreement between the measured and simulated flows in all calibration events. Table 2 presents the percent errors between measured and simulated hydraulic and water quality variables for each calibration event. As shown in Table 2, the parameters were calibrated to ensure that the errors are less than 10%. The calibrated δ is 0.68, and the values of a and b remain at 0.05 and 0.08, respectively.

The degradation of m_{soil} was identified in the model calibration. Furthermore, m_{soil} can be described as a function of the operation time (t_o) since Event 1. The relationship between m_{soil} and t_o developed based on the model calibration was expressed as Eq. 13. The calibrated relationship was then applied in the model validation to determine m_{soil} of each validation event.

[13] $$m_{soil} = -0.006t_o + 34.79 \qquad (R^2 = 0.98)$$

where t_o is in the unit of day.

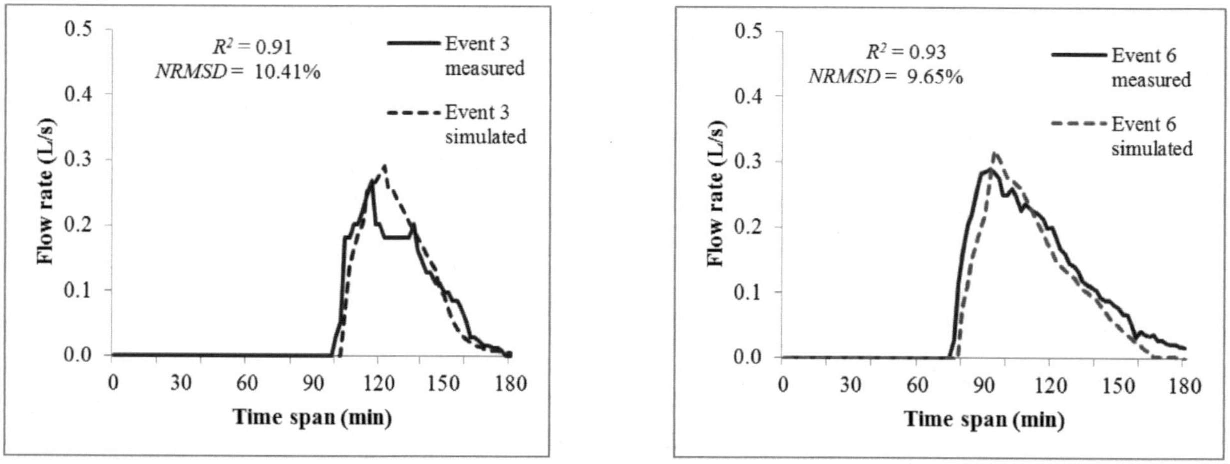

Figure 3. Measured and simulated outflow hydrographs of Events 3 and 6 in the model calibration

Table 2. Percent errors between measured and simulated variables in the model calibration

Event	Parameter	Error (%)	Event	Parameter	Error (%)	Event	Parameter	Error (%)
	Q_p (L/s)	5.13		Q_p (L/s)	7.78		Q_p (L/s)	6.03
	t_p (min)	6.25		t_p (min)	5.17		t_p (min)	8.00
1	V_{out} (L)	1.92	3	V_{out} (L)	4.52	5	V_{out} (L)	-8.79
	VR (%)	-0.01		VR (%)	-0.35		VR (%)	0.87
	R_{TSS} (%)	-2.90		R_{TSS} (%)	-6.34		R_{TSS} (%)	-3.84
	Q_p (L/s)	10.00		Q_p (L/s)	10.00			
	t_p (min)	2.17		t_p (min)	8.11			
6	V_{out} (L)	-9.70	8	V_{out} (L)	-9.64			
	VR (%)	1.13		VR (%)	0.62			
	R_{TSS} (%)	-6.28		R_{TSS} (%)	-3.10			

4.2 Model validation and discussion

The m_{soil} of Events 2, 4 and 7 is estimated to be 34.7%, 34.4% and 32.8%, respectively, based on the calibrated relationship between m_{soil} and t_o (Eq. 13). Figure 4 presents the measured and simulated outflow hydrographs of Event 4 as an example in the model validation. The percent errors between measured and simulated hydraulic and water quality variables are shown in Table 3. The calculated errors of all hydraulic and water quality variables are all less than 10%. In the model validation, R^2 ranges from 0.77 to 0.95, while *NRMSD* varies in the range between 8.48% and 15.64%.

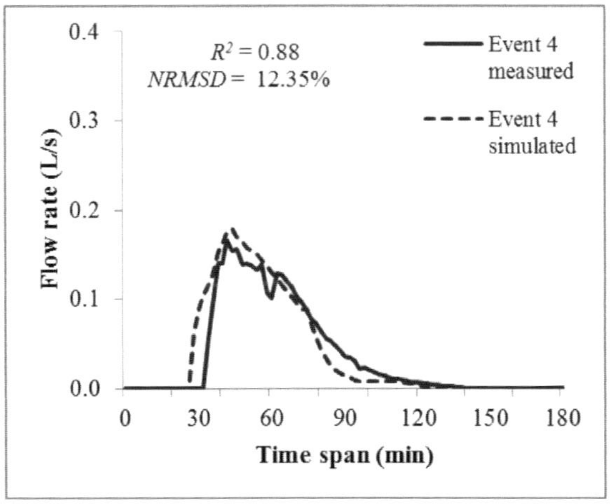

Figure 4. Measured and simulated outflow hydrographs of Event 4 in the model validation

Table 3. Percent errors between measured and simulated variables in the model validation

Parameter	Error (%) in Event 2	Error (%) in Event 4	Error (%) in Event 7
Q_p (L/s)	-4.76	7.83	5.85
t_p (min)	8.57	9.52	-9.26
V_{out} (L)	-8.37	6.06	-5.96
VR (%)	0.27	-0.31	0.30
R_{TSS} (%)	-5.48	-6.74	-4.05

The model validation shows that the calibrated model can simulate each validation event and is especially capable of capturing the temporal variation of the hydraulic performance of the bioretention cell. As shown in the outflow characteristics in Table 1, the hydraulic performance of the cell, in general, tends to degrade with time, as Q_p and V_{out}, and consequently VR increase whereas t_p decreases with time. The results might be ascribed to the decrease of m_{soil} with t_o as illustrated in the model calibration results shown in Figure 5. However, the recovery of the hydraulic performance was observed in the last two events, Events 7 and 8. The results from both the model calibration and validation clearly show that the proposed model effectively predicted the variation of hydraulic variables, which can be ascribed to the effects of the combination of the porosity of the soil layer and the antecedent moisture levels of each simulated event.

For a bioretention cell, physical properties such as porosity and antecedent moisture level, are not stationary over time. The developed model successfully captured the dynamics of hydraulic performance of the cell. This result strongly suggests that the model is applicable for modeling long-term hydraulic performance of bioretention cells. The model, however, currently applies the model calibration process to establish the influence of the porosity of the soil layer on operation time. In addition, the field measurements provide information on the moisture level prior to an event. In order to make the model more generalized, a model is required to capture the temporal variation of the porosity of the soil layer and the antecedent soil moisture levels and combine it with the developed model. The model was developed based on field observations in 100-year storm events, thus, model verification under different magnitude events is recommended. Furthermore, several municipalities define the level of service for TSS at a specified particle size range; therefore, further model improvements to account for PSD is also recommended.

5 CONCLUSION

This paper developed a physically-based model for bioretention cells. The results proved its applicability by applying it to simulate both hydraulic and water quality (TSS removal) performance of a bioretention cell in Calgary. More importantly, the proposed model successfully captured the temporal variation, especially of the hydraulic performance, which is characterized by peak flow, time to peak, and volume reduction. Thus the model is of practical value in aiding in the engineering design of bioretention. In order to generalize the model for wider application, further research is recommended on several issues, such as modeling temporal variation of porosity and antecedent moisture levels, further verifying for other magnitude storm events, and taking in account PSD when modeling TSS removal.

ACKNOWLEDGEMENT

The authors would like to thank Mitacs Elevate program and Westhoff Engineering Resources Inc. for the financial support. The authors also would like to acknowledge the contribution of Dr. Usman Khan (York University) during his MSc at the University of Calgary in data collection.

REFERENCES

Akan, A.O. 2013. Preliminary design aid for bioretention filters. *Journal of Hydrologic Engineering*, **18**(3): 318-323.

Braga, A., Horst, M. and Traver, R. G. 2007. Temperature effects on the infiltration rate though an infiltration basin BMP. *Journal of Irrigation and Drainage Engineering*, **133**(6), 593-601.

Brown, R.A., Skaggs, R.W. and Hunt, W.F. 2013. Calibration and validation of DRAINMOD to model bioretention hydrology. *Journal of Hydrology*, **486**: 430-442.

Davis, A.P. 2008. Field performance of bioretention: Hydrology impacts. *Journal of Hydrologic Engineering*, **13**(2): 90-95.

Davis, A.P., Hunt, W.F., Traver, R.G. and Clar, M. 2009. Bioretention technology: an overview of current practices and future needs. *Journal of Environmental Engineering*, **135**(3): 109-117.

Greater Vancouver Sewerage & Drainage District. 2012. Stormwater source control design guidelines. Vancouver (Canada.)

Hatt, B.E., Fletcher, T.D. and Deletic, A. 2009. Hydrologic and pollutant removal performance of biofiltration systems at the field scale. *Journal of Hydrology*, **365**(3-4): 310-321.

He, Z. and Davis, A.P. 2010. Process modeling of storm-water flow in a bioretention cell. *Journal of Irrigation and Drainage Engineering*, **137**(3): 121-131.

Hsieh, C.H. and Davis, A.P. 2005. Evaluation and optimization of bioretention media for treatment of urban storm water runoff. *Journal of Environmental Engineering*, **131**(11): 1521-1531.

Huang, J., He, J., Valeo, C. and Chu, A. 2016. Temporal evolution modeling of hydraulic and water quality performance of permeable pavements. *Journal of Hydrology*, **533**: 15-27.

Hunt, W.F., Smith, J.T., Jadlocki, S.J., Hathaway, J.M. and Eubanks, P.R. 2008. Pollutant removal and peak flow mitigation by a bioretention cell in urban Charlotte, N. C. *Journal of Environmental Engineering*, **134**(5): 403-408.

Khan, U.T., Valeo, C., Chu, A. and van Duin, B. 2012a. Bioretention cell efficacy in cold climates: Part 1—hydrologic performance. *Canadian Journal of Civil Engineering*, **39**(11): 1210-1221.

Khan, U.T., Valeo, C., Chu, A. and van Duin, B. 2012b. Bioretention cell efficacy in cold climates: Part 2—water quality performance. *Canadian Journal of Civil Engineering*, **39**(11): 1222-1233.

Khan, U.T., Valeo, C., Chu, A. and He, J. 2013. A Data Driven Approach to Bioretention Cell Performance: Prediction and Design. *Water*, **5**(1): 13-28.

Lee, J.H. and Bang, K.W. 2000. Characterization of urban stormwater runoff. *Water Research*, **34**(6): 1773-1780.

Li, H. and Davis, A.P. 2008. Urban particle capture in bioretention media. II: Theory and model development. *Journal of Environmental Engineering*, **134**(6): 419-432.

Muthanna, T.M., Viklander, M. and Thorolfsson, S.T. 2008. Seasonal climatic effects on the hydrology of a rain garden. *Hydrological Processes*, **22**(11): 1640-1649.

Palhegyi, G.E. 2009. Modeling and sizing bioretention using flow duration control. *Journal of Hydrologic Engineering*, **15**(6): 417-425.

Roy-Poirier, A., Champagne, P. and Filion, Y. 2010. Review of bioretention system research and design: past, present and future. *Journal of Environmental Engineering*, **136**(9): 878-889.

The City of Calgary. 2011. Stormwater management and design manual. Wastewater and Drainage.

Toronto and Region Conservation Authority. 2010. Low impact development stormwater management planning and design guide. Toronto (Canada.)

Trowsdale, S.A. and Simcock, R. 2011. Urban stormwater treatment using bioretention. *Journal of Hydrology*, **397**(3): 167-174.

Wong, T.H., Fletcher, T.D., Duncan, H.P. and Jenkins, G.A. (2006). Modelling urban stormwater - a unified approach. *Ecological Engineering*, **27**(1): 58-70.

Wu, Y. 1994. An analysis of constant-pressure filtration. *Chemical Engineering Science*, **49**(6): 831-836.

Yao, K.M., Habibian, M.T. and O'Melia, C.R. 1971. Water and waste water filtration: concepts and applications. *Environmental Science and Technology*, **5**(11): 1105-1112.

Leadership in Sustainable Infrastructure
Leadership en Infrastructures Durables

Vancouver, Canada
May 31 – June 3, 2017/ *Mai 31 – Juin 3, 2017*

MODELLING ICE DYNAMICS IN THE UPPER SAGUENAY FJORD

A. Cornett [1,5], D. Watson [2], H. Babaei [3], M. Sayed [4]
[1, 2, 3, 4] National Research Council, Ottawa, Ontario, Canada
[5] Andrew.Cornett@nrc-cnrc.gc.ca

ABSTRACT

A numerical model of ice dynamics developed at the National Research Council of Canada has been set-up and applied to predict the dynamics of the ice cover in the upper Saguenay Fjord subject to forcing by ground level winds and near surface water currents. The model has been used to predict the evolution of ice pressure, ice concentration and ice thickness at sites being considered for development of new port facilities. Predictions for eleven different hypothetical scenarios were produced, each corresponding to a different combination of wind speed and direction, tide, freshwater inflow, initial ice thickness, and initial ice concentration. The scenarios include both extreme conditions with very low probability of occurrence, as well as more common conditions. The set-up and application of the ice dynamics model as well as the sensitivity of the predicted ice cover dynamics to variations in initial ice concentration, initial ice thickness, wind direction and wind speed is described and discussed.

Keywords: Saguenay Fjord, ice dynamics, ice pressure, ice thickness, ice concentration, numerical model

1. INTRODUCTION

The Saguenay Fjord extends for 110 km between the communities of Chicoutimi, where the fjord transitions to the Saguenay River, and Tadoussac, where the fjord meets the St Lawrence Estuary, see Figure 1. The Saguenay Fjord is notable for its complex and strongly stratified flows, with freshwater at the surface flowing over denser seawater at lower depths. During winter a stable ice cover forms over the upper part of the fjord, and commercial ships accessing port facilities in Chicoutimi and Port Alfred are escorted by ice breaking vessels operated by the Canadian Coast Guard. Ice breaking vessels are normally deployed before mid-March to initiate the Spring break-up and assist with ice clearing operations.

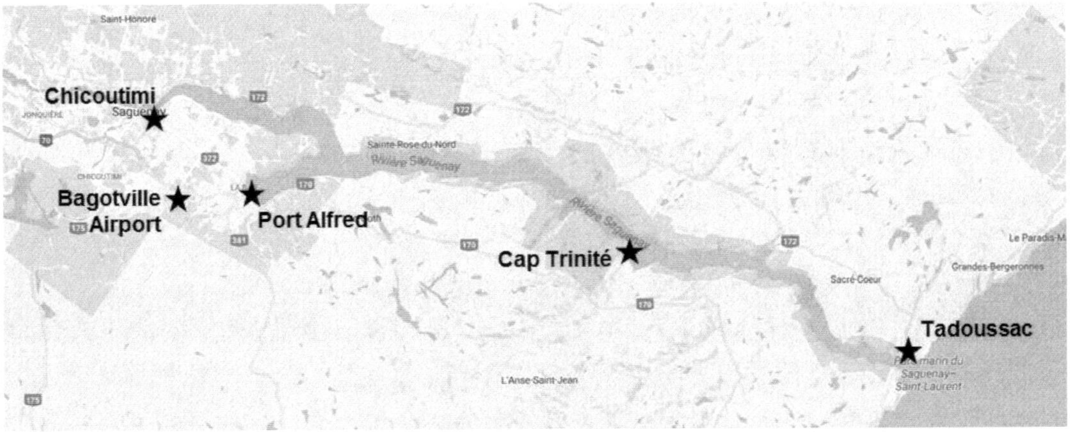

Figure 1. Map of the Saguenay Fjord region (source: Google).

A numerical model of ice dynamics previously developed at the National Research Council Canada has been setup and applied to support the planning of future port facilities in the upper part of the fjord. The model predicts the deformation of an ice cover over time and space due to forcing by wind and surface current. The wind and current forcing can be steady and uniform or unsteady and spatially varying. Initial conditions are specified by prescribing the initial ice concentration and initial ice thickness throughout the computational domain. Since the horizontal length scales of interest are much larger than the ice thickness, a depth-averaged formulation is employed, whereby the stresses and velocities are averaged over ice thickness. A parameterization of ice thickness redistribution is introduced in order to relate ridge (and ice rubble) formation and lead opening to the strain rates. Thus, the numerical solution keeps track of the evolution of deformed ice thickness and ridging. Further details on the ice dynamics model are given in Sayed *et al.* (2002) and Kubat *et al.* (2011, 2015).

In this study the ice dynamics model was used to predict the temporal and spatial evolution of ice concentration, ice thickness and ice pressure throughout the region for various hypothetical scenarios, each comprised of an initial ice condition combined with different wind and current forcings. The current forcing was obtained from a 3D numerical model of stratified hydrodynamics in the fjord, while the wind forcing was either derived from analysis of historical wind measurements or obtained from a numerical wind model.

Review and Characterization of Ice Conditions

A brief review of available information on ice conditions in the upper Saguenay Fjord was undertaken to characterize the range of local ice conditions and inform the ice dynamics modelling. Useful information was found in a report by Arctec Canada Ltd. (ACL, 1980) and in Urgales *et al.* (2002), Bourgault *et al.* (2011), and De Vernal *et al.* (2011).

Landsat satellite images for the period of 1972-2016, accessed through U.S. Geological Survey (USGS) online repository, were reviewed. Two examples showing the dramatic difference in the extent of the ice cover before and after the spring break-up are presented in Figure 2. The narrow shipping routes cleared by ice breaking vessels leading to/from two existing port facilities can be seen in Figure 2a.

Ice observation charts published by the Canadian Ice Service (CIS) were also obtained and reviewed. Two examples showing conditions in mid-March are shown in Figure 3. Various sub-regions are defined, each designated by a unique letter (A, B, etc.), and the concentration, stage of development (age) and form (floe size) of the ice in each sub-region is specified through a corresponding Egg code. Further guidance on interpreting ice charts can be found on the Canadian Ice Service website.

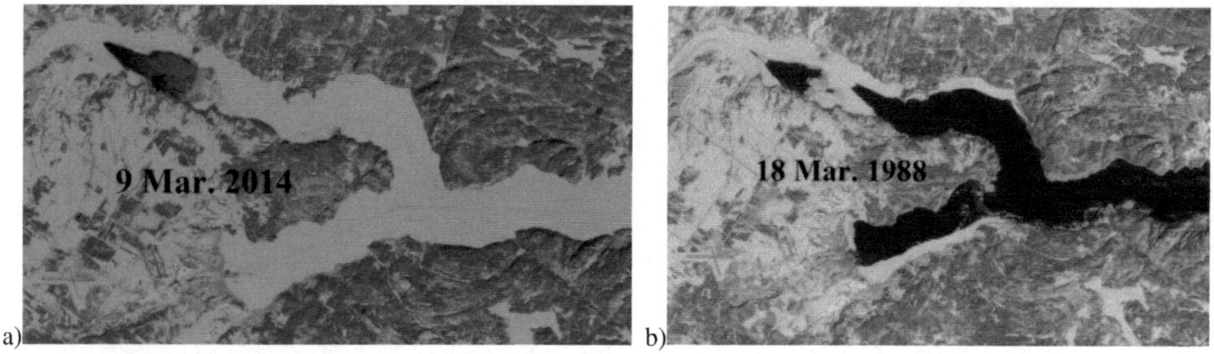

Figure 2. Landsat images showing typical ice cover (a) before and (b) after Spring Breakup.

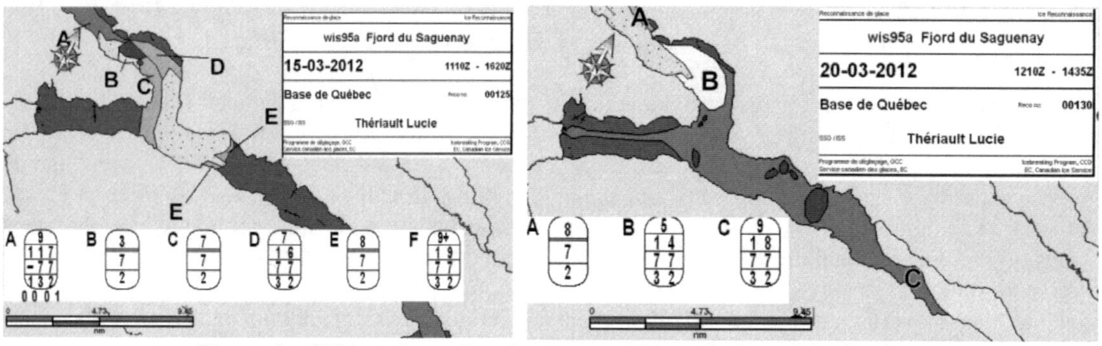

Figure 3. CIS ice charts showing ice conditions during mid-March.

Data on ice thickness acquired under the Ice Thickness Collection Program run from 1947-2002 by Environment Canada was also obtained and analyzed. Two hundred nine historical ice thickness measurements from three sites in the upper fjord are plotted in Figure 4a, while Figure 4b shows two exceedance probability curves derived from the data. Considering only observed seasonal maximums, the median thickness is 48 cm while the maximum is 70 cm. It is noted that thicker ice may well have occurred over the period but was not measured.

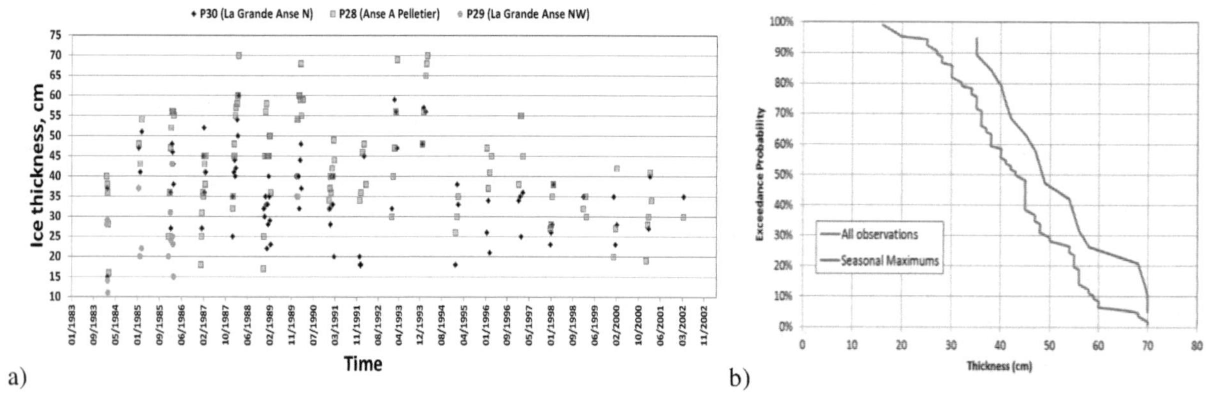

Figure 4. Ice thickness data measured in the upper fjord from 1984 – 2002.

Following Gosink *et al.* (1986) thermodynamic thickness growth of ice sheets can be estimated by:

[1] $h = 3.5\times10^{-2}\times CFDD^{0.5}\times\alpha$

in which h and $CFDD$ denote ice thickness in meters and cumulative freezing degree days in degrees Celsius (°C). The coefficient α, a positive number less than 1, depends mainly on local wind and snow conditions. Although this equation was originally developed to estimate ice thickness growth for lakes, it is thought to be applicable to streams when current speeds are smaller than approximately 0.6 m/s, which is the case for a large part of the upper Saguenay Fjord. Using some of the observed ice thickness values in Figure 4 for calibration yields a best fit value of $\alpha = 0.503$. Consideration of temperature records recorded at the Bagotville airport over the period from 1950 to 2015 yields a peak seasonal value of *CFDD* = 1,833 °C (for the 1958-1959 season). With this peak *CFDD* and assuming $\alpha = 0.503$ in Equation 1, a maximum ice thickness of 0.75 m can be estimated over the 65 year period.

Hydrodynamic Conditions

Detailed information on near surface currents throughout the upper fjord was required to force the ice dynamics model. As noted previously, the Saguenay Fjord is notable for its complex and strongly stratified flows, with freshwater at the surface flowing over denser seawater at lower depths. Moreover, the hydrodynamics are sensitive to the tide range and phase (ebb or flood) as well as the freshwater discharge entering the upper fjord from the Saguenay River.

The tides throughout the fjord are semi-diurnal, rising and falling twice daily. The tide range at Tadoussac, located near the mouth of the fjord, varies from ~2 m during Neap tides up to ~6 m during Spring tides. Typical tide ranges at Port Alfred, located in the upper fjord, are slightly higher, fluctuating from ~2.5 m up to ~6.5 m.

Data on the Saguenay River discharge at Shipshaw for the period 1972 - 2012 was obtained and analyzed. Based on this data, the average freshwater inflow is 1,310 m^3/s, while the inflow exceeded 1% of the time is 6,493 m^3/s. The inflow during the month of March when the ice-cover breaks up was also investigated. The mean inflow during March was found to be 1,244 m^3/s, while the maximum recorded inflow during March is 1,907 m^3/s.

A three-dimensional hydrodynamics model based on the Telemac-3D solver was set up to simulate the stratified hydrodynamics throughout the fjord. Open water conditions (without ice cover) were assumed for all of the hydrodynamic simulations. Further information on the development, calibration and application of the 3D hydrodynamics model together with a discussion of the simulation results is given in Cornett *et al.* (2017). Figure 5 shows the modelled distribution of near-surface current speed and direction across the upper fjord for one time step.

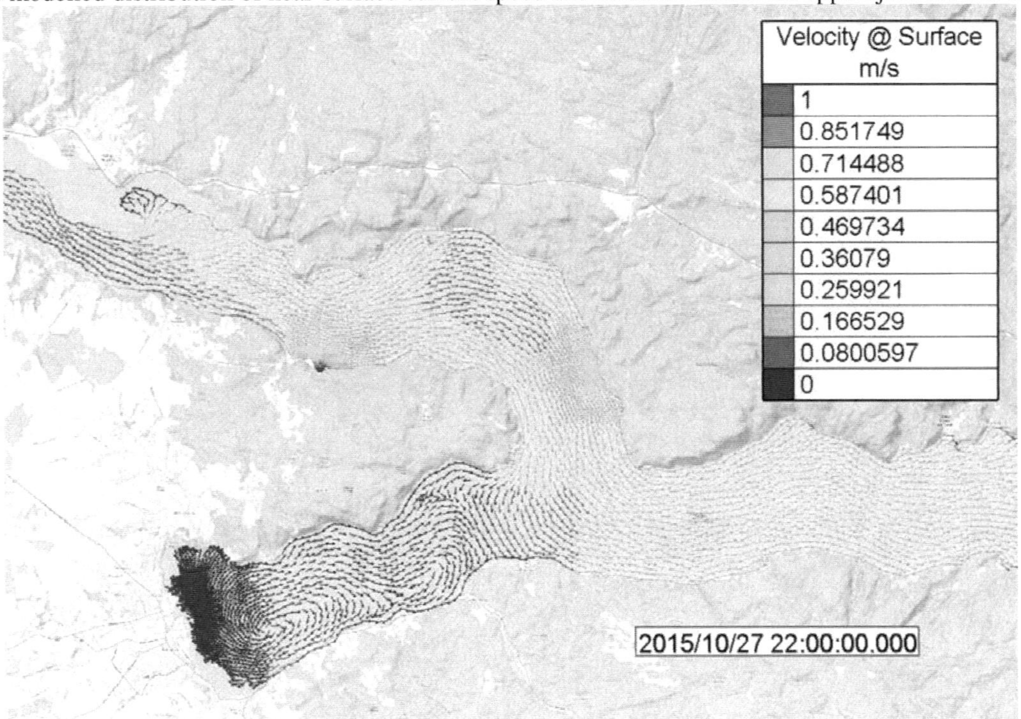

Figure 5. Modelled distribution of near surface current speed and direction for one time step.

Wind Conditions

Information on ground level winds in the region was also required to force the ice dynamics model. Two approaches were used to obtain this information: a) review and analysis of historical wind data recorded at Bagotville airport and several other sites in the region; and b) numerical simulation of wind fields.

Summary wind statistics derived from the Bagotville airport data are presented in Table 1. The two most common wind directions are west followed by east, while the strongest winds tend to approach from the west, north-west and east directions.

The Anemoscope wind simulation model, described in NRC (2006), was used to estimate the spatial distribution of ground level wind (speed and direction) for 16 extreme wind scenarios or atmospheric states. Anemoscope embodies the WEST wind mapping system described in Pinard *et al.* (2005) and Yu *et al.* (2006). Results for one such simulation are presented in Figure 6, for the case of an extreme wind approaching from ESE.

Table 1. Summary wind statistics (occurrence frequency in %) for Bagotville Airport (source: Environment Canada).

		Direction								
		NE	**E**	**SE**	**S**	**SW**	**W**	**NW**	**N**	**All**
Speed in km/hr	**< 1**									8.9
	1-10	1.6	8.4	4.7	3.2	3.9	7.0	2.6	1.3	32.6
	11-20	0.5	10.2	1.4	0.6	2.7	14.2	4.1	0.4	34.2
	21-30	0.1	5.3	0.4	0.1	0.9	7.7	2.5	0.1	17.1
	31-40	0.0	1.5	0.1	0.0	0.2	3.0	0.9	0.0	5.6
	41+	0.0	0.2	0.0	0.0	0.0	1.1	0.2	0.0	1.5
	All	2.2	25.6	6.6	3.9	7.6	32.9	10.4	1.8	100.0
	Avg. speed	9.0	15.7	9.0	7.6	12.1	18.6	17.9	8.8	14.2

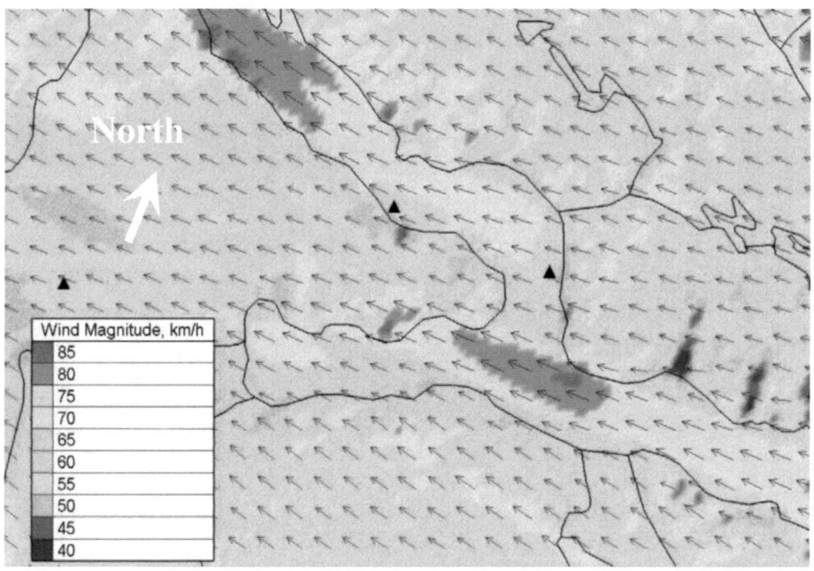

Figure 6. Modelled distribution of near ground wind speed and direction, extreme wind from ESE.

2. SCENARIO DEFINITION

An ice dynamics model previously developed at NRC was applied to investigate ice dynamics in the upper fjord for eleven hypothetical scenarios. The environmental and initial conditions for each scenario are summarized below in Table 2. All eleven scenarios assume a Spring (large) tide and begin when the initial water level in the upper fjord is high. Hence, the water level falls during the first 6 hours of each simulation, it then rises between the 6th and 12th hours. The narrow ice-free shipping corridors evident in Figure 2a were included in the initial condition for all scenarios.

Scenarios 1 to 4 have a high freshwater inflow of 1,907 m^3/s with an initial ice thickness of 70 cm and an initial ice concentration of 95%. The 1,907 m^3/s inflow is the maximum freshwater inflow recorded during the month of March. Scenario 1 has no wind (zero speed) while Scenarios 2 to 4 use spatially varying wind fields calculated by AnemoScope (example in Figure 6). It is worth noting that the extreme winds were sustained and held constant for the full duration of each ice dynamics simulation, typically six hours or more. While it is acknowledged that this represents an unrealistic scenario, the effect of extreme winds acting for shorter durations can be observed by examining intermediary simulation outputs (i.e. outputs after one hour).

Scenarios 5 to 11 all assume a freshwater inflow of 1,244 m^3/s, corresponding to the average freshwater inflow to the upper fjord during March, when the Spring break-up occurs. Initial ice thickness is either 43 cm (the average ice thickness reported in measurements taken over 18 seasons) or 70 cm (close to the thickest ice observed over 18 seasons). Initial ice concentration is either 95%, representing a full and virtually complete ice cover – i.e. stable conditions before the start of Spring break-up, or 70%, representing a situation with ice floes of various sizes and 30% open water. At the start of each simulation, both the ice thickness and ice concentration were assumed to be evenly distributed across the non-land portion of the computational domain, except for the narrow shipping corridors, where open water was assumed.

The initial distribution of ice thickness and ice concentration for Scenario 6 are shown in Figure 7. In this figure and other similar ones to follow, six different colours are used to denote ice concentration over the range from < 1/10 (low) to > 9/10 (high). Similarly, ice thickness variations over the range from < 20 cm to > 100 cm are denoted using six different colours. In both figures, warm colours denote more adverse ice conditions (thicker ice or higher concentrations) while colder colours denote more benign ice conditions. Conventional black wind barb symbols are used to denote the wind speed and direction, while small black vectors are used to indicate the water current speed and direction.

The 260° (from the west) wind direction modelled in Scenarios 5 to 8 is approximately shore normal to the eastern shore (labelled in Figure 7). Two wind speeds were modelled for this direction: 19 km/hr and 35 km/hr. The 19 km/hr wind corresponds to the average speed of the westerly winds (based on observed wind statistics at Bagotville), while the 35 km/hr speed is a more extreme condition that occurs less frequently. Again, it is worth noting that the winds were held constant for the full duration of each simulation.

The 22.5° (NNE) wind direction modelled in Scenarios 9 to 11 is approximately shore normal to the southern shore (labelled in Figure 7). Two wind speeds are modelled for this direction: 9 km/hr and 25 km/hr. The 9 km/hr speed is the average speed of the winds from NNE (based on observed wind statistics at Bagotville), while the 25 km/hr speed is a more extreme condition that occurs less frequently. These winds were sustained for the full duration of each simulation.

Table 2. Summary of initial ice conditions and environmental conditions for Scenarios 1 – 11.

Scenario	Tide range	Initial water level	Freshwater inflow (cms)	Current field	Wind speed @ Bagotville (km/hr)	Wind direction (deg N)	Wind field	Initial ice concentration (%)	Intial ice thickness (cm)
1	Spring	High	1907	Spatially and time varying	0	-	Spatially varying	95	70
2	Spring	High	1907	Spatially and time varying	80 [1]	262 (W) [1]	Spatially varying	95	70
3	Spring	High	1907	Spatially and time varying	53 [2]	30 (NNE) [2]	Spatially varying	95	70
4	Spring	High	1907	Spatially and time varying	61 [3]	90 (E) [3]	Spatially varying	95	70
5	Spring	High	1244	Spatially and time varying	19	260 (W)	Spatially uniform	95	43
6	Spring	High	1244	Spatially and time varying	35	260 (W)	Spatially uniform	95	43
7	Spring	High	1244	Spatially and time varying	35	260 (W)	Spatially uniform	95	70
8	Spring	High	1244	Spatially and time varying	35	260 (W)	Spatially uniform	70	43
9	Spring	High	1244	Spatially and time varying	9	22.5 (NNE)	Spatially uniform	95	43
10	Spring	High	1244	Spatially and time varying	25	22.5 (NNE)	Spatially uniform	95	43
11	Spring	High	1244	Spatially and time varying	25	22.5 (NNE)	Spatially uniform	70	43

1 Wind field derived from Climate State ANU1D293C14M (AnemoScope Simulation)
2 Wind field derived from Climate State ANU1D068C14M (AnemoScope Simulation)
3 Wind field derived from Climate State ANU1D135C14M (AnemoScope Simulation)

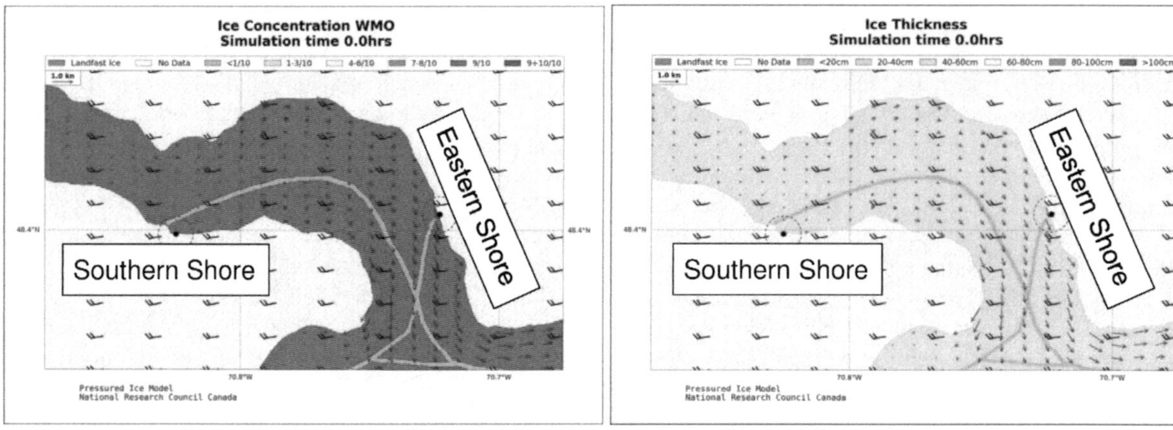

Figure 7. Initial distribution of ice concentration and ice thickness for Scenario 6.

3. ICE DYNAMICS MODEL SET UP

A rectangular computational grid extending 15 km in the east-west direction by 9 km in the north-south direction was set up to cover the region of interest, see Figure 8. The grid cell size chosen was 25 meters which is required to properly model the ice-free corridors that are roughly 125 m wide. The final grid is composed of 600 cells in the east-west direction and 360 cells in the south-north direction.

The ice dynamics model was forced using both ground-level wind and surface water current prescribed across the computational domain. A simple approach with no active coupling between the hydrodynamics and ice dynamics was followed in this study. The influence of the hydrodynamics on the ice cover was modelled, but the effect of the ice cover on the hydrodynamics was neglected. Similarly, the influence of the wind on the ice cover was modelled, but the effect of the ice cover on the wind (expected to be small) was neglected.

Numerical predictions of surface water currents that varied over both space and time were extracted from the output files of 3D hydrodynamic simulations that had been pre-computed (example in Figure 5). The water current was specified on an unstructured triangular grid, and linear interpolation was used to obtain the water current speed and direction for each cell of the ice dynamics grid at every time step. It is acknowledged that open water conditions (without ice cover) were assumed in the hydrodynamic simulations, hence the influence of the ice cover on the hydrodynamics was neglected. There is no doubt that this simplification introduces some uncertainly into the final results.

Only steady wind fields (that do not vary with time) were considered in this study, and in all cases the wind was assumed to remain steady throughout the ice dynamics simulation. For the cases where spatially varying ground-level wind fields were computed using the Anemoscope wind model, linear interpolation was used to obtain wind speed and direction at each cell of the ice dynamics grid. For other cases where spatially uniform ground-level winds were assumed, a uniform wind speed and direction was applied at each grid cell.

Test simulations were carried out to ensure that the model set-up could appropriately simulate the ice dynamics over the area of interest. The grid extent, grid cell size and simulation time step were all varied until stable simulation outputs were achieved. The final computational time step for the ice dynamics model was 5 s, the hydrodynamic forcing was updated every 30 minutes, and model outputs were stored every 15 minutes.

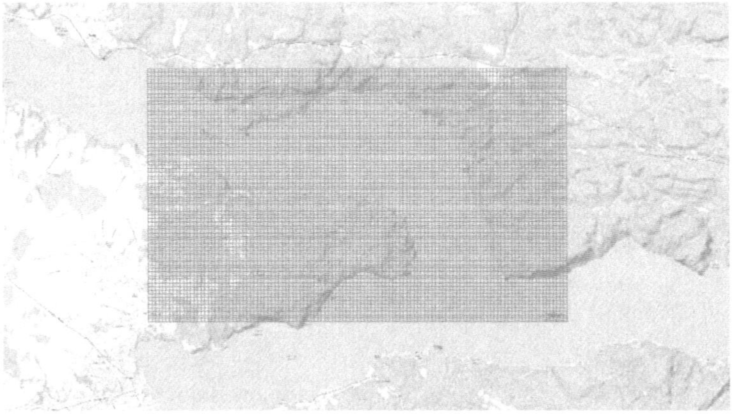

Figure 8. 15 km x 9 km ice dynamics model grid with 25 m x 25 m resolution.

4. RESULTS AND DISCUSSION

Each simulation predicts the evolution of the distributions of ice thickness, ice concentration and the stresses (or pressure) within the ice cover for the prescribed initial conditions and environmental forcing. We note that the pressure values resulting from the present simulations correspond to compaction and deformation processes that take place along length scales of 100s of meters. While they are indicative of the severity of ice conditions that may be manifested in ridging, rafting and pile-up of ice, they should not be used to estimate ice forces on a wharf or a vessel. The ice pressure can be characterized in terms of relative severity as *no pressure* (< 6 kN/m), *light* (6 to12 kN/m), *moderate* (12 to 24 kN/m), *strong* (24 to 40 kN/m) or *very strong* (> 40 kN/m), where ice pressure is expressed as a force per unit width. These pressure values are based on experience with the effect of pressure levels on the ability of vessels to navigate through an ice cover (Kubat *et al.*, 2015).

We present here in some detail the results for two scenarios. One of these (Scenario 6) produces pronounced ice action against the eastern shore, where construction of a new wharf is being considered. The other (Scenario 10) produces ice action against the southern shore where port facilities are currently located.

The initial evolution of ice pressure over the first hour of Scenario 6 is illustrated in Figure 9. In this scenario an ice cover with an initial uniform thickness of 43 cm, and uniform concentration of 95% is subject to a 35 km/hr wind approaching from west. The hydrodynamic conditions correspond to a falling Spring tide during March. For these conditions, the ice dynamics are dominated by the 35 km/hr wind from the west, which is sufficient to mobilize the ice cover, which deforms and compacts against the eastern shore. Under these strong winds, the effects of hydrodynamic forcing on the ice cover are relatively minor compared to the wind stress. The ice-free channels close early in the simulation as the wind pushes the ice eastwards. The ice cover deforms progressively over the first five hours until reaching a near steady state condition thereafter. The modelled distributions of ice concentration, thickness and pressure after five hours are shown in Figure 10. These results show that the ice conditions along the eastern shore become relatively severe, with concentration reaching 100% and thickness reaching up to 80 cm (initially 43 cm) combined with strong pressure. While ice conditions along the eastern shore worsen, the western and southern parts of the domain become largely free of ice.

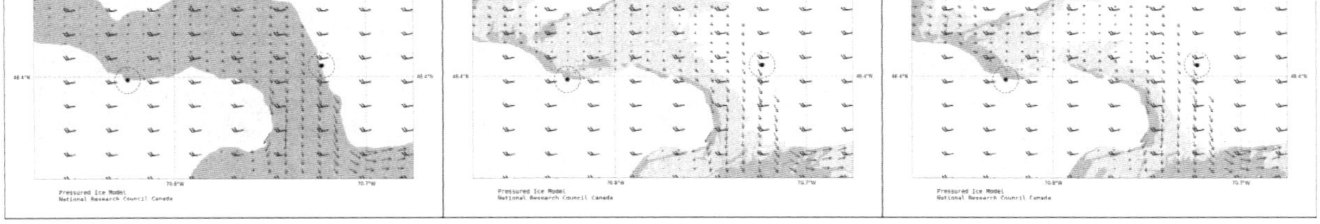

Figure 9. Initial evolution of ice pressure severity for Scenario 6 (35 km/hr wind from west).

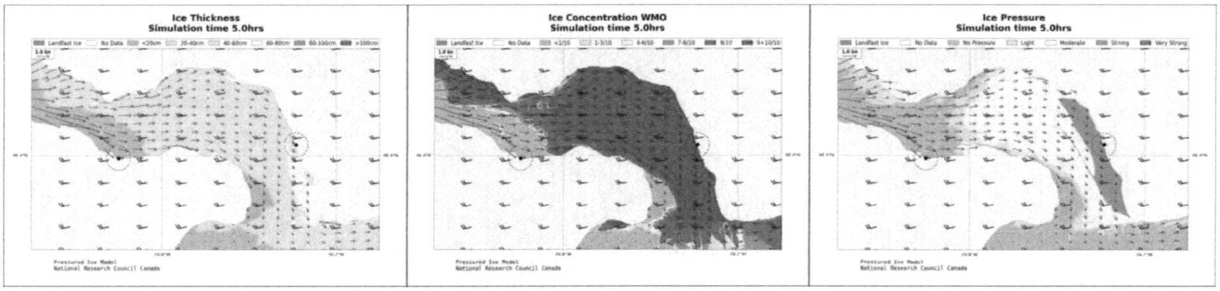

Figure 10. Distribution of ice cover thickness, concentration and pressure after 5 hours for Scenario 6.

In Scenario 10, the same initial ice cover considered in Scenario 6 is exposed to the same hydrodynamic forcing, this time combined with a 25 km/h wind approaching from the north-northeast direction. The initial evolution of ice pressure for this case is illustrated in Figure 11, while the distributions of ice thickness, concentration and pressure after 5 hours are shown in Figure 12. For these conditions the ice dynamics is again dominated by the wind forcing: the ice cover rapidly deforms and compacts against the southern shore in response to the wind stress. The ice cover deforms progressively over the first ~5 hours, becoming nearly stable thereafter. The simulation results indicate that the ice conditions along the southern shore grow increasingly severe, with concentration reaching 100% and thickness exceeding 80 cm (initially 43 cm) combined with strong to very strong pressure. In contrast, the northern and eastern shores become largely ice free.

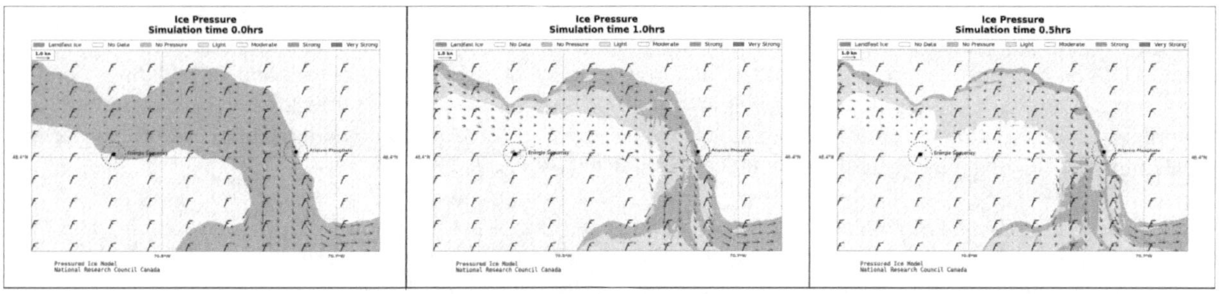

Figure 11. Initial evolution of ice pressure severity for Scenario 10 (25 km/hr wind from north-northeast).

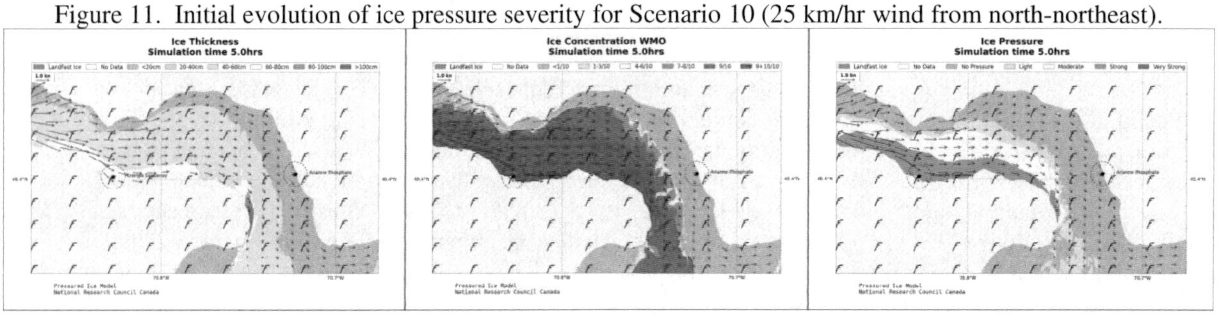

Figure 12. Distribution of ice cover thickness, concentration and pressure after 5 hours for Scenario 10.

The average ice thickness and average ice pressure forecast at the 5 hr mark of each scenario is presented graphically in Figure 13 for two locations, one near the center of the Southern shore and the other near the center of the Eastern shore. According to the simulation results, Scenarios 5 to 8, which all consider wind from the west, produce moderate to severe ice conditions along the eastern shore. The severity of ice conditions in this location is greatest for scenario 7 in which relatively thick 70 cm ice is modelled together with a relatively strong 35 km/hr wind. Thicker ice and/or stronger winds produce more severe ice conditions, as expected. For Scenario 8, which combines average ice thickness (43 cm) with reduced initial ice concentration (70%), the build-up of ice pressure is delayed in comparison with scenarios 5 to 7 in which higher initial ice concentration (95%) was modelled. Thus, the build-up of pressure along the eastern shore after a specific intermediate duration, say 3 hours, is significantly reduced for Scenario 8 compared to the others. However, the peak pressure reached at the end of the simulation (12 hours) remains similar, it just takes longer for the pressure in the ice cover to reach these levels when the initial concentration is lower.

Winds from the north-northeast were modelled in Scenarios 9 – 11, resulting in appreciable ice action against the southern shore. The ice build-up along the southern shore was most severe for Scenario 10 (stronger wind combined with high initial ice concentration) and least severe for Scenario 11 (stronger wind combined with lower initial ice concentration). Results for Scenario 9 (weaker wind combined with high initial ice concentration) indicate a medium ice severity when compared against the other two cases. The build-up of ice pressure against the southern shore occurred much more slowly in Scenario 11, when a lower initial ice concentration was modelled.

It should be mentioned that the results of this study should be associated with a moderate degree of uncertainty due to the many assumptions, simplifications and limitations inherent in the methodology followed to obtain them. Despite this important caveat, the simulation results show that wind conditions are the primary factor in mobilizing and deforming the ice cover leading to moderate or severe ice conditions along the shore of the upper Saguenay Fjord. Moderate and strong sustained winds acting against either shoreline can be expected to increase the severity of ice conditions. In contrast to wind action, water currents appear to play a minor role in the ice dynamics for this region. This conclusion is corroborated by the results of Scenario 1, in which wind forcing was excluded (calm conditions). Only the time-varying water current acted on the ice cover in Scenario 1. Simulation outputs for this scenario indicate that the ice cover remains stationary with no detectable movement; the navigation channels remain intact and ice-free, and the ice pressures in the ice cover remain light throughout the region.

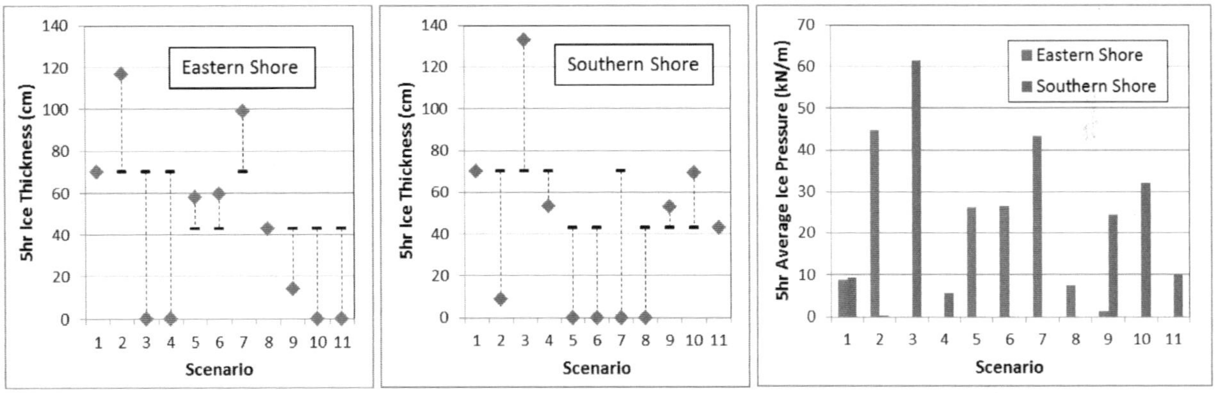

Figure 13. Average Ice thickness and pressure near Southern and Eastern shores for Scenarios 1-11, after 5 hours.

5. CONCLUSION

A numerical model of ice dynamics previously developed at the National Research Council Canada has been setup and applied to support the planning of future port facilities in the upper part of the fjord. The ice dynamics model has been used to predict the evolution of ice pressure, ice concentration and ice thickness in response to forcing by ground level winds and surface water currents. The hydrodynamic forcing was obtained from a newly developed 3D numerical model of stratified hydrodynamics in the fjord, while the wind forcing was either derived from analysis of historical wind measurements or obtained from a numerical wind model. Results for eleven hypothetical scenarios covering both non-extreme and extreme conditions have been obtained.

The simulation results suggest that, aside from the action of ice clearing ships, wind conditions are the primary factor in mobilizing and deforming the ice cover in the upper Saguenay Fjord. The hydrodynamic forcing is a secondary factor in comparison to the wind, due to the relatively low water current speeds that prevail over most of the study region. Moderate or severe sustained winds from northerly directions can mobilize the ice cover, which deforms and compacts as it moves southwards, creating moderate to severe ice conditions along the southern shore. Similarly, strong or extreme winds approaching from westerly directions are able to mobilize the ice cover, which deforms and compacts as it moves eastwards, creating moderate to severe ice conditions along the eastern shore. The numerical simulations suggest that the severity of the ice conditions along either shore intensify with increasing initial ice thickness and/or increasing wind speed. Lower wind speeds and/or thinner ice tend to produce more moderate near-shore ice conditions, as expected. Lower initial ice concentration tends to delay the build-up of ice pressure and thickness near both shores.

The NRC ice dynamics model is an important tool for forecasting ice conditions that may influence navigation safety, port operations and the development of new port facilities at higher latitudes where ice covered waters persist, such as the Saguenay Fjord.

6. REFERENCES

Arctec Canada Ltd. (ACL), 1980. *Ice Conditions at a Proposed Dock Site at Grande Anse, Saguenay River.* Report to the SNC Group.

Bourgault, D., Galbraith, P., Winkler, G., 2011. Exploratory Observations of Winter Oceanographic Conditions in the Saguenay Fjord. *J. ATMOSPHERE-OCEAN* 50 (1) 2012, 17–30.

Cornett, A., Pilechi, V., Cousineau, J., 2017. Modelling stratified hydrodynamics in the Saguenay Fjord with Telemac-3D. *Proc. 23rd Canadian Hydrotechnical Engineering Conf.*, CSCE, Vancouver, Canada.

De Vernal, A., St-Onge, G., Gilbert, D., 2011. Oceanography and Quaternary geology of the St. Lawrence Estuary and the Saguenay Fjord. *IOP Conf. Series: Earth and Environmental Science* 14 (2011).

Gosink, J., Marcotte, N., Muller, A., and Osterkamp, T.E., 1986. Thermal Regime of Lakes and Rivers: Chapter 4 of *River and Lake Ice Engineering*, Edited by Larsen P., Ashton, G., Water Resources Publications, Colorado.

Hervouet, J-M, 2007. *Hydrodynamics of Free Surface Flows: Modelling with the Finite Element Method.* Wiley Scientific. 360 p.

Kubat, I., Watson, D. and Sayed, M., 2011. Characterization of Pressured Ice Threat to Shipping, *Proc. 21st Int. Conf. on Port and Ocean Engineering under Arctic Conditions (POAC)*, 2011, Montreal, Canada, pp 11-136.

Kubat, I, Sayed, M., and Lamontagne, P., 2015. Analysis of vessel besetting over the Gulf of St. Lawrence and the Strait of Belle Isle, winter 2013-2014, *Proc. 23rd Int. Conf. on Ports and Ocean Engineering under Arctic Conditions (POAC)*, June 14-18, Trondheim, Norway.

National Research Council Canada (NRC), 2006. *AnemoScope© Reference Guide.* 55 pages.

Sayed, M., Carrieres, T., Tran, H. and Savage, S.B., 2002. Development of an operational ice dynamics model for the Canadian Ice Service. *Proc. 2002 Int. Offshore and Polar Eng. Conf., ISOPE*, Kitakyushu, Japan, pp 841-848.

Urgeles, R., Locat, J., Lee, H., Martin, F., 2002. The Saguenay Fjord, Quebec, Canada: Integrating marine geotechnical and geophysical data for spatial seismic slope stability and hazard assessment. J Marine Geology 185 (2002) 319-340.

Leadership in Sustainable Infrastructure
Leadership en Infrastructures Durables

Vancouver, Canada
May 31 – June 3, 2017/ *Mai 31 – Juin 3, 2017*

MODELLING STRATIFIED FLOWS IN THE SAGUENAY FJORD WITH TELEMAC-3D

Andrew Cornett [1], Abolghasem Pilechi [2], and Julien Cousineau [3]
[1, 2, 3] Ocean, Coastal & River Engineering, National Research Council Canada, Ottawa, Ontario, Canada
[4] Andrew.Cornett@nrc-cnrc.gc.ca

ABSTRACT

The Saguenay Fjord, located on the north shore of the St. Lawrence Estuary in the Province of Quebec, Canada, is notable for its complex and strongly stratified flows, with freshwater at the surface flowing over and mixing with denser seawater at lower depths. A new three-dimensional numerical model based on the TELEMAC-3D solver has been developed to simulate the stratified flows in the Saguenay Fjord and provide useful predictions of the surface currents in the upper part of the fjord for use in a parallel study of ice cover dynamics. The development, calibration and validation of the new model is presented and discussed, as are results from several simulations. The simulation results help clarify the influences of tide range, tide phase and the freshwater discharge entering the fjord on the surface currents in the upper fjord. This study demonstrates that the TELEMAC-3D solver can be applied with a modest level of effort to obtain reasonable and useful estimates of hydrodynamics in a fjord or estuary with complex stratified flows.

Keywords: Saguenay Fjord, hydrodynamics, numerical model, stratified flow, Telemac

1. INTRODUCTION

The Saguenay Fjord extends for 110 km between the communities of Chicoutimi, where the fjord transitions to the Saguenay River, and Tadoussac, where the fjord meets the St Lawrence Estuary, see Figure 1. The fjord varies in width from approximately 1.5 to 4 km, while the depth varies from ~15m near Tadoussac to over 275 m near Cap Trinité. The Saguenay Fjord is notable for its complex and strongly stratified flows, with freshwater at the surface flowing over and mixing with denser seawater at lower depths. The sketch presented in Figure 2 depicts the more buoyant freshwater moving downstream near the surface over the denser seawater at lower depths. Over 90% of the water volume in the fjord is estimated to be salt water. During winter a stable ice cover forms over the upper part of the fjord, and commercial ships accessing port facilities in Chicoutimi and Port Alfred are escorted by ice breaking vessels operated by the Canadian Coast Guard.

A three-dimensional numerical model of the hydrodynamics in the Saguenay Fjord has been developed to support the planning of future port facilities and improved navigation safety. In particular, surface currents predicted by the new hydrodynamic model have been used to force an ice dynamics model that in turn gives predictions of ice concentration, ice thickness and ice pressure throughout the region for a range of hypothetical scenarios. The development of the 3D hydrodynamic model is described herein, while the set-up and application of the ice dynamics model is discussed in Cornett *et al.* (2017).

The new hydrodynamic model is based on the TELEMAC-3D solver, which employs finite-element methods to solve the Navier-Stokes equations, in non-hydrostatic form, over a computational domain subject to initial conditions and time-varying boundary conditions. In two-dimensions (plan) the computational domain is represented by an unstructured mesh of triangular elements. Discretization in the third dimension (depth) is achieved by defining a constant number of sigma layers, whose thickness varies with space and time depending on the local water depth. In its final iteration the entire 110 km long fjord was discretized into 21 sigma layers, with 22,800 triangular elements defined in each layer.

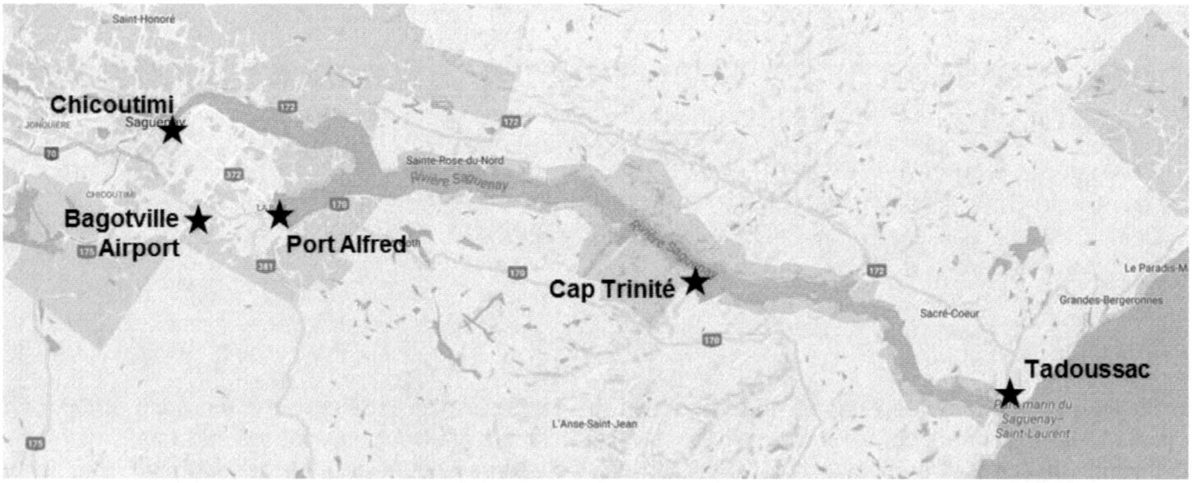

Figure 1. Map of the Saguenay Fjord region (source: Google).

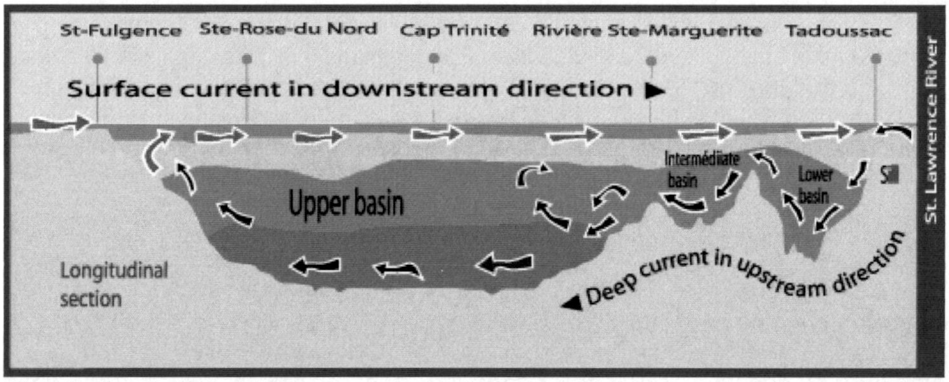

Figure 2. Sketch of general circulation pattern in the Saguenay Fjord (source: http://www.virtualmuseum.ca).

2. HYDRODYNAMIC CONDITIONS

Available information on tides, freshwater inflows to the fjord, water currents and salinity was reviewed, analyzed and referenced in setting up and calibrating the 3D hydrodynamic model. Some of the most important field measurement information is summarized in what follows. The present research was also informed by previous investigations into the hydrodynamic, oceanographic and geologic conditions of the Saguenay Fjord, including those reported in Urgales *et al.* (2002), Bourgault *et al.* (2011), and De Vernal *et al.* (2011).

Freshwater Inflow

A flow-duration curve for the total upstream freshwater flow into the Saguenay Fjord was generated based on analysis of measured discharge data from five sources, including the outflow from the dam at Shipshaw and four other important tributaries. The freshwater flow data was provided by Division Énergie Électrique of Rio Tinto Alcan and the Water Survey of Canada. Figure 3a shows the temporal variation of the Saguenay River discharge at

Shipshaw from 1972 to 2001, while Figure 3b shows the flow-duration curve obtained from analysis of the flow data. Based on this analysis, the average inflow is 1,310 m^3/s while the inflow exceeded 1% of the time is 6,493 m^3/s. The inflow during the month of March when the ice-cover breaks up was also investigated; the mean freshwater inflow during March is 1,244 m^3/s while the maximum recorded inflow during March is 1,907 m^3/s.

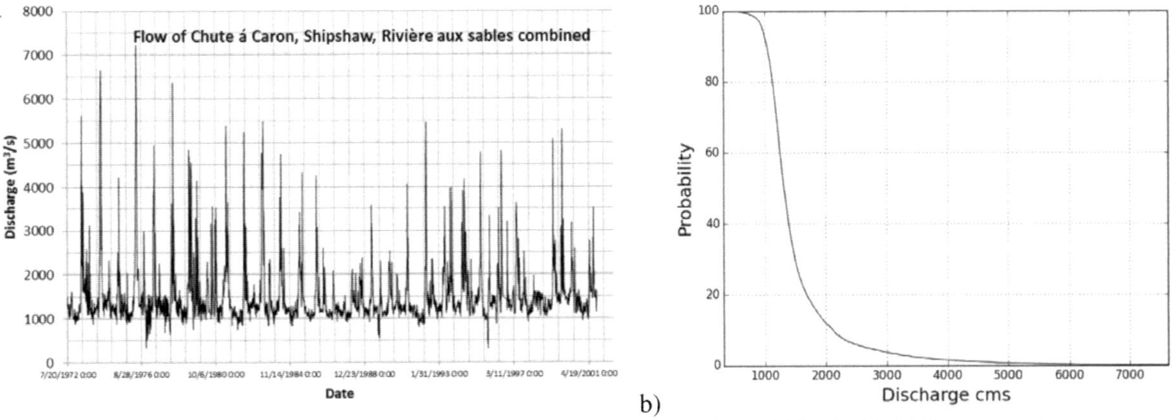

a) b)

Figure 3. Freshwater inflow: a) Saguenay River discharge at Shipshaw, 1972-2001; b) flow-duration curve.

Tides

Information on the tides at Port Alfred and Tadoussac was obtained from Fisheries and Oceans Canada and reviewed, see Figure 4. The tides in the region are semi-diurnal, rising and falling twice daily. The tides at Port Alfred, located in the upper fjord, are generally between 0.5 to 1.0 m larger than the tides at Tadoussac, located at the lower end of the fjord where the fjord joins the St. Lawrence Estuary. The Spring tide range at Port Alfred can exceed 6 m while the Neap tide range can be less than 2.5 m.

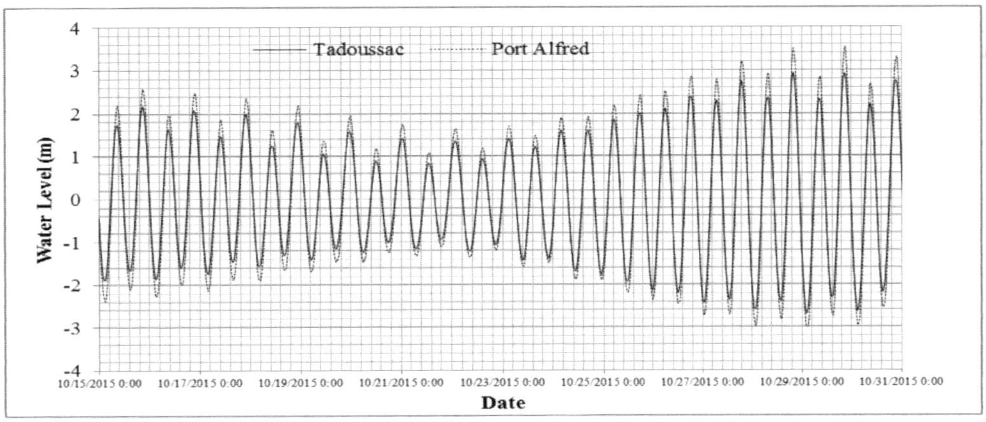

Figure 4. Observed tides at Tadoussac and Port Alfred, 15-30 October 2015.

Water Current

The currents near the Grande Anse wharf were measured (by others) on two occasions using an Acoustic Doppler Current Profiler (ADCP) unit. A mobile survey was conducted on 23 November 2015 wherein the ADCP unit was mounted to a boat and used to measure water currents along six shore-perpendicular transects. Data was collected for a ~3 minute duration at roughly five stations along each transect.

A stationary upward-looking ADCP unit was deployed in ~17 m water depth near the Grande Anse wharf from 27–29 October, 2015. The raw ADCP data from the stationary deployment was analyzed to estimate the speed and direction of the near-surface currents at this site over the deployment, accounting for the vertical fluctuation of the free surface. The raw ADCP data and the estimated speed of the surface currents are shown in Figure 5. It is noted that the estimated surface velocity may not be entirely accurate due to difficulties measuring velocities near the free surface, where reflections can produce erroneous readings, and uncertainties related to the variable density of the water in the fjord.

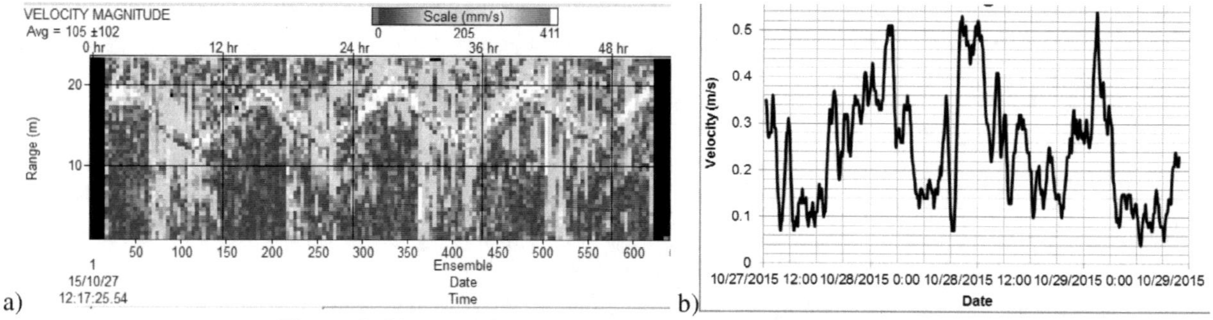

Figure 5. Observed flow velocity near the Grande Anse wharf:
a) raw ADCP data from stationary deployment; b) estimated surface current speed.

Salinity

Twenty-five salinity profiles (describing the variation of salinity with depth) measured in the upper fjord (by others) at various times and locations were reviewed and analyzed. The salinity profiles coincide with times when the freshwater inflow to the fjord was low (~1,100 m^3/s), moderate (~1,600 m^3/s) and high (>5,000 m^3/s). Three typical salinity profiles indicative of each freshwater inflow condition are presented in Figure 6. All the profiles show that the water column in the fjord is strongly stratified, with lower-density freshwater or near freshwater (salinity < 10 PPT) at the surface, and higher-density seawater (salinity > 25 PPT) at lower depths. The thickness of the freshwater layer and the gradient of the transition from freshwater to seawater varies with time and location. In general, the thickness of the freshwater layer varies from a minimum of approximately 3 m when the freshwater discharge is low, up to 15 m when the freshwater discharge is exceptionally high.

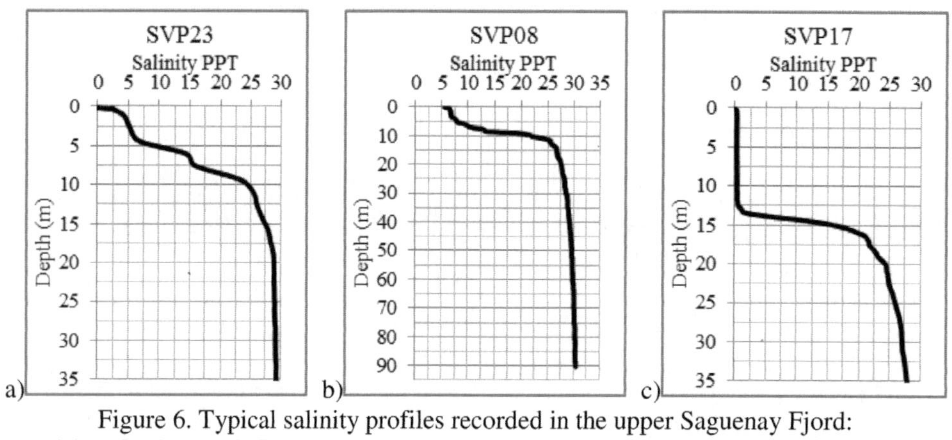

Figure 6. Typical salinity profiles recorded in the upper Saguenay Fjord:
a) low freshwater inflow; b) medium freshwater inflow; c) high freshwater inflow.

Bathymetry

Information on the bathymetry throughout the region was assembled from several sources, including navigation charts published by the Canadian Hydrographic Service (CHS) as well as high-resolution bathymetric surveys covering several smaller sub-regions. The information from these sources was referenced to a common datum and merged to form a single digital elevation model. Figure 7 shows the resulting bathymetry for a portion of the upper fjord. The upper fjord is generally deep, with water depths less than 30 m prevailing only very near the shoreline, and at the head of the fjord near Shipshaw.

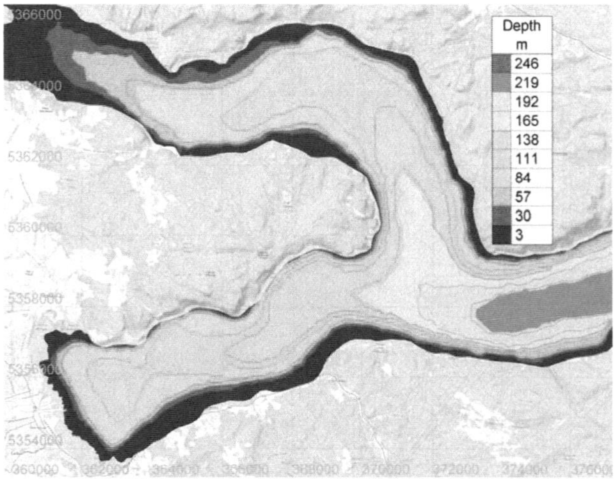

Figure 7. Bathymetry for a portion of the upper Saguenay Fjord.

3. NUMERICAL MODEL SETUP

Telemac Modelling System

A three-dimensional numerical model based on the TELEMAC-3D solver from the TELEMAC System (v7.0) was developed to simulate the hydrodynamics in the Saguenay Fjord. The model employs finite-element methods to solve the Navier-Stokes equations, in non-hydrostatic form, over a computational domain subject to initial conditions and time-varying boundary conditions. In two-dimensions (plan) the computational domain is represented by an unstructured mesh of triangular elements. Discretization in the third dimension (depth) is achieved by defining a constant number of layers, whose thickness can vary with space and time depending on the local water depth. The unstructured grid allows the size of the elements to vary in space: a fine mesh can be used to improve accuracy in areas of interest and in areas with high velocities and/or strong spatial gradients, while a coarse mesh can be used in less important areas in order to reduce computation effort and cost. Further details on the TELEMAC system are provided in Hervouet (2007). The TELEMAC System was originally created by the Laboratoire national d'hydraulique et environment d'Électricité de France (EDF) but is now managed by a consortium of core organizations under the OpenTELEMAC-MASCARET umbrella (see www.opentelemac.org).

Computational Domain and Grid

The new model extends from a point ~700 m downstream from the Chute-á-Caron dam at Shipshaw down to the St-Lawrence Estuary. The model domain includes the entire Saguenay Fjord and small portion of the St-Lawrence River, see Figure 8. The configuration of the triangular mesh in four sub-regions (A – D) is also shown in this figure. Some shallow tidal flats in the upper part of the fjord were excluded in order to simplify and speed-up the numerical computations. The exclusion of the tidal flats is believed to have little influence on the water levels and currents in the region of interest.

To minimize numerical errors and computational cost, the total number of grid cells and the distribution of the cell sizes were chosen based on a sensitivity analysis. The final mesh has 22,838 triangular elements and 12,456 nodes in each layer. The horizontal resolution varies from 15 m near the upstream boundary, where velocities are greatest, to ~1,000 m in the St. Lawrence River. The horizontal resolution near two reference sites of interest (B and C in Figure 8) was refined to ~40 m.

Initially, the water column throughout the domain was discretized into 11 layers. Results from these initial simulations suggested that additional layers would be required in order to simulate the strong density stratification that exists in the fjord. In the final set-up, the water column throughout the domain is discretized into 21 layers. The layers are unevenly distributed throughout the water depth: a finer layer thickness was specified in the upper part of the water column near the free surface, a medium layer thickness was used near the bottom, while the largest layer thickness was specified over the middle portion of the water column, away from the upper and lower boundaries. It is noted that the thickness of each layer varies with space and time based on the local water depth, which depends on the local bathymetry and the local free surface elevation.

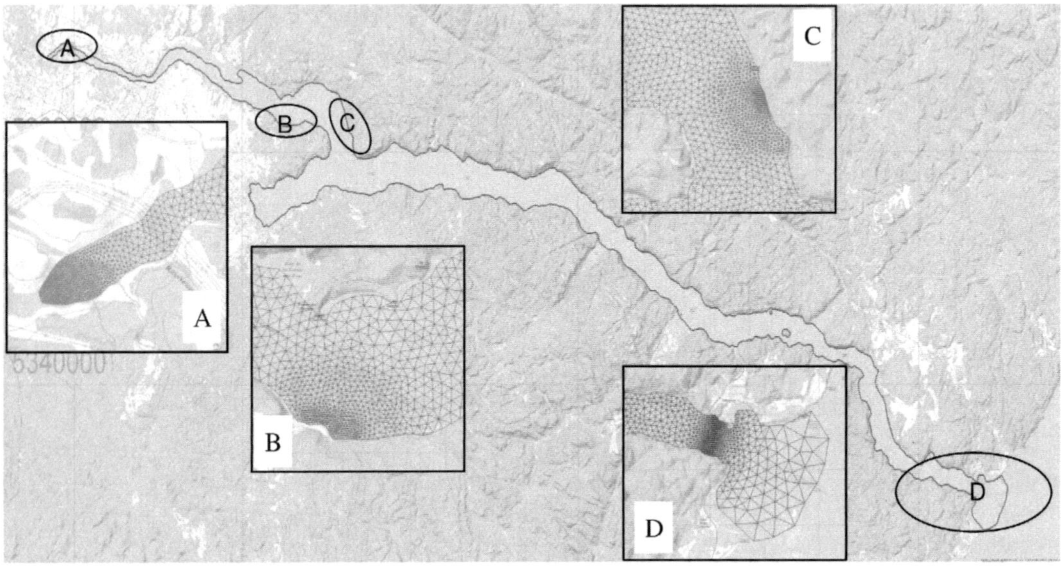

Figure 8. Computational domain (red outline) and triangular mesh with variable horizontal resolution.

Boundary and Initial Conditions

The model was set-up with two dynamic boundaries. Time-varying water levels were prescribed at the downstream boundary in the St-Lawrence Estuary (D in Figure 8), while a steady inflow (discharge) of freshwater was specified at the upstream boundary near Shipshaw (A in Figure 8). The water level boundary condition was developed from analysis of tide data for Tadoussac and other nearby tide stations in the St. Lawrence Estuary. The steady upstream discharges were based on analysis of the freshwater inflows discussed above (see Figure 3).

The hydrodynamic model was initially run for a one-month warm-up period to establish realistic equilibrium conditions for the density stratification and salinity gradient throughout the domain. The one-month warm-up simulation was conducted using a steady up-stream freshwater inflow of 2,000 m^3/s combined with the tides during May 2014. The quasi-equilibrium condition obtained at the end of the one-month warm-up simulation is depicted in Figure 9. This figure shows the spatial distribution of surface salinity computed over the entire domain, together with the salinity profiles computed for six stations along the fjord. In the model, the surface salinity varies gradually from 0 PPT (pure freshwater) at location A (near the upstream boundary) to 30 PPT (pure seawater) at location G (near the downstream boundary), as expected. The surface salinity at location B (upstream from Grande Anse) is less than ~3 PPT, while the surface salinity at location C (downstream from Grande Anse) is ~12 PPT. The modelled salinity profile at location A shows only freshwater, as expected, while the modelled salinity profile at location G in the St. Lawrence shows pure seawater, again as expected. The modelled salinity profiles at locations B and C show a strongly stratified water column comprised of a thin layer of near freshwater at the surface over a much thicker layer of near seawater at depth, and is generally consistent with the form of the measured salinity profiles for this region (three examples in Figure 6). The salinity/density stratification predicted by the model is considerably less pronounced in the lower parts of the fjord, below location D.

The equilibrium condition obtained at the end of the one-month warm-up simulation was used as the initial condition for subsequent model runs conducted primarily to estimate surface currents in the upper fjord for different scenarios.

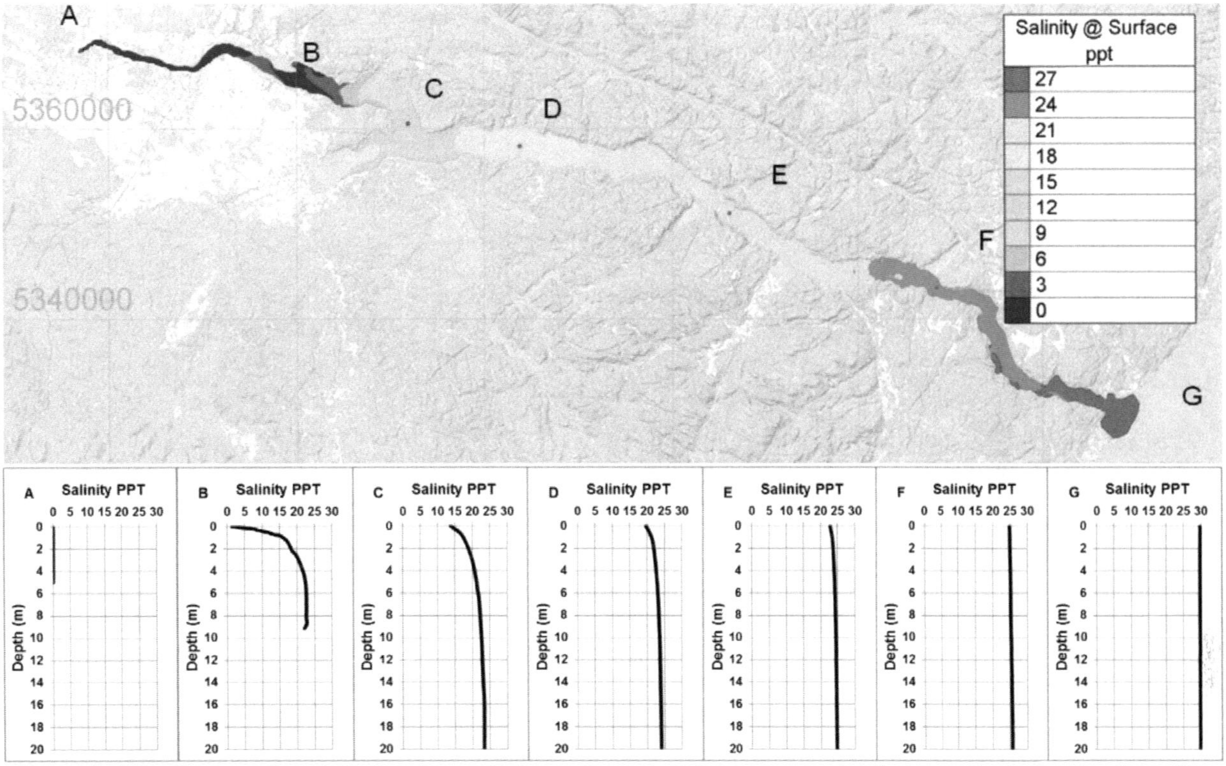

Figure 9. Distribution of modelled surface salinity at the end of the warm-up simulation and salinity-depth profiles for selected locations A-G.

Calibration and Validation

The hydrodynamic model was initially calibrated to ensure that it provided reliable predictions of the water level fluctuations at Port Alfred. The model was calibrated through an iterative process of varying the bottom roughness until the water-surface elevations predicted by the model were in close agreement with tide level predictions for Port Alfred issued by Fisheries and Oceans Canada. In the end, a single Manning's roughness coefficient value of 0.04 was used throughout the computational domain. Using the optimized roughness coefficient, both the magnitude and the timing of the high and low tides at Port-Alfred were reasonably well predicted by the model.

Next, the surface currents predicted by the model were compared with the surface current estimated from data collected during the stationary ADCP deployment. The horizontal eddy viscosity specified in the model was adjusted to achieve reasonable agreement between the modelled surface currents and those estimated from measurements. In the end, a horizontal eddy viscosity of 20 m^2/s was selected for use in the model.

Although the agreement between modelled and 'measured' water currents was not perfect, it was confirmed that the model provides a reasonable prediction of the speed of the surface current and the timing of its fluctuations at the ADCP deployment location. The model indicates that the current speed near the Grande Anse wharf is highest during times of ebbing tide, when the tidal current reinforces the freshwater outflow at the surface, and lowest during times of flooding tide, when the tidal current opposes the freshwater outflow at the surface.

The numerical simulation of the density stratification in the fjord was found to be sensitive to several factors, including the vertical turbulence model, the number of layers, and vertical distribution of the layers throughout the water column. Better results were obtained when the number of layers was doubled from 11 to 21, and the layers were concentrated in the upper part of the water column where the vertical gradient in salinity was greatest, instead of being evenly distributed with depth. Several alternative vertical turbulence formulations were investigated and assessed, but in the end, the "Nezu and Nakagawa" mixing length model (Nezu and Nakagawa, 1993) was selected as the most appropriate model from those available within TELEMAC-3D.

Extensive comparison between measured and modelled salinity profiles indicates that the model generally over-predicts the salinity near the surface and under-predicts the salinity at lower depths, relative to the measured data. The freshwater layer predicted by the model is generally thinner than indicated by the salinity profile data. These observations suggest that the degree of vertical mixing may be somewhat over-estimated in the model. It is noted that the hydrodynamic model was developed to provide a reasonable prediction of surface currents for various scenarios, and was not intended to provide a complete and full description of the salinity/density stratification and mixing over the water column and throughout the fjord.

The numerical model was validated by comparing model output with different data sets and for different periods than those used during calibration. The results of the validation were judged to be satisfactory and adequate for the model's intended purpose.

4. RESULTS AND DISCUSSION

Once the final model configuration and set-up was established, the model was used to simulate the water levels and three-dimensional flows in the fjord for four different scenarios. The scenarios were selected to include both typical and extreme conditions during the month of March, when the ice cover breaks up and moves out of the fjord, as well as conditions during the peak of the spring freshet when freshwater inflows are highest. Conditions on 27 October 2015, when the stationary ADCP was deployed near the Grande Anse wharf, were also modelled. The specifications for the four scenarios are summarized in Table 1. In all cases, the model was run for a 15-day period that included both a full Spring tide and a full Neap tide. The end state of the one-month warm-up simulation (see Figure 9) was adopted as the initial condition of salinity for each scenario.

Scenario 1 represents conditions at the peak of the average annual freshet, when freshwater inflows to the fjord are ~4,297 m^3/s. The distribution of surface currents over the upper fjord predicted by the model for Scenario 1 is shown in Figure 10, while Figure 11 shows the modelled currents at mid-depth.

Table 1 : Specifications for selected scenarios

Scenario	Tide signal	Freshwater inflow (m^3/s)	Remark
1	15-30 October 2015	4297	Average annual maximum freshwater inflow
2	15-30 October 2015	1907	Maximum freshwater inflow during March
3	15-30 October 2015	1244	Average freshwater inflow during March
4	15-30 October 2015	1120	Freshwater inflow on Oct 27, 2015

The simulation results suggest that the surface currents and circulation in the upper fjord are dependent on the tide range, the phase of the tide (ebb or flood) and the discharge of freshwater entering the upper fjord. The strongest surface currents tend to occur during Spring (large) tides when the water level is falling (ebb tide) and when the freshwater inflow discharge is high. The surface current tends to weaken at times when the freshwater inflow is lower, when the tide is rising (flooding), and when the tide range is smaller. In many locations the surface current continues to flow downstream (towards the lower fjord) even during times of rising (flood) tide.

The currents and circulation below the surface tend to be quite different. Near mid-depth, the flows are influenced more strongly by the tides and less by the fresh-water inflow, such that the flow direction generally reverses every ~6.4hrs. The horizontal circulation at depths near the interface between the freshwater and the salt wedge tends to feature many eddies and gyres of varying sizes; possibly a manifestation of mixing between the freshwater and saline flows.

It should be noted that in reality the surface currents may also be influenced by the local wind conditions and the state of the ice cover (if any); however, these factors were not explored in the present study. It should also be noted that smaller eddies or gyres may form close to shore in small bays that were not resolved by the present hydrodynamic model and therefore are not included in the simulation results presented here.

Probability curves for modelled surface current speed at two reference sites (B, C) are presented in Figure 12. The reference site locations are shown in Figure 8. Scenarios 3 and 4 produced very similar results, as expected, considering that the initial conditions and tide signal forcing were identical and the upstream freshwater inflows were very similar in these two cases. Surface current speeds for Scenario 2 were slightly larger than for Scenarios 3 and 4; a result that can be attributed to the greater freshwater inflow (1,907 m^3/s in Scenario 2 compared to ~1,200 m^3/s in Scenarios 3 and 4). The surface current speeds throughout the region are consistently greater for Scenario 1; in which a freshwater inflow of 4,297 m^3/s was modelled.

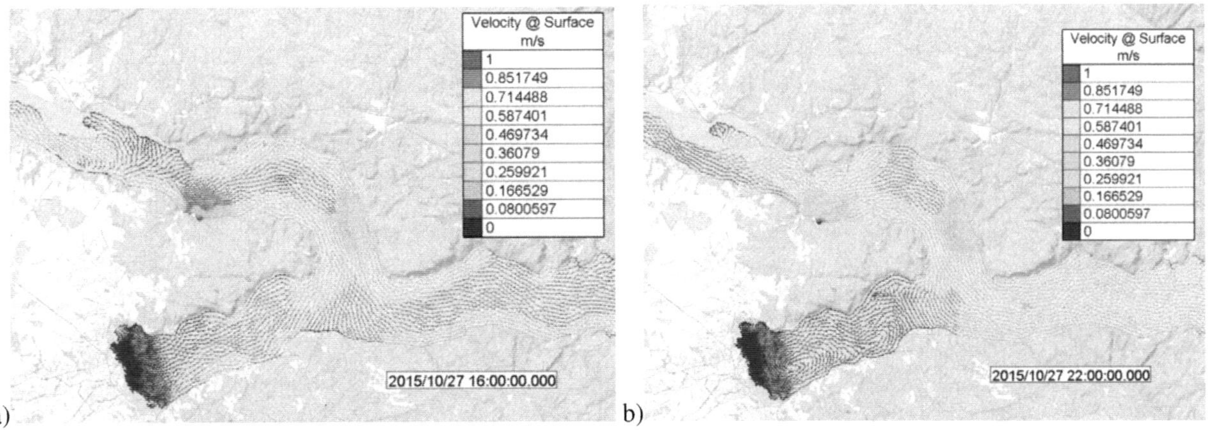

a) b)

Figure 10. Modelled surface currents in the upper fjord, Scenario 1: a) Spring flood tide; b) Spring ebb tide.

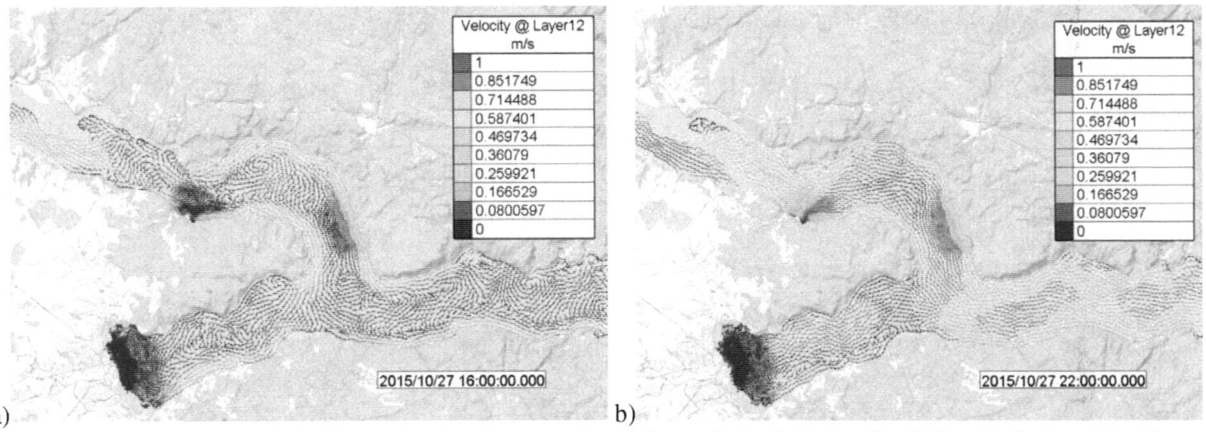

a) b)

Figure 11. Modelled mid-depth currents in the upper fjord, Scenario 1: a) Spring flood tide; b) Spring ebb tide.

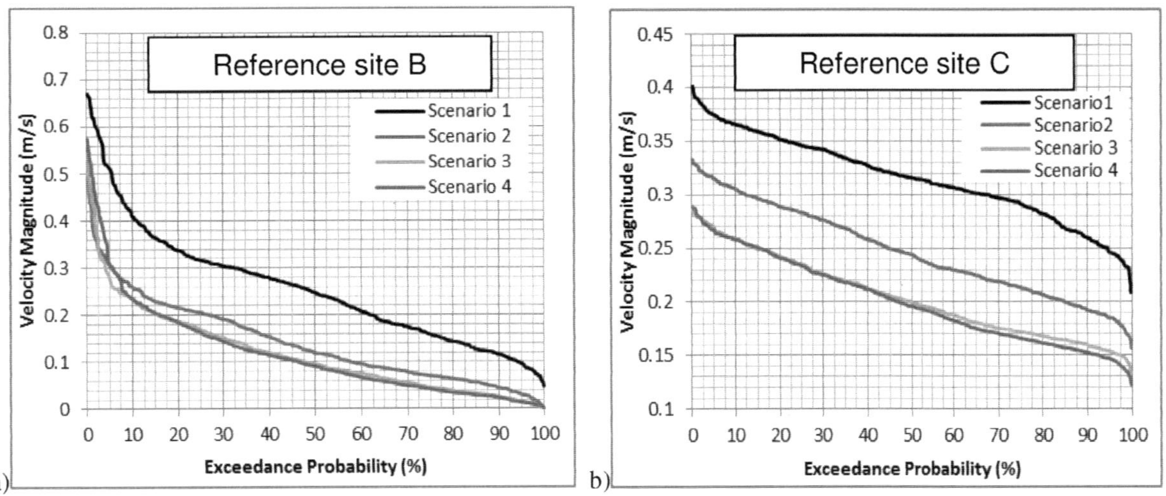

a) b)

Figure 12. Comparison of exceedance probability curves for surface current speed at two reference sites, Scenarios 1 – 4.

5. CONCLUSION

The Saguenay Fjord, located on the north shore of the St. Lawrence Estuary in the Province of Quebec, Canada, is a highly complex and dynamic body of water where freshwater and seawater combine, mixed by the tides. The complex and highly stratified flows in the fjord are particularly challenging to model numerically.

A three-dimensional numerical model based on the TELEMAC-3D solver has been developed to simulate the stratified flows in the Saguenay Fjord and provide useful predictions of the surface currents in the upper part of the fjord for use in a parallel study of ice cover dynamics. In the final version of the model, the 110 km long fjord is discretized into 21 layers, with 22,838 triangular elements in each layer. The model is forced by prescribing a fluctuating water level at the St. Lawrence Estuary, combined with a steady freshwater inflow entering near Shipshaw. A quasi steady state initial condition for the salinity and density stratification throughout the fjord was developed by conducting a warm-up simulation with one month duration. The model has been calibrated and validated by comparing simulation outputs with field observations of fluctuating water levels, near-surface currents, and salinity profiles. A reasonable level of agreement between the various measured and modelled quantities was achieved, considering the complexity of the hydrodynamic processes involved and the relatively short amount of time that could be dedicated to the effort. Useful predictions were generated for four hypothetical 14-day long scenarios, each corresponding to a different freshwater inflow. The simulation results help clarify the influences of tide range, tide phase and the freshwater discharge entering the fjord on the surface currents in the upper fjord. Due to the complexity of the flows and mixing in the fjord, and the numerous assumptions and approximations inherent in the simulating these flows in a numerical model, the numerical simulations of near-surface currents that have been obtained should be associated with a moderate degree of uncertainty.

This study has demonstrated that the TELEMAC-3D solver can be applied with a modest level of effort to obtain reasonable and useful estimates of hydrodynamics in a fjord or estuary where freshwater and seawater combine. It is noted that the presence of an ice cover and the potential effects of an ice cover on the flows and surface currents in the fjord were not included in these numerical simulations, nor were the potential effects of winds blowing over the water surface. The simulation results are therefore truly valid only for open-water conditions in the absence of strong winds.

6. REFERENCES

Bourgault, D., Galbraith, P., Winkler, G., 2011. Exploratory Observations of Winter Oceanographic Conditions in the Saguenay Fjord. *J. ATMOSPHERE-OCEAN* 50 (1) 2012, 17–30.

Cornett, A., Watson, D., Babaei, H., Sayed, M., 2017. Modelling Ice Dynamics in the Upper Saguenay Fjord. *Proc. 23rd Canadian Hydrotechnical Engineering Conf.,* CSCE, Vancouver, Canada.

De Vernal, A., St-Onge, G., Gilbert, D., 2011. Oceanography and Quaternary geology of the St. Lawrence Estuary and the Saguenay Fjord. *IOP Conf. Series: Earth and Environmental Science* 14 (2011).

Hervouet, J-M, 2007. *Hydrodynamics of Free Surface Flows: Modelling with the Finite Element Method*. Wiley Scientific. 360 p.

Nezu, I., Nakagawa, H., 1993. *Turbulence in Open-Channel Flows*. Int. Assoc. for Hydraulics Research Monograph, A. A. Balkema, Rotterdam, the Netherlands.

Urgeles, R., Locat, J., Lee, H., Martin, F., 2002. The Saguenay Fjord, Quebec, Canada: integrating marine geotechnical and geophysical data for spatial seismic slope stability and hazard assessment. *J. Marine Geology* 185 (2002) 319-340.

Leadership in Sustainable Infrastructure
Leadership en Infrastructures Durables

Vancouver, Canada
May 31 – June 3, 2017/ *Mai 31 – Juin 3, 2017*

MODELLING STREAMFLOW AND SEDIMENT EXPORT IN POTHOLE-DOMINATED, COLD-CLIMATE PRAIRIE WATERSHEDS USING THE SOIL AND WATER ASSESSMENT TOOL (SWAT)

Mekonnen, Balew A.[1,3], Mazurek, Kerry A.[2], and Putz, Gordon[2]
[1] Guelph University, Canada
[2] University of Saskatchewan, Canada
[3] mekonnen@uoguelph.ca

Abstract: The Canadian prairies are dominated by millions of depressions, or potholes, that have a significant impact on streamflow generation in the region. It has been difficult to incorporate the dynamic storage, or "fill and spill", processes of these depressions in hydrologic models of prairie watersheds. Additionally, the region has a cold-climate, where snowmelt and soil thaw processes impact the generation of streamflow, sediment export, and non-point source pollutants. This paper discusses improvements to the hydrological model SWAT (Soil and Water Assessment Tool) for modelling streamflow and sediment export in two pothole-dominated watersheds in Saskatchewan: the Assiniboine and Qu'Appelle watersheds. The Pond module of SWAT was modified to incorporate dynamic storage in the potholes using a probability distribution to assess how many of the depressions would be spilling and therefore contributing to streamflow. Soil erodibility coefficients were adjusted seasonally to improve estimates of sediment export. Improved performance for simulating streamflow was seen over the existing SWAT Pond module for simulating landscape storage. Sediment export estimates improved when using seasonally adjusted erodibility coefficients over annual sediment erodibility coefficients.

1 INTRODUCTION

1.1 Background

There is much interest in predicting streamflow and water quality for streams in the prairie region of Canada, both for development of capabilities to better assess potential flooding and for developing strategies for overcoming water quality challenges locally or nationally. Modelling the hydrology of the region has been a challenge; it is dominated by millions of depressions, called potholes and sloughs, which significantly influence runoff generation in the region (Eulis *et al.* 1999; Hayashi *et al.* 2004). For example, Godwin and Martin (1975) estimated that about 67 % of the Assiniboine Watershed at Kamsack is intercepted by potholes. When not full, such as would occur in dry periods, the ponds intercept runoff and an area can become non-contributing to streamflow. When full, the ponds can interconnect and do contribute to streamflow. Due to the "fill-and-spill" characteristics of the ponds, the contributing areas for streams in the prairie region are dynamic (Shaw *et al.* 2011).

To begin to incorporate the dynamic storage effects into hydrological models for the prairie region, researchers have attempted to describe in detail the physical processes occurring for individual depressions in their models (e.g. Su et al. 2000; Pomeroy et al. 2007; Fang and Pomeroy 2008; Shook et al. 2013; Chu et al. 2013). Due to this physical detail, their work has tended to be for smaller scale watersheds (or at plot scale). For example, Su et al. (2000), Pomeroy et al. (2007) and Fang and Pomeroy (2008) tried to simulate the water budget for an individual wetland. Shook et al. (2013) used a fully-distributed hydrological model for three small-scale watersheds to model the fluxes from the depressions. Chu et al. (2013) used a physically-based distributed model to simulate the pothole water balance on plot scale.

Conversely, Abedini (1998) used the probability distributed models that were used to describe soil moisture storage (e.g. Moore, 1985) for describing landscape depression storage and its heterogeneity. Using this approach for laboratory and plot scale measurements, Abedini (1998) simulated streamflow from surfaces dominated by depressions. M.A. Mekonnen et al. (2014) also implemented a probability distributed model for depression storage in the prairie region within the Modélisation Environmentale Communautaire - Surface and Hydrology (MESH) model. They simulated runoff from depressions for a portion of Saskatchewan's Assiniboine Watershed (an area of 1939 km^2). However, they did not attempt to find the parameters of the probability distribution used to describe storage capacity from the topography of the watershed and could not validate their model due to data limitations.

This paper reports on a study to incorporate a probability distribution to describe landscape depression storage heterogeneity into SWAT (the Soil and Water Assessment Tool) in attempt to improve streamflow simulation results for two Canadian prairie watersheds in Saskatchewan: the Assiniboine and the Qu'Apelle watersheds. Both are large watersheds where the land use is predominantly for agriculture. Additionally, the modified SWAT model that includes this probability distributed approach was used to model sediment export in the watersheds and these results are also reported. Among the reasons the SWAT model was chosen for this study were its development for agriculture-dominated watersheds, its wide application, the free access to its source code (Neitsch et al. 2011), and computational efficiency that affords its use to very large watersheds.

1.2 SWAT Model Description

The Soil and Water Assessment Tool (SWAT) is a continuous, semi-distributed hydrological model. The version used for the work described herein was SWAT2009. SWAT simulates the different hydrological processes of a watershed and can estimate water, sediment, and pollutant yields (Arnold *et al.* 1998). In SWAT, a watershed is partitioned into sub-basins that are further grouped into what are called "Hydrological Response Units" (Arnold *et al.* 1998). These HRU's are non-spatially specific lumped areas within a sub-basin that are comprised of unique combinations of land cover, soil type, and slope (Neitsch *et al.* 2011).

Water balance computations and hydrological process simulations are performed at the HRU level and either the modified Curve Number Method (SCS: USDA Soil Conservation Service 1972) or the Green-Ampt method (Green and Ampt 1911) is used to compute surface runoff. Potential evapotranspiration is estimated using one of three methods: Hargreaves (Hargreaves *et al.* 1985); Priestley-Taylor (Priestley and Taylor 1972), or Penman-Monteith (Monteith 1965). The actual evaporation from soils and plants is calculated by the method presented in Ritchie (1972). For the estimation of baseflow, groundwater is partitioned into a two-aquifer system, which represent its shallow and deep components (Arnold *et al.* 1998). To find the water yield to the main channel within a sub-basin, the contributions from each HRU within the sub-basin are summed. The water is then routed to the outlet of the sub-basin using either the variable storage coefficient method (Williams 1969) or the Muskingum routing method (Cunge 1969).

With respect to the cold-climate features of SWAT, for snowmelt prediction a temperature-index method is used (Neitsch et al. 2011). Other modifications include a seasonally variable snowmelt rate (Fontaine et al. 2002), division of each sub-basin into 10 elevation bands (Fontaine et al. 2002), and modification of the curve number value for frozen soil conditions for enhanced surface runoff and reduced infiltration (Tolston and Shoemaker 2007).

Soil erosion within SWAT is quantified for each HRU using the modified Universal Soil Loss Equation (MUSLE) (Williams 1975) and the estimated surface runoff. The sediment export on a given day, S, in metric tons is given as:

[1] $S=11.8(QqA_{hru})^{0.56}K \cdot LS \cdot C \cdot P \cdot CFRG$

where Q is the surface runoff volume (mm H_2O/ha); q is the peak rate of runoff (m^3/s); A_{hru} is the HRU's area (ha); K is the soil erodibility factor; LS is a topographic factor; C is a cover and management factor; P is a supporting practice factor; and CFRG is coarse fragment factor. The soil erodibility factor in SWAT2009 is an annualized value (it does not vary with season). Before reaching the main channel of a sub-basin, sediment that is mobilized can be lagged and routed through grassed waterways, vegetative filter strips, and wetlands. Sediment is then moved through the main channel considering deposition and erosion processes. The equations used to describe these processes are reviewed in Mekonnen et al. (2016b).

1.3 Previous Uses of SWAT in Depression-Dominated Prairie Watersheds

SWAT has been used to model streamflow in the prairie region in several previous studies (e.g. Sophocleous et al. 1999; Shrestha et al. 2011; Almendinger et al. 2012), for which either the landscape depressions were neglected or treated as a lumped storage component. In their work, Shrestha et al. (2011) assumed the entire watershed area was contributing to the watershed outlet for all events. Conversely, Sophocleous et al. (1999) treated the depressions as if they were always non-contributing. In Almendinger et al. (2012) depressions were aggregated as a single, lumped storage per sub-basin using SWAT's surface depression module. The latter approach appeared to improve streamflow simulations and could provide some simulation of the dynamic storage in the ponds, but did not bring in considerations of the heterogeneity in storage capacity of the ponds. Studies where incorporation of heterogeneity was attempted include Wang et al. (2008), who developed a routine to simulate the processes occurring for each individual depression. However, because of computational demands with this type of approach, it was considered unsuitable for the simulations of the very large watersheds that were the focus of the work herein.

2 STUDY WATERSHEDS

The first study watershed was the Assiniboine River Basin, which drains areas in Eastern Saskatchewan and Western Manitoba and terminates at the Red River in Winnipeg. Figure 1 shows a map of the watershed. The Saskatchewan portion of the watershed is 17,300 km² in area. For the simulations, data from a gauging station on the Assiniboine River was used (Kamsack Gauging Station – Water Survey of Canada Station No. 05MD004 at 51°33′53″N latitude and 101°54′48″ W longitude). Measurements of the daily flow and daily sediment load were available at this station, which were obtained from the Environment Canada HYDAT hydrometric database. At the Kamsack gauging station, the station has a gross drainage area of 13,000 km². However, only 4320 km² of the area is considered contributing by the definition of Godwin and Martin (1975), which considered what areas of the watershed would be contributing in a 1:2 year storm. In the northwest and southwest region of the watershed, the elevation is about 718 m and near the Kamsack gauging station the elevation is about 428 m. The mean annual precipitation is 450 mm per year, while the mean annual temperature is about 1°C (Saskatchewan Watershed Authority (SWA) 2005). It is estimated that 63 % of the total streamflow in the watershed is generated from snowmelt runoff in April and May (SWA 2005). The land use in the watershed is predominantly agriculture, with 62 % of the watershed in annual crops, 25 % in pasture and range grass, and 11 % in forest (Ltholf *et al.* 2008). Soils in the watershed are predominantly characterized as Black Chernozemic soils (70 %) (SWA 2005).

The second watershed studied was the Moose Jaw River watershed, which is within the Qu'Appelle watershed (Fig. 1). The Moose Jaw River is a major tributary of the Qu'Appelle River. For the simulations, daily flow and daily sediment data from the gauging station near Burdick (50°24′1.2″ N latitude and 105°23′52.3″W longitude; Station No. 05JE006) were used. The watershed has a gross drainage area of 9230 km²; however, for the 1:2 year return period storm, only 3470 km² is considered to be contributing to streamflow (Godwin and Martin 1975). The elevation of the watershed is about 877 m in its southwest region and 395 m in its central region near Burdick. The mean annual precipitation is 365 mm with an annual average air temperature of 4°C (Environment Canada 2009). The predominant contribution to streamflow is snowmelt, which occurs in early spring. The watershed is also dominated by agriculture with 71 % in annual crops, 21 % in pasture and range grass, and 4 % in forest (Olthof *et al.* 2008). There are heavy clay soils in the east of the watershed and gravelly sandy soils in its western portions (SWA 2005).

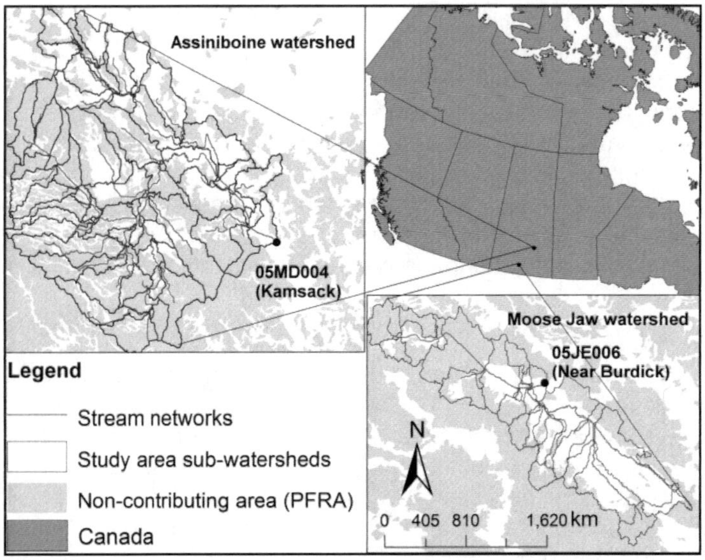

Figure 1: Map showing the studied Moose Jaw River and Assiniboine watersheds

3 MODIFICATIONS TO SWAT

3.1 Probability-Distribution Approach to Depression Storage

For the development of the probability distribution to describe storage in the depressions in the study watersheds, terrain analysis of DEM (Digital Elevation Model) data was performed using the ArcGIS 10 tool called "ArcHydro Tools" (Environmental Systems Research Institute 2011). Within ArcHydro Tools is a depression analysis routine. This was used to identify each landscape depression in the study watersheds and the associated area that drains to it, as well as its surface area (when full) and volume. Probability distributions were then fitted to the data in terms of the storage capacities of the depressions; storage capacity is defined as the volume of a depression divided by its surface area. It was found that both the exponential and Pareto distributions fit the data reasonably well. However, the exponential distribution was chosen because the distribution has fewer parameters and this is more computationally efficient to model. A comparison of the distributions to the storage capacity data for the study watersheds and the equation for the probability distribution are given in Mekonnen *et al.* (2016b).

For an individual depression, the water balance is described as follows. If the inputs of water to the depression exceed that which can be stored, runoff from the depression is equal to the inputs of water to the depression minus any losses and the storage in the depression. The inputs of water to a depression include precipitation and runoff from upland areas contributing to the depression. Losses from a depression include evapotranspiration and seepage. That stored within the depression is equal to the water storage capacity of the depression minus the water initially stored in the depression. For the work herein, all terms are written in terms of volume per surface area. If the inputs of water do not exceed the sum of the losses and capacity of the depression to store the water, there is no runoff.

The probability distributed model keeps track of the percentage of the depressions that are full to their capacity. With precipitation, the precipitation is distributed evenly to the potholes. Those that are full to their capacity contribute runoff. Those that are not full, collect precipitation until they are full per the water balance described above. This probability distributed model was coded into the SWAT Pond module. The details of the mathematics of this model and its integration into SWAT are given in Mekonnen *et al.* (2016a,b).

3.2 Seasonally Adjusted Soil Erodibility Factors

Research conducted by McConkey *et al.* (1997) indicated for the nearby Swift Current watershed, soil erodibility varied by "season". They divided the year into four periods to define these seasons: Period 1 was November 1 to March 15; Period 2 was March 16 to March 31; Period 3 was April 1 to April 30; and

Period 4 was May 1 to October 31. These periods corresponded to varied conditions of runoff and whether the soil was frozen or partially-frozen. In Period 1, the ground is frozen and runoff is due to snowmelt. In Period 2, the soil is partially-frozen and runoff is due to snowmelt. In Period 3, the soil is either thawed if not covered by snow or partially-thawed if covered by snow. Finally, in Period 4, the soil is thawed and there is no snow. In field plot tests, McConkey *et al.* (1997) observed varied soil erodibility coefficients for the different periods.

To test whether the study watersheds were showing variations in erodibility during these periods, the observed daily sediment load was plotted against observed daily streamflow for each of the listed periods, as given in Mekonnen *et al.* (2016b). It was seen there is greatly varied slopes in this relationship between periods indicating that seasonal erodibility values should be considered. As such, soil erodibility values in SWAT were modified to vary K (see Eq. 1) for each of the listed periods. For this work, the ratio of the soil erodibility coefficient to its typical annualized value for each period suggested by McConkey *et al.* (1997) were used to adjust the annualized erodibility coefficient for the soils given in SWAT.

4 INPUT DATA REQUIREMENTS

The spatial data required for use in the model include land cover, topography, and soils data. The land cover data were obtained from the GeoBase Canada (2007) and was prepared through vectorization of raster-thematic data originating from classified Landsat 5 and Landsat 7 ortho-images. The land-cover data were available at a scale of 1:250,000. The DEM of the case study basins were obtained from the GeoBase Canada website at a scale of 1:50,000. Detailed soil data at a resolution of 1:1,000,000 along with soil properties used in the SWAT model (version 2009) were obtained from the Agriculture and Agri-Food Canada database (Soil Landscapes of Canada Working Group 2007).

Gridded daily temperature (minimum and maximum) and precipitation data was used as the meteorological input data for the SWAT model (version 2009). The gridded climate dataset for Canada (Hutchinson *et al.* 2009) was obtained from Agriculture and Agri-Food Canada. This dataset covers south of 60°N latitude in Canada for the period 1961–2003, and was prepared through interpolation of observations from Environment Canada using a thin-plate smoothing spline-surface fitting method at a 10 km spatial resolution. Choi *et al.* (2009) demonstrated the suitability of gridded climate data to calibrate a hydrologic model in a prairie environment.

5 MODEL CALIBRATION AND VALIDATION

Model calibration was carried out using the shuffled complex evolution-uncertainty analysis algorithm (SCE-UA). Flow parameters were adjusted first to best represent flow processes. Then sediment calibration parameters were optimized while keeping the flow parameters fixed. Details of the model calibration procedures and reasons for this two-step approach are given in Mekonnen *et al.* 2016b. Observed daily discharge data from the Assiniboine River at Kamsack and Moose Jaw River near Burdick hydrometric stations were used for the flow calibration. Four years of flow data (1992–1995) were used for model calibration, and another four years of flow data (1996–1999) for model validation at the Kamsack gauging station. Similarly, five years of flow data (1992–1997) were used for model calibration, and an additional five years of flow data (1998–2002) for model validation at the Moose Jaw River near Burdick gauging station. Furthermore, two years of model warm-up period were allowed prior to model calibration for both watersheds to reduce uncertainty associated with initial conditions.

For the calibration for sediment, the challenge for the study watersheds was the limited frequency of sediment load measurements and period of coverage. For instance, Water Survey of Canada had terminated sediment data collection in the study watersheds in 1983. Consequently, sediment data are available only in the period of 1970–1983, and limited to spring and summer time. Calibration and validation were performed by comparing the simulated sediment load with observations corresponding to the dates when observation data were available (mostly available for the spring and summer periods). A total of 980 observations of sediment loading over four years (1972–1975) and 979 observations over four years (1976–1979) were used for calibration and validation, respectively, for the Assiniboine River watershed (Kamsack gauging station). Similarly, sediment data consisting of 1,238 observations over the years 1972–1977 were used to calibrate the model for the Moose Jaw River watershed. Additional sediment data consisting of 385 observations over the years 1978–1983 were used to validate the model for the Moose Jaw River watershed. Prior to the model calibration and validation periods, two years of model input data and warm-up period were allowed to reduce uncertainty associated with initial conditions. Therefore, because of data limitations, the flow and sediment calibrations were performed over different periods. A similar methodology had to be implemented by Santhi *et al.* (2006) while calibrating SWAT for a large river basin in the United States (Bosque River Watershed). Thus, the model's performance at simulating streamflow was assessed for the sediment calibration and validation periods. Results show that the model had good performance during these periods.

Performance of the model was judged by graphical means and several statistical parameters including the Nash-Sutcliffe efficiency (NSE) and the root mean square error (RMSE). The Nash-Sutcliffe Efficiency, NSE, is:

$$\mathrm{NSE} = 1 - \frac{\sum_{i=1}^{n} (\widehat{Q}_i - Q_i)^2}{\sum_{i=1}^{n} (\bar{Q} - Q_i)^2} \qquad [2]$$

where n is the number of data points; $\widehat{Q}_i$, is the simulated value at time i; Q_i is the observed value at time i; and $\bar{Q}$ is the averaged observed streamflow. According to van Liew et al. (2003), a model can be rated as "good" for an NSE greater than 0.75, satisfactory for an NSE value between 0.36 and 0.75, and unsatisfactory if NSE less than or equal to 0.36. The root mean square error is calculated by:

$$\mathrm{RSME} = \sqrt{\sum_{i=1}^{n} (Q_i - \widehat{Q}_i)^2} \qquad [4]$$

For the model, the calibrated parameters for streamflow were the curve number, soil evaporation compensation factor, surface runoff lag coefficient, baseflow factor, snowfall temperature, snowmelt base temperature, maximum melt factor, minimum melt factor, snowpack temperature lag factor, areal snow coverage threshold at 100 %, areal snow coverage threshold at 50 %, depression maximum storage capacity, and Manning's n for the main channel. The calibrated parameters for predictions of sediment yield in the study watersheds are the peak rate adjustment factor for sediment, linear parameter for maximum sediment re-entrained, exponent parameter for sediment re-entrained, USLE support practice factor, channel erodibility factor, and channel cover factor. Details of the calibration results are given in Mekonnen *et al.* 2016b.

6 RESULTS AND DISCUSSION

6.1 Streamflow Simulations

First, the probability distributed approach (SWAT-PDLD) was evaluated for the two study watersheds for the prediction of streamflow. The results from SWAT-PDLD were compared to those from SWAT assuming no depressions in the watershed and SWAT with the lumped approach in the existing POND module. Figures 2 and 3 give the simulated *vs.* observed data for the calibration and validation periods for the average monthly streamflows during the calibration and validation periods, with the associated statistical parameters given in Table 1. Additional daily streamflow simulation results and duration curves are given in Mekonnen *et al.* (2016a).

It is seen that the application of SWAT considering no-depressions (all drainage areas contributing) gave poor model performance, where SWAT consistently overestimated streamflow. The assumption of lumping all depressions into one for each sub-basin, which is done in the SWAT Pond Module, did significantly improve model performance as compared to the no depression case. However, this modelling approach tended to underestimate peak flows, which was also observed by Wang *et al.* (2008). Further improvement in modelling results did occur with the probability distributed approach to storage, as evidenced by increased NSE values and reduced RMSE as shown in Table 1.

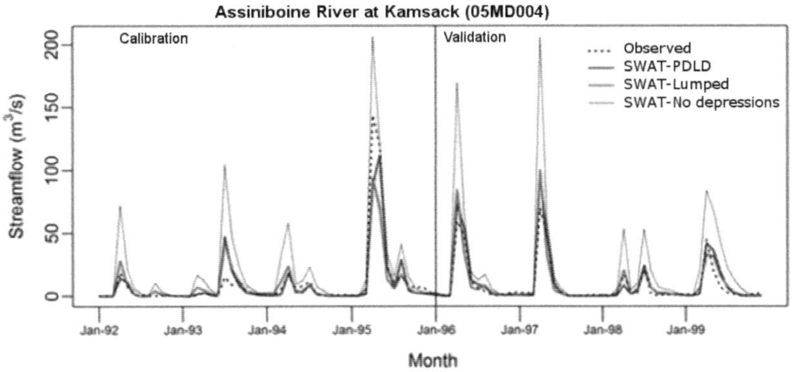

Figure 2: Average monthly hydrographs for Assiniboine River at Kamsack during the calibration and validation periods for the three SWAT configurations (adapted from Mekonnen *et al.* 2016a)

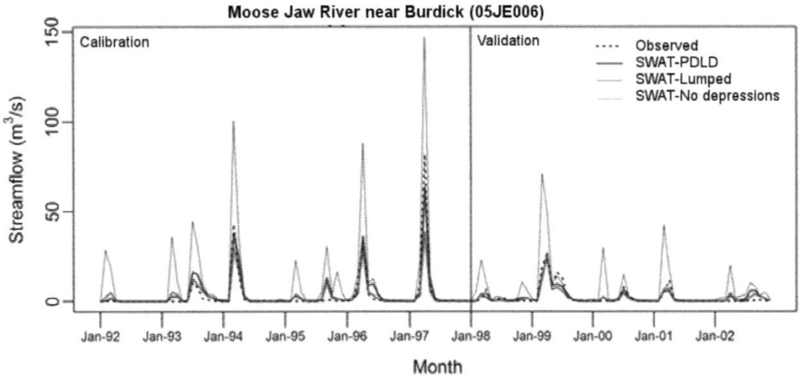

Figure 3: Average monthly hydrographs for Moose Jaw River at Burdick during the calibration and validation periods for the three test SWAT configurations (adapted from Mekonnen *et al.* 2016a)

Table 1: Statistical performance measures for monthly streamflow simulations

	Model Performance of Monthly Streamflow: Calibration (Validation)			
Model	Location	NSE	RMSE	NSE Model Performance Rating
SWAT - No depressions	Kamsack	0.37(-2.0)	26.93 (28.98)	Unsatisfactory
	Burdick	-0.80 (-2.10)	15.58 (12.77)	Unsatisfactory
SWAT lumped	Kamsack	0.81 (0.80)	10.82 (7.40)	Good
	Burdick	0.76 (0.84)	5.74 (2.94)	Good
SWAT with PDLD	Kamsack	0.86 (0.89)	8.84 (5.60)	Good
	Burdick	0.90 (0.89)	3.67 (2.46)	Good

6.2 Sediment Export Simulations with the Probability Distribution Approach for Landscape Depression Storage

The SWAT model with the probability approach to storage was then used to test the model for sediment export simulation for the study watersheds. The model was tested in two cases: (1) using annual values of soil erodibility; and (2) using seasonally adjusted erodibility values as defined above. The graphical results of the simulations for the Assiniboine River at Kamsack for these cases are given in Figures 4 and 5. The model performance results for simulating the mean daily sediment year (in tons per day) is given in Table 2 for both watersheds. The results show improved sediment export prediction using the seasonal soil erodibility values, however sediment export was still under predicted. This is especially true for the prediction of spring peak values. It is thought that some of this under prediction by the model may be due to unaccounted for management practices in these agricultural watersheds. The model assumed a cover crop in agricultural areas in all years. However, it has been common practice to leave some agricultural fields as fallow in some years (van Kooten et al. 1989) and this is known to result in increased erosion (van Kooten and Furtan 1987).

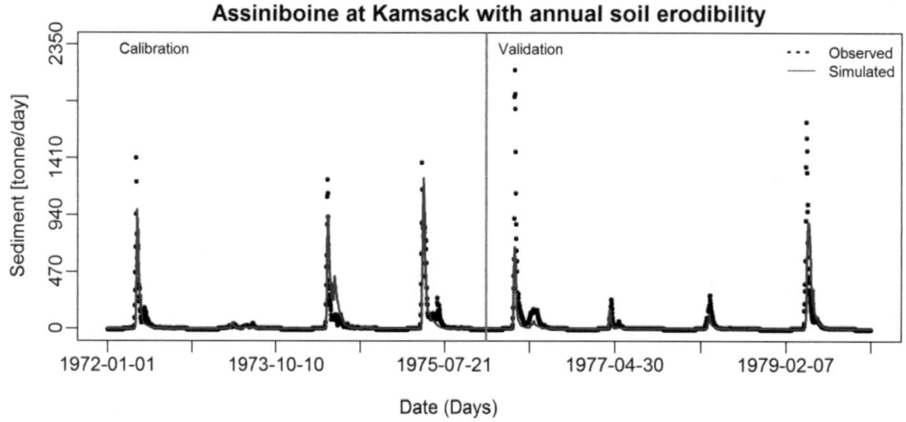

Figure 4: Observed and simulated daily sediment export for the Assiniboine River at Kamsack with SWAT-PDLD and an annual value of soil erodibility (K) (adapted from Mekonnen et al. 2016b)

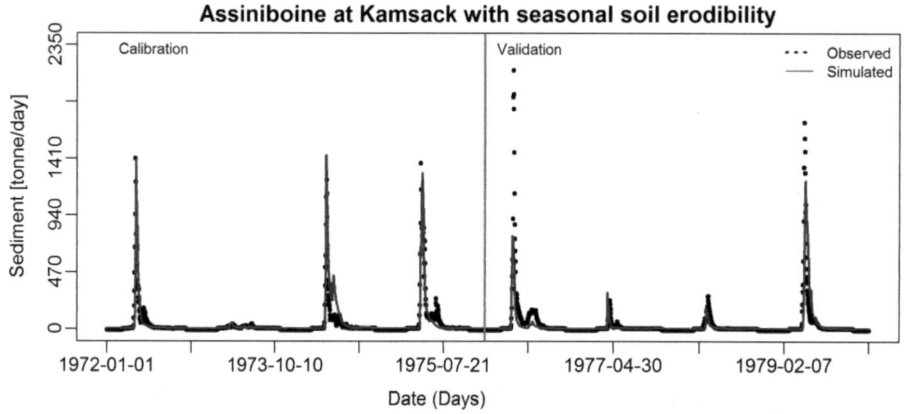

Figure 5: Observed and simulated daily sediment export for the Assiniboine River at Kamsack with SWAT-PDLD and a seasonally varying value of soil erodibility (K) (adapted from Mekonnen et al. 2016b)

Table 2: Sediment export simulation results as compared to observed values (*No observed data)

				Mean daily sediment yield (tons per day)	
Season	Months	Location	Observed	Model (annual K)	Model (seasonal K)
Fall	September, October,	Kamsack	0.94	0.76	0.77
	and November	Burdick	0.73	0.46	0.48
Winter	December, January,	Kamsack	*	0.10	0.10
	and February	Burdick	*	0.03	0.03
Spring	March, April,	Kamsack	127.4	102.3	120.0
	and May	Burdick	750.3	279.8	486.9
Summer	June, July	Kamsack	22.9	13.6	13.8
	and August	Burdick	26.9	12.9	15.5

7 CONCLUSIONS

Canadian prairie region watersheds are dominated by depression storage in millions of potholes of varied size that have a significant impact on streamflow generation in the region. In modelling streamflow within the region using SWAT, it was found that incorporation of dynamic contributions to streamflow from depression storage, herein using a probability distributed approach, significantly improves modelling results. Further, in modelling sediment export in these watersheds, seasonal adjustments for soil erodibility were found also to be important, as erodibility changes when the soil is frozen, partially-frozen or thawed.

REFERENCES

Abedini, M.J. 1998. *On Depression Storage, its Modelling and Scale*. Ph.D. Thesis, Department of Water Resources Engineering, University of Guelph, Guelph, Canada.

Almendinger, J.E, Murphy, M.S., and Ulrich, J.S. 2012. Use of the Soil and Water Assessment Tool to Scale Sediment Delivery from Field to Watershed in an Agricultural Landscape with Topographic Depressions." *J. Environ. Qual.,* **43**(1), 9-17.

Arnold, J.G., Srinivasan, R., Muttiah, R.S., and Williams, J.R. 1998. Large-Area Hydrologic Modeling and Assessment: Part I Model Development." *J. American Water Resour. Assoc.,* **34**(1): 73-89.

Chu, X., Yang, J., Chi, Y., Zhang, J. 2013. Dynamic Puddle Delineation and Modeling of Puddle-to-Puddle Filling-Spilling-Merging-Splitting Overland Flow Processes. *Water Resour. Res.*, **49**(6): 3825-3829.

Choi, W, Kim, S.J., Rasmussen, P.F., and Moore, A.R. 2009. Use of the North American Regional Reanalysis for Hydrological Modelling in Manitoba. *Can. Water Resour. J.,* **34**(1): 17-36.

Cunge, J.A. 1969. On the Subject of a Flood Propagation Method (Muskingum method). *J. Hydraulic Research,* **7**(2): 205-230.

Environment Canada. 2009. Canadian climate normals 1971-2000(online). Available at: 573 http://www.climate.weatheroffice.ec.gc.ca/climate_normals/index_e.html. (Retrieved: 10 Dec 2011).

Environmental Systems Research Institute. 2011. Arc Hydro Tools: Tutorial, v2. Redlands, CA.

Euliss, N.H. Jr., Mushet, D.M., and Wrubleski, D.A. 1999. "Wetlands of the Prairie Pothole Region: Invertebrate Species Composition, Ecology, and Management." In *Invertebrates in Freshwater Wetlands of North America: Ecology and Management*, John Wiley & Sons, New York.

Fang, X., Pomeroy, J.W. 2008. Drought impacts on Canadian prairie wetland snow hydrology. *Hydrological Processes*, 22(15): 2858-2873.

Fontaine, T.A., Cruickshank, T.S., Arnold, J.G., and Hotchkiss, R.H. 2002. Development of a Snowfall-Snowmelt Routine for Mountainous Terrain for the Soil Water Assessment Toil (SWAT). *J. Hydrol.*, **262**(1-4): 209-223.

GeoBase Canada. 2007. *Canadian Digital Elevation Data, Level 1 Production Specifications*. Natural Resources Canada, Canadian Centre for Topographic Information, Quebec, Canada.

Godwin, R.B., and Martin, F.R. 1975. Calculation of Gross and Effective Drainage Areas for the Prairie Provinces. *Canadian Hydrology Symposium – 1975 Proceedings*, Winnipeg, Manitoba, NRC, Canada.

Green, W.H., and Ampt, G.A. 1911. Studies on Soil Physics-I: The Flow of Air and Water through Soils. *J. Agric. Sci.* **4**: 11-24.

Hargreaves, G.L., Hargreaves, G.H., and Riley, J.P. 1985. Agricultural Benefits for Senegal River Basin. *J. Irrig. Drain. Eng.,* **111**(2):113-124.

Hayashi, M., Quinton, W.L., Pietroniro, A., Gibson, J.J. 2004. Hydrologic Functions of Wetlands in a Discontinuous Permafrost Basin Indicated by Isotropic and Chemical Signatures. *J.Hydrol.*, **296**: 81-97.

Hutchinson, M.F., McKenney, D.W., Lawrence, K., Pedlar, J.H., Hopkinson, R.F., Milewska, E., Papadopol, P. 2009. Development and Testing of Canada-Wide Interpolated Spatial Models of Daily Minimum–Maximum Temperature and Precipitation for 1961-2003. *J. Appl. Meteorol. Clim,* **48**: 725-741

McConkey, B. G., Nicholaichuk, W., Steppuhn, H., and Reimer, C.D. 1997. Sediment Yield and Seasonal Soil Erodibility for Semiarid Cropland in Western Canada." *Can. J. Soil Sci.,* **77**: 33-40.

Mekonnen, B.A., Mazurek, K.A., Putz, G. 2016a. Incorporating Landscape Depression Heterogeneity into the Soil and Water Assessment Tool (SWAT) using a Probability Distribution, *Hydrol. Process.* **30**(13): 2373-2389.

Mekonnen, B.A., Mazurek, K.A., and Putz, G. 2016b. Sediment Export Modeling in Cold-Climate Prairie Watersheds", *Journal of Hydrologic Engineering*, 21(5): 05016005 1-14.

Mekonnen, M.A., Wheater, H.S., Ireson, A.M., Spence, C., Davison, B., and Pietroniro, A. 2014. Towards an Improved Land Surface Scheme for Prairie Landscapes." *J. Hydrol.,* 511, 105-116.
Monteith, J.L. 1965. Evaporation and the Environment in the State and Movement of Water in Living Organisms. *Proc. Soc. Experimental. Biol., 19th Symposium*, Swansea, UK.
Moore, R.J. 1985. The Probability-Distributed Principle and Runoff Production at Point and Basin Scales. *Hydrological Sciences Journal,* **30**(2): 273-297.
Neitsch, S.L., Arnold, J.G., Kiniry, J.R., and Williams, J.R. 2011. *Soil and Water Assessment Tool Theoretical Documentation: Version 2005.* Blackland Research Center, Temple, TX
Olthof, I., Latifovic, R., and Pouliot, D. 2008. *Circa-2000 Northern Land Cover of Canada.* Earth Sciences Sector, Canada Centre for Remote Sensing, Natural Resources Canada.
Pomeroy, J.W., Gray, D.M., Brown, T., Hedstrom, N.M., Quinton, W., Granger, R.J., and Carey, S. 2007. The Cold Regions Hydrological Model, a Platform for Basing Process Representation and Model Structure on Physical Evidence, *Hydrological Processes,* **21**: 2650-2667.
Priestley, C.H.B., and Taylor, R.J. 1972. On the Assessment of Surface Heat Flux and Evaporation using Large-Scale Parameters." *Mon. Weather Rev.,* **100**: 81-92.
Ritchie, J.T. 1972. A Model for Predicting Evaporation from a Row Crop with Incomplete Cover. *Water Resources Research,* **8**: 1204-1213.
Santhi, C., Srinivasan, R., Arnold, J.G., and Williams, J.R. 2006. A Modeling Approach to Evaluate the Impacts of Water Quality Management Plans Implemented in a Watershed in Texas. *Environmental Modelling & Software,* **21**: 1141-1157.
Saskatchewan Watershed Authority. 2005. Background report: Assiniboine River watershed. Regina, SK.
Shaw, D.A., van der Kamp, G., Conly, F.M., Pietroniro, A., and Martz, L. 2011. The Fill-Spill Hydrology of Prairie Wetland Complexes During Drought and Deluge. *Hydrological Processes,* **26**:3147-3156.
Shook, K., Pomeroy, J.W., Spence, C., Boychuk, L. 2013. Storage Dynamics Simulations in Prairie Wetland Hydrology Models: Evaluation and Parameterization. *Hydrol. Processes*: **27**(13): 1875-1889.
Shrestha, R.R., Dibike, Y.B., Prowse, T.D. 2011. Modeling Climate Change Impacts on Hydrology and Nutrient Loading in the Upper Assiniboine Catchment. *J. Am. Water Resour Assoc.,* **48**(1): 74-89.
Soil Conservation Service. 1972. Section 4: Hydrology, *National Engineering Handbook,* SCS, USA.
Soil Landscapes of Canada Working Group. 2007. Soil Landscapes of Canada v3.1.1. Ag. & Agri-Food Canada, Digital Map & Database 1:1 Million Scale, http://sis.agr.gc.ca/cansis/nsdb/slc/v3.1.1/intro.html
Sophocleous, M.A., Koelliker, J.K., Govindaraju, R.S., Birdie, T. Ramireddygari, S.R. Perkins, S.P. 1999. Intergrated numerical modeling for basin-wide water management: The case of the Rattlesnake Creek Basin in South-Central Kansas. *J. Hydrol.,* **214**: 179-196.
Su, M., Stolte, W.J., van der Kamp, G. 2000. Modelling Canadian Prairie Wetland Hydrology using a Semi-Distributed Flow Model. *Hydrological Processes*, **14**: 2405-2422.
Tolson, B.A., and Shoemaker, C.A. 2007. Cannonsville Reservoir Watershed SWAT2000 Model Development, Calibration, and Validation. *J. Hydrol.* **337**(1-2): 68-86.
van Kooten, G.C., and Furtan, W.H. 1987. A review of issues pertaining to soil deterioration in Canada. *Can. J. Agr. Econ,* **35**: 33-54.
van Kooten, G.C., Weisensel, W.P., and de Jong, E. 1989. Estimating the costs of soil erosion in Saskatchewan. *Can. J. Agr. Econ.,* **37**: 63-75.
van Liew, M.W., Arnold, J.G., Garbrecht, J.D. 2003. Hydrologic Simulation on Agricultural Watersheds Choosing Between Two Models. Trans. ASABE, **46**(6): 1539-1551.
Wang, X., Yang, W, Melesse, A.M., 2008. Using Hydrologic Equivalent Wetland Concept within SWAT to Estimate Streamflow in Watersheds with Numerous Wetland. *Trans. ASAE*, **51**(1): 55-72.
Williams, J.R. 1975. Sediment-Yield Prediction with Universal Equation using Runoff Energy Factor., *Proceedings of the sediment-yield workshop*, USDA Sedimentation Lab., Oxford, MS.
Williams, J.R. 1969. Flood Routing with Variable Travel Time or Variable Storage Coefficients." *Trans. ASAE,* **12**(1): 100-103.

Leadership in Sustainable Infrastructure
Leadership en Infrastructures Durables

Vancouver, Canada
May 31 – June 3, 2017/ *Mai 31 – Juin 3, 2017*

MODELLING SUPERCRITICAL FLOW-INDUCED SCOUR AROUND STRUCTURES

Sanaz. Mehrzad[1] , Colin D. Rennie[2], Ioan Nistor[3]
[1.]PhD Student, University of Ottawa, Canada
[2,3.] Professor, University of Ottawa, Canada
[4] rmehr026@uottawa.ca

Abstract: The present study investigates the scour mechanism and vortex structure around a square structure due to inland-propagating tsunami-like bores on a dry or flooded horizontal mobile bed. A series of hydraulic bores were simulated using a dam-break wave generated by the release of water impounded behind a rapidly-opening swing gate. A novel video-recording system was used to monitor the evolution of the scour and vortex structure. Image processing allowed tracking the time and spatial evolution of the scour and the observation of sediment movement around the structure. It was found that the scour mechanism associated with dam break waves differs from that of a steady flow. The influence of bed condition (fixed versus mobile) and (dry versus wet) on bore characteristics, and how they might affect the scouring process will be presented and discussed.

1 INTRODUCTION AND RESEARCH NEEDS

Field evidence indicates that destructive tsunamis cause substantial coastal sediment mobilization (Chen et al, 2016). Measurements collected from multiple tsunami events, such as the 1992 Nicaragua Tsunami and the 2004 Indian Ocean Tsunami, have recorded substantial evidence of scour around damaged buildings and bridge foundations (Yeh and Li, 2008). The field surveys noted that scouring was one of the primary causes of coastal structural damage (Chen et al, 2016).Tsunami waves have been observed with wave heights of 5–10 m. The inundating waves will be inevitably accompanied by very high flow velocities when they penetrate inland. The high flow velocities produce high bed shear stresses and large amounts of sediment movement over large areas, resulting in substantial beach erosion and scouring around a large number of structures (Li et al, 2012). Yeh and Li (2008) observed local scour at the seaward corner of a schoolhouse in Kalapakkom, India. The inundation depth was 0.95 m above the building's floor and the run up height was 4.1 m. The scour depth was approximately 1.5 m with a horizontal span of 5 m. They also observed an undermined patio in Devanaanpattinam, India, where the observed runup height was 3.0 m. The undermining resulted from swift channelized flows during the drawdown of the tsunami wave. During the 1993 Okushiri Tsunami attack, at the entrance of Okushiri Port, Japan, a scour depth of 4 m was created between breakwaters, causing the breakwater to overturn as a result of foundation failure. Generally, the first tsunami wave propagates over dry lands (dry bed condition) while the subsequent waves may intrude on the flooded coastline (wet bed condition) before full recession of the first wave. Previous research have shown that the water layer presence on the bed could influence the tsunami bore characteristics (Stansby et al 1998; Douglas and Nistor 2014). Laboratory experiments related to scour around structures caused by tsunamis are limited. Tonkin et al (2003) examined scour around a cylindrical structure located on a sloped wet sandy beach. Nakamura et al (2008) focused on scour around a square structure located on a horizontal wet sand bed. Both studies underline that such a tsunami-induced scour hole, occurring in such transient flow conditions, exhibited significantly different mechanisms than observed in steady current and consistent short-wave field studies. Laboratory waves employed in these studies were either solitary waves or long waves. At present, general scientific consensus indicates that solitary waves tend to be less representative of actual tsunamis due to their short period. This is particularly important for studies focusing on scour, where flow duration is a critical variable. The use of hydraulic bores generated from dam break waves have been shown to better replicate the temporal features of tsunami-induced flooding (Chanson. 2006). Using dam-break flow conditions, this study investigated the scour mechanism and vortex structure around the square structure due to inland-propagating tsunami bores on a dry and flooded horizontal bed. Additionally, characteristics of the bore, including run up, stream-wise velocity and bore front velocity for different initial downstream conditions (dry bed vs wet bed) will be presented.

2 EXPERIMENTAL SET UP

2.1 Flume and instrumentation

The hydrodynamic boundary conditions for these experiments were simulated using a dam-break method in a flume located in Hydraulics Laboratory at the University of Ottawa (Canada). The flume was 30 m long, 1.5 m wide and 0.8 m deep. The dam-break wave was generated by releasing water impounded behind a rapidly-opening swinging gate which was equipped with a lock and release mechanism. The sediment bed section was 3.30 m long and 1.5 m wide, delimited at each end by false floors as shown in Figure 1.

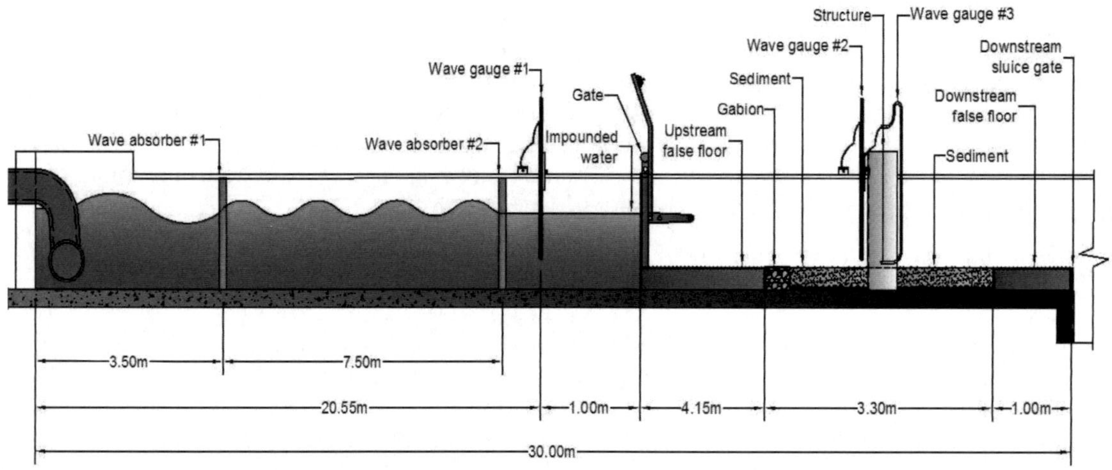

Figure 1: (a) Experiment setup, side view, (not to scale)

Two aluminum false floors with a height of 0.2 m were installed before and after the sediment bed, downstream of the gate, to provide formwork for the sediment section. A sluice gate was installed at the end of the downstream false floor to be able to create a wet bed initial condition. The upstream and downstream false floor had a length of 4.15 m and 1 m, respectively. The upstream false floor was covered with a layer of glued sand with a uniform grain size of 1 mm, in order to provide surface roughness, limiting the influence of the transition between the fixed and moveable bed. A grid system of 20 x 20 cm spacing was also painted on the top surface of the upstream false floor for measurement purposes.

The base of the sediment bed was composed of a layer of 1 cm thick coarse gravel covered by a geotextile sheet. The base layer drained water that infiltrated from below the false floor, to prevent the water from influencing the sediment bed. The sediment section was filled with 1 mm uniform sand. Directly downstream of the false floor, a 15 cm wide gabion was installed across the flume. The gabion was constructed with a wire box filled with coarse gravel, approximately 3 cm thick. The gabion was used to minimize the inevitable local scour that would occur at the interface of false floor and sediment section.

Three wave absorbers were installed in order to attenuate secondary waves within the reservoir and prevent reflective waves-induced in water level fluctuation at the test section.

Three WG-50 capacitance type wave gauges with an accuracy of 0.4% were used to record the time-history of the water surface elevation at different locations, using a sampling rate of 30 Hz. Two wave gauges were positioned as follows: one on the front face and the other one on the side face of the structure. The third wave gauge was placed at a distance of 1.0 m behind the swing gate, inside the reservoir. This wave gauge was used to record the time-history of the water surface elevation within the reservoir over the duration of the test. Two video cameras, a GoPro Hero™ Black and an IO Industrial Flare™ high speed video camera, were used during the experimental investigations a) to track the bore front as it propagated downstream towards the structure, and b) to record the evolution the of scour and vortex structure. A 0.2 m x 0.2 m square structure built of Plexiglas was used in the experiment. The structure was installed in the center of the sediment section. The square structure was transparent to allow for monitoring the scouring process by means of the high speed camera which was placed at the top of the structure. The monitoring of the scour was achieved with the help of a mirror inclined at 45° inside the structure as shown in Figure 2.

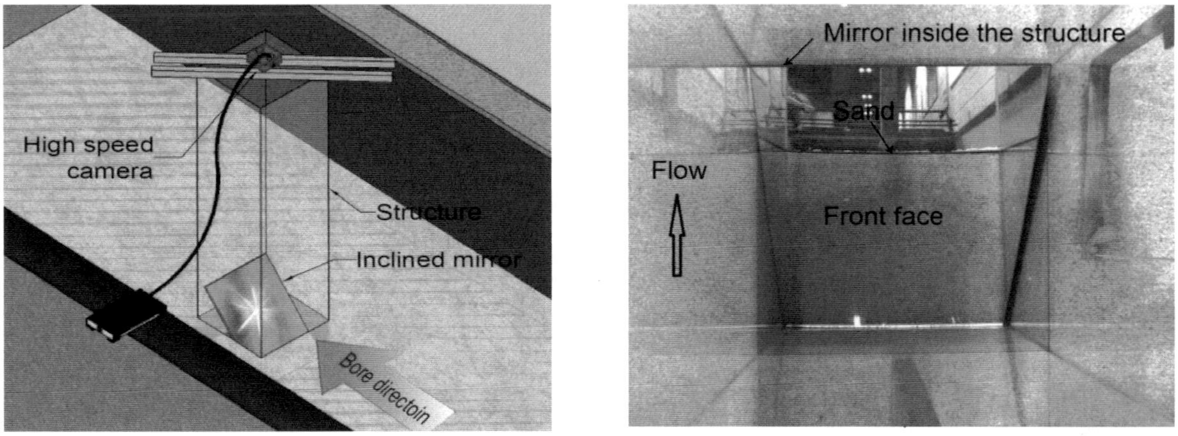

a) b)

Figure 2: (a) Experiment setup to monitor the scour process; (b) Reflected (looking from above-mirror inside) view of the structure's front face.

Turbulent velocity measurements were conducted with an Acoustic Doppler Velocimetry (ADV) using the Nortek Vectrino with a side-looking head equipped with four receivers using a sampling rate of 25 Hz (installed at x = 5.15 m from the gate).

2.2 Experimental Procedure

Before each experiment, the sediment bed was levelled and the reservoir was filled with water to the desired impounded water depth. The swinging gate was then manually unlocked and rapidly opened. The bore then advanced over the false floor and finally over the sediment bed section, while all measuring instruments were recording. The opening time of the gate was recorded to calculate the non-dimensional removal period $t\,(g/h)^{1/2}$ .The expression should be less than the square root of two to achieve the ideal dam-break condition (Lauber and Hager, 1997). To investigate the effect of bed condition on bore velocity and scour depth, two initial conditions (wet versus dry conditions) were considered downstream of the gate. A schematic illustration of the experiment is presented in Figure 3, where h_u and h_d are the initial water depths (m) upstream and downstream of the gate, respectively. All the tests were conducted with an initial impoundment depth of h_u = 0.25 m, while downstream water depth h_d varied between tests (0 m, 0.025 m, 0.05 m, or 0.125 m). Since the impoundment water depths were kept constant, tests are named according to downstream water depths as dry bed, 0.025 m wet bed, 0.05 m wet bed and 0.125 m wet bed, respectively. For the experiment, a 1:40 Froude scaling ratio was used, resulting in h_u = 0.25 m corresponding to a prototype height of 10 m.

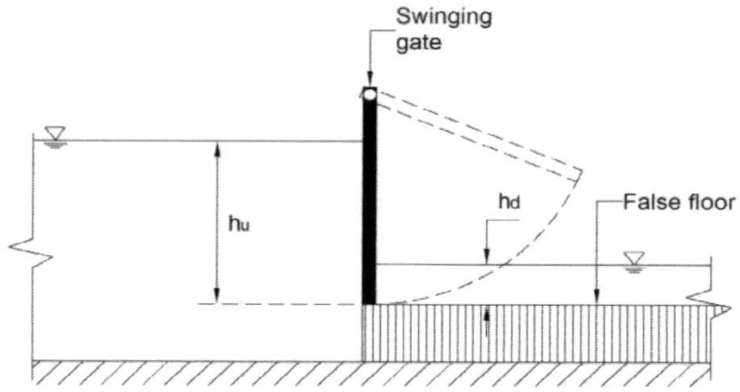

Figure 3: schematic illustration of the experiment

3 VALIDATION OF THE MODEL

A comparison between a dam-break analytical solution and the experimentally obtained bore profile is presented here. Because of the limited number of wave gauges, a number of identical tests were repeated with the same experimental configuration by installing wave gauges in sequential steps of 1 m to cover the entire flume. Repeated tests performed with the wave gauges at the same location showed a high degree of reproducibility in terms of wave arrival time and evolution of water depth. Figure 4 presents the comparison between the dam-break analytical solution and the experimentally obtained bore profile on the fixed floor before arrival of the bore on the sediment section. The x-axis corresponds to the direction along the flume length and h represents the water surface elevation. Chanson's (2006) solution considers the bottom friction in the dominant wave tip region, using the diffusive wave equation and Darcy-Weisbach friction factor (*f*). The value of *f* was taken as 0.03, described as a smooth surface. As shown in Figure 4, a good agreement can be observed between the experimental bore profiles and Chanson's solution.

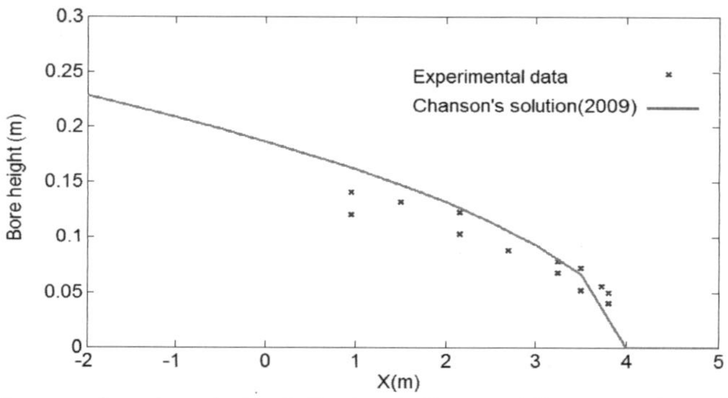

Figure 4: Comparison between dam-break analytical solution and the experimentally obtained bore profile for h_u = 0.25 m, h_d = 0 m at t = 2.48 sec

4 RESULT AND DISCUSSION

4.1 Scour mechanism

Figure 5 shows the measured results for the initial condition of h_u = 25 cm and h_d = 0 cm (dry bed condition). The black line represents the horizontal velocity measured 50cm upstream of the structure. The green and red lines show the scour depth and water surface elevation time-histories, respectively, measured at the front face of the structure. It can be seen that stream-wise velocities were captured after 2 s. The ADV measurements were noisy for the first 2 seconds of advancing bore as a result of air bubble entrainment and cavitation during the first few seconds after the bore arrival. The graph shows that significant scour depth occurs over a short period of time after the initial impact of the bore.

The short duration and turbulent nature of the bore induced a rapid scouring process. The initial bore front had a vertical front profile, and the maximum bore depth was attained within 8 seconds. The maximum scour depth occurred 13 seconds after the arrival of the bore front.

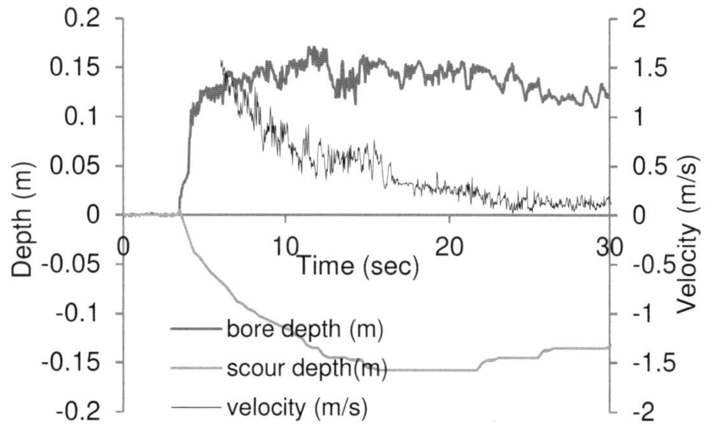

Figure 5: Water surface elevation (red line) and scour depth time history (green line) measured at the front face of the structure and horizontal velocity (black line) measured 50cm upstream of the structure for $h_u = 0.25$ m, $h_d = 0$ m

The critical shear stress for initiation of sediment motion was calculated from the Shields diagram in order to compute critical shear velocity. For the 1 mm grain size, the calculated critical velocity was 0.28 m/s. Figure 5 shows that the highest scour rate coincided with the largest flow velocities greater than the critical velocity. As the flow velocity decreased, the scouring pace slowed and the suspended sediment started to settle, filling the scour hole.

4.2 Effect of bed condition

Generally, the first tsunami wave propagates in a form of turbulent hydraulic bore over coastal areas that are in a dry bed condition, while the second or third tsunami waves may advance into the previously inundated inland area, which is referred to as the "wet bed condition". It is reported that the presence of a still water layer (h_d) can significantly influence the bore propagation characteristics. In this section, the bore propagation characteristics and the resulting scour profile on both wet and dry bed conditions are compared and discussed.

4.2.1 Bore front velocity (U)

Figure 6 presents bore front average velocity as it propagated to the downstream end of the flume. The propagation of the bore front along the flume was tracked using the high-speed video camera. The velocity was calculated using the time sequence of the video recordings and the distance between the two gridlines. This velocity was assumed to represent the average velocity for that section. In general, for both fixed and movable beds, the highest velocities were recorded at locations closest to the gate, as a result of the bed friction. The bore front velocity gradually decreased as the bore propagated to the downstream end of the flume.

On the fixed bed, maximum instantaneous velocities reached as high as 2.26 m/s for the dry bed. The bore front velocities were 2.15 m/s, 1.84 m/s and 1.75 m/s for the 0.025 m wet bed, 0.05 m wet bed and 0.125 m wet bed, respectively. The bore front velocity decreased as the downstream water depth increased. The presence of the still water layer caused a resistance to the propagation of the incoming bore, inducing energy dissipation in the bore front. The higher downstream water depth resulted in lower total head and subsequent less kinetic energy and lower velocity. Therefore, the fastest bore front velocity occurred in the dry bed condition.

Conversely, the velocity measurement on the mobile bed showed different trends from that of a fixed bed. The bore front velocity in wet bed conditions with a thin layer of water was faster than observed in dry permeable bed. This was likely a result of water infiltrating into the bed under the dry condition, given the porosity and permeability of the sediment bed. In other words, energy dissipation due to seepage is presumed to originate from the interface between the fluid and the sediment bed. This counters some previous results in other studies discussed in Section 1. Chanson (2003) and Stansby et al (1998) observed that the speed of the bore on a dry (fixed) bed is higher than that on a wet bed. However, for both of their experimental conditions, the bed was solid and hydraulically smooth, with no sediment (zero porosity). Therefore, the present results indicate that the conclusion of higher bore front velocity moving over a dry fixed bed cannot be expanded to the case with a sediment bed.

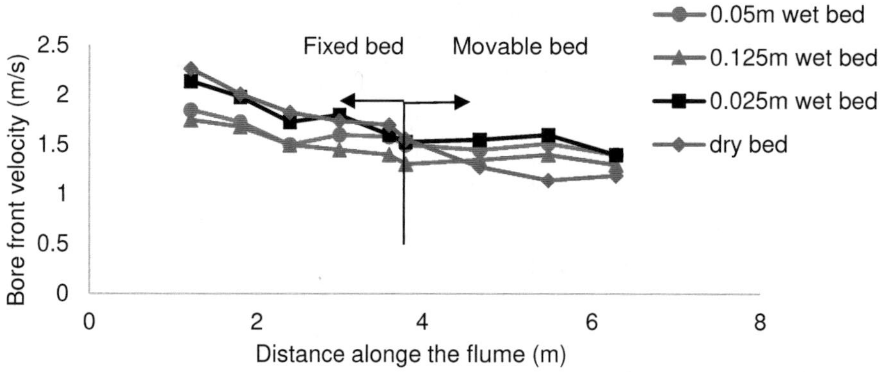

Figure 6: Bore front velocity along the flume for all initial conditions

4.2.2 Run up and stream-wise velocity of bore propagation

Comparisons of the time-history of water surface elevation at the upstream face of the structure and stream-wise velocity are shown for different initial conditions in Figure 7. It can be seen that the run-up height resulting from the initial impact was influenced by the bed condition. The run-up height reached the highest elevation (0.175 m) for the case of bore propagating over a thin layer of water (h_d =0. 025 m). The maximum run up heights were 0.17 m, 0.15 m and 0.11 m for h_d =0 m, h_d =0.05 m and h_d =0.125 m, respectively. Therefore, it can be seen that splash height in wet bed condition with thin layer of water was slightly greater than that of the dry bed condition.

Comparing the results in wet bed condition, a consistent reduction in the run up height was observed with increasing downstream water depths. The maximum run up height for the dry condition was 0.17 m, which occurred 7.9 s after the arrival of the bore front; while maximum bore depth was attained within less than a second for wet bed conditions. This also contrasted other studies, such as Stansby et al (1998) and Douglas and Nistor (2014), which observed that maximum bore heights occurred within 0.5 s after the initial impact for bores propagating over dry smooth solid beds. This delay in maximum run up height again could be caused by permeability of the mobile bed under dry conditions. Moreover, a delay in the arrival of the bore front was observed as the downstream depth increased, which shows a similar trend as the bore front velocity. For the wet bed conditions, a relatively vertical bore front was measured with respect to the dry bed condition. Observations showed steeper and deeper bore front with higher air entrainment in the bore front for wet bed conditions as the downstream water depth increased. To further analyze the effects of bed condition on propagation characteristics, a time-history of stream-wise velocity, measured upstream of the structure, is also given in Figure 7 for all initial downstream water depths. In all cases, the ADV sample volume was set such that it would be at or just below the bore water surface. As noted above, the stream-wise velocity measurements began 2 s after arrival of the bore. According to Figure 6, the bore front velocities at the location of the model structure were 1.14 m/s, 1.6 m/s, 1.51 m/s and 1.4 m/s, while from Figure 7 the maximum stream-wise velocities were 1.57 m/s,1.45m/s, 1.005 m/s and 0.554 for h_d=0 m, h_d=0.025 m, h_d=0.05 m and h_d=0.125 m, respectively. It appears that the lowest bore front velocity occurred in the dry bed condition, whereas the dry bed run had the highest measured stream-wise velocity 2.53 seconds after bore arrival. It appears that saturation of the initially dry bed allowed for increased velocity following passage of the bore front. Moreover, these results show reductions in the stream-wise velocity for increasing initial downstream depths, similar to the trend for bore front velocity.

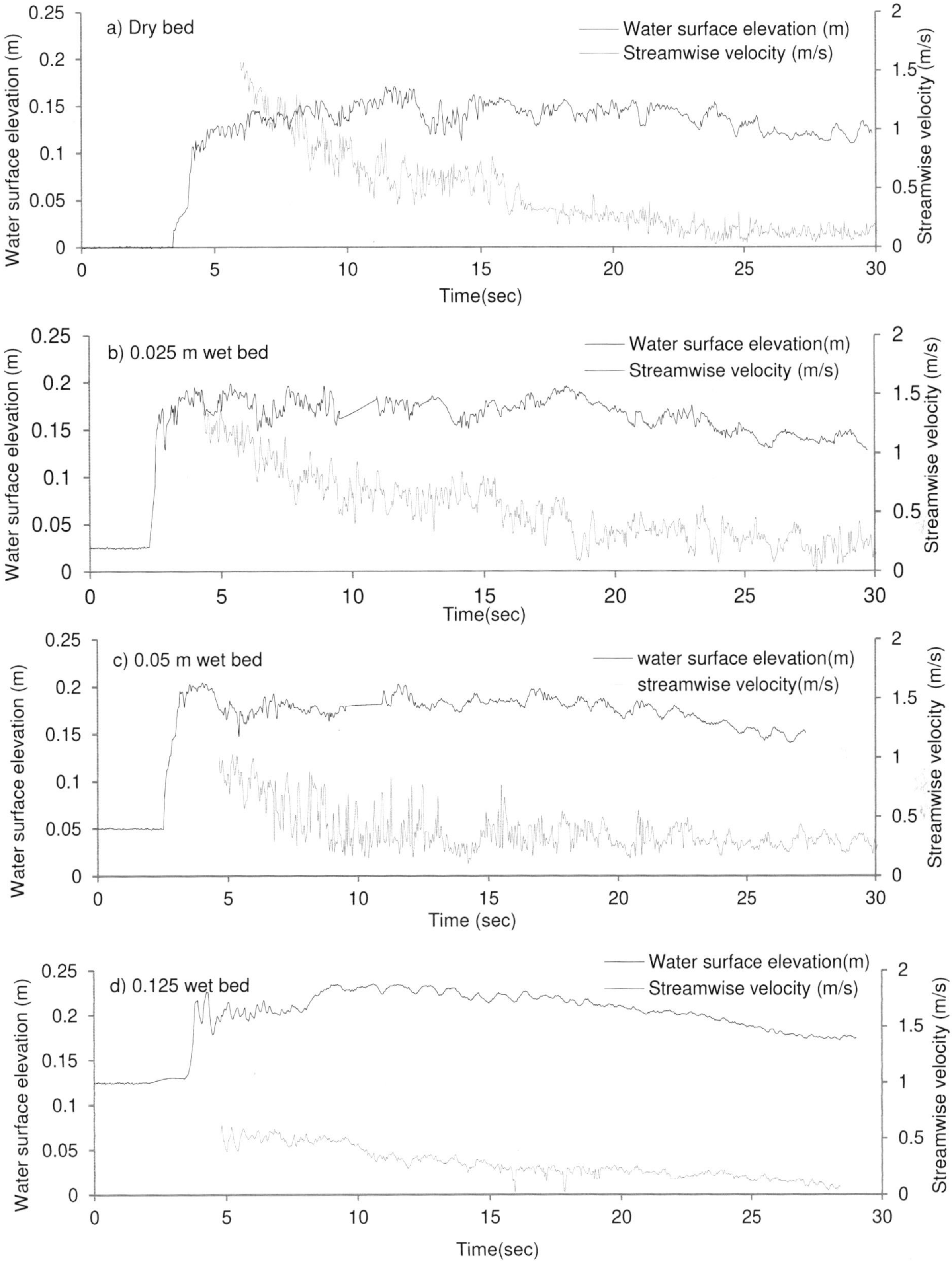

Figure 7: measured water surface elevation (black line) and measured streamwise velocity (red line) for a) h_d = 0 m, b) h_d = 0.025 m , c) h_d= 0.05 m and d) h_d = 0.125 m

4.2.3 Scour depth

Figure 8 demonstrates time-history of scour depth for all the experimental tests. For better comparison of all the tests, time zero refers to the bore arrival time at the structure, i.e., when the scouring process would be initiated. The magnitude of the scour depth for the 0.025 m wet bed was 0.16 m, which was slightly larger than that generated for the dry bed test (0.15 m scour depth). This would imply that the second tsunami wave can be even more destructive than the first wave propagating on dry land. The scour depths were respectively 0.12 m and 0.03 m for the 5 cm and 12.5 initial water depths, reductions of 40% and 140% compared to the 2.5 cm initial water depth case. Therefore, similar to bore front velocity and stream-wise velocity, a substantial reduction in scour depth was observed with increasing the downstream initial water depth. The presence of the initial still water depth weakened strength of the down-flow at the structure face and the bore became less effective at displacing the sediment grains around the structure. In contrast, Arneson et al (2012) reported that the scour capacity of a steady flow field reduces as the flow depth decreases. The down-flow at the pier face becomes less well developed because it has a shortened length over which to develop. The vorticity of the horseshoe vortex weakens as the down-flow weakens, and the wake vortex also weakens due to the increased importance of bed friction in a shallower depth. There is a noticeable difference between the time that it takes for the maximum scour depth to form in a dry bed compared to that of wet bed conditions The maximum scour depth in the dry bed condition was attained 14 s following the bore arrival, while for the wet bed tests it was attained within 7, 7 and 4.6 s. This was likely because under the dry bed condition the bore reached the structure before the bed was saturated completely.

Lastly, it can be seen that the final scour depth was less than the maximum scour depth. Therefore, using the observed final scour depth as a design criterion for the foundation of coastal structures is not recommended.

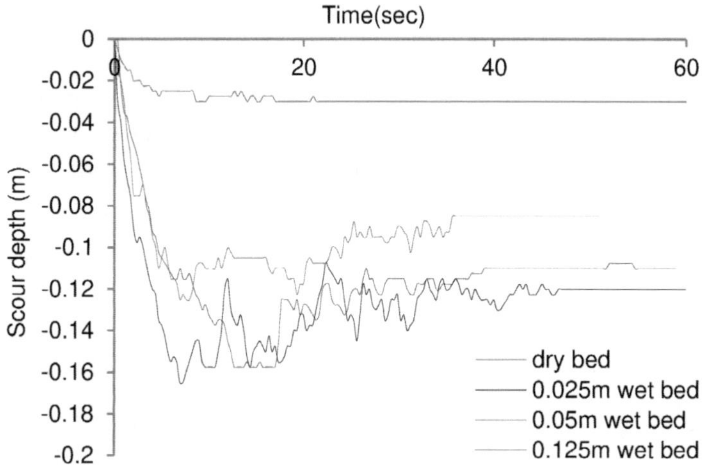

Figure 8: scour depth time history for different downstream initial conditions

5 CONCLUSIONS

In this study, the scouring mechanism around a square structure due to a tsunami-like, inland propagating bore and the influence of bed condition (dry vs wet) and (fixed vs movable bed) on bore characteristics and the resulting scour has been investigated experimentally. Based on the analysis of the recorded data, the following conclusions have been drawn:

- The short duration and very turbulent nature of the bores induced a rapid scouring process. Significant scour depths were reached over a short period of time at the front of the structure.
- The importance of the downstream depth on bore development and its propagation characteristics is shown. The existence of the still water layer caused a resistance to the propagation of the incoming bore and resulted in energy dissipation of the bore front.

- The present results indicate that the previous observation of higher bore front velocity moving over a dry fixed bed cannot be expanded to sediment bed cases. As a result of water infiltrating into the bed, the bore front velocity was less for the dry permeable bed case than wet bed cases with downstream water depth.
- Similar to bore front velocity and stream-wise velocity, a substantial reduction in scour depth was observed with increasing downstream initial water depth.

6 REFERENCES

Arneson L.A, Zevenbergen L.W., Lagasse P.F., Clopper P.E. (2012). Evaluating scour at bridges fifth edition. *Hydraulic Engineering Circular*, **18:** FHWA-HIF-12-003

Chanson, H., Aoki, S., & Maruyama, M. (2003). An Experimental Study of Tsunami Runup on Dry and Wet Horizontal Coastlines. *Science of Tsunami Hazards*, **20**(5), 278-293

Chanson, H. (2004). *Environmental Hydraulics of Open Channel Flows*. Elsevier Butterworth Heinemann,Oxford, UK, 483 pages (ISBN 0 7506 6165 8).

Chanson, H. (2006). Tsunami surges on dry coastal plains: application of dam break wave equations. Coast. Eng. J. 48 (04), 355–370.

Chen, J., Jiang, C., Yang, W. et al. (2016). Laboratory study on protection of tsunami-induced scour by offshore breakwaters. *Nat Hazards,* **81**: 1229. doi:10.1007/s11069-015-2131-x

Douglas, S., Nistor, I. (2015). On the effect of bed condition on the development of tsunami-induced loading on structures using OpenFOAM. Natural Hazards, 76(2), 1335-1356

Lauber, G., & Hager, W. H. (1997). Experiments to dam break wave: Horizontal channel. *Journal of Hydraulic Research*, **36**(3), 291-307

Li L, Huang Z, Qui Q, Natawidjaja DH, Sieh K. (2012). Tsunami-induced coastal change: scenario studies for Painan, West Sumatra, Indonesia. Earth Planets Space **64**:799–816

Nakamura, T., Yasuki, M., & Mizutani, N. (2008). Tsunami Scour Around a Square Structure. *Coastal Engineering Journal*, **50**(2), 206-246

Stansby, P. K., Chegini, A., & Barnes, T. C. (1998). The initial stages of dam-break flow. *Journal of Fluid Mechanics*, **374**, 407-424.

Tonkin, S., Yeh, H., Fuminori, K., & Shinji, S. (2003). Tsunami scour around a cylinder. *Journal of Fluid Mechanics*, **496**, 165-192

Yeh, H., Robertson, I., and Preuss, J., 2005, Development of Design Guidelines for Structures that Serve as Tsunami Vertical Evacuation Sites, Open File Report 2005-4, Washington Division of Geology and Earth Resources, State of Washington (contract 52-AB-NR- 200051), Olympia, Washington

Yeh, H and Li, W (2008). "Tsunami Scour and Sedimentation", *Fourth International Conference on Scour and Erosion*, pp. 95-106

Leadership in Sustainable Infrastructure
Leadership en Infrastructures Durables

Vancouver, Canada
May 31 – June 3, 2017/ *Mai 31 – Juin 3, 2017*

MODELLING THE GENERATION AND PROPAGATION OF LANDSLIDE-GENERATED WAVES

J.A. Vasquez [1,2]
[1].Northwest Hydraulic Consultants Ltd. (NHC), North Vancouver, British Columbia, Canada
[2].jvasquez@nhcweb.com

ABSTRACT

Sub-aerial landslides in water bodies such as bays, fjords, lakes and reservoirs can generate water waves tens or even hundreds of meters high, with potentially devastating effects on infrastructure and human safety. British Columbia has experienced several historic landslide-generated wave events, such as the 20-m high Haney slide wave in the Fraser River (1880); the 25-m high Attachie slide wave in the Peace River (1973) and the recent 35-m high Chehalis Lake wave event (2007). Using the Telemac-2D hydrodynamic model, it is demonstrated that wave propagation results are very sensitive to the mesh resolution and numerical scheme applied to solve the flow equations, typically resulting in excessive wave dissipation (i.e. amplitude under-prediction). It is also shown that correct modelling of these waves requires reproducing the solid slide impacting the water body at high velocity, which is possible using the computational fluid dynamics (CFD) model Flow-3D. In order to accurately model in Flow-3D the wave propagation over distances larger than 10 times the water depth away from the wave generation region, it is recommended to use second order numerical schemes to solve the momentum advection equations.

Keywords: landslide-generated waves; CFD; numerical modeling; Flow-3D; Telemac-2D.

1. INTRODUCTION

1.1 Landslide-generated waves and their impacts

Landslides impacting water bodies can generate large impulse water waves, also known as tsunami waves. The landslides can occur completely underwater (sub-aqueous) or above water (sub-aerial), the latter being usually the most relevant because slides falling at high velocity can generate extremely high waves (mega-tsunamis). Figure 1 shows the three phases of sub-aerial landslide-generated waves (Heller et al. 2009): 1) wave generation caused by the impact of slide with the water body; 2) wave propagation radially away from the point of impact accompanied by attenuation of its amplitude; and 3) wave run-up against the shoreline and dam in case of reservoirs (with potential dam overtopping or dam breach).

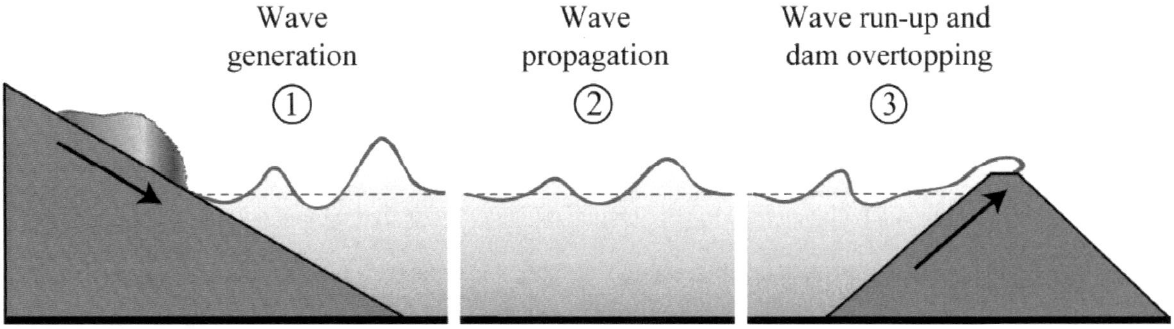

Figure 1. The three phases of landslide-generated waves (Heller et al. 2009).

Table 1 provides some examples of landslide-generated wave events recorded around the world. The largest mega-tsunami wave ever recorded reached 524 m high in Lituya Bay, Alaska (Miller 1960; Fritz et al. 2009). Another well-known case is the 240 m high tsunami wave that overtopped the Vajont Dam in Italy, killing more than 2000 people (Flacker and Eyzaguirre 1979; Wikipedia 2016). The Chungar landslide in Peru caused a 30 m high wave that flooded and destroyed a mining camp, killing an estimated 400 to 600 people (Flacker and Eyzaguirre 1979; Slingerland and Voight 1982). In addition to bays, fiords, lakes, rivers and reservoirs (Table 1); tailings dams and abandoned mining pits flooded with water are also quite susceptible to experience large landslide-generated wave events (Vasquez and de Lima 2006; NHC 2012, 2013, 2016).

Table 1. Examples of historic landslide-generated wave events (Flacker and Eyzaguirre 1979; NHC 1983b; SFU 2011; Wikipedia 2016; Vasquez and de Lima 2016).

Year	Name and location	Slide volume ($\times10^6$ m^3)	Wave height (m)	Casualties
2008	Three Gorges, China	0.4	32	
2007	Chehalis Lake, BC	3	38	-
2002	Safuna Lake, Peru	20	100	-
1980	Spirit Lake, WA	>2000	260	
1973	Attachie, Peace River, BC	15	25	-
1971	Chungar, Peru	0.1	30	400-600
1963	Vajont Dam, Italy	240	260	2000
1958	Lituya Bay, Alaska	30	524	2
1936	Loen Lake, Norway	1	70	73
1934	Tafjord, Norway	1.5	62	44
1905	Loen Lake, Norway	0.4	40	61
1880	Haney, Fraser River, BC	1	20	1
1792	Shimabara, Japan	535	100	15000+

The province of British Columbia (BC) in Canada has also experienced several landslide-generated wave events. One of the oldest events recorded was the 1880 Haney Slide on the north bank of the Fraser River, 40 km east of Vancouver, when about 1 million m^3 of glacial-marine clay collapsed and moved almost one kilometre across the Fraser River, partially blocking its flow. The wave height was estimated to reach about 20 m high, killing one farmer on the opposite south bank. Fifteen kilometres upstream the wave had attenuated to about 3 m high (SFU 2011; Clague and Turner 2006). The Peace River and tributaries near Fort St. John, where the 60 m high Site C dam is currently under construction, have experience several historic landslides (Van Esch 2012), such as the Attachie slide in 1973 that generated a 25 m high wave in the Peace River (NHC 1983b). More recently in 2007, a large slide entering Chehalis Lake generated a wave that ran up to 38 m in the opposite shore, destroying three empty summer campgrounds. In the Columbia River, the construction of the Revelstoke and Mica hydroelectric dams has increased water levels along the toes of nearly 1000 km of steep mountain slopes, where dozens of large potential rock slides have been identified (Watson et al. 2007), including Downie Slide and Checkerboard, where physical model studies (NHC 1976, 2005) showed that maximum waves heights could potentially exceed 10 m.

Studying and predicting the relevant characteristics of landslide-generated waves for engineering applications is a complex multidisciplinary problem. Geotechnical engineers assess the potential features of the slide, such as location, dimensions, sliding angle and speed, and provide that information to hydrotechnical engineers, who are tasked with predicting the wave hydrodynamics caused by the slide impacting on the water body (Figure 1). Predicted wave information can then be used to determine potential flood extents (impacts lines) around the shores of the water body; assess the risk of dam overtopping; compute outflow hydrographs of water spilling out of the water body and route the floodwave downstream; compute wave forces on structures such as bridge piers; amongst others. Such engineering studies typical involve analyzing several scenarios in which the landslide features or initial water elevations are varied. Mitigation measures can involve geotechnical measures (e.g. slope stabilization, continuous monitoring); protection against wave impact; relocation of people and infrastructure; raising dams or dykes; lowering of water levels; early warning systems; amongst others.

1.2 Predicting wave hydrodynamics

There are three available methods for assessing wave hydrodynamics: analytical equations (Slingerland and Voight 1982; Hughes and Berry 1991; Heller et al. 2009); physical models (NHC 1976, 1983a, 1983b, 2005; Ataie-Ashtani and Nik-Khah 2008; Spheric 2015); and numerical models (Raney and Butler 1975; Basu et al. 2009a,b; Biscarini 2010; Heller et al. 2016; Vasquez and de Lima 2016). Analytical methods are well suited for slide and water bodies of simple geometry, but are less reliable for complex cases, which is why physical modelling has been traditionally applied for important engineering projects; for example, Site C (NHC 1983b) and Revelstoke (NHC 1976, 2005) hydroelectric projects in BC. Numerical models, especially computational fluid dynamics (CFD) models, are now an attractive cost-effective alternative to physical model; but our present knowledge of their real capabilities and limitations is not as mature. The objective of this paper is to provide further insight into the advantages and limitations of numerical modelling for assessing the hydrodynamics of landslide-generated waves. A topic of particular interest is the possibility of numerical dissipation for modelling wave propagation over long distances, which if present in a numerical model would cause the model to underpredict wave amplitude.

Vasquez and de Lima (2016) already proved that the CFD code Flow-3D can accurately reproduce wave generation, propagation and runup in a small basin, intended to represent a flooded mine pit. Although wave patterns were highly three-dimensional and complex, Flow-3D predicted wave runup heights within ±10% of observations made in a physical model, which is quite good for practical applications. However, the length of the basin was only 4 times the water depth and numerical dissipation was likely not issue. Therefore, the need for further verification using the additional experimental data in a long flume (Spheric 2015, Heller et al. 2016), with length in excess of 35 times the water depth where numerical dissipation could be an issue.

The paper is organized as follows. Section 2 provides a brief summary introduction to the numerical modelling of landslide-generated waves. Section 3 describes validation of the Flow-3D CFD model using Heller et al. (2016) experimental data in a long flume. The possibility of modelling wave generation by imposing a known inflow wave boundary condition is explored in Section 4, where the influence of numerical scheme and mesh resolution on wave generation is also illustrated. Summary and conclusions are presented in Section 5. Finally, references are listed at the end in Section 6.

2. NUMERICAL MODELLING

2.1 Wave generation modelling

Conventional depth-averaged hydrodynamic models such as River2D, Telemac-2D and similar could in principle be able to simulate the propagation and runup phases of landslide-generated waves (Figure 1). However, since they lack a mechanism to simulate the transfer of energy and momentum from the solid slide to water, hydrodynamic models cannot simulate the wave generation phase. A very early attempt to deal with this problem was investigated by Raney and Butler (1975), who developed a 2D vertical plane model to simulate the generation and propagation of landslide-generated waves. The moving slide was modelled as a transient bottom deformation which transferred pressure and viscous forces to the liquid by the use of empirical drag equations. The calibrated model reproduced well observations made in a physical model. Although Raney and Butler (1975) semi-empirical approach seemed promising, I was not applied further; perhaps in part because wave generation can now be explicitly solved by computational fluid dynamic (CFD) models.

2.2 CFD modelling

CFD models solve the 3D mathematical flow equations to compute the fluid velocity vector and pressure at each cell of a computational mesh. Almost all CFD models (e.g. Fluent, Star-CCM+, CFX, OpenFOAM) use body-fitted meshes that conform to the external boundaries of solids present in the computational domain. In this type of models, wave generation can be easily modelled if the slide is assumed as another fluid interacting with water (e.g Biscarini 2010). However, if the slide is modelled as a rigid non-deformable solid these models run into problems because the body-fitted mesh needs to be remeshed at every time step while the slide is moving, which can lead to numerical stabilities if mesh deformation is excessive. A possible way around is the use of overlapping or chimera meshes, which are small body-fitted meshes attached to the solid that move together inside a larger fixed body-fitted mesh. Another way is to avoid the use of body-fitted meshes altogether, as done by the CFD code Flow-3D.

2.3 Flow-3D

Flow-3D (www.Flow3D.com) uses a simple rectangular orthogonal mesh (Cartesian grid with rectangular prismatic cells). The solid is interpolated into the grid using the Fractional Area-Volume Obstacle Representation (FAVOR) method, in which a solid is allowed to cut through a cell and its location is recorded, not by moving the edges of the cell to conform to the solid as in body-fitted meshes, but in terms of the fractional face areas and fractional volume of the cell that are not covered by the solid. In this way, the grid and the solid remain independent of each other and it is easy to modify the solid while keeping the grid unchanged. This makes it easier for a moving solid, like a falling landslide, to move through a grid that remains invariant during the simulation (Das et al. 2009a; Basu et al. 2009a,b; Vasquez and de Lima 2016). Finally, another way to model landslide-generated waves without worrying about meshing issues is to use mesh-free numerical methods such as smoothed particle hydrodynamics (SPH).

2.4 SPH modelling

In SPH modelling, the fluid and solid are represented by discrete particles that interact with each other; but without the use of a computational mesh. SPH has been successfully applied to simulate wave generation and propagation (Das et al. 2009b; Heller et al. 2016). SPH is a very promising technique subject to very intense research, but at the moment it not as mature as CFD modelling for practical engineering applications.

3. FLOW-3D VALIDATION

3.1 Experimental data

Heller et al. (2016) conducted experiments of landslide-generated waves using the experimental setup shown in Figure 2. The experiment considered here was conducted in a two-dimensional arrangement, conducted within a straight flume 0.6 m wide, almost the same as the slide width such that waves propagate only along the 10 m long axis of the flume. The experiment with an initial water depth h = 0.24 m is considered here for Flow-3D validation (Figure 2).

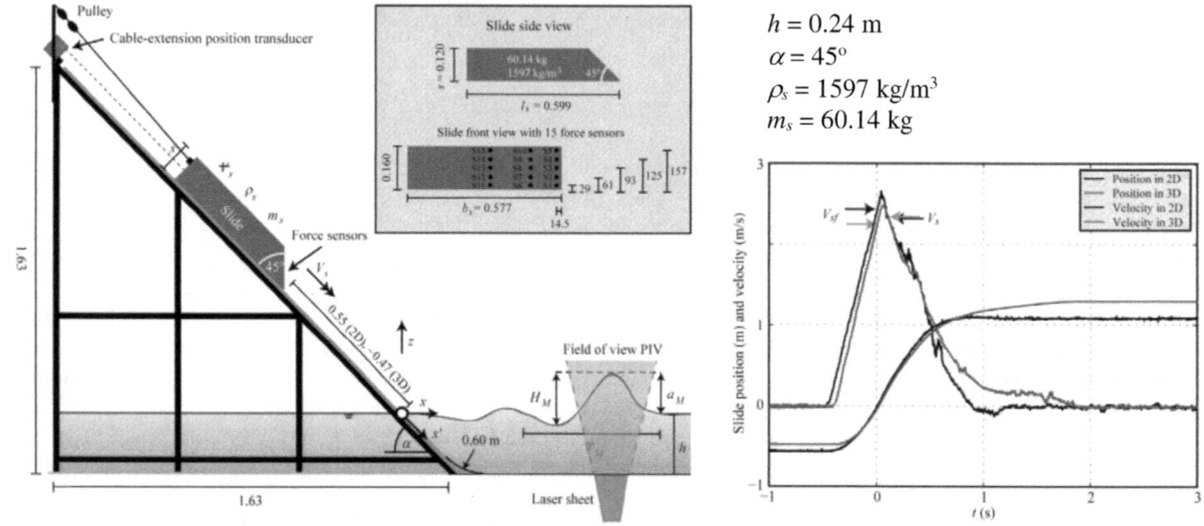

Figure 2. Experimental setup of landslide-generated wave experiment (Heller et al. 2016, Spheric 2015).

Water levels in time were measured at 7 gauges located at distances between 3 and 35 times the water depth from the sloping bank (Figure 2). Figure 3 shows the wave height above the initial water surface, measured at the 7 gauges. The slide generates a solitary wave traveling away from the impact zone. The water surface elevation is most disturbed closer to the impact zone, with the water level not only rising above, but also being depressed below the initial surface. However, at a distance 35*h* (8.4 m) the wave exhibits almost perfect symmetrical solitary wave shape, with no disturbance below the water surface. Notably, there was little dissipation in the wave amplitude as it propagated away from the slide area.

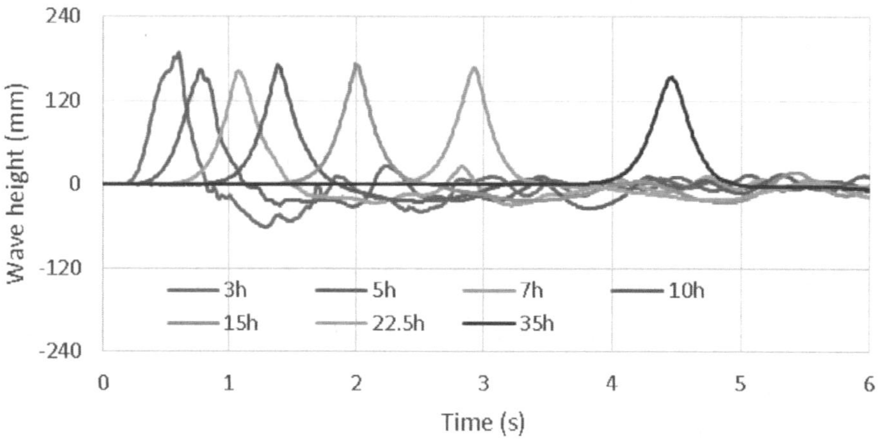

Figure 3. Wave height in time measured at 7 gauges along the experimental flume (h = 0.24 m)

3.2 Flow-3D Setup

Modelling the slide movement in Flow-3D version 11.2 requires providing the spatial components of the both the linear and angular velocity of the moving slide. Heller et al. (2016) provide information on the "position" of the slide in time (Figure 2); whose exact meaning is not clear, as it could represent the distance along the curved surface of the slide slope or the distance along the cable-extension position transducer shown in Figure 2. For Flow-3D application, the experimental velocity magnitude (Figure 2) was decomposed by trial-and-error into the horizontal (V_x) and vertical (V_z) components, plus an angular velocity (ω_y) that approximated the velocity plot in Figure 2 assuming the slide followed the curved bottom trajectory. The geometry of the slide and circular bottom transition were provided by Spheric (2015). Initial water depth was assumed as h = 0.24 m with a hydrostatic pressure distribution. Friction was neglected. Turbulence was modelled using the RNG k-epsilon turbulence model. Preliminary 3D tests showed no improvement, so all runs were conducted in a 2D vertical plane. Total simulation time was 5.5 s. The domain was discretized using uniform grid cell sizes of 10 and 5 mm. Only the results with the fine 5 mm grid, which is considered mesh independent, are reported here. Flow-3D can solve the flow momentum equations using first order (FO) and second order (SO) numerical schemes; both of which were tested to assess their influence on the results, as discussed below.

3.3 Flow-3D results

Figure 4. Time sequence showing trajectory of slide in Flow-3D shows the trajectory of the falling slide assumed in the simulation. The slide impacts the water surface at 2.4 m/s, causing water to be displaced and gain kinetic energy, which leads to the generation and propagation of a single wave downstream.

3.3.1 First Order Momentum Advection

The results of Flow-3D using the default first order momentum advection numerical scheme are shown in Figure 5. The results are quite good near the slide area, up to a distance around 15*h* (3.6 m). However, beyond that distance the wave's shape deviates from the observed shape (Figure 5) becoming asymmetrical, longer and flatter.

3.3.2 Second Order Momentum Advection

In order to reduce numerical dissipation, another test was performed using the second order momentum advection as shown in Figure 6. Now the shape of the wave remains symmetrical as in the experiment, but the peak heights near the slide have decreased relative to the experiment (Figure 3) and previous results with the FO scheme (Figure 5).

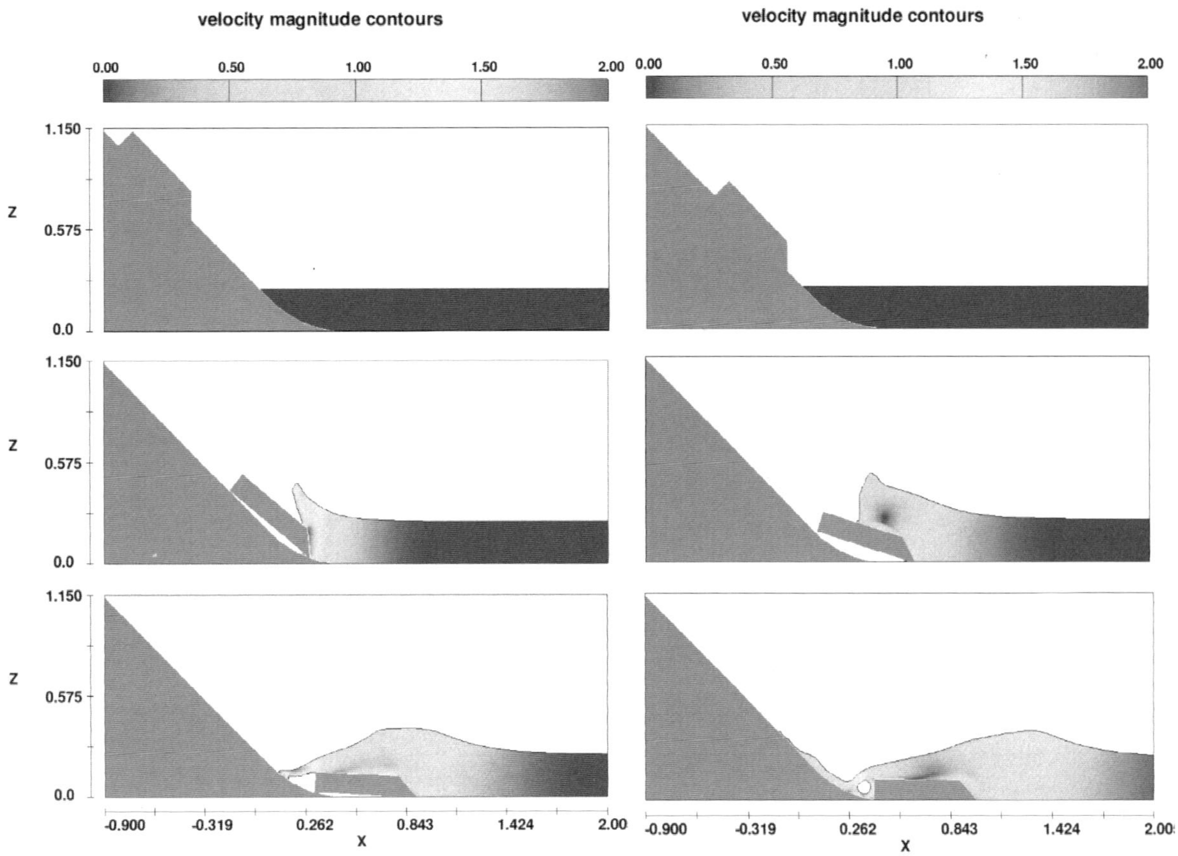

Figure 4. Time sequence showing trajectory of slide in Flow-3D

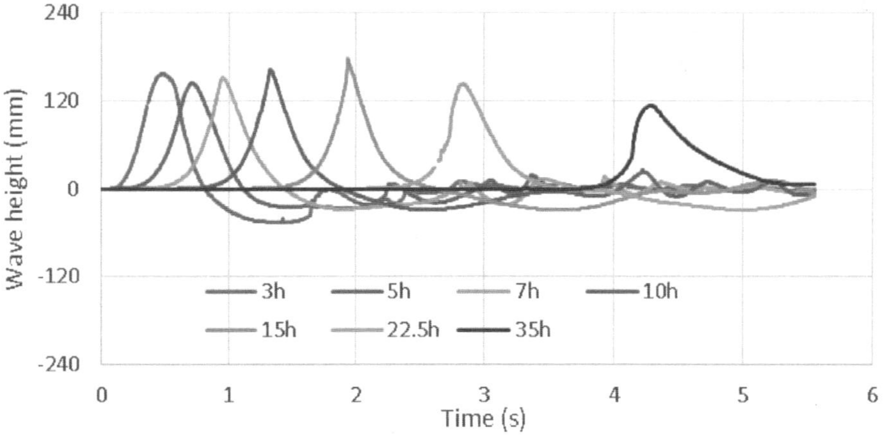

Figure 5. Flow-3D results using the default first order momentum advection numerical scheme

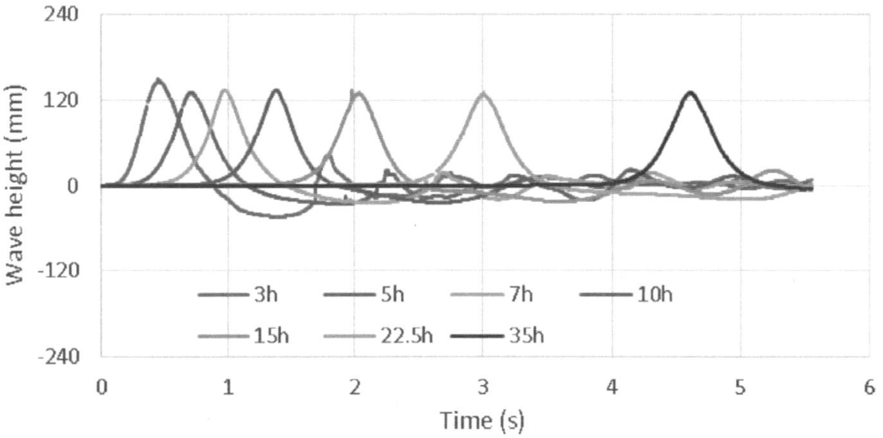

Figure 6. Flow-3D results using the second order momentum advection numerical scheme

4. WAVE PROPAGATION WITH IMPOSED BOUNDARY CONDITION

4.1 Hybrid approach?

Because conventional hydrodynamic models cannot simulate wave generation explicitly (phase 1 in Figure 1), they cannot be applied for landslide-generated wave modelling. However, when modelling the propagation of a sub-aqueous ocean tsunami using Telemac-2D, Horsburgh et al. (2008) did not simulate explicitly the tsunami wave generation, but instead assumed the initial shape of the tsunami wave and impose it as boundary condition in Telemac-2D to model wave propagation in the ocean. This raises the question of whether a similar approach could be applied for sub-aerial landslides. That is, a hybrid approach in which analytical wave generation equations are used to estimate the initial wave amplitude and length, and then that information is imposed as inflow boundary condition in a conventional hydrodynamic model to simulate the wave generation phase. The appeal of such an approach is that hydrodynamic models are more widely available and faster to run than CFD models.

In order to test this hypothesis, additional numerical models of Heller et al (2016) flume were developed using both River2D and Telemac-2D, but starting at $x = 3h = 0.72$ m and using a node spacing of 5 mm, similar to that of Flow-3D. The water levels measured in the experiment at $x = 3h$ were applied as inflow boundary condition. For comparison, as similar Flow-3D model was also developed and ran. Figure 7 shows the experimental results at $x = 35h$, compared with the numerical results of the various models. Telemac-2D offers several numerical schemes to choose from (see

Figure 8); two of them which provided the highest wave amplitudes, the Harten-Lax-van Leer-Contact (HLLC) and Second Order Kinetic (SOK) schemes are shown in Figure 7.

All the numerical tests underpredicted the wave amplitude observed in the experiment (Figure 7). This is not completely unexpected, because the information on momentum transferred by the slide to water during the wave generation phase is not included in the simple time-varying water level used as inflow boundary condition. However, the high asymmetry of the predicted wave is harder to explain, considering that the imposed inflow wave at $x = 3h$ (Figure 3) is rather symmetrical. Although imposing a known water level inflow boundary condition to represent wave generation could be an acceptable approach for modelling sub-aqueous landslide-generated waves (Horsburgh et al. 2008); the numerical results shown in Figure 7 suggest this may not be true for the highly energetic sub-aerial landslide-generated waves, as the shape of the propagating wave gets significantly distorted for all numerical models tested.

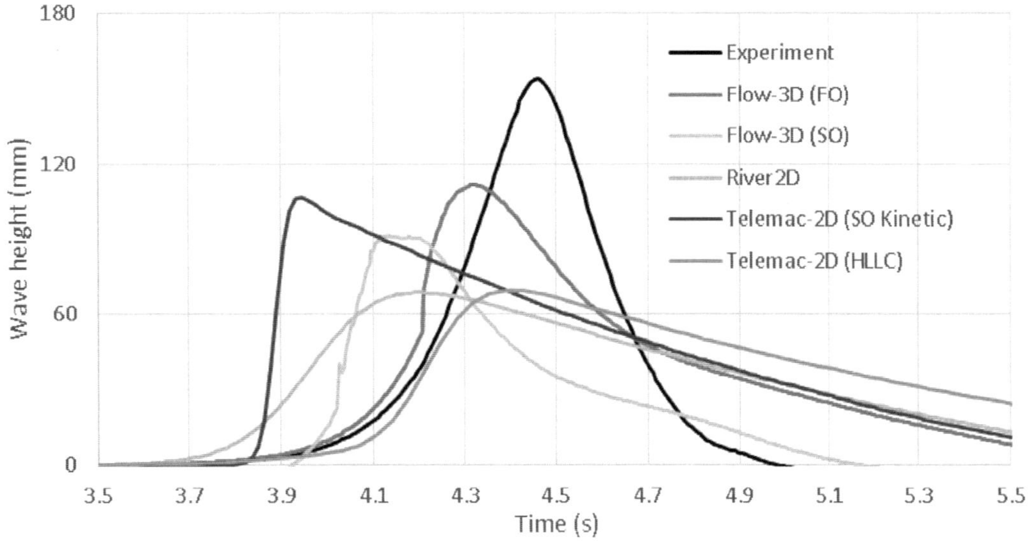

Figure 7. Wave height at $x = 35h$ predicted by various models assuming that wave at $x = 3h$ is known and imposed as inflow boundary conditions

4.2 Example in full scale reservoir

In order to illustrate the challenges of wave propagation modelling in a more realistic setting, a hypothetical example is presented here applying Telemac-2D in large $h = 42$ m deep reservoir. In this example, a single squared-sinusoidal wave with an amplitude of 18 m and period of 15 s is introduced at one shore of the reservoir. The wave propagates radially from its point of origin (slide location) with its amplitude decaying with distance as the wave propagates. The maximum wave amplitude predicted by Telemac-2D at an arbitrary point located 1.3 km away from the origin (a distance about $30h$) is reported in Figure 8 (for reference, when Flow-3D was applied for this case using a 5 m mesh and SO scheme, the maximum wave amplitude predicted was 5.5 m). Several simulations were performed in which the numerical scheme used to solve the equations or the mesh resolution were varied, while all other factors remained constant. The initial mesh with 30 m node spacing was refined first to 15 m and then again to 7.5 m.

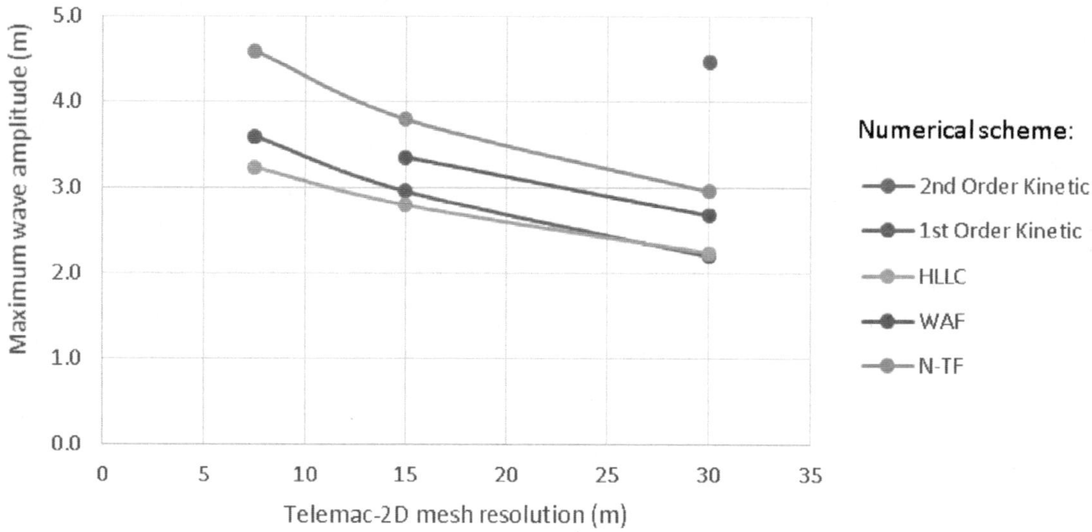

Figure 8. Influence of mesh resolution and numerical schemes used by Telemac-2D model on predicted wave maximum wave amplitude at point 30*h* away from hypothetical landslide location.

Figure 9 shows that predicted maximum wave amplitudes can be quite sensitive to the mesh resolution and adopted numerical scheme. Wave amplitude increased as the mesh was refined, regardless of the numerical scheme used, probably due to excessive numerical dissipation in the coarser meshes. The details of each numerical scheme used by Telemac-2D are not very relevant, but the important message here is that different approximations in the way the equations are solved lead to significantly different results when modelling wave propagation far from the point of wave generation. Since most numerical schemes tend to suffer from excessive dissipation (i.e. tend to underpredict wave amplitude), it could be argued that the SOK numerical scheme, which predicted the highest amplitudes for the 30 m mesh is perhaps less dissipative and hence more accurate (SOK also predicted the highest amplitude of Telemac-2D schemes tested in previous case, see Figure 7). But unfortunately, SOK became unstable in the finer 15 and 7.5 m meshes and results could not be obtained. The same was true for the WAF numerical scheme in the 7.5 m mesh. This illustrates another usual problem with numerical models, higher order or sophisticated schemes (which usually are more accurate) tend to be more problematic and cannot be applied in all practical cases.

5. SUMMARY AND CONCLUSIONS

Landslides falling into water bodies such as bays, fjords, rivers, lakes, mine pits and reservoirs have been observed to generate waves that can reach amplitudes tens of meters high (many of them in British Columbia); which can have devastating effects on human life and infrastructure. Therefore the need for reliable ways to predict wave amplitudes. In the case of numerical models, their predictions of maximum wave amplitudes are quite sensitive to the mesh resolution and numerical scheme chosen to solve the flow equations. For practical applications that require a high level of confidence in the predicted wave dynamics (e.g. significant risk to life and infrastructure), physical model is recommended. In lieu of physical modelling, CFD models with proven capabilities to explicitly simulate all the phases of landslide-generated waves (generation, propagation and runup), such as Flow-3D should be used.

This paper provides further evidence that Flow-3D can reproduce quite well the wave generation and propagation dynamics of sub-aerial landslide-generated waves caused by highly energetic slides falling in water. But in contrast with previous research that looked at wave generation and wave propagation in short flumes (Das et al. 2009a; Basu et al. 2009a,b; Vasquez and de Lima 2016), this paper extends those results to longer flumes where the choice of numerical scheme to solve the momentum equations can play a more important role in the prediction of wave propagation. In the case of Flow-3D, it was found that the First Order momentum advection scheme works well for wave generation and propagation near the landslide region. However, for distance beyond 10 times the water depth, the Second Order scheme is recommended for modelling wave propagation.

An alternative approach to model wave propagation using depth-averaged flow models is to assume the initial wave amplitude near the slide region as known and impose it as inflow boundary condition in the flow model. Although such an approach may work for sub-aqueous ocean tsunami waves (Horsburgh et al. 2008), it remains unproven for more energetic sub-aerial landslide-generated waves and its application is not recommended until fully validated. In general, the use of numerical models that have not been properly validated using experimental or field data specific to landslide-generated wave phenomena is not recommended.

6. REFERENCES

Ataie-Ashtani, B.; Nik-Khah, A. 2008. Impulsive waves caused by subaerial landslides. *Environmental Fluid Mechanics*, 8 (3): 263-280.

Biscarini, C. 2010. Computational fluid dynamics modelling of landslide generated water waves. *Landslides*, 7 (2): 117-124.

Basu, D., Green, S., Das, K., Janetzke, R.; Stamatakos, J. 2009a. Navier-Stokes simulations of surface waves generated by submarine landslides: effects of slide geometry and turbulence. 2009 *Society of Petroleum Engineers Americas Environmental & Safety Conference*, Santo Antonio, Texas.

Basu, D., Green, S., Das, K., Janetzke, R., and Stamatakos, J. 2009b. Numerical simulations of surface waves generated by subaerial landslide at Lituya Bay, Alaska. *28th International Conference on Ocean, Offshore and Artic Engineering*, Honolulu, Hawaii.

Clague, J.; Turner, B. 2006. *Vancouver, City on the Edge: Living with a Dynamic Geological Landscape.* Tricouni Press Ltd., Vancouver, BC.

Das, K., Green, S., Basu, D., Janetzke, R., and Stamatakos, J. 2009a. Effect of slide deformation and geometry on waves generated by submarine landslides: A numerical investigation. *2009 Offshore Technology Conference*, Houston, Texas, EEUU.

Das, K., Janetzke, R., Basu, D., Green, S., and Stamatakos, J. 2009b. Numerical simulations of Tsunami wave generation by submarine and aerial landslides using RANS and SPH. *28th International Conference on Ocean, Offshore and Artic Engineering*, Honolulu, Hawaii.

Flacker, G. and Eyzaguirre, V.R. 1979. Rock Avalanche and Wave at Chungar, Peru. *Rockslides and Avalanches, Vol. 2 Engineering Sites - Developments in Geotechnical Engineering*. Elservier Scientific Publishing Company, the Netherlands.

Fritz, H.M., Mohammed, F.; Yoo, J. 2009. Lituya Bay Landslide Impact Generated Mega-Tsunami 50th Anniversary. *Pure Applied Geophysics*, 166 (1-2): 153-175.

Heller, V., Hager, W. and Minor, H-E. 2009. *Landslide-generated waves in reservoirs- Basics and Computation.* ETH – Universidad de Zurich.

Heller, V., Bruggemann, M., Spinneken, J.; Rogers, B.D. 2016. Composite modelling of subaerial landslide-tsunami in different water body geometries and novel insight into slide and wave kinematics. *Coastal Engineering*, Vol. 109 (2016): 20-41.

Horsburgh, K. J., C. Wilson, B. J. Baptie, A. Cooper, D. Cresswell, R. M. W. Musson, L. Ottemöller, S., Richardson, and S. L. Sargeant 2008. Impact of a Lisbon-type tsunami on the U.K. coastline and the implications for tsunami propagation over broad continental shelves. *Journal of Geophysical Research: Oceans.* 13 (C4).

Hughes, B. R.; Berry, B.A. 1991. Landslide-generated waves. *Dam Safety* 1991.

Miller, D.J. 1960. *Giant Waves in Lituya Bay, Alaska*. Geological Survey Professional Paper 354-C.

NHC 1976. *Hydraulic Model Study of Wave from Downie Slide*. Report prepared by Northwest Hydraulic Consultants Ltd. (NHC) for British Columbia Hydro and Power Authority, August 1976.

NHC 1983a. *Review of Landslide Generated Wave Analysis*, Report prepared by Northwest Hydraulic Consultants Ltd. (NHC) for British Columbia Hydro and Power Authority, May 1983.

NHC 1983b. *Peace River, Site C Project, Model of Landslide Waves at Attachie and Bear Flat Areas*. Report prepared by Northwest Hydraulic Consultants Ltd. (NHC) for BC Hydro, June 1983.

NHC 2005. *Checkerboard Creek Landslide Generated Wave – Physical Model Study*. Report prepared by Northwest Hydraulic Consultants Ltd. (NHC) for British Columbia Hydro and Power Authority, August 2005.

NHC 2012. *Análise Hidráulica de Movimentos de Massa nos Taludes da Mina de Águas Claras – CFD Simulation and Results*. Technical Memo prepared by Northwest Hydraulic Consultants Ltd. (NHC) for POTAMOS Engenharia e Hidrologia Ltda., July 2012.

NHC 2013. *Questa Mine Hydraulic Modelling of Landslide-Generated Waves*. Report prepared by Northwest Hydraulic Consultants Ltd. (NHC) for Golder Associates Inc., October 2013.

NHC 2016. *WSAP Mine Wave Hazard Assessment*. Report prepared by Northwest Hydraulic Consultants Ltd. (NHC) for Piteau Associates, April 2016.

Raney, D.C and Butler, H.L. 1975. *A Numerical Model for Prediction the Effects of Landslide-Generated Water Waves*. Hydraulics Laboratory, US Army Engineering Waterways Experiment Station. Report H-75-1.

SFU 2011. *Risky Ground*. Newsletter of Center for Natural Hazards, September 2011, Fall Edition. Simon Fraser University, Canada.

Slingerland, R. and Voight, B. 1982. Evaluating Hazard of Landslide-Induced Water Waves. *Journal of the Waterway, Port, Coastal and Ocean Division*, ASCE, Vol. 108, No. WW4, November 1982, pp. 504-512.

Spheric 2015. *Test 11: Subaerial landslide-tsunami generation with a rigid slide in a channel (2D) and basin (3D)*. https://wiki.manchester.ac.uk/spheric/index.php/Test11, last modified on 27 August 2015. Accessed on 5 February 2016.

Van Esch, K.J.B. 2012. *Failure Behaviour of Bedrock and Overburden Landslides of the Peace River Valley near Fort St. John, British Columbia.* Master's Thesis. University of British Columbia.

Vasquez, J.A.; de Lima, G. 2016. Modelación CFD de Ondas Tsunami en Reservorios, Lagos y Minas Causadas por Deslizamientos de Laderas. Proceedings of the *27th IAHR Latin American Conference*, Lima, Peru

Watson, A. D., Psutka, J. F., Stewart, T. W., and Moore, D. P. 2007. Investigations and monitoring of rock slopes at Checkerboard Creek and Little Chief Slide. In *1st Canada-US Rock Mechanics Symposium*. American Rock Mechanics Association, Vancouver, Canada.

Wikipedia 2016. Megatsunami. https://en.wikipedia.org/wiki/Megatsunami.

Leadership in Sustainable Infrastructure
Leadership en Infrastructures Durables

Vancouver, Canada
May 31 – June 3, 2017/ *Mai 31 – Juin 3, 2017*

NUMERICAL MODELING OF THE FLOW HYDRODYNAMICS AND FLOATING DEBRIS PATHWAYS AROUND THE BEQUIA ISLAND

Philippe April LeQuéré [1,4], Ioan Nistor [2], and Ronald Townsend [3]
[1, 2, 3] University of Ottawa, Canada
[4] papri010@uottawa.ca

ABSTRACT

To accommodate an increasing number of tourists visiting Bequia, the largest island of Saint-Vincent and the Grenadines, the local government constructed an airport, through a major coastline land-reclamation project. However, due to the prevailing ocean current patterns in the area, an inlet created on the East side of the new airport is prone to trapping significant amounts of ocean-borne debris. This litter accumulation creates a health risk to local fishermen who clean their daily catch using water from the inlet. It is proposed to install a rock-fill groyne structure on the eastward side of the new inlet to address this problem. The utilisation of a coast-line groyne in this case is somewhat unorthodox, as the latter is normally employed to mitigate against coastal erosion.
The goal of this study is to optimise the groyne design with the assistance of a 3D numerical model. The 'Delft3D' open-source model (WAVE and FLOW modules) was selected to examine the effects of different orientations and lengths of the proposed groyne on the movements of floating debris. Included in the initial phase of the study was a field investigation to collect certain data which was necessary for model calibration and validation. This involves the use of an Acoustic Doppler Current Profiler (ADCP) to measure local shore bathymetry and also current velocities over a range of tidal cycles. This paper reports on the preliminary results of this study.

Keywords: - Bequia, numerical model, Delft3D, hydrodynamic, tide-induced current, groyne

1. INTRODUCTION

This paper focuses on the numerical modeling exercise undertaken of the coastal region of the Paget Farm community on the island of Bequia, Saint-Vincent and the Grenadines. This particular region is being studied in order to address the problem of floating debris accumulation that is occurring in the Jeff Gregg inlet. This new coastal inlet was created in the process of constructing the new J.F. Mitchell airport on the island in 1990. Because the airport is sited on new land reclaimed from the ocean, the resulting modified coastline in this area has changed the local coastal hydrodynamics and ocean bed morphology. These changes have been found to be largely responsible for the aforementioned accumulation of floating ocean-borne debris in the Jeff Gregg inlet, which is located immediately east of the airport. The prevalent winds and wave directions are mostly from east to west in the area and hence the resulting ocean surface currents tend to capture and transport marine debris (sargassum, Paget Farm community litter, etc) into the Jeff Gregg inlet

Figure 1 Paget Farm Coast Map (Tour du monde en images. 2015)

The goal of the study is to redirect the prevalent surface current in this area (and with it the related path of debris) away from inlet entrance and towards the main (bypass) ocean current. This will be accomplished by building a rubble mound groyne just east of the inlet. Optimizing the design of this structure will require a better understanding of the existing coastal hydrodynamics and the related debris movements in the area. This will be accomplished with the aid of a dedicated 3-D mathematical model of the ocean region in question. The model selected for this exercise is the Delft3D numerical model, which is a software that is particularly well-suited for studying complex coastal hydrodynamics. Very little information is available about this remote region. Accordingly, a field campaign was initiated to gather the necessary information to build and calibrate the model.

2. THE DELFT3D MODEL

As previously stated, a computational model of the coastal area in question was developed using the open source of Delft 3D. This software is capable of simulating the coastal hydrodynamics using a module called FLOW, by solving an approximation of the Navier-Stokes equations for incompressible shallow water flow (Deltares, Delft3D-FLOW. 2011) (Trouw et al. 2012) (Visser, 2002). FLOW can interpret the effects of wind, tide and other defined inputs and outputs on water. However, it cannot simulate the effect of waves. The waves are therefore modeled as forcing factors on the surface grid layer into the FLOW simulation. These factors are calculated by the WAVE module, which will be presented later. Figure 2 illustrates the computational grid used in the FLOW module, superposed onto Bequia's map:

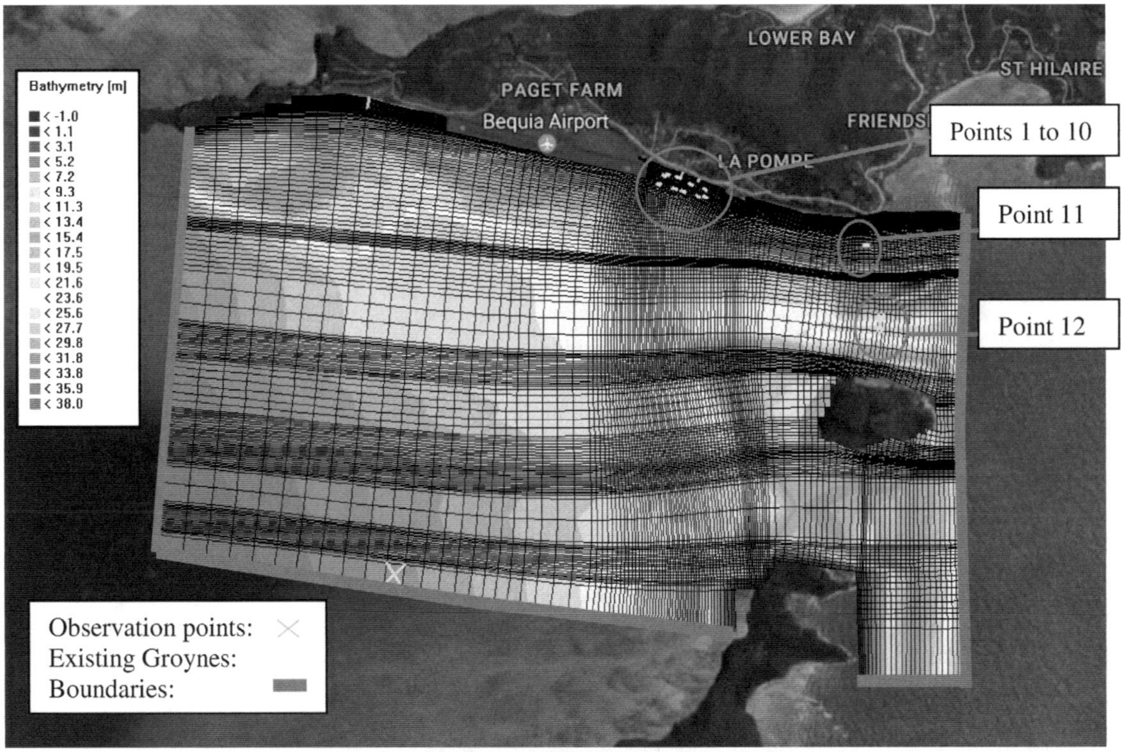

Figure 2: FLOW Module Grid

Shown in yellow in the legend, are existing groynes that were found along this particular section of the coastline. The offshore boundaries (shown in red) were defined as a sinusoidal function that mimics the effect of the tide at the boundaries. A total of 14 Eulerian points, called observations points, were also defined at the approximate locations where velocity measurements were taken in the field (see the Data Collection section). Points 1 to 10 were used for roughness calibration while points 11 and 12 were used for tide-induced current calibration. Two additional points at the south and east boundaries were used for the monitoring of the tide at corresponding locations to ensure that the correct variation of the tide cycle was simulated. The grid is a 3D grid with 121 cells in the X-direction, 153 in the Y-direction and 6 in the Z-direction, for a total of 111,078 cells. The horizontal grid was refined around the islands within the grid domain because of the effect it had on the tide-induced current. The vertical cells get smaller when closer to they are water surface or to the sand bottom. They are expressed as percentages of the water depth, causing the cells to also be smaller when closer to the shore.

The wave simulation is done with the WAVE module and the calculated forces are then communicated to FLOW for the hydrodynamic simulation. These wave forces are calculated using a specific wave model called SWAN (Simulating Waves Nearshore) (Deltares, Delft3D-WAVE. 2011) (Visser. 2002). This model uses the linear wave theory to derive the necessary equations.

3. FIELD MEASUREMENTS

In order to run Delft3D-FLOW with Delft3D-WAVE modules, the following information is necessary: depth of the whole area of interest, characteristics of the tide, bottom roughness, wind and waves. A good understanding of the mechanisms that generate currents is also very useful. Details on current velocities at multiple locations under known conditions are also necessary to calibrate the numerical model.

Forecasts for wind, wave and tide for the Friendship Bay, which is located two kilometers east of the airport, were obtained on the website www.wisuki.com The daily field data observations undertaken by the authors over two separate field campaigns of data collection were used for the calibration of the model and some of the historical values will be also used for the design of the groyne. The historical values indicate that the wind's nautical direction is always between 70 and 100 degrees of North and have a typical velocity of 10 m/s. The waves are also always coming from the same direction with a typical swell significant height between 1 and 2 meters. The tide in the area is semi-diurnal with maximum amplitude of 0.7m. As well, a navigation map of the area was employed, which provides approximate bathymetry of the region. This map was useful for the coarser regions of the model grid.

However, it was not precise enough for those areas closer to the shore. To address this deficiency, two field campaigns were conducted to collect the necessary data. The first necessary data were detailed bathymetry measurements which were carried out along the near-shore region. An acoustic Doppler current profiler (ADCP) was used to collect this information. This device can be used to generate a bathymetric map using its transducers to get the depth (Z coordinate) and an integrated GPS unit to get X and Y coordinates as well as correcting the depth with elevation variations caused by waves and changes of tide. Over the course of the campaign, a total of 23400 x-y-z points were collected and were used to construct the model's grid. Continuous bathymetry measurements in the study area were obtained using the ADCP unit tethered to a boat. Figure 3 shows the area that was surveyed in these exercises.

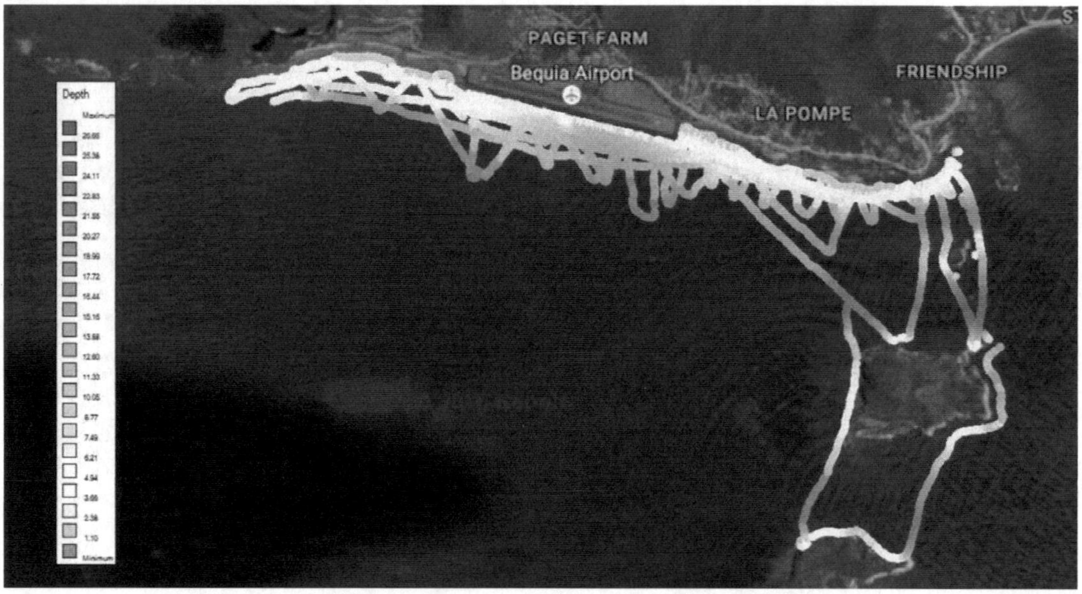

Figure 3: Field Measurements – Schematic of the Depth Transects

Eulerian point velocities in the water column were also collected to assist in the calibration of the model. Using the ADCP, velocity can be measured using the transducers that compare the reflection of sound waves on small particles in the water column between two time steps. This provides point velocities through the entire water column. The boat was anchored at a particular location which was then identified using GPS. GPS was also used to correct velocities by adjustments that reflected movements of the boat. In order to use these points in the calibration of the model, the velocities were averaged over time to account for fluctuations due to the waves and the movement of the boat. Since the software used (ADCP post processing program) could not separate the velocity in the Z direction at any given time step, the velocity was also averaged over the whole water column. A total of 29 points were collected. Table 1 summarizes some of the points that were collected on August 29th 2016.

Time	Depth (m)	Velocity X (m/s)	Velocity Y (m/s)	Time of Measurement (s)	UTM(m)	UTM(m)
7:03:02	2.16	-0.055	-0.002	80	689243.04	1436328.50
7:05:11	2.66	-0.038	0.017	64	689238.87	1436322.48
7:10:53	2.08	-0.068	-0.036	77	689319.94	1436308.85
7:15:16	3.28	-0.118	-0.038	79	689397.92	1436262.99
7:19:13	3.29	-0.120	-0.034	74	689477.33	1436238.94
7:23:15	5.00	-0.170	-0.057	103	689485.18	1436189.92
7:27:31	5.67	-0.162	-0.059	85	689438.65	1436193.31
7:31:56	5.53	-0.144	-0.051	88	689351.88	1436223.49
7:35:46	5.70	-0.154	-0.038	75	689286.46	1436243.29
7:39:44	5.96	-0.127	0.004	83	689182.92	1436269.85

Table 1: Example of Velocity Points

Each point has a location defined in the UTM (Universal Transverse Mercator coordinate system) which helped trace back its exact position on the numerical model grid. The velocities at those points were converted into X and Y components for an easier comparison with the model's results.
However, two important ingredients for the model construction were still missing, namely, the bottom roughness and the tide-induced current. As the bottom roughness can vary on a case- by- case basis, calibration is necessary. Also, during the first site visit, it was observed that there was a strong current between the islands located east of the airport and that its direction changes with the reversal of tide. At flood tide, the current moves from east to west, but with an ebb tide, the current moves from west to east. Knowing that the current direction changes depending on the instant of the tide, the latter is the most likely driving factor for the current phenomenon. The way that this was simulated in the model was by defining the east boundary as a harmonic water level and the west boundary as a harmonic total discharge. This is described in more detail in the calibration of the tide-induced current section. However, no data is available for the velocity of this current under any given tide condition. Therefore, a calibration using the measured depth-averaged velocities at the given points is needed in order to calibrate the boundary condition. Figure 4 illustrates the direction and the location of the tide-induced current:

Figure 4: Tide-Induced Current

The movements of floating debris were observed during the velocity measurements just east of the inlet. It was noted that the debris were moving approximatively parallel to the shore and were moving westward even when the current was moving eastward. This indicates that the tide-induced current isn't the main driving factor of the surface current. Due to time constraints, the measurements were limited. If more time had been available, a tracking of floating debris, using GPS placed on floaters released at different times and sea conditions, could have extended the information available to validate the model. Also, a time series of the depth-averaged velocity over a full tide cycle at a location where the tide-induced current creates high velocities (such as point 11 or 12 in Figure 2) would have been preferable to create a better numerical estimation of the phenomenon.

4. CALIBRATION OF THE MODEL

Because data on the bottom roughness and the tide-induced currents are missing, the model needs to be calibrated. It was decided to separate the calibration by adjusting two model parameters: one for the tide-induced current and one for the bottom roughness.

Calibration of the Tide-Induced Current

The tide-induced current is modeled by defining the east boundary as a harmonic water level and the west boundary as a harmonic total discharge. Delft3D has a function called harmonic boundary where a defined value for water level (or total discharge in this case) varies following a sinusoidal function. The boundaries were defined as such because of some instability problems experienced when the two boundaries were defined as water level. The east

boundary will be modeled following the tide's amplitude and speed (degree/h) at the moment that the point velocities were measured (see Table 2). The west boundary has three components that need to be controlled: the total discharge for the west boundary and that for the south boundary and the lag between the east boundary and the west boundary harmonic cycles. The total discharge was defined by the authors as a factor of the tide amplitude so it that varies proportionally when the tide amplitude is changed.
As the calibration process, by adjusting the bottom roughness, is intrinsically related to the calibration of the tide-induced current and vice versa, a value for the bottom roughness has to be assumed to first obtain a preliminary value for the boundaries' parameters. For simplicity, the bottom roughness value for the calibration of the tide-induced current was assumed to be the same in the whole region. From scuba dives performed by the authors in the area, it was observed that, except near the shore, the ocean bottom in the area is covered mostly by sand. Samples of the sand were collected and analysed. The resulting grain-size distribution curves provided representative D_{50} – values for the three samples collected. The Manning's roughness value for the ocean bed was then calculated to be n = 0.018, using the following formula: (California Departement of Water Resources. 1970)

[1] $n \approx 0.039 * D_{50}^{\frac{1}{6}}$

The required data on wave conditions, tide amplitude and wind conditions were obtained on the web site www.wisuki.com, where daily information on the local weather is continuously updated. As point velocities were recorded over two different days, under different weather conditions, two calculations were required. The following table summarizes the tide, wave and wind conditions for those two days:

	Date	August 21st	August 22nd
Wave condition	Significant height (m)	0.8	0.8
	Direction (d)	90	120
	Period (s)	8	8
Wind Condition	Speed (m/s)	5	5
	Direction (d)	90	90
Tide condition	Time/Water Level	2:30 +0m	3:07 +0m
		8:38 +0.4m	9:24 +0.4m
		15:16 +0.1m	16:30 +0m
		20:23 +0.3m	21:27 +0.3m

Table 2: Information required for the model's Tide Lag Calibration (wisuki. 2016)

The tide-induced current was calibrated using the velocity measured on those two days at approximatively the same location which is between Bequia and Middle Cay and between Middle Cay and Petit Nevis (Points 11 and 12 in Figure 2). Figure 5 summarizes the results of the calibration:

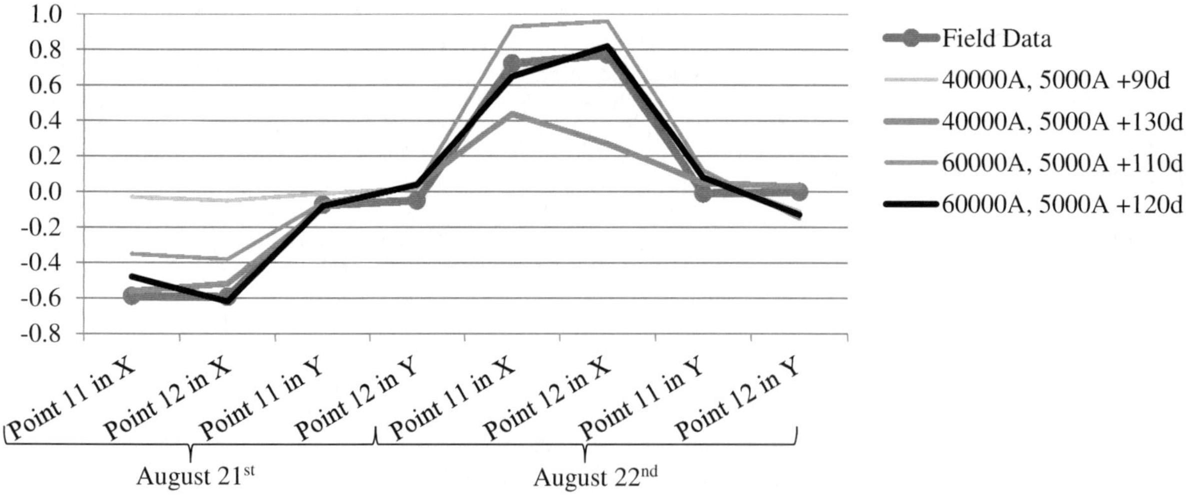

Figure 5: Tide Calibration Results

After analysing the data, it was observed that a total discharge for the west and south boundaries, of 60000 times the tide amplitude and 5000 times the tide amplitude respectively, represents most accurately the tide-induced current. A lag between the definitions of the two boundaries of 120 degrees was also found to be more appropriate. A root mean square error of 0.083 m/s was calculated. The main source of that error stems from the fact that the measured velocities at points 11 and 12 under the two conditions are similar and almost identical when considering velocities in the Y-direction. In the numerical model, point 11 exhibits lower velocities in the X-direction while the velocities in the Y-direction indicate that the currents are not parallel. This can probably be attributed to the uniform conditions for the eastern boundary.

Calibration of the Bottom Roughness

As soon as the tide lag was correctly reproduced, the second element used in the calibration is the bottom roughness. When diving in the area for visual confirmation of the nature of the bottom terrain, it was observed that the bottom near the coast (within 100m of the shore) consisted of small reefs and sand patches. This fact was also observed via satellite imagery. As there is no precise information on a Manning's roughness for that type of bottom, the bottom roughness adopted for this region will have to rely on a calibration process. As a reference, Manning's roughness was estimated as high as 0.22 for some reef systems in earlier similar-type studies (Cialone et al. 2007). For the rest of the grid, the roughness was assumed to be constant with a Manning's n = 0.018, a value that was used to verify tide lag calibration. At a specific distance, the bottom roughness was assumed uniform with a Manning's n = 0.018 and the roughness was then increased close to shore as needed. As all the measurements used for this calibration were taken on the same day, only one sea condition was used in the model. Table 3 summarizes the wave, tide and wind information inputted into the model:

	Date	August 29th
Wave condition	Significant wave height (m)	1.2
	Direction (d)	90
	Wave period (s)	7
Wind Condition	Speed (m/s)	7
	Direction (d)	90
Tide condition	Time/Water level	3:14 +0.2m
		8:23 0m
		15:03 0.5m
		23:03 0m

Table 3: Model Information for Roughness Calibration (wisuki. 2016)

As the data was collected between 7:00AM and 7:40AM on August 29th 2016, only the ebb tide between 3:14AM and 8:23AM will be simulated. The points used for the calibration are shown in Figure 2 and marked from 1 to 10. All these points are very close to the location of the intended groyne. The following table summarizes the results of the calibration for bottom roughness:

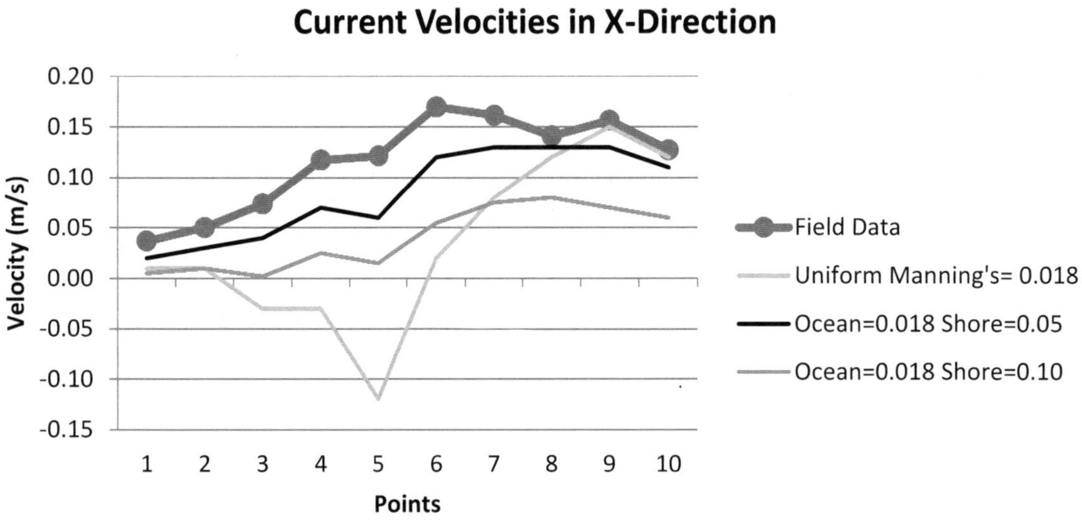

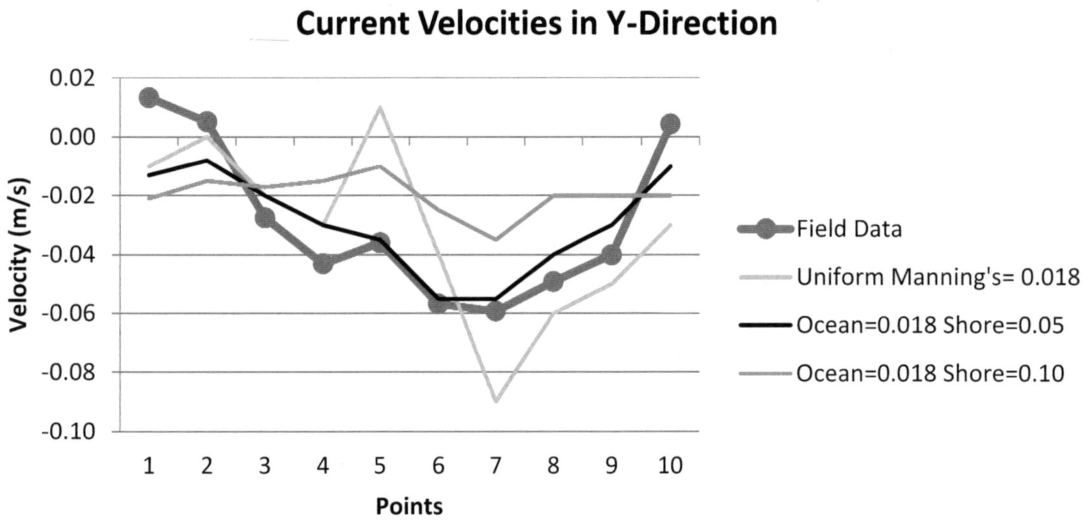

Figure 6: Bottom roughness Calibration

After analysing the data, it was observed that when the bottom roughness for the first 100m of shore is increased to a Manning's coefficient of 0.05, while keeping the rest of the grid at 0.018, the model was most accurately representing the field measurements. The calculated root mean square error is 0.0264 m/s. This difference is acceptable since it is within the mean averaging error of 0.048 m/s for the data set. Also, the model is representing correctly the trend of the measured velocities. This gives a good indication that it will accurately model the movement of the floating debris. Knowing that the shore is composed of a mix of small reefs and sand, it is in line with previous studies that have also found that a reef bottom will have a higher roughness. (Cialone et al. 2007) Previous authors have also suggested similar Manning's coefficient for reefs systems in shallow water. (Rosman et al. 2011) (Prager. 1991)

5. VALIDATION OF THE MODEL

Validation of the model was done by tracking the movement of floating debris and testing if the surface current along the shore will push the debris toward the inlet. This is done by releasing floating buoys, defined as drogues by Delft3D. It was observed that using the default value of JONSWAP's bottom friction in Delft3D (0.067 m^2s^{-3}), the debris are moving offshore perpendicularly to the shore. This observation is contrary to what was observed during the field campaigns. So, different values of JONSWAP were tried and it seems that a lower value is more appropriate for this situation of 'shallow water with reefs' bottom. This confirms the newest findings that the assumed value of JONSWAP for wind driven waves is too high. (Cialone et al. 2007) Figure 7 illustrates the movement of floating debris released at 3 locations every hour during a full tide cycle.

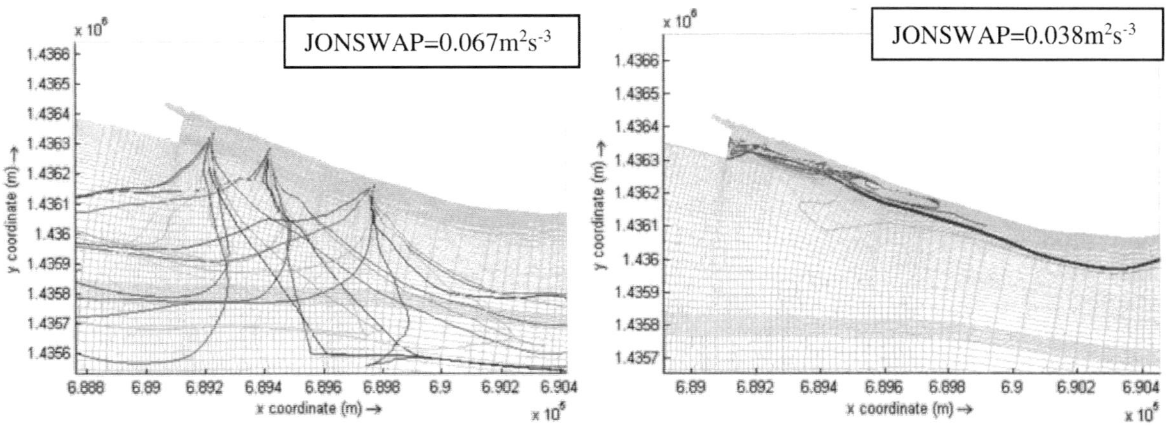

Figure 7: Floating Debris with Different JONSWAP

A JONSWAP parameter of 0.038 m^2s^{-3} will be used as it was proposed as a more appropriate value for this situation and represents better what was observed. (Vledder. 2010) However, the value of JONSWAP can't be adjusted further since no actual floating debris tracks were measured in the field. The change of JONSWAP has had little effect on the depth-averaged velocity calibration and the values presented for the calibration are for the reduced JONSWAP parameter.

6. CONCLUSIONS AND WORK IN PROGRESS

A 3D hydrodynamic model Delft3D was established for the coastal region of Paget Farm on Bequia Island in St. Vincent and Grenadines, to study the wave- and tide-induced hydrodynamics of the area relevant to the movement of floating litter along the coastline. The tide-induced current was accurately reproduced. The validation of the model has confirmed findings that JONSWAP's parameter for wind driven waves is too high. Now that the calibration process is completed, the model will be used to find the optimal size and orientation of a proposed new groyne which will have a triple role: (1) protect the local coastal area against erosion, (2) provide additional landing ground for fishing boats and (3) redirect the current (and litter) away from the entrance to the inlet.

The groyne will be represented in the model as a series of 'thin dams'. (Deltares, Delft3D-FLOW. 2011) (Visser. 2002) These are cells that are defined as 'always dry'; therefore they will be excluded from the computation. Since there is no exchange of water through those cells, the velocity around these cells will be zero. This will modify the distribution of velocities around the area of the inlet and thus redirect the influx of litter away from the airport. Several scenarios will be studied in combination with assessing the influence of the groyne configuration. The final selected configuration will be further modeled to investigate the morphologic impact of the groyne. This will evaluate the assumption that a groyne has the potential to reduce the present coastal erosion eastward of the airport.

7. REFERENCES

Booij, N., Ris, R. C. and Holthuijsen L. H. 1999. A third-generation wave model for coastal regions. *Journal of Geophysical Research, Vol. 104*, p. 7649-7666.

California Departement of Water Resources. 1970. *Determination of the Manning Coefficient From Measured Bed Roughness in Natural Channels,* Washington: United States Government Printing Office.

Chatzirodou, A. and Karunarathna, H. 2014. Impacts of Tidal Energy Extraction of Sea Bed Morphology. *34th Coastal Engineering Proceedings*, Seoul, Korea.

Cialone, M. A., McKee Smith J. 2007. *Wave Transformation Modeling with Bottom Friction Applied to Southeast Oahu Reefs*, US Army Engineer Research and Development Center, Vicksburg, Mississippi.

Deltares. 2011. *Delft3D-FLOW User Manual*, Deltares, Delft, Netherlands.

Deltares. 2011. Delft3D-WAVE *User Manual*, Deltares, Delft, Netherlands.

Elias, E. 1999. The Egmond Model, Calibration, Validation and Evaluation of Delft3D-MOR with Field Measurements. Delft Hydraulics. Delft, Netherlands

Mitchell, S. J. 2016. *Bequia Airport - an essential lifeline with an even brighter future*. Consulted on March 25th 2016, http://www.iwnsvg.com/2016/01/28/bequia-airport-an-essential-lifeline-with-an-even-brighter-future/

Prager, E. J. 1991. Numerical simulation of circulation in a Caribbean-type backreef lagoon. *Coral Reefs,* 10 (4): 177-182.

Rosman, J. H. 2011. A framework for understanding drag parameterizations for coral reefs. *Journal of Geophysical Research*, 33 (8): 1-15.

Tour du monde en images. 2015. *Saint-Vincent-et-*Grenadines, Consulted on January 17th 2017, http://www.jump-voyage.com/carte-saint-vincent-et-grenadines/vacances-saint-vincent-et-grenadines-1/

Trouw, K., Zimmermann, N., Mathys, M., Delgado, R. and Roelvink, R. 2012. Numerical Modelling of Hydrodynamics and Sediment Transport in the Surf Zone: a Sensitivity Study with Different Types od Numerical Models. *Proceedings of the International Conference on Coastal Engineering*, Santander, Spain

U.S. Geological Survey. 2009. *Measuring Discharge with Acoustic Doppler Current Profilers from a Moving Boat*. Reston, Virginia: U.S. Departement of the Interior.

Visser, R. 2002. Morphological modelling in the vicinity of groynes, an extended application of Delft3D-RAM including impact, Delft University of Technology, Delft, Netherlands.

Vledder, v. G. 2010. Revisiting the JONSWAP Bottom Friction Formulation. *Coastal Engineering Proceedings*, Shanghai, China, 32 : 41 in waves.

wisuki. 2016. *Statistics, Friendship Bay*. Consulted on August 22nd to August 29th, 2016, http://wisuki.com/statistics/925/friendship-bay?a_wi=4&wi_m=0&a_wa=4&wa_m=0&temp=monthly&rain=quantity

Leadership in Sustainable Infrastructure
Leadership en Infrastructures Durables

Vancouver, Canada
May 31 – June 3, 2017/ *Mai 31 – Juin 3, 2017*

NUMERICAL MODELLING OF ROCK-WEIR TYPE NATURE-LIKE FISHPASSES

Baki, Abul Basar M. [1, 6], Zhu, David Z. [2], Harwood, Andrew [3], Lewis, Adam [4], and Healey, Katie [5]
[1] Water Resources Engineer/Scientist, Ecofish Research Ltd., Vancouver, BC, CANADA.
[2] Professor, Department of Civil and Environmental Engineering, University of Alberta, Edmonton, AB, CANADA.
[3] Fisheries Scientist, Ecofish Research Ltd., Vancouver, BC, CANADA.
[4] Fisheries Biologist, Ecofish Research Ltd., Vancouver, BC, CANADA.
[5] Senior Analyst, Ecofish Research Ltd., Vancouver, BC, CANADA.
[6] baki@ualberta.ca

ABSTRACT

In recent years many migration barriers have been replaced by nature-like fishpasses to mitigate the effects of human development and habitat fragmentation on fish. This research study numerically investigated the complex flow characteristics in rock-weir type nature-like fishpasses for different structure geometries and channel characteristics to optimize design. Initially, the numerical model was validated with physical model results and good agreement was achieved. As part of this research, the study described herein examined the flow characteristics of rock-weir fishpasses with and without passage slots/notches. We investigated physical characteristics within rock-weirs with and without notches to determine how they affect the hydraulics within the structures, focusing on water surface profiles, local velocity fields, volumetric energy dissipation rates, fish resting zones, and fish passage performance. It is hoped that the results of this study will advance our knowledge on flow in a rock-weir fishpass and be useful to fishpass designers and fish biologists.

Keywords: - Computational fluid dynamics, hydraulics, energy dissipation, resting zone, and rock-weir fishpass.

1. INTRODUCTION

The fragmentation of the longitudinal river corridor by different hydraulic structures (e.g., weirs, dams, and culverts) represents a major global human impact on running waters (Jungwirth 1998). This fragmentation has a negative effect on fish movement and dispersal (Yamamoto et al. 2004). To mitigate these effects of human development and habitat fragmentation on fish, considerable effort has been expended in recent years to provide passage at migration barriers via construction of nature-like fishpasses (NLFs) (DVWK 2002). NLF designs generally fall into two categories: pool-weir type and rock-ramp type. Pool-weir fishpasses are constructed using a series of weirs and pools in a stepped fashion, rather than having a single large drop (Katopodis and Williams 2011). Pool-weir type structures are designed in consideration of fish passage goals and the physical setting; common structures include rock-weirs, step-pools, and crossbar block ramps (Baki et al. 2017a). Rock-weirs are composed of boulders in rows at regular intervals to form a series of pools, which provide resting zones for migrating fish (Baudoin et al. 2014). The rock-ramp fishpass consists of a long sloping channel interspersed with boulders providing resting places for fish swimming upstream (Baki et al. 2014; 2015; 2016). They are typically constructed of boulders, cobble, and other natural materials to create diverse physical and hydraulic conditions providing efficient passage to multiple species including migratory and resident fish assemblages (Turek et al. 2016).

Generally, a NLF is designed to function by providing passable conditions over a range of flows from the 95% (low) to 5% (high) flow exceedance during periods when migrating fish are normally present at the site (NMFS 2008). The target/design flow regime, determines the size of boulder that will be used in the rock-weir design: boulder size is typically uniform if the NLF is designed for a constant flow (Baki et al. 2017a; 2017b; Kupferschmidt 2015), whereas two different sizes of boulders are used if multiple design flows are contemplated (DVWK 2002; Wang and Hartlieb 2011; Oertel and Schlenkhoff 2012; Baudoin et al. 2014; Turek et al. 2016; Baki et al. 2017c). The use of both small and large boulders provides small passage openings, hereafter called notches, which guarantee a minimum water depth over the weirs and maintain connectivity between pools at low flows to allow fish to pass (Oertel and Schlenkhoff 2012). Typically, fish preferentially pass through weir openings rather than over weir crests (Turek et al. 2016). With increasing flow rates, the water overflows the larger sized boulders and the flow regime changes from a transitional flow regime to a streaming flow regime (Baki et al. 2017a).

Depending on specific project goals, researchers have proposed various configurations of pool-weir type NLF structures. Recently, several laboratory experiments, numerical modelling, and field studies have been carried out to examine the local flow patterns associated with rock/boulder-weirs (Thomas et al. 2000; Rosgen 2001; DVWK 2002; Haro et al. 2008; and Meneghetti 2009). Thomas et al. (2000) and Rosgen (2001) proposed conceptual designs for rock-vortex structures, DVWK (2002) provided some guidelines for crossbar block ramps, Baudoin et al. (2014) proposed pre-assessment procedures for rock-weirs, and Wang and Hartlieb (2011) and Oertel and Schlenkhoff (2012) investigated flow characteristics for crossbar block ramps using physical and numerical models. Recently, Turek et al. (2016) reported some guidelines for rock-weirs with passage notches specifically designed for the target fish species. Cox (2005) found that available guidelines and literature related to pool-weirs were scarce and consistently lacked investigation of hydraulic properties and/or performance.

The studies conducted to date have usually focused on the hydraulics associated with a single or limited range of rock-weir geometries. None of these efforts have linked the physical/numerical processes associated with weir performance to variations in weir structure geometry. Due to the lack of reliable design guidance for the rock-weir fishpasses, this research study attempts to develop a physical/numerical model matrix to investigate the physical characteristics within rock-weirs and how variations in structure geometry affect local hydraulics. This study is part of a NSERC IRDF research project on rock-weir type NLFs, which was undertaken at Ecofish Research Ltd. in collaboration with the University of Alberta. The objective of this research study was to examine the complex flow characteristics and to optimize the design and construction of rock-weir type NLFs, so that they can be used to pass fish at run of river hydroelectric projects. Part of the outcome of this research, the numerical modeling results for the NLFs, is presented in this paper.

2. NUMERICAL MODELLING APPROACH

The numerical model was based on an experimental rock-weir fishpass, installed in a rectangular flume 0.92 m wide, 0.61 m tall, and 8.89 m long at the T. Blench Hydraulics Laboratory of the University of Alberta (more details in Kupferschmidt 2015). The model domain was 9 m in length consisting of six rock-weirs and five pools (
Figure **1**), with a constant bed slope of 3%. Weirs were arranged in a V-shape with the V facing in the upstream direction having an arm angle (θ) of 70 degrees, plan view angle of departure from bankline (
Figure **1**a). The domain included two different types: Type (I) - without notches and Type (II) - with notches. In (I), the rock-weirs consisted of a single size of boulder having 14 cm in diameter (*D*) and weir height of d = 12.5 cm from the bed, meaning 1.5 cm was embedded in the channel bed (
Figure **1**a). In (II), the boulders in the weirs consisted of two different sizes: larger boulders 14 cm in diameter and smaller boulders 7 cm in diameter (bringing the weirs to the height of 12.5 cm and 5.5 cm

above the bed, 1.5 cm was embedded in the channel bed), resulting in an opening height of 7 cm in the passage notches (
Figure **1**a&b). Simulations were conducted varying the flow through the fishpasses (Table 1) from 0.03 m^3s^{-1} to 0.15 m^3s^{-1} in Series-A and -B.

The numerical simulations were conducted using a commercial software ANSYS-CFX. This CFD model solves the three-dimensional Reynolds-averaged Navier-Stokes equations using the finite volume approach. The standard k - ε turbulence model was used to determine the turbulent viscosity. The performance of the numerical model was validated with a series of experimental observations for water depth and velocity, and good agreement was achieved. A complete outline of the model, model development, corresponding boundary conditions, and model validation is provided in Baki et al. (2017a).

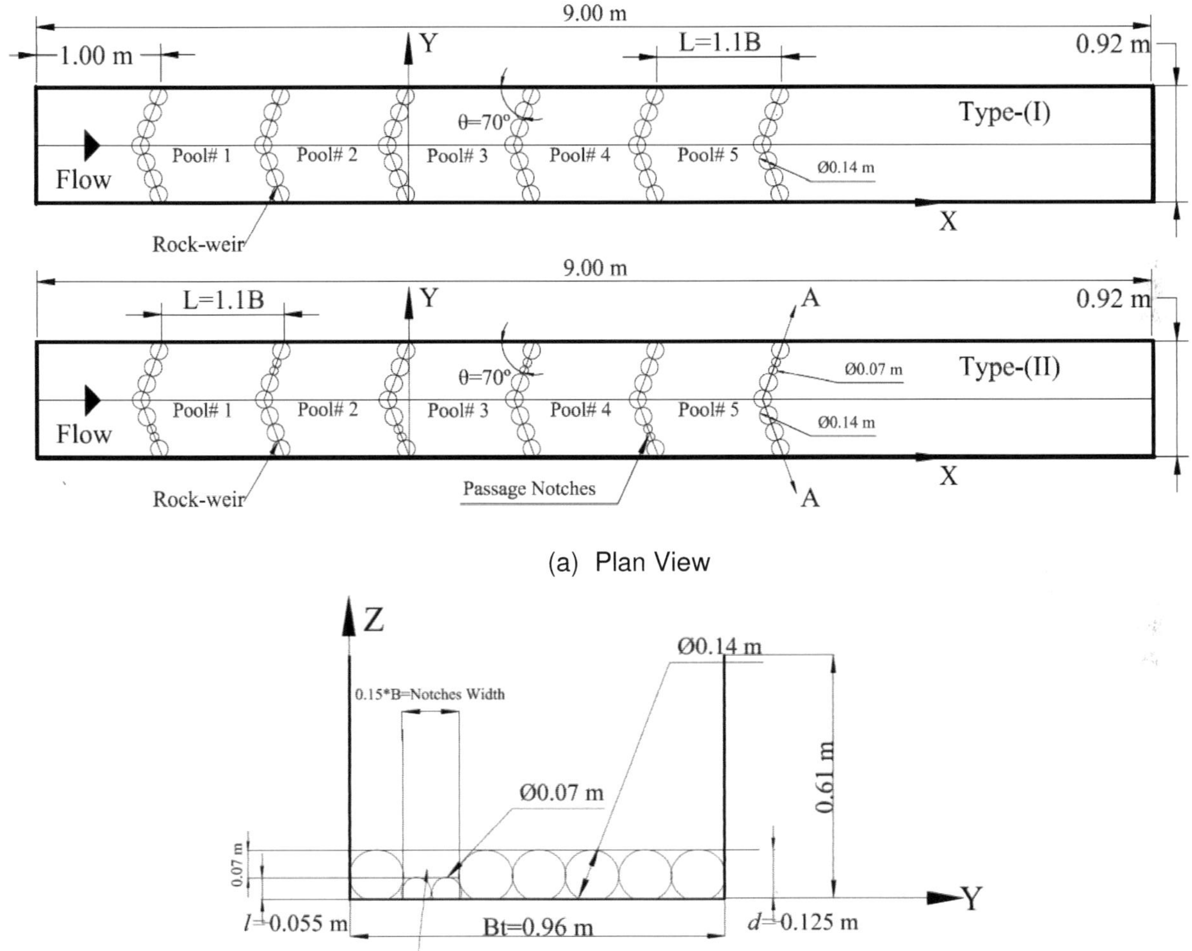

(b) Cross-sectional View (A-A)

Figure 1: (a) Plan view of CFD model domain for the Type: (I) without notches and (II) with notches, and (b) cross-sectional view of A-A in (II).

3. RESULTS AND DISCUSSIONS

1.1 Water surface profiles

The dimensionless longitudinal water surface profiles along the channel center line under the without and with notch alternatives are plotted in Figure 2 for different flow conditions in Series-A and -B. For both series from 0.03 to 0.06 m^3s^{-1} (A1-4, B1-4), a short hydraulic jump appeared immediately downstream of the weir crest and then a horizontal water surface existed. For 0.085 to 0.15 m^3s^{-1} (A6-9, B6-9), a long hydraulic jump and oscillating water surface appeared. Evaluation of the flow fields, especially the water surface profiles and hydraulic jump for 0.03 to 0.06 m^3s^{-1} compared to 0.085 to 0.15 m^3s^{-1}, confirmed two different flow regimes: weir flow regime and transitional flow regime (Chanson 1994; Dust and Wohl 2012; Baki et al. 2017a).

Table 1: Descriptions of all Simulations.

Simulation	Flow (m^3/s)	Bed Slope (%)	Boulders Diameter (cm)	C_g * (coefficient)	θ ^ (degrees)	Pool Spacing, L	Type
A1	0.030	3	14	0	70	1.1B	(I) Without Notches
A2	0.040						
A3	0.050						
A4	0.060						
A5	0.075						
A6	0.085						
A7	0.100						
A8	0.120						
A9	0.150						
B1	0.030	3	14, 7	0.15	70	1.1B	(II) With Notches
B2	0.040						
B3	0.050						
B4	0.060						
B5	0.075						
B6	0.090						
B7	0.110						
B8	0.130						
B9	0.150						

*Passage notches width = $C_g \times B$, where B is the channel width.

^Arm angle (θ), plan view angle of departure from bankline.

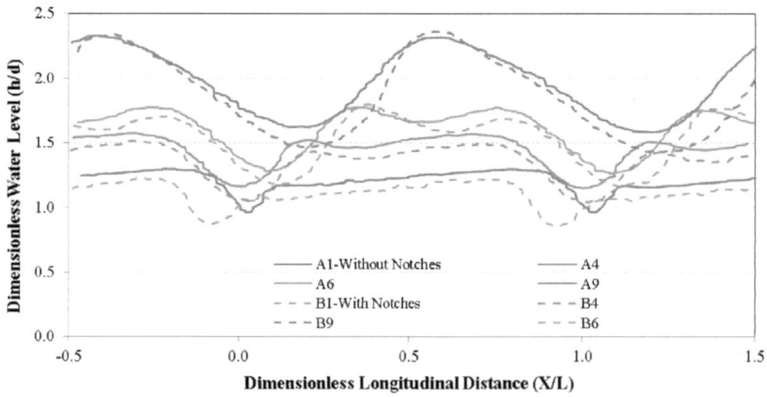

Figure 2: The dimensionless water-surface profiles along the center line of the flume in Series-A (A1= 0.03 m^3s^{-1}, A4=0.06 m^3s^{-1}, A6= 0.085 m^3s^{-1}, and A9= 0.15 m^3s^{-1}) and -B (B1= 0.03 m^3s^{-1}, B4=0.06 m^3s^{-1}, B6= 0.09 m^3s^{-1}, and B9= 0.15 m^3s^{-1}).

In general, the water level for the rock-weirs with notches is less than the water level without notches for each flow condition. The water level in Series-A without notches varied from 15.1 to 24.6 cm and in Series-B with notches 13.8 to 23.4 cm, and was ~7% higher for Series-B with notches. This result is expected as the passage notches provide additional openings to direct flow into the pool, reducing the obstruction of flow and decreasing the pool water depth due to less backwatering.

1.2 Flow pattern and velocity field

The local flow fields are important in the design of rock-weirs because they can affect fish passage and overall structure performance (Baki et al. 2017a). This study investigated the local flow patterns in a pool on the horizontal plane parallel to the bed at z = 0.5H for different flow conditions in Series-A (without notches) and -B (with notches) (Figure 3 and Figure 4, respectively), where *H* is pool averaged water depth along the center line of the channel.

In Figure 3, immediately downstream of the weir crest in the Series-A trials, the alignment of flow is directed towards the center of the downstream pool, which is similar to that observed in step-pool structures (Thomas et al. 2000). As a result, it prevents flow being directed towards the outer banks, directing scour away from the banks towards the centre of the channel and limiting bank erosion (Thomas et al. 2000). As the flow approaches the upstream side of the weir, the alignment of flow changes direction from center line and diverges towards the channel banks. The flow is spread over the weir instead of concentrated in the center line of the weir. The flow alignment in a pool, towards center line and then the channel banks, generates flow re-circulation zones along both channel banks (Figure 3). The two dimensional (transverse and longitudinal directions) extent of the high velocity zone and re-circulation zone in the pool increases with increasing flow rates, resulting in the reduction of pool low velocity zones (discussed later). In general the flow pattern in Series-B with notches is the same as that in Series-A (Figure 4).

The magnitude and extent of the high velocity zone, and re-circulation zone in the pool, varied from without notches to with notches. The maximum velocity (U_{max}) in all simulations occurred in the location of notches (Figure 4), whereas all simulations without notches U_{max} occurred along the channel center line (Figure 3). Moreover, the magnitude of U_{max} in Series-B with notches was ~20% greater than in series-A without notches due to a reduced backwatering effect. The flow recirculation zone and the slow velocity area in Series-B were larger than in Series-A due to flow being concentrated to one side of the pool.

The maximum velocity reduction, η, was calculated to quantify the effects of notches for different flow condition as follows:

[1] $\eta = (V_o - U_{max})/(V_0*100)$

where, V_0 is the pre-structure velocity referring to the normal depth. The results confirmed that the maximum velocity reduction η decreased from 45% to 37% when the flow rate increased from 0.03 m^3s^{-1} to 0.15 m^3s^{-1} in Series-A. Similarly in Series-B, η decreased from 30% to 25% when the flow rate varied from 0.03 m^3s^{-1} to 0.15 m^3s^{-1}. Therefore, a significant reduction of flow velocity occurred in rock-weirs without notches, which was larger than the reduction in velocity in weirs with notches.

1.1 Slow velocity zones and energy dissipation rate

The availability of fish resting areas in a pool (i.e., areas of slow velocity) was examined by calculating the relative cumulative frequency distributions of dimensionless velocities (U/U_{max}) in a horizontal plane parallel to the bed at z= 0.5H (Figure 6). The results of this analysis are independent of discharge (Wang et al. 2010); results are presented only for 0.06 m^3s^{-1} simulations in Series-A (without notches) and -B (with notches). The frequency distribution curves provide an estimate of the average percentage of areas on a plane parallel to the bed in a pool for U/U_{max} from 0 to 1. Following Baki et al. (2017b) the average percentage of areas where the velocity is lower than $0.4U_{max}$ was 30% for without notches and 46% for with notches at a flow rate 0.06 m^3s^{-1}. Therefore, the slow velocity zones under $0.4U_{max}$ are significantly altered by the opening in the weir (notches), with the areas of slow velocity in a pool increased 1.5 times in a pool with notches compared to a pool without notches (Q = 0.06 m^3s^{-1}).

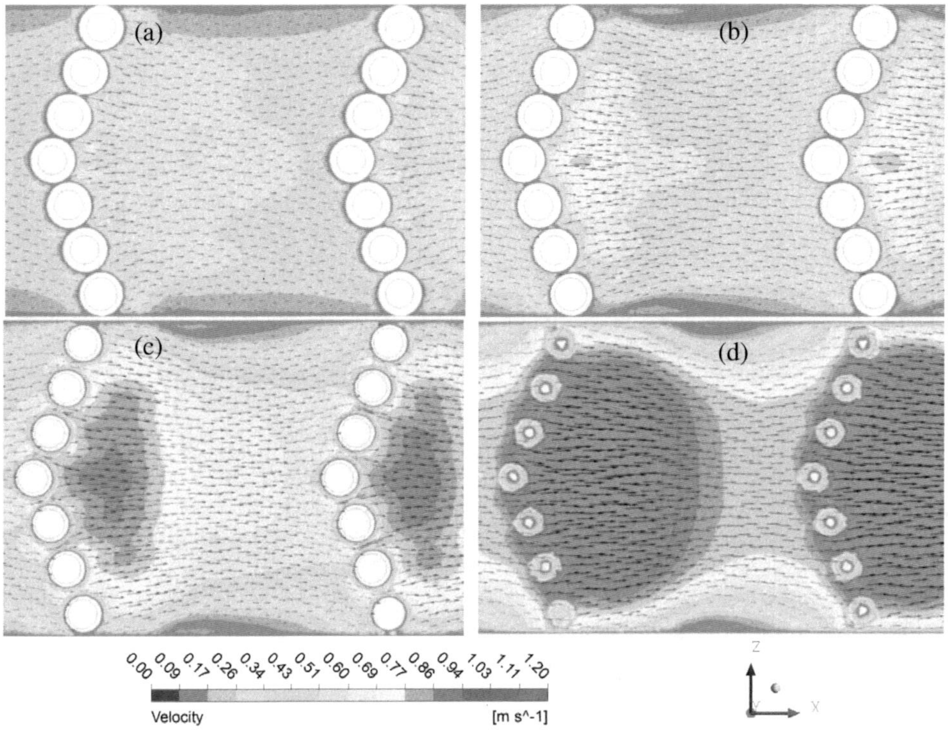

Figure 3: Spatial distributions of velocity magnitude with directions on horizontal plane parallel to bed at z = 0.5H for different flow rates in series-A (without notches): (a) 0.03 m^3s^{-1}, (b) 0.06 m^3s^{-1}, (c) 0.085 m^3s^{-1}, and (d) 0.15 m^3s^{-1} (not to scale).

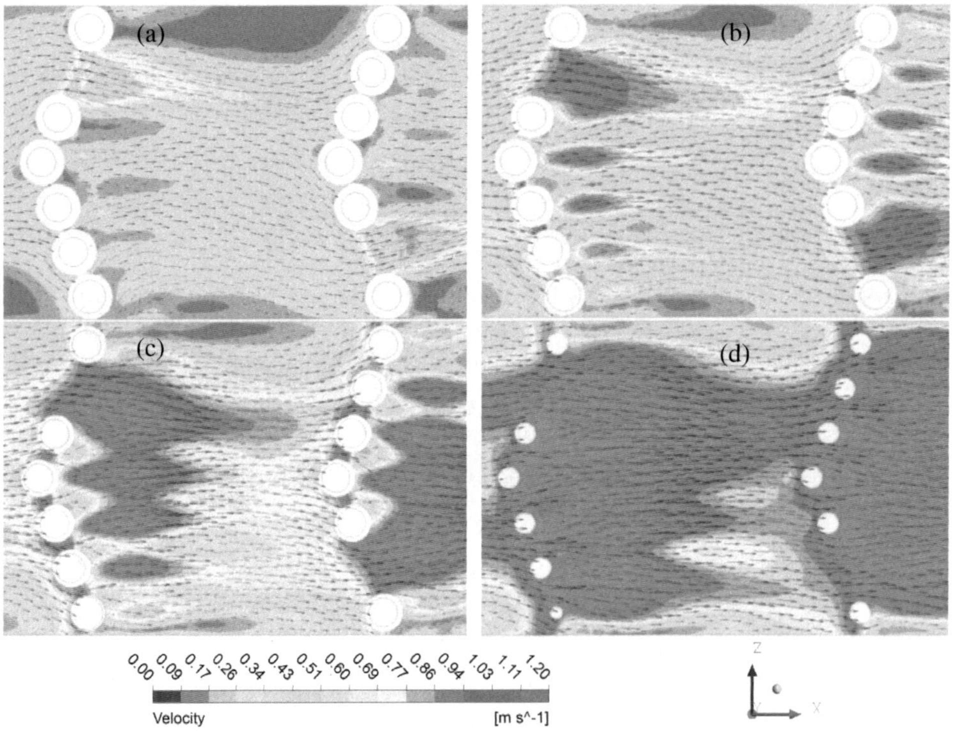

Figure 4: Spatial distributions of velocity magnitude with directions on horizontal plane parallel to bed at z = 0.5H for different flow rates in series-B (with notches): (a) 0.03 m^3s^{-1}, (b) 0.06 m^3s^{-1}, (c) 0.09 m^3s^{-1}, and (d) 0.15 m^3s^{-1} (not to scale).

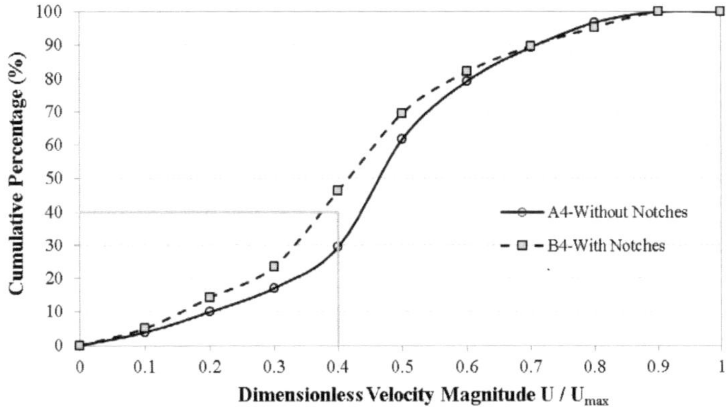

Figure 5: The relative cumulative frequency distribution curves of U/U_{max} on horizontal plane parallel to bed at z = 0.5H in the pool at 0.06 m^3s^{-1} (reference lines highlight the area for U/U_{max} < 0.4 and Cumulative Percentage > 40%).

Pool type fishpasses are designed to cause energy dissipation and to provide lower current velocity and higher water depth in the downstream pool section while maintaining acceptably low levels of turbulence (Towler et al. 2015). The average volumetric energy dissipation rate (E) in a pool is calculated using the following basic formula:

[2] $E = (\rho g Q \Delta h) / V_p$

where, Δh is the difference in water level between two pools (mid-point at each pool), V_p is the volume of pool excluding the volume of the rock-weir,ρ density of water, and g is the gravitational acceleration. The relationship between E and discharge per unit channel width (specific discharge, q) is presented for all series of simulations in Figure 7. E generally increases with specific discharge; which is consistent with the findings of Yagci (2010), who found E increased with increasing discharge at a constant bed slope for a pool-weir fishpass. In Series-A without notches, E increased from 63 W/m^3 to 198 W/m^3 as specific discharge increased from 0.033 m^2s^{-1} to 0.163 m^2s^{-1}. Similarly, E increased from 76 W/m^3 to 232 W/m^3 as specific discharge increased from 0.033 m^2s^{-1} to 0.163 m^2s^{-1} in Series-B with notches. On average, E increased approximately 15% from the simulation in series-A without notches to with notches (Series-B).

The average volumetric energy dissipation rates in Series-A and -B were typically within the energy dissipation criteria of 150 W/m^3 and 200 W/m^3 under specific discharge of 0.09 m^2s^{-1} (except simulation B8 and B9, Figure 7). Larinier (2008) recommended that the average volumetric energy dissipation rate should not exceed 200 W/m^3 for large salmon and sea trout, and 150 W/m^3 for smaller shad and riverine species. For a small stream on a 1:1 Froude model scale, these results could be applied *in-situ*. For a full-scale structure (prototype), where the model-to-prototype scale is 1:4, the average volumetric energy dissipation rates for the Series-A (without notches) were below the upper limit of 200 W/m^3 at a specific discharge of 0.054 m^2s^{-1} or less. In Series-B (with notches), *E* was below the upper limit of 200 W/m^3 at a specific discharge of 0.043 m^2s^{-1} or less. The average volumetric energy dissipation rates were mostly above the lower criteria of 150 W/m^3 in all series of simulations, except the A1 simulation.

1.1 Fish swimming performance

The effectiveness of fishpass hydraulics depends on both fish swimming speed and endurance (i.e., the length of time during which the fish can maintain a specific speed) (Puertas et al. 2012). This study analyzed swimming performance of salmonids or cyprinids fish species for a full scale prototype structure (where the model-to-prototype scale is 1:4) based on the velocity fatigue restriction as recommended by Puertas et al. (2012). The restriction states that the maximum flow velocity cannot exceed the fish burst speed (assumed to be 10 body lengths per second, or 10 L_B/s), i.e., $U_{max} \leq$ burst speed. This restriction limits the value of $S_0 \times L$ (bed slope × pool spacing) using dimensionless maximum velocity,

[3] $U^*_{max} = U_{max} / \mathrm{sqrt}(2gS_0L)$

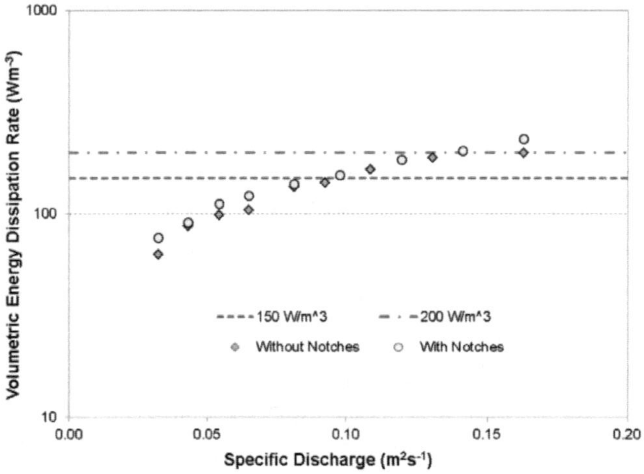

Figure 6: Relationship between average volumetric energy dissipation rate *E* and flow rate per unit channel width (specific discharge) for all simulations in Series-A and -B.

The effort that the fish must exert to traverse the fishpass was evaluated by comparing the maximum velocity $U_{max} \leq 2.5$ ms^{-1} (equivalent to the burst swimming speed of 250 mm salmonids or cyprinids) and allowable pool spacing $L \geq B$ (at least channel width) in Figure 7. In Series-A without notches, the maximum flow velocities in the simulations at flow rate ≤ 3.2 m^3s^{-1} in prototype (equivalent to 0.10 m^3s^{-1} in the model) achieved the above requirements (Figure 7). In Series-B with notches, the maximum flow velocities in all simulations at a flow rate of 1.92 m^3s^{-1} (equivalent to 0.06 m^3s^{-1} in the model) or less in the prototype achieve the above requirements (Figure 7). The increased flow velocity in the fishpass with notches did not produce hydraulic conditions compatible with the fish swimming performance of salmonids or cyprinids at higher flow rates ($Q > 1.92$ m^3s^{-1}).

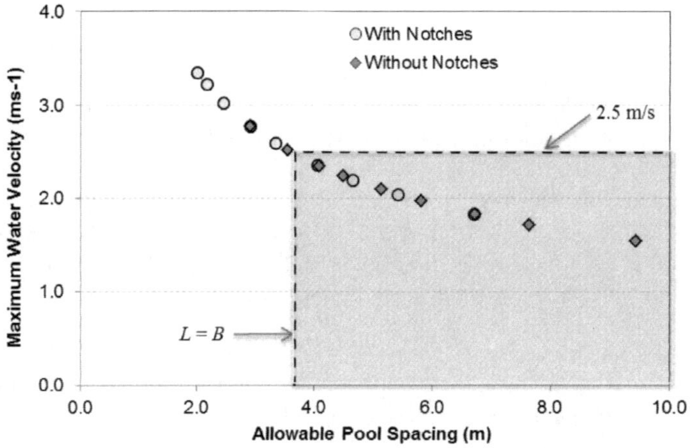

Figure 7: Relationship between maximum water velocity and allowable pool spacing at a prototype scale for 250 mm body length of salmonids or cyprinids. The dashed lines show $U_{max} \leq 2.5$ ms-1 (horizontal) and $L \geq B$ (vertical).

4. CONCLUSIONS

This study numerically investigated the hydraulic characteristics associated with rock-weirs with and without passage notches. The simulated water surface profiles confirmed two distinct flow regimes present within the structures: weir flow and transitional flow. When notches were added to the rock weir, the pool average water depth decreased by approximately 7% and the maximum velocity increased by about 20%. As water flows over the rock-weir, flow is directed towards the center of the downstream pool, then started to change direction from center line to channel banks when approaching the next weir, distributing flow across the weir. The magnitude and extent of the high velocity zone, and re-circulation zone in the pool varied between the trials with and without notches. The slow velocity zones which fish require for resting in a pool increased 1.5 times in a pool with notches compared to a pool without notches. On the other hand, the average volumetric energy dissipation rates in a pool with notches increased approximately 15% over that in a pool without notches. From the perspective of fish swimming performance, both rock-weirs with and without notches ensured flow fields low enough to be overcome by 250 mm long salmonids or cyprinids at flow rates of 1.92 m^3s^{-1} or less in a full scale prototype structure (1:4). The results of this research study have improved our understanding of flow fields associated with rock-weir NLFs and improved design procedures for designing rock-weir NLFs.

ACKNOWLEDGEMENT

This research was made possible through grants from the Natural Sciences and Engineering Research Council (NSERC) of Canada and Ecofish Research Ltd.

5. REFERENCES

Baki, A. B. M., Zhu, D. Z., and Rajaratnam, N. 2014. Mean Flow Characteristics in a Rock-Ramp-Type Fish Pass. *Journal of Hydraulic Engineering*, 140(2): 156–168.

Baki, A. B. M., Zhu, D. Z., and Rajaratnam, N. 2015. Turbulence Characteristics in a Rock-Ramp-Type Fish Pass. *Journal of Hydraulic Engineering*, 141(2): 04014075.

Baki, A. B. M., Zhu, D. Z., and Rajaratnam, N. 2016. Flow Simulations in a Rock-Ramp-Type Fish Pass. *Journal of Hydraulic Engineering*, 142(10): 04014075.

Baki, A. B. M., Zhu, D. Z., Harwood, A., Lewis, A., and Healey, K. 2017a. Rock-Weir Fishpass. I: Flow Regimes and Hydraulic. Submitted to *Journal of Ecohydraulics.*

Baki, A. B. M., Zhu, D. Z., Harwood, A., Lewis, A., and Healey, K. 2017b. Rock-Weir Fishpass. II: Design Evaluation and Considerations. Submitted to *Journal of Ecohydraulics.*

Baki, A. B. M., Zhu, D. Z., Harwood, A., Lewis, A., and Healey, K. 2017c. Hydraulic Design Aspects of Rock-Weir Fishpasses for Habitat Connectivity. To be submitted to *Journal of Ecohydraulics.*

Baudoin, J. M., Burgun, V., Chanseau, M., Larinier, M., Ovidio, M., Sremski, W., Steinbach, P., and Voegtle, B. 2014. *Assessing the Passage of Obstacles by Fish.* Concepts, design and application. Onema, Paris, France.

Chanson, H. 1994. Hydraulics of Nappe Flow Regime Above Stepped Chutes and Spillways. *Aust. Civ. Struct. Eng. Trans.*, CE36(1): 69–76.

Cox, A. 2005. A Study of In-Stream Rehabilitation Structures in Sand-Bed Channels. M.S. Thesis, Colorado State University, Department of Civil Engineering, Fort Collins, Colorado.

Dust, D. and Wohl, E. 2012. Characteristics of the Hydraulics at Natural Step Crests in Step-Pool Streams via Weir Flow Concepts. *Water Resources Research*, 48, W09542.

DVWK 2002. *Fish Passes-Design, Dimensions and Monitoring.* Published by the Food and Agriculture Organization of the United Nations in arrangement with German Association for Water Resources and Land Improvement as DVWK-Merkblatt.

Haro, A., Franklin, A., Castro-Santos, T., and Noreika, J. 2008. Design and Evaluation of Nature-Like Fishways for Passage of Northeastern Diadromous Fishes. Final report, Submitted to NOAA National Marine Fisheries Service Office of Habitat Conservation.

Jungwirth, M. 1998. River Continuum and Fish Migration- Going Beyond the Longitudinal River Corridor in Understanding Ecological Integrity. In: Jungwirth et al. (ed.) Fish Migration and Fish Bypasses. Fishing News Books, Oxford, 19–32.

Katopodis, C. and Williams, J. G. 2011. The Development of Fish Passage Research in a Historical Context. *Ecological Engineering*, 48: 8-18.

Kupferschmidt, C.P.L. 2015. Ecohydraulics of Nature-Like Fishways and Applications in Arctic Aquatic Ecosystem Connectivity. M.S. Thesis, University of Alberta, Edmonton, Alberta.

Larinier, M. 2008. Fish Passage Experience at Small-Scale Hydro-Electric Power Plants in France. *Hydrobiologia.* 609:97-108.

Meneghetti, A. M. 2009. Stage discharge relationships for U-, W-, and A-weirs. M.S. Thesis, Colorado State University, Department of Civil Engineering, Fort Collins, Colorado.

NMFS (National Marine Fisheries Service). 2008. Anadromous Salmonid Passage Facility Design. NMFS, Northwest Region, Portland, Oregon.

Oertel, M., Schlenkhoff, A. 2012. Crossbar Block Ramps: Flow Regimes, Energy Dissipation, Friction Factors, and Drag Forces. *Journal of Hydraulic Engineering.* 138(5):440–448.

Puertas, J., Cea, L., Bermudez, M., Pena, L., Rodriguez, A., Rabunal, J. R., Balairon, L., Lara, A., and Aramburu, E. 2012. Computer Application for the Analysis and Design of Vertical Slot Fishways in Accordance with the Requirements of the Target Species. *Ecological Engineering.* 48:51-60.

Rosgen, D. L. 2001. The Cross Vane, W-weir and J-hook Structures: Their Description, Design and Application for Stream Stabilization and River Restoration. *Proceedings of the Wetland Engineering and River Restoration Conference*, ASCE, Reno, Nevada, USA.

Thomas, D., Abt, S., Mussetter, R., and Harvey, M. 2000. A Design Procedure for Sizing Step-Pool Structures. *Proceedings of the joint conference on water resource engineering and water resources planning and management*, ASCE, Minnesota, USA.

Towler, B., Mulligan, K., Haro, A. 2015. Derivation and Application of the Energy Dissipation Factor in the Design of Fishways. *Ecological Engineering.* 83(10):208-217.

Turek, J., Haro, A., Towler, B. 2016. Federal Interagency Naturelike Fishway Passage Design Guidelines for Atlantic Coast Diadromous Fishes. Interagency Technical Memorandum for USGS.

Wang, R.W., David, L., and Larinier, M. 2010. Contribution of Experimental Fluid Mechanics to the Design of Vertical Slot Fish Passes. *Kwol. And Manag. of Aquat. Ecosyst.* 396(02): 1-21.

Wang, R. W., Hartlieb, A. 2011. Experimental and Field Approach to the Hydraulics of Nature-Like Pool-Type Fish Migration Facilities. *Knowledge and Management of Aquatic Ecosystems.* 400:05.

Yagci, O. 2010. Hydraulic Aspects of Pool-Weir Fishways as Ecologically Friendly Water Structure. *Ecological Engineering.* 36:36-46.

Yamamoto, S., Morita, K., Koizumi, I., and Maekawa, K. 2004. Genetic Differentiation of White-Spotted Charr (Salvelinus Leucomaenis) Populations after Habitat Fragmentation: Spatial–Temporal Changes in Gene Fequencies. *Conserv Genet*, 5(4):529–538.

Leadership in Sustainable Infrastructure
Leadership en Infrastructures Durables

Vancouver, Canada
May 31 – June 3, 2017/ *Mai 31 – Juin 3, 2017*

OPTICAL REMOTE SENSING TECHNIQUE FOR THE GENERATION OF MEANDERING RIVER CHANNEL TOPOGRAPHY AND SEDIMENT SIZE

Harada, Sho.[1,4], Li, S. Samuel[2] and Lapointe, Michel[3]
[1] Concordia University, Canada
[2] Concordia University, Canada
[3] McGill University, Canada
[4] shoharada10@gmail.com

Abstract: Traditional cross-section survey techniques are costly, time-consuming, and difficult to implement, particularly in inaccessible areas. Practical limitation in the spacing and frequency of survey points restricts ground-based survey to reach scale. This paper demonstrates the potential of sizing bed materials within complex shallow channels, using high-resolution multispectral and stereo images. The demonstration uses a 13-km long reach of meandering, alluvial river (the Goulais River in Ontario). Fluvial remote sensing provides complimentary alternative to field survey, in which a watershed-scale, synoptic view of the river may be acquired. The presented technique generates a river topography model (RTM) by combining water depth map (bed to free surface) and the dry surface elevation map (free surface and above). The former is generated from the depth-to-brightness ratio that is empirically estimated by correlating available field survey points to the digital numbers of the image (excluding the bed elevation, z). The latter is generated through a photogrammetric analysis of stereo images. A challenge arises in combining the maps when the image used to derive the dry surface elevation is at a higher stage than the image used to derive the water depth. This is corrected by applying constant vertical displacement. The resulting RTM is a continuous digital terrain that encompasses the channel bed, floodplains, and the dry terrain features. Qualitative observations of the RTM indicate a correct placement of geomorphic features (pool-riffle system, point bars) for the river. The RTM is used as a model domain for depth-averaged simulations to estimate the bed shear stress corresponding to the formative discharge. The bed material sizes derived from the simulations compare well with field observations. The method presented offers a promising complimentary tool for river analysis.

1 Introduction

In the fields of fluvial sciences, particularly in the ecology department but not limited to, it is often desirable to have a synoptic view of a river in assessing the nature of the system. In the context of sediment transport and channel morphology, a river must be analyzed from the headwaters to the estuary to characterize the sediment caliber distribution, sediment transport rate/capacity, and the resulting channel aggradation/degradation (Church 2002). From an ecology stand point, identification of the major determinants of biota composition, i.e. the sedimentary links (Davey and Lapointe 2007), the various in-stream habitat zones and their longitudinal and lateral connectivity (Bunn and Arthington 2002) require a large-scale understanding of the river.

Despite the need for large-scale assessment of rivers, terrestrial survey methods have been limited in spatial extent. Conventional field surveys (i.e. cross-sectional measurements, sonar, patch sampling) are time consuming, costly, and logistically constrained. In addition, sequential data measurements are prone to error due to the changing environmental variables (imagine making a measurement downstream of a peaking dam over a day). Consequently, there is considerable difficulty in achieving survey with sufficient spatial extent, spatial density, and frequency of resurvey. (Review of literature show that) There is an increasing trend in supplementing aerial and satellite data to overcome the above-mentioned constraints set by traditional data collection methods.

Since the launch of LANDSAT 1 in 1972, the first of many earth-observing satellites, availability and accessibility to remote sensing data has improved exponentially due to the number of earth-observing satellites launched and the improvements in data storage and transfer technology (J. B. Campbell 2002). Application of remote sensing in aquatic environments initiated in the coastal sciences domain with the primary aim to remotely obtain costal bathymetry. In the past two decades, the river science community has adapted techniques developed for the coastal analysis to drive research in fluvial information extraction from remotely sensed data (i.e. bathymetry, substrate types, turbidity, flow characteristics, in-stream habitat, large woody debris, water temperature). Three major techniques for extracting water depth from remotely sensed data are briefly reviewed.

Physically based technique is based on physics of optics and employ atmospheric and subaquatic radiative transfer models to resolve the components of detected radiance to isolate the bottom-reflected radiance (Bukata 1995). Legleiter et al. (2004) simulated aquatic radiative transfer using the Hydrolight to determine the effect of water depth, substrate type, suspended sediment concentration and turbulence on the bottom reflectance in the Soda Butt-Pebble Creek confluence, US. Legleiter et al. (2009) use the optimal band ratio analysis algorithm and radiative transfer model to retrieve bathymetry for the Soda Butt Creek and the Lamar River, WY, USA at a r^2 of 0.79-0.98. The physically based techniques have shown the capacity to be accurate, yet the complexity of the radiative transfer models seems to deter user. In addition, accuracy and precision of depth estimates vary depending on the channel morphology, thus it is not advisable to rely solely on the technique (Marcus 2012).

Correlation technique employs a regression between the brightness of the image (pixels) and the field depth measurement at the same location to estimate the attenuation coefficient of the water column as a function of depth. It is the simplest of the techniques and most commonly used. Gilvear et al. (1995) correlated field depth measurement to bins of linearly stretched spectral histogram to create a depth map of the Faith River, AK, USA, with a spatial resolution of 1.3m. Winterbottom and Gilvear (1997) used reflectance-measured depth linear function proposed by Lyzenga (1981) for the three bands of the Multi Spectral Scanner (band 5, 6, 8) and aerial panchromatic image to map the depth of the River Tummel, Scotland, with a spatial resolution of 2m and 0.8m and r^2 of 0.67 and 0.55 respectively. Fonstad and Marcus (2005) developed and demonstrated the Hydraulically Assisted Bathymetry model that combined the linear regression of depth-brightness ratio and open channel hydraulics to determine the bathymetry of the Brazos River, TX, USA, with a spatial resolution of 1m and a r^2 of 0.77. Gilvear et al. (2007) regressed field measured reflectance, eliminating the dependence of bottom substrate on the bottom reflectance, and image brightness to obtain a reflectance-depth ratio of the Fourth Estuary, UK, with spatial resolution of 0.66m and r^2 (of the regression) of 0.81-0.87. Lejot et al. (2007) applied the depth-brightness ratio to very high resolution aerial images obtained by paraglider drone to reconstruct the bathymetry of the Rhone River, France with spatial resolution of 14 cm and r^2 of 0.86.

Photogrammetric technique is based on stereoscopy and utilizes a pair of stereo images and stereo matching algorithm (identifying identical ground points on the two images) to reconstruct the water depth. Westaway et al. (2001) used unsupervised stereo matching algorithm of ERDAS Imaging Software with high-resolution aerial photo scans of the North Ashburton River, New Zealand, to retrieve bathymetry with spatial resolution of 37.1 cm and mean errors of 36 cm. Lane et al. (2010) combined photogrammetric technique and depth-brightness modelling in the inundated zones to reconstruct the South Saskatchewan River, Canada, with spatial resolution of 4 cm and mean error of -25 cm-18 cm. Photogrammetric technique requires image resolution in the centimeter ranges, since submerged stereo points can only be match at a spatial resolution fine enough to resolve the texture of the channel bottom. Consequently, such technique often requires custom low flying surveys (i.e. planes and drones), which limit the temporal resolution of the site image (availability of historical records) and increase cost.

In this study, depth-brightness correlation technique was used to reconstruct the water depth map from 0.5 m Geoeye-1 multispectral image. The dry land digital elevation model (DEM) was reconstructed by the vendor using a pair of 0.5 m stereo images from Pleiades-A1. The depth map and the dry land DEM were combined to create a river topography map (RTM), a channel bottom-and-up digital representation of the riverine landscape. As an example of RTM application, digitally reconstructed river was used as a model domain to simulate a bankfull flow in a 2-D hydrodynamic model. Subsequently the simulated bed shear stress was compared against the field sampled sediment caliber in the riffle zone to assess the feasibility of predicting sediment caliber distribution from an image-derived RTM.

In the following, preliminary results of the study is presented. The study site is described in section 2. Next field survey, RTM reconstruction, bankfull discharge estimation and hydrodynamic modelling methodologies are described in section 3. Subsequently, geoprocessing and computational results are presented in section 4. Lastly, conclusion will be drawn in section 5.

2 Study Site

The Goulais River is an eastern tributary of Lake Superior, located 30km north of Salute Ste. Marie, Ontario. It drains an area of approximately 2000 km^2 into the Goulais Bay. It is an unregulated alluvial river, with well-developed meanders for many kilometers upstream of its mouth. The study site is a 13-km reach situated 3.5 km downstream from the township of Searchmont. Its characteristic pool-riffle system is accompanied by alternating point bars and mid channel bars. There are numerous meander cutoffs evident by clearly identifiable oxbow lakes in the satellite images. The floodplain is densely forested with much of the banks covered by tree canopy. During late summer, the site measures as much as 78 cm of flow depth. The channel width ranges between 20 m and 100 m and width transitions occur smoothly.

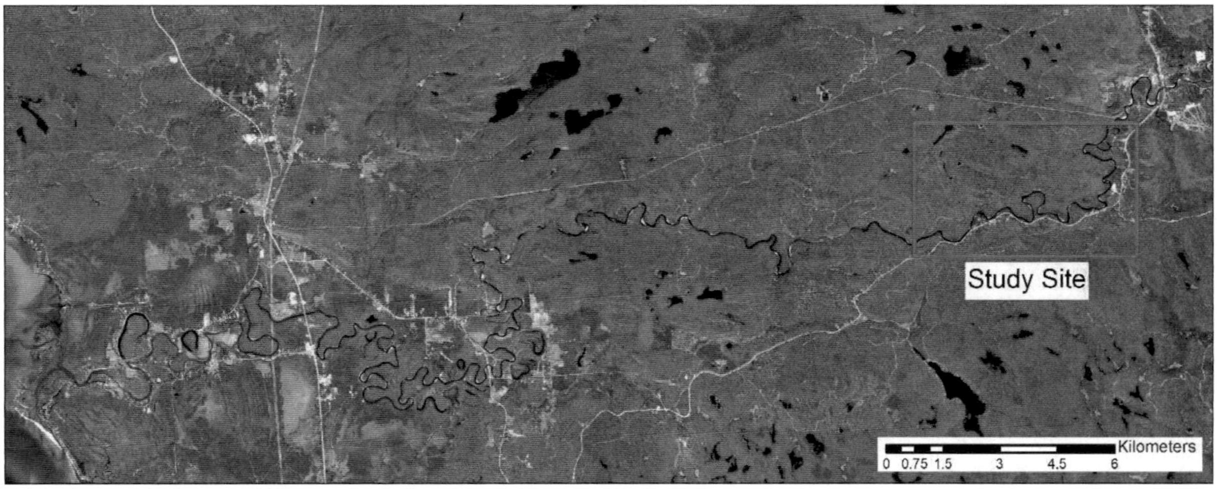

Figure 1. Satellite image of the Goulais River; the river flows from the town of Searchmont (top right), the study site (red box), and drains to the Goulais Bay (bottom left)

3 Methodologies

3.1 Field Measurement

The field survey was conducted by the biologists of Universite de Montreal as a part of the NSERC-HydroNet project on August 14-16, 2011 and July 17-30, 2012. Depth and grain size were measured in 35 rectangular patches (60m x 5m oriented in the streamwise direction).

Depth measurements were collected at 10 different locations for each patch, distributed in a 4 middle- 3 left- 3 right configuration. Grain class distribution was assessed visually at 10 different 50 cm x 50 cm quadrat for each patch, where trained personnel called out the percent grain class based on the Wentworth grain scale. Median diameter, D_{50}, was measured by collecting medial axial grain width for 50 randomly selected sediments in the patch.

3.2 Depth – band ratio correlation to determine water depth

The image-derived depth map was constructed by a fluvial geomorphology PhD student at McGill University. The 4-band 0.5-m Geoeye-1 satellite image was captured on July 7, 2012 with a discharge of 2.59 m^3/s (hence forth referred to as low flow) measured at the Searchmont gauging station 22km upstream. The field measured depth was regressed against logarithm of band ratio (band 3 [red; 655-690μm] over band 2 [green; 510-580μm]) to yield a linear relationship. The band ratio was chosen to maximize the difference in radiation absorbing properties of water while maintaining adequate penetration to avoid signal saturation (Hugue, et al. 2016). 5 by 5 kernel filter was applied to smooth out the band ratio raster prior to applying the linear relationship to the rest of the submerged pixels. Superficially deep pixels on the shaded side banks were excluded and reinterpolated using minimum curvature spline technique based on 12 neighboring points.

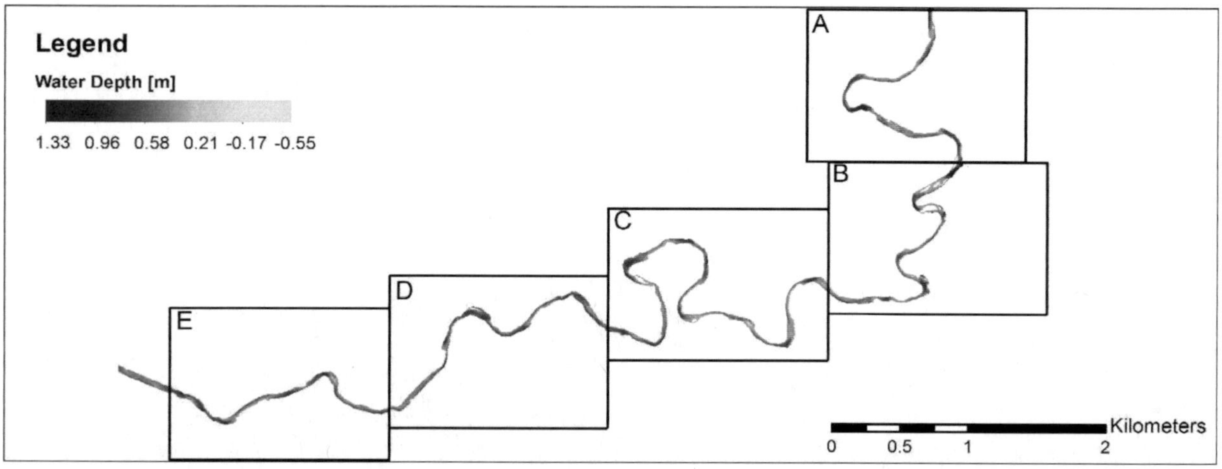

Figure 2. Water depth map of the Goulais River study reach derived from depth – band ratio correlation. The sub-sections shown in Figure 5 are marked. Flow from sub-section A towards E

3.3 Photogrammetry to determine the surface elevation

The 2.5m dry land DEM was reconstructed from 0.5 m stereo images captured by Pleiades-A1. The images were captured on June 14, 2016 with a discharge of 15.01 m3/s (hence forth referred as high flow) measured at the upstream gauging station. The DEM consists of pixel elevations of high flow free surface and the ground topography above it.

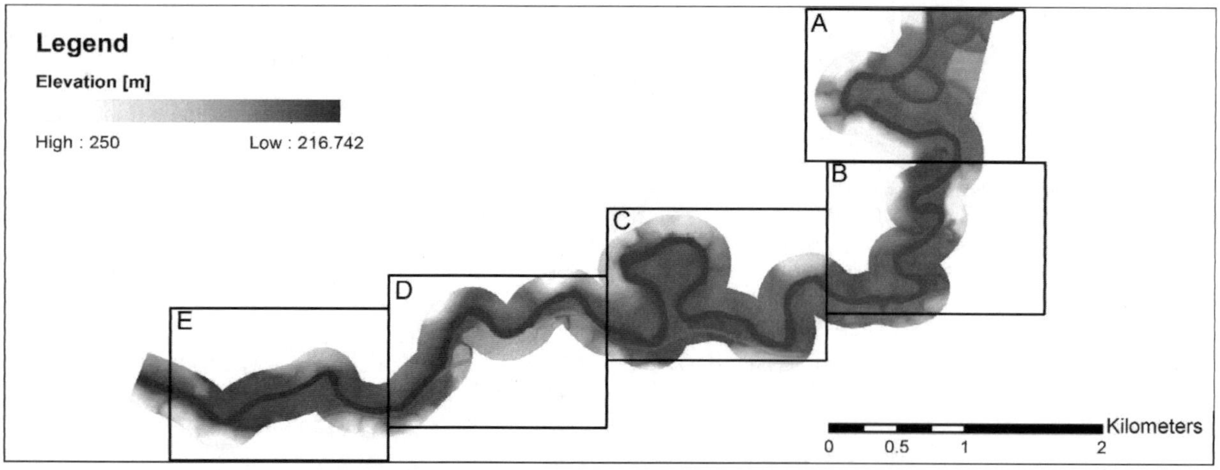

Figure 3. Dry land DEM of the Goulais River study reach derived from photogrammetry. The sub-sections shown in Figure 5 are marked. Flow from sub-section A towards E

3.4 Combining the depth map and the DEM

The water depth map and dry DEM was combined to generate the RTM, which include the submerged channel and the dry land topography. In the process, it has been assumed that no significant channel morphology has taken place between the 4-year gap of the image acquisitions. Here an issue arises due to the mismatched discharge of the images, resulting in a gap, Δh, between the free surfaces that varies in space (streamwise direction).

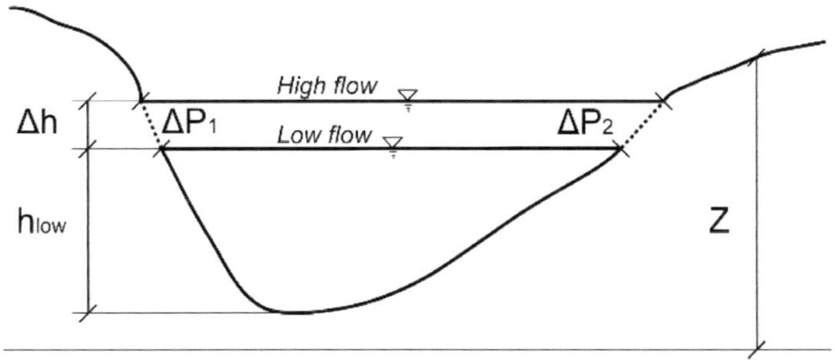

Figure 4. Schematic representation of the missing depth and wetted perimeter between the high flow (dry land DEM) and low flow (water depth map)

Determination of Δh using hydraulic principles has proven to be difficult due to the imperfections in the generated water depth map. In particular, the locust of deepest points in cross sections do not produce a smooth thawleg commonly observed in natural channels. This is mainly due to the persistence of superficially deep sections present in the shaded side banks despite the corrective measure taken. In this study, it was assumed that Δh would have small impact on the eventual bankfull flow simulation results because: 1. For a reach with average channel width of 45 m, the difference in stage between low flow (2.59 m^3/s) and high flow (15.01 m^3/s) should be insignificant; 2. Δh compared to the flow depth at the bankfull discharge (134 m^3/s) should be insignificantly small. Based on the above reasoning, a constant value of 50 cm was applied throughout the reach. The missing wetted perimeter, ΔP, was approximated by discretizing the channel into cross sections at 2 m interval and applying a linear interpolation between the upper edges of low flow and high flow free surfaces.

3.5 Delineation of bankfull depth and bankfull flow

The bankfull discharge was estimated through ensemble of statistical and graphical method. Frequency analysis was conducted on 47-years of peak annual discharge data obtained from the Searchmont gauging station. Peak discharge with a 1.6-year recurrence was determined based on the general consensus of the peak annual discharge of 1-2 year recurrence matching the channel forming bankfull discharge (Petit and Pauquet 1997, Emmett and Wolman 2001, Castro and Jackson 2001).

Subsequently, 9 plots of width-depth ratio against stage was constructed for cross sections in the vicinity of the downstream boundary. The stages of multiple cross sections corresponding to the minimum width-depth ratio was averaged to determine the bankfull stage. Assuming normal flow, the bankfull discharge was determines using the Manning's equation with n=0.0425, Sf=So. The average of the two methods yielded a bankfull discharge of 134 m^3/s.

3.6 Hydrodynamic Simulation

The present study deals with 2D depth-averaged open channel flow. The model equations are solved with finite element solver, Open-TELEMAC 2D.

The depth-averaged model equations are based on the principles of continuity (eq. [1]) and momentum (eq. [2], [3]) as

[1] $$\frac{\partial h}{\partial t} + u_1 \cdot \vec{\nabla}(h) + h\,div\left(\overrightarrow{u_1}\right) = 0$$

[2] $$\frac{\partial u_i}{\partial t} + \vec{u_i} \cdot \vec{\nabla}(u_i) = -g\frac{\partial Z}{\partial x_i} + \frac{1}{h}div(h\upsilon_t \vec{\nabla} u_i) + F_i^f$$

[3] $$F_i^f = -\frac{u_i}{\cos(\alpha)}\frac{gn^2}{h^{4/3}}\sqrt{u_1^2 + u_2^2}$$

where subscript $i = 1$ and $i = 2$ denotes the streamwise and lateral component of the variable, h is the flow depth; u is the velocity component; Z is the bed elevation; υ_t is the turbulent eddy viscosity; F_i^f is the friction force; $\frac{1}{\cos(\alpha)}$ is the norm of the vector normal to the bottom; g is the gravitational acceleration; n is the Manning's roughness coefficient. Depth-averaged $k-\varepsilon$ model was used as turbulence closure. Turbulent viscosity is estimated as a function of turbulent kinetic energy and rate of turbulent dissipation.

Table 1. Summary of hydrodynamics simulation parameters

	Location	Type	Value
Boundary Condition	Inlet	Constant discharge	134 m^3/s
	Outlet	Rating curve[a]	
	Lateral	Free velocity	
	Channel Bottom	Constant Manning's n[b]	0.0425
Initial Condition	Entire Domain	Constant free surface elevation	223 m (1 m/7 m above inlet/outlet)
		Constant velocity	0 m/s
	Inlet	Turbulen quantities gradient	$\frac{\partial k}{\partial h} = 0$, $\frac{\partial \varepsilon}{\partial h} = 0$
Time Step		Constant time step	0.25 s
Simulation Time			5 hrs
CPU Time			25 hrs
Convergence		Minimum residual	< 1E-5

[a] Constructed from low, high and bankfull flow stages. Higher flow stages estimated from normal flow Manning's equation

[b] Based on (Chow, 1959) for major streams with top with at flood stage >100ft

The active channel for the high flow plus a buffer zone of 5 m was discretized with 2 m unstructured mesh (approximately 50 cross channel node count and 2 m along channel interval). In the floodplain, the mesh sized is increased to an edge length of 100m at a growth rate ratio of 1.2 away from the main channel. The details of the simulation are summarized in table 1. By monitoring the 7 cross sectional discharge time series, a steady-state solution was reached at 5 hours simulation time.

3.7 Comparison of simulated bed shear stress and measured sediment caliber

The computed bed shear stress is related to sediment particle diameter based on critical entrainment shear stress for an incipient motion problem. Accordingly, the principles and techniques for rigid-boundary hydraulics are respected. Assuming a homogenous patch, Shield's experiments suggests a nondimensional number (Shield's stress) as

[4] $$\frac{\tau_c}{(\rho_s - \rho)D_s} = \frac{b}{a}$$

where τ_c is the critical shear stress; ρ_s is the sediment density; ρ is the fluid density; D_s is the sediment particle diameter; a is the particle exposure coefficient, b is the static friction coefficient. Shield demonstrated that for gravel and coarser sediments (D_s>5 mm), the Shields stress reaches a constant value of approximately 0.06, suggesting a proportional relationship between critical shear stress and sediment particle diameter. For the smaller sediments, Shield's experimental plot (Shield 1936) was used to graphically extract the corresponding critical shear stress.

4 Results and Discussion

4.1 Reconstructed RTM

In this study the accuracy of the RTM was not verified against detailed field survey due to a lack of available data and resources to carry out a survey of such scale. However, visual inspection of the model conforms with many of the prominent morphological features expected in a meandering river. Series of pool-riffle system can be seen throughout the reach, with a wavelength of approximately 10 channel width. The majority of the thalweg shifts toward the outer bank and high gradient outer bank slopes are observed in the pools. Between two pools, a lateral shift in the location of lowest bed elevation can be observed from one bank to the another, where the centerline is crossed approximately at the intermediary riffle zone. The high gradient side banks of the low flow depth map match with the locations of terraces (remnant of ancient floodplain left behind during the glacial retreat), illustrated by steep banks followed by gently sloping tread.

4.2 Estimation of sediment caliber from shear stress

The sediment caliber in riffle zones were successfully predicted with an r^2 value of 0.90. On the contrary, majority of the predictions in the glides, pools, and runs were overestimated. The author argues that this is the case because the bed shear stress in the riffle zone remain relatively high even at discharge lower than the bankfull discharge. As bankfull discharge events, by definition, occur approximately every 1-2 years, the interim lower flows bring about sediment transport that cause the sediment caliber distribution to deviate from the distribution just after the bankfull event. This effect is most prominent in the pools, and to a lesser degree in the proximal and distal ends, due to the reduced velocity (reduced bed shear stress) during low flows, leading to vertical accretion of finer sediments (sand and silt). This causes the previously mobilized larger sediments during the bankfull event to be covered by the finer sediments. In addition, one to one relationship is not expected to hold between sediment caliber and bankfull shear stress in the non-riffle zone due to the complex 3-D flow patterns of helicoidal flow, which cannot be accounted for in 2-D simulation. The cork-screw like secondary flow acts as a mechanism to sort sediments in the lateral plane.

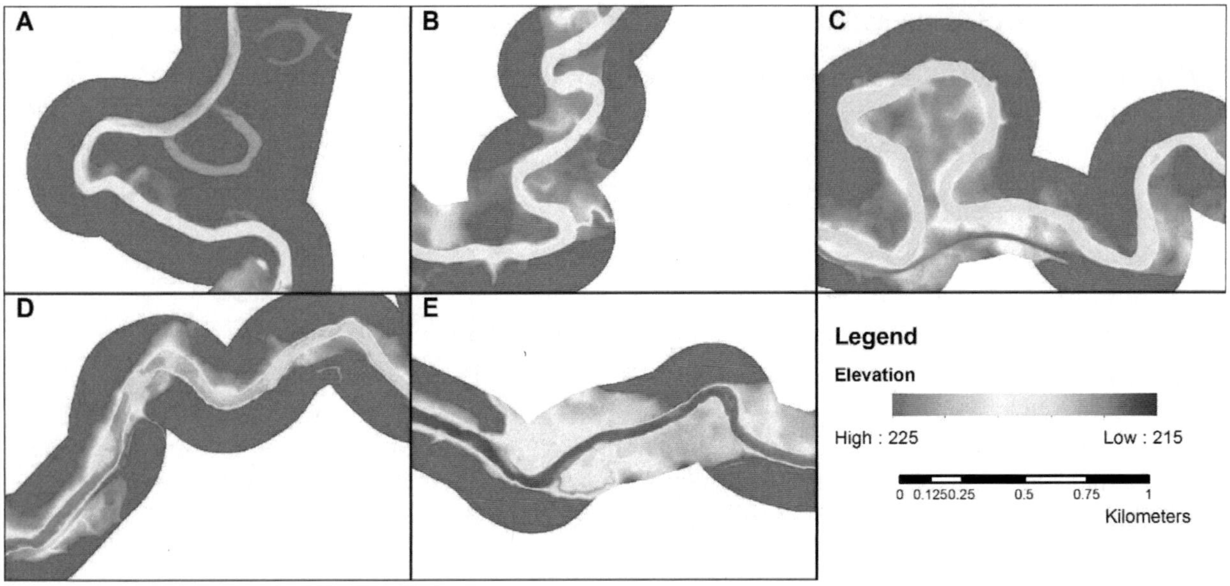

Figure 5. River terrain model of sub-sections of the Goulais River study site

The underprediction of the sediment caliber is most likely due to misrepresentation of colluvium (sediments transported to a location independent of the river flow, i.e. rainwash, sediments falling from side banks, etc.) as alluvium (sediments transported to a location by fluvial transport) during the field survey. Colluvium is supply limited, meaning its loading on the river is independent of the hydrodynamics of the river. Since the biologists did not discern the two types of sediments, it is hard to conclude with certainty. For the underestimated patches in the left bank, particularly in those situated adjacent to the road, measurements may have coincided with locations of riprap, where larger boulders were introduced artificially to prevent lateral erosion.

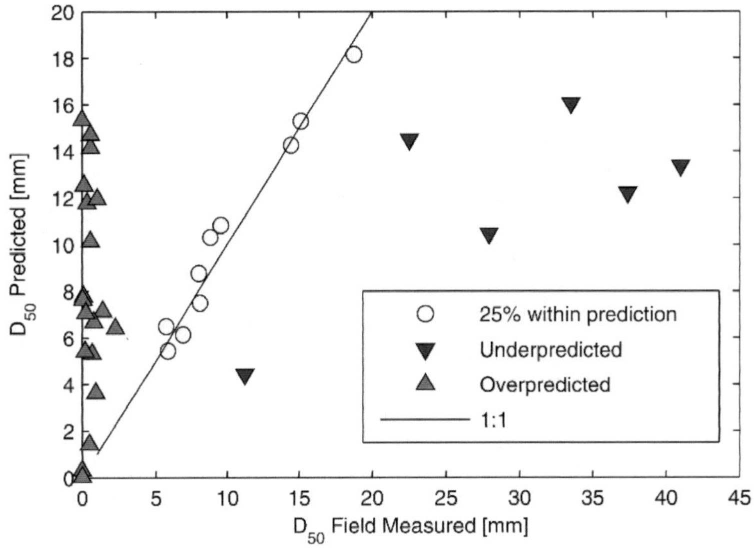

Figure 6. Comparison of field measured and predicted sediment caliber

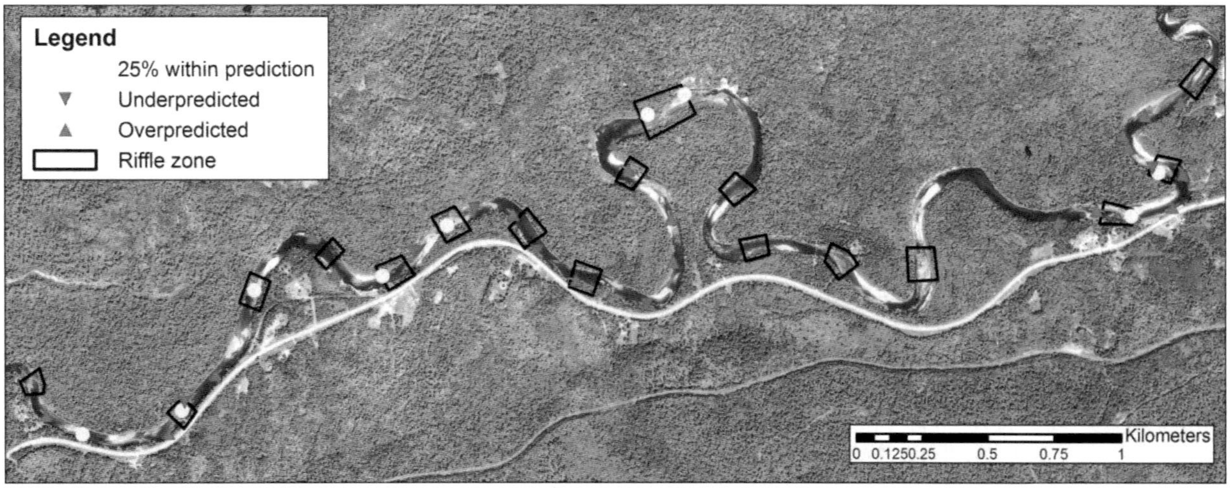

Figure 7. Spatial representation of the prediction performance

5 Conclusion

This paper reports digital reconstruction of channel bed, floodplain and dry land topography using minimal field measurement and multispectral and stereo pair satellite imagery. Depth-band ratio (red/green) regression was used to estimate the bathymetry from the image pixels. The depth map and the dry land DEM was combine with an addition of constant depth discrepancy to account for the difference in discharges. The RTM was used as a model domain to simulate a bankull flow event in a 2D hydrodynamic model. The following conclusions have been reached:

- The simulated bed shear stresses in the riffle zones showed strong positive correlation with the field measured sediment caliber. The bed shear stress in the riffle zone is large enough during the interim low flow seasons to maintain the sediment distribution to be comparable to that of immediately after a bankfull event.
- The wellness of correlation between the predicted and measured sediment caliber confirms that, in a bankfull event study, the choice of depth discrepancy between the low and high flows is insensitive to the resulting shear stress distribution.
- Overprediction of sediment caliber in the pools and their neighbors suggests the necessity for more sophisticated technique to account for the complex hydrodynamics and the multi-dimensional interactions of the sediments and the flow.

Acknowledgement

Financial support from National Sciences and Engineering Research Council of Canada through Discovery Grant held by S. Li is acknowledged.

References

Bukata, R. P. 1995. "Optical Properties and Remote Sensing of Inland and Coastal Waters." 54. Boca Raton, FL: CRC Press.
Bunn, S., and A. Arthington. 2002. "Basic principles and ecological consequences of altered flow regimes for aquatic biodiversity." Environmental Management 492-507.
Campbell, J. B. 2002. Introduction to Remote Sensing. New York: Gilford Press.
Campbell, J. 2002. "Introduction to Remote Sensing." 46-48. New York: Guilford Press.
Castro, J. M., and P. L. Jackson. 2001. "Bankfull discharge reccurence intervals and regional hydraulic geometry relationships: patters in the Pacific Northwest, USA." Journal of American Water Resources Assiciation 1249-1262.

Church, M. 2002. "Geomorphic thresholds in riverinee landscapes." Freshwater Biology 541-557.
Davey, C., and M. Lapointe. 2007. "Sedimentary links and stapial organization of Atlantic salmon (Salmo salar) spawmomg habitat in a Canadian Shield river." Geomorphology 82-96.
Emmett, W. W., and M. G. Wolman. 2001. "Effective discharge and gravel-bed rivers." Earth Surface Processes and Landforms 1369-1380.
Fonstad, M. A., and W. A. Marcus. 2005. "Remotesensing of stream depths with hydraulically assisted bathymetry (HAB) models." Geomorphology 320-339.
Gilvear, D. J., P. Hunter, and T. Higgins. 2007. "An experimental approach to the measurement of the effect of water depth and substrate on optical and near infra-red reflectance: a field-based assessment of the feasibility of mapping submerged instream habitat." International Journal of Remote Sensing 2241-2256.
Gilvear, D. J., T. M. Waters, and A. M. Milner. 1995. "Image analysis of aerial photograph to quantiify changes in channel morphology and instream habitat following placer mining in interior Alaska." Freshwater Biology 389-398.
Hugue, F., M. Lapointe, B. C. Eaton, and A. Lepoutre. 2016. "Satellite-based remote sensing of running water habitats at large riverscape." Geomorphology 353-369.
Lane, S. N., P. E. Widdison, R. E. Thomas, P. J. Ashworth, J. L. Best, I. A. Lunt, G. H. Samrook Smith, and C. J. Simpson. 2010. "Quantification of braided river channel change using archival digital image analysis." Earth Surface Processes and Landforms.
Legleiter, C. J., D. A. Roberts, and R. L. Lawrence. 2009. "Spectrally based remote sensing of river bathymetry." Earth Surface Processes and Landforms 1039-1059.
Legleiter, C. J., D. A. Roberts, W. A. Marcus, and M. A. Fonstad. 2004. "Passive optical remote sesing of river channel morphology and in-stream habitat: Physical basis and feasibility." Remote Sensing of Environment 493-510.
Lyzenga, D. R. 1981. "Remote sensing of bottom reflecrance and water attenuation parameters in shallow water using aircraft and LANDSAD data." International Journal of Remote Sensing 71-82.
Marcus, W. A. 2012. "Remote sensing of hydraulic environment in grave-bed rivers." In Gravel-bed Rivers: Processes, Tools, Environments, by M. Church, B. M. Pasccak and A. G. Roy, 269. Singapore: Wiley-Blackwell.
Petit, F., and A. Pauquet. 1997. "Bankfull discharge reccurence interval in gravel-bed rivers." Earth Surface Processes and Landforms 685-693.
Shield, A. 1936. "Application of similarity principles and turbulence research to bed-load movement." Hydrodynamics Laboratory publication.
Westaway, R. M., S. N. Lane, and D. M. Hick. 2001. "Remote sensing of clear-water, shallow, grave-bed rivers using digital photogrammetry." Photogrammetric Engineering & Remote Sensing 1271-1281.
Winterbottom, S. J., and D. J. Gilvear. 1997. "Quantification of channel bed morphology in gravel-bed rivers using airborne multispectral imagery and aerial photograph." Regulated Rivers: Research & Management 489-499.

Leadership in Sustainable Infrastructure
Leadership en Infrastructures Durables

Vancouver, Canada
May 31 – June 3, 2017/ *Mai 31 – Juin 3, 2017*

QUANTITATIVE ASSESSMENT OF RIVER ICE GENERATION USING SHORE-BASED PHOTOGRAMMETRIC TECHNIQUES

Ansari, Saber[1 3], Rennie, Colin D.[1], Seidou, Ousmane[1], Malenchak, Jarrod[2], and Ghareh Aghaji Zare, Soheil[1]
[1] Department of Civil Engineering, University of Ottawa, Canada, K1N 6N5
[2]Section Head, Ice, and Environmental Engineering, Department of Water Resources Engineering, Manitoba Hydro, Winnipeg, Canada
[3] sansa048@uottawa.ca

Abstract: Ice cover has great influence on hydraulics and morphology of rivers in high latitude areas of the northern hemisphere. Ice generation rate dictates the ice cover behavior and propagation during the river freeze-up. However our understanding of the process is limited, due to insufficient observation data sets. The collection of reliable hydraulic data under the ice cover conditions is difficult, dangerous, or even impossible using traditional methods. In this study, a time series of images captured by a digital time-lapse camera was used to estimate the ice generation rate at two locations along the Lower Nelson River (LNR), an open water patch and on a border ice leading edge. It was observed that ice periodically formed and was advected out of the persistent open water patch. The image time series were geo-rectified and then ice generation in the patch was estimated based on observed changes in ice volume. The volume of thermal ice generation in the open water patch and on the shores of an island was estimated using the automated image analysis algorithm. Application of this algorithm in similar studies and acquiring more comprehensive data will lead us to more precise equations for estimation of river ice production.

1 Introduction:

Ice-covered lakes and rivers influence living conditions in countries of colder regions. Canadian lakes and rivers are exposed to extreme cold weather that promotes river ice cover formation in most areas. Although ice cover may be beneficial for the economy of the region, such as the recreational industry, it also may lead to various adverse effects such as loss of marine navigation, ice jamming floods and frost heaving damage to hydraulic structures. River flow characteristics are influenced by the presence of river ice in the whole process of pre-formation, formation and deterioration. Several researchers have studied the influence of river ice cover on river hydraulics, morphology and sediment transport such as Ghareh Aghaji Zare, et al. (2016a) and Guo et al. (2016). Moreover, ice cover phenomena can also influence water quality and mixing ability (Ranjie & Huimin 1987; P.A. Chambers, et. al. 1993; Tang et al., 2016). Despite these previous studies, still much remains to be understood. Studying river ice cover processes in field locations is expensive and can be dangerous. Therefore scientists are interested in monitoring techniques in which a vast set of data is acquirable with less effort and resources.

In many research studies of different fields, photogrammetry has been introduced as an art and technology of data acquisition. Oblique terrestrial imagery however has been employed as an easy to use, efficient and economic method of monitoring and data gathering in earth sciences. Several studies have used terrestrial photogrammetry for data acquisition in various fields, such as glaciology, land cover investigations, watershed management, transportation, river engineering and sediment transport studies. Monitoring of river ice processes both in lakes and rivers, and both remote and shore-based imagery have been used for different purposes such as navigation and ice jam related floods (Jędrzychowski & Kujawski 2014).

1.1 Research area:

For an improved understanding of the river ice processes, a comprehensive research investigation has been conducted since 2012 on Nelson River, Manitoba, Canada. The Nelson River is one of the largest rivers across North America, with a length of 2575 km and a drainage area of about 892300 km^2. This large river has great importance for the economy of the region considering its tremendous potential in hydropower generation and fisheries. Nelson River drains Lake Winnipeg to Hudson's Bay through several lakes such as Sipiwesk, Split and Stevens. The study reach is on the Lower Nelson River, LNR, a regulated semi-alluvial river. This research was conducted between Limestone Generation Station (LGS) and Gillam Island, the last 102 km of the river (Figure1).

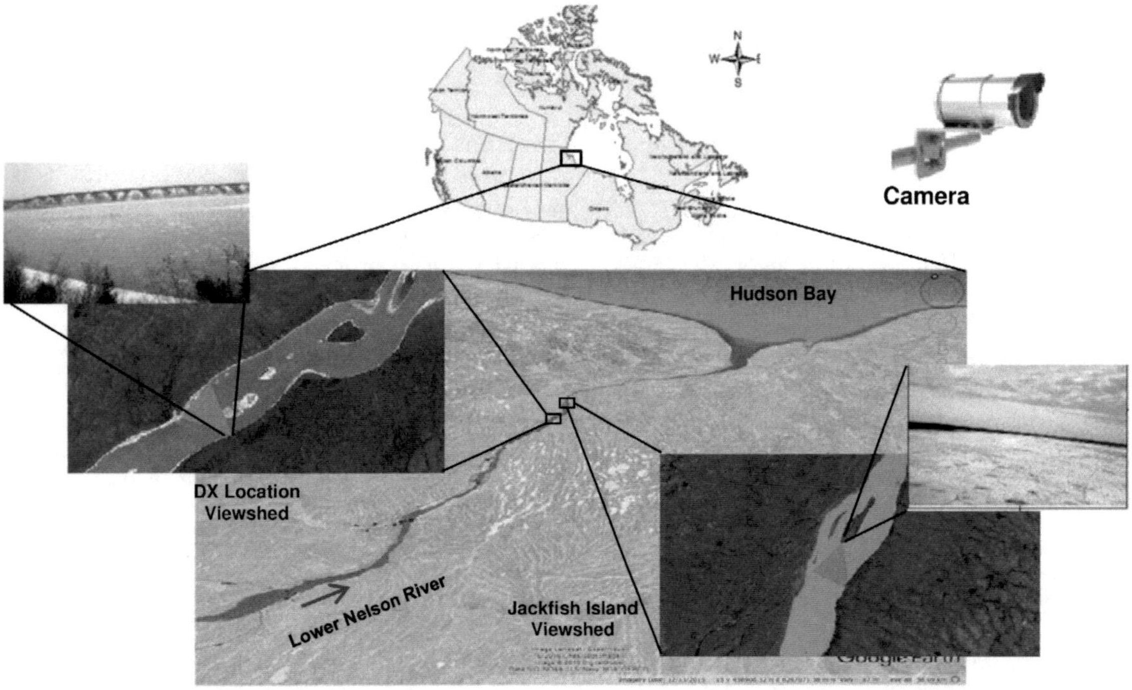

Figure 1: Study area locations and mounted camera along LNR; DX, and Jackfish Island (respectively, 65 km and 50 km upstream of Gillam Island)

1.2 Methodology:

This study included long-term hydraulic data gathering and shore-based monitoring of the river ice processes, during four consecutive winters. A set of bottom mounted instruments including a 1200-kHz Teledyne RD Instrument Acoustic Doppler Current Profiler (ADCP) and a 546-kHz Shallow Water Ice Profiling Sonar (SWIPS), was employed for hydraulic data acquisition of the river. Moreover, two shore mounted high resolution cameras were also used for monitoring purposes. The monitoring system was used for validation of the bottom mounted measurements and to obtain spatial and temporal distribution of the river ice cover. The upward looking ADCP and SWIPS instruments provided comprehensive data sets of the river ice cover thickness along with water elevation and river flow velocity. On the other hand, to extract quantified data from the time-series images taken at two study sites along the LNR an automated image processing algorithm was also developed (Ansari et al., in revision).

The robust image analysis technique is able to extract and quantify spatial and temporal distribution of the river ice cover. Extracted results of this algorithm can be used for quantification of important ice cover characteristics. The image processing algorithm developed in MATLAB consists of five distinct steps: 1) Preprocessing; in this step initial processing of the images was performed to read date, time, and temperature data recorded on each image. 2) Image registration; this step was developed to rectify the discrepancies caused by camera movements. 3) Image rectification; a currently existing method (Corripio et al. 2004; Härer et al. 2013) was modified for georectification of the oblique images. 4) Target detection; a modified version of the edge sensitive variational image thresholding technique (Ray & Saha 2006) was used to detect ice cover in its different types. 5) Calculations; this step was developed to calculate river ice cover spatial and temporal information.

A brief description of the developed method along with its quantified results to study border ice and thermal ice growth are addressed herein. Application of this method in shore-based monitoring of river ice cover studies will help better understanding of the phenomenon and the continuous advancement of river ice models.

2 Apparatus and data:

Continuous data recording was performed at two locations along the LNR, "DX" and "Jackfish Island" (Figure 1). A Teledyne RD Instruments 1200 kHz Sentinel Acoustic Doppler Current Profiler (ADCP) along with an ASL 546 kHz Multi-Functional Acoustic water column profiler (MF-AWCP) were bottom-mounted at both locations (Ghareh Aghaji Zare et al. 2014). Shore-based monitoring was performed using two Campbell Scientific digital cameras. The CC5MPX weighs about 1.06 kg, with a built-in clock and thermometer. The optical zoom lens of the camera is (4 to 12 mm) and the field of view of the camera is 27° to 80°. The cameras were programmed to take images with the quality of 1280×720 pixels and with a time lapse of every 30 or 60 minutes. Field measurements of the Jackfish Island site performed for three winters starting at December 2012 (~ 6200 images), whereas the DX site measurements were performed for winter 2012 (~ 2300 images). Figure 1 shows two sample pictures taken at the two study locations. The number of the acquired images indicates the importance of developing an automated image analysis algorithm.

3 Automated image analysis algorithm:

An automated image analysis algorithm was developed to extract data from the time series of shore-based images. This algorithm consists of five distinct subroutines as explained here.

3.1 Preprocessing:

The preprocessing step deals with three different tasks prior to proceeding with the main body of the image analysis algorithm. These steps include 1) saving recorded data on each image such as: date, time and temperature. 2) Quality screening of the pictures to eliminate the poor quality images from being processed. 3) Quality increase of the images, by the application of the Median filter (Ansari et al. in revision).

3.2 Image registration:

Due to the installation of the cameras on tree trunks, they were subject to some minor movements induced by wind. An image registration subroutine was added to the algorithm to deal with the errors induced by these movements. In this step, the Canny edge detection method (Canny,1986), as well as Hough transform method (Hough, 1962) were used to detect a matching part in all images. Each image was then transformed to the reference image using the method introduced by Goshtasby (1988).

3.3 Image rectification:

To be able to extract quantified information from the images, the image rectification subroutine was developed to assign correct and real world coordinates to the corresponding pixels within the images. There have been numerous efforts to address the image rectification problem in the literature(e.g., Corripio 2004; Bourgault 2008; Härer et al. 2013; Messerli & Grinsted 2015). In this algorithm, an existing method of image rectification developed by Corripio 2004 was employed to perform the rectification step. This method applies the camera's intrinsic and extrinsic characteristics and is independent of the Ground Control Points (GCPs) (Härer et al. 2013). In this method pixels of the camera field of view Digital Elevation Model (DEM) are assigned to the pixels of the images. A similar method has also been used in a study by (Härer et al. 2013) to monitor snow cover patterns. In the present algorithm, this method was modified to account for water surface elevation fluctuations using the water elevation measurements (Ansari et al. in revision).

3.4 Target detection:

Prior to the final calculation, river ice cover detection is the most important and challenging step. The main goal of this method is the automatic detection of different kinds of river ice cover. In computer vision, target detection is performed using different image thresholding techniques. In image thresholding techniques the image is segmented into several regions of interest. In the developed algorithm the edge sensitive image thresholding method developed by Ray and Saha (2007) was modified to account for altering intensity of pixels for ice and water at different times of the day (Ansari et al. in revision).

3.5 Ice characteristics calculations:

The final step of the algorithm is to draw boundaries around the detected river ice cover floes and margins to calculate the area, thickness, and volume of the detected river ice cover. Once the preprocessing, image registration, rectification and target detection steps are performed successfully, other subroutines can also be added to the main code and in the final calculations to extract quantified desired information. Figure 2 shows the developed pipeline of the image analysis algorithm.

4 Results:

4.1 Model verification:

The main sources of error in this algorithm are the image *georectification* step and the *ice detection* subroutine. Verification of the results acquired by this method was performed distinctly for rectification and detection steps. Ground Control Points taken on the stable ice cover were used to verify the georectification method. Results show that the average distance Root Mean Square Error (RMSE) of the rectified reference images was less than 30 meters. The *target detection* algorithm was verified by manual detection of the river ice cover. The manually detected boundaries of ice cover were then rectified to obtain the area. Comparison of the results extracted by the algorithm and manual detection method led to error estimation. The target detection RMSE of the analyzed images using the developed algorithm ranged from 20 to 253 square meters over the entire viewshed of 492408 m^2 at Jackfish Island location.

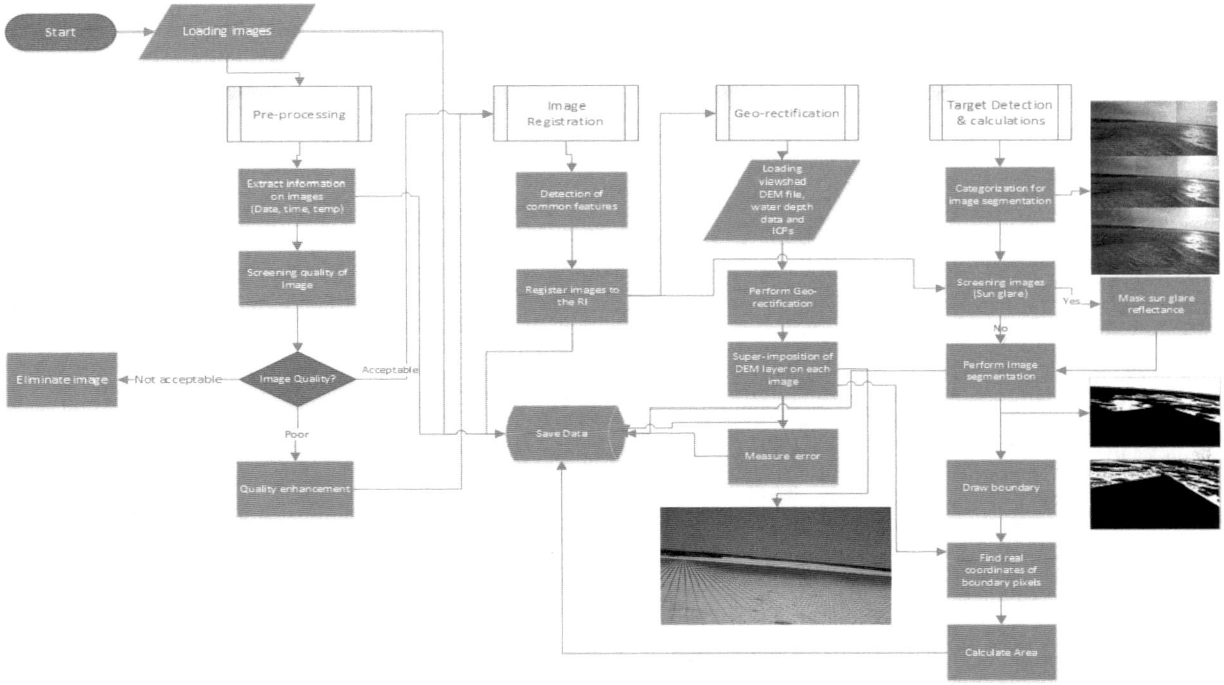

Figure 2: Steps of the developed image analysis pipeline

4.2 Thermal growth of river ice:

One of the major specifications of the locations chosen for field measurements was the occurrence of three main stages of river ice cover processes (freeze-up, consolidation and break-up). Frazil ice generation is the first stage of the freeze-up and river ice cover formation process in LNR. Ice formation occurs in various shapes and types, mainly related to the flow intensity. Border ice and skim ice that form at sections with lower flow intensity are the first types of ice forming in this river. Other than border ice, frazil ice particles that are formed in supercooled water are the primary elements of river ice cover processes. Frazil ice particles adhere to one another growing and forming frazil ice floes. On the other hand frazil particles that are present throughout the cross section of the flow can attach to the objects on the river bottom and form anchor ice. According to the recorded temperature data, air temperature at LNR drops consistently below zero in late October, that is when border ice growth and frazil ice generation initiates.

The analysis of the time series of images taken at two locations of "DX" and "Jackfish Island" presented in (Ansari et al. in revision), led to the investigations over the thermal growth of river ice cover at two locations with different conditions. DX site was located at a steeper section. Flow intensity was higher at this location causing an open area to emerge during the monitoring period. This open water patch was subject to periodical formation of thermal ice. On the other hand, a leading edge of river ice cover detected at the Jackfish Island site was observed to grow gradually starting from mid-January. The thermal growth of river ice cover in the open water patch at DX site and the gradual propagation of river ice cover leading edge at Jackfish Island site that has been quantified using the automated image analysis algorithm are described in the following sections.

4.3 DX location:

The DX site is located about 35 km downstream of Limestone Generation Station (LGS) and 65 km upstream of Gillam Island. The river at this location has a steeper bed compared to the Jackfish Island site. The high flow intensity at this location forms an area with an open water patch. However, thermal ice formation in the open area periodically covers the open water patch during the winter. Field measurement and monitoring at this location was conducted for one winter (2012-13). Unfortunately, the bottom mount equipment was lost (presumably due to high flow intensity); however, acquired time-series images were used. Observations show that frazil ice generation began as early as mid-October. It was also observed that complete ice cover formed on January 30th 2013. Nevertheless, due to high flow intensity, complete cover at this location did not last for long and an open water patch appeared just the day after (Figure 3). The open water area grew with time (see figure 4) until reaching to an equilibrium. Figure 4 shows the growth and shrinkage of the DX open water patch in terms of area percentage of river ice cover. Air temperature and river discharge are also shown in this figure. Flow discharge at the cross section was calculated based on released discharge time shifting at LGS. It was observed that in addition to the frazil ice production at this section that flowed to the downstream; thermal ice periodically formed at the borders of the open area. The thermal growth of river ice on the borders of the open area is the main cause of the calculated open water patch area fluctuations.

a) b)

Figure 3: a) Aerial photo, b) shore-based image at DX location

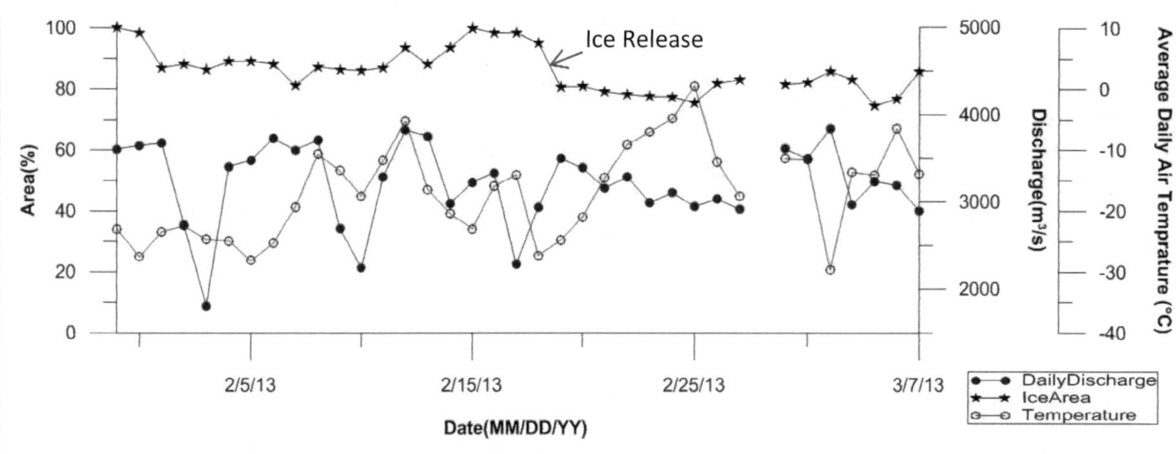

Figure 4: Extracted river ice cover area, water discharge and average daily air temperature at DX location along LNR for the period January 30, 2013 to March 7, 2013

Thermal growth of river ice cover led to the production of smooth ice with different texture detectable by the automated image analysis algorithm (Figure 3-b). Thermal ice periodically was produced in the open water area and it was advected out of camera field of view after its detachment from the ice cover body. A major detachment of the river ice cover within the open water patch is shown in Figure 4.

In order to estimate the volume of thermal ice production in the open water patch from the observed ice area, ice thickness was calculated using Stefan's ice cover growth equation. This method uses the cumulative Freezing Degree Days in degrees Celsius (FDD) to calculate the thermal ice cover thickness. In Stefan's equation as presented below (Stefan,1889), α is an empirical coefficient depending on the site, k_i is the thermal conductivity of river ice (2.24 W m^{-1} $^oC^{-1}$), ρ_i is the density of ice (916 kg m^{-3}) and λ is the latent heat of freezing (3.34×10^5 J kg^{-1}) of ice.

$$[1]\ h_i = \alpha \sqrt{\frac{2k_i}{\rho_i \lambda} FDD}$$

This formula has been used in several other studies to estimate and simulate thermal river ice cover growth. Stefan's formula can be simplified to the following equation, in this case ß is site dependent and based on Michel 1971 it varies between 1.3 ~ 1.7 ($^oC^{-1/2}$.$Day^{-1/2}$).

$$[2]\ h_i = \beta \sqrt{FDD}$$

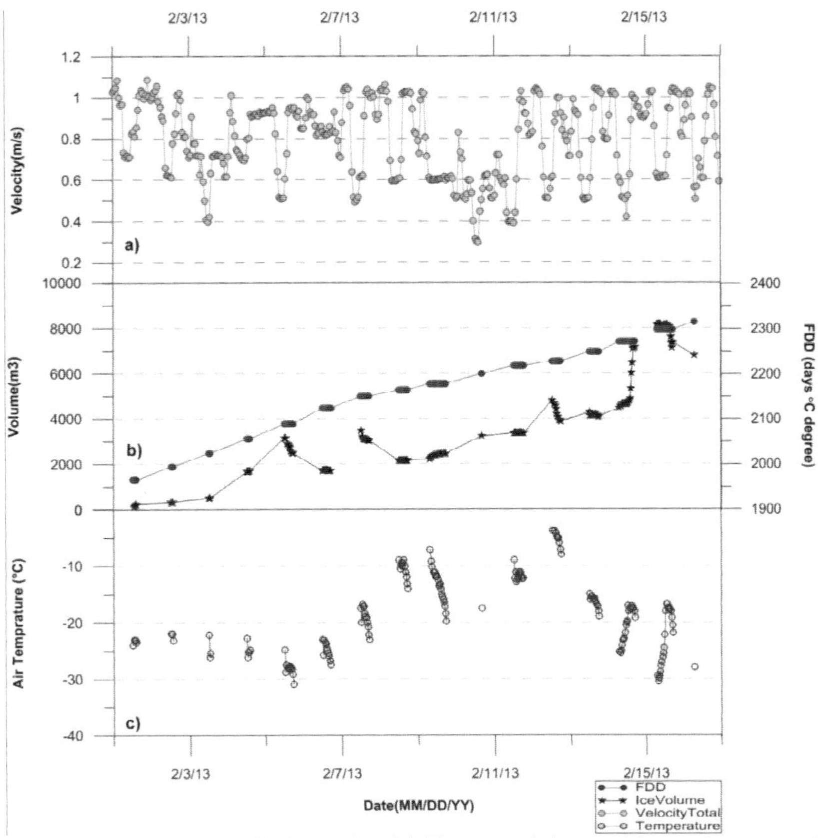

Figure 5: a) Average water velocity at DX location b) Extracted thermal ice volume production in an open water patch, and Freezing Degree Days (FDD) c) air temperature at DX location along LNR (February 2013)

Figure 5 shows for the DX location the hourly distribution of Freezing Degree Days (FDD), generated thermal ice and section averaged water velocity (discharge divided by section area). The developed image analysis algorithm was used to detect the thermal river ice cover and calculate its area, while the thickness of the generated river ice cover was calculated using Stefan's formula. These results were used to estimate the thermal river ice cover volume. The hourly results show that more important than air temperature, the increase in water velocity led to thermal ice cover deterioration due to detachment of thermal ice from borders of the open water patch.

4.4 Jackfish Island:

The relatively mild slope of the river at this location allowed a comprehensive study for three winters (2012 to 2015). The total monitoring area at this location was around 492408 m^2. While border ice growth and open water patch were monitored in the first two years, in the last winter ice break up was the primary goal of the research. Detailed results of this study are presented in (Ansari et al. in revision). In this section extracted results of the border ice growth at the Jackfish Island location are explained. Figure 6 shows a view of Jackfish Island with the formed border ice. Terrestrial monitoring at this location started in the second week of November 2013 and continued until March 2014. Figure 7 shows the average air temperature along with the area percentage covered by river ice and water surface. As is shown in this figure the algorithm was able to calculate and quantify the amount of river ice cover as river ice floes and border ice formed at the front edges of Jackfish Island (Figure 7). In Figure 8 it is demonstrated that the border ice cover increase at the shores of the island has a direct relation with temperature drop and increase in freezing degree days, although comprehensive statistical investigation is needed to verify this relation.

a) b)

Figure 6: Jackfish Island site. a) taken on November 2013, b) taken on January 2014

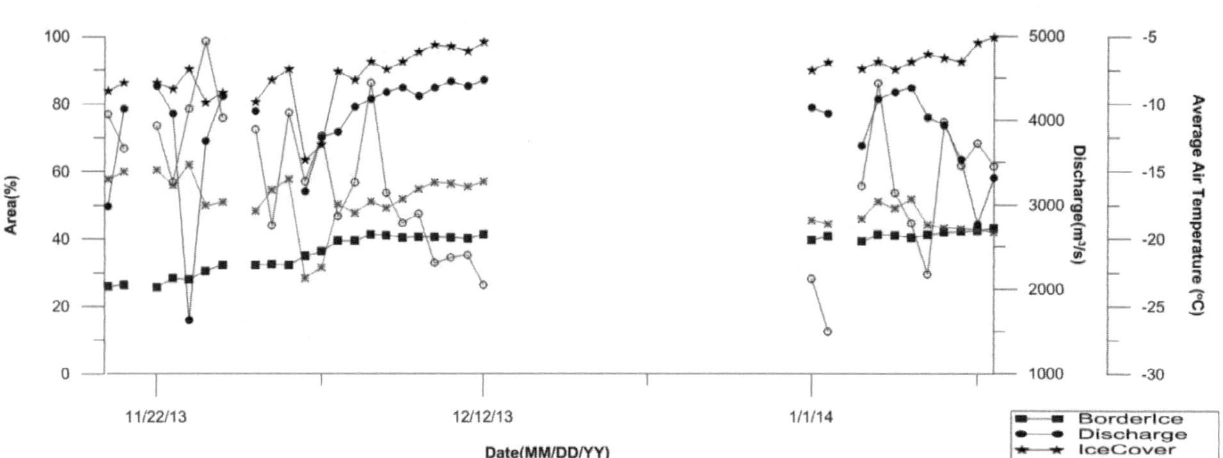

Figure 7: Water discharge, river ice cover area and air temperature quantified extracted results from time series of the Jackfish Island images (Duration of November 2013 to February 2014)

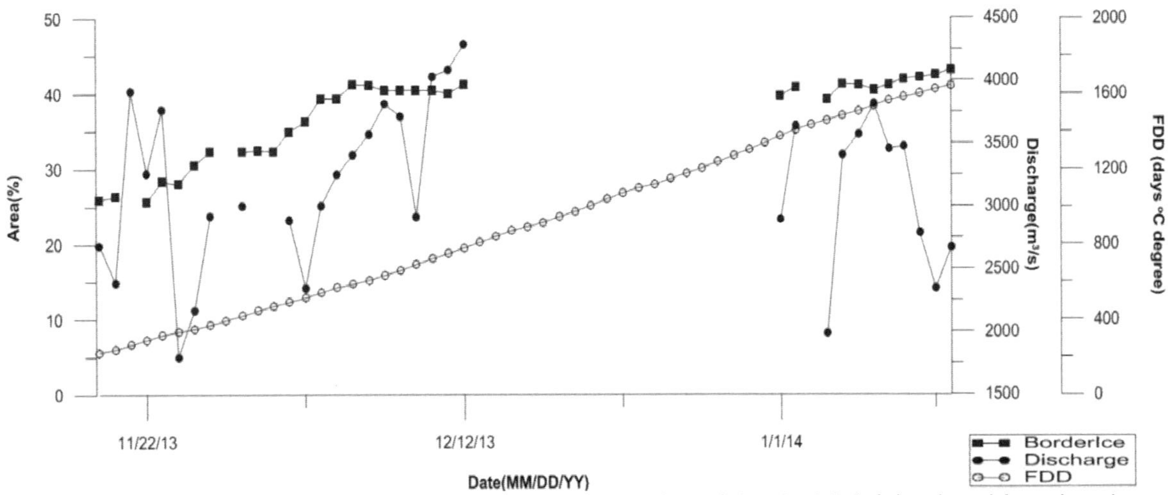

Figure 8: Water discharge, border ice growth at leading edge of the Jackfish Island and freezing degree days at this location (Duration of November 2013 to February 2014)

5 Conclusions:

An automated image analysis algorithm was developed to extract quantified data from the time series images of the shore-based terrestrial monitoring. This algorithm consists of five subroutines as: "*preprocessing*", "*image registration*", "*image rectification*", "*target detection*" and "*final calculations*". While the two steps of pre-processing and image registration are considered as the pre-processing steps of the image analysis algorithm, the image rectification subroutine is used to assign correct real word coordinates to the pixels of the images using the DEM map and camera intrinsic and extrinsic characteristics such as Field of View (FOV). Some modifications were added to this method to consider the water surface fluctuations in the river. For the "target detection" algorithm a categorization method was added to a variational thresholding technique developed by (Ray & Saha 2006). The categorization algorithm increased the applicability of the algorithm for detection of water and ice cover surfaces from time series images.

At DX location it was observed that the rate of thermal river ice production on the borders of an open water patch can be quantified using the developed algorithm. Periodical production and advection of the thermal ice cover in the open water patch shows the importance of water flow intensity in river ice cover formation. On the other hand, at Jackfish Island the ice floes and the border ice cover at the upstream island edge were monitored. It was shown that the border ice growth is mainly due to the decrease of air temperature. The leading edge of the thermal ice that is formed on the borders of this island grows until reaching to the other sections to form a stable ice cover across the river. The unique ability of the developed algorithm to extract quantified data from images demonstrates its suitability for automated analysis of shore-based terrestrial imagery.

Acknowledgements

The authors would like to thank Nathan Lambkin and his crew from Manitoba Hydro for all their help and contributions during field measurements, including deployment and retrieval of instruments. This work was funded by Manitoba Hydro (R&D grant), Hatch corporation Ltd., and the Natural Sciences and Engineering Research Council of Canada (CRD grant).

References

Ansari, S., Rennie, C.D., Seidou, O., Malenchak, J., Zare, S.G., (In Revision) Automated Monitoring of River Ice Processes using Shore-based Imagery. *Cold Regions Science and Technology.*

Corripio, J.G., Durand, Y., Guyomarc'h, G., Mérindol, L., Lecorps, D. and Pugliése, P.,, 2004. Land-based remote sensing of snow for the validation of a snow transport model. *Cold Regions Science and Technology*, 39(2-3), pp.93–104. http://dx.doi.org/10.1016/j.coldregions.2004.03.007

Corripio, J.G., Javier G., 2004. Snow surface albedo estimation using terrestrial photography. *International Journal of Remote Sensing*, 25(24), pp.5705–5729. http://dx.doi.org/10.1080/01431160410001709002

Ghareh Aghaji Zare, S., Asce, S.M., et al., 2016. Boundary Shear Stress in an Ice-Covered River during Breakup. *American Society of Civil Engineers* , 142(4), pp.1–14. DOI: 10 .1061/(ASCE)HY.1943-7900.0001081

Ghareh Aghaji Zare, S. Moore S., Rennie C.D., Seidou, O., Ahmari, H., 2014. Influence of river ice break-up on stream hydraulics and sediment transport. *River Flow 2014* – Schleiss et al. (Eds), pp.951–959.

Goshtasby, Ardeshir, "Piecewise linear mapping functions for image registration" Pattern Recognition, Vol. 19, 1986, pp. 459-466. doi:10.1016/0031-3203(86)90044-0

Goshtasby, Ardeshir, "Image registration by local approximation methods" Image and Vision Computing, Vol. 6, 1988, pp. 255-261. doi:10.1016/0262-8856(88)90016-9

Guo, X., Yang, K., Fu, H., Xia, Q., Wang, T. and Yang, S., 2016. Ice processes modeling during reverse water transfer of open canals: a case study. *Journal of Hydro-environment Research.* http://dx.doi.org/10.1016/j.jher.2016.03.002

Hough, P.V.C. Method and means for recognizing complex patterns, U.S. Patent 3,069,654, Dec. 18, 1962

Härer, S. Bernhardt, M., Corripio, J. G., & Schulz, K., 2013. PRACTISE – Photo Rectification And ClassificaTIon SoftwarE (V.1.0). *Geosci. Model Dev, 6*, pp.837–848.doi:10.5194/gmd-6-837-2013

Jędrzychowski, K. & Kujawski, A., 2014. Method of image analysis in the process of assessment of ice occurrences. Zeszyty Naukowe/Akademia Morska w Szczecinie, 37(109), pp.45–49.

Kaufmann, V., 2012. The evolution of rock glacier monitoring using terrestrial photogrammetry: The example of Ausseres hochebenka

Ranjie, H.O.U. & Huimin, L.I., 1987. Modelling of BOD-DO dynamics in an ice-covered river in Northern China. Pegman Journals Ltd , 21(3), pp.247–251.

Ray, N. & Saha, B.N., 2006. Edge sensitive variational image thresholding. Proceedings - *International Conference on Image Processing*, ICIP, 6. DOI: 10.1109/ICIP.2007.4379515

Tang, G. et al., 2016. Modelling and Analysis of Hydrodynamics and Water Quality for Rivers in the *Northern Cold Region of China. International Journal of Environmental Research and Public Health*, 13(4), p.408. Available at: http://www.mdpi.com/1660-4601/13/4/408. doi:10.3390/ijerph13040408

Leadership in Sustainable Infrastructure
Leadership en Infrastructures Durables

Vancouver, Canada
May 31 – June 3, 2017/ *Mai 31 – Juin 3, 2017*

QUANTITATIVE STEEP CREEK RISK ASSESSMENT, DISTRICT OF NORTH VANCOUVER, BRITISH COLUMBIA

Kris Holm[1,3], Matthias Jakob[1], and Hamish Weatherly[1] Fiona Dercole[2], Steve Bridger[2]
[1] BGC Engineering Inc., 500-980 Howe Street, Vancouver, British Columbia, Canada
[2] District of North Vancouver, 355 West Queens Road, North Vancouver, British Columbia, Canada
[3] kholm@bgcengineering.ca

Abstract: The District of North Vancouver (DNV) has a long history of managing geohazards. Starting in the 1990s, and updated approximately every 10 years, DNV has retained geotechnical and geoscientist consultants to assess debris geohazard (i.e., debris floods and debris flows) risks and make recommendations for reducing risk to tolerable levels. In 2015-2016, we completed comprehensive flood, debris flood- and debris-flow risk assessments for 35 steep creeks within the District using a variety of custom-tailored methods described in this contribution. While most creeks' headwaters are in forested and largely undeveloped terrain, the lower reaches flow through municipal areas containing over 20,000 buildings and a network of roads, utilities, and stormwater management infrastructure. The objectives of the assessment were to assess debris geohazards including their frequency, magnitude, extent, and potential to result in blockage and overflow of DNV stormwater management infrastructure; estimate the risk posed by these hazards to buildings and persons within buildings, prioritize locations for risk reduction planning; and develop risk control options and costs. Based on the results of the assessment, DNV staff are developing a 10-year work plan that will be integrated with the District's asset management, GIS-based inspection program, climate change adaptation and hazard mitigation plans. The work presented herein is an example of a pro-active science-based creek management program aiming to optimize funds for public safety and economic risk reduction.

1 Introduction

Following a damaging debris flood on Kilmer Creek in November 2014, the DNV requested a district-wide geohazard risk and risk control options assessment of 35 steep creeks that could block stormwater management assets and/or damage buildings and infrastructure (Figure 1). Of these creeks, this paper is focused on risk assessment of 25 "urban" creeks defined as being accessible by road and serviced by DNV's stormwater management infrastructure. The remainder of the creeks are in rural areas in the northeast part of the District, which are accessible by boat only. The objectives of this study were to: (i) assess debris geohazards including their frequency, magnitude, extent, and potential to result in blockage and overflow of DNV stormwater management infrastructure; (ii) numerically model geohazard scenarios and create maps showing the estimated extent and potential severity of impact to buildings and infrastructure; (iii) estimate safety and economic risk posed by these hazards to buildings and persons within buildings; (iv) prioritize locations for risk reduction planning based on results of the risk assessment; and (v) propose risk control measures and costs to reduce these risks to levels considered tolerable by the DNV.

The creeks assessed are defined as "steep creeks" subject to floods, debris floods or debris flows, with the potential to cause economic damages and loss, and/or pose risk to life for persons within buildings. For the purposes of this paper, a channel is considered "steep" when the channel slope is equal to or exceeds approximately 3° (~ 5%). Stormwater management infrastructure assessed on these creeks include culverts and stormwater mains owned by the DNV with potential for debris blockage and/or physical damage. Culverts in upper watershed areas that are not owned by the DNV, but that affect watershed boundaries and associated creek flows into the DNV, were also assessed. However, hazard scenarios and risks associated with blockage of these culverts were not assessed in detail. Of the 25 urban creeks assessed, 14 are classified as having potential to produce debris floods, 2 to debris flows, and 9 that are unlikely to produce either debris flows or debris floods.

Table 1 lists the types of elements at risk and types of risks or impacts assessed. Specifically, "safety risk" considers risk to life for persons within buildings, while "economic risk" considers direct damages to buildings, roads and culverts due to debris impact. Assessment of drainage infrastructure focuses on identifying requirements and costs to upgrade assets as part of debris risk management. Assessment of other types of infrastructure was limited to identifying their location in relation to areas impacted by debris hazard scenarios.

Table 1: Elements at risk

Element at Risk	Type of Risk or Impact Assessed
Persons within Buildings	Quantitative estimation of individual and group risk to life for persons located within buildings.
Buildings	Quantitative estimation of damage to buildings due to debris impact expressed as a proportion of appraised building value and direct damage cost, and as an annualized damage cost. Identification of facilities considered critical by DNV for function during an emergency that are located within areas potentially subject to debris geohazard impact. These include schools, police stations, or fire stations.
Roads	Estimation of spatial extent and intensity of impact by debris geohazard scenarios.
Culverts	Estimation of event return periods likely to result in blockage by debris for a given culvert.

2 Study Area Physiography

The DNV is a 160 km^2 district municipality bounded by Burrard Inlet to the south, Capilano River to the west, and Indian Arm to the east (Figure 1). The developed southern portion of the District is flanked by the North Shore Mountains, which rise up to 1,449 masl and contain steep, mostly forested slopes overlain by colluvium and till. In addition to the larger Capilano, Lynn and Seymour rivers, numerous steep creeks originate in undeveloped, smaller (1-15 km^2) watersheds and flow through stormwater management infrastructure downstream. Figure 1 shows the watershed boundaries for the steep creeks assessed, which are affected in some areas by drainage alterations caused by road construction and logging.

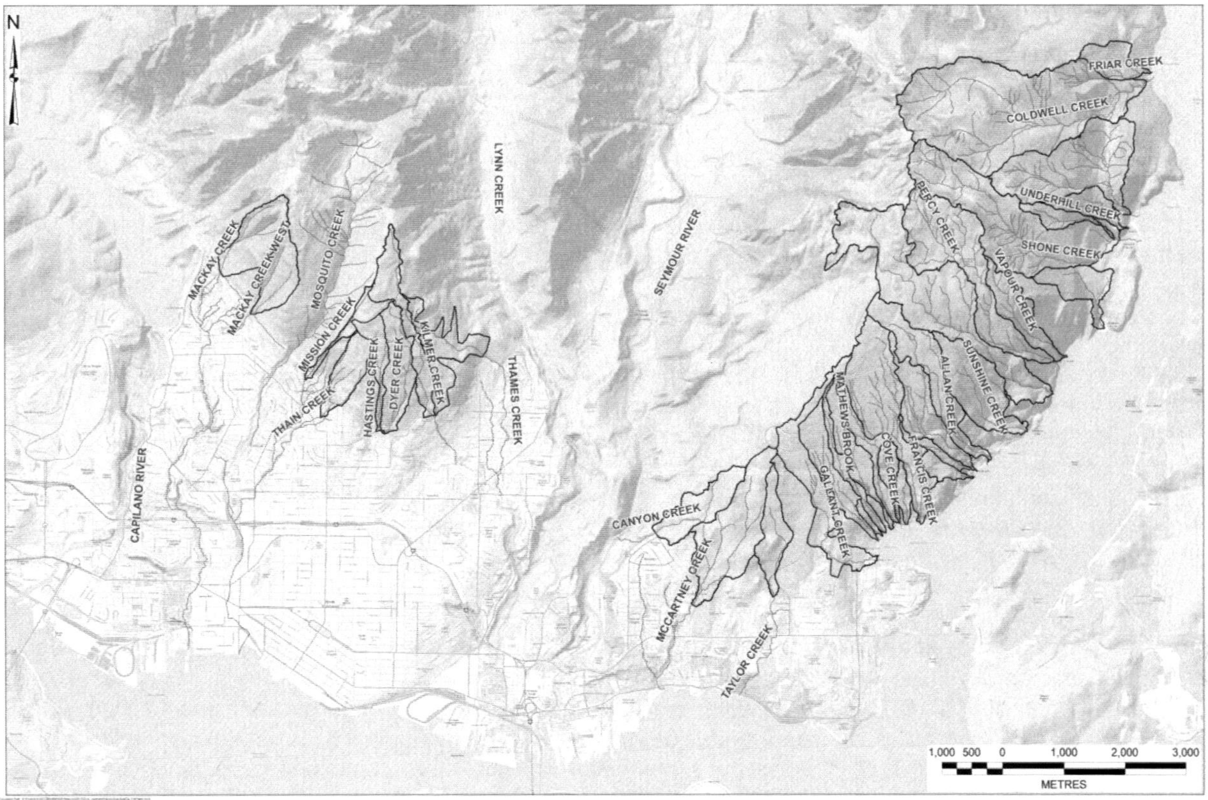

Figure 1: Study area and creek watershed boundaries upstream of development interface

3 Infrastructure

DNV buildings, drainage management works and transportation infrastructure were considered in the assessment of debris geohazard impact. All of these assets are classified by the DNV using a unique identifier (Asset ID). Other types of infrastructure, such as power infrastructure (transmission lines) or oil/gas infrastructure, were not assessed.

The DNV also maintains a geospatial database of building locations and characteristics throughout the District, including building type, assessed value, estimated replacement value, and population. The building inventory was developed from a combination of DNV information and windshield surveys[1] completed by Carlos Ventura and colleagues at the UBC Earthquake Research Facility as part of an earthquake risk assessment (Journeay et al. 2015). All data are associated with a unique Building ID and geospatially related to building footprints in GIS.

Drainage infrastructure within the current DNV asset inventory includes 371 culverts, 7 natural hazard mitigation structures, and approximately 350 kilometres of storm mains with replacement costs totalling $296 million (DNV 2015). An additional 86 culverts also exist within the DNV that are owned by others (e.g., the Province of BC). Of these, a total of 93 culverts and stormwater mains were considered to have credible potential for debris blockage and/or that are located in creek sections subject to debris hazard processes, and were assessed in detail.

Transportation infrastructure within the DNV and applicable to this study includes roadways and pedestrian or vehicle bridges. Pedestrian and vehicle bridges were reviewed where they intersect with a study creek upstream of development or within the area modelled. However, pedestrian bridges were not prioritized for

---

[1] Windshield surveys are systematic observations made from a vehicle.

evaluation of risk reduction alternatives. No vehicle bridges were identified as having credible potential for debris blockage.

4 Methods

During hazard characterization, the creeks are assessed with respect to geographic location, development and local infrastructure, and geomorphic and hydrological characteristics. The process involves compilation and review of previous assessments on the creeks in conjunction with desktop analyses and field investigation. The details of each component are described in the following sections.

4.1 Previous Studies

The initial phase of the assessment included a thorough review of previous studies conducted within the DNV that focused on debris geohazard and risk estimation, stormwater management and culvert hydraulics, infrastructure vulnerability, and that provide records of previous water conveyance failures. The outcomes of the review included a compilation of material that remains current and applicable to the current assessment, identification of data gaps and limitations to be addressed by the current assessment and identification of data gaps and limitations that lie outside the scope of work but affect the methodology or expected results of the study.

The steep creeks around DNV have a documented history of flood, debris flood and debris flow events. Therefore, previous assessments of clear-water flood, debris flood, debris flow, landslide and other slope movement features were reviewed as documented by Kerr Wood Leidal (1995, 2003a-k, 2011), Northwest Hydraulic Consultants (2010), the Ministry of Environment (1995), and BGC Engineering Inc. (2009, 2010, 2011, 2013). DNV records, news reports and field observations, including a detailed description of the November 2014 event on Kilmer Creek, were also reviewed. The November 2014 records were of critical importance in that they were used to calibrate the hazard analysis and debris-flood modelling described here-in. Calibration of event frequency was also provided by a review of information kept by DNV maintenance personnel of culvert inlets that have historically required maintenance for debris blockage during storm events.

4.2 Desktop Study

Terrain analysis formed the initial stage of hazard characterization. The analysis involved delineation of creek and creek tributaries, delineation of watershed boundaries and mapping of geomorphic features as described in the sections below. All terrain analysis was completed using 2013 LiDAR data provided by DNV and 1 m topographic contours.

Initial creek delineations were provided by DNV and modified based on terrain analysis coupled with field observations. Modifications to the creek delineations were completed where the initial version did not extend to the upstream headwaters, where notable tributaries identified during field inspections had not been delineated, or where anthropogenic modification of the landscape had altered the drainage pattern.

Watersheds were delineated within ArcGIS using watershed boundaries calculated using Global Mapper v15.2 and BASINS 4.1 (United States Environmental Protection Agency (EPA) 2013). This work was supplemented by field observations of local and regional drainage patterns and reference to previous investigations of drainage along roadways or other infrastructure. In particular, a drainage assessment was completed in 2007 along Old Grouse Mountain Highway to facilitate culvert sizing (BGC 2007). The delineated drainage patterns identified during the assessment were included in the present assessment.

4.3 Field Investigation

The majority of study creeks visited were traversed from the development interface (e.g., boundary of residential development with the undeveloped upper watersheds) upstream to the point at which the creek had little potential to contribute to sediment transport to development. In many instances, creeks were hiked along the full length to the headwaters in order to delineate the entire creek. Select creeks were traversed through development in order to characterize the potential for blockage of culverts and attendant flooding potential.

The primary objective of field investigation for the hazard component of this study was to collect parameters supporting debris volume and frequency estimation, as well as hydraulic analyses at culvert crossings and observations on infrastructure maintenance. These parameters included channel hydraulic characteristics (gradient, width, and bankfull height), channel bed characteristics (grainsize distribution, sediment availability and bedrock exposure), cross sections where high water marks could be reconstructed, sediment sources (bank erosion, rockfall, localized slope instability), culvert properties (dimensions, gradient, freeboard and condition), debris control structure dimensions, and pedestrian footbridge channel geometry.

In addition to the creeks, major roadways affecting regional drainage patterns were visited to check and refine the local watershed boundaries. This includes Mt. Seymour Road, Indian River Drive and Old Grouse Mountain Hwy. On Percy Creek, which is subject to debris flows, four test pits were mechanically excavated on the fan to use stratigraphic information for volumetric determination and collect organic samples for radiocarbon dating. Tree core samples were also collected for dendrochronological analysis.

4.4 Hydrogeomorphic Process Assignment

The study creeks are subject to hydrogeomorphic processes whose dominant driver is water with varying sediment concentrations; these include clear water flood, debris flood, and debris flow processes (Jakob et al. 2015). Identifying the dominant hydrogeomorphic process on a creek is important for detailed assessment of flow magnitude and behaviour, selection of parameters for numerical modelling of flows, selection of criteria to estimate vulnerability, and associated risk and mitigation design. A dominant hydrogeomorphic process type was assigned to each study creek based on fieldwork and desktop study that included review of previous work, the geomorphology of fans (where existing), channel deposits and their associated watersheds, channel gradients, field observations, and records of previous events.

4.5 Frequency-Magnitude Analysis

Frequency-magnitude (F-M) estimations form the core of any hazard assessment because they combine the findings from frequency and magnitude analyses in a format suitable for numerical analysis. Frequency and magnitude of hydrogeomorphic events are inversely related. The higher the event frequency, the lower its magnitude and vice versa. In short, the rarer an event, the larger it will be. Hazard magnitude is expressed as the peak flow and the total volume of sediment mobilized in the specific hazard event.

4.5.1 Frequency Estimation

Frequency analysis determines how often hydrogeomorphic events occur, on average. Frequency can be expressed either as a return period or an annual probability of occurrence. Frequencies of hydrogeomorphic events were established through flood frequency analysis, determination of the critical threshold for full bed mobilization, radiocarbon dating of organics in test trenches (Percy Creek) and dendrochronology (Percy Creek).

4.5.2 Magnitude Estimation

The objective of the magnitude analysis is to estimate debris volumes and peak discharges of past hydrogeomorphic events. The results form inputs to numerical modelling and assessment of debris blockage at culverts and stormwater mains. Table 2 summarizes the approaches used to estimate debris volumes for creeks subject to debris floods and debris flows, which have the potential to carry much higher sediment loads than clear-water floods.

Table 2: Methods employed for estimation of debris magnitudes

Application	Method
Flood and debris flood prone streams with a mobile bed but not steep enough to produce debris flows.	• Determine the shear stress threshold (i.e., critical shear stress) for bed mobilization. • Use average channel dimensions and Manning's equation to determine the discharge that corresponds to the critical shear stress. • Use a hydrograph associated with a specific return period to calculate the amount of time that the flow exceeded the discharge threshold. • Use Rickenmann's (2001) sediment transport equation to calculate sediment discharge based on stream power. • Calculate sediment volume based on the estimated sediment discharge rate multiplied by the duration over which the critical shear stress occurs. • Use of regionally-derived frequency-magnitude curves for debris floods as a comparative tool with previously employed methods.
Percy Creek debris flows	• Excavation of test pits to collect samples of organic material for radiocarbon dating and collection of tree core samples for dendrochronological analysis. • Reconstruct yield rates along the creek to estimate the total volume of erodible sediment available. • Add estimated sediment volumes from point sources along the creek. • Use of regionally-derived frequency-magnitude curves for debris flows.
Debris flood, debris flow hybrids	• Use appropriate combination of methods from above.
Dam outbreak floods	• Determination of peak discharge and outflow hydrographs from the failure of a landslide dam using the model BREACH (Fread 1991).

While based on the best available data, estimates of debris-flow or debris-flood magnitudes contain uncertainties. Older deposits can be eroded or reworked and are therefore often difficult to distinguish unambiguously from one another. Creek-side development also creates practical limits to subsurface investigation of past deposits. Hydromorphic processes also involve hazard mechanisms that cannot be quantified by flow modelling, such as bank erosion.

Frequency and magnitude information was combined to produce frequency-magnitude relationships for each debris-flood and debris-flow prone creek.

4.6 Hazard Scenarios

Hazard scenarios are hypothetical scenarios where flows occur outside the normal creek channel in areas where they can impact development or infrastructure. The scenarios quantify the extent and "intensity", or destructive potential, of a flood, debris flood or debris flow and are used to estimate building damages and risk to life. Hazard scenarios do not encompass every possible flow or avulsion scenario that could occur but define representative events at a range of return periods that are suitable for risk estimation. These return periods are based on the results of frequency-magnitude analysis.

For each creek, we estimated the return period where culverts and stormwater mains are anticipated to overflow either due to capacity exceedance or blockage by sediment or organic material. Judgment was required in some cases where a series of culverts or storm mains were expected to block during a representative hazard event. Flow magnitudes and expected blockage scenarios then formed the inputs for flow modelling. For Percy Creek, representative debris flow hazard scenarios were based on estimated debris-flow frequency-magnitude relationships for return period classes up to 3000 years including avulsion scenarios and estimated representative avulsion scenarios. Debris-flood prone creeks were assessed up to return periods of 200 years, which was considered the upper limit of the flood frequency analysis. Many of the debris-floods creek in the DNV are also supply-limited, placing an upper threshold on sediment mobilization.

4.7 Hazard Scenario Modelling and Mapping

Having established an F-M curve for each creek, numerical flow routing models were then used to estimate the extent, velocity and depth of flow for each hazard scenario on a given creek. For debris flood modeling, the commercially available two-dimensional hydraulic model FLO-2D (2007) was used to estimate debris-flood velocity, depth and the extent of inundation for each debris-flood hazard scenario. Culverts or stormwater mains were blocked in the model for each scenario. Estimated capacities of culverts or stormwater mains not assumed to be blocked were also input into the model. For debris flow modeling on Percy Creek, the three-dimensional numerical model *DAN3D* was used (McDougall and Hungr 2004). The model grid cell outputs were then imported into GIS, overlaid on base maps, and used to interpret hazard intensity maps.

Hazard intensity maps were developed for each scenario from the modeling that show the destructive potential, of a flood, debris flood or debris flow. Model results were displayed using a "flow intensity index" (I_{DF}) and flow depth. The flow intensity index is calculated as flow depth multiplied by the square of flow velocity (Jakob et al. 2012). It is not appropriate for estimating damages associated with low velocity flooding (e.g., approximately < 1 m/s), where values of I_{DF} will approach zero irrespective of flood depth. Below an intensity threshold of $I_{DF}<1$, flood vulnerabilities for buildings were estimated based on assumed flood depths.

5 Risk Assessment

Risk assessment involves estimation of the likelihood that a debris flood or debris flow scenario will occur, impact elements at risk, and cause particular types and severities of consequences. We estimated individual and group safety risk (loss of life risk) for persons within buildings, and economic risk associated with building damages due to debris impact. The primary objective of the safety risk assessment is to identify cases where the estimated risk level exceeds risk tolerance thresholds defined by the DNV. Following consideration of safety risk, the economic risk assessment can support further prioritization of risk reduction options.

Both quantitative risk assessment (QRA) and semi-quantitative risk assessment (semi-QRA) methods were used to estimate risk. QRA methods were used to estimate safety and economic risk for each creek system as a whole, considering all hazard scenarios and culvert blockages modelled for a given creek. The results support risk reduction prioritization for each creek (e.g., should creek "X" be higher priority, from a risk perspective, than creek "Y"). Semi-QRA methods were used to assign relative risk ratings to individual culverts or storm water mains. The economic risk rating for individual culverts addressed the question, "what is the probability that a particular culvert blocks and results in some level of economic consequences?". The results of semi-QRA support risk reduction prioritization for individual culverts (e.g., should culvert "X" be higher priority, from a risk perspective, than culvert "Y").

5.1 Safety Risk

Safety risk was quantitatively estimated for each creek from two perspectives: risk to individuals and groups. Individual safety risk considered the risk to a particular individual exposed to a hazard, and is independent of the number of persons exposed to risk. Group safety considered the collective risk to all individuals exposed to hazard, and is proportional to the number of persons exposed to risk. In both cases, safety risk estimates considered the spectrum of hazard scenarios assessed for a given creek. Hazard scenarios were considered as having a credible (non-negligible) risk to life only where flows exceeded a minimum intensity threshold ($I_{DF} > 1$), or where buildings were identified as particularly vulnerable to impact.

Individual risk estimate results were compared to geohazard tolerance criteria adopted by the DNV in 2009. For risk to groups, estimated risks were compared to group risk tolerance criteria formally adopted in Hong Kong (GEO 1998) and informally applied in Australia (AGS 2007) and DNV. Group risk tolerance thresholds based on criteria adopted in Hong Kong (GEO 1998) are shown on an F-N Curve in Figure 1.

5.2 Economic Risk – Creek Systems

Economic risk estimates for each creek were based on the hazard scenarios and considered direct damages to buildings. The results were reported as a direct damage cost for each scenario and as an annualized figure. The annualized cost was calculated by multiplying the hazard scenario probability by the estimated damage cost for a given scenario, and then summing the results for all scenarios.

Building damage cost estimates were based on vulnerability criteria relating damage levels to flows with a certain level of intensity or destructive power. Damage was measured as a proportion of the building replacement cost or as an absolute cost. Annualized damage cost was calculated by interpolating a damage curve from cost estimates for individual events at a given probability of occurrence.

Building damages associated with low intensity flows ($I_{DF} < 1$) were assumed to include inundation by water and sediment. Damages for such flows were estimated using flood stage-damage curves developed for Alberta Environment and Parks (AEP) following the damaging floods in southwestern Alberta in June 2013 (IBI Group 2015).

Building damages associated with higher intensity flows ($I_{DF} > 1$) have the potential to cause structural building damage due to dynamic impact pressure, and were considered to have credible potential to cause loss of life. Building damages levels for higher intensity flows were assigned using intensity-damage criteria developed from judgement with reference to Jakob et al. (2011).

5.3 Economic Risk – Culverts

Table 3 displays the matrix used to determine relative economic risk ratings for individual culverts. The hazard rating in the matrix corresponds to the culvert blockage rating, and the consequence rating to estimated direct building damage costs downstream of the culvert.

Table 3: Economic risk matrix

Hazard Rating (Probability Hazard Scenario Occurs and Impacts Elements at Risk)			Economic Risk Rating				
Classification	Hazard Scenario Probability	Culvert Overflow Rating (Years)					
Very Low	0.001-0.0003	n/a	1	1	2	3	4
Low	0.003-0.001	>200	1	2	3	4	5
Moderate	0.01-0.003	200	2	3	4	5	6
High	0.03-0.01	50	3	4	5	6	7
Very High	0.1-0.03	20	4	5	6	7	7
Consequence Rating	Indices		Very Low	Low	Moderate	High	Very High
	Direct Damage Cost ($M)		<0.1	<0.5	0.5-1	1-10	>10

6 Results

6.1 Safety Risk

Of the 25 creeks assessed in detail, a total of 5 single family residential buildings on three creeks were subject to estimated risk levels exceeding DNV's individual safety risk tolerance standard for existing development (>1×10^{-4} risk of fatality per year). An additional 13 single family residential buildings on various creeks were estimated as subject to risk levels exceeding DNV's individual safety risk tolerance standard for proposed development (>1×10^{-5} risk of fatality per year). Figure 1 displays estimated group risk for Percy Creek on an F-N curve, the only urban creek where the best-estimate of group risk falls in the unacceptable range. The lower, upper and best-estimates shown on the graph (where the dashed line represents the bounds) reflect the range used to estimate vulnerability.

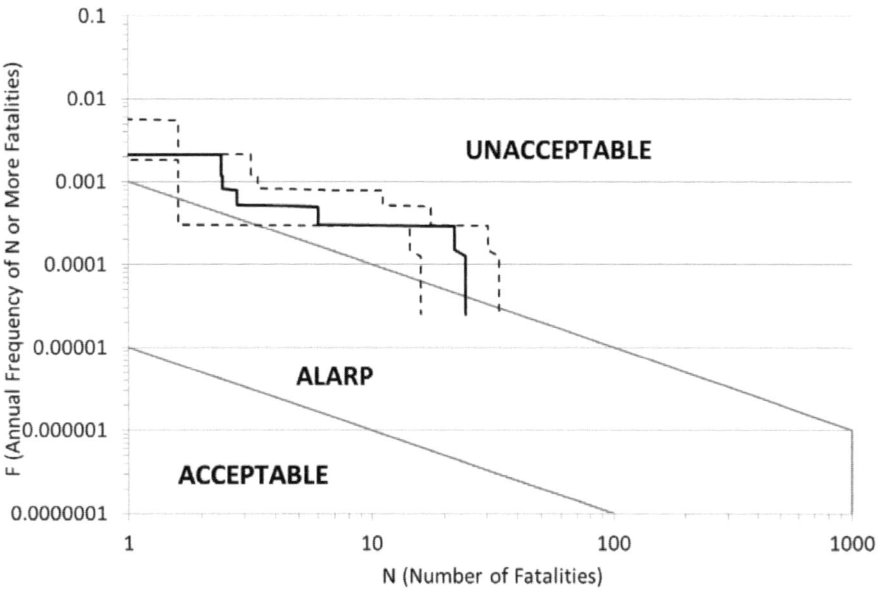

Figure 1: F-N Curve for Percy Creek

6.2 Economic Risk

Estimated damage costs for each hazard scenario and creek ranged from \$10,000 on Francis Creek to over \$3M on Mackay Creek, with annualized damage costs ranging from negligible to about \$100,000. Because the annualized cost is proportional to event frequency, highly damaging but rare events may have a lower annualized cost than more frequent but less damaging events. Semi-quantitative risk ratings assigned to each stormwater drainage asset ranged from 2 to 7 and support prioritization of risk control measures along a given creek.

7 Conclusions

This study provided debris geohazard risk assessment and conceptual debris risk control options for creeks within the DNV. The study includes 35 "steep" creeks (creeks with channel gradients >5%) prone to flood, debris-flood or debris-flow processes that could cause economic damages or pose risk to life for persons within buildings.

For stormwater drainage assets with credible potential for debris blockage, we estimated the return period where culverts and stormwater mains are anticipated to overflow either due to capacity exceedance or blockage by sediment or organic material. We developed and modelled representative hazard scenarios and estimated safety and economic risk associated with these scenarios. An interactive geohazard asset management application, the "Debris Hazard Info Tool" (DNVHIT), was developed to display the results and supporting data for this work.

Based on the results of the hazard assessment, we assessed safety and economic risk at both a creek and individual asset level of detail. Risk estimates for each creek support prioritization of overall risk reduction planning. Relative risk ratings at the asset level support risk reduction prioritization for each stormwater management asset, and quantify the level of risk reduction achieved by implementing the risk control measures. Based on the results of the assessment, DNV staff are developing a 10-year work plan that will be integrated with the District's asset management, GIS-based inspection program, climate change adaptation and hazard mitigation plans.

Acknowledgements

This work was motivated and funded by the District of North Vancouver. The authors would like to thank Michael Porter for his detailed review of the original assessment.

References

Australian Geomechanics Society (AGS). 2007. A national landslide risk management framework for Australia, 42(1).

BGC Engineering Inc. 2007. Old Grouse Mountain Highway – Culvert Sizing. *Memo prepared for* Hedberg and Associates Consulting Ltd.

BGC Engineering Inc. 2009. North Vancouver Debris Flow and Debris Flood Quantitative Risk Assessment: Update. Report prepared for District of North Vancouver, September 22, 2009.

BGC Engineering. 2010. Mosquito Creek Debris Flood Hazard Assessment. Final report prepared for Natural Resources Canada dated March 31, 2010.

BGC Engineering. 2011. Quantitative Risk and Mitigation Option Assessment. Final report prepared for District of North Vancouver dated January 6, 2011.

BGC Engineering Inc. 2013. Mosquito Creek Post-Mitigation Quantitative Risk Assessment. Report prepared for District of North Vancouver, dated October 2, 2013.

DNV 2015. Asset Management Plan, Drainage. Prepared by the District of North Vancouver, 27 pp.

Fread DL (1985, revised 1991) BREACH: An Erosion Model for Earthen Dam Failures. NWS Report, National Oceanic and Atmospheric Administration, Silver Spring, Maryland.

Hong Kong Geotechnical Engineering Office (GEO) 1998. Guide to Slope Maintenance (Geoguide 5). Geotechnical Engineering Office, Hong Kong.

IBI Group 2015. Provincial Flood Damage Assessment. Prepared for Government of Alberta ESRD - Resilience and Mitigation, dated February 2015.Intergovernmental Panel on Climate Change (IPCC). 2014. Climate Change 2014: Impacts, Adaptation, and Vulnerability. Part A: Global and Sectoral Aspects. Contribution of Working Group II to the Fifth Assessment Report of the Intergovernmental Panel on Climate Change.

Jakob, M., Stein, D. and Ulmi, M. 2012. Vulnerability of buildings to debris flow impact. Natural Hazards 60:241–261.

Journeay, J.M., Dercole, F., Mason, D., Weston, M., Prieto, J.A., Wagner, C.L., Hastings, N.L., Chang, S.E., Lotze, A., and Ventura, C.E. 2015. A profile of earthquake risk for the District of North Vancouver, British Columbia. Geological Survey of Canada Open File 7677, 223 pp.

McDougall S, Hungr O. 2004. A model for the analysis of rapid landlside motion across three-dimensional terrain. Can Geotech J 41(6):1084–1097 doi:10.1139/t04-052

Ministry of Environment, Lands and Parks. 1995. Design Brief on the Floodplain Mapping Study. *Prepared for the* Water Management Branch, Province of British Columbia. January, 1995.

Northwest Hydraulic Consultants (NHC). 2010. Flood Assessment Study, North Vancouver. *Prepared for* Natural Resources Canada and the District of North Vancouver.

Natural Resources Canada (NRCAN), 2015. A profile of earthquake risk for the District of North Vancouver, British Columbia; by Journeay, J M; Dercole, F; Mason, D; Westin, M; Prieto, J A; Wagner, C L; Hastings,

N L; Chang, S E; Lotze, A; Ventura, C E; Geological Survey of Canada, Open File 7677, 2015; p. 145-150.

Rickenmann D. 2001. Comparison of bed load transport in torrents and gravel bed streams. *Water Resources Research* **37**(12): 3295-3305.

United States Environmental Protection Agency. 2013. BASINS [computer program]. Version 4.1. US EPA, Washington, DC.

Leadership in Sustainable Infrastructure
Leadership en Infrastructures Durables

Vancouver, Canada
May 31 – June 3, 2017/ *Mai 31 – Juin 3, 2017*

RECONSTRUCTION OF AREA-CAPACITY CURVES FOR ALBERTA LAKES

Islam, Zahidul[1, 3] and Seneka, Michael[2]
[1, 2] Alberta Environment and Parks, Canada
[3] Zahidul.Islam@gov.ab.ca

Abstract: Area-Capacity curves represent the relationship of surface area and volume of a lake or reservoir to elevation. Area-Capacity curves are essential to the development of water balance models for lakes; more specifically, they provide science-based information on surface water availability at any given time. This information can be vital to the execution of informed management decisions, such as the permitting of water withdrawals from lakes in such a way as to protect ecosystems while supporting reliable, quality water supplies that contribute to a sustainable economy. Currently, there are three major sources of area/capacity information in Alberta: the *Atlas of Alberta Lakes* (Mitchell 1990); Alberta Environment and Parks (AEP) archived files of historical bathymetric surveys and lake area/capacity calculations; and digital lake bathymetry from the Alberta Geological Survey (AGS). The Atlas of Alberta Lakes provides Area-Capacity curves for about 80 lakes in Alberta. However, these curves are only available on paper or in digital image format and are not usable unless digitized to retrieve the area/capacity information at various elevations. The AEP archived files contain manual calculations of area/capacity based on original bathymetry surveys for about 60 lakes and are not immediately accessible to the public. AGS data provide lakes bathymetry (GIS files) on their website, but not the actual information on area/capacity. In the current study, an attempt has been made to combine these sources and reconstruct Area-Capacity curves for numerous Alberta lakes. Digital images of Area-Capacity curves from the Atlas of Alberta Lakes were manually digitized, and higher order polynomial equations were fitted to develop Area-Capacity curves. Lakes for which area/capacity calculations are not available but digital bathymetry data could be acquired through the AGS, an ArcGIS based tool was developed to construct Area-Capacity curves. A mathematical extrapolation technique was applied to extend the Area-Capacity curves beyond the original surveyed water level.

1 INTRODUCTION

Area-Capacity curves represent the relationship of surface area and volume of a lake or reservoir to elevation (Figure 1). These curves are essential to the development of water balance models for lakes; more specifically, they provide science-based information on surface water availability at any given time. This information can be vital to the execution of informed management decisions, such as the permitting of water withdrawals from lakes in such a way as to protect ecosystems while supporting reliable, quality water supplies that contribute to a sustainable economy.

Currently, there are three major sources of area/capacity information in Alberta:

I. Atlas of Alberta Lakes (the Atlas): The Atlas of Alberta Lakes (Mitchell 1990) is a collection of information on 100 of Alberta's most studied lakes. It provides information on location, drainage basin characterises, water quantity, water quality, and biological characterises of lakes. The Atlas provides Area-Capacity Curves for about 80 lakes in Alberta. However, these curves are only available on paper or in digital image format and are not usable unless digitized to retrieve the area/capacity information at various elevations.

II. Alberta Environment and Parks (AEP) archived files: AEP has a collection of historical information on lake bathymetric surveys. The AEP archived files contain manual calculations of area/capacity (Area-Capacity Table) based on original bathymetry surveys for about 60 lakes. However, this information is not immediately accessible to the public.

III. Alberta Geological Survey (AGS) digital bathymetry: AGS has a collection of digital bathymetry, comprised of Geographic Information System (GIS) data of about 169 lakes (Alberta Geological Survey 2016). This dataset has been generated by capturing by digitizing the historic hydrographic survey maps and then digitally modelled to produce a continuous surface for each lakebed. However, AGS data provide lakes bathymetry, but not the actual information on area/capacity.

In the current study, an attempt has been made to combine these sources and reconstruct Area-Capacity curves for numerous Alberta lakes.

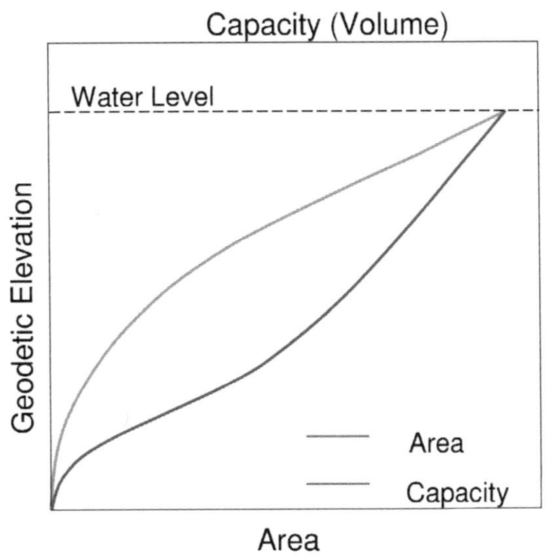

Figure 1: Schematic Diagram of Area-Capacity Curves

2 DATA AND CHALLENGES

The Atlas is publicly accessible digitally through the University of Alberta website (Mitchell 1990). First, the digital images (JPEG Format) of Area-Capacity curves of about 80 lakes are downloaded from the website. Then, an open source screen digitizing software Plot Digitizer (SOURCEFORGE.NET 2015) is used to manually digitize those digital images of Area-Capacity curves. Plot Digitizer is a Java program used to digitize functional X-Y type scatter or line plots from a scanned image of a plot (SOURCEFORGE.NET 2015). The advantage of this screen digitizing is that enough points can be collected from the digital images to capture the curvature of Area-Capacity curves. However, this process has human errors which can lead to irregular shapes, especially when they are zoomed in.

The AEP archived files contains Area-Capacity calculations based on original bathymetry contours. Lake surface area and lake volume are then calculated at regular elevation intervals (from the lake bottom to the surveyed water level) from those bathymetry contours. However, often these calculations have fewer points compared to the depth of a lake.

The AGS digital lake bathymetry data is a combination of multiple data sources (Alberta Geological Survey 2016). All available bathymetry and related information for lakes were collected and hard copy maps digitized where necessary. The data were validated against more recent data (Shuttle Radar Topography Mission 'SRTM' imagery and Indian Remote Sensing 'IRS' imagery) and corrected where necessary. ESRI ArcGIS 3D Analyst could be used to calculate lake surface area and volume (capacity) at various elevations to construct Area-Capacity curves from this digital lake bathymetry dataset. However, often Area-Capacity curves constructed through this process do not match with existing Area-Capacity curves collected from other sources, whenever available (e.g., the Atlas, AEP archived files).

Moreover, most of the Area-Capacity curves available through historical sources (e.g., the Atlas or AEP Archived files) are based on bathymetry surveys conducted in 1960s/1970s/1980s. As the calculation of lake surface area and volume (capacity) are limited by water level information available at the time of the survey, sometimes these curves are unable to provide information on area/capacity at a water level above the surveyed water level. Hence, a mathematical extrapolation is often required to estimate area/capacity above the surveyed water level.

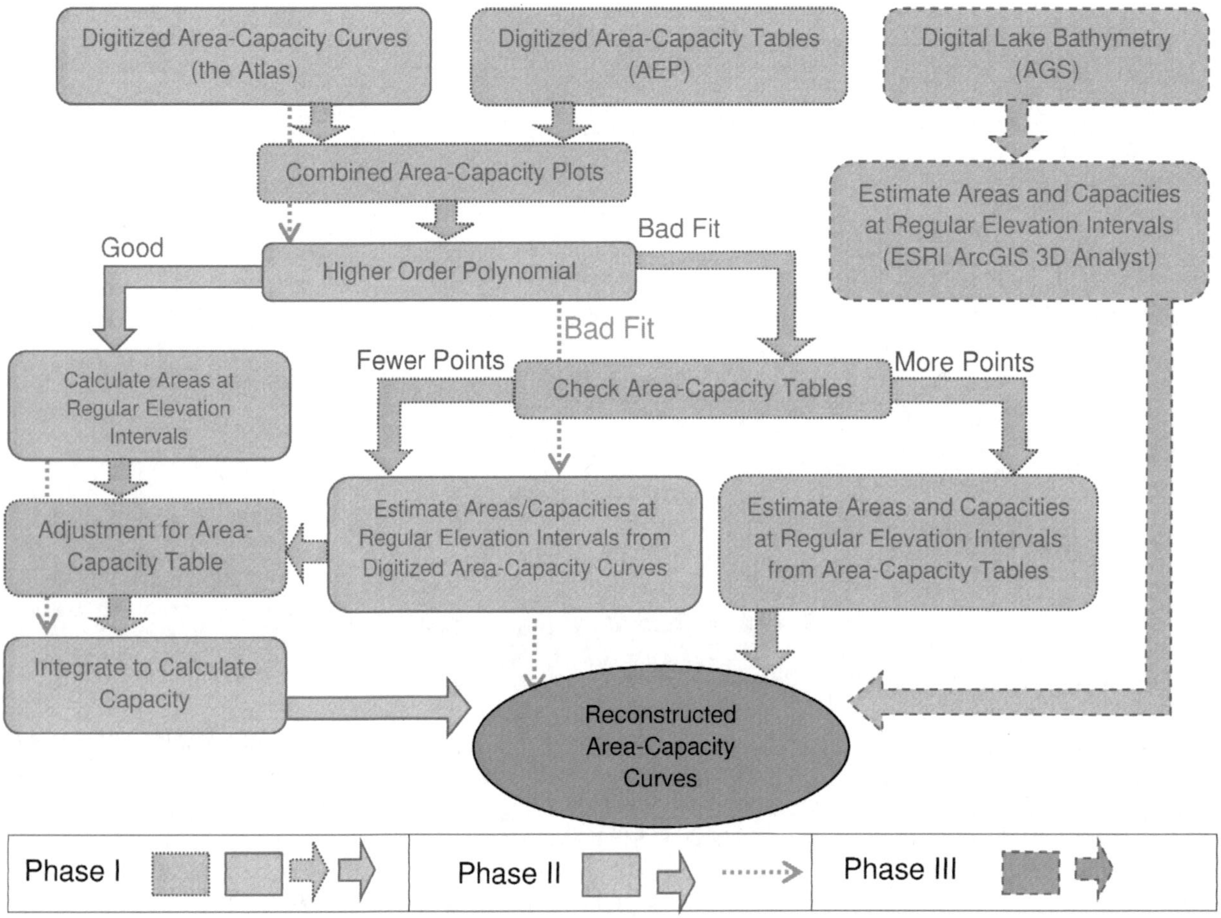

Figure 2: Reconstruction Flowchart for Phase I, II, and III

3 METHODOLOGY

Based on data availabilities and challenges discussed in the aforementioned sections, three different phases have been planned to reconstruct Area-Capacity curves. The following subsections will demonstrate the reconstruction methodologies of these three phases:

Phase I: Lakes have "Area-Capacity Curves" (through the Atlas) and "Area-Capacity Tables" (through AEP Archived files)

Phase II: Lakes have "Area-Capacity Curves" (through the Atlas) but no "Area-Capacity Tables" (through AEP Archived files)

Phase III: Lakes have no Area-Capacity curves/tables; have AGS lake bathymetry data

3.1 Phase I Methodology

Figure 2 shows a flowchart of Area-Capacity curves reconstruction methodology for all Phases. The purple & blue rectangular boxes and purple & blue wide arrows show steps for the Phase I reconstruction process (i.e. for the lakes which has "Area-capacity Curves" available through the Atlas as well as "Area-Capacity Tables" available through AEP archived files). The step by step methodology is also given below with the demonstration of the steps for the Marie Lake and Crimson Lake of Alberta in Figure 3:

Step 1: The digital images of Area-Capacity Curves are manually digitized using the Plot Digitizer.

Step 2: The Area-Capacity Tables are collected (if available) from AEP archived files, digitized and plotted.

Step 3: Area-Capacity plots from both sources (Step 1 & Step 2) are combined (Figure 3a) for each lakes.

Step 4: Based on the combined points, a higher order polynomial (usually 6th order) is fitted to represent the analytical relationship between lake surface area and lake depth (with respect of the surveyed water level) (Figure 3b).

Step 5a: If the polynomial shows a good fit between lake surface area and depth (Figure 3b), then the polynomial equation is used to reconstruct lake surface area at regular elevation intervals from the lake bottom to the lake surface (surveyed water level).

It was observed that, often the reconstructed lake surface areas calculated from the polynomial equation do not match with the values from the Area-Capacity Tables. The Area-Capacity Tables from AEP archived files are based on original bathymetric contours and ideally the reconstructed curves should follow these points. Hence, an adjustment of reconstructed curves was necessary. Following steps were applied to adjust the reconstructed curves so that they follow the data points based on original bathymetry:

I. The polynomial equation developed in **Step 4** is used to estimate lake surface area at the elevations where original bathymetric contours are available (points of the Area-Capacity Tables based from AEP archived files).

II. Area Adjustment Factors are calculated by dividing the lake surface areas found from the Area-Capacity Table by the lake surface areas found in **Step 5a (I)** (for the same elevation or lake depth). Note, near the lake bottom, these factors could provide unrealistic higher values (dividing a small number with another small number). In order to avoid this to happen, these factors were capped at 1.5.

III. The Area Adjustment Factors were then interpolated at regular elevation intervals (as mentioned in **Step 5a**).

IV. The reconstructed lake surface areas were then adjusted by multiplying the lake surface areas from **Step 5a** (i.e., using the polynomial equation) by the Area Adjustment Factors.

Finally, the capacities (volumes) at regular elevation intervals are then calculated by numerically integrating the adjusted reconstructed lake surface areas using trapezoidal rule.

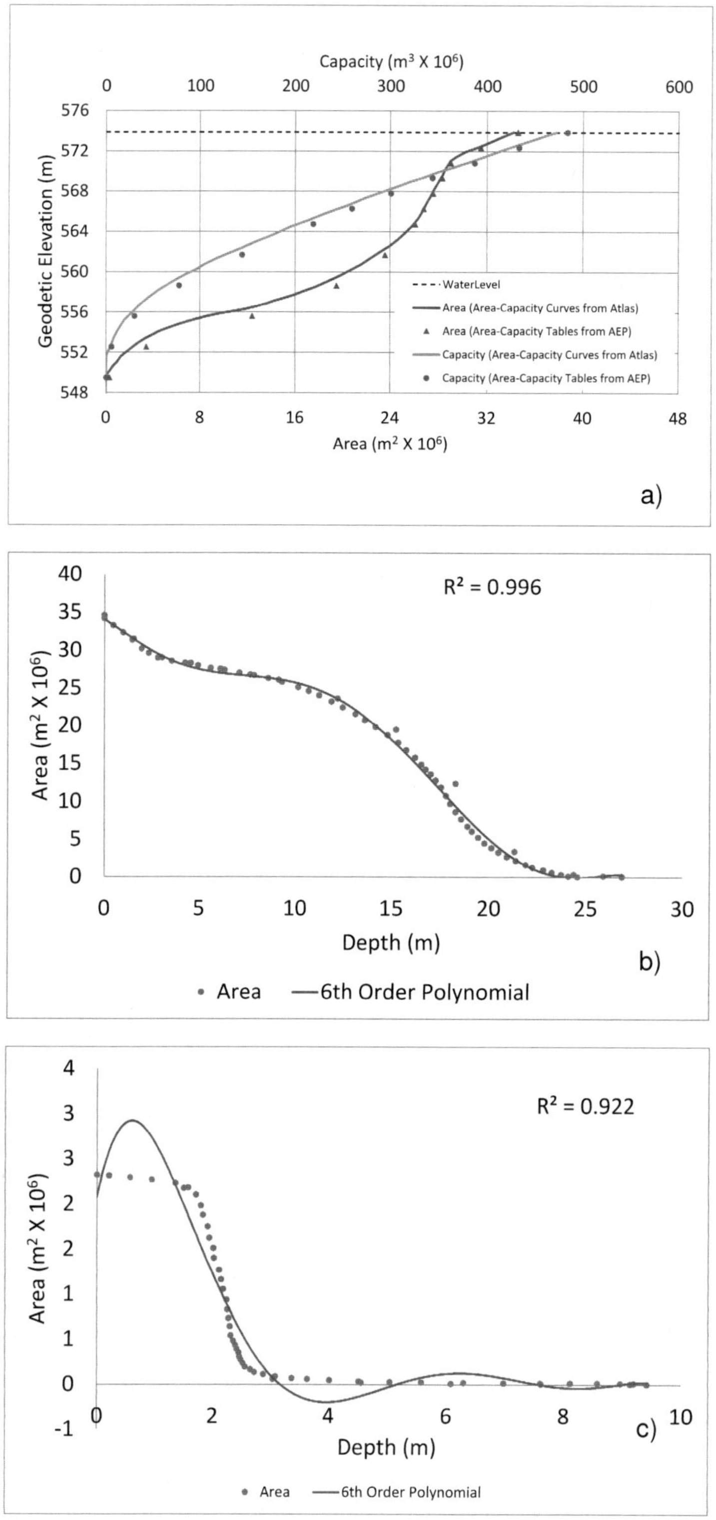

Figure 3: a) The combined (from the Atlas and from the AEP archive) Area-Capacity plots for the Marie Lake b) 6th Order Area-Depth Polynomial for the Marie Lake (example of good fit), c) 6th Order Area-Depth Polynomial for the Crimson Lake (example of bad fit)

Step 5b: If the polynomial does not show a good fit between lake surface area and depth (Figure 3c), then the Area-Capacity curves are reconstructed either from the Area-Capacity Tables (AEP) or from the digitized Area-Capacity Curves (the Atlas).

If the Area-Capacity Table has sufficient points (to maintain the curvature), the Table is used to linearly interpolate areas/capacities at regular elevation intervals (from lake bottom to lake water level).

If the Area-Capacity Table does not have sufficient points, the digitized Area-Capacity Curves are used to linearly interpolate areas/capacities at regular elevation intervals. However, in order to make sure that the reconstructed curves developed through this interpolation represent calculated areas/capacity from the original bathymetric contours (points from the Area-Capacity Tables), adjustments are necessary. Following steps were applied for the adjustments:

I. From the digitized Area-Capacity Curves, linear interpolation is used to estimate lake surface area and capacity at the elevations where original bathymetric contours are available (points of the Area-Capacity Tables).

II. Area Adjustment Factors and Capacity Adjustment Factors were calculated by dividing the lake surface areas and capacities found from the Area-Capacity Table by the lake surface areas and capacities found in **Step 5b (I),** respectively. Note, near the lake bottom these factors could provide unrealistic higher values (dividing a small number with another small number). In order to avoid this to happen, they (factors) were capped at 1.5.

III. The Area Adjustment Factors and Capacity Adjustment Factors were then interpolated at regular elevation intervals.

IV. The reconstructed lake surface areas and capacities were then adjusted by multiplying the lake surface areas and capacities from **Step 5b** by the Area Adjustment Factors and Capacity Adjustment Factors, respectively.

3.2 Phase II Methodology

The blue rectangular boxes, blue wide arrows, and green arrows in Figure 2 show steps for the Phase II reconstruction process (i.e. for the lakes which has "Area-capacity Curves" available through the Atlas **but no** "Area-Capacity Tables" are available through AEP archived files). The step by step methodology is also given below:

Step 1: The digital images of Area-Capacity Curves are manually digitized using the Plot Digitizer.

Step 2: Based on the digitized points, a higher order polynomial is fitted to represent the analytical relationship between lake surface areas and lake depths (with respect of the surveyed water level)

Step 3a: If the polynomial shows a good fit between lake surface areas and depths, then the polynomial equation is used to reconstruct lake surface areas at regular elevation intervals from the (lake bottom to the lake surface, or the surveyed water level). The lake capacities (volumes) at those elevations are then calculated by numerically integrating the reconstructed lake surface areas using trapezoidal rule.

Step 3b: If the polynomial does not show a good fit between lake surface area and depth, then the digitized Area-Capacity Curves are used to linearly interpolate areas/capacities at regular elevation intervals.

3.3 Phase III Methodology

For the lakes where no information of area/capacity is found, however digital bathymetry data is available through AGS, the following methodology is applied to construct Area-Capacity curves:

Step 1: ESRI ArcGIS© is used to collect information on lake bottom elevation from lake bathymetric DEM. It was assumed that the lowest elevation of the bathymetry grid represent the lake bottom.

Step 2: Lake surface areas and volumes at regular elevation intervals (from lake bottom to lake water level) are calculated using the "Surface Volume" tool under 3D Analyst toolset of ESRI ArcGIS. Surface Volume is a Geoprocessing tool that calculates the area and volume of a surface above or below a given reference plane (ESRI 2017). An ArcGIS tool was developed to make the process automated by providing lake bottom elevation, water level, and the elevation increments.

Step 3: Lake surface areas and volumes are plotted against the geodetic elevations to construct Area-Capacity curves.

3.4 Mathematical Extrapolation of Reconstructed Area-Capacity curves

Lake area/capacity calculations are limited by water level information available during the hydrographic survey. Anytime lake level increases from the surveyed lake level, these curves are unable to provide information on area/capacity at a water level above the surveyed water level. Figure 4 shows an example (Baptiste Lake) of such case where the lake level during the hydrographic survey (conducted in January 1979) is 578.52 meter. However, the lake levels fluctuated within 577.84 to 579.20 meter over a period of 38 years (1972-2009). The historical Area-Capacity Curves/Tables is unable to provide a full range of information on lake area and volume for this case.

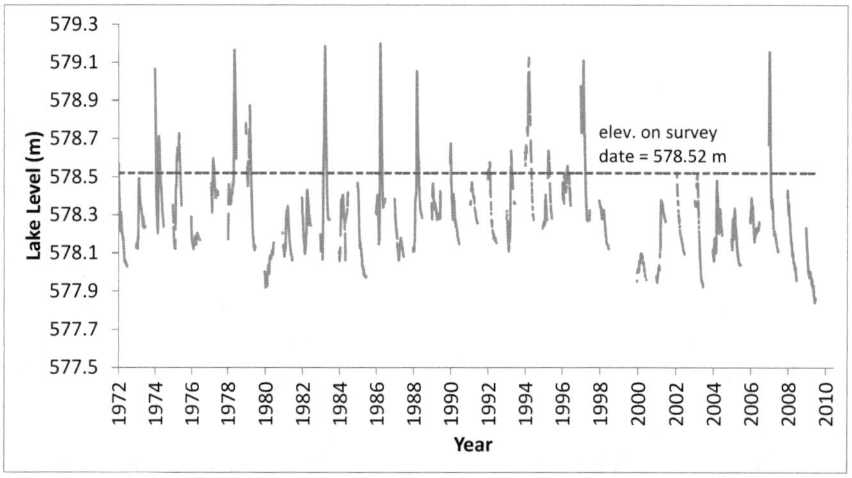

Figure 4: Historical water levels of Baptiste Lake, Alberta. The red horizontal dashed line shows lake level (578.52 meter) during the hydrographic survey conducted in January 1979

Based on the observation of lake levels of about 80 lakes, following assumptions are considered to mathematically extrapolate the reconstructed Area-Capacity curves above the surveyed water level:

I. Volume-Depth relationship of a lake at the higher elevations (nearer the surface water level portion of a lake) is more realistic to extrapolate for the elevations higher than the surveyed lake level.

II. The top 1/4th depth of a lake is defined as the "near water level portion" and is chosen the development of mathematical extrapolation.

III. Area/Capacity curves should not be extended beyond 1.5 meter above the surveyed water level.

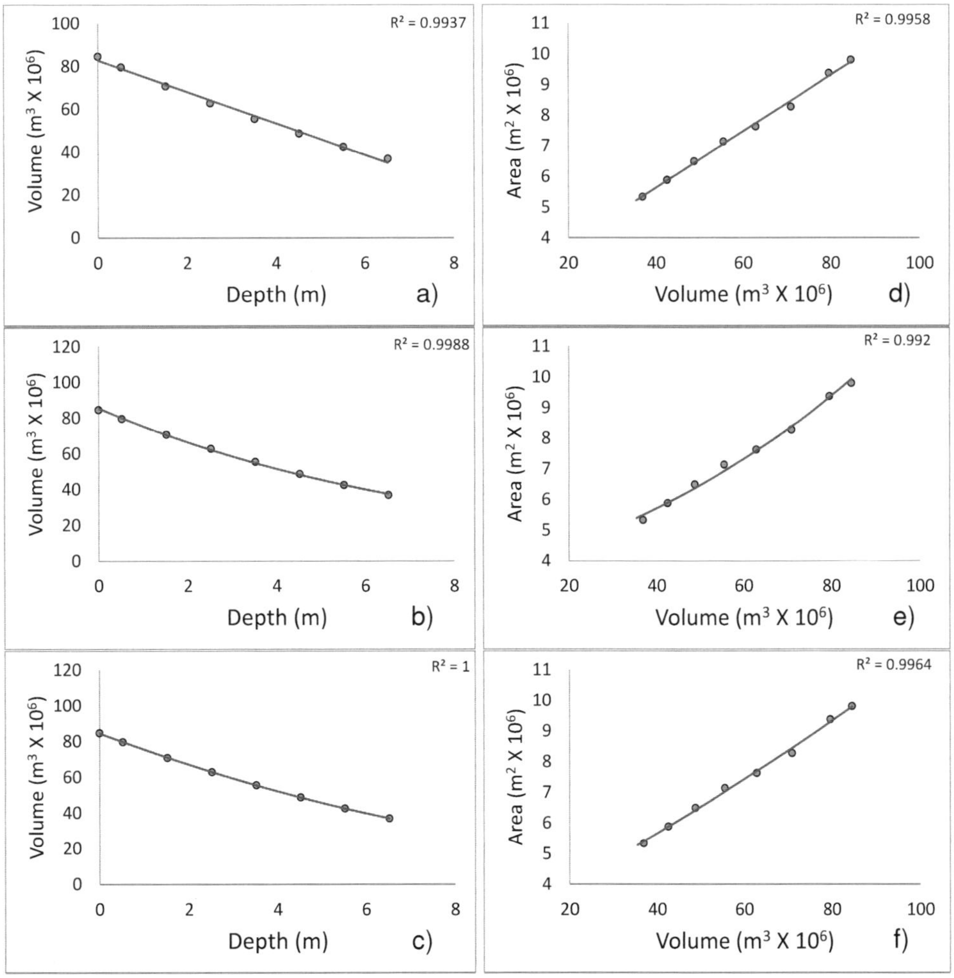

Figure 5: Analytical models for near-surface water level portion Volume-Depth and Area-Volume relationships of Baptiste Lake: Linear (**a** & **d**), Exponential (**b** & **e**), and 2nd Order Polynomials (**c** & **f**)

The step by step methodology is given below:

Step 1: For the near-surface water level potion of a lake (top 1/4th depth) the "Volume-Depth" and the "Area-Volume" analytical relationship are developed. Three types of analytical models are used for each of those two relationships: linear, exponential and 2nd order polynomial. Figure 5 shows different analytical models for these two relationships of the Baptiste Lake. Note, in some cases these analytical models developed for the top 1/4th portion of the lake do not produce a good relationship. Manual adjustment of the "near water level portion" depth is necessary for those cases to develop a representative analytical relationship (See Table 1 for details).

Step 2: The "Volume-Depth" analytical model, which provides the best Coefficient of Determination (R^2), is then chosen to extrapolate lake volumes at regular elevation intervals above the surveyed water level (not exceeding 1.5 meter above the surveyed water level).

Step 3: A Volume Adjustment Factor is calculated by dividing the lake volume at surveyed water level found from the reconstructed Area-Capacity curves by the lake volume at surveyed water level found from the chosen analytical model (**Step 2).** The lake volumes above the surveyed water levels estimated by the analytical model are then adjusted by multiplying with the Volume Adjustment Factor to avoid any discontinuity.

Step 4: The "Area-Volume" analytical model, which provides the best Coefficient of Determination (R^2), is then chosen to extrapolate the lake surface areas for the extrapolated and adjusted lake volumes from **Step 3**.

Step 5: An Area Adjustment Factor is calculated by dividing the lake surface area at surveyed water level found from the reconstructed Area-Capacity curves by the lake surface area at surveyed water level found from the chosen analytical model (**Step 4).** The lake surface areas above the surveyed water levels estimated by the analytical model are then adjusted by multiplying with the Area Adjustment Factor to avoid any discontinuity.

Note, as most of the reservoirs are surveyed up to their full supply level, no mathematical extrapolation is considered /necessary for reservoirs.

4 RESULTS AND DISCUSSIONS

Appendix A shows a list of Alberta lakes for which the reconstruction process is completed. The table also includes information on the reconstruction method and which portion of the lake considered in extrapolation of Area-Capacity curves above the surveyed water level. This is an ongoing process and preliminary data generated from this project need to be reviewed to ensure the quality of information. Once the Quality Assurance and Quality Control (QA/QC) is performed, the data will potentially be shared with public. Reconstructed Area-Capacity curves for numerous lakes have already been applied for the development of generic shapes of Alberta lakes to support water policy development (Islam and Seneka 2016).

Acknowledgements

The Authors would like to thank Taisiya Chuchvaha and Chen Xu of AEP for their support in this project.

References

Alberta Geological Survey. 2016. *Alberta Lake Bathymetry Data.* http://ags.aer.ca/alberta-lake-bathymetry-data.htm (accessed February 2016).

ESRI .2017. *ArcGIS Pro Tool Reference.* http://pro.arcgis.com/en/pro-app/tool-reference/main/arcgis-pro-tool-reference.htm (accessed February 2016).

Islam, Z. and Seneka, M. 2016. Development of Generic Shapes of Alberta Lakes to Support Water Policy Development. *Proceedings of Canadian Society for Civil Engineering*: GEN-147-1 to11.

Mitchell, P. 1991. *Atlas of Alberta Lakes.* http://albertalakes.ualberta.ca/ (accessed February 2017).

SOURCEFORGE.NET. 2015. *Plot Digitizer.* http://plotdigitizer.sourceforge.net/ (accessed February 2016).

Appendix A: List of Alberta lakes for which reconstructions of Area-Capacity curves are completed

Name	Reconstruction Phase	Reconstruction Method	Extrapolation Depth[1]
Amisk Lake	Phase I	Polynomial[2]	25%
Baptiste Lake	Phase I	Interpolation (Tables)[3]	25%
Battle Lake	Phase I	Polynomial	25%
Beauvais Lake	Phase I	Interpolation (Curves)[4]	19%
Bonnie Lake	Phase I	Interpolation (Curves)	29%
Buffalo Lake	Phase I	Polynomial	31%
Chestermere Lake	Phase I	Polynomial	30%
Coal Lake	Phase I	Interpolation (Curves)	4%
Cold Lake	Phase I	Interpolation (Tables)	25%
Crimson Lake	Phase I	Interpolation (Curves)	5%
Dillberry Lake	Phase I	Polynomial	28%
Ethel Lake	Phase I	Polynomial	20%
Figure Eight Lake	Phase I	Polynomial	6%
Garner Lake	Phase I	Polynomial	25%
Gregoire Lake	Phase I	Polynomial	25%
Halfmoon Lake	Phase I	Polynomial	25%
Hasse Lake	Phase I	Polynomial	25%
Hastings Lake	Phase I	Polynomial	25%
Hubbles Lake	Phase I	Polynomial	25%
Iosegun Lake	Phase I	Polynomial	25%
Island Lake	Phase I	Interpolation (Tables)	25%
Isle Lake	Phase I	Interpolation (Curves)	25%
Kananaskis Lakes	Phase I	Polynomial	25%
Lac La Nonne	Phase I	Polynomial	25%
Lac St. Cyr	Phase I	Interpolation (Tables)	25%
Lac Ste. Anne	Phase I	Polynomial	25%
Lake Eden	Phase I	Polynomial	25%
Lessard Lake	Phase I	Interpolation (Tables)	25%
Lesser Slave Lake	Phase I	Polynomial	25%
Little Fish Lake	Phase I	Interpolation (Curves)	25%
Long Lake (near Athabasca)	Phase I	Polynomial	25%
Mann Lakes	Phase I	Polynomial	25%
Marie Lake	Phase I	Polynomial	25%
McLeod Lake	Phase I	Interpolation (Tables)	18%
Miquelon Lake	Phase I	Polynomial	25%
Moose Lake	Phase I	Interpolation (Tables)	25%
Muriel Lake	Phase I	Polynomial	25%
Nakamun Lake	Phase I	Polynomial	25%
Narrow Lake	Phase I	Interpolation (Curves)	25%
North Buck Lake	Phase I	Interpolation (Curves)	11%
Pine Lake	Phase I	Interpolation (Tables)	25%
Pinehurst Lake	Phase I	Polynomial	25%
Rock Lake	Phase I	Polynomial	25%
Sandy Lake	Phase I	Polynomial	15%
Skeleton Lake	Phase I	Interpolation (Tables)	25%

Name	Reconstruction Phase	Reconstruction Method	Extrapolation Depth[1]
Smoke Lake	Phase I	Interpolation (Tables)	25%
Spring Lake	Phase I	Polynomial	25%
Steele Lake	Phase I	Polynomial	25%
Sylvan Lake	Phase I	Polynomial	25%
Thunder Lake	Phase I	Interpolation (Curves)	25%
Touchwood Lake	Phase I	Polynomial	25%
Twin Lake	Phase I	Polynomial	25%
Utikuma Lake	Phase I	Polynomial	25%
Wizard Lake	Phase I	Polynomial	25%
Wolf Lake	Phase I	Interpolation (Tables)	25%
Beaver Lake	Phase II	Interpolation (Curves)	25%
Blood Indian Creek Reservoir	Phase II	Interpolation (Curves)	N/A
Buck Lake	Phase II	Polynomial	25%
Chain Lakes Reservoir	Phase II	Interpolation (Curves)	N/A
Cooking Lake	Phase II	Polynomial	6%
Crawling Valley Reservoir	Phase II	Interpolation (Curves)	N/A
Elkwater Lake	Phase II	Interpolation (Curves)	7%
Gleniffer Lake	Phase II	Polynomial	N/A
Glenmore Reservoir	Phase II	Interpolation (Curves)	N/A
Gull Lake	Phase II	Interpolation (Curves)	25%
Lac La Biche	Phase II	Polynomial	25%
Little Bow Lake Reservoir	Phase II	Interpolation (Curves)	N/A
McGregor Lake	Phase II	Interpolation (Curves)	N/A
Pigeon Lake	Phase II	Polynomial	25%
Reesor Lake	Phase II	Polynomial	N/A
Sturgeon Lake	Phase II	Interpolation (Curves)	16%
Tucker Lake	Phase II	Polynomial	25%
Wabamun Lake	Phase II	Polynomial	25%
Christina Lake	Phase III	Surface Volume[5]	N/A
Bear Lake	Phase III	Surface Volume	N/A
Red Deer Lake	Phase III	Surface Volume	N/A
Seibert Lake	Phase III	Surface Volume	N/A
Jackfish Lake	Phase III	Surface Volume	N/A

[1] Top portion of the depth considered in mathematical extrapolation (As most of the reservoirs are surveyed up to their full supply level, no mathematical extrapolation is considered for reservoirs)

[2] 6th Order Polynomials were fitted

[3] Area-Capacity Tables (AEP archived files) is used to linearly interpolate areas/capacities at regular elevation intervals

[4] Area-Capacity Curves (the Atlas) is used to linearly interpolate areas/capacities at regular elevation intervals

[5] Surface Volume tool under 3D Analyst toolset of ESRI ArcGIS is used to calculate area/capacity from AGS digital bathymetry

Leadership in Sustainable Infrastructure
Leadership en Infrastructures Durables

Vancouver, Canada
May 31 – June 3, 2017/ *Mai 31 – Juin 3, 2017*

RISK ASSESSMENT OF WASTE WATER PROJECTS IN EGYPT

Elsanabary, Mohamed H.[1,5], Nema, Amira S.[2], Tolba, Ehab R.[3] and Hassan, Hassan M.[4]
[1]Assistant professor, Civil Engineering Department, Port Said University, Egypt
[2]Civil engineer, Projects Sector, Damietta Holding Company of Potable and Waste Water, Egypt
[3]Associate professor, Civil Engineering Department, Port Said University, Egypt
[4]Professor, Civil Engineering Department, Port Said University, Egypt
[5] mohamed.elsanabary@eng.psu.edu.eg

ABSTRACT: Waste water projects are dynamic sector which affect human's life and should meet the basic needs of individuals. Through the last decade, Egyptian Government has increased its intention to support waste water projects. One of the challenges facing waste water projects in Egypt is how to assess risks that affecting this kind of projects. These risks cause losses and consequently lead to an increase in cost, time delays and lack of project quality. The purpose of this paper is to identify risk factors that affect waste water projects in Egypt and assess the impacts of such factors in the project's cost and time. The research study area was taken in Damietta, Egypt through five years' time frame from 2011 to 2015. Eight major risk factors were identified to study their impact on the cost and time of waste water projects. The study claims that cost and time related risks are most likely to occur and have the major influence on replacement and renovation of waste water projects. Furthermore, waste water projects that are totally constructed, new in nature, have low risk impact score. Defective design risk is ranked as the first of the 8 major risk factors and affecting 50% of the waste water projects. The study divided risks in waste water projects into two main categories; technical risks such as defective design and managerial risks such as permits and information unavailability. Finally, in waste water projects, the study concluded that the technical risks are more frequent than managerial risks.

1. INTRODUCTION

Waste water service measures the social and healthy progress of urban/rural communities. On a global level, problems of the waste water projects are the largest source of pollution. Such problems could include: population growth, lack of waste water capacity estimate and inaccurate planning especially in the developing countries. According to the United Nations' Human Development Report, there is a remarkably absence of effective national policies especially in waste water services more than potable water. Also, there is a strong relationship between poor access to drinking water and waste water service and health aspects.

Egypt is one of the early countries that constructed waste water projects in the Middle East. In 1914, sewage main department has been established. The sewage sector was managed by many authorities, but the lack of coordination between authorities caused lots of problems to appear. In 2004, the holding company for potable and waste water was established as an independent entity affiliated to the Ministry of Housing. The holding company is responsible for potable water and waste water sectors. Currently, this vital sector is facing many challenges. The most important challenges are increasing in population and industrial and agricultural activities. Waste water projects can be extremely complex and full with uncertainty. Waste water projects can be categorized into two main categories; First, new construction projects which include: Establishing new waste water networks and Create new treatment plants. Second, replacement and renovation projects which include: Increasing the network capacity or improve the network quality and Increase the network capacity or improve the performance.

In year 2006/2007, the total number of replacement and renovation projects, which listed according to the Egyptian government five year plan, is 222 projects as the year target all over the country. Only 4 projects were completed with 3.8% of the target and 25 projects have been completed, partially, through 2006/2007 then delivered in 2007/2008. It was found that other projects about 193 projects are delayed in the implementation more than one year and some of them delayed to 2011/2012. Risks in waste water projects affect cost and project time. Consequently, that leads to delay benefit of the projects and therefor the unsafe disposal of wastes, increasing pollution and disease. The large construction companies that granted the implementation of waste water projects which might hire some subcontractors. In many cases the subcontractors are eligible to work in those major projects which need physical potential and human experience. Accordingly, that would lead to wasting a lot of money and time without any use. The most critical risk is the lack of clear vision of the project from the beginning which causes modification of project drawings and maps, in addition to other modifications that appear during the work in the field of the project.

Risks come from many sources as incorrect design, project team, site problems, etc. Moreover, the size and complexity of waste water projects increases the risks. Risk and uncertainty cause potential damage consequences to the construction of waste water projects. Cohen et al. (2004) defined risk as the potential for complications and problems with respect to the completion of a project and the achievement of project goals. The Uncertain future event or condition, with occurrence rate greater than 0% but less than 100%, has an effect on, at least, one of the project objectives, schedule, cost and quality. Furthermore, the impact or consequences of this future event must be unexpected or unplanned. It is well accepted that risk can be, effectively, managed to mitigate its adverse impacts on project objectives even if it is inevitable in all project stages. It is important to face these uncertain risks by assessing their impacts on the project objectives.
Managing risks involves identifying, assessing and prioritizing risks. This will occur by monitoring, controlling, and applying managerial resources with coordinated and economical effort. Managing risks minimizes the probability and/or impact of unfortunate events and maximizes the realization of project objectives (Hubbard, 2009). Effective risk management may lead the project manager to several benefits such as identification of favorable alternative course of actions, increased confidence in achieving project objectives, improved chances of success, reduced surprises, more precise estimates (through reduced uncertainty) and reduced duplication of effort (through team awareness of risk control actions). Risk analysis and management continue to be a major feature of the project management. Risk analysis is dealing effectively with uncertainty and unexpected events to achieve project success. The main objectives of this research are:

I. Defining waste water projects in Egypt
II. Explaining waste water projects types
III. Identifying risks that facing this type of projects
IV. Assessing the risk impacts as time increase and cost overrun in waste water projects.

2. DATA DESCRIPTION

There are different types of risks associated with the construction activities. These risks might be physical, environmental, design, logistics, financial, legal, political, construction and management risks. The study area was taken in Damietta, Egypt through five years' time frame from 2011 to 2015. By studying and reviewing all the archived documents of the 222 projects, Table 1 shows that there are 33 types of risk factors affecting potable and waste water projects through 2011 to 2015 which causing delay in project time and cost overrun. Risk can be repeated in the same project more than one time during the various stages of implementation. Most of these risks are depending on each other and affecting on each other. For example, incorrect design risk could lead to inaccurate quantities in the project and mean more project cost and time due to both risks incorrect design and inaccurate quantities.

Table1: Classifications of risk factors that affect the potable and waste water projects with the code of each factor

Risk Class	Risk Factor	Code
Design risk	Defective design (incorrect)	R1
	Rush design	R2
	Awarding the design to unqualified designers	R3
	Not coordinated design (structural, mechanical and electrical)	R4
	Lack of consistency between bill of quantities, drawings and specifications	R5
	Inaccurate quantities	R6
Environmental risk	Difficulty to access the site	R7
	Adverse weather conditions	R8
	Acts of God	R9
Physical risk	Supplying invalid materials	R10
	Change the productive capacity of labor and machinery	R11
	Unavailable labor, materials and equipment	R12
	Occurrence of accidents because of poor safety procedures	R13
Logistics risk	Inaccurate project program	R14
	Poor communications between the home and field offices	R15
	High competition in bids	R16
	Undefined scope of working	R17
Financial risk	Delayed payments on contract	R18
	Financial failure of the contractor	R19
	Exchange rate fluctuation affecting material and equipment price	R20
	Monopolizing of materials due to closure and other unexpected political conditions	R21
Legal risk	Difficulty to get permits	R22
	Legal disputes during the construction phase among the parties of the contract	R23
	Delayed disputes resolutions	R24
	Ambiguity of work legislations	R25
Construction risk	Gaps between the Implementation and the specifications due to misunderstanding of drawings and specifications	R26
	Undocumented change orders	R27
	Design changes	R28
	Actual quantities different from the contract quantities	R29
Management risk	Poor communication between involved parties	R30
	Changes in management ways	R31
	Information unavailability	R32
	Resource management	R33

Waste water projects are one of the infrastructure projects. The life cycle of waste water projects from the origin started by construction then followed by operation and maintenance, replacement and renovation, operation and maintenance then finally demolition. Table 2 shows that replacement and renovation of waste water projects has the largest annually share more than new construction. New construction is more complex and need more detailed studies. Also, new construction should provide land spaces to establish treatment plants and networks. In 2013, waste water projects were 100 projects with new construction projects present 2% of waste water projects while replacement and renovation present 98% of waste water projects.

Studying risk impacts in waste water projects requires assessment of risk facing these types of projects. From the archived documents of Damietta holding company for potable and waste water from 2011 to 2015, the average number of implemented projects was about 155 projects annually. The average number of projects exposed to risk during the same period was about 105 projects annually. Through the 5- years study period, waste water project had 440 projects with replacement and renovation of waste water projects represented 429 projects and new construction projects represented 11 projects. About

68% of projects were exposed to risk. Waste water project have the most shares in the holding company projects. The number of risky projects was 275 projects i.e. 55 projects exposed to risk annually and most of them were replacement and renovation projects.

Table 2: Total annually number of projects and number of waste water projects (new construction and renovation) in Damietta holding company from 2010to 2015

Year	Total number	Waste water projects	New construction projects	Replacement and renovation
2010 / 2011	150	80	1	79
2011 / 2012	110	60	2	58
2012 /2013	180	100	2	98
2013 / 2014	210	130	3	127
2014 / 2015	120	70	3	67

Risk facing waste water projects can be divided into two main categories:

- Technical risks: affecting the project site and depends on design and drawings, they can present in defective design, gaps between implementation and specification, design change and not coordinated design
- Managerial risks: depend on the factors that affect the projects from the company side, such as company policies, communication between individuals and the relation between the company and other authorities. Managerial risk is representing in Poor communication between involved parties, Changes in management style, Lack of data and difficulty to get permits.

3. METHODOLOGY

Because if the dependency of risk factors which represented in Table 1, to overcome this issue a statistical technique called the principal component analysis (PCA) is used to classify the risk factors into categories. PCA is a statistical technique which used to analyze the inter-relationships among a large number of variables and explain these variables in terms of a smaller number of variables (Elsanabary and Gan, 2014). Principal component analysis (PCA) is a multivariate technique that analyzes a data table in which observations are described by several inter-correlated quantitative dependent variable (Abdi, 2010). PCA is useful for finding clusters of related variables and thus ideal for reducing a large number of variables into a more easily understood framework (Badu, 2009).The goals of PCA are, (Abdi, 2010):

- Extract the most important information from the data
- Compress the size of the data set by keeping only this important information.
- Simplify the description of the data set.
- Analyze the structure of the observations and the variables.

There are 8 major risks resulted from PCA as shown in Table 1. These risks are sorted in a descending order with respect to their effectiveness:

- Defective design (incorrect),
- Not coordinated design (structural, mechanical and electrical),
- Difficulty to get permits as roads or traffic permits,
- Gaps between the Implementation and the specifications due to misunderstanding of drawings and specifications,
- Design changes,
- Poor communication between involved parties,

- Changes in management ways,
- Information unavailability.

4. RESULT DISCUSSION

From the PCA results, the 8 risk factors that affect waste water projects with high percentage from 2010 to 2015 are shown in Table 1. Table 3 shows that "defective design" risk factor was affecting , in average, 49% of the total waste water projects annually, started with 68% in 2010 and ended with 36% in 2015. The "Not coordinated design" risk factor affected was affecting, in average, 22% of the total waste water projects annually, started with 23% in 2010 and ended with 9% in 2015. The affected ratio decreased from 2011 to 2015, but the total number of risky projects sill high and hampered the utilization of the project.

The total number of waste water projects in 2010 is close to that of 2015. In 2011, the total number of waste water projects was 80 projects with 55 projects affected by the risks. In 2015, the total number of waste water projects was 70 projects with 40 projects affected by the risks. The percentage decreased but the risk still affecting the projects with high percentage.

Table 3: Approximate occurrence percentage of each of the 8 risk factor in waste water projects (2010 - 2015)

Year	Total no. of projects	Risky project	Defective design	Not coordinated design	Difficulty to get permits	Gaps between Implementation and specifications	Design changes	Poor communication between involved parties	Changes in management ways	Information unavailability
2010/2011	80	69%	68%	23%	23%	44%	23%	24%	18%	19%
2011/2012	60	83%	55%	30%	23%	33%	25%	12%	11%	12%
2012/2013	100	60%	50%	30%	11%	35%	27%	27%	36%	10%
2013/2014	130	54%	38%	19%	8%	19%	10%	15%	46%	7%
2014/2015	70	57%	36%	9%	6%	21%	17%	21%	19%	4%

Major risks caused cost overrun and delaying time in waste water projects are shown in Table 4. For example, defective design affected 54 projects of 80 projects in 2010/2011; it caused 1.2 million Egyptian Pounds (LE) cost overrun and 4.8 years delaying time. In 2012/2013 gaps between the implementation and specifications affected 35 projects out of 100 projects; it caused 1.5 million LE cost overrun and 2.4 years delaying time. From Table.3, "Defective design" risk is the most effective risk of major risks; it affected between 25 and 54 of waste water projects annually from 2010 to 2015. "Defective design" risk caused cost overruns about 1.14 million LE annually and 4 years delaying time in average. "Defective design" risk is depending on the efficiency of design team or the consultant.

Table 4: Cost overrun and delaying time caused by major risks through study years (2010 - 2015)

Year	Risk factors	Defective design	Not coordinated design	Difficulty to get permits	Gaps between Implementation and specifications	Design changes	Poor communication between involved parties	Changes in management ways	Information unavailability	Total
2010/2011	No. of projects	54	18	18	35	27	18	16	15	
	Cost x 10^6 LE	1.2	0.12	0.12	1.2	0.9	0.3	0.1	0.3	4.24
	Time (years)	4.8	3	1.8	3	1.8	1.2	0.9	0.9	17.4
2011/2012	No. of projects	33	18	14	20	15	7	6	7	
	Cost x 10^6 LE	0.9	0.06	0.09	0.6	0.6	0.12	0.06	0.18	2.61
	Time (years)	3.6	2.4	1.8	1.8	1.2	0.9	0.6	0.6	12.9
2012/2013	No. of projects	50	30	11	35	27	27	18	10	
	Cost x 10^6 LE	1.2	0.15	0.09	1.5	1.2	0.3	0.15	0.18	4.77
	Time (years)	4.2	3.6	1.2	2.4	1.8	1.8	1.2	0.6	16.8
2013/2014	No. of projects	50	25	10	25	13	20	18	9	
	Cost x 10^6 LE	1.5	0.12	0.12	0.6	0.6	0.3	0.12	0.15	3.51
	Time (years)	4.2	3	2.4	2.4	1.2	1.2	0.9	0.6	15.9
2014/2015	No. of projects	25	6	4	15	12	15	7	3	
	Cost x 10^6 LE	0.9	0	0.06	0.3	0.6	0.3	0.06	0.12	2.34
	Time (years)	3	1.8	1.2	2.4	1.2	0.9	0.6	0.6	11.7

4.1. Risk Impact on Project Cost

To improve controlling of cost variation for constructing waste water projects in Egypt; one must identify and recognize the influence of the main factors affecting it (Aziz, 2013). By collecting data from the company document archive, the rate of increasing cost in waste water projects due to the 8 major risk factors is about 3.5 million LE, in average, annually, as shown in Figure1. The rate of cost overrun was varying irregularly through the study years. The highest cost overrun due to risk factors was 4.77million LE during 2012 /2013, about 60% of waste water projects were affected by risk. In 2014 / 2015 the percentage of projects that were affected by risk was 57%, which caused cost overrun of about 2.34 million LE. During 2012 / 2013 an increase in risky projects percentage was noticed, by 3% difference, from 2014 / 2015 which is quite small difference while the cost overrun increased by 2.43 million LE more than that of 2014 / 2015. The cost overrun does not depend on the number of projects only, but also depends on the cost of the projects and the degree of risk affecting. Projects with expensive budget are exposed to more cost overruns due to more risk exposure.

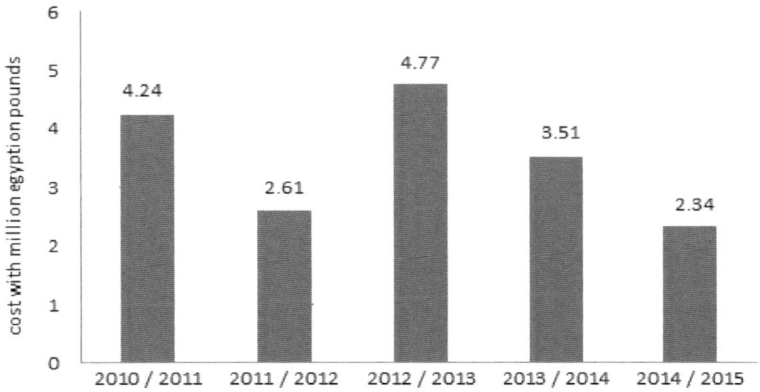

Figure1: Cost overrun due to the 8 risk factors from 2010 to 2015

It is noticed from Table 5 that the technical risks, such as "defective design" risk, have a high impact on the projects cost. Through the study period, "Defective design", "Gaps between the implementation and specification" and "Design changes" risks caused cost overrun of about 1.14, 0.84 and 0.78 million LE annually, respectively. However, managerial risks as "Poor communication between parties" risk increased cost overrun by 0.26 million LE annually and considered the largest cost overrun due to managerial risks. Therefore, technical risks cause more cost overrun than the managerial risks in waste water projects.

Table 5: Total cost overruns as an impact of each risk factor from 2010 to 2015

Risk factors	Total increase in cost (million LE)
Defective design	5.7
Gaps between the Implementation and specifications	4.2
Design changes	3.9
Poor communication between involved parties	1.32
Information unavailability	0.93
Changes in management ways	0.49
Difficulty to get permits	0.48
Not coordinated design	0.45

4.2. Risk Impact on Project Time

The rate of delaying time of waste water projects annually due to the 8 major risk factors is about 15 years as shown in Figure 2. The rate of delaying time is changing irregularly through the study period. The longest delayed time due to the risk factors was 17.4 years in 2010 /2011 where about 69% of waste water projects were affected by risk. During 2011 / 2012 the delayed time due to the risk factors was 12.9 years where about 83% of waste water projects were affected by risk. Year 2011 / 2012 increased in risky projects percentage from 2010 /2011 with 14% increase, but still 2010 / 2011 has the largest time delaying in waste water projects due to risk factors. The time delaying is not depending on the number of projects only but also depending on the number of risks affecting on the project.

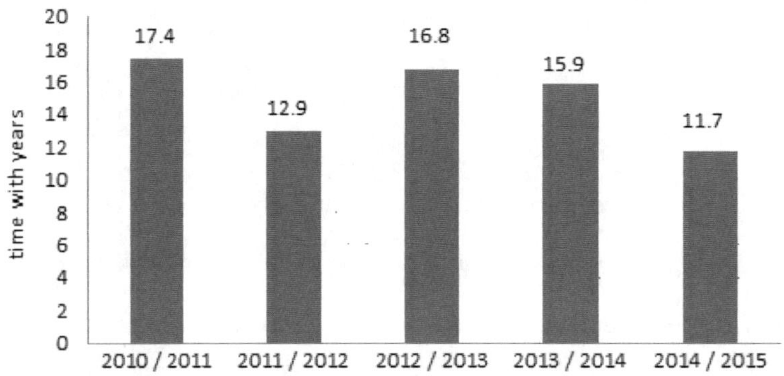

Figure 2: Time delay due to the 8 risk factors from 2010 to 2015It is noticed from Table 6 that the technical risks, such as "defective design" risk, have a time delay of 19.8 year during the study period. Through the study period, "Not coordinated design", "Gaps between the implementation and specification" and "Design changes" risks caused delaying time by 13.8, 12 and 7.2 years, respectively. However, managerial risks as "Difficulty to get permits" risk caused delaying time by 8.4 years from 2010 to 2015 and considered the highest time delaying due to managerial risks. Therefore, managerial risks cause more time delaying than the technical risks in waste water projects.

Table 6: Total delayed time as an impact of each risk factor from 2010 to 2015

Risk factors	**Total delay in time (Years)**
Defective design	19.8
Not coordinated design	13.8
Gaps between Implementation and specifications	12
Difficulty to get permits	8.4
Design changes	7.2
Poor communication between involved parties	6
Changes in management styles	4.2
Data unavailability	3.3

5. CONCLUSIONS AND SUMMARY

Waste water projects are more sensitive to risk during the implementation stage more than other types of projects. Replacement and renovation type of waste water projects is more affected by risk than the new construction type. The replacement and renovation is used to increase the capacity of the network or the station. The risk impact in waste water projects appears in cost overrun and the project time delay.

The study area was taken in Damietta, Egypt through five years' time frame from 2011 to 2015. By studying and reviewing all the archived documents of the 222 projects. Out of 33 risk factors, 8 major risks factors were identified based on the principal component analysis. "Defective design" risk has the highest effect between the 8 major risk factors in cost and projects time. "Defective design" risk caused costs overrun of about 1.14 million LE yearly and 4 years delaying time in waste water projects. "Defective design" risk is depending on the design team or the consultant. "Defective design" risk doesn't depend on any external factor as; site nature, environment condition and management system. The company should study the risk factor and find solution to decrease its occurrence.

The 8 major risk factors classified to technical risks and managerial risks. Technical risks have major impact on both project time and cost, while managerial risks have more impact on the project time than project cost. Risk impact, not only depend on the number of projects affected by the risk but also depend on several factors such as projects priority, project cost and the degree of risk affecting the project. Risk affects differently in two projects with same condition, budget and time. Projects with high priority and expensive budget are exposed to more impact due to risk.

Technical risks are depending on site problems, such as design and drawing misunderstanding. Technical risks are impact both the cost and project time. Decision maker in the company should give more attention to the design team or choose responsible consultant for the design and site works. Managerial risks are depending on the company strategy in management as the way of giving and receiving orders between projects parties and changing management way. Managerial risks are also depending on relationship between company and other governmental authorities. Managerial risks have major impact on project time. The decision maker in Damietta holding company should review the management way, providing facilities to decrease the effect of managerial risks on both cost and time.

Finally, the Egyptian government should give more attention to waste water projects and increase the approved budget for them. Government should prepare effective plans to cover the rural areas with waste water service. The replacement and renovation of waste water projects are considered the most frequently projects between the company projects. There are some studies must be done to avoid the risk impacts in replacement and renovation type including:

- Cadastral and contour maps to determine the size of the required served area
- Detailed maps of the diameters and depths of the existing network
- Identifying regression lines and expulsion lines.
- Identifying the obstructions such as gas pipes or electrical cables.
- Considering the expected population growth in the future to determine the expansion in stations and network.

6. REFERENCES

Abdel-Shafy, H., and Aly, R. 2002. Water issue in Egypt: Resources, pollution and protection endeavors. Navigation, 49(3.1), 4-6.

Abdi, H., and Williams, L. J. 2010. Principal component analysis. Wiley Interdisciplinary Reviews: Computational Statistics, 2(4), 433-459.

Arias, J. C. 2011.Review of Risk Management Methods.Volume 4-Number 1-January 2011-Semiannual Publication, 4(1), 59.

Aziz, R. F. 2013. Factors causing cost variation for constructing wastewater projects in Egypt. Alexandria engineering journal, 52(1), 51-66

Bakr, A. F., El Hagla, K., and Rawash, A. N. A. 2012. Heuristic approach for risk assessment modeling: EPCCM application (Engineer Procure Construct Contract Management). Alexandria Engineering Journal, 51(4), 305-323.

Bedford, T., and Cooke, R. 2001.Probabilistic risk analysis: foundations and methods. Cambridge University Press.

Berg, H. P. 2010. Risk management: procedures, methods and experiences. Risk Manage, 1, 79-95.

Cohen, M. W., and Palmer, G. R. 2004. Project risk identification and management, AACE International

Elsanabary, M. H., and Gan, T. Y. 2014. Weekly streamflow forecasting using a statistical disaggregation model for the Upper Blue Nile basin, Ethiopia. Journal of Hydrologic Engineering, 20(5), 04014064.

Hillson, D. 2003. Effective opportunity management for projects: Exploiting positive risk. CRC Press

Hillson, D. 2006. Integrated risk management as a framework for organizational success. In PMI Global Congress proceedings.

Hubbard, D. W. 2009. The failure of risk management: Why it's broken and how to fix it. John Wiley & Sons.Transaction, pp. I013-15. Retrieved September, 10, 2006.

Leadership in Sustainable Infrastructure

Leadership en Infrastructures Durables

Vancouver, Canada

May 31 – June 3, 2017/ *Mai 31 – Juin 3, 2017*

SCOUR IN COMPLEX BRIDGE PIERS: FRASER AND PADMA RIVERS

Jose A. Vasquez[1,2], Kara I. Hurtig[1], Matt S. Gellis[1], Andre Zimmerman[1], Bruce W. Walsh[1].

[1] Northwest Hydraulic Consultants Ltd. (NHC), North Vancouver, British Columbia, Canada

[2] jvasquez@nhcweb.com

ABSTRACT

The construction of complex bridge piers in large sand-bed rivers has become increasingly common. These piers are made by driving or boring a group of piles deep into the riverbed and then connecting them by a pile cap on top, over which the stem of the pier supporting the deck is located. In contrast with conventional footings, the pile cap is located high above the riverbed and close to the water surface, functioning also as protection against ship collision. The combination of several piles, pile cap and stem above the riverbed gives the pier a complex geometry, which does not easily fit with the simple geometry commonly assumed by most scour prediction equations; hence the need for mobile-bed physical modelling in order to determine scour depths for design purposes. We report the results of several complex piers scour experiments carried out at Northwest Hydraulic Consultants' laboratory including the Golden Ears Bridge, Port Mann Bridge and Pattullo Bridge in the Fraser River, British Columbia and the Padma River Bridge in Bangladesh. These pier tests encompass a wide range of conditions such as vertical and inclined piles (diameters from 1.8 to 3.0 m), rectangular, octagonal and dumbbell-shaped pile caps (pile cap lengths between 18 and 60 m), and flow discharges ranging from 2-year to 100-year floods. Maximum observed local scour depths varied between 14 and 22 m.

Keywords: Bridge; pier scour; complex pier; physical modelling; Fraser River.

1. INTRODUCTION

It is well-known that the most common cause of bridge failure is due to the erosion of riverbed material surrounding the foundations of piers and abutments (Arneson et al. 2012). The local erosion or scour of solid piers, like the cylindrical pier shown in Figure 1.a has been thoroughly studied and is well understood. In the horizontal plane, water flowing around the cylindrical pier typically detaches and creates wake vortices behind it, while in the vertical plane a downflow current generated at the upstream nose of the pier directs water towards the bed. The combination of downflow current and flow around the pier creates a complex three-dimensional horseshoe vortex, which removes bed sediment and generates a scour hole around the pier (Figure 1.a). The Transportation Association of Canada (TAC 2004) provides empirical equations to estimate scour depths in solid piers of different shapes. However, in large sand bed rivers bridge piers typically have complex shapes consisting of several clearly definable components (Figure 1.b), for which TAC (2004) only provides some interim guidance.

Complex piers (Figure 1.b) are typically built by driving or boring a group of piles deep into the riverbed and then connecting them by a pile cap on top, over which the stem or column of the pier supporting the deck is located (Figure 2). In the United States, the Federal Highway Administration (Arneson et al. 2012) and the Florida Department of Transportation (FDOT 2005) provide empirical methods to estimate scour in complex piers. The basic simplifiying assumption in these methods is that scour of each individual component (column/stem, pile cap and pile group) can be computed independently and then summed together to find the total local scour in the complex pier, in some way neglecting the possible interaction of the flow structure between the components, especially the downflow current (Moreno et al. 2014).

a) Solid pier

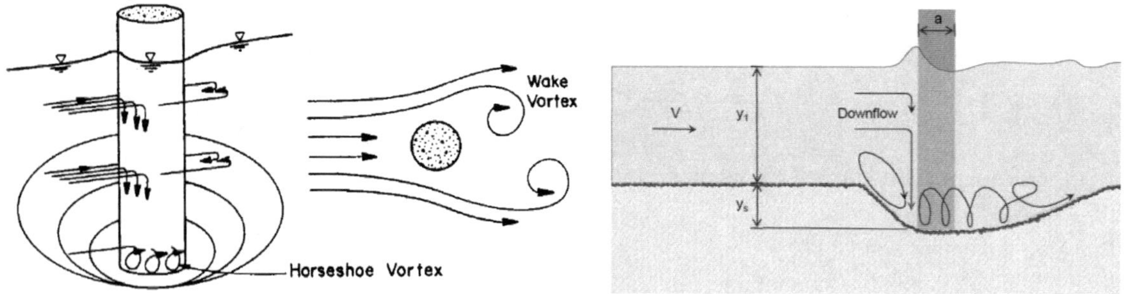

b) Complex pier

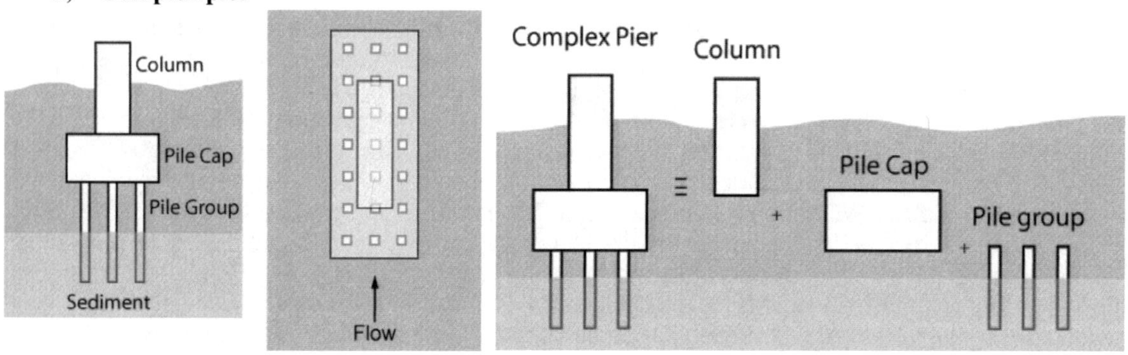

Figure 1. (a) Scour in single solid cylindrical pier (Arneson et al. 2012); and (b) scour-producing components of complex pier (FDOT 2005).

Although complex piers are sensible and cost effective from a structural standpoint, their shape can present a challenge for hydrotechnical engineers responsible for estimating design scour depths at these structures (FDOT 2005). Figure 2.a illustrates the contrast between the old solid pier of the Port Mann Bridge in the Fraser River and its complex pier replacement built downstream. The old solid pier had a hydrodynamic shape in the form of an elongated ellipse aligned with the main flow direction. The complex pier replacement is rectangular in shape and aligned with the bridge deck, not the flow, which further complicates the application of empirical scour equations.

a) Port Mann Bridge **b) Pitt River Bridge**

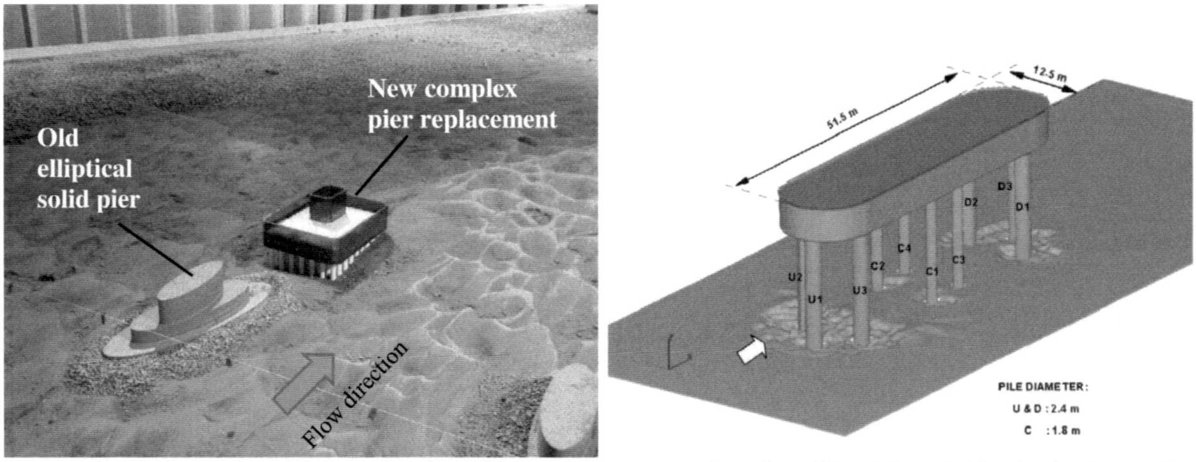

Figure 2. (a) Physical model showing old solid pier and new complex pier of Port Mann Bridge in the Fraser River (NHC 2009a); and (b) initial scour around the complex pier of Pitt River Bridge computed using Flow-3D (Vasquez and Walsh 2009).

As shown in Table 1, complex piers are becoming the norm in recent bridge crossings built during the past decade in the Fraser River and the Pitt River in Metro Vancouver, BC (Figure 3). The same is true in other large sand-bed rivers, such as the Padma River in Bangladesh, where the largest river crossing in the world is currently under construction. The apparent popularity of complex piers in large rivers is not due to any particular hydrotechnical advantage, but it is probably more related to advances in pile driving technology and the risk of soil liquefaction during seismic events (Hannigan et al. 2016). In the navigable Fraser and Padma Rivers, the pile cap is located at the water surface to provide protection against ship collision. Therefore, the column or pier stem above the pile cap remains dry and can be disregarded as a scour-producing component.

Table 1. Examples of complex pier dimensions in the Fraser, Pitt and Padma Rivers (NHC 2006, 2009a,b; Vasquez 2007; Vasquez and Walsh 2009)

Bridge	Completed	Pile cap length (m)	Pile cap width (m)	Pile diameter (m)	Number of piles (-)	Figure
Golden Ears Bridge	2009	44.2	18.8	2.4	12	Figure 4
Pitt River Bridge	2009	51.5	12.5	2.4/1.8	6/4	Figure 2.b
Port Mann Bridge	2012	41.4	32.4	1.8	63	Figure 2.a
Padma River Bridge	Construction	16.1	16.1	3.0	6	Figure 6
New Pattullo Bridge	Proposed	26.5	17.5	1.8	24	-

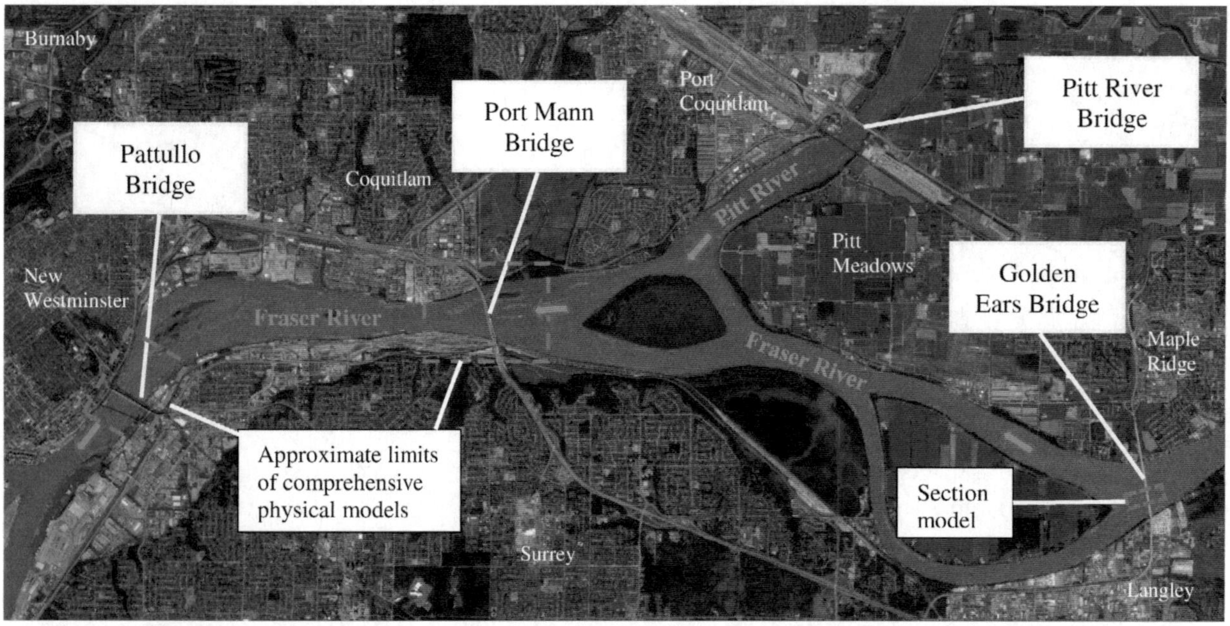

Figure 3. Google Earth image of Fraser River showing location of bridge crossings mentioned in Table 1 and approximate limits of movable-bed physical models used to study local pier scour.

The purpose of this paper is to describe some observations made in physical mobile-bed models of local scour in complex piers in the Fraser River and Padma River.

2. SCOUR MODELLING FOR COMPLEX PIERS

Although numerical models have been successfully applied to compute scour in solid cylindrical piers since the 1990s (Olsen and Melaaen 1993); the same is not true for complex piers due to their shape and size. Figure 2.b shows an early attempt by Vasquez and Walsh (2009) to simulate scour in the complex pier of the Pitt River Bridge using the computational fluid dynamics (CFD) code Flow-3D version 9. Flow-3D correctly predicted the flow interaction between the pile groups and the initial scour development, but the computational time needed to compute the maximum equilibrium scour was excessive. Also, in tests with a simpler 3-pile group under tidal conditions (Vasquez and Walsh 2009) the CFD model showed a tendency to under-predict the maximum scour depth. At present, CFD models are still not capable of accurately and efficiently predicting local scour in complex piers, leaving physical modelling as the only viable option. However, numerical models still play an important role in providing boundary conditions to physical models, such that the size of the physical model can be reduced.

Physical models intended to study local pier scour can be classified as comprehensive models if they include the entire bridge crossing and piers, or section (flume) models if they only include a single pier. Comprehensive models are needed in complex crossings such as when more than one bridge is present, which was the case for the Port Mann and Pattullo Bridge replacements (NHC 2009a; Figure 2.a and Figure 3). The main disadvantage of mobile-bed comprehensive models is their large size and complex operation. Section models are simpler to operate, less expensive and are appropriate when the pier is not influenced by other nearby structures. Water depth and velocity in a section model can be changed to represent local flow conditions at individual piers. Similarly, the pier orientation can be rotated to simulate various angles of attack. This approach was used to model complex piers of the Golden Ears Bridge (NHC 2006; Vasquez et al. 2007; Figure 3; Figure 4) and Padma River Bridge (Figure 7); in both cases the local boundary conditions to the section model were provided by a two-dimensional flow model.

One important challenge when modelling sand-bed rivers using reduced-scale physical models is the impossibility of scaling sand grain sizes geometrically. For example, the 0.35 mm sand grain in the Fraser River, if scaled geometrically in a 1 to 100 reduced-scale physical model will require 0.0035 mm silt in the model, which is cohesive and hence behaves differently from sand. For that reason, using sand in a physical model may be preferred although it may be more representative of a gravel-bed river. Fortunately, local scour is mostly dependent on the pier geometry and not so much on the grain size. However, the general bed mobility and sediment transport in the model becomes very low and bed features such as dunes do not fully develop. One option is to use light-weight material in the model to represent sand in the prototype, but this is complicated and requires considerable experience (Shen 1990, Vollmers 1990). The hydraulics laboratories in Northwest Hydraulic Consultants (NHC) have been using crushed walnut shell over the years, an industrial abrasive with relative submerged density of 0.3 (compared to 1.65 for natural sand) to model sand bed rivers. Crushed walnut shell allows the formation of realistic dune patterns (Figure 2.a), which is particularly important when studying the stability of riprap in the riverbed used as scour protection in piers and pipeline crossings (Vasquez et al. 2007). All the results presented here come from mobile bed models that used crushed walnut shell to model sand.

3. FRASER RIVER

3.1 Setting

The Fraser River is the longest river in British Columbia, draining a 250,000 km^2 watershed and flowing for 1,375 km into the Strait of Georgia at the city of Vancouver. The Fraser's mean annual discharge at its mouth is 3,550 m^3/s and it discharges 20 million tons of sediment into the Pacific Ocean. Near the town of Mission, about 85 km from the ocean the river transitions from gravel to sand. The sand-bed reach continues for 50 km down to the head of the alluvial delta at New Westminster, below Port Mann Bridge (Figure 3). From there the Fraser delta flows for 35 km to the ocean (McLean et al. 1999). The median sand grain size in the riverbed is 0.35 mm, and dunes typically range from 3 m to 5 m high in the lower reaches.

The flood of record in the Fraser occurred in 1894 and was estimated to reach 17,200 m^3/s at Mission (McLean et al. 1999) and 21,150 m^3/s at Pattulo Bridge (Figure 3). This flood of record is commonly used for testing local pier scour in physical models. Because the Fraser River below Mission is tidally influenced and the river planform morphology exhibits several bends, islands, bars, contractions and expansions (Figure 3), defining boundary conditions for physical models would be challenging if not for the use of large-scale numerical models. A one-dimensional (1D) unsteady flow model of the entire Fraser River below Mission is used to provide boundary conditions to local two-dimensional (2D) models. The 2D model, which is typically at least 10 km long and includes the entire extent of the physical model, is then used to provide the velocity distribution across the entrance section of the physical model (headbox) and water levels at the downstream section; in this way reducing the size and cost of the physical model. In some instances and when available, ADCP field velocity data can also be used to provide inflow boundary conditions.

3.2 Golden Ears Bridge

Golden Ears Bridge, which is located 54 km upstream from the sea, was completed in 2009 to link the municipalities of Surrey and Langley on the south side of the Fraser River with Maple Ridge and Pitt Meadows on the north side (Figure 3). The 6-lane bridge is a 968 m long cable-stayed structure supported by 4 towers, spaced 242 m apart. The towers are H-shaped with two legs; each leg transfers load to a hexagonal pile cap and from there to the ground through 6 cylindrical piles. The two hexagonal pile caps are connected by a rectangular element, making a single dumbbell-shaped structure at each tower (Table 1, Figure 4).

An 11-km long reach of the Fraser River around the bridge crossing was modelled with the 2D flow model River2D to compute the local flow conditions (depth, velocity and angle of attack) at each pier (Vasquez et al. 2007), which was used as a boundary condition for a 1:100 scaled section model of an individual pier. The most adverse flow conditions that led to the maximum scour depth of 14 m (Figure 4), were for a flow depth of 13 m, flow velocity of 2.5 m/s and angle of attack of 6 degrees. Due to the skewed alignment (angle of attack), maximum scour depth occurred at a downstream pile directly attacked by the flow.

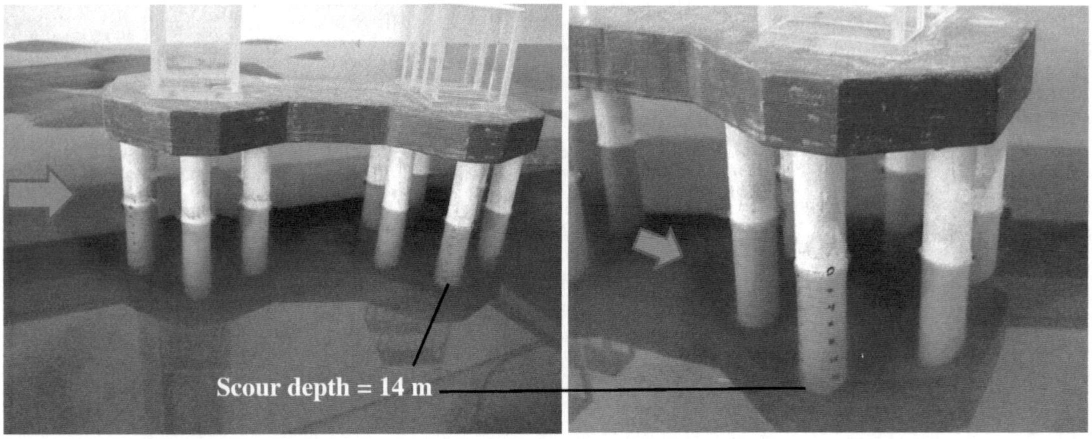

Figure 4. Physical model of Golden Ears Pier showing maximum scour depth of 14 m at downstream pile (drained model with water surface matching initial bed level elevation)

3.3 Port Mann Bridge

The Port Mann Bridge is a 10-lane cable-stayed bridge in the Trans-Canada Highway that opened to traffic in 2012, connecting Coquitlam to Surrey over the Fraser River (Figure 3). It is currently the second longest cable-stayed bridge in North America, handling a traffic volume in excess of 100,000 trips per day. The main free span between towers is 470 m long. The new bridge replaced a steel arch bridge that was located upstream until fully removed in 2015. Pier N1 is the main in-channel pier and initially located downstream from an existing pier of the old bridge (Table 1, Figure 2.a). Pre-construction bed levels around pier N1 were around El. -11 m.

A 1:75 scale comprehensive physical model of Port Mann Bridge plus a 2 km long reach of the Fraser River (Figure 3) was built to study local pier scour during the construction phase, when the old and new bridge structures were present, and during normal operation with the old bridge removed (NHC 2009a). The riprap scour protection was also studied in the physical model. Boundary conditions were provided by a 12-km long River2D model. Approximate local flow conditions at Pier N1 were 14 m water depth and 2.4 m/s flow velocity. Results from the physical model showed that if left unprotected (i.e. without riprap), local scour will cause bed levels at Pier N1 to reach El. -28 m (Figure 5), equivalent to 17 m of scour depth.

Figure 5. Comprehensive physical model of Port Mann showing scour at Pier N1 (NHC 2009a)

3.4 Pattullo Bridge Replacement

The existing Pattullo Bridge opened to traffic in 1937 to link New Westminster and Surrey over the Fraser River (Figure 3). It is located right downstream from the New Westminster Rail Bridge. A new bridge replacement has been proposed for Pattullo Bridge, to be located upstream of the Rail Bridge and is currently being studied in a 1:80 comprehensive physical model. Preliminary testing of two bridge concepts that include a complex bridge pier design has resulted in maximum local scour of 18 m for an unprotected pier (NHC 2017).

4. PADMA RIVER BRIDGE

4.1 The Padma River

The Padma River forms by the confluence of the Brahmaputra (Jamuna) and Ganges Rivers in Bangladesh, both originating in the Himalayas. The Padma River flows for 100 km until its confluence with the Lower Meghna River, which discharges 90 km downstream into the Bay of Bengal in the Indian Ocean. The mean annual flow of the Padma River, around 30,000 m^3/s, is the third largest in the world. Bankfull discharges vary between 60,000 and 90,000 m^3/s with an approximate mean value of 76,000 m^3/s. Considering the effects of climate change, the 100-year and 500-year floods have been estimated at 148,000 m^3/s and 160,000 m^3/s respectively (McLean et al. 2012).

The Padma River is one of the most morphodynamically active rivers in the world; over the years its planform geometry changes almost cyclically between straight, meandering and braided morphologies. Channel widths have been observed to vary between 1.5 and 10 km. Annual sediment loads have been estimated to be around 1 billion tonne/year; bed material is fine 0.11 mm sand (McLean et al. 2012).

4.2 The Padma River Bridge

At present, a 6.15 km long bridge is under construction over the Padma River, 35 km southwest of Dhaka. The bridge is made of 41 spans, 150 m long between piers. The complex pier has 6 inclined piles each of 3 m diameter, raked at 1 horizontal to 6 vertical in circular arrangement (Table 1). Natural scour depths in the area near the bridge section have been observed to reach El. -50 m. Three mobile-bed physical models were used to study different scour aspects (Vasquez et al. 2012). A comprehensive physical model was used to study scour around the river training works (RTW); a section model was built to study scour along the south RTW; while another section model was used to study local pier scour, which is reported here. Four pier options, shown in Figure 6 and summarized in Table 2, were tested in the 1:80 scale section model under various flow conditions and angles of attack. Bed levels in the section model were read using a laser scanner, examples are shown in Figure 7.

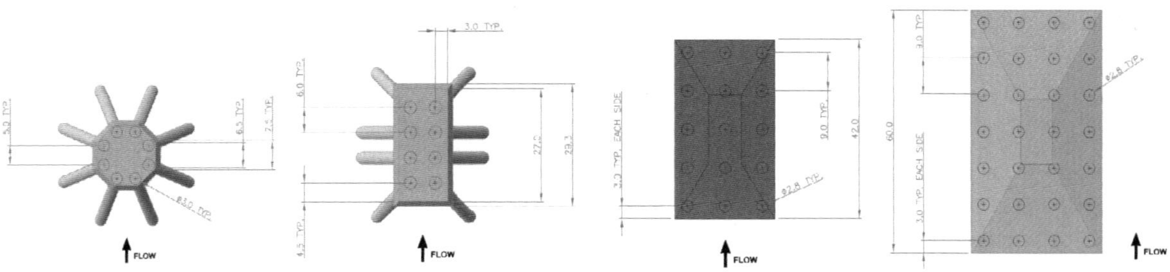

S8 **R8** **B15** **B28**

Figure 6. Plan views of four pier options for Padma Bridge tested in physical model (NHC 2009b)

4.3 Pier scour section model

In order to take into account the effects of general bed scour in the Padma River during floods, bed levels in the section models were set to El. -21 m for the 2-year general scour and El. -50 m for the 100-year general scour, corresponding water levels were El. 5.7 m and El. 7.3 m, respectively. The scour around individual piles varied considerably, but was always deepest in the vicinity of the farthest downstream piles (Figure 7). The location of the maximum scour consistently occurred in the same piles regardless of the bed at either El. -21 m or El. -50 m.

Table 2. Maximum scour depths for single pile and four pier options of the Padma Bridge measured in physical section model (NHC 2009b)

Pier	Pile group layout	Pile cap width (m)	Pile cap length (m)	Pile diameter (m)	Scour depth (m)
S1*	1 single pile without pile cap	-	-		7
R8	8 inclined piles in 2 × 4 grid	14.3	29.3	3.0	17
B15	15 vertical piles in 3 × 5 grid	24.0	42.0	2.8	21
B28	28 vertical piles in 4 × 7 grid	33.0	60.0	2.8	22
S8	8 inclined piles in circular pattern	18.1	18.1	3.0	16

*Used for reference

4.3.1 Pier options

For the **S8 pier** design, the physical model results indicate 14 to 16 m of local scour in the vicinity of the farthest downstream piles, and 11 to 13 m as an average depth among all 8 piles. Whether the starting bed elevation was at El. -21 m or El. -50 m had no significant effect. Also attack angles of 0, 11.25 and 22.5 degrees produced closely similar local scour depths.

For the **R8 pier** design with attack angles of 0 and 22.5 degrees, local scour depths of 13 m and 17 m, respectively were observed in the vicinity of the farthest downstream pile, while the average depths among all piles were 11 m and 12 m respectively. Scour was clearly deeper at the 22.5 degrees attack angle.

In comparison with the S8 and R8 raked-pile arrangements, the **B15 and B28** vertical-pile arrangements generated significantly more scour. The maximum local scour depth at an individual pile ranged between 18 and 21 m for the B15 design and exceeded 22 m for the B28 design. The average scour depth was between 16 to 17 m for the B15 design and about 20 m for the B28 design. With the B15 design, an attack angle of 11.25 degrees produced about 2 m more scour than angles of 0 or 22.5 degrees. With the B28 design, attack angles of 0 and 22.5 degrees produced the same depth of scour.

Based on these observations, it seems that piles arranged in circular or ring arrangements such as S8, given their symmetrical layout are rather insensitive to the angle of attack compared to piles in rectangular arrangements.

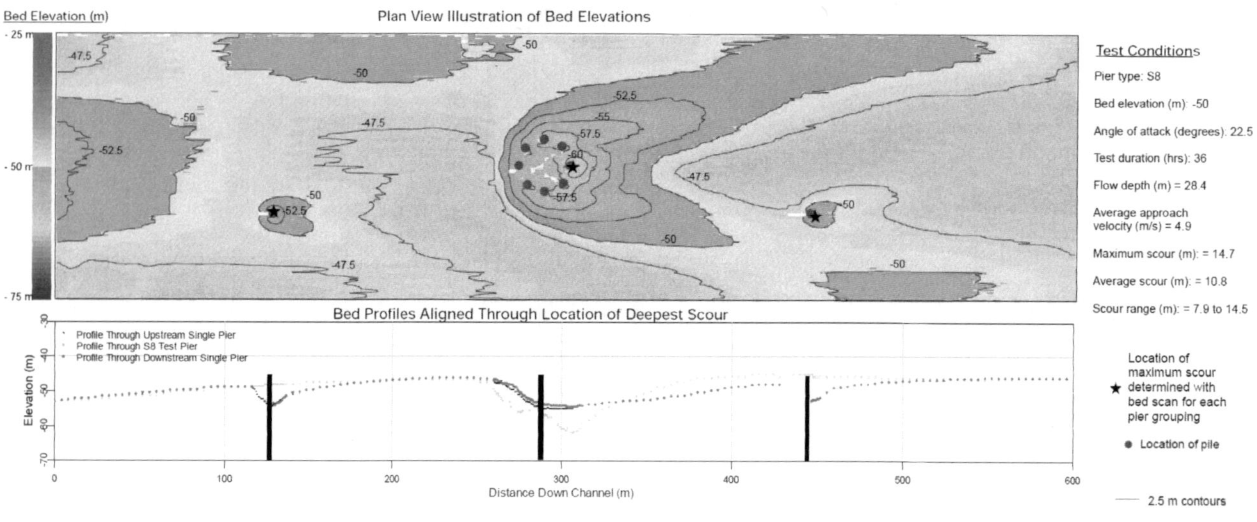

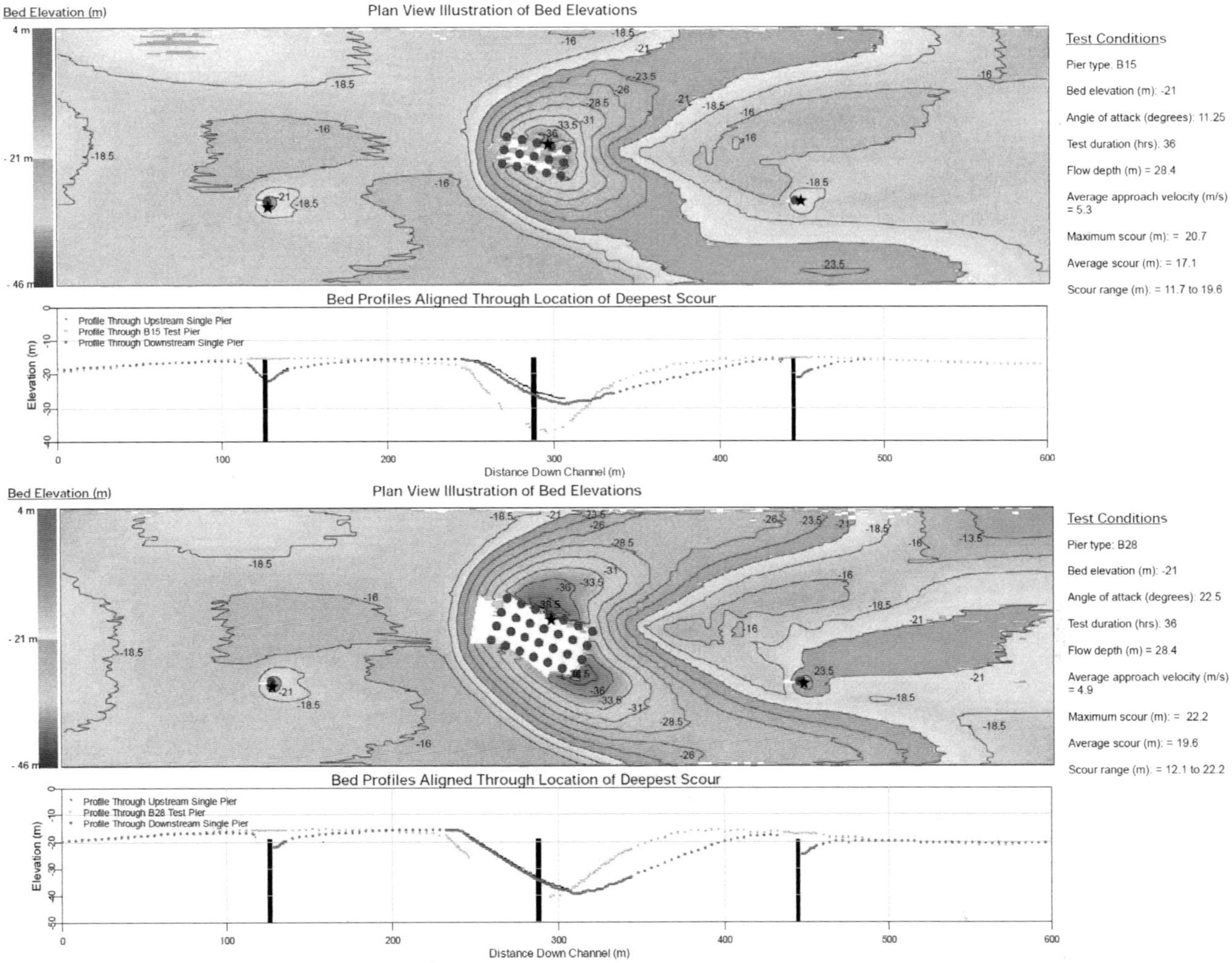

Figure 7. Examples of bed level measurements made by laser scanner for pier options S8, B15 and B28.

4.3.2 Raked piles in Pier S8

Raked piles driven on an inclination from vertical are used when large static loads are expected (Hannigan et al. 2016). However, the effect of raked piles on complex pier scour has not been investigated in great detail, and the available empirical equations assume vertical piles. During the S8 physical model tests, it became evident that the rake on the downstream piles could increase local scour. Scour reached 16 m at the downstream piles compared with 9 m at the most upstream pile (Figure 8.a).

It can be hypothesised that the vertical angle of the pile affects how the downflow current dives into the bed, increasing or decreasing scour. Figure 8.b shows this effects in solid pier, which seems to agree with the observations for Pier S8.

The real pier geometry currently under construction in the Padma River was not tested in the physical model, but it is a 6-pile arrangement somewhat similar in shape to pier S8 and it is expected that scour depths would be similar.

a) Pier S8 scour **b) Effect of pier inclination on scour depth (TAC 2004)**

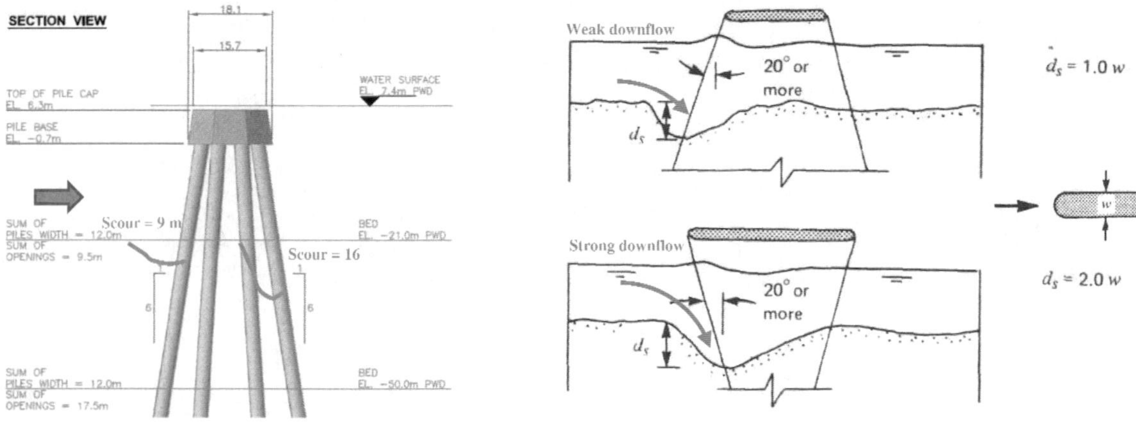

Figure 8. Influence of pier face inclination on downflow current and scour depth that could explain scour observed in complex pier S8 with raked piles.

5. SUMMARY AND CONCLUSIONS

The construction of complex piers in bridge structures has become increasingly common, as demonstrated by the recent bridge projects in the Fraser and Pitt rivers in BC, and Padma River in Bangladesh (Table 1). In contrast with conventional solid piers, complex piers are made of several piles connected on top by a pile cap over which the pier column is located (Figure 1). Given their complex geometry, empirical equations are less reliable to estimate maximum scour depth and physical modelling is required to model local scour in important crossings.

Because complex piers are seldom aligned with the flow, maximum scour depths typically tend to occur in downstream piles. Maximum scour depths in the various complex piers that were modelled ranged between 14 and 22 m. In relative terms, maximum scour depths varied between 4.3 and 10.0 times the pile diameter, which is much larger than the typical range of 1.5 to 2.7 times observed in a single solid cylindrical pile (Figure 1a). This increased pile scour in complex piers can be attributed to the grouping effect of several surrounding piles plus the pressurized flow effect under the pile cap.

Scour in complex piers with rectangular pile arrangements seem to be more sensitive to the angle of attack than piers where piles are arranged in more symmetrical circular (ring) arrangements.

Scour at the downstream piles of raked piles groups can be considerably deeper, which may be attributable to the vertical angle of the piles affecting downflow currents.

6. REFERENCES

Arneson, L.A., Zevenbergen, L.W., Lagasse, P.F.,and Clopper, P.E. 2012. *Evaluating Scour at Bridges – Fifth Edition. Hydraulic Engineering Circular No. 18*, Rep. No. FHWA-HIF-12-003 HEC-18, U.S. Department of Transportation, Federal Highway Administration, Washington, DC; April 2012.

FDOT 2005. *Bridge Scour Manual.* Florida Department of Transportation. May 2005.

Hannigan, P.J., Rausche, F., Likins, G.E., Robinson, B.R., and Becker, M.L. 2016. *Geotechnical Engineering Circular No. 12 – Volume I Design and Construction of Driven Pile Foundations*. National Highway Institute, U.S. Department of Transportation, Federal Highway Administration, Washington, DC.

McLean, D.G., Church, M. and Tassone, B. 1999. Sediment Transport along Lower Fraser River: 1 Measurements and Hydraulic Computations. *Water Resources Research*, 35(8): 2533-2548. August 1999.

McLean, D.G., Vasquez, J.A., Oberhagemann, K. and Sarker, M.H. 2012. Padma River Morphodynamics near Pardma River Bridge. *International Conference on Fluvial Hydraulics, River Flow 2012*, Vol. (1): 741-747. San José, Costa Rica. Sep. 5-7, 2012.

Moreno, M., Maia, R., Couto, L. and Cardoso, A., 2014. Contribution of complex pier components on local scour depth. In *3rd IAHR Europe Congress*, Portugal.
NHC 2006. *Golden Ears Bridge, Hydraulic Modelling*, Final Report prepared by Northwest Hydraulic Consultants Ltd. (NHC) for Golden Crossing Group.
NHC 2009a. *Port Mann Bridge Hydrotechnical Assessment Construction and Ecohydraulics*. Final Report prepared by Northwest Hydraulic Consultants Ltd. (NHC) for Kiewit-Flatiron General Partnership, May 2009.
NHC 2009b. *Padma Bridge Project – Pier Scour Physical Model Study*. Final Report prepared by Northwest Hydraulic Consultants Ltd. (NHC) for Maunsell Ltd., December 2009.
NHC 2017. *Pattullo Bridge Replacement Project, Physical Model Dtage 1 Reference Concept – Preliminary Tests Results Summary*. Technical Memorandum prepared by Northwest Hydraulic Consultants Ltd. (NHC) for Parsons Corporation, January 2016.
Olsen, N.R. and Melaaen, M.C., 1993. Three-dimensional calculation of scour around cylinders. *Journal of Hydraulic Engineering*, 119(9):1048-1054.
Shen, H.W. (1990). Preface to NATO Workshop on Movable Bed Physical Models. Proceedings of the *NATO Advanced Research Workshop on Movable Bed Physical Models*. Delft, The Netherlands. August 18-21, 1987.
TAC 2004. *Guide to Bridge Hydraulics - Second Edition*, Transportation Association of Canada. Thomas Telford Publishing.
Vasquez, J.A., McLean, D.G., and Walsh, B.W. 2007. Modelling Scour and Riprap Protection in Golden Ears Bridge. Proceedings of the *18th Canadian Hydrotechnical Conference*, Canadian Society for Civil Engineering, Winnipeg, Manitoba; Aug. 22-24, 2007.
Vasquez, J.A., and Walsh, B.W. 2009. CFD simulation of local scour in complex piers under tidal flow. 33rd *IAHR World Congress*, International Association for Hydraulic Research, Vancouver, BC; Aug. 9-14, 2009.
Vasquez, J.A., McLean, D.G., O'Connor, V.F., and Zimmermann, A. 2012. Hydraulic Modelling for the Padma River Bridge. *International Conference on Fluvial Hydraulics*, River Flow 2012, Vol. (2): 1227-1233. San José, Costa Rica. Sep. 5-7, 2012.
Vollmers, H.J. (1990). Examples for Using Sand and Lightweight Material in Movable Bed Model. Proceedings of the *NATO Advanced Research Workshop on Movable Bed Physical Models*. Delft. August 18-21, 1987.

Leadership in Sustainable Infrastructure
Leadership en Infrastructures Durables

Vancouver, Canada
May 31 – June 3, 2017/ *Mai 31 – Juin 3, 2017*

SCS STORM TYPE SELECTION FOR ESTIMATING DESIGN FLOWS IN BRITISH COLUMBIA

Millar, Robert Ph.D., P.Eng/P.Geo.[1,4]

[1] Senior Hydrotechnical/Water Resources Engineer, Golder Associates Ltd., Suite 200 - 2920 Virtual Way, Vancouver, BC, V5M 0C4 Canada
[4] Robert_Millar@Golder.com

Abstract: The US Soil Conservations Service (SCS) methodology for estimating design peak flows in small watersheds is widely used for design of water management infrastructure and has been incorporated in the HEC-HMS computer software. A key component of the methodology is selection of the appropriate design storm. Four synthetic 24-hour rainfall temporal distributions, or storm types (Type 1, 1A, 2 and 3), have been developed and are options within HEC-HMS. Within the USA, the geographical boundaries of the four storm types have been determined and the appropriate storm type can be readily selected. However, there is no established methodology for determining the appropriate storm type for projects located outside the USA, or whether these storm types are even applicable. A methodology is presented that allows for a rational approach to determining the appropriate SCS storm type for a project that is located outside of the USA. The approach requires depth-duration-frequency (DDF) table values. The storm-selection methodology is applied to three climate stations in British Columbia that have Environment Canada DDF values: Vancouver YVR, Williams Lake, and Stewart A stations. For the Stewart A station, none of the SCS storm types are appropriate, and a site-specific temporal distribution is developed using the alternating-block methodology to provide a user-defined hyetograph for input to HEC-HMS

1 Introduction

The US Soil Conservation Service[1] (SCS) methodology (USDA, 1986) for estimating design peak flows in small watersheds is widely used for design of water management infrastructure in the USA and globally. The SCS method has been incorporated in the HEC-HMS computer software (USACE, 2009). A key component of the methodology is selection of the appropriate design storm. Four synthetic 24-hour rainfall temporal distributions, or storm types (Type 1, 1A, 2 and 3) have been developed and are options within HEC-HMS. Peak flow estimates are sensitive to the choice of storm type. For a typical small watershed of a few km^2, peak flow estimates assuming a Type 2 storm may be three or more times greater than derived using a Type 1A storm.

Within the USA, the geographical boundaries of the four storm types have been determined and the appropriate storm type can be readily selected. However, there is no established methodology for

[1] Now the Natural Resources Conservation Service (NRCS)

determining the appropriate storm type for projects located outside the USA, or whether these storm types are even applicable.

This paper presents a methodology for determining selecting the appropriate SCS storm type outside of the USA and is applied to examples from British Columbia.

2 SCS Storm Types

Four synthetic storm types that provide 24-hour rainfall distributions have been developed for the lower 48 states of USA, and for Alaska and Hawaii (SCS, 1986). The cumulative 24-hour rainfall distributions for SCS Storm Types 1, 1A, and 2 are presented in Figure 1. Storm Type 3 is similar to the Type 2 distribution, and applies to eastern and south-eastern USA. The Type 3 storm is not considered further in this paper.

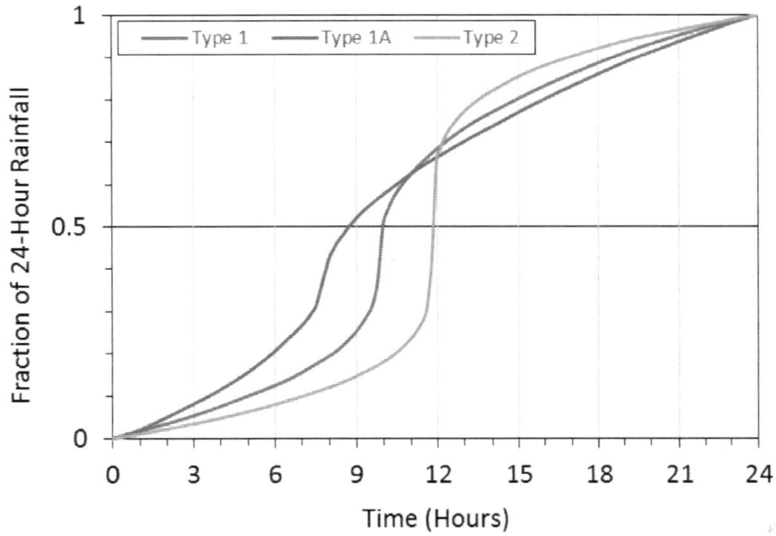

Figure 1: Cumulative 24-hour Rainfall Distributions for SCS Storm Types 1, 1A, and 2 (after SCS, 1986)

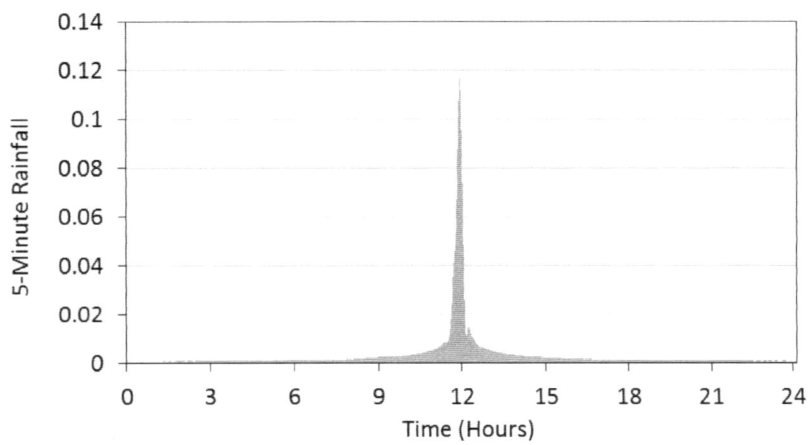

Figure 2: Temporal Rainfall Distribution for the Type 2 Storm in 5-minute increments as a proportion of the total 24-hour Rainfall. The peak 5-minute interval captures 11.7% of the total 24-hour rainfall.

The rainfall distributions represent synthetic design storms and do not reflect actual storm events. The storms begin with low intensity, increase to a maximum, with intensity decreasing later in the storm. Figure 2 shows the incremental distribution of the Type 2 storm in 5-minute increments. The peak intensity is near the mid-point of the storm, with the maximum 5-minute rainfall depth equal to 11.7% of the total 24-hour depth.

2.1 Geographic Distribution

The geographic distribution of the storm types is limited to the USA, and there is no guidance for selecting the appropriate storm type in British Columbia (BC). The geographic distribution of the storm types in western North America adjacent to BC is shown in Figure 3. Within BC, the distribution suggests that Type 1A storm would be appropriate for the south coastal and the adjacent south-west interior, Type 2 within the interior of BC, and Type 1 for north coast and areas adjacent to Alaskan pan handle.

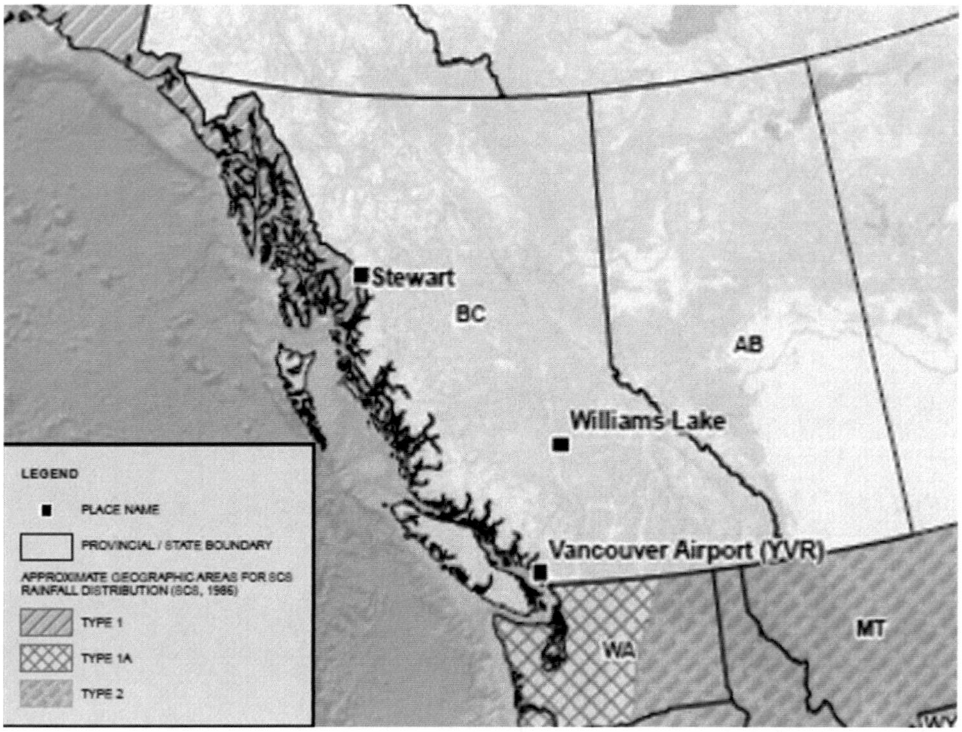

Figure 3: Approximate Geographic Boundaries for SCS Storm Types in Western US adjacent to British Columbia (after SCS, 1986)

2.2 Depth-Duration

An alternate way of characterising the SCS storm types developed here is to extract the depth-duration (DD) values. Maximum rainfall depths as a proportion of the 24-hour total have been determined for Type 1, Type 1A, Type 2 storms for selected durations from 5 minutes through 24-hours (Table 1). For example, the maximum 5-minute rainfall depth for the Type 2 storm is 11.7% of the total 24-hour depth (see also Figure 2). The values in Table 1 can be readily scaled to the total 24-hour depth.

The values in Table 1 form the basis for selecting the appropriate storm type by comparing with station depth-duration-frequency (DDF) values. In BC, short duration intensity-duration-frequency (IDF) datasets including DDF data are available for 121 stations in British Columbia as part of the Environment Canada

Engineering Climate Datasets program[2]. Data from Vancouver YVR, Williams Lake, and the Stewart A climate stations are used to demonstrate the proposed methodology for storm type selection. The location of these climate stations in shown in Figure 3.

Table 1: Maximum Rainfall Depth-Duration Values as a Proportion of the Total 24-hour Rainfall

	Rainfall Depth-Proportion of 24-hour		
Duration (mins)	Type 1	Type 1A	Type 2
5	0.064	0.020	0.117
10	0.116	0.040	0.213
15	0.153	0.060	0.279
20	0.179	0.079	0.327
30	0.216	0.115	0.386
60	0.281	0.172	0.454
120	0.370	0.252	0.538
240	0.492	0.372	0.639
360	0.578	0.468	0.707
720	0.761	0.687	0.841
1440	1.0	1.0	1.0

3 Application to BC Climate Stations

The DDF values (mm) for Vancouver YVR, Williams Lake, and Stewart A stations are plotted with scaled values from Table 1 (Figures 4 to 6). The storm depth-duration curves are scaled with the 1:100-year 24-hour rainfall depth from Environment Canada. Comparing the depth-duration curves for the three storm types with the Environment Canada DDF data allows the most appropriate storm type to be selected.

- For Vancouver YVR (Figure 4), the Environment Canada DDF values correspond closely to the Type 1 storm distribution. However, the geographic distribution of storm types (Figure 3) would suggest that Type 1A, not Type 1, would be applicable to Vancouver YVR.
- For Williams Lake (Figure 5), the Environment Canada DDF values correspond closely to the Type 2 storm distribution, which is consistent with the geographic distribution of storm types (Figure 3).
- For Stewart A (Figure 6), the DDF values do not correspond to any of the SCS storm distributions, whereas the geographic distribution of storm types (Figure 3) would suggest that the Type 1 distribution would be appropriate.

For Stewart A station, additional analysis is provided below to develop a site-specific storm distribution.

---

[2] Can be downloaded from ftp://ftp.tor.ec.gc.ca/Pub/Engineering_Climate_Dataset/IDF/

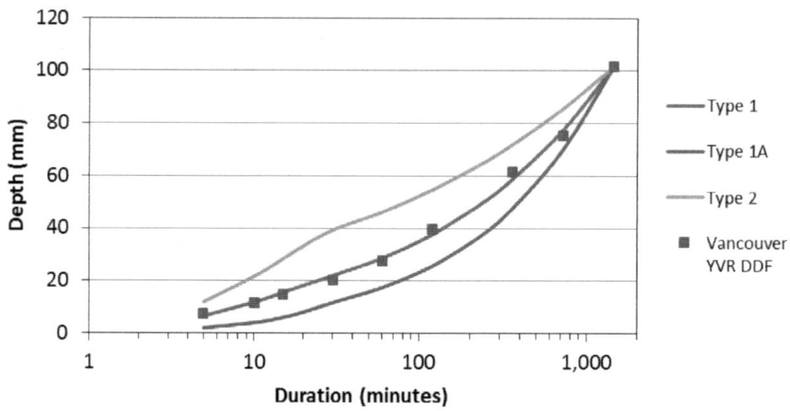

Figure 4: Vancouver YVR 1:100-year Depth-Duration plot. The 1:100-year 24-hour depth is 101.7 mm. The Vancouver YVR DDF values correspond closely to the Type 1 distribution.

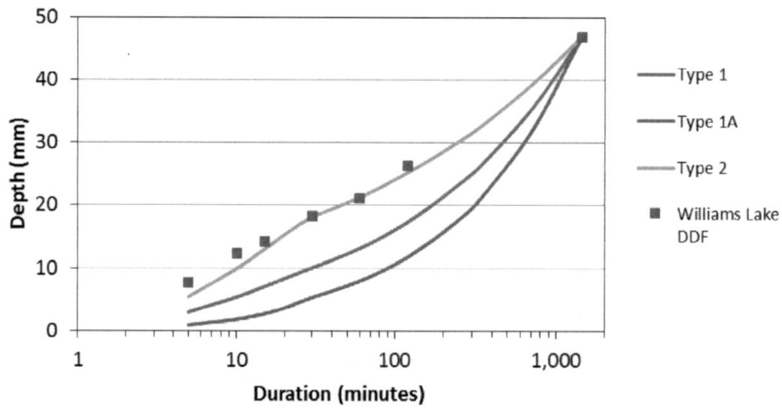

Figure 5: Williams Lake 1:100-year Depth-Duration plot. The 1:100-year 24-hour depth is 46.9 mm. The Williams Lake DDF values correspond closely to the Type 2 distribution.

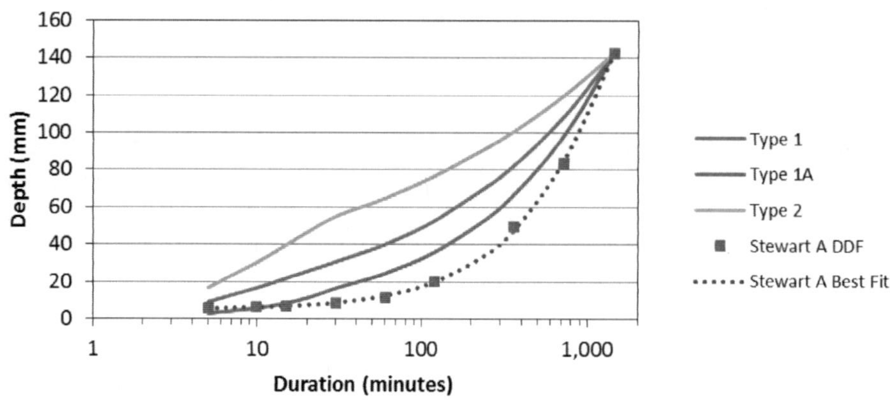

Figure 6: Stewart A 1:100-year Depth-Duration plot. The 1:100-year 24-hour depth is 142.5 mm. The William Lake DDF values do not correspond to an SCS distribution, and a polynomial has been fit to the data (Equation 1).

3.1 Development of a Site-Specific Rainfall Distribution

The DDF values for Stewart A station (Figure 6) do not correspond to any of the standard SCS Storm Types. An assumption that the Type 1 storm is applicable based on geographic distribution (Figure 3) would potentially have significant implications for hydrologic design. For example, the Environment Canada 30-minute 1:100-year depth is 8.5 mm, whereas the corresponding value from the Type 1 storm is 30.7mm, or 3.6 times greater. Using the Type 1 storm distribution for hydrologic design could therefore significantly overestimate the design flow. An alternate approach is presented below to derive a site-specific storm rainfall distribution (hyetograph) using the Environment Canada DDF values.

The process for deriving a site-specific storm requires two steps:

1. Fit a curve (polynomial) to the Environment Canada DDF values;
2. Develop a 5-minute hyetograph using the “alternating-block” methodology (Chow et al., 1988).

A second order polynomial was fit to the Steward A DDF values (Figure 6, $r^2 = 0.9997$):

[1] $D(T) = AT^2 + BT + C$

Where D = depth (mm) for duration T (mins), and A, B, and C are empirical coefficients. Other equation types can be used as appropriate.

The alternating block methodology is used to develop a design hyetograph from DDF values. Here, the cumulative rainfall depth $D(T)$ is calculated for every 5 minute interval up to 24-hours using Equation [1]. The design storm is centered on the peak 5-minute rainfall depth at 12 hours, which corresponds to the 5 minute DDF value (5.7 mm). For each additional 5-minute time interval, or block, the incremental rainfall depth is added successively on alternating sides of the centre so that for any duration, the total depth is given by Equation [1]. The resultant design storm consists of a central peak, with remaining blocks arranged in descending order alternately either side of the central block. Close inspection of the peak of the Type 2 Storm incremental rainfall hyetograph (Figure 2) indicates that an alternating-block methodology has been used.

For Stewart A (Figure 7), the resultant cumulative rainfall distribution exhibits a generally constant accumulation over the 24-hour period, with a central “jump”, which represents the maximum 5-minute rainfall accumulation (5.7 mm). The corresponding incremental hyetograph (Figure 8) consists of a prominent 5-minute central peak (5.7 mm), and decreasing incremental rainfall depths from 0.63 mm to 0.32 mm for the remaining 5-minute intervals. The distribution is substantially different from the SCS storms types. While the derived Stewart distribution and hyetograph might appear unusual, they are consistent with the Environment Canada DDF values (Figure 6).

Within HEC-HMS there is an option for a “Specified Hyetograph” to allow the derived hyetograph to be used as input.

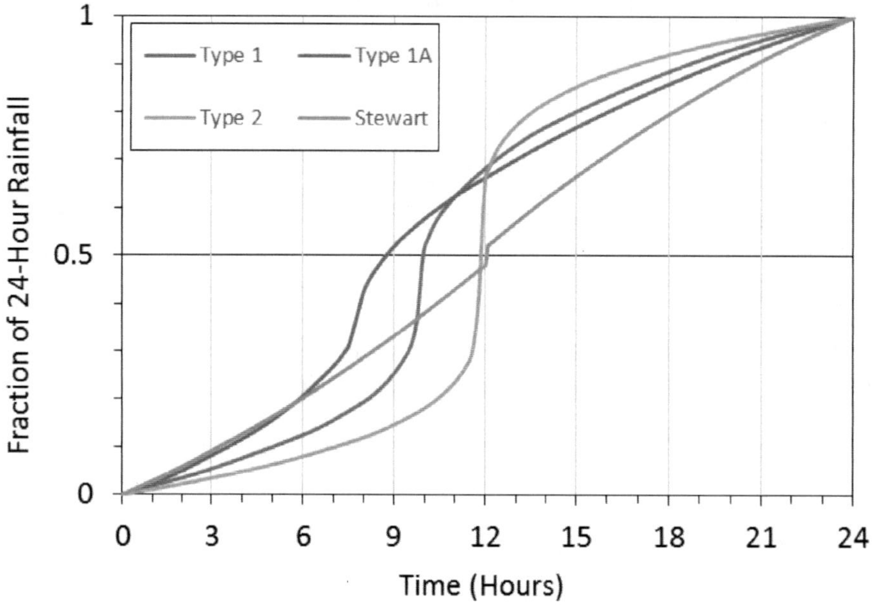

Figure 7: Derived Cumulative 24-hour Rainfall Distribution for Stewart, BC.

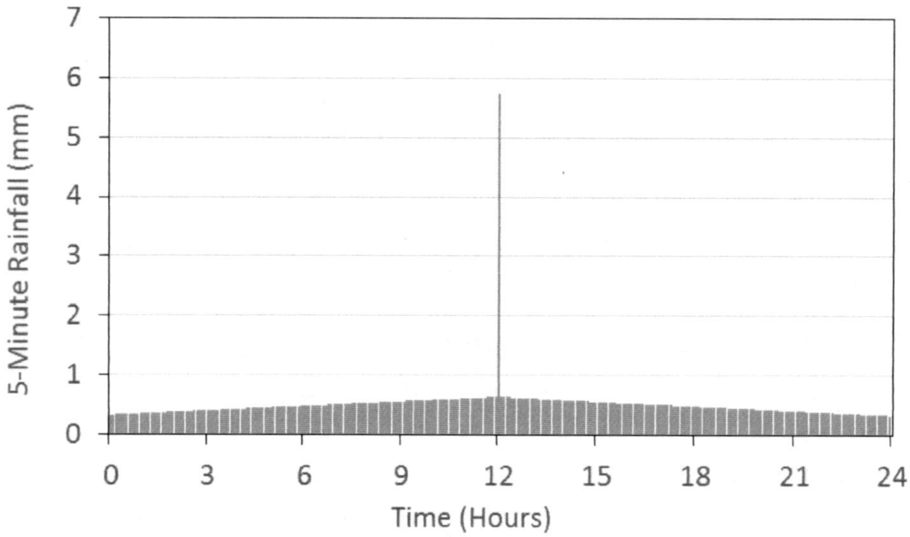

Figure 8: Derived Incremental 24-hour Hyetograph for Stewart, BC.

4 Summary and Conclusions

The analysis of Environment Canada DDF data from 3 climate stations in BC indicate that in general the appropriate SCS storm type (SCS, 1986) in BC cannot reliably be inferred from the geographic distribution of storm types within the USA. Furthermore, DDF data from a specific station in BC may not conform to any of the standard storm types.

A methodology is proposed using the "alternating-block" method to derive site-specific design storm hyetographs, which can be input to HEC-HMS as a "Specified Hyetograph".

References

Chow, V.T., Maidment, D.R., and Mays, L.W. 1988. *Applied Hydrology.* McGraw Hill, 572 p.

US Army Corps of Engineers (USACE), 2009. *HEC-HMS Hydrologic Modeling System. Version 3.4* (computer software). Hydrologic Engineering Center, Davis CA.

US Department of Agriculture (USDA), 1986. *Urban Hydrology for Small Watersheds*, Technical Release 55, (TR-55).

Leadership in Sustainable Infrastructure
Leadership en Infrastructures Durables

Vancouver, Canada
May 31 – June 3, 2017/ *Mai 31 – Juin 3, 2017*

SIMULATION OF ICE AND SEDIMENT PROCESSES IN AN ALLUVIAL STREAM

Ghareh Aghaji Zare, Soheil[1,8]; Seidou, Ousmane[2]; Groeneveld, Joe[3]; Rennie, Colin D.[4]; Ahsan, Rajib[5]; Melindy, Sandy[6]; and Malenchak, Jarrod[7]
[1,3,5,6] Hatch Ltd, Canada
[2,4] University of Ottawa, Canada
[7] Manitoba Hydro, Canada
[8] Soheil.Zare@hatch.com

Abstract: The flow regime of streams in the northern hemisphere can be largely affected by ice formation during the winter period. The ice cover can affect not only the river hydraulics of a particular reach but can also influence important sedimentation processes. Although the effects that an ice cover may have on flow hydraulics are reasonably well understood, our understanding of its impact on sedimentation processes is much less developed. The sparse availability of continuous and comprehensive information on fluvial behavior during the winter period has contributed to this lack of understanding. Numerical models are one means by which to bridge this knowledge gap and improve our understanding of the complex interaction between ice covers, river hydraulics and sediment transport. Several ice simulation models have been developed and applied over the years. Of these, ICESIM has been widely (and successfully) applied in studies of several hydroelectric projects across Canada. The robust algorithms of the model have been successfully calibrated and utilized to simulate the river hydraulic characteristics and ice cover formation. The ICESIM model was originally developed by Acres International Limited (now Hatch) in 1973 for studies of the Nelson River hydroelectric plants and since then it has been continuously advanced and improved. In a recent series of improvements, the model was converted (from *Fortran*) to run on a *Matlab* platform, and a sediment transport simulation module was also added. The resulting model, ICESIMAT, is a one-dimensional, steady-state model capable of the simulation of ice cover formation and sediment transport in a river. In a recent study, ICESIMAT was applied to simulate ice cover formation and sediment transport on the Lower Nelson River in northern Manitoba, Canada. The results of the numerical simulations were validated by *in-situ* measurements performed during several field campaigns at the study site. This paper summarizes the model capabilities and the results of these investigations.

1 Introduction

The role that an ice cover plays in rivers of the northern hemisphere makes it a subject of importance for all industrial projects being conducted in these areas. Rivers in Canada, amongst other countries located in cold regions, are significantly affected by the presence of ice cover during winter. River ice introduces a dynamic boundary to the top of the river and alters flow magnitudes and patterns. Though these variations are temporary, the significant scale of the phenomenon often causes substantial impacts on the environment, economy and public safety.

The presence of an ice cover has some important implications for hydroelectric plants, too. The variation in flow and stream conveyance capacity is the main source of instability in the operation of hydroelectric generating stations during winter. Despite the fact that ice cover impacts are relatively well recognized, there are still gaps in understanding river ice processes and their consequences on hydraulic behaviour of the affected streams. This is mainly due to the difficulty associated with conducting comprehensive field investigations during the ice-covered period. Therefore, efforts have been made to overcome the shortage in this area using new field measurement techniques, as well as other approaches such as physical or numerical modeling of the ice cover processes.

1.1 Ice Processes in Natural Streams

Ice processes in rivers can be categorized into three main stages: freeze-up, stabilization/consolidation and break-up. Each of these stages includes several sub-stages accompanied by various ice features. The complex process of river ice formation and evolution is the result of the combination of different meteorological, hydrologic, mechanical and thermal factors. However, the dominating element in the ice cover processes is river hydraulics. River hydraulics plays a reciprocal role in river ice formation. In other words, while it is the main element in ice cover formation, river hydraulics is also greatly affected by the formation of the ice cover.

During the formation of an ice cover, the water temperature begins to drop at the interface of the water surface and the supercooled air above it. Frazil ice particles are the product of the heat transfer from the "warmer" water to the "colder" air. Hereafter, the flow intensity defines the type of ice being formed. In areas with low flow intensity, frazil crystals form at the flow edges, bond to each other and then expand laterally into the stream. This process creates the *skim ice* which thickens thermally and eventually forms *border ice*.

In more turbulent areas, frazil ice particles formed at the surface are distributed to various depths before traveling downstream. The adherent nature of frazil ice causes it to either accumulate on objects exposed to the flow (river bed material, vegetation, flow intakes, etc.) and form *anchor ice*, or to agglomerate and float at the surface in the form of *ice pans*. While traveling downstream, ice pans grow in size which leads to an increase in the concentration of river ice at the surface. At narrower cross sections (contractions due to change in morphology or border ice formation), the moving ice floes start to congest and bridge an ice arch at the cross section. The formation of an *ice bridge* acts as a barrier to the incoming floes, and ice pans start to accumulate at the bridge location. This process initiates the formation of an ice cover which then progresses upstream by the accumulation of incoming floes through a process called *juxtaposition.*

External forces such as water thrust and wind drag, along with internal forces due to the weight of the cover, mechanically thicken the ice through a process called *shoving*. Shoving causes the leading edge of the ice cover to retreat in order to provide the necessary strength against the imposed forces. When the ice cover reaches an equilibrium thickness that is capable of withstanding the above-mentioned forces, the ice cover progression continues. The thickness and strength of the resulting ice cover are dependent on several climatic parameters such as temperature, precipitation and solar radiation. Each of these elements can favorably or adversely affect the cover stability. Climatic factors affect the ice cover mainly by altering the thermal budget through the cover. This defines the heat transfer rate from the underside of the cover to the air in contact with the ice surface.

Variation in the thermodynamic balance of the ice during the spring triggers another important ice cover stage known as *thermal break-up*. Notably, the thermal break-up can start even before the air temperature starts to rise. The ice cover melting process is initiated due to incoming solar radiation which introduces another source of heat to the thermodynamic system of the river. Thermal break-up causes the cover to weaken, consequently reducing the mechanical resistance against the flow force (i.e. hydraulic thrust) and initiating ice cover destabilization through the process of *mechanical break-up*. During the break-up period, the cover starts to decay and retreat through a series of mechanical cover breakings accompanied by a retreat of the leading ice edge and open water emerging along the river reach. The process continues until the cover is completely removed and an open-water regime returns. Some of the most drastic ice cover features, such as an *ice jam* formation can take place during the break-up period.

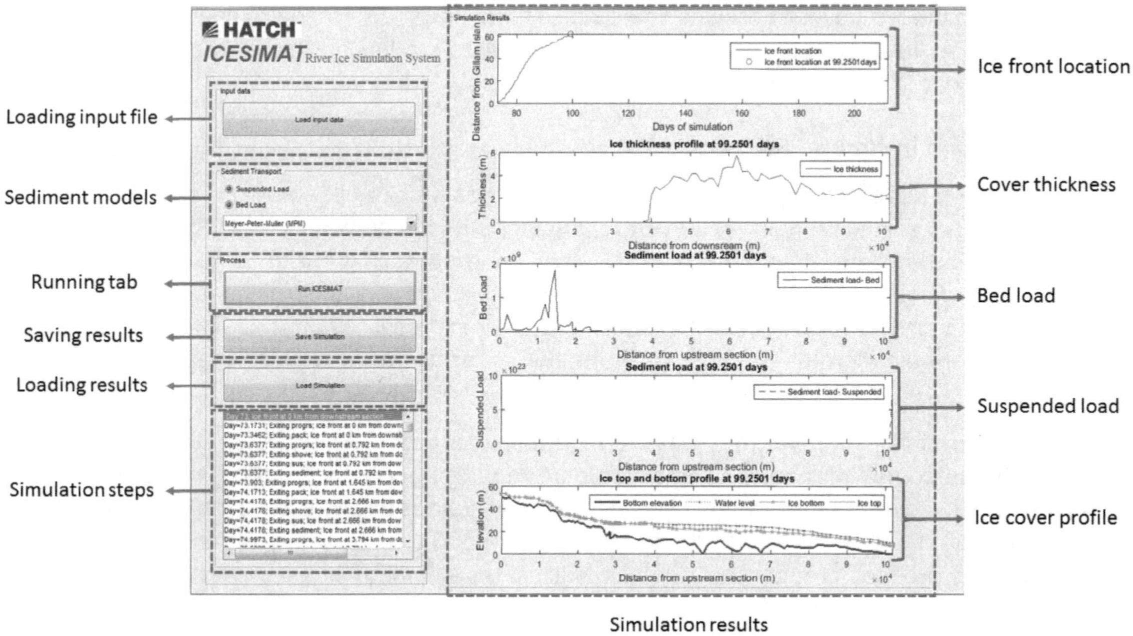

Figure 1 ICESIMAT Graphical User Interface (GUI)

1.2 River Ice and Channel Morphology

The effect of river ice on channel morphology peaks during the break-up period when the extreme river hydraulic condition is capable of causing significant erosional and depositional features. Several researchers such as Prowse and Culp (2003), Milburn and Prowse (2002), Milburn and Krishnapan (2002), and Turcotte and Morse (2012) studied the morphological effects of the ice cover on rivers.

The sediment transport process consists of three main components: initiation of motion, transfer, and deposition, all of which are determined by local flow strength and sediment properties. Channel morphology dictates the spatial variation of flow strength and sediment properties, and thus potential scouring zones. As a simplified description, sediment tends to be eroded from areas where flow converges and is deposited at locations where flow diverges. An ice cover presence complicates this process by adding an upper boundary to the river, which is often as rough as the lower boundary, and temporarily changes the river sediment transport capacity compared to free surface conditions. Consequently, sediment transport tends to decrease under a floating ice cover due to reduced bed shear (Sayre and Song, 1979; Ettema et al., 2000). Conversely, an ice cover and/or ice jam can locally constrict the flow and thereby causes local scour (Neill, 1976; Sui et al., 2006), which can locally increase sediment transport (Sui et al., 2000). Furthermore, local flow concentration by river ice has been observed to cause direct geomorphic effects including thalweg shifting and cut-off of meander loops (Ettema and Zabilansky, 2004).

During spring ice break-up, increased freshet flow combined with flow surges (Jasek, 2003) due to temporary ice damming results in increased bed shear and sediment transport, and break-up flows can be the dominant channel forming flow (Beltaos, 2007). However, as mentioned, this critical process remains poorly studied largely because of the difficulty and danger of accessing rivers during the spring break-up condition. Heaving of ice on river banks can also scour the river bank (Smith, 1979; Doyle, 1988) which can both remove sediment from the bank and inhibit the growth of woody vegetation and thereby decrease bank strength which ultimately results in a wider channel (Smith, 1979; Hey and Thorne, 1986). It appears that only Prowse (1993), Milburn and Prowse (2002), and Beltaos and Prowse (2001) have reported sediment transport measurements during ice break-up. Notably, Prowse and Milburn described a field campaign on the Liard River, Northwest Territories, in which suspended sediment samples were collected from a

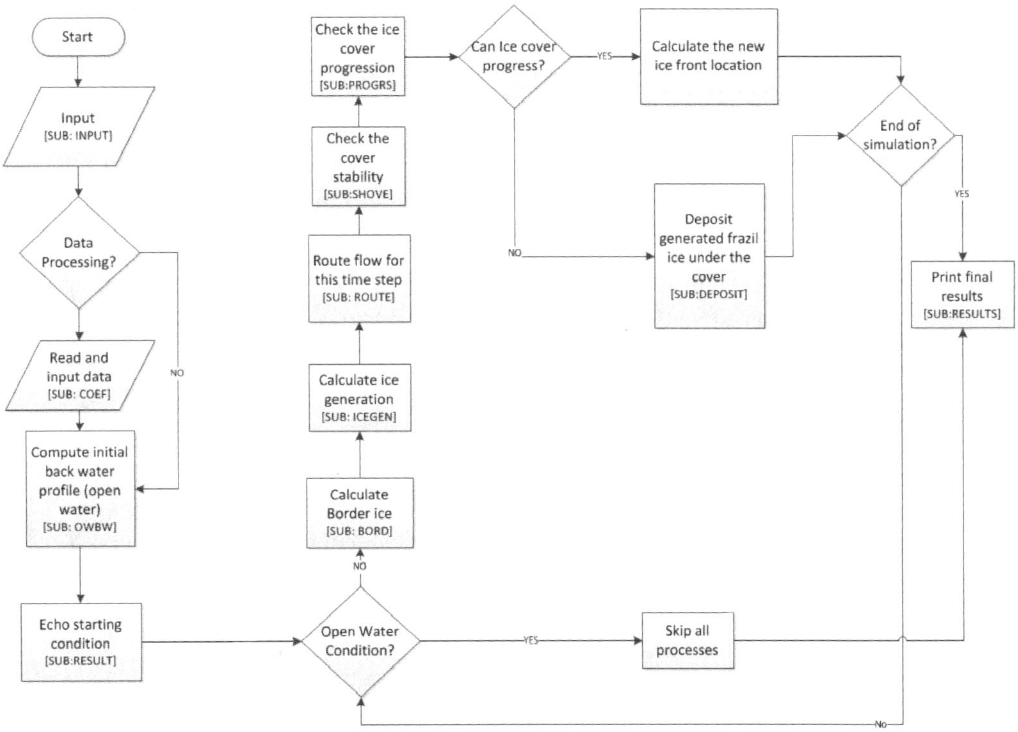

Figure 2 Flowchart of river ice simulation in ICESIMAT/ICESIM

helicopter during the break-up. They observed a 30-fold increase in suspended sediment concentration during the break-up.

River Ice and Sediment Transport Modeling

The ICESIMAT model is programmed and developed in *Matlab* based on the Hatch river ice simulation model (ICESIM). ICESIMAT is designed to simulate the river ice cover formation, river hydraulics and sediment transport under the variety of ice cover stages. The model includes a multifunctional Graphical User Interface (GUI) to provide a direct user interaction with the model through graphical manipulation of model parameters. Simulation results are also simultaneously post processed and presented at each simulation time step. The simulation in total, or each individual time step, can also be loaded into the model through the GUI. ICESIMAT GUI and its components are shown in Figure 1.

ICESIMAT and its predecessor ICESIM mostly share the same type of required information to set up and perform the simulation. However, the ICESIMAT needs an additional set of geomorphological information to conduct the sediment transport modeling.

The necessary information for simulation of the river ice cover formation include daily air temperature, river discharge, observed water level and observed ice front location throughout the winter. Furthermore, ICESIMAT requires information about bed material at each cross section in order to estimate sediment load during the simulation. Figure 2 shows the river ice simulation flowchart in ICESIM/ICESIMAT.

1.3 Numerical Simulation of River Ice Processes

The major physical processes of ice cover formation are simulated in discrete time steps in ICESIMAT. The model is formulated to estimate the parameters that affect water stage and ice cover characteristics in the river. Amongst the numerous factors and features of river ice processes, important processes such as volumetric ice generation, ice cover frontal progression, ice floe transport and deposition, cover erosion, border ice formation and mechanical cover thickening, and ice front retreat are well characterized and modeled through several specific subroutines and algorithms.

As illustrated in Figure 2, the simulation of the river ice process starts with the estimation of frazil ice generation in the river. ICESIMAT calculates the frazil ice generation rate using the equation below.

$$[1]\ V_f = \frac{L_0 B_0}{\rho_i \lambda_t} C_0 (T_w - T_a)$$

In this equation, V_f defines the volumetric generation of frazil ice (m³), $L_0 B_0$ is the upstream open water area (m²), C_0 is the heat transfer coefficient (W/m² °C), λ_t is latent heat of fusion of ice (J/kg), ρ_i is the density of ice (kg/m³) and T_w andT_a are water and air temperatures (°C), respectively.

The generated volume of ice then starts to accumulate at the bridging location(s) and forms the leading edge. ICESIMAT uses the following equation developed by Mitchel (1971) to calculate the ice front thickness:

$$[2]\ h_j = \frac{V_u^{\,2}}{[2g\left(1-\frac{\rho_i}{\rho_w}\right)(1-P_j)]}$$

In this equation, h_j represents the ice thickness (m), V_u is the water velocity under ice (m/s), ρ_i and ρ_w are ice and water densities (kg/m³) and P_j is the ice cover porosity (dimensionless). Due to the difficulty of direct measurement of water velocity under the ice cover, ICESIMAT calculates the water velocity as:

$$[3]\ V_u = \frac{V}{\left(\frac{R_c - h_j}{R_c}\right)}$$

In which R_c is equal to the mean hydraulic radius (m) and V is the water velocity (m/s) immediately upstream of the ice edge.

The main criterion to define whether the frontal edge would progress or not is the Froude number at the leading edge location. Assuming the constant values of $P_j = 0.73$ and ρ_i=920 kg/m³, the Froude number at the ice cover leading edge can be represented as:

$$[4]\ F_r = \frac{V}{\sqrt{gy}} = 208\sqrt{\frac{h_j}{y}}\,(1 - \frac{h_j}{y})$$

The Froude number limit for the ice cover leading edge to progress upstream is about 0.12; for values greater than the limit, the incoming ice floes start to underturn and travel below the cover to reach the point where the velocity decreases and allows the floes to deposit under the existing ice cover.

According to Michel (1971), frazil ice starts to deposit under the ice cover at velocities equal to or less than 0.3 m/s. Deposited ice particles, as well as consolidated ice cover, start to be eroded at velocities between 1.5 m/s to 2 m/s. These mobilized ice particles then start to be conveyed until they reach a point where the velocity permits the floes to be re-deposited, otherwise they exit the simulated reach at the downstream model boundary.

The process of ice generation, deposition/erosion, consolidation and frontal progression continues until the ice covers the entire river or reaches a point where it is hydraulically impossible for the cover to progress any further (i.e., due to a drastic elevation change or at a very fast flowing cross section).

1.4 Simulation of Sediment Transport Under the Ice Cover

ICESIMAT is capable of simulating sediment load in ice-covered rivers both in the form of suspended and bed loads. Suspended sediment calculation is based on the estimation of non-cohesive suspended materials using the method developed by vanRijn (1984). Bed load, however, can be calculated using several approaches available for this purpose. ICESIMAT includes Meyer-Peter-Muller (1948), Einstein (1950), Einstein-Brown (1950), Yalin (1963), Bagnold (1980), Modified Ackers-White (1982), Parker (1982, 2000), vanRijn (1984) and Wilcock-Crowe (2003) models to estimate the sediment bed load in the ice-covered river. The appropriate model to be applied in the calculations can be selected by the user during model set-up (Figure 1).

2 Field Investigations

Field investigations of this study were conducted on the Lower Nelson River (LNR) in northern Manitoba. The LNR is a regulated semi-alluvial river that drains Lake Winnipeg and flows into Hudson's Bay through the northern Manitoban shield. The study site of this research project was located upstream of Jackfish Island in the middle of the reach between the Limestone Generating Station (LGS) and Gillam Island at the mouth of the bay. The river stretches about 102 km along this section and flows fairly straight and fast. River banks are relatively steep along this reach. The elevation of the river thalweg undergoes two sets of elevation drops in the reach with an average slope of 0.0011 for the upper portion (about 28 km long) and 0.00018 for the remaining lower portion.

Ice processes in the LNR start with ice bridging at the lower boundary of the reach. Ice cover then propagates upstream toward the LGS and the leading ice edge stalls at Sundance Rapids about 3 km downstream of the LGS. The ice cover processes normally take place between mid-November to mid-May.

River hydraulics and ice cover variations were studied using acoustic techniques in this research. Autonomously recording acoustic instruments including a 1200 kHz Acoustic Doppler Current Profiler (ADCP) and 546 kHz Shallow Water Ice Profiling Sonar (SWIPS) were utilized to monitor the interactions between the ice cover and river hydraulics and sediment transport. Instruments were installed on a bottom moored aluminum mount and deployed in the river during three winter field campaigns. Recorded information was then analyzed and interpreted to evaluate the mentioned effects.

2.1 Estimation of Ice Cover Thickness Using Acoustic Data

Ice thickness can be calculated according to the data measured by the instruments in a time series format of river ice draft. The three equations corresponding to calculation of ice thickness are provided below.

Ice draft, *d*, can be calculated as the difference between the simultaneous water level and elevation of the bottom of the ice, which is derived from the peak in the vertical profile of backscatter intensity. The measured acoustic range of the underside of the cover is corrected for transducer tilt (θ).

[5] $d = H - \beta . D \cos\theta$

In which β is the correction for the actual sound speed to the nominal value.

Water depth, *H*, can be determined using the measured bottom pressure P_{btm} , which should be adjusted according to the atmospheric pressure P_{atm}, measured at a nearby weather station:

[6] $H = \frac{P_{btm} - P_{atm}}{\rho_w g}$

in this equation, ρ_w is the water density

Hence, for a fairly flat ice cover, the ice thickness, T_{ice}, can be calculated as:

[7] $T_{ice} = d . \frac{\rho_w}{\rho_{ice}}$

3 Results and Analysis

3.1 Ice Cover Formation

Figure 3 (left) shows the ice cover simulation results in the study area during the three main stages of formation, stabilization, and removal. As shown in the figure, the ice cover propagates upstream from Gillam Island (downstream boundary of the simulation) through the processes of juxtaposition. Juxtaposition relocates the leading edge of the ice cover upstream until the cover progresses to its most upstream

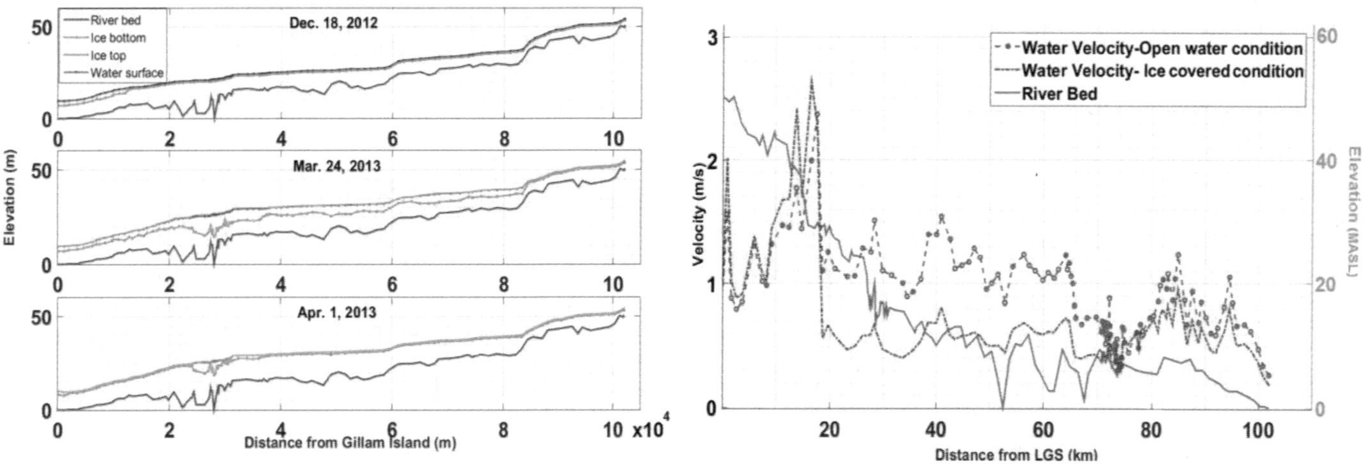

Figure 3 ICESIMAT results showing the ice-covered condition in three stages of progression, stabilization and removal (Left); Calculated flow velocity magnitude during open-water and ice-covered conditions (Right)

hydraulic extent. Subsequently, the generated ice volume passes under the leading edge (underturning) and deposits in downstream cross sections where the hydraulic condition meets the deposition limits. The process of propagation is presented in Figure 3 (left, upper plot).

Ice deposition under the existing cover, as well as the cover adjustment through the mechanical thickening, forms the stable ice cover over the study reach as shown in Figure 3 (left, middle plot). Once the ice cover is stable, it thickens and thins according to the thermal and hydraulic parameters of the flow, but is likely to be stable in terms of thickness and propagation extent if no drastic variation in temperature or flow occurs.

At the start of the break-up period, ice cover decays and retreats as previously described. Therefore, the cover state and condition differ according to the new thermal and hydraulic condition. The ice cover removal can be noticed from the difference between the cover condition at two days of model simulation presented in the middle and lower plots of Figure 3 (left).

Estimated ice cover thickness profiles during both the early and late stages of the winter 2012/2013 simulation are represented in Figure 4 (left). As shown, the ice cover experienced an increase in thickness twice during the winter. This is likely due to the mechanical thickening of the cover during the winter.

Figure 4 (right) shows the ice thickness variation in three locations along the reach calculated by the model. As illustrated in this figure, ice cover thickness varies drastically at different locations along the river reach. This is mainly due to the different parameters involved in ice cover formation at various locations. Hence, depending on how dynamically or statically the cover forms, it may evolve to different cover thicknesses to maintain the force balance on the body of the cover. Some interesting features can be addressed in this figure and are worthy of a more detailed discussion.

One interesting feature is the similarity in the start of ice cover formation between the presented locations, although the cover at each location forms at a different rate. This is mainly due to border ice growth and the approach that is taken in the model to calculate the border ice formation through the simulation. ICESIMAT uses different approaches to simulate the border ice growth including the Newbury (1967) method as well as the border ice factor (the second approach is applied in this study). Through this approach, the border ice growth is calculated using the border ice factor which relates the border ice growth to the Accumulative Freezing Degree-Days (AFDD) at the study reach. The AFDD is assumed to be uniform along the study reach and, consequently, applying the border ice factor results in the formation of border ice at the same rate and thickness for all the simulated river cross sections. However, the Newbury method leads to the formation of border ice at a different magnitude and rate

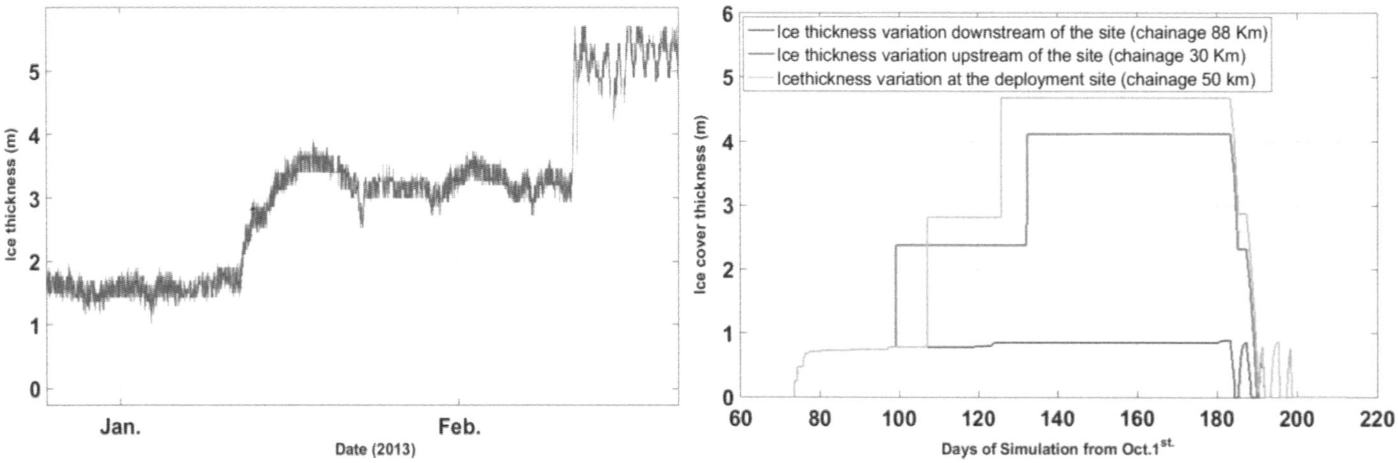

Figure 5 River ice thickness; field estimation (Left), Simulated (Right)

between the cross sections. Selecting the method of border ice simulation is based on the direct observations or experience / interpretations of the user from the study site, which results in some variability between different simulation cases.

Another important feature that can be observed in the cover thickness results is the significant difference between the ice cover thickness values upstream and downstream of the deployment site. As described before, the simulated reach of the LNR experiences two sets of different bed slopes before reaching the estuary at the mouth of Hudson Bay. Hence, different hydraulic conditions affect the cover, causing it to form different cover thicknesses. Due to the relative similarity in ice thickness between the deployment location and upstream site, the cover decayed at almost the same rate in both places. However, the downstream cross section experienced a completely different decay process than its upstream counterpart both in terms of timing and value.

The other important feature shown in the figure is the formation and removal of mechanically thickened ice covers in different locations presented in the figure, during the break-up period. These mechanically-thickened ice covers are shown graphically in the form of ice cover thickness peaks before the complete removal occurs. It should be noted that the ICESIMAT model is not capable of simulating an ice-jam formation during the freeze-up and break-up periods. Therefore, the calculated ice cover thickness during the break-up period is based on the equilibrium cover thickness calculation. This is the reason why even the timing and duration of the peaks vary while the ice thickness values are almost identical. This happens because ICESIMAT currently simulates the ice cover removal process using a series of equilibrium, steady ice cover formation processes at sites where open water emerges while the ice cover is still present in adjacent cross sections. As a result, the cover will only thicken enough to resist the thrust forces at the time of simulation based on the ice volume available from upstream cross sections.

As it can be seen, the model simulated the trend of ice cover thickness variation, but there is a slight difference in the thickness values. The difference between the observed and simulated ice cover thickness values can be explained according to the nature of the ICESIMAT model. Estimated thickness values are more representative of the local thickness whereas the ICESIMAT estimation, considering the 1D nature of the model, is basically a cross-sectional average value. Hence, considering the assumption of the uniform cover condition in ICESIMAT calculations for each cross section, local variations in ice cover thickness are not simulated in the model at the current stage.

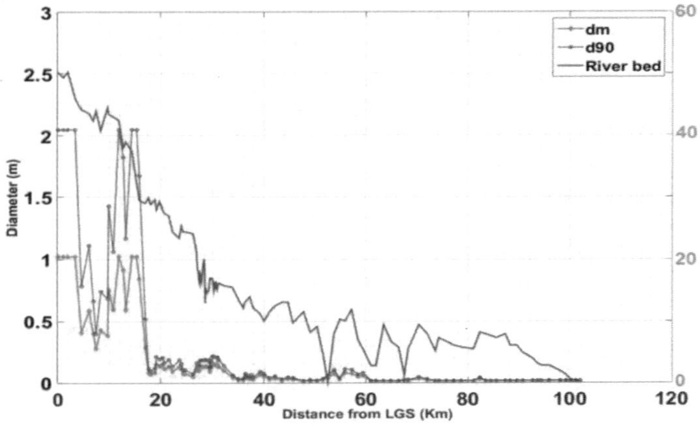

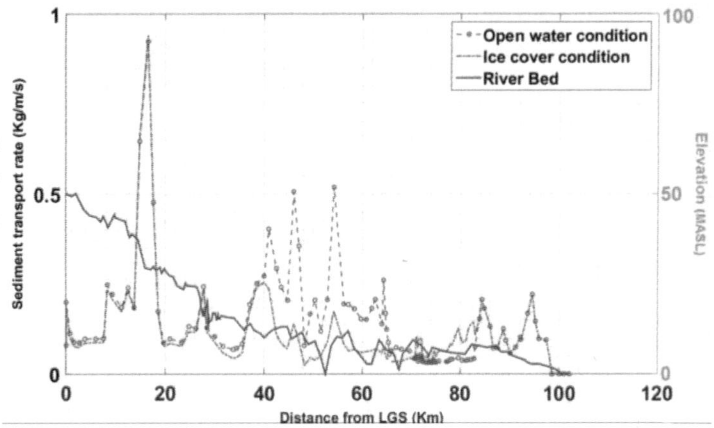

Figure 6 Bed material size distribution along the study reach (Left); Simulated sediment transport rate during the open-water and ice-covered periods (Right)

3.2 Water Velocity

River hydraulic conditions play an important role in defining the stage and condition of the ice cover. This is particularly important in ice cover modeling since the model calculates the ice cover thickness based on the water velocity and elevation at each cross section. Figure 3 (right) compares flow velocity magnitude in all simulated cross sections during two stages of open-water and ice-covered conditions.

As shown in the figure, the ICESIMAT results show the flow velocity magnitude during the ice-covered period to be lower than the open-water period. These results confirm the model consistency with the theory in terms of ice cover effects on flow velocity magnitudes and also match the field observations of the ice cover condition (also refer to Figure 4) and water stage (not shown here).

The lower velocity magnitudes of the model simulations can be explained by the extra resistance added to the stream at the top of the water column in places where the ice cover formed. However, higher simulated water velocities on upstream cross sections, beyond the cover extent, can be described as the results of higher water stages (i.e., greater hydraulic gradient occurred due to the backwater effect from the downstream ice covered cross sections). Notably, the hydraulic simulation of ICESIMAT during the open-water stages was also compared with HEC-RAS during the model calibration process (not shown here), and the results were found to be almost identical.

3.3 Sediment Transport

According to the variation in water velocity (i.e., shear stress), the sediment transport rate estimation in the river is expected to be different between ice cover and open water conditions. As previously discussed, lower water velocities and shear stresses during the stable ice cover period cause the sediment transport rate to decrease, except for the break-up period when the initiation and transport rate significantly increase due to the highly dynamic stages of flow and ice cover.

Figure 6 shows the bed material distribution (left) and sediment transport rate estimation (right) for the study reach during both the open-water condition and the ice-covered period. Comparing the results, it is clear that the sediment transport occurred at a lower rate at the study reach during the ice-covered period. There are two important points to note in terms of sediment transport rate, as discussed below.

1. The calculated sediment rate is based on the vanRijn method for suspended loads and the Wilcock-Crowe method for bed load. According to the included models, it can be concluded that the current stage of ICESIMAT is developed for non-cohesive suspended sediment materials, though it can calculate the bed load using almost all available bed load models. This is mainly due to the fact that

the sediment load in the LNR, which ICESIMAT was originally developed for, is mostly in the form of bed load.

2. The ICESIMAT is limited to a single model application for the calculation of bed load over the study reach even though the cross sections might be different in terms of bed material size and distribution. As shown in Figure 6 (left), the mean particle diameter, d_m, and 90th percentile diameter, d_{90}, of the bed material change from the scale of meters to millimeters (i.e., from boulder/large gravel to sand particles or smaller). Therefore, the selection of a sediment transport model should be performed cautiously to ensure the model simulations are accurate and consistent with observed data on sediment load. However, similar to all numerical simulation models for river hydraulics and sediment transport, it is important to select an appropriate approach for estimation of sediment load based on the nature of the flow and morphological characteristics of the channel.

4 Conclusion

Considering the capabilities and limitations of ICESIMAT, its performance was found to be acceptable in terms of simulation of river hydraulics and sediment transport under an ice cover. Further improvements to upgrade the model capabilities are under consideration to account for the appropriate suspended and bed load calculation approaches according to the varying bed material and size distribution. The performance of the ice simulation model will evolve as it continues to be used to simulate real-world river ice scenarios. Potential future developments in this field will include model enhancements to simulate ice jam formation and removal, as well as the hydraulic features associated with it.

References

Ashton, G.D., 2010. River Lake Ice Engineering, Water Resources Publications, LLC, Highlands Ranch, Colorado, USA.

Beltaos, S., 2007. *River Ice Break up Processes: Recent advances and Future Directions*, Canadian Journal of Civil Engineering, 34,703-716.

Benke, A.C. and Cushing, C.E., 2005. Rivers of North America, Elsevier Academic Press, Burlington, Massachusetts, USA.

Breland, A. and Carson, R.W. 1995. *Studies and Field Tests of Ice Cover Effects on the Yukon River,* Proceedings of Canadian Electrical Association, March 1995.

Carson, R.W. and Groeneveld J.L., 1997. *Evolution of the ICESIM Model*, Proceeding of the 9th Workshop on River Ice, Fredericton, NB.

Carson, R.W. 1991, *Proceedings of Ice Cover Evolution Modules for RIVICE*, Proceedings of the 6th Workshop on Hydraulics of River Ice, Ottawa, ON.

Carson, R.W. 1982. *Experience with River Ice at the Limestone Site*, Proceedings of Workshop on Hydraulic Resistance of River Ice, June 1982.

Carson, R.W. and Jonassen, L.P., 1979. *River Diversion During Construction of Limestone Generating Station*, Proceedings of the Fourth National CSCE Hydrotechnical Conference, May 1979.

Carson, R.W. and Simonsen, C.P., 1977. *Ice Processes During Construction of Limestone Generating Station*, Proceedings of the Third National CSCE Hydrotechnical Conference, May 1977.

Davar, K.S., Belatos, S., and Pratte, B., 1996. *A Primer on Hydraulics of Ice Covered Rivers*, Canadian Committee on River Ice Processes and the Environment, British Columbia, Canada.

Doyle, P.F., 1988. *Damage from a Sudden River Ice Break-up*, Canadian J. Civil Eng.,15: 609-615.

Ettema, R. and Zabilensky, L., 2004. *Ice Influence on Channel Stability: Insights from Missouri's Fort Peck Reach*, J. Hydr. Eng., ASCE, 130, 279-292.

Judge. D.G., Lavender, S.T., and Carson, R.W., and Ismail, S., 1997. *Headpond Ice Jams - Where Will They Occur?* Proceeding of the 9th Workshop on River Ice, Federiction, NB.

Turcotte.B and Morse,B, 2011. *Sediment Transport in Ice-affected Rivers*, Journal of Hydrology, 409(1): 561-577.

Leadership in Sustainable Infrastructure
Leadership en Infrastructures Durables

Vancouver, Canada
May 31 – June 3, 2017/ *Mai 31 – Juin 3, 2017*

SIMULATION OF NEGATIVELY BUOYANT FOUNTAINS USING DATA MINING METHODOLOGY

Niyousha Mohammadidinani [1], and Amir H. Azimi, P.Eng [2]
[1] Graduate student, Dept. Civil Eng., Lakehead University, Thunder Bay, ON, CA
[2] Associate Professor, Dept. Civil Eng., Lakehead University, Thunder Bay, ON, CA
[3] azimi@lakeheadu.ca

ABSTRACT

Data mining and boundary visualization techniques were employed to model the hydrodynamics of negatively buoyant fountains. Experimental and numerical results of two different fountain types of axisymmetric and plane fountains were selected from the literature for numerical simulation. Fountain characteristics such as penetration height y_m, fountain width x_w, and the thickness of the temperature layer ΔT were considered. Experimental studies in the literature indicated that the fountain characteristics can be correlated with non-dimensional parameters such as Froude, Reynolds, and Prandtl numbers. All proposed empirical models from the literature are nonlinear functions of Froude and Reynolds numbers. This nonlinearity causes a considerable prediction error. The M5P model tree was used for prediction of fountain characteristics such as penetration height and penetration width of negatively buoyant fountains. The selected model restructures non-linear correlations into a tree of linear models. It was found that all model trees accurately predict fountain parameters with maximum 3% error for y_m and x_w. Different fountain shapes were formed based on variations of Froude and Reynolds numbers. Regime classification was performed using boundary visualization. Different classifiers in Weka software were tested for boundary visualization to define regime boundaries. It was found that the Logistic classifier can properly define the boundary of different fountain flow regimes.

Keywords: Data mining, Weka, Negatively buoyant jets, fountains, modeling

1. INTRODUCTION

Fountain structure forms when a dense fluid is injected upward into another fluid of different density. In this condition the buoyancy force opposes the momentum of the flow. The injected fluid penetrates a distance into the ambient fluid before stagnating and falling back around itself. The characteristics of the fountain flow, such as the penetration height and mixing strength, depend on the Froude and Reynolds numbers (Williamson et al., 2008). Round fountain structure can be classified using Froude, Reynolds and Prandtl numbers as

$$[1] \quad \mathrm{Fr} = \frac{u_o}{\sqrt{R_o g'}}$$

$$[2] \quad \mathrm{Re} = \frac{u_o R_o}{\nu}$$

$$[3] \quad \Pr = \frac{\nu}{k}$$

where R_o is the radius of the nozzle, u_o is the characteristic velocity, g' (i.e., $g'=g(\rho_o-\rho_a)/\rho_a$) is the reduced gravity between the fountain source with a density of ρ_o and the ambient fluid with a density of ρ_a, g is acceleration due to gravity, ν is the kinematic viscosity of the fluid at the fountain source and k is the thermal diffusivity.

Fountain flow are categorized as weak and forced fountains. Weak fountains are generally steady and symmetric (Lin and Armfield, 2000a) and forced fountains are unsteady with stronger mixing capability with the ambient fluid (Turner 1966, Friedman et al., 2007). For a fountain with a weak discharge rate, the discharge momentum is less than the negative buoyancy and the flow is characterized when Fr $\leq$1.0. Forced fountains occur for Fr$\geq$3 (Kaye & Hunt 2006, Lin and Armfield, 2010). Transition fountains are classified for 1<Fr<3. Weak fountains occur in the replenishment of cold water in solar ponds, in the melting of magma chamber roofs as well as many other environmental and industrial settings. Weak fountains are further classified to as axisymmetric and plane fountain (Lin and Armfield, 2000a). The main differences between axisymmetric and plane fountains are that the latter penetrate to a greater height, have a greater spread and take longer to achieve a steady state (Williamson et al., 2008, Lin and Armfield, 2000b).

In this paper, experimental data for weak negatively buoyant fountains were modeled using data mining methodology. Experimental data of Williamson et al. (2008) and numerical simulation of Lin and Armfield (2000c) were used for data mining simulation. Williamson et al. (2008) performed experiments by injecting salt water into fresh water. The saline water is fed from a header tank to the base of the fresh-water tank as shown in Figure 1. The water is injected from a sudden start and maintained at a constant flow rate throughout the experiment. The fountain inlet flow rate, inlet pipe diameter and the salinity of the inlet fluid were varied to cover a wide range of Reynolds and Froude numbers. The volume flow rates varied from 0.06 to 20 cm^3/s. Nozzle diameters ranging from 0.54 mm to 4.80 mm were used. The density ratio range was varied with $0.004<(\rho_o-\rho_a)/\rho_a<0.16$. The kinematic viscosity of saline water was varied in the range of $1.01\times10^{-6}<\nu<1.4\times10^{-6}$ m^2/s. The flow was recorded with two digital cameras at a rate of 15 frames per second. For very weak fountains ($0.2\leq Fr\leq1.0$) a vertical cylindrical container considered for axisymmetric fountains whereas for a plane fountain it was a rectangular container. Numerical data of Lin and Armfield (2000c) have been carried out for $0.0025\leq Fr\leq0.2$ with Re=200 and Pr=7 to study the effects of Froude number. The thickness of the temperature layer ΔT was determined from temperature profile and it normalized with the nozzle radius.

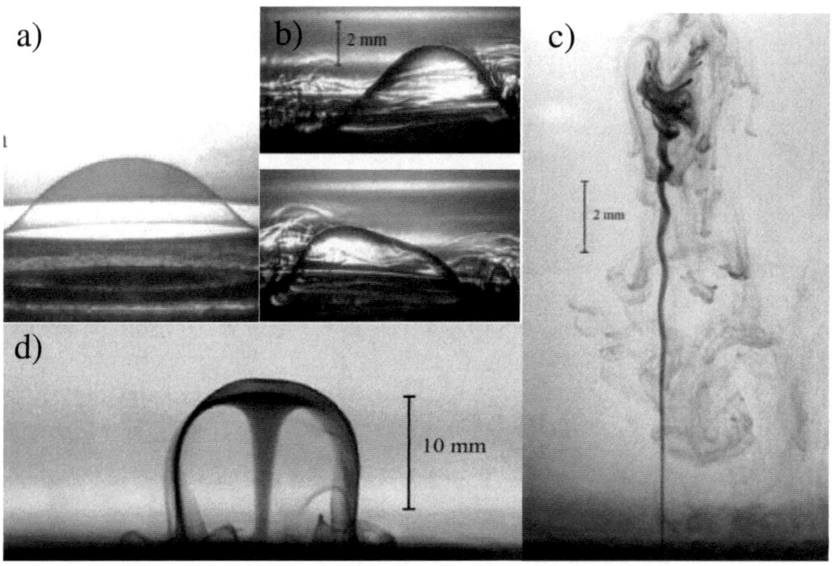

Figure1. Visualization of axisymmetric fountain at different flow conditions; a) weak steady axisymmetric fountain (Re =21 and Fr =0.71); b) flapping fountain (Re=72 and Fr=1); c) sinuous behaviour (Re =105 and Fr =105); d) fountain behaviour (Re =26 and Fr =26). Images were taken from (Williamson et al., 2008).

2. MODELLING BACKGROUND

Knowledge Discovery in Data (KDD) or data mining is a computational process to extract meaningful patterns and rules from large number of variables and data using Weka software. Weka software written in Java language and it has a collection of machine learning algorithms to extract the relationship between different parameters. The classifiers in Weka enable to prepare classification, regression algorithms and estimate the accuracy of the proposed model. Different classifiers were implemented in Weka for data classification named as Bayesian, Trees, Rules, Functions and Lazy classifiers (Witten and Frank, 2005). Attributes in classifiers are categorized into numeric and nominal. Numeric attributes are variables such as Froude, Reynolds and Prandtl numbers and nominal attributes can be defined to be certain such as different shapes and regimes of negatively buoyant fountains.

Weka software used the model tree algorithm for flow classification. The main goal of the model tree approach is the process of dividing complex problems into smaller problems (Bhattacharya et al., 2007). Model trees are accurate, understandable, and easy to train and it can be employed as a robust method for classification, prediction and dealing with missing data (Witten and Frank, 2005; Jung et al., 2010). One of the most famous approaches of model tree simulation is M5 algorithm which was initially introduced by Quinlan (1992). Two main procedures are involved in the algorithm as building the tree and inferring knowledge from it (Jung et al., 2010; Etemad-Shahidi et al., 2010). M5P is an improved version of the M5 algorithm which is proposed by Wang and Witten (1997). The new version has a similar structure to the M5 algorithm which produces easier trees and effectively deals with missing values and enumerated attributes (Jung et al., 2010; Etemad-Shahidi et al., 2010).

The M5P algorithm generally consists of three main steps as building the tree, pruning the tree and smoothing it. One distinct advantage of this mechanism is more transparent than other machine learning algorithms such as artificial neural network (ANN), k nearest neighbouring (kNN) and logistic regression. Therefore, it is easy to follow a tree structure to understand how a decision has been made (Pedrycz and Sosnowski, 2001). Other advantages of model trees are that they are more accurate than ANN, easy to train and robust when dealing with missing data (Witten and Frank, 2005). Figure 2 is an example of regime classification of negatively buoyant fountain from experimental study of (Williamson et al., 2008) based on variations of Reynolds and Froude numbers.

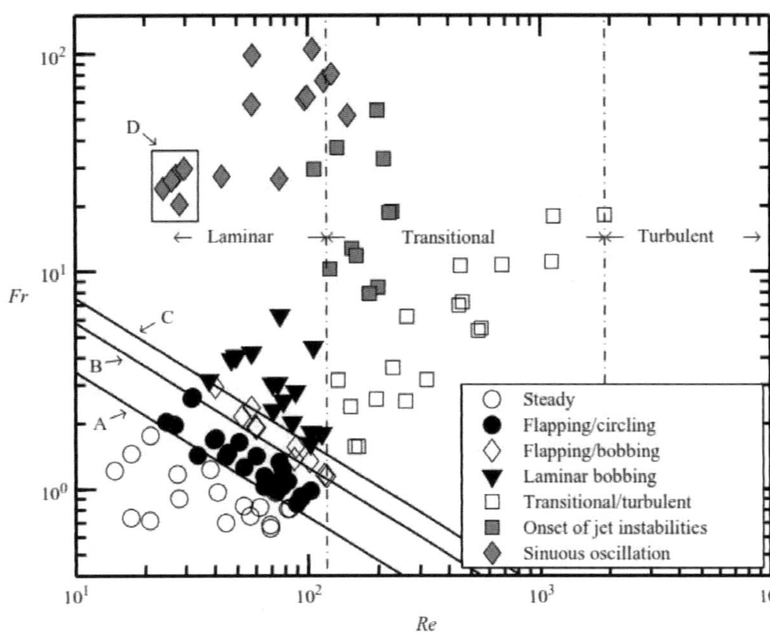

Figure 2. Regime classification of fountain behaviour with non-dimensional parameters of Re and Fr. Solid lines $FrRe^{2/3}=C$ where the constant C is: A=16; B=27; C=35. (From Williamson et al., 2008).

3. RESULTS AND DISCUSSION

3.1 Fountain Classification using Boundary Visualization

Boundary visualization technique in Weka software was employed for regime classification of negatively buoyant fountains. Many classifiers are available in Weka software such as Bayesian, Trees, Rules, Functions and Lazy classifiers (Witten and Frank, 2005). It is important to select a proper classifier for boundary visualization. Figure 3 shows two boundary visualization maps of data points in the study of Williamson et al. (2008). Figure 3a shows the boundary visualization using one of Trees classifiers (i.e., J48 classifier). Comparison of the boundary visualization with J48 and experimental data (see Figure 2) indicates that the J48 classifier in not able to properly define regime boundaries. Many classifiers from different categories in Weka software were tested to find the best match between experimental regime map and boundary visualization. Figure 3b shows the boundary visualization results using function/logistic classifier. As can be seen, the classifier properly defined the boundaries of each fountain regimes.

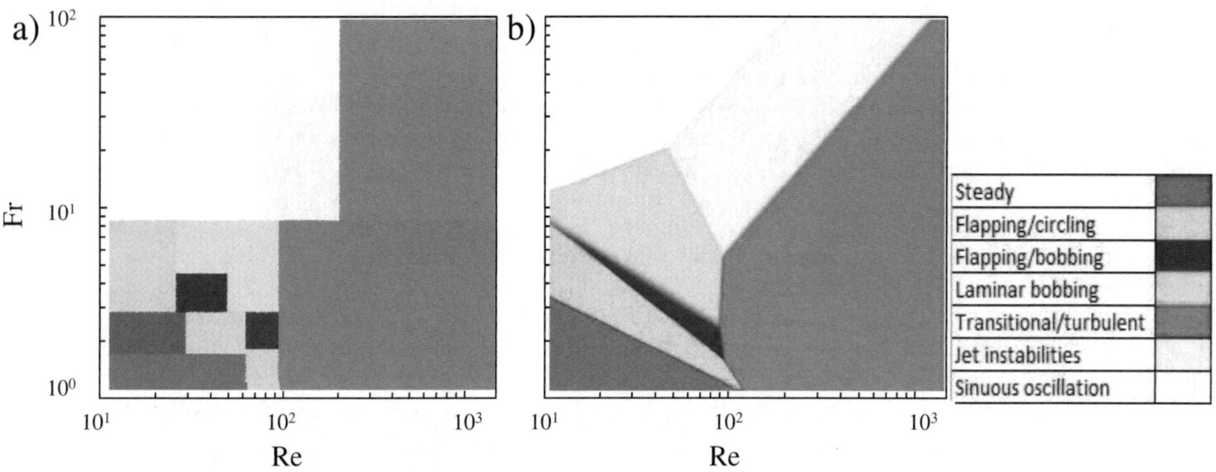

Figure 3. Boundary visualization map of fountain behaviour with Re and Fr numbers by Weka software, a) Tree-J48 classifier, b) Function-Logistic classifier.

3.2 Model Tree and Fountain Characteristics

The M5P model was used for correlating the fountain's characteristics with the mentioned non-dimensional parameters. Fountain height y_m, fountain width x_w and the thickness of temperature layer ΔT were correlated with Re, Pr, and Fr. The number of branches in model tree indicate the complexity of correlation between measured parameters and non-dimensional numbers. Figure 4 shows the model tress of y_m, x_w and ΔT based on non-dimensional parameters. As can be seen in Figure 4, linear modeling of the fountain height is more complex than fountain width since M5P defined five linear models for y_m whereas for x_w and ΔT, two linear equations were defined. As can be seen in Figure 4a, three linear models (LM1, LM4, and LM5) were classified based on Fr. For $Fr \leq 0.078$, the proposed linear models were classified based of the value of Re. As can be seen in Figure 4a, Prandtl number had no impact on regime classification and Reynolds number effect became important for very low values of Froude number. Coefficients of linear models (i.e., η_1, η_2, η_3 and C) for prediction of y_m, x_w, and ΔT for both axisymmetric and plane fountains are listed in Tables 1, 2, and 3.

As can be seen in Table 1, y_m is correlated with Fr and Re in all five linear equations whereas for x_w, linear equations are a function of Froude number (see Table 2). Correlation of the thickness of temperature layer ΔT with non-dimensional numbers indicates that for $Re \leq 150$, the thickness of temperature layer is only a function of Re whereas for $Re > 150$, ΔT correlates with Re, Fr, and Pr.

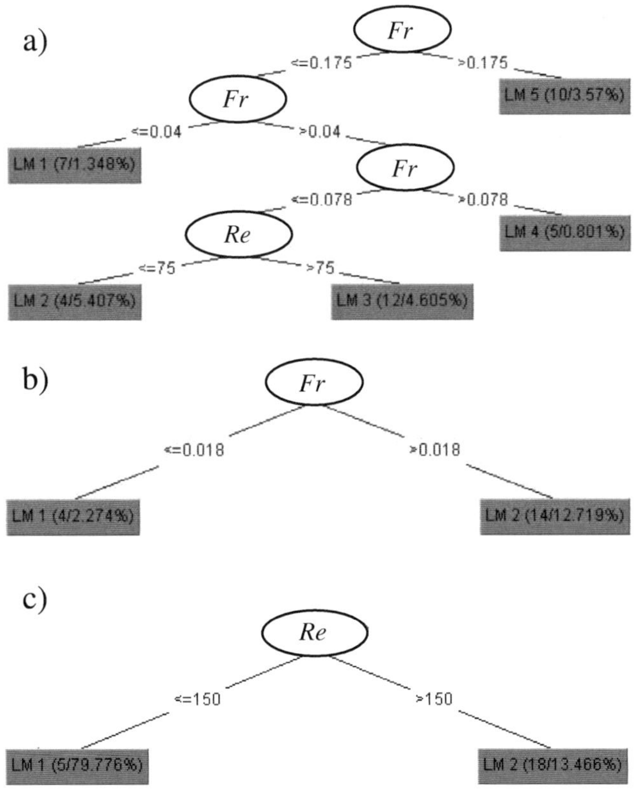

Figure 4. Structure of model trees constructed by M5P and linear models based on numerical data of Lin and Armfield (2000c); a) Model tree for y_m, b) Model tree for x_w, c) Model tree for ΔT.

Table 1. Performance of M5P classification model in form of series of linear models to predict y_m.

Parameter	Linear model	η1	η2	C
y_m (axisymmetric)	LM1	-0.0001	2.1335	0.085
	LM2	-0.0002	1.3624	0.1334
	LM3	-0.0001	1.3624	0.1277
	LM4	-0.0001	1.703	0.1107
	LM5	-0.0001	1.227	0.1695
y_m (plane)	LM1	-0.0001	3.0256	0.1242
	LM2	-0.0002	2.0304	0.1893
	LM3	-0.0001	2.0304	0.1817
	LM4	-0.0001	2.5476	0.1538
	LM5	-0.0001	1.9695	0.2225

* LM=(η_1Re)+(η_2Fr)+C

Table 2. Performance of M5P classification model in form of series of linear models to predict x_w.

Parameter	Linear model	η1	C
x_w (axisymmetric)	LM1	1.1913	1.087
	LM2	0.8869	1.1001
x_w (plane)	LM1	2.4042	1.1203
	LM2	1.7612	1.1499

* LM=(η_1Fr)+C

Table 3. Performance of M5P classification model in form of series of linear models to predict ΔT.

Parameter	Linear model	η1	η2	η3	C
ΔT (axisymmetric)	LM1	-0.0011	0	0	0.2411
	LM2	-0.0002	-0.0102	0.2668	0.1874
ΔT (plane)	LM1	-0.0015	0	0	0.3069
	LM2	-0.0002	-0.0137	0.357	0.2397

* $LM=(\eta_1 Re)+(\eta_2 Pr)+(\eta_3 Fr)+C$

Performance of M5P model tree to simulate fountain height of both axisymmetric and plane fountains is shown in Figure 5. Comparison of the M5P predictions with numerical data of Lin and Armfield (2000c) indicated that the M5P model is able to accurately predict the fountain height with ±0.5%. The M5P model over-predicts y_m for fountains with very small Fr values (i.e., LM1). Figure 6 shows the performance of M5P model for fountain width for both negatively buoyant axisymmetric and plane fountains. As can be seen, the M5P model over-predicts the experimental data by 2%.

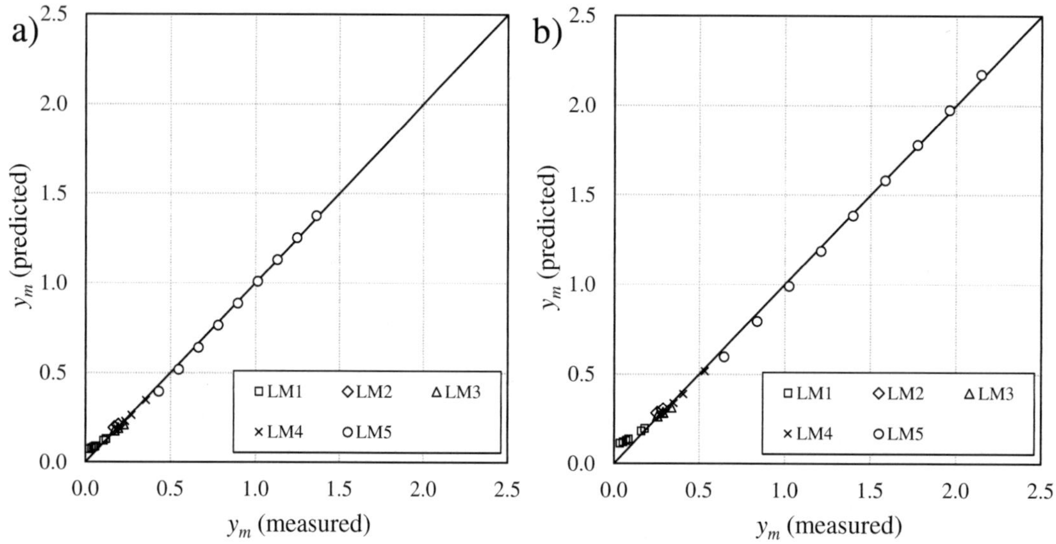

Figure 5. Performance of M5P model tree to simulate fountain height; a) axisymmetric fountain, b) plane fountain.

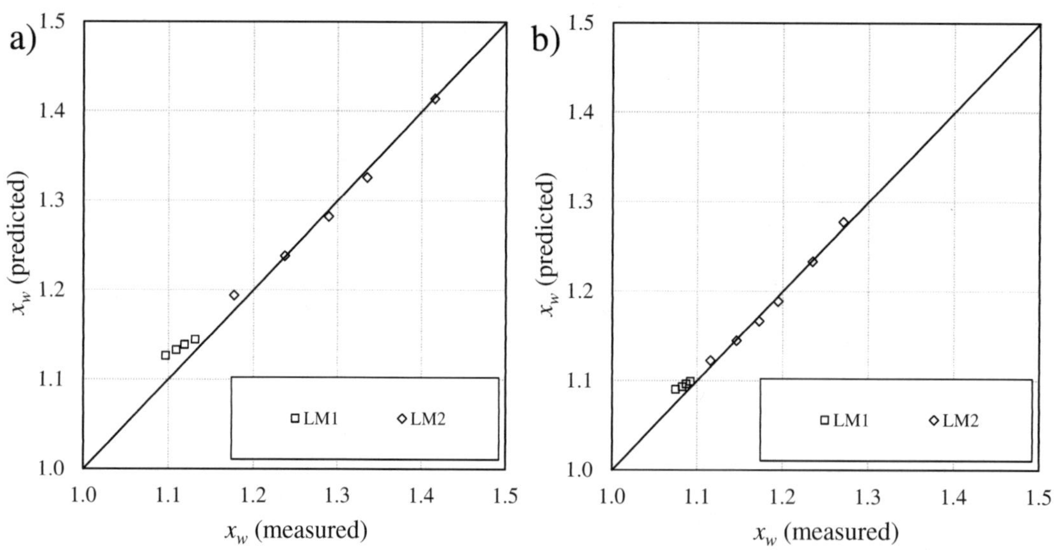

Figure 6. Performance of M5P model tree to simulate fountain width; a) axisymmetric fountain, b) plane fountain.

More data scatter were found on correlation between model predictions and experimental data for ΔT since the thickness of the temperature layer is a function of all non-dimensional parameters (i.e., Re, Fr, Pr) in this study. Lin and Armfield (2000c) shows that the thickness of the temperature layer is a function of Re, Fr, and Pr as

$$\Delta T \approx \frac{\mathrm{Fr}^{2/3}}{(\mathrm{Re\,Pr})^{2/3}} \quad [4]$$

Figure 7 shows the performance of M5P model on prediction of ΔT for both negatively buoyant axisymmetric and plane fountains. The average prediction uncertainty for LM2 was 12%±15%.

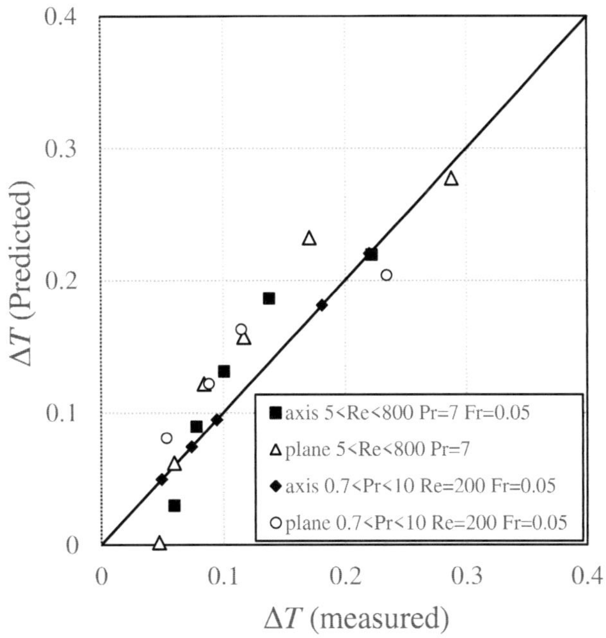

Figure 7. Performance of M5P model tree to simulate the thickness of the temperature layer, ΔT.

4. SUMMARY AND CONCLUSION

Data mining methodology was employed for regime classification and modeling the main characteristics of negatively buoyant fountains. For model validation, data points for both axisymmetric and plane fountains were used from experimental studies of Williamson et al. (2008) and numerical modeling of Lin and Armfield (2000c). Non-dimensional parameters such as Reynolds, Froude, and Prandtl numbers were defined as the numeric data entry for data mining using Weka software. Regime classes such as steady, flapping, bobbing, laminar bobbing, transitional, onset of jet instability, and sinuous oscillation were selected as nominal data entry. Boundary visualization technique in Weka software was shown that the accuracy of the regime map is directly related to the type of classifier. It was found that the function/logistic classifier can accurately determine the regime map.

Three fountain parameters named as fountain height y_m, fountain width x_w and the thickness of the temperature layer ΔT were selected to study the performance of a data mining model (i.e., Weka) to simulate fountain characteristics. A robust model tree (i.e., M5P) was selected amongst other model tree algorithms in Weka software to predict fountain characteristics. Each model tree provided different numbers of linear equation to cover the entire range of data. It was found that the M5P model can accurately simulate y_m and x_w with 0.5% variations. Higher uncertainty level of prediction to model the thickness of the temperature layer ΔT was found due to complexity of the linear model. It was found that the M5P model can predict the thickness of the temperature layer with12%±15% accuracy.

5. REFERENCES

Bhattacharya, B., Price, R. K., and Solomatine, D. P. 2007. Machine learning approach to modeling sediment transport. J. Hydraul. Eng. 133(4), 440–450.

Etemad-Shahidi, A., Taghipour, M. 2010. Predicting longitudinal dispersion coefficient in natural streams using M5 model tree. J. Hydraulic Eng. 138(6), 542–554.

Friedman, P. D., Vadokoot, V. D., Meyer, W. J. and Carey, S. 2007 Instability threshold of a negatively buoyant fountain. Exps. Fluids 42, 751–759.

Jung, N. C., Popescu, I., Keldeman, P., Solomatine, D. P., and Price, R. K. 2010. Application of model trees and other machine learning techniques for algal growth prediction in Yongdam reservoir, Republic of Korea. J. Hydroinf. 12(3), 262–274.

Kaye, N.B.&Hunt,G.R.2006 Weak fountains. J. F. Mech. 558, 319–328.

Lin, W. and Armfield, S. W. 2000a. Direct simulation of weak axisymmetric fountains in homogeneous fluid. J. Fluid Mech. 403, 67–88.

Lin, W. & Armfield, S. W. 2000b. Direct simulation of weak laminar plane fountains in a homogeneous fluid. Intl J. Heat Mass Transfer 43, 3013–3026.

Lin, W. & Armfield, S. W. 2000c. Very weak axisymmetric fountains in a homogeneous fluid. Heat Transfer A 38, 377–396.

Lin, W. and Armfield, S. W. 2010. Very weak fountains in homogeneous fluid. Numerical Heat Transfer, Part A, 38:377-396.

Pedrycz, W., and Sosnowski, Z. A. 2001. The design of decision trees in the frame work of granular data and their application to software quality models. Fuzzy Sets Syst. 123, 271–290.

Quinlan, J. R. 1992. Learning with continuous classes. Proc., 5thAustralian Joint Conf. on Artificial Intelligence, World Scientific, Singapore, 343–348.

Turne r, J. S. 1966 Jets and plumes with negative or reversing buoyancy. J. Fluid Mech. 26, 779–792.

Wang, Y., and Witten, I. H. 1997. Induction of model trees for predicting continues classes. Proc. of the Poster Papers of the European Conf. on Machine Learning, 1997, Univ. of Economics, Faculty of Informatics and Statistics, Prague, Czech Republic.

Williamson, N., Srinarayana, N., Armfiled, S. W., McBain, G. D. and Lin, W 2008. Low- Reynolds number fountain behaviour. J. Fluid Mech. 608, 297-317.

Witten, I. H., Frank, E. 2005. Data mining-Practical machine learning tools and techniques.Morgan Kaufmann, San Francisco.

Leadership in Sustainable Infrastructure
Leadership en Infrastructures Durables

Extended Abstract

Vancouver, Canada
May 31 – June 3, 2017/ *Mai 31 – Juin 3, 2017*

SOURCING AND CHANNELLING INFORMATION FLOWS FOR HYDROLOGICAL PREDICTION

Weijs, Steven V.[1,2], Foroozand, Hossein[1] , Kumar, Akhil[1] and Galindo, Luis C.[1]
[1] Civil Engineering, UBC, Canada
[2] steven.weijs@civil.ubc.ca

1 The challenge of hydrological prediction

Hydrological engineering and science is tasked with providing predictions of the behaviour of various components of the hydrological cycle on various time scales. Since hydrological systems are vastly more complex than we could capture with the data we can normally collect, we are stuck in a situation of predicting from partial knowledge, and uncertainty is inevitable. Choices and trade-offs have to be constantly made between different options for monitoring network layouts (variables, locations, times, scales), as well as choices between model approaches and different model complexities (data-driven/physically based, lumped / distributed).

2 Theoretical framework: Information Theory

Information theory gives us a number of measures of information, which are based in probabilities. Due to the fundamental derivations from first principles (Shannon, 1948), these measures allow formulation of laws and principles that govern the flow of information as a fundamental measurable quantity, evaluated in "bits". These measures, such as Shannon entropy, mutual information, relative entropy, and transfer entropy, allow us to look into contributions and interactions of various sources of information, without restricting assumptions on linearity of relations or Gaussianity of distributions. Diverse applications in hydrology have been pursued (Ruddell and Kumar, 2009; Alfonso et al, 2010; Nearing et al 2015).

3 The information-centered approach to hydrology and water resources management

In this presentation, we will give an overview of some recent work on the intersection of information theory and hydrology, done in the "hydro-info-theory" research group at UBC. The overarching aim of the research is to investigate approaches to increase the amount of useful information that our water management decisions are based on. Examples of elements that contribute to this aim are 1) better quantification of uncertainties (example on discharge); 2) sourcing hydrological information in more economic ways, that trade accuracy for spatial coverage (example on snow cover); 3) improving uptake of information into models by using an information-based approach to calibration and complexity control.

4 Discussion: The information-centered approach to hydrology and water resources management

We argue that elements 1 and 3, where the raw data (and thus its information content) is fixed, are questions of channeling information and should be lead by purely information-based and epistemic considerations, while element 2, which represents the choice of what information to collect, is a question of sourcing information, and should be guided by utility-based and economic considerations (Weijs et al, 2010).

5 Conclusion and Recommendations

Studying the dynamics of information in the process of learning from hydrological data can provide a useful avenue to inform choices that are essentially about optimizing the processing of the information in that data into predictions and decisions.

There are still many open questions surrounding the correct application of information theory to this field, which include dealing with numerical issues, false data, and translation of concepts and results from information theory in other fields to applications in engineering hydrology. Although many translations are possible, the challenge is to find those where the fundamental properties of information retain their significance and meaning in the application. Interdisciplinary exchange is key to advance these ideas.

References

Alfonso, L.; Lobbrecht, A. and Price, R. 2010. Information theory--based approach for location of monitoring water level gauges in polders. *Water Resources Research,* **46**(3):W03528

Nearing, G. S. and Gupta, H. V. 2015. The quantity and quality of information in hydrologic models *Water Resources Research,* **51** (1): 524-538

Ruddell, B. L. and Kumar, P. 2009. Ecohydrologic process networks: 1. Identification *Water Resources Research,* **45**(3): W03419

Shannon, C. E. 1948. A mathematical theory of communication. *Bell System Technical J.,* **27**(3), 379-423

Weijs, S. V.; Schoups, G. and van de Giesen, N. 2010. Why hydrological predictions should be evaluated using information theory. *Hydrology and Earth System Sciences,* **14**(12): 2545-2558

Leadership in Sustainable Infrastructure

Leadership en Infrastructures Durables

Vancouver, Canada

May 31 – June 3, 2017/ *Mai 31 – Juin 3, 2017*

STATISTICAL MODELING OF EXTREME RAINFALL PROCESSES (SMEXRAIN): A DECISION SUPPORT TOOL FOR CONSTRUCTING INTENSITY-DURATION-FREQUENCY RELATIONS FOR URBAN WATER INFRASTRUCTURE DESIGIN

Nguyen, Truong-Huy[1,2], and Nguyen, Van-Thanh-Van[1]
[1] McGill University, Canada
[2] huy.nguyen5@mail.mcgill.ca

Abstract: In recent years, it has been recognized that society has become more vulnerable to extreme storm events. Many studies have been carried out to investigate the variability of these extreme storms. Of particular interest for urban water infrastructure design is the investigation of the probability of occurrence of the extreme rainfalls using frequency analysis method. Many probability distributions have been proposed to model the extreme rainfall processes. However, there is no general agreement as to which distribution should be used. Consequently, in practice, a number of popular distributions are often selected and their descriptive and predictive abilities are then investigated and compared. This task requires a significant investment of time to analyze a large amount of available rainfall data of different time scales and record lengths for a number of different sites. This paper presents therefore the development of a decision-support tool for statistical modeling of extreme rainfall processes (SMExRain). The proposed tool can assist stakeholders and decision-makers in the selection of the most suitable distribution(s) for accurate estimation of the extreme rainfalls. More specifically, the SMExRain can be used to evaluate the descriptive and predictive abilities of ten commonly-used probability models, Beta-K, Beta-P, Generalized Extreme Value, Generalized Logistic, Generalized Normal, Generalized Pareto, Gumbel, Log-Pearson Type III, Pearson Type III, and Wakeby, for their accuracy and robustness in the estimation of annual maximum rainfalls (AMRs). Results of a numerical application using daily and sub-daily AMR data available from a network of 84 stations in Ontario have indicated the feasibility and accuracy of the proposed decision support tool SMExRain.

1 INTRODUCTION

The design and management of various water infrastructure systems require an adequate knowledge of extreme rainfall events of high return periods and for different durations (from several minutes to hours or days). In most cases, these extreme rainfalls corresponding to the desired return periods could not be extracted directly from the observed data since the available historical rainfall record is usually too short. In current engineering practice, the estimated design rainfall values are accomplished based on statistical frequency analyses of annual maximum rainfall data where available rainfall records of adequate lengths could be used to estimate the parameters of a selected probability distribution (Chow 1964, Kite 1977). The selected distribution model is then used to estimate rainfall intensities corresponding to return periods greater than or less than those of the recorded storm events. Accurate estimation of extreme rainfall could help alleviate the damage caused by these extreme storms and it can help achieve more efficient design and management of water infrastructure systems.

In general, selection of an appropriate distribution to representing annual extreme rainfalls is the most difficult and time-consuming task since there are many probability models available in the literature as well as in the design guidelines from different countries (Chow 1964, Kite 1977, Stedinger et al. 1993, Hosking and Wallis 1997, Rao and Hamed 2000, WMO 2009). Hence, the choice of a suitable model still remains as one of the major problems in engineering practice since there is no general agreement as to which distribution, or distributions, that should be used for the frequency analysis of extreme rainfalls. The selection of an appropriate model, thus, depends mainly on the characteristics of the available rainfall data at a particular site. Therefore, it is necessary to evaluate many available distributions in order to find a suitable model that could provide accurate extreme rainfall estimates for a given location.

In view of the above issues, the main objective of the present study is to propose a decision-support tool (hereafter referred to as SMExRain – Statistical Modelling of Extreme Rainfall processes) that can be used to identify the most appropriate distribution(s) for estimating accurately the extreme rainfalls for design purposes. The proposed tool can provide a systematic and objective assessment of both the descriptive and predictive abilities of various distributions in the estimation of the extreme rainfalls using different graphical and statistical assessment criteria. In addition, this tool is highly efficient and convenient for the analysis of a large database of extreme rainfall data for different durations at a given location as well as for a large number of sites. The structure of the SMExRain and the procedure for identifying the best distribution is described in Section 2. An illustrative application of this decision-support tool using daily and sub-daily annual maximum rainfall data for Ontario region is presented in Section 3. Results of this numerical application have indicated the feasibility and accuracy of the proposed SMExRain. Finally, it can be noted that the results of the extreme rainfall frequency analyses using the SMExRain can be displayed in the popular form of Intensity-Duration-Frequency (IDF) relations for a given location of interest.

2 THE DECISION-SUPPORT TOOL: SMEXRAIN

2.1 General Description

The decision-support tool SMExRain has been coded in Matlab environment and equipped with a user-friendly ribbon interface. This software runs independently without the need of installation of a Matlab version. However, as a standalone application compiled from the Matlab environment, SMExRain requires the installation of the Matlab Compiler Runtime (MCR) v9.0 corresponding to the Matlab R2015b version (Mathworks 2016). The MCR v9.0 is free of charge and can be easily downloaded from the Matlab's website. The structure of SMExRain is depicted in Figure 1a.

Regarding the data screening and preliminary analysis, SMExRain provides users with many commonly used statistical criteria, including maximum, minimum, mean, standard deviation, skewness, and kurtosis of the input extreme data series. In addition to these numerical values, it also provides users with many useful graphs for statistical analyses, including the histogram plot for empirical probability density function analysis, the time series plot for trend analysis, and the boxplot for outlier detection.

For selecting a best-fit probability distribution, various numerical and graphical criteria could be employed to evaluate the best fit of the model to the selected data. This descriptive ability assessment includes the use of the popular L-moment ratio diagram (Hosking and Wallis 1997), the application of different statistical goodness-of-fit tests, and the use of various graphical displays such as probability plots and quantile-quantile plots to determine whether the fitted distributions are consistent with the given set of observations (Stedinger et al. 1993). In addition, the SMExRain provides necessary tools for evaluating the predictive ability of a model. This assessment is important to assess the capability of the selected model for describing accurately future extreme rainfall events. Often in selecting a particular distribution, one may be tempted to select a distribution with a high number of parameters. Generally, these distributions could provide a better fit to the existing data than those models with a fewer number of parameters, but they could be quite rigid and consequently are not able to provide an accurate estimation of future extreme rainfall events that are beyond the range of the available data. Nonparametric data resampling schemes such as bootstrap (Efron and Tibshirani 1994) is hence included in the SMExRain to permit an evaluation of the predictive ability of each distribution model. Detailed description of probability distributions, parameter estimation methods, as well as the graphical and numerical criteria to assess the descriptive and predictive abilities of a distribution are presented in the following sections. Notice also that, for convenience, the SMExRain has been designed to allow users to perform the assessment and comparison of up to 12 probability distributions simultaneously rather than to evaluate a single distribution at a time. Depending upon the number of distributions selected for assessment as well as the number of generated bootstrap samples considered, the total simulation process can be completed within a short period (from seconds to minutes on a laptop computer).

2.2 Probability Distributions and Parameter Estimation Methods

SMExRain includes common two-to-five parameter probability distributions that have been selected based on their popularity in hydrologic frequency analyses: Beta-K (BEK), Beta-P (BEP), Generalized Extreme Value (GEV), Generalized Normal (GNO), Generalized Logistic (GLO), Generalized Pareto (GPA), Gumbel (GUM), Log-Pearson Type III (LP3), Pearson Type III (PE3), and Wakeby (WAK) distributions. Other special cases of these distributions, such as exponential (EXP) and normal (NOM) were also included in the software (Chow 1964, Kite 1977, Stedinger et al. 1993, Hosking and Wallis 1997, Rao and Hamed 2000, WMO 2009).

Regarding the estimation of the distribution parameters, the SMexRain software include some common estimation procedures such as the method of moments, the maximum likelihood method, the method of L-moments (Stedinger et al. 1993, Rao and Hamed 2000), and the method of non-central moments (NCMs) (Nguyen et al. 2002). These methods differ in the weights they give to different elements in the selected data set. The maximum likelihood method yields asymptotically optimal estimators of the parameters for some distributions; however, it often involves tedious computation, and it is very sensitive to the computational techniques considered. The L-moments estimators are unbiased and discriminate the behavior of skewed data, which is ideal for parameter estimation of hydrologic data. They are more robust than conventional moments to outliers in the data and sometimes yield more efficient parameter estimates than the maximum likelihood estimates (Stedinger et al. 1993, Hosking and Wallis 1997). The method of NCMs has been shown to be able to consider some scale-invariance property of the NCMs of extreme rainfall data for different durations (Nguyen et al. 2002, Nguyen and Nguyen 2008, Nguyen et al. 2007). More specifically, in the SMExRain, the method of L-moments is used for all distributions (Hosking and Wallis 1997) except for the BEK and BEP models that are estimated by the method of maximum likelihood (Mielke and Johnson 1974). It is noted that the method of L-moments could be used for BEK (Murshed et al. 2011), but it is not preferable since the estimation procedure is more complicated and more tedious than the method of maximum likelihood and the results are almost the same. GEV parameters are estimated by both the L-moments (denotes as GEV) and non-central moments (denotes as GEV*) methods.

2.3 Goodness-of-Fit Tests for Assessing the Descriptive Ability of a Distribution

To visually assess the adequacy of a fitted distribution, SMExRain provides probability plots or quantile-quantile (Q-Q) plots as a means of comparing observed to the fitted (or estimated) data (see Figure 1b). To support the decision-making, SMExRain is equiped with many commonly-used empirical plotting position (EPP) formulas described in literature, for example, Bloom's, Cunnane's, Weibull's formulas and so on (Nguyen et al. 1989, Helsel and Hirsch 2002). In addition, it also provides a general EPP formula which can be customized by users.

In addition to the visual assessment, SMExRain also provides different numerical assessment criteria to indicate the accuracy of the best fit. Four popular criteria are included in the SMExRain, including the root mean square error (RMSE), the relative root mean square error (RRMSE), the maximum absolute error (MAE), and the correlation coefficient (CC) as described in the following equations:

$$[1]\ \mathrm{RMSE}=\left\{\sum\frac{(x_i-y_i)^2}{(n-m)}\right\}^{1/2}$$

[2] $\text{RRMSE} = \left[\frac{1}{(n-m)}\sum\left\{\frac{(x_i - y_i)}{x_i}\right\}^2\right]^{1/2}$

[3] $\text{MAE} = \max(|x_i - y_i|)$

[4] $\text{CC} = \frac{\sum\{(x_i - \bar{x})(y_i - \bar{y})\}}{\{\sum(x_i - \bar{x})^2 \sum(y_i - \bar{y})^2\}^{1/2}}$

where $x_i, i = 1,2,\ldots,n$ are the observed values and $y_i, i = 1,2,\ldots,n$, are the values estimated from an assumed probability distribution for the same probability level; m is the number of distribution parameters; $\bar{x}$ and $\bar{y}$ denote the average value of the observations and estimated quantiles, respectively.

For ease of computation and assessment the descriptive ability of a probability distribution, a ranking scheme has also been developed to judge the overall goodness-of-fit of each distribution by comparing the four categories of test criteria mentioned above. Rankings are assigned to each distribution for every test category according to the relative magnitude of the statistical test results. A distribution with the lowest RMSE, RRMSE, MAE or highest CC would be given the rank of 1. In case of a tie, average ranks are assigned to those distributions. It should be pointed out that equal ranks could also be used in case of ties with almost identical results. SMExRain has not been equipped with the case of equal ranks; however, since Microsoft Excel has supported this function, users could estimate the equal ranks using the Excel built-in function after exporting the results from SMExRain without any difficulty.

2.4 Bootstrap Method for Assessing the Predictive Ability of a Distribution

In general, it is expected that models with more parameters could fit better to the observed data; however, their parameter estimates could be more complex; and more importantly, their prediction abilities may not be better than those models with fewer parameters. Additionally, the performance of a distributions in extrapolating beyond the data is often of primary interest in real applications. To better evaluate the performance of various distributions at predicting extreme right-tail data, a bootstrap method was used in the SMExRain. This is a nonparametric approach to quantify the estimation uncertainty with statistical sampling procedure that yields multiple synthetic samples of the same sizes as the original observations (Efron and Tibshirani 1994). It is found that the distribution of sample statistics computed from the bootstrap samples is a good representation of the respective distribution of the observed statistics (Vogel 1995).

In order to generate bootstrap samples, a portion size n of the original dataset size N, for instance, n= 50% or half, must be extracted first. SMExRain provides users with two options: common validation and cross validation. In the former option, users can select the first or second halves to bootstrap. In the latter option, a portion of the selected size n is extracted with the starting point selected randomly. After extracting the sample of size n, users can enter the number of bootstrap samples to be generated, for example, hundreds to thousands samples etc. The default value in SMExRain is 1000 samples for stable results and efficient computation costs. Each candidate distribution is then fitted to the generated bootstrap samples and is extrapolated to estimate the right-tail quantiles corresponding to the k largest (k=4 by default) observed rainfall amounts in the full data set (see Figure 1c). The variability in the estimation of these extrapolated quantiles is presented in SMExRain in the form of modified boxplots by default. However, users can also switch to the standard boxplots (Helsel and Hirsch, 2002). The standard boxplot generally portrays the median, interquartile range, and outliers of the investigated data set. In this paper, the modified boxplots are used to show the robustness of each distribution predictive ability. The middle line of a modified box is the sample mean, the box height is twice standard deviation, the upper and lower whisker extend to the maximum and minimum value of the sample respectively (see Figure 1c). Large box widths or long whiskers imply high uncertainty in the estimation of these extreme values. If the observed values fall outside the box, then the distribution fitted to the bootstrap samples has overestimated or underestimated the true values and is therefore not commendable. It is important to know that, rather than considering each distribution at a time, SMExRain allows user to compare the predictive performance of up to 12 probability distributions simultaneously using the same generated bootstrap samples to ensure a fair comparison.

2.5 Construction of IDF relations

In standard engineering practice, the results of at-site rainfall frequency analyses are often summarized and presented in the form of depth-duration-frequency (DDF) relations or intensity-duration-frequency (IDF) relations for each rain-gauge site with adequate rainfall records. In SMExRain, these IDF or DDF curves are available in both tabular and graphical forms for estimated rainfall intensities (or depths) at investigated durations (generally from five minutes to one day) and for return periods of interests (generally from two to a hundred years). Depending upon the choice of an empirical regression equation representing IDF relations, the coefficients are also computed for the ease of use and of interpolation of rainfall intensities at unobserved durations. SMExRain supports many popular regression equations (WMO 2009, Green et al. 2016) in both real-space (with two or three coefficients) and log-space (with polynomial up to order 6). The coefficients are estimated based on the least-square technique.

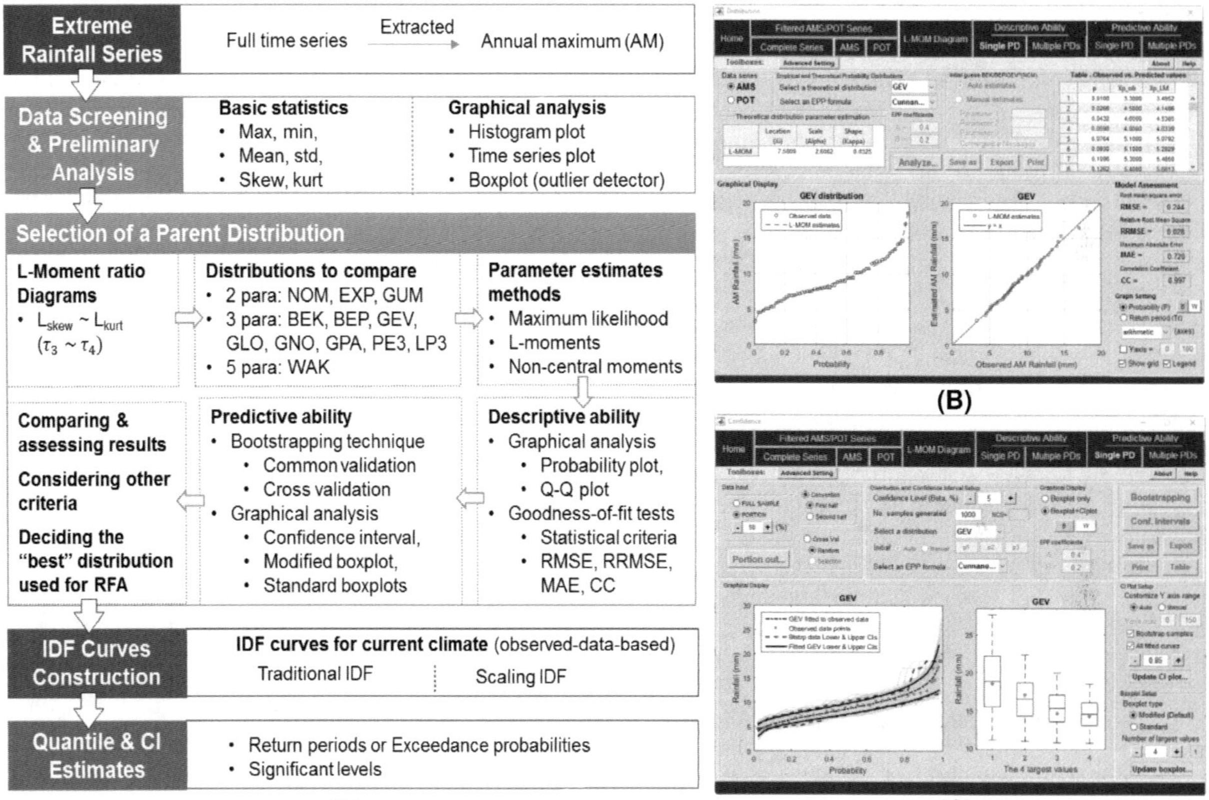

Figure 1: (a) SMExRain structure and functions, (b) SMExRain user interface for the descriptive ability test, and (c) for the predictive ability test

3 ILLUSTRATIVE APPLICATION USING ONTARIO DATA

3.1 Database

For purposes of illustrative application, this paper presents the utilization of SMExRain to identify the "best" distribution(s) for representing the distribution of sub-daily and daily annual maximum rainfalls in Ontario. A total of 252 datasets for three different rainfall durations (5 minutes, 1 hour, and 24 hours) from a network of 84 stations were used. The selection of these stations relied on the quality of the data, the adequate length of available historical extreme rainfall records (containing at least 20 years), and the spatial distribution of the rain-gauges to represent the different climatic conditions in this study area.

3.2 Decision-Support Process

3.2.1 Descriptive Ability Assessment Results

The Q-Q plots of all 252 AMS shows that all distributions can describe well the left-tail and central parts. The right-tail parts, however, are less well described and there are no obvious trends. These values can be accurately estimated, over-estimated, or under-estimated by any of the 11 candidate distributions. For purposes of illustration, only 1-hour AMS from one of the longest record and highest data quality stations – Toronto Int. Airport, is presented here as shown in Figure 2. From the visual standpoint, all distributions seem to perform well in this case, except the BEK and GPA distributions. However, the significance of the differences between the remaining models is difficult to judge merely based on the graphical display, as the differences are minor. A more objective evaluation using numerical comparison criteria is thus necessary.

Rankings of the 11 candidates at each of the 84 stations based on the four statistical criteria were performed. Ranking from number 1 to 11 indicates the gradual decrease from the best to the worst distributions. On the basis of these goodness-of-fit numerical comparison results, it was found that no unique distribution ranked consistently best for all locations and for all three selected rainfall durations. The overall ranks of each distribution were obtained for every test category by summing the individual point rank at each location. The rank sum results of the 84 stations of three durations for different record lengths (i.e. stations contain at least 40 years, 30 years, and 20 years), presented in Figure 3, show that WAK model outperforms the others in describing the distribution of daily and sub-daily AMS. The GEV, GNO, and PE3 models also performed well overall and their scores are close to each other. This can be expected since these models are advocated for use in frequency analyses of hydrologic extreme variables by many previous studies. It is also noticed that the PE3 model performed slightly better than GEV and GNO models for 5-min duration data. However, for data set of longer durations – 1-hour and 24-hour, GEV and GNO are slightly better (see Figure 3). The GLO, LP3, and GEV* distributions moderately performed and they stand among middle positions (see Figure 3). It is also interesting to notice that if only RRMSE criterion is considered, LP3 is the best candidate for data of all durations.

After assessing how well each distribution fit to the overall data sets, the focus is on the right tail region of the distribution since this is the region of importance to engineering design and planning applications. The degree of fit on the right tail of the distributions was visually examined using the quantile-quantile plots to gain further appreciation of the overall fit of the distributions. From the visual standpoint, there was very little to choose from between the various distributions for representing the data used in this study since a very small variability was found in the comparison results for different locations as shown, for instance, in Figure 2 – Toronto Int. Airport

3.2.2 Predictive Ability Assessment Results

The sampling characteristics of extrapolated right-tail quantiles were investigated using the bootstrap procedure for all stations containing at least 30-year record length (i.e. 47 stations). In this study, one thousand bootstrap samples of size equal to half (i.e. 50%) of the actual sample size were drawn with replacement from the observations for 21 stations containing at least 40 years of record. For 26 stations contain between 30-40 years of record, approximate two third (i.e. 65%) of the actual sample size were used in order to make sure that the generated bootstrap samples contain at least 20 years of record. Each candidate distribution was then fitted to the generated bootstrap samples and used to extrapolate the right-tail quantiles corresponding to the four largest observed rainfall amounts in the full data set. The distribution of the extrapolated amounts obtained was presented in the form of the modified box plots. The performance of each distribution in estimating the four largest values was evaluated.

The modified boxplots of 141 AMS show that, generally, the Beta-K, Beta-P gave consistently the worst performance with large sampling variation and bias for all three rainfall durations. For instance, Figure 4 shows the modified boxplots of extrapolated right-tail bootstrap data for 5-min AMS at Toronto Int. Airport. Unlike the BEK and BEP models, the modified boxplots for the WAK model do not show large box widths; however, they reveal long upper whiskers (see Figure 4 as an example). In addition, results of modified boxplots reveal that, the LP3 model produced larger box widths than other remaining distributions, yet it was not as poorly performed as BEK, BEP or WAK (see Figure 4). Although the Gumbel distribution exhibited the lowest sample variation in most cases, it tends to overestimate or underestimate the observed values most frequently. The GEV, GEV*, GLO, GNO, GPA, and PE3 distributions produced satisfactory results at most stations where the box enclosed the observed right-tail values with a reasonable whisker spread and correlation with the observed values. In particular, the GEV, GNO, and PE3 produced almost identical results. Occurrences of over- or under-estimation of largest rainfall amounts did occur for all distributions at several locations.

3.3 Decision-Making Process

In general, it is observed that no one distribution performed the best at every station for each category. This could be due to the strong spatial variabilty of rainfall characteristics within this study region. While it is difficult to provide a clear physical interpretation of the regional variability of the probability distribution parameters, one is still able to rely on the proposed tool to identify the GEV, GNO, and PE3 as the best distributions for a large number of cases considered. Furthermore, it is easy to recognize distributions that perform less satisfactory, it is more difficult to identify the overall best distribution. These three models could be thus used alternately for the frequency analysis of annual extreme rainfalls for a given site as shown for instance in Figure 5 for Toronto Int. Airport station. The difference in extreme design rainfall estimates produced by the three distributions is also further investigated for all stations containing at least 30-year records. Results reveal that the estimated values for return periods within the twice sample lengths are almost identical for the three distributions. However, the GEV model tends to provide slightly higher values for high return periods, while the PE3 model tends to give slightly higher values for low return periods. The three models, therefore, could be used interchangeably in constructing IDF relations and estimating extreme design rainfalls for Ontario region. Nonetheless, if only one probability model is preferred for the entire region, other criteria should be thus considered in the choice of an appropriate distribution. For instance, the GEV model is based on a more solid theoretical basis than the other two distributions because it was derived from the statistical theory of extreme random variables. Therefore, the GEV could be considered as the most suitable distribution if only a unique probability model is required for describing the distribution of annual maximum rainfalls for this study area.

4 CONCLUSIONS

A decision-support tool (SMExRain) has been developed for evaluating systematically the performance of various commonly-used probability distributions in hydrologic frequency analyses in order to identify the most suitable model for representing the distribution of extreme rainfalls for a study region of interest. Based on a number of graphical and numerical criteria, and being equipped with a user-friendly ribbon interface, this tool can be used to assess in an efficient manner the descriptive and predictive abilities of each distribution for a large database of extreme rainfall data of different durations at a given location as well as for a large number of sites. More specifically, it is relied on the results of four goodness-of-fit tests, including root mean square error (RMSE), root mean square relative error (RRMSE), maximum absolute error (MAE), and correlation coefficient (CC), and is also supported by the visual comparison of the probability plots, quantile-quantile plots and the modified and standard box plots from the bootstrap samplings. Furthermore, SMExRain can provide IDF relations for the current climate (i.e. based on historical data) in both tabular and graphical forms for a given location of interest. These IDF relations can also be presented in mathematical formulas (in real or log-space) for the convenience of computation and application in practice.

The proposed SMExRain tool has been tested using extreme rainfall data for other regions in Canada. In this paper, it has been successfully applied to identify the best probability distributions that could provide accurate annual maximum rainfall estimates for the selected Ontario region. In particular, it was found that, among the eleven selected candidates, the GEV, GNO and PE3 provided the most consistent and the best performance for extreme rainfall data for different durations and for a number of locations in this study area. The estimated rainfall values for return periods within the twice sample lengths are almost identical for these three distributions. However, the GEV model tends to provide slightly higher values for high return periods, while the PE3 model tends to give slightly higher values for low return periods. The three models, therefore, could be used interchangeably in constructing IDF relations and for estimating extreme design rainfalls.

Nonetheless, for practical application purposes and for safety in urban water infrastructure design and management, the GEV could be preferable to the GNO and PE3 because it has a more solid theoretical basis and it produces higher estimates (or quantiles) for high return periods as compared to the others.

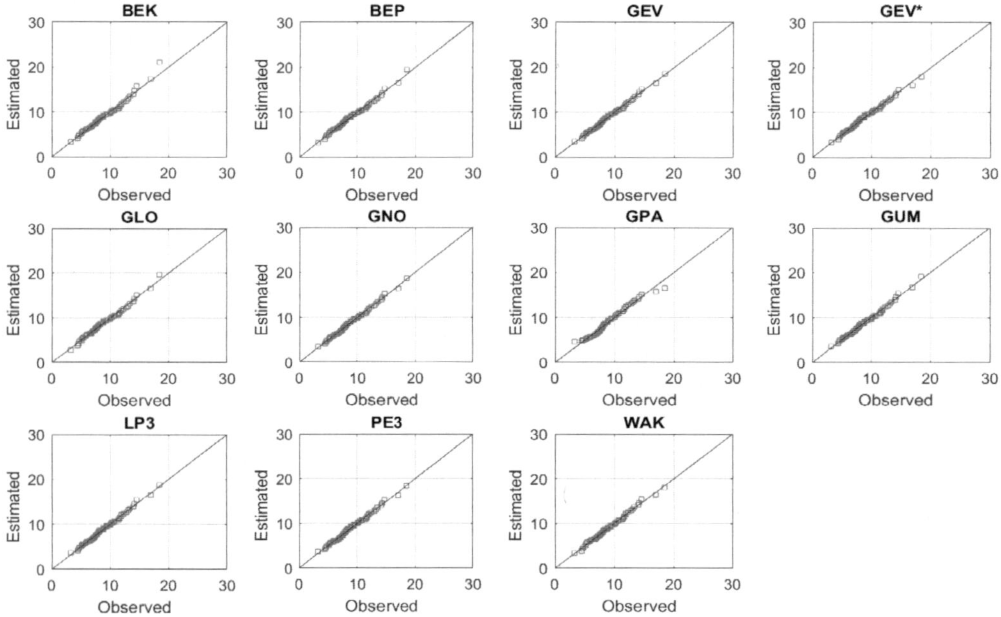

Figure 2: Q-Q plots for distributions fitted to 5-min AMS at Toronto Int. Airport station

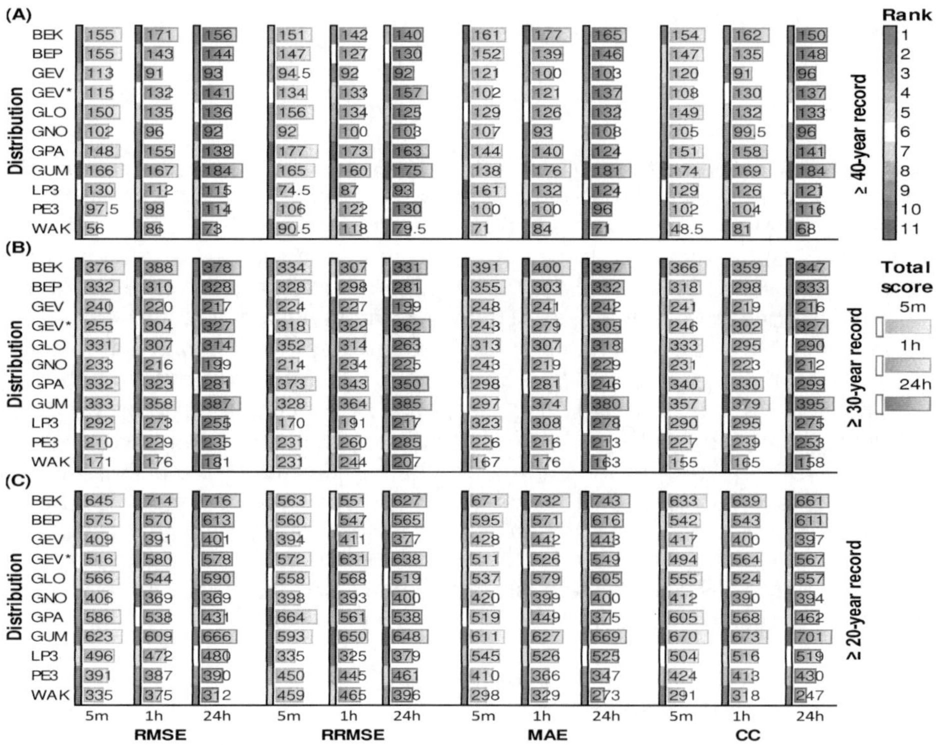

Figure 3: Overall rank for all stations containing at least (A) 40-year, (B) 30-year, and (C) 20-year records based on the four statistical tests for all three durations of 5-min, 1-hour, and 24-hour AMS (The lowest scores or the shortest bar indicates the best distribution)

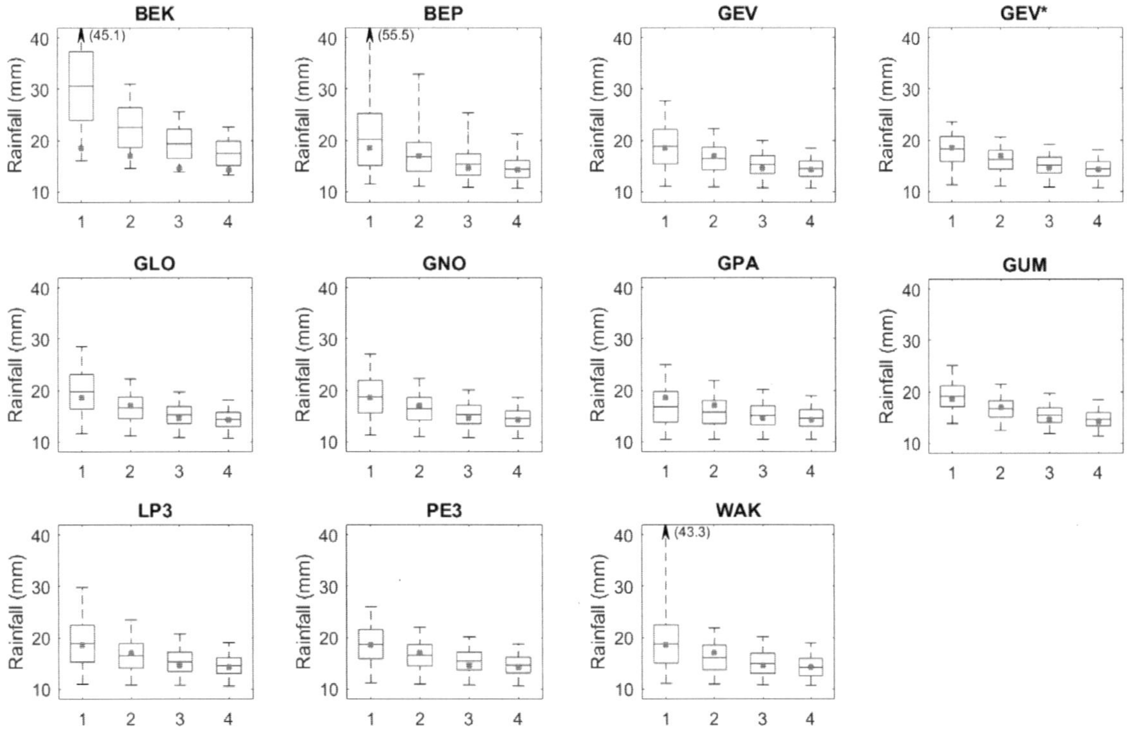

Figure 4: Boxplots of extrapolated right-tail bootstrap data for 5-min AMS at Toronto Int. Airport

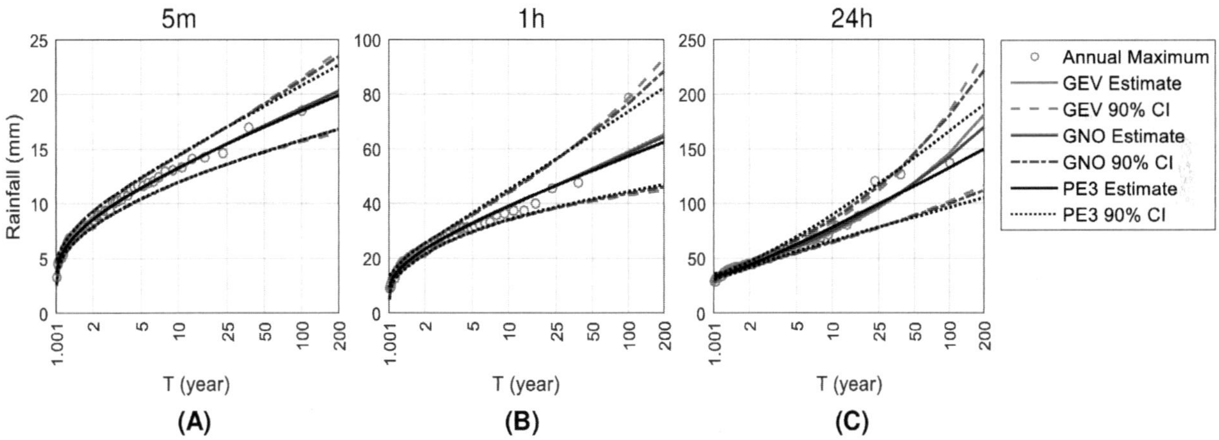

Figure 5: Frequency curves (solid lines) and 90% confidence limits (90% CI, dashed lines) of (a) 5-minute, (b) 1-hour, and (c) 24-hour annual maximum rainfalls (circle markers) at Toronto Int. Airport station using the top three distributions – GEV, GNO, and PE3

ACKNOWLEDGEMENTS

The authors would like to acknowledge the funding provided by the Faculty of Engineering at McGill University and the NSERC FloodNet Strategic Network for this project.

REFERENCES

Chow, V.T. 1964. *Handbook of Applied Hydrology*. McGraw-Hill, New York, USA.

Efron, B. and Tibshirani, R.J. 1994. *An introduction to the bootstrap*. CRC Press, New York, USA.

Environment Canada. 2014. *Engineering Climate Dataset.* Rainfall Intensity-Duration-Frequency (IDF) Tables and Graphs V2.3. http://climate.weather.gc.ca/prods_servs/engineering_e.html. Accessed on Jan 2016.

Green, J., Johnson, F., Beesley, C., The, C. 2016. *Design Rainfall. Book 2, Chapter 2 in Australian Rainfall and Runoff - A Guide to Flood Estimation*, Commonwealth of Australia.

Helsel, D.R. and Hirsch, R.M. 2002. *Statistical Methods in Water Resources Techniques of Water Resources Investigations, Book 4, chapter A3*. U.S. Geological Survey, US.

Hosking, J.R.M. and Wallis, J.R. 1997. *Regional Frequency Analysis: An Approach Based on L-Moments.* Cambridge University Press, Cambridge, UK.

Kite, G.W. 1977. *Frequency and Risk Analyses in Hydrology*. Water Resources Publications, Fort Colling, Colorado, USA.

Mathworks. 2016. *Matlab Compiler Runtime (MCR).* https://www.mathworks.com/products/compiler/mcr/. Accessed on Jan 2016.

Mielke, P.W. and Johnson, E.S. 1974. Some generalized beta distributions of the second kind having desirable application features in hydrology and meteorology. *Water Resources Research,* **10** (2): 223-226.

Murshed, M.S., Kim, S., Park, J.-S. 2011. Beta-κ distribution and its application to hydrologic events. *Stochastic Environmental Research and Risk Assessment*, **25**(7): 897-911.

Nguyen, V-T-V., Nguyen, T-D., Ashkar, F. 2002. Regional frequency analysis of extreme rainfalls. *Water Science and Technology Journal*, **45**(2): 75-81.

Nguyen, V-T-V., Inna, N., Bobee, B. 1989. New Plotting-Position Formula for Pearson Type-Iii Distribution. *Journal of Hydraulic Engineering-Asce*, **115**(6): 709-730.

Nguyen, V-T-V. and Nguyen, T-D. 2008. A spatial-temporal statistical downscaling approach to estimation of extreme precipitations for climate-related impact studies at a local site, *World Environmental and Water Resources Congress 2008*: Honolulu, HI.

Nguyen, V-T-V., Nguyen, T.D., Cung, A. 2007. A statistical approach to downscaling of sub-daily extreme rainfall processes for climate-related impact studies in urban areas, In: Fang, H.H.P., Lee, J.H.W. (Eds.), *Water Science and Technology: Water Supply*, pp. 183-192.

Rao, A.R. and Hamed, K.H. 2000. *Flood frequency analysis.* CRC Press, Boca Raton, London.

Stedinger, J.R., Vogel, R.M., Foufoula-Georgiou, E., 1993. *Frequency Analysis of Extreme Events, chapter 18 in Handbook of Hydrology* edited by D. A. Maidment. McGraw-Hill, New York, USA.

Vogel, R.M. 1995. Recent advances and themes in hydrology. *Reviews of Geophysics*, **33**(S2): 933.

WMO. 2009. *Guide to hydrological practices, volume II: Management of water resources and application of hydrological practices.* 6th edition, World Meteorological Organization, Geneva, Switzerland

Leadership in Sustainable Infrastructure

Leadership en Infrastructures Durables

Vancouver, Canada

May 31 – June 3, 2017/ *Mai 31 – Juin 3, 2017*

TESTING EVOLUTIONARY ALGORITHMS FOR OPTIMIZATION OF WATER DISTRIBUTION NETWORKS

Moosavian, Naser [1,3], and Lence, Barbara[2]

[1] Graduate Research Assistant, Department of Civil Engineering, University of British Columbia, Canada

[2] Professor, Department of Civil Engineering, University of British Columbia, Canada

[3] naser.moosavian@alumni.ubc.ca

Abstract: Water distribution networks (WDNs) are one of the most important elements in the urban infrastructure system and require large investment for construction. Design of such networks is classified as a large combinatorial discrete non-linear optimization problem. The main concerns associated with optimization of water distribution networks are related to the nonlinearity of the discharge-head loss relationships for pipes and the discrete nature of pipe sizes. This paper compares different techniques, all based on evolutionary algorithms (EAs), which yield optimal solutions for design of water distribution networks. The fundamental concept of EAs is that they search for the global optimum with populations of solutions, rather than by improving a single solution, as in Newton-based and other search methods. EAs start with an initial population and use different operators on these populations to change their composition and improve their performance over repeated iterations, or generations. In this paper, six EAs are applied for design of two benchmark pipe networks, the Two-Loop and Hanoi networks, and one real water distribution system for the City of Farhadgerd, Iran. Results show that as the size of the network increases, the soccer league competition algorithm increasingly becomes the most efficient among these algorithms, and consistently converges to the global optimum.

1 INTRODUCTION

EAs are becoming increasingly popular for their use in solving engineering decision problems and in practical applications such as calibration, because they: (i) are based on rather simple concepts and are easy to implement; (ii) do not require gradient information; (iii) can consider and bypass local optima; and (iv) can be utilized in a wide range of problems covering different disciplines. Generally, the main structure and searching process of different EAs are similar, however, their operators may vary. Over the last three decades, many heuristic optimization techniques have been successfully used to identify water network designs, see, e.g., applications of genetic algorithms (Murphy and Simpson 1992); simulated annealing (Cunha and Sousa 2001); harmony search (Geem 2006); shuffled frog leaping (Eusuff and Lansey 2003); ant colony optimization (Maier et al. 2003); particle swarm optimization (Suribabu and Neelakantan 2006); cross entropy (Perelman and Ostfeld 2007); scatter search (Lin et al. 2007); differential evolution (Vasan and Simonovic 2010); self-adaptive differential evolution (Zheng et al. 2013); soccer league competition (Moosavian and Roodsari 2014); and improved genetic algorithms (Bi et al. 2015).

Because these heuristic algorithms were originally developed to address specific engineering problems, there is no guarantee that the global optimum may be found or that the method will be efficient in the solution of specific problems, such as the design of water distribution systems. Wolpert and Macready (1997) cite the No-Free-Lunch (NFL) theorem, assert that no one optimization algorithm may be suited for solving all kinds of optimization problems, and underscore the need for new algorithms that may address a wide range of problems or improve on efforts to reach global optima. In the area of water distribution systems, some comparisons of EA algorithms have been undertaken (Zheng et al. 2013; Moosavian and Roodsari 2014; Bi et al. 2015), though these studies focus on a select few methods. This paper presents a rigorous comparison of six EAs in application to the optimum design of WDNs, and evaluates the algorithms in terms of the best solution obtained, the speed of convergence, and the numbers of function evaluations.

In the optimization of water distribution systems, the primary objective is to minimize the total cost of the network. This cost is a function of pipe size diameters and these pipe diameters in turn affect the network pressures. The main constraint of the problem is to satisfy minimum required pressure at all network nodes. In applications of EAs, at each evaluation of the objective function, hydraulic analysis of the network must be conducted, which exacts a high computational burden, and is in fact the most computationally intensive portion of the optimization process. Therefore, the optimal solution, for a fixed number of function evaluations, is a suitable comparator for EAs. In this paper, six meta-heuristic algorithms are compared in terms of the WDN design selected for three example networks, under different stopping criteria based on the allowable number of function evaluations. To guard against selection of local optima, for each criterion and each pipe network, these algorithms are implemented 20 times. The algorithms include the: genetic algorithm (GA); harmony search (HS); differential evolution (DE); particle swarm optimization (PSO); artificial bee colony (ABC); and soccer league competition (SLC) meta-heuristics. In the following sections, the general WDN optimization model is introduced, the algorithms are briefly described, their application to three example networks are assessed, and conclusions are drawn from this comparison.

2 PROBLEM FORMULATION

A water distribution system is a collection of many components such as pipes, reservoirs, pumps and valves which are connected in order to provide water to consumers. The optimal design of such a system can be defined as the best combination of component sizes and component settings (e.g., pipe size diameters, pump types, pump locations and maximum power, and reservoir storage volumes) that gives the minimum cost for the given layout of network, such that hydraulic laws governing continuity of flow and energy are maintained and constraints on quantities and pressures at the consumer nodes are fulfilled. In this paper, water distribution system design is formulated as a least-cost optimization problem with the selection of pipe sizes as the decision variables, while pipe layout and its connectivity, nodal demand, and minimum pressure head requirements, are imposed. The optimization problem is stated mathematically as:

[1] $$Min\ C=\sum_{k=1}^{np} c_k(D_k)\times L_k$$

where $c_k(D_k)\times L_k$ = the cost of pipe k with length L_k and diameter D_k, and np = the number of pipes in the network. This objective function is minimized under the following constraints:

Flow continuity at nodes

For each node, flow continuity must be satisfied,

[2] $$\sum Q_{in}-\sum Q_{out}=q_j, \qquad \forall k\epsilon nn$$

where q_j = demand at node j (meters^3/second); nn = number of nodes; and Q_{in} and Q_{out} = flow into and out of node j (meters^3/second), respectively.

Energy conservation in loops

The total head loss around a closed pipe loop should be equal to zero, or the head loss along a loop between two fixed head reservoirs should be equal to the difference in water level of the reservoirs:

[3] $$\sum_{k \in loop\, l} hf_k = \Delta H, \qquad \forall l \epsilon nl$$

where ΔH = difference between nodal pressures at both ends of a path (meters), and $\Delta H = 0$, if the path is closed; nl = number of loops; and hf_k = head loss due to friction in the pipe *k* (meters) which is obtained from following equation:

[4] $$hf_k = H_i - H_j = R_k Q_k^n$$

where H_i and H_j = the nodal heads at the start node and the end node of the pipe (meters); R_k = the resistance coefficient of the *k*th pipe with flow rate Q_k (second/ meters^2); and *n* = a constant depending on the head loss equation, and is 1.852 for the most common expression for head loss, the Hazen-Williams head loss formulation.

Minimum pressure at nodes

For each junction node in the network, the pressure head should be greater than the prescribed minimum pressure head:

[5] $$H_j \geq H_j^{min}, \forall\, j \in nn$$

where H_j = the pressure head at node *j* (meters); *nn* = the number of nodes; and H_j^{min} = the minimum required pressure head (meters).

In this work, the Global Gradient Algorithm, GGA (Todini and Pilati 1988), in a MATLAB environment, was applied to conduct the hydraulic analysis of network. GGA satisfies the continuity and energy conservation equations (Eq 2-4), while calculating the pressure head H_j at each junction node and the discharge Q_k in each pipe.

Pipe size availability

The diameter of the pipes should be available from a commercial size set:

[6] $$D_k = \{D(1), D(2), ..., D(K)\}, \quad \forall k \in np$$

where K = the number of candidate diameters.

3 EVOLUTIONARY ALGORITHMS (EAs)

Generally, EAs are implemented in five main steps, including: (i) creation of a random initial population; (ii) evaluation of the cost function or fitness; (iii) selection of two or more solution vectors for the evolution process; (iv) implementation of an evolution strategy with different operators; and (v) selection of good solutions, and replacement into the population. Differences in EAs most often include variations in how they evolve new solution vectors, how they select these solutions from the population, and how they replace solutions into the updated population. In this section, a summary of the different EAs compared in this paperis presented.

3.1 Genetic Algorithm (GA)

Implementation of the GA (Holland 1972) begins with identification of a random set, or population, of solutions (chromosomes). The cost, or fitness, of each solution is determined by evaluation of the objective function and compared to determine those solutions, or population members, that are allowed to evolve. Typically the comparison process employs tournament selection, which involves a series of competitions, or "tournaments," among pairs of solutions selected at random from the population. The winner of each tournament is the solution with the best fitness. These winners are selected to undergo evolution with crossover and mutation operations, probabilistic-based mechanisms, through which new solutions are constructed. The new solutions are then evaluated and may be used to further evolve the population, if they provide better fitness (i.e., improved objective function value), than other population members. The process is continued for a large number of generations or until no further improvement in the bestobjective function is obtained.

3.2 Harmony search (HS)

The HS algorithm (Geem, 2006) is conceptualized based on the musical process of searching for a 'perfect state' of harmony, such as in jazz improvisation. Jazz improvisation seeks a best state (fantastic harmony) determined by aesthetic estimation, and analogously HS seeks a best state (global optimum) determined by evaluating the objective function. Aesthetic estimation is evaluated by the set of pitches played by each instrument, and in HS the objective function evaluation is performed by the set of values assigned by each decision variable. The harmony quality is enhanced with practice, and in HS the solution quality is enhanced over iterations. Each new harmony or solution vector is generated based on three rules: memory consideration, pitch adjustment, and random selection. If this new harmony is better than the existing worst harmony in memory, the new harmony is included in memory and the worst harmony is excluded from memory. This procedure is repeated until what is defined as a fantastic harmony is found.

3.3 Differential Evolution (DE)

DE (Price et al. 2005) is an improved version of GA. Similar to the GA, there are three important operators involved in the DE algorithm including the mutation, crossover, and selection operators. The main difference between GA and DE is that GA relies on its crossover operator to exchange information among solutions, while DE primarily relies on its mutation operator to form new solution vectors. Mutation is based on the fitness difference of randomly sampled pairs of solutions in the population. DE automatically adapts the mutation increments (i.e., step size of allowable variable changes), based on the stage of the evolutionary process. At the beginning of the evolution process, the mutation operator favors exploration and as evolution progresses, it favors exploitation. The algorithm uses a uniform crossover operator that can take child vector parameters from one parent more often than from the other. By using components of existing population members to construct trial vectors, the crossover operator efficiently shuffles information about successful combinations, enabling the search for an optimum to focus on the most promising area of the solution space.

3.4 Particle swarm optimization (PSO)

PSO (Kennedy and Eberhart 1995) is an evolutionary optimization algorithm which originated as a simulation of a simplified social system such as birds flocking and fish schooling. Similar to GA, PSO is also population-based and searches for optimal solutions by updating generations. However, unlike GA, PSO possesses no evolution operators. Instead, PSO relies on the exchange of information between individuals, or particles, of the population, or swarm. In order to select fitter particles for further populations, each particle adjusts its trajectory towards its current position, towards its previous best position and towards the current best position attained by any other member in its neighborhood. Compared with GA, PSO presents the advantage of being conceptually very simple and fast. However, its main disadvantage is the risk of a premature search convergence.

3.5 Artificial Bee Colony (ABC)

In the ABC algorithm (Karaboga 2005), the position of a food source represents a possible solution to the optimization problem and the nectar amount of a food source corresponds to the quality (fitness) of the associated solution. Search processes start based on three kinds of bees: employed bees; onlooker bees; and scout bees. The number of the employed and onlooker bees is equal to the number of solutions considered in the population. The employed bee whose food source has been exhausted by the bee colony becomes a scout bee. Each cycle of the search procedure consists of the following actions: employed bees are sent onto their food sources and their nectar amounts are evaluated; employed bees share the nectar information of food sources with the observer bees, and based on this information, onlooker bees select among food source regions, and evaluate the nectar amount of the selected food sources. This process continues iteratively, and iff a solution represented by a food source is not improved by a pre-determined number of iterations, then the food source is abandoned and the employed bee for that source is converted to a scout, and is sent randomly onto possible new food sources. In this case, a new solution vector is randomly created and replaces the abandoned solution.

3.6 Soccer League Competition (SLC) algorithm

The SLC algorithm (Moosavian and Roodsari 2014) is inspired from professional soccer leagues. It involves different teams, or collections of solutions, where each solution is a team member, and a number of operators that act on the team members to perform an effective search for finding the near optimal solution. SLC mimics matches between teams and determines the winners based on their relative power, where the winner of a match has a higher probability of increasing its power for future matches. After each match, all players (decision vectors) on the winning team undergo operations that change their decisions, and thus their power (i.e., strengthening or weakening each player), producing modified team members. Finally, the original strength of each player on this winning team is compared with its modified strength and the player, i.e., decision vector, with the greater strength, is allowed to play on the team in the next match.

An iteration, or round, is defined as a set of matches that allows each team, or modified team, to play with each other team in the league. Thereafter, the next round is played with these new teams. The user specifies the stopping criteria to be either based on a limit of the number of rounds undertaken, or the number of function evaluations made.

4 APPLICATIONS TO WDN DESIGN

The performance of the six above-mentioned EAs is tested for three water distribution networks, the Two-Loop and Hanoi benchmark networks, and the WDN for the City of Farhadgerd, Iran.

For consistency in comparison, three stopping critera are defined as: (i) the maximum number of function evaluations is set to 1000. Given this criteria, the initial speed of convergence and performance of the algorithms may be evaluated; (ii) the maximum number of function evaluations is set to 10000. Here, the performance of the algorithms is evaluated for a more mature evolution, compared with (i); and (iii) the maximum number of function evaluations is set to 40000, and the evolution of the algorithms is considered to be fully mature. For each of these criteria, all algorithms are executed 20 times, i.e., 20 sets of initial populations of solutions are generated based on 20 different random number sequences.

Statistical analyses of the results for all algorithms are performed and are presented as box plots for the 20 different executions of the algorithm. All of the computations are implemented in the MATLAB programming language environment with an Intel(R) Core(TM) 2Duo CPU P8700 @ 2.53GHz and 4.00 GB RAM.

4.1 Two-loop benchmark network

This network, shown in Figure 1, is a hypothetical benchmark which has seven nodes and eight pipes with two loops, and is fed by a reservoir with a 210-m fixed head (Alperovits and Shamir 1977). The pipes are all 1,000 m long with a Hazen-Williams coefficient of 130. The minimum pressure limitation is 30 m above ground level for all nodes. There are 14 available commercial pipe diameters. The degree of candidate diameter is an indicator of the size and complexity of the network problem, and is defined as the number of candidate diameters divided by the number of pipes. For this case, the number of candidate diameters is equivalent to $14 / 8 = 1.75$. Geem (2006) determines the minimum cost of this network to be $419,000 US.

The performance of the algorithms, in terms of the minimum total cost obtained, under the three stopping criteria, is shown in the box plots in Figures 2, 3, and 4, for the 1000-, 100000, and 40000-function evaluation stopping criteria, respectively. The results show that, for the first stopping criterion, the PSO alsgorithm exhibits the best convergence speed, and over the 20 executions of the algorithms, its minimum and mean cost is $423,000 and $435,000 US, respectively. Of the remaining algorithms, the GA and HS algorithmsresult in the lowest and second lowest mean and minimum costs, respectively, and the DE algorithm exhibits the worst performance among all algorithms.

As the stopping criterion is relaxed to 10000 function evaluations, the performance of the ABC, SLC, HS, and DE algorithms improves significantly, and each of these algorithms achieve the global optimal minimum cost. Here, the ABC, DE, and SLC algorithms have the least standard deviation, in this order. These small standard deviation values show that the algorithms perform the search process consistently, for the breadth of different initial random populations. In contrast, the PSO algorithm also achieves the global optimum, but its average performance over 20 executions is relatively poor.

Finally, for the stopping criterion of 40000 function evaluations, the box plot shown in Figure 4 indicates that the high-performance algorithms are ABC, DE, SLC, and HS, in decreasing order of performance.

4.2 Hanoi benchmark network

The Hanoi network, shown in Figure 5, consists of 32 nodes, 34 pipes with 3 loops, and is fed by gravity from a reservoir with a 100-m fixed head. All pipes have a Hazen-Williams coefficient of 130 and the minimum head limitation at all nodes is 30 m above ground level (Alperovits and Shamir 1977). There are six possible pipe diameters and 34 pipes in the system, thus the degree of candidate diameter is equivalent to *6 / 34 =0.1765*. The global optimum solution has the total cost of $6.1 million US (Alperovits and Shamir 1977).

Optimization results for this network are provided in Figures 6, 7, and 8 for the 1000-, 10000-, and 40000-function evaluation stopping criteria, respectively. For the first criterion, the SLC algorithm exhibits the highest performance. with a the minimum and mean least cost equal to $7.0 and $8.3 million, respectively. It also has the smallest standard deviation. The PSO and GA algorithm obtain minimum costs of $11 and $8.4 million US, respectively. For the 10000-function evaluation criterion, the SLC also exhibits the best performance, and has a minimum and mean least cost of $6.1 and $6.3 million US, respectively. The DE algorithm has the smallest standard deviation, but does not find the lowest cost overall. The GA and DE algorithms are the second and third best algorithms, respectively. A similar trend is exhibited for the 40000-function evaluation criterion, where the SLC and DE algorithms exhibit the highest performance, in this order, and PSO exhibits the worst performance. This indicates that the convergence properties of the PSO algorithm do not improve when the number of iterations increases. A comparison of the Hanoi and Two-Loop benchmark networks indicates that as the degree of candidate diameter decreases, the convergence potential of the SLC algorithm improves significantly.

4.3 Farhadgerd network

The Farhadgerd WDN serves a population of approximately 8200 in the town of Farhadgerd, Iran, a residential community near the regional capital, Mashhad. The network shown in Figure 9, comprises 64 pipes and 53 nodes, has one reservoir with a head of 510 m, and a minimum pressure requirement of 20 m at all nodes. There are nine possible pipe diameters and thus the degree of candidate diameter for this network is equivalent to *9 / 64 =0.14*. The recorded optimal cost for Farhadgerd is $120 million US.

Optimization results for this network are provided in Figures 10, 11, and 12 for the 1000-, 10000-, and 40000-function evaluation stopping criteria, respectively. For the first criterion, the SLC surpasses the other algorithms with a minimum and mean least cost of $210 and $240 million US, respectively. The ABC algorithm exhibits the worst performance by far with respect to the others under this criterion. Among the other algorithms, the mean result of the PSO algorithm indicates that it performs second-best, with a minimum and mean least cost of $260 and $290 million US, respectively. For the 10000-function evaluation criterion, the SLC algorithm again out-performs all other algorithms, with a minimum and mean least cost of $120 and $160 million US, respectively. The GA algorithm performs second-best, under this criterion. The DE algorithm results are comparable with other algorithms, and exhibit the smallest standard deviation, however, it is unable to find a global optimum solution. These findings are similar for the cases with a stopping criterion of 40000 function evauations. The minimum and maximum least cost solutions obtained by the SLC algorithm are $120 and $140 million US, respectively, which are less than all solutions obtained by other algorithms. This confirms that for the range of networks investigated in this paper, the performance of the SLC improves as the degree of candidate diameter decreases, though the DE algorithm has the lowest standard deviation.

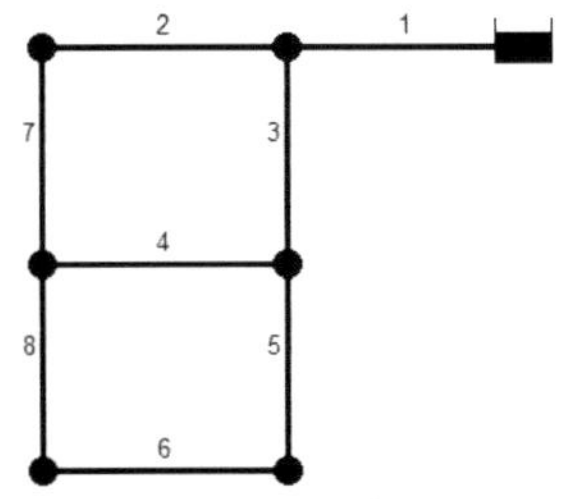

Figure 1: Two-Loop benchmark network

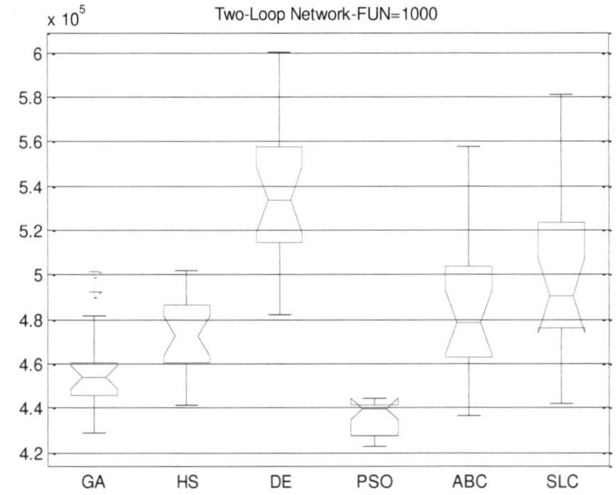

Figure 2: Box-plot of least cost solutions for the Two-Loop benchmark network: 1000 function evaluations

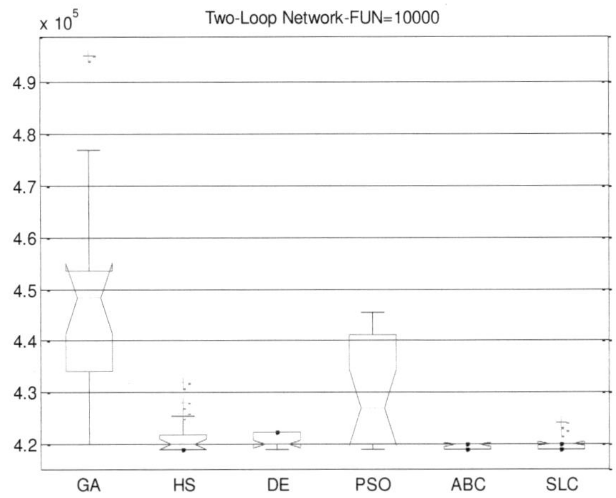

Figure 3: Box-plot of least cost solutions for the Two-Loop benchmark network: 10000 function evaluations

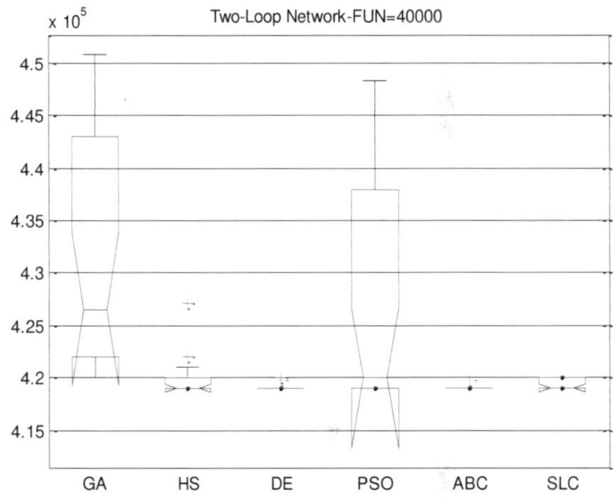

Figure 4: Box-plot of least cost solutions for the Two-Loop benchmark network: 40000 function evaluations

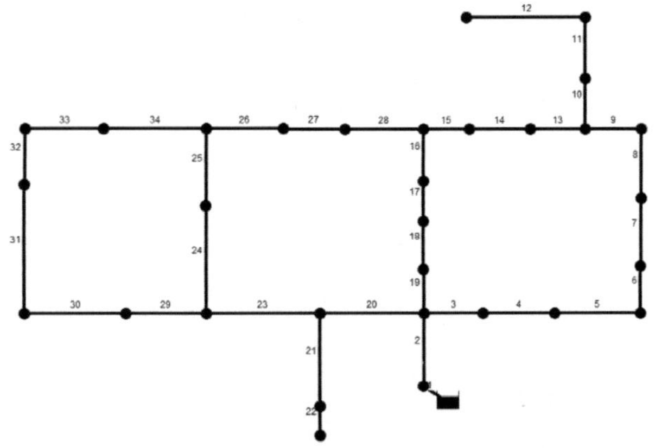

Figure 5: Hanoi benchmark network

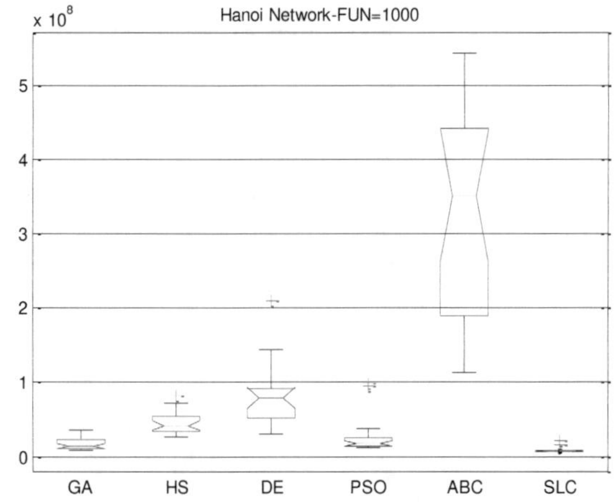

Figure 6: Box-plot of least cost solutions for the Hanoi benchmark network: 1000 function evaluations

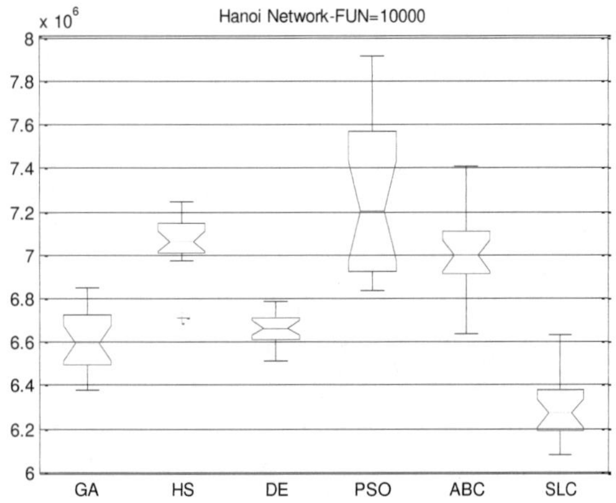

Figure 7: Box-plot of least cost solutions for the Hanoi benchmark network: 10000 function evaluations

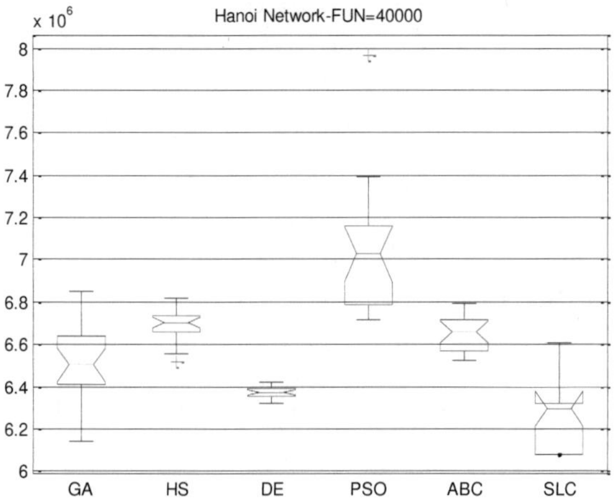

Figure 8: Box-plot of least cost solutions for the Hanoi benchmark network: 40000 function evaluations

5 CONCLUSION

In this paper, six EAs are evaluated for their performance in finding the optimal least-cost design of water distribution networks, under three pre-defined stopping criteria based on function evaluations. The performance is shown to be dependent on the stopping criteria, and on the degree of candidate diameter, which is an indication of the size and complexity of the network. For larger benchmark networks, the SLC, ABC, and DE algorithms result in better performance. For the Farhadgerd WDN, with a large number of pipes, and small candidate diameter, the SLC algorithm exhibits the best performance in terms of mean and minimum total cost.

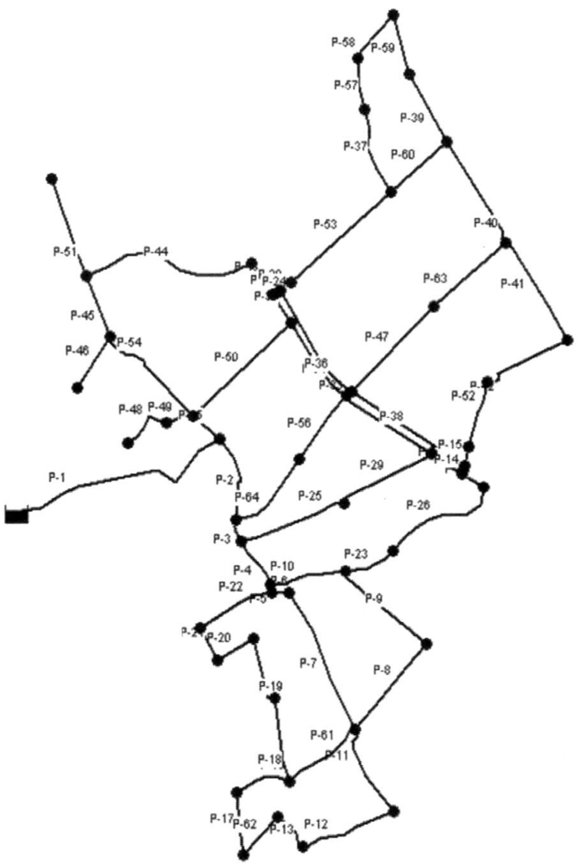

Figure 9: City of Farhadgerd network

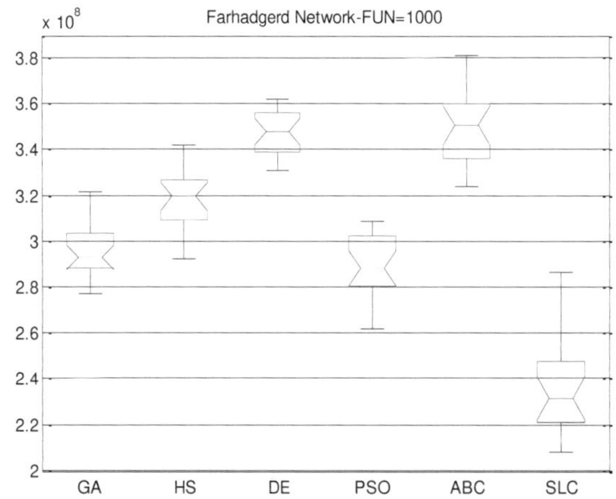

Figure 10: Box-plot of least cost solutions for the City of Farhadgerd network: 1000 function evaluations

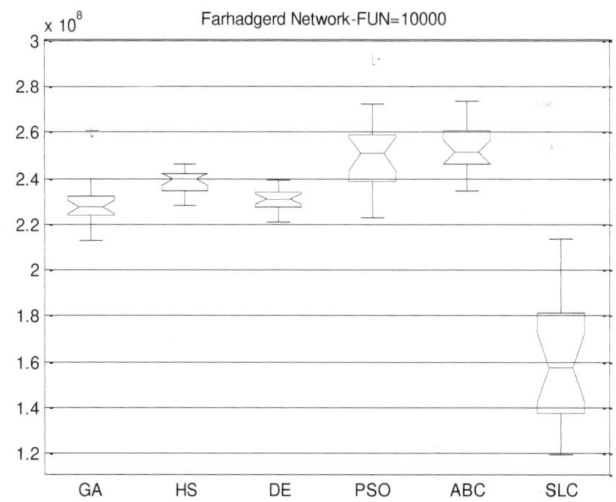

Figure 11: Box-plot of least cost solutions for the City of Farhadgerd network: 10000 function evaluations

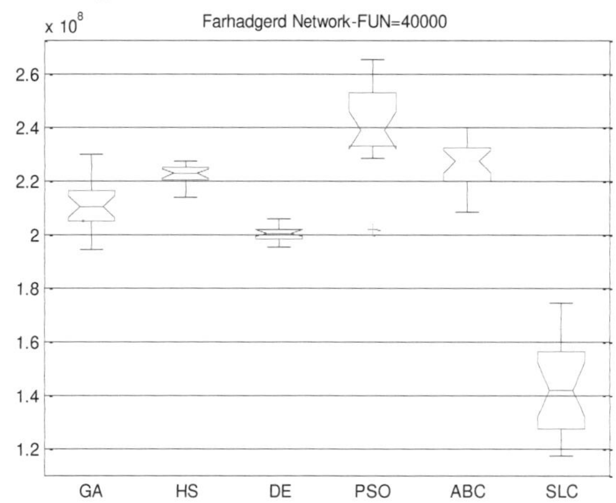

Figure 12: Box-plot of least cost solutions for the City of Farhadgerd network: 40000 function evaluations

6 REFERENCES

Alperovitz, E. and Shamir, U., 1977. Design of Optimal Water Distribution Systems. *Water Resour. Res.,* **13**(6): 885–900.

Bi. W., Dandy, G.C. and Maier, H.R. 2015. Improved Genetic Algorithm Optimization of Water Distribution System Design by Incorporating Domain Knowledge, *Environmental Modelling & Software,* **69**: 370–381.

Cunha, M.C. and Sousa, J. 2001. Hydraulic Infrastructures Design Using Simulated Annealing, *J. Infrastruct. Syst.,* **7**(1): 32–38.

Holland, J. H. 1975. Adaptation in Natural and Artificial Systems: An Introductory Analysis with Applications to Biology, Control, and Artificial Intelligence. Ann Arbor, MI: University of Michigan Press, 1975.

Eusuff, M.M. and Lansey, K.E. 2003. Optimisation of Water Distribution Network Design Using Shuffled Frog Leaping Algorithm, *J. Water Resour. Plan. Manag.,* **129**(3): 210–225.

Geem, Z.W. 2006. Optimal Cost Design of Water Distribution Networks Using Harmony Search, *Eng. Optim.*, **38**(3): 259–280.

Karaboga, D. 2005. An Idea Based On Honey Bee Swarm For Numerical Optimization, Technical Report-TR06, Erciyes University, Engineering Faculty, Computer Engineering Department, 2005.

Kennedy, J. and Eberhart, R.C. 1995. Particle Swarm Optimization, *in: Proceedings of the 1995 IEEE International Conference on Neural Networks*, **4**: 1942–1948.

Lin, M.D., Liu, Y.H., Liu, G.F. and Chu, C.W. 2007. Scatter Search Heuristic For Least-Cost Design of Water Distribution Networks, *Eng. Optim.,* **39** (7): 857–876.

Maier, H.R., Simpson, A.R., Zecchin, A.C., Foong, W.K., Phang, K.Y., Seah, H.Y. and Tan, C.L. 2003. Ant Colony Optimization For The Design of Water Distribution Systems, *J, Water Resour. Plan. Manag.,* **129** (3): 200–209.

Moosavian, N and Roodsari, B.K. 2014. Soccer League Competition Algorithm: A Novel Meta-Heuristic Algorithm For Optimal Design of Water Distribution Networks, *Swarm and Evolutionary Computation*, **17**: 14-24.

Murphy, L.J. and Simpson, A.R. 1992. Pipe Optimization Using Genetic Algorithms, Research Rep. No. **R93**, Dept. of Civil Engineering, Univ. of Adelaide, Adelaide, Australia.

Perelman, L. and Ostfeld, A. 2007. An Adaptive Heuristic Cross-Entropy Algorithm For Optimal Design of Water Distribution Systems, *Eng. Optim.,* **39** (4): 413–428.

Price, K.V., Storn, R.M. and Lampinen, J.A. 2005. (Eds.), Differential Evolution: A Practical Approach to Global Optimization, Springer Natural Computing Series.

Suribabu, C.R. and Neelakantan, T.R. 2006. Design of Water Distribution Networks Using Particle Swarm Optimization, *Urban Water J.,* **3** (2): 111–120.

Todini, E., and Pilati, S.1988. A Gradient Algorithm for the Analysis of Pipe Network, *Computer Applications in Water Wupply, **1**, System Analysis and Simulation,* B. Coulbeck and O. Chu-Hou, eds., Wiley, London, 1–20.

Vasan, A. and Simonovic, S. 2010. Optimization of Water Distribution Network Design Using Differential Evolution, *J. Water Resour. Plan. Manag.,* **136** (2): 279–287.

Wolpert, D.H. and Marcready W.G. 1997. No Free Lunch Theorems For Optimization, *IEEE Transactions On Evolutionary Computation*, **1**(1): 67-82.

Zheng, F., Zecchin, A.C. and Simpson, A.R. 2013. Self-Adaptive Differential Evolution Algorithm Applied to Water Distribution System Optimization, *J. Comput. Civil Eng.,* ASCE, **27**(2): 148–158.

Leadership in Sustainable Infrastructure
Leadership en Infrastructures Durables

Vancouver, Canada
May 31 – June 3, 2017/ *Mai 31 – Juin 3, 2017*

THE DEVIL IS IN THE DETAILS: ASSUMPTIONS AND REALITY COLLIDE IN URBAN DIKE BREACH MODELLING

David Roche[1,3], Alisson Seuarz[1], David Sellars[1], and David Roulston[2]
[1] Kerr Wood Leidal Associates Ltd., Canada
[2] District of Squamish, Canada
[3] droche@kwl.ca

Abstract: Two-dimensional hydrodynamic modelling is a preferred tool for simulating complex outflow from dike breaches in urban areas. As always, engineers must ensure that results are based on an appropriate suite of assumptions and processes. Too much detail increases engineering costs while too little can lead to misleading model results and inefficient risk mitigation decisions. The costs of such decisions can exceed the modelling budget by orders of magnitude as risk management strategies are implemented. As part of the District of Squamish Integrated Flood Hazard Management Plan (IFHMP), a detailed dike breach model was developed for the Squamish River floodplain. The model considers the interdependence between architecture and preferential flow pathways as well as alternative representations of floodplain structures, compatibility of climate change and development assumptions, the iterative nature of floodproofing and flood risk, and the potential for breaches occurring anywhere along the developed floodplain's 20 km perimeter. To further complicate matters, Squamish's sea level rise planning area and river dike breach zone overlap within the downtown core. The IFHMP explores options for a sea dike and examines the effects of intentionally breaching the proposed sea dike to reduce floodplain inundation during a river dike breach. Model results highlight some significant changes from the (now decades-old) 1D modelling. Squamish faced difficult decisions about long-term floodplain land use and upgrades to the dike system. Detailed modelling supported the adoption of new structural, non-structural and policy measures to manage flood risk as part of a sustainable community plan.

1 INTRODUCTION

In 2014, the District of Squamish (District) retained a multidisciplinary consulting team led by Kerr Wood Leidal Associates Ltd. (KWL) to prepare an Integrated Flood Hazard Management Plan (IFHMP). The three-year project is summarized in a series of four reports that provide Squamish with an updated suite of tools for managing flood risk (KWL, 2017a, b, c; KWL and Arlington Group, forthcoming). The scope of work included two-dimensional (2D) hydraulic modelling of the Squamish River and Mamquam River floodplains under dike breach conditions. The 2D modelling described herein superseded steady-state 1D modelling completed as part of the District's 1994 Flood Hazard Management Plan (Klohn Leonoff, 1994).

The objectives of the 2D dike breach modelling were to better understand how the hazards of a dike breach would affect and interact with floodplain development, and to support a discussion of mitigation opportunities. A key modelling deliverable was the generation of new Flood Construction Levels (FCLs) appropriate for development and redevelopment within the dike-protected floodplain areas, as required by BC provincial guidelines (BC MWLAP, 2004). Other modelling deliverables included inundation depth maps, flood velocity maps, and physical hazard maps based on the United Kingdom's Hazard Rating concept (i.e., identifying the most critical combination of depth and velocity).

A total of eight dike breach scenarios spread across two floodplain areas provided a baseline indication of hazard and consequence using a simplified assessment procedure that focussed on a subset of potential hazards. Model scenarios were selected to reflect the conditions most likely to govern flood risk management decisions, considering flood magnitude, planning horizon, and breach location.

The District was particularly interested in moving beyond a simple hazard assessment toward a risk-based approach. Although available funding did not support a comprehensive Quantitative Risk Assessment (QRA), model results were used to generate economic consequence calculations using the HAZUS-MH platform. Social and economic consequences were assessed using new GIS algorithms developed as part of the IFHMP. The complete suite of consequence results can be incorporated into a comprehensive QRA framework as more modelling is completed under future projects. The model itself will be updated periodically to evaluate development applications that propose to vary existing zoning or land use.

The most important consideration for the model was to balance simplifying assumptions needed to meet scope and budget constraints against the need for detail in representing complex floodplain behaviours. Failure to achieve an appropriate representation could leave the District and its constituents exposed to unacceptable risks or facing millions of dollars in costs for excessive mitigation over the lifespan of the IFHMP. This paper describes the balance between simplicity and detail that was achieved for the IFHMP.

2 MODEL DOMAIN AND SETUP

The Mamquam River divides the developed Squamish River floodplain into an "upper" (north) floodplain area and a "lower" (south) floodplain area (Figure 1). Each floodplain has an extensive urban footprint, and is protected by an independent dike system. This allows dike breaches on the upper and lower floodplains to be modelled separately.

Dikes at the downstream (south) end of the upper floodplain stop overland flow from returning to the river, and internal drainage works would be quickly overwhelmed or blocked during a dike breach event. Inflow from a dike breach is therefore trapped in the floodplain until it rises enough to overtop the confining dike. The result is extensive internal ponding (a "bathtub" effect) and water levels that exceed those in the adjacent river. A proposed sea dike around downtown Squamish will create a similar situation for the lower floodplain south of the Mamquam River, as shown in Figure 2 below.

The IFHMP Background Report (KWL, 2017a) updates pre-existing peak flow hydrology and 1D modelling of the Squamish River and Mamquam River. These updates become upstream boundary conditions for the 2D dike breach model. The report also describes a coastal flood analysis that provides downstream boundary conditions.

Modelling software had to be commercially available and able to integrate the pre-existing 1D river model with an unstructured, high-resolution 2D representation of the floodplain. DHI's MikeFlood software package was selected and run on a desktop PC with quad-core Intel i7 CPU at 2.8 GHz supported by 16 GB RAM. Two-dimensional calculations were executed using an NVIDIA Titan GPU. Simulation time for a single high-precision dike breach simulation varied from about one to several days, and would have been impractical without GPU parallelization.

Model topography was developed from a composite of high-resolution 2013 LiDAR, previous ground-level surveys of the river channels and dike crest, and near-shore coastal bathymetry data obtained from Canadian Hydrographic Services. The model incorporated cadastral, zoning, growth, roadway, and economic data drawn from the District's GIS system. Digitized roof polygons were also provided for existing buildings within the floodplain. High-level information concerning future land use for Squamish Nation Reserves within the floodplain was obtained directly from the Squamish Nation.

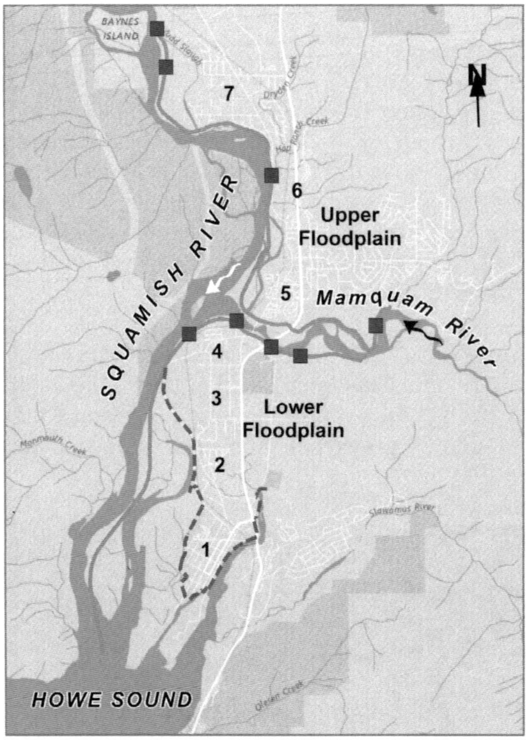

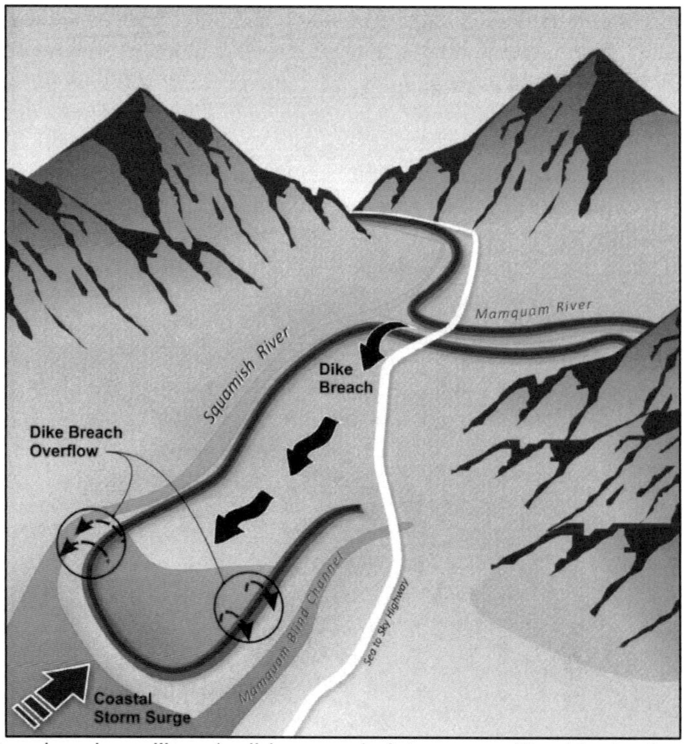

Figure 1 (Left): Squamish River floodplain showing river dikes (solid orange), future sea dike (dashed brown), and modelled breach locations (blue squares). Key at-risk areas include Downtown (1), Dentville (2), Industrial Park (3), North Yards (4), Garibaldi Estates (5), Eagle Run (6) and Brackendale (7)

Figure 2 (Right): Conceptual Illustration of Internal Ponding due to Upstream Dike Breach

Bed resistance uses a generalized representation of the floodplain's complex, spatially-varied roughness characteristics. Characteristic roughness values are assigned to polygons simulating city blocks (or large rural lots) and interstitial roads across the floodplain. Roughness values were weighted by governing land use. Since no calibration data are available, representative values for each land use were selected based on experience and engineering judgement. Initial water levels were set equal to the ground surface (for dry areas) and to a typical ponded elevation (for sloughs and ditches).

3 MESH DEVELOPMENT

The model's two-dimensional flexible mesh uses represents topography with a network of small triangular elements. Maximum default element areas of 500 m^2 provided the best balance between model efficiency and resolution. Size was reduced to 100 m^2 along the dike perimeter to improve flow exchange between 1D and 2D model components.

During a dike breach, flow concentration is expected along highways, local road networks, and other linear lower-resistance corridors. This is particularly important for densely developed areas where buildings, fences, and other obstructions limit conveyance through developed lots. Allowing mesh elements to span these linear corridors results in spatial averaging of elevation and roughness that in turn creates irregular and discontinuous conveyance capacity. This makes it extremely difficult to identify areas of elevated hazard within the model output. To provide more realistic floodplain hydraulics, linear features (dikes, railways, roads, etc.) were assigned a maximum element size of 10 - 20 m^2.

Buildings within a 2D model act as resistance elements to obstruct flow but also store water and thereby influence attenuation. For the IFHMP, consideration was given to ignoring buildings, excluding building footprints from the mesh, or modelling footprints with an elevated Manning's n value. Syme (2008) notes that blocking flow at the upstream building face while leaving the downstream side open can provide realistic results for both conveyance and storage; however, this more complex approach is not feasible when dealing with thousands of structures and multi-directional flow patterns.

After careful consideration, excluding building footprints from the mesh was selected as the most appropriate option for the Squamish IFHMP. This decision was based on the critical importance of conveyance and the availability of GIS roof polygons for existing floodplain structures. Implications of this assumption for floodplain storage and attenuation were checked and confirmed to be relatively limited. Building polygons were simplified and incorporated directly into the model's built-in mesh generator. Simplification was kept to the minimum required to achieve a stable mesh.

The final lower floodplain model contains about 184,000 individual mesh elements. The larger domain of the upper model has nearly 256,000 individual mesh elements. This is considered a relatively high resolution for 2D hydraulic modelling given the required scale, current tools and data limitations, and the processing capability of a single desktop computer. Testing clearly demonstrated the importance of the high-resolution mesh in simulating realistic floodplain behaviour (Figure 3).

Figure 3: Detailed representation of the floodplain (left) supports a more nuanced analysis of maximum Hazard Rating (depth x velocity, right), particularly where flow is concentrated along the road network

4 FUTURE CONDITIONS

Flood hazards are not constant in time, and may increase as a result of climate change. Guidance from provincial regulators requires that BC professionals consider 1 m of Sea Level Rise (SLR) by Year 2100 and 2 m by Year 2200 (BC MFLNRO, forthcoming), as well as a minimum +10% adjustment to peak discharges (APEGBC, 2012).

The consequences of a given flood hazard can also increase as a community expands or intensifies its development of floodplain areas. Changes can affect both new elements at risk (i.e., new infrastructure and populations) and existing elements (e.g. as new development reduces conveyance and raises upstream water levels). Failure to consider and plan for future conditions can leave today's developments exposed to unacceptable risk at the end of their service life.

As a rapidly-growing community in a high-hazard area, Squamish faces expected future changes in both hazard and consequence. The IFHMP dike breach model was adapted to explore potential flood risks under assumed Year 2100 conditions. Results may be sensitive to the resulting assumptions; however, the goal is to establish reasonable (rather than accurate) future conditions and considerable uncertainty can be tolerated. Sensitivity analysis would be appropriate but was beyond the project scope.

4.1 Year 2100 Flood Hazards

In the absence of a historical trend, river hydrographs were scaled up by 10% (APEGBC, 2012), and a constant 1 m was added to the coastal tide series to represent SLR (BC MFLNRO, forthcoming). Coastal storm surge, wind field and tide harmonics were assumed to maintain stationarity. Year 2100 wave conditions were assessed for the IFHMP but not incorporated into the dike breach model.

4.2 Year 2100 Floodplain Assumptions

Key Year 2100 floodplain modifications include floodproofing, allowances for infill development, and floodways. Floodproofing and infill development assumptions are based on District zoning data and high-level land use information provided by the Squamish Nation. The three types of modification are discussed individually below.

4.2.1 Floodproofing

There is an intuitive feedback relationship between future water levels and future floodproofing assumptions. The assessment process is typically iterative, since it is difficult to estimate with precision how the model results will respond to changes to the floodplain surface. Multiple simulations are usually completed to explore mitigation options and achieve an optimum balance between inflow, outflow, floodproofing, floodplain storage (attenuation), and conveyance.

Since an iterative approach to establishing future flood levels was beyond the project scope, FCLs from the 1994 Flood Hazard Management Plan (Klohn Leonoff, 1994) were used to approximate an initial suite of Year 2100 floodproofing targets. The implicit assumption is one of convergence in a single iteration (i.e., updating floodproofing assumptions would not have a significant effect on results). This approach results in considerable uncertainty where IFHMP FCL recommendations differ from the modelled conditions, but was the best approach possible given project funding and schedule constraints.

A multidisciplinary team of planners and engineers developed a suite of zoning-specific floodproofing allowances. The allowances target the 1994 FHMP FCLs but also recognize expected land use constraints. For example, large rural lots were not raised in the model, since the footprint of floodproofing fill relative to lot size would be small and unlikely have a significant effect on conveyance.

Floodproofing assumptions were applied in GIS based on the most representative zoning class for each city block. A buffer was retained at the outer perimeter of each block to allow the mesh generator to smoothly interpolate between Year 2100 fill assumptions and surrounding 2013 topography.

4.2.2 Infill Development

Infill development will increase the footprint of buildings on both occupied and vacant lots within each city block. These changes will reduce the effective conveyance of the floodplain. It is impossible to accurately predict how much and where infill development will occur. In accordance with District development objectives and to provide a conservative representation, each city block is assumed to infill to the maximum allowable footprint for its corresponding land use.

The percentage of each city block occupied by existing building footprint (if any) was determined and compared to the maximum allowed under the bylaw. Existing buildings are represented explicitly in the mesh. Potential increases in building footprint were represented by pro-rating each city block's roughness value assuming that the additional building footprint has a Manning's n value of 0.4 (Syme, 2008). The nature of infill development means that it is likely to be adequately served by existing roads, so no changes were required for the road network.

4.2.3 Floodways

In the IFHMP context, a floodway is a dedicated or *ad hoc* corridor that provides a continuous and unobstructed flow path during a dike breach scenario. Properly-planned floodways with sufficient capacity can help mitigate rate of rise and maximum water level in upstream areas, although they may create elevated hazards (due to higher velocities) along the floodway itself. Floodways also play an essential role in situations where gravity drainage works remain functional, or an intentional outlet dike breach could relieve an internal ponding condition.

The effectiveness of a floodway can only be confirmed through hydraulic modelling. The IFHMP incorporates floodways by maintaining present-day ground elevations along selected corridors where floodproofing fill would otherwise be applied.

5 MODEL SCENARIOS

Several factors must be considered when defining an appropriate portfolio of scenarios for dike breach modelling: floodplain area affected (e.g., upper or lower floodplain for the IFHMP), breach location and geometry, magnitude of concurrent river flood, magnitude of concurrent coastal flood, time horizon for modelling, and treatment of downstream dikes. Each of these factors is discussed individually below.

5.1 Concurrent Flood Conditions

A dike breach model is driven by a concurrent flood condition. Most often, the concurrent flood is assumed capable of overtopping, eroding, undermining, destabilizing or otherwise initiating a dike failure. The probability of hydraulically-induced failure is usually considered low (but not necessarily zero) for floods at or below the dike's design condition, and approaches unity when an earthfill dike is overtopped. The probability of a dike breach failure is therefore the product of the probability of the concurrent flood and the conditional probability of a dike breaching during that flood.

When an analysis is limited to a subset of potential hazard scenarios, flood scenarios must be carefully selected to provide a context-appropriate characterization of the risk. Because IFHMP model results would be used to generate FCLs, the study prioritized 1 in 200 Annual Exceedance Probability (AEP) design conditions for the concurrent river flood. Experience with seepage, piping, and compromised freeboard during past flood events suggests that the existing dikes have a non-zero probability of failure at similar discharges. Future work is expected to explore dike breach consequences under more extreme flood conditions.

Coastal flooding was assessed separately at the 1 in 200 AEP level; however, river and coastal floods at Squamish are not strongly correlated. Review of historical streamflow and metocean records suggests that the 1 in 10 AEP coastal flood at Howe Sound is an appropriate concurrent boundary condition for 1 in 200 AEP river flood simulations.

5.2 Breach Location

Early modelling showed that dike breaches were required at eight locations (shown in Figure 1) to adequately characterize the floodplain response. Each breach was positioned where experience suggests there could be an elevated risk of dike failure (landside sloughs, impinging flow, penetrations, etc.). Quantitative characterization of failure probabilities (e.g., using reliability functions in a Monte Carlo analysis) can be useful but is effort-intensive and beyond the scope of the IFHMP.

Breaches on the upper floodplain are hydraulically straightforward and strategically located to identify and characterize the floodplain's governing hydraulic behaviours and flow directions. Multiple breaches were required on the lower floodplain to capture preferential north-south flow along corridors defined by linear infrastructure (e.g., Highway 99, CN Rail mainline).

Each of the eight scenarios involves a single dike breach occurring at a unique location. Multiple dike breaches are possible, particularly as water levels approach widespread overtopping conditions; however, multiple breaches would tend to govern rate of initial rise more than ultimate water level due to the widespread internal ponding that occurs on the diked Squamish River floodplain. A detailed assessment of multiple-breach scenarios was beyond the scope of the IFHMP.

5.3 Breach Mechanics and Geometry

Dike breaches are relatively uncommon and their mechanics can depend significantly on situational factors such as breach mechanism, dike fill materials, construction standards, and local hydraulics. Lacking resources for quantitative characterization, the IFHMP relied on dike breach information provided in the Fraser Basin Council's 2004 floodplain mapping guidelines (WMC, 2004). Maximum dike breach widths of 200 m and 100 m are assumed for the Squamish River and Mamquam River, respectively. Breach growth is based on a historical relationship for sand dikes.

The model is expected to be relatively insensitive to the initial incision (assumed to progress from dike crest to floodplain in 0.25 h) and side slopes (assumed 1H:1V side slopes). The initiation mechanism for each breach is not specified and the breach mechanics are considered representative of a range of failure processes. All failures commence on the rising limb of the river hydrograph so that the breach is already well-developed when hydrograph peaks.

5.4 Time Horizon

Ideally, the IFHMP would evaluate both existing and future conditions. However, all eight available model scenarios were required to provide an adequate description of floodplain response for a single river flood. The need to craft stable long-term policy meant that the IFHMP had to prioritize future conditions. The focus on Year 2100 conditions means that IFHMP model results may not provide an accurate picture of existing or short-term risks under current development and mitigation conditions.

5.5 Outlet Dike Breaches

The District's emergency plan would be activated in response to any major river flood that could result in a dike breach. Advance planning can make provision for intentionally breaching downstream dikes to mitigate internal ponding and avoid a "bathtub" effect. High downstream river levels outside the Squamish River and Mamquam River dikes would limit the effectiveness of this approach for the upper floodplain, but the weaker correlation between coastal and river flood conditions make intentional breaches a more promising option for the future sea dike.

Outlet dike breaches have the predictable effect of reducing inundation depths but increasing velocity. Supplemental modelling was completed to assess the effect of adding intentional sea dike breaches into a river dike breach scenario (KWL, 2015). Increases in velocity and Hazard Rating were relatively modest and more than offset by significant decreases in maximum water level.

6 POST PROCESSING AND RESULTS

Post-processing algorithms played a key role in producing IFHMP technical deliverables. Specific post-processing tasks included compilation of model results, calculations to account for potential dike breaches at other locations, flood hazard map production, and consequence assessment. Each of these topics is discussed below.

6.1 Compilation of Model Results

Results from each of KWL's eight scenarios represent one possible flood scenario resulting from a dike breach at a specific location. Flood risk management policy must consider all possible outcomes. To be useful in a planning context, model results must be generalized and spatially comprehensive. The eight IFHMP dike breach scenarios were chosen and modelled specifically to support this process.

Post-processing combined scenario results into envelope surfaces of water level, flow velocity, and Hazard Rating by extracting the highest value for each mesh element. Results were then interpolated from point shapefiles (based on mesh element locations) to a continuous raster using standard GIS tools.

The resulting envelope surfaces provide an acceptable representation of generalized floodplain behaviour that is independent of dike breach location.

6.2 Breach Zone Calculations

The generalized results discussed above may not be the governing hazard in areas close to the dike where a local breach could generate higher velocities and inundation depths. Different agencies have different names for this area of local effect; the IFHMP refers to this area as the "dike breach zone".

Because a breach could theoretically occur at any location along the dike, local effects within the dike breach zone must be approximated and incorporated into the final composite results. Key breach characteristics include water depth in breach, Hazard Rating in breach, and breach zone width. The IFHMP relates these hydraulic characteristics to maximum head, defined as the difference between a representative floodplain elevation and the corresponding maximum water surface elevation for the 1 in 200 AEP flood without dike breaches.

Maximum head can be calculated at any point along the dike profile and used to extrapolate potential breach characteristics. The IFHMP calculated potential breach water depth, Hazard Rating, and breach zone width at 10 m intervals along the ±20 km length of river dike. A spatially-distributed estimate of local effects was obtained by interpolating between breach characteristics at the dike and generalized floodplain results at the breach zone limit. Figure 4 shows an example of the locally-elevated results for Hazard Rating along the dike. Although there is considerable uncertainty in the breach zone results, the result is far more useful for mitigation and response planning than results based exclusively on a small number of assumed breach locations.

6.3 Final Map Production

Final water surface elevation, water velocity and HR rasters were produced by merging the dike breach zone results and coastal flood conditions with the envelope surface of 2D model results. Inundation maps were created by subtracting the 1 m resolution topographic raster from the final water surface elevation (Figures 5 and 6). Areas potentially subject to other hazards (debris flow, avulsion, overland flow, etc.) are indicated on the final IFHMP maps to highlight the need for further assessment. Internal drainage hazards can be incorporated once the District completes its planned Integrated Stormwater Management studies.

Considerable discussion was devoted to the concept of appropriate freeboard for floodplain FCLs. Freeboard (typically 0.6 m) is a required part of conventional FCL calculations in British Columbia, but there is no accepted standard for dike-protected areas. Due to the conservative suite of modelling assumptions, the comprehensive nature of the modelling, and the challenges associated with implementing future-oriented FCLs, no additional allowance for freeboard was applied within the diked Squamish River / Mamquam River floodplain.

The final flood hazard maps are composite envelope results that encompass the potential for a single breach occurring at any location along the Squamish River and Mamquam River dikes. This is the tool required for floodplain-level mitigation planning, which must consider and prioritize the hazards and consequences of multiple dike breaches. Weighting the hazard and consequence results by probability (i.e., as part of a risk assessment) would be beneficial but is beyond the scope of the IFHMP.

6.4 Consequence Assessment

The final hazard maps (flood level, depth, velocity and HR) served as inputs for a GIS-based consequence assessment. The inundation extents shown in Figures 5 and 6 were the basis for a $447 million lower-bound estimate of economic damages. Results also indicate that a dike breach could damage or destroy as many as 1,400 buildings and generate nearly 40,000 tonnes of debris. Over 50% of the population of Squamish could be displaced.

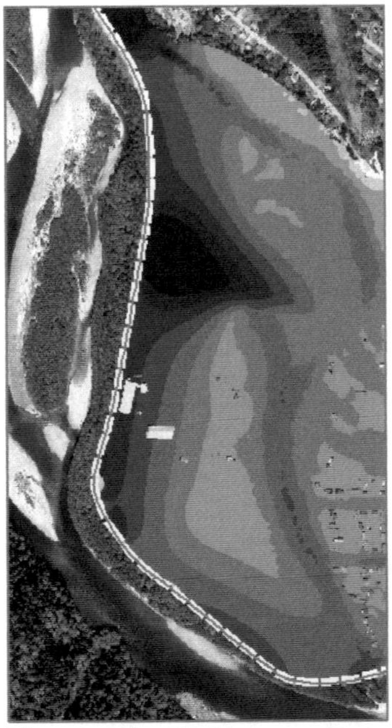

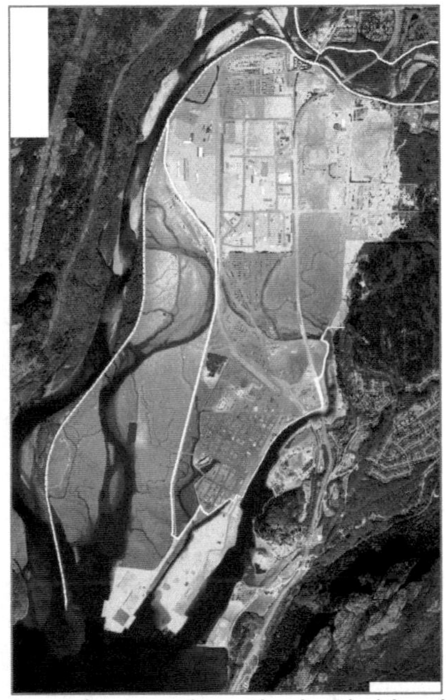

Figure 4 (Left): Map excerpt showing elevated Hazard Rating for dike breach zone along the land side of the river dike. Darker colours represent higher hazards

Figure 5 (Centre): Map excerpt showing dike breach inundation extents for the upper floodplain

Figure 6 (Right): Map excerpt showing dike breach inundation extents for the lower floodplain

Social consequences were assessed using a new GIS-based approach developed as part of the IFHMP. The process considers vulnerability, hazard intensity, and duration to assess displacement of residents, disruption of employment, and interruption of important community services. Maximum intensities for social consequence are governed by inundation of critical community facilities such as schools, wastewater treatment infrastructure and fire halls. The approach was also adapted to map environmental consequences at sensitive habitat areas and storage sites for potentially dangerous materials.

7 PROJECT OUTCOMES

The dike breach modelling and parallel assessments of other flood hazards formed the basis for a detailed flood risk mitigation strategy that includes over 100 specific management tools. The strategy incorporates elements of protection (diking), accommodation (floodproofing, appropriate land use, and designation of floodways), avoiding new risks, managed retreat of key infrastructure, and selectively accepting risk where it brings significant benefits for the community. Tools are grouped into seven categories: land use planning, development and building controls, structural flood protection, watershed and river management, public education, emergency planning, and flood insurance. Some of the most important aspects of the strategy include adopting a higher standard for high-consequence dike protection works, limiting densification in the highest-hazard parts of the floodplain, and directing growth to lower-risk areas of the community. Key deliverables included a prioritized list of structural flood protection upgrades, significant revisions to Official Community Plan hazard policies, the District's first Floodplain Bylaw, and new Development Permit Area guidelines for flood hazard lands.

A careful balance of simplifying assumptions and dike breach model complexity allowed the IFHMP to support both general and detailed outcomes. The importance of simplifying assumptions is clear from the complex combinations of hazard and consequence that could be usefully considered in risk mitigation discussions. At the same time, model complexity was essential for producing realistic results. The chosen level of complexity resulted in both direct and indirect benefits for the study; for example, the identification of severe hazard conditions on planned evacuation routes, the potential benefits of intentionally breaching a downstream dike, development of new algorithms to simulate local effects in the dike breach zone, and establishment of long-term FCLs that will not become "moving targets". In addition, future work can leverage the detailed model to evaluate specific redevelopment proposals or produce a Quantitative Risk Assessment.

Not all dike breach assessments will require the Squamish IFHMP's level of detail. The appropriate level of complexity will depend on project-specific factors: technical flood hazard, available data, project constraints and desired outcomes. Some issues – such as the consideration of multiple breach locations and the local breach zone – must be addressed to produce accurate results, while others – such as the incorporation of future development assumptions – may need to balance desired outcomes with more practical considerations. It is hoped that detailed studies like the IFHMP can help decision-makers reach effective decisions about costs, benefits and the most appropriate level of detail for their specific situation.

ACKNOWLEDGEMENTS

This work would not have been possible without the dedicated assistance of District of Squamish planning and engineering staff and the support of Mayor and Council. Technical advancements described herein are the result of a multidisciplinary effort that included engineers from KWL, SNC-Lavalin Inc., and Thurber Engineering Ltd., planners from Arlington Group Planning + Architecture, and environmental scientists from Cascade Environmental Resource Group. Graham Farstad of Arlington Group provided specific support for assumptions about future infill development. Outcomes were shaped in part by input from the District's flood management partner Skwxwú7mesh Úxwumixw (Squamish Nation), a stakeholders' Technical Working Group, and the Squamish community.

REFERENCES

APEGBC (Association of Professional Engineers and Geoscientists of British Columbia). 2012. *Professional Practice Guidelines-Legislated Flood Assessments in a Changing Climate in BC*, v1.1. APEGBC, Burnaby, BC, Canada.

BC MFLNRO (British Columbia Ministry of Forests, Lands and Natural Resource Operations). Forthcoming. *Amendment - Section 3.5 and 3.6 – Flood Hazard Area Land Use Management Guidelines.* BC Ministry of Forests, Lands and Natural Resource Operations, Water Management Branch, Victoria, BC, Canada.

BC MWLAP (British Columbia Ministry of Water, Land and Air Protection). 2004. *Flood Hazard Area Land Use Management Guidelines.* Amended August 17, 2011. BC Ministry of Forests, Lands and Natural Resource Operations, Water Management Branch, Victoria, BC, Canada.

Klohn Leonoff. 1994. *Flood Hazard Management Plan.* Klohn Leonoff file no. PB 6166 0101. Richmond, BC, Canada.

KWL (Kerr Wood Leidal Associates Ltd.). 2016. *Intentional Sea Dike Breach Assessment – Revision 2.* Technical Memorandum dated February 16, 2016. KWL file no. 463.278-300. Burnaby, BC, Canada.

KWL (Kerr Wood Leidal Associates Ltd.). 2017a. *Integrated Flood Hazard Management Plan - Background Report.* KWL file no. 0463.278-300. Burnaby, BC, Canada.

KWL (Kerr Wood Leidal Associates Ltd.). 2017b. *Integrated Flood Hazard Management Plan - Coastal Flood Risk Mitigation Options.* KWL file no. 0463.278-300. Burnaby, BC, Canada.

KWL (Kerr Wood Leidal Associates Ltd.). 2017c. *Integrated Flood Hazard Management Plan – River Flood Risk Mitigation Options.* KWL file no. 0463.278-300. Burnaby, BC, Canada.

KWL (Kerr Wood Leidal Associates Ltd.). Forthcoming. *Integrated Flood Hazard Management Plan.* KWL file no. 0463.278-300. Burnaby, BC, Canada.

Syme, W.J. 2008. Flooding in Urban Areas - 2D Modelling Approaches for Buildings and Fences. *9th National Conference on Hydraulics in Water Engineering*, Engineers Australia, Darwin, NT, Australia.

WMC (Water Management Consultants). 2004. *Floodplain Mapping Guidelines and Specifications.* Water Management Consultants, Richmond, BC, Canada.

Leadership in Sustainable Infrastructure
Leadership en Infrastructures Durables

Vancouver, Canada
May 31 – June 3, 2017/ *Mai 31 – Juin 3, 2017*

THE IMPORTANCE OF GEOMORPHOLOGY IN HYDRAULIC ENGINEERING: THE CONTRIBUTION OF ROLF KELLERHALS

David G. McLean [1,2]
[1] Northwest Hydraulic Consultants Ltd., Canada
[2] dmclean@nhcweb.com

ABSTRACT

Dr. Rolf Kellerhals was an inspiration to young engineers and geomorphologists in western Canada during his career that spanned the decades of the 1970s, 1980s and 1990s. As a specialist in river engineering, he emphasized the importance of utilizing the tools of the geomorphologist to improve our understanding of river processes and our predictions on how rivers respond to engineering projects. Much of his work related to assessing the effects of hydro power projects and diversions on river stability and morphology. He demonstrated how interpretative methods could provide independent predictions from standard hydraulic engineering methods such as hydraulic modelling and how in many cases the interpretative methods were the most reliable tools that were available.

This paper reviews the contribution made by Dr. Kellerhals using case histories and examples from past projects. It also shows that his message is still very relevant today, where so much emphasis is on the use of complex morphodynamic models that often have limited validation testing over the long time scales that are required to assess project impacts.

Keywords: - (River engineering, fluvial geomorphology, sedimentation)

1. INTRODUCTION

1.1 The Role of Geomorphology in Hydraulic Engineering

In 1954 Emory Lane, a civil engineer with the US Bureau of Reclamation, explained the importance of using geomorphic concepts for predicting the response of alluvial channels to engineering works such as dams, diversions and channelization measures. Lane (1954) stated, "a knowledge of fluvial morphology will enable the hydraulic engineer to predict the changes and to reduce the detrimental effects to a minimum and take advantage of favorable effects". Lane introduced engineers to the concept of equilibrium in natural stream systems to show how changing a parameter such as discharge (Q) or sediment load (G) would affect other channel parameters such as slope (S) or sediment grain size (D) in order to restore long-term equilibrium. This is illustrated in the simple stream balance relation below (Eq. 1):

$$[1] \quad GD \propto QS$$

For example, if a stream is flowing in a condition of equilibrium and its sediment load is decreased, equilibrium can be restored if the water discharge or slope of the channel is decreased sufficiently, or if the size of the sediment is increased (the channel becomes armoured).

1.2 Progress in Canada

During the 1960s and 1970s a number of civil engineers in western Canada embraced Lane's advice and integrated geomorphology with classical hydraulics to develop the multi-disciplinary practice of river engineering that is commonly applied today. Much of this development took place at the University of Alberta in Edmonton and various provincial agencies, where applied research was conducted under the auspices of the Research Council of Alberta. This highly creative group of engineers who advanced the subject of river engineering included Prof. Tom Blench, Charles R. Neill, Dr. Victor J. Galay, Dr. Dale Bray and Dr. Rolf Kellerhals. However, adopting this approach into general practice has been relatively slow, as evidenced by the introductory comments on assessing river processes in Kellerhals, Church and Bray, 1976:

"*Rivers have fascinated generations of hydraulic engineers with their variety of form and behavior. The large potential benefits of successful river engineering works, combined with the dire consequences of failures, have provided one of the earliest and greatest challenges to the profession. Today this challenge continues to be met in research on river-related processes and by systematic collection of river data. Much progress has been made in both respects but it is the writer's contention that the bias of most engineers towards readily quantifiable topics has led to a serious gap in this work, the neglect of interpretive work on river-related landforms. Lane made a similar plea 20 years ago and although that paper is often quoted, its message seems to have been largely ignored.*"

1.3 Contribution of Rolf Kellerhals

Rolf Kellerhals' career spanned a period of over 30 years and involved both research and consulting from his office in Heriot Bay on Quadra Island, British Columbia. He received his MSc. Degree in Civil Engineering from the University of Alberta in Edmonton and a PhD in the Interdisciplinary Hydrology Program from the University of British Columbia in Vancouver. He was the editor of the first Canadian Hydraulics Conference, which was held in Edmonton in May 1973 and was a widely regarded authority in both the engineering and geoscience community. During his career, he collaborated with many experts in the field of geomorphology, including Prof. Michael Church and Mike Miles, and helped to expand the role of geoscience as a field of practice. He passed away in August, 2016.

1.4 Purpose

This paper reviews some of the key concepts that were embedded in Rolf Kellerhals' work and highlights its relevance to present engineering practice. It is important to emphasize that although he often used the tools and methods of a geomorphologist, his writing and conference presentations were often directed at the engineering community, emphasizing the importance of using geomorphic methods as an independent check or confirmation of traditional engineering methods based on hydraulics and semi-theoretical sediment transport equations. In this respect, he shared and extended the insights provided in Lane's earlier work. Therefore, it is fitting to re-emphasize his message in the setting of an engineering conference.

2 APPLYING FLUVIAL GEOMORPHOLOGY TO HYDRAULIC ENGINEERTING PROBLEMS

2.1 Need for Improved Understanding of River Behaviour

The 1960s and 1970s were a time of unprecedented natural resource development in western Canada, including hydroelectric development as well as expansion of roads, rail lines and pipelines routes that crossed some of the most challenging rivers in North America. The available hydraulic engineering methods for predicting river processes such as scour and erosion and for assessing the channel response to large engineering projects such as dams and water diversions were very limited. Part of the problem is that alluvial river processes are complex and it is difficult to formulate realistic models that can accurately represent physical processes such as turbulence, sediment transport and erosion, given the variability and range of sediments that compose the bed and banks of most streams. Furthermore, most of the practical experience and empirically-based engineering methods (such as Regime Theory) had been developed on sand bed rivers (Lacey, 1929) from parts of the world having a very different physiography and geological setting than the recently glaciated, steep, mountainous gravel bed rivers of western Canada. Consequently, there was a recognized need to improve predictive methods for design and project assessment.

This paper reviews four areas of research where significant contributions were made by Rolf Kellerhals:

1. Critical importance of collecting field data and long term observations to support predictions of future river response. More generally, he emphasized throughout his career the importance of long-term observations and data collection as a critical component for making sound water resource management decisions.
2. Systematic description of the hydraulic and geomorphic characteristics of gravel bed rivers, including development of equations for predicting stable channel dimensions.
3. Development of methods for classifying rivers and identifying the factors that govern their morphology and channel stability.
4. Assessment of the downstream effects of large engineering projects such as dams and water diversions on channel morphology. Dr. Kellerhals emphasized that environmental impacts from projects could not be properly evaluated unless the physical changes to the habitat were accurately estimated.

2.2 Hydraulic and Geomorphic Characteristics of Gravel Bed Rivers

So-called "Regime Theory" equations were developed in the 1920s to predict stable channel dimensions on sand bed canals transporting low sediment loads in the Indian sub-continent (Lacey, 1929). Other semi-theoretical investigations had been carried out in hydraulic laboratories. However, most of this work was restricted to sand bed channels and was of limited use when applied to most mountainous environments in western Canada. In fact, in the early 1960's very little systematic investigations of river hydraulics and channel processes on the steep, laterally unstable, gravel bed rivers of western Canada. Systematic observations and characterization of gravel bed rivers and river processes was pioneered by the combined efforts of several researchers and provincial agencies in Alberta starting in the 1960s and 1970s.

Kellerhals (1969) analyzed the hydraulics of coarse gravel bed channels in natural channels that were near the threshold of sediment transport. These so-called "threshold" channels had adopted to past flow conditions by adjusting their dimensions and slope until the bed load transport rate in the channel was negligible. Subsequently, Kellerhals, Neill and Bray (1972) characterized the hydraulic and geomorphic characteristics of alluvial mobile bed channels at 110 reaches in Alberta. These field studies encompassed the entire province and represented a major undertaking that to the author's knowledge has not been replicated. The information was subsequently used to develop empirical equations to predict stable channel dimensions of gravel bed rivers (Bray, 1973). The primary data has continued to be valuable for developing more general, semi-theoretical stable channel equations (Parker, 1978). Kellerhals and Church (1989) combined the Alberta river reach data with data from sand bed rivers and canals and as well as from large rivers from Asia and South America to illustrate the consistency of these channel scaling relations. Figure 1 is a version of this relation, as presented in Church (2015). The plots show that channels can be thought of as dynamic models of each other. However, the width to depth ratio of alluvial channels increases with increasing channel forming discharge, so that the scaling is distorted (the channels are not geometrically similar).

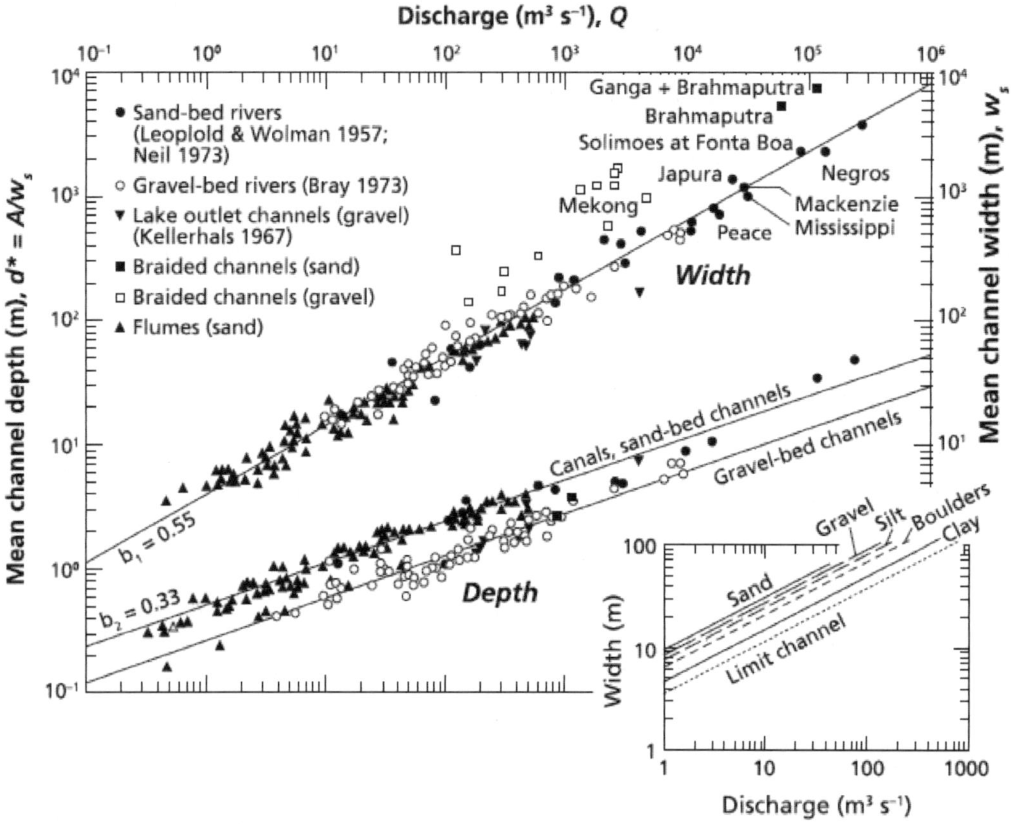

Figure 1: Scaling relationships for river channel geometry (reproduced from Church, 2015).

2.3 River Classification and Interpretation of River Processes

Geologists have developed a wide range of systems and methods for classifying and characterizing rivers. However, Kellerhals, Church and Bray (1976) were more concerned with using river classification as a diagnostic tool to understand the key factors that governed the river's morphology in order to make engineering predictions about its stability and future pattern of channel changes. Kellerhals, Church and Bray (1976) commented that the characteristics of a river channel will change wherever a change occurs in any of the conditions that govern fluvial morphology. Reasonably detailed knowledge of these conditions is therefore an essential prerequisite to the proper planning of any engineering interference with a river. The main factors that govern the morphological characteristics of a river were considered to be:

- The supply of sediment and water;
- The nature of the materials that the river flows through, and
- The geological setting and geological history of the region.

The role of glacial history in affecting river processes was generally not appreciated by engineers. Kellerhals, Church and Bray (1976) pointed out that the end of the last major glaciation (about 10,000 years ago near the 49th Parallel) remains a significant factor for understanding present river behavior. During much of the period since then, the rivers have been moving, sorting and re-depositing the major slug of glacially derived sediments; today many rivers flow over lag deposits of glacial till or outwash that they are not now competent to move.

The classification system was intended as a practical tool for engineers in the field and used readily available information such as historical air photos and maps. The basis for the method and type of information that was collected are generally consistent with the systematic evaluation of river regime described in Neill and Galay (1967). Figure 2 illustrates the approach for classifying channel plan form features.

The end point for the classification assessment was not to just describe the river morphology but to make quantitative predictions about future river behavior. In particular, it was emphasized that geomorphic-based methods should, whenever possible, be used to provide independent checks on traditional engineering calculations. The following examples were cited:

- Sediment transport computations can sometimes be checked against growth rates of alluvial fans and deltas or against point bar progression rates.
- Maximum flood levels can be checked with evidence of silt deposition, ice scour trim lines, or evidence contained in vegetation.
- Local scour at natural constrictions or spurs is still one of the most reliable indicators of potential scour at bridge abutments or pipeline crossings.

It was emphasized that if geomorphological evidence disagrees with computed values, most prudent engineers will give it precedence.

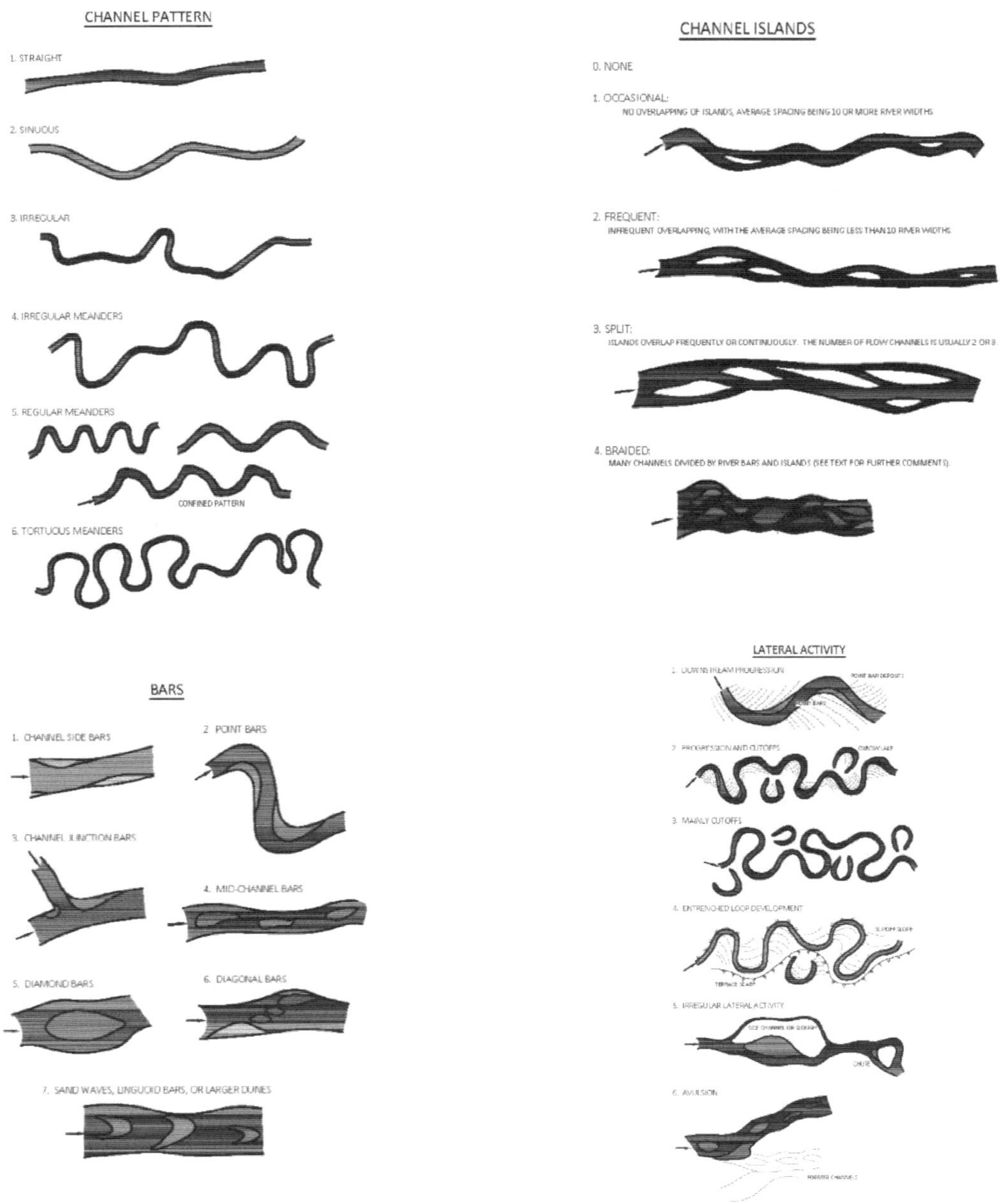

Figure 2: Classification of plan form features of river channels (from Kellerhals, Church and Bray, 1976).

Kellerhals and Church (1990) described a more specific case of classification applied to assessing hazards on alluvial fans, including high gradient fans subject to debris flows and debris floods.

2.4 Assessing Downstream Effects of Dams and River Diversions

Lane's application of dynamic equilibrium (Eq. 1) provides a tool to predict the expected change to a river's longitudinal profile in response to flow regulation or reduction in sediment supply downstream of a dam. However, it does not describe how the plan form of a river will respond. Mollard (1973) presented a conceptual model which related channel plan form to factors such as sediment supply and river discharge. Kellerhals and Church (1989) adapted this model and illustrated how it could be used for predicting channel plan form response to engineering works (Figure 3).

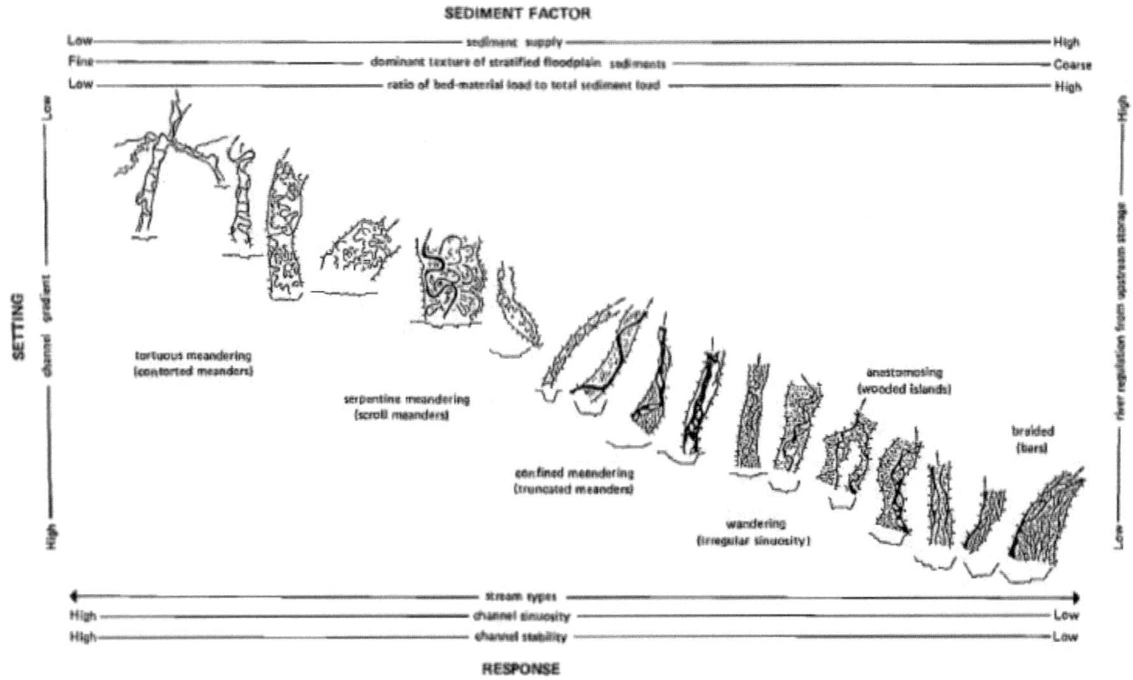

Figure 3: Conceptual model of the relation between channel plan forms types (Kellerhals and Church, 1989)

Figure 2 and Figure 3 provide simple tools for assessing the impact of engineering projects such as dams or river diversions on downstream river morphology. Kellerhals and Miles (1996) concluded that although there was a pressing need for predictive models, these simple relations remained the state of the art for assessing the relation between channel morphology and the imposed driving factors of discharge and sediment supply. The main reason for this unsatisfactory situation was attributed to our poor understanding of sediment transport processes in a three dimensional environment. This was because the engineering community has been absorbed with two-dimensional sediment transport in laboratory flumes and in some artificially trained and straightened European rivers. When the results of this work are misapplied to the more normal three-dimensional river situations (where channel width and alignment are free to adjust), they often lead to wrong answers. In some cases, even the predicted direction of change has been wrong (Kellerhals and Miles, 1996).

Inter-basin diversions provide an example of an engineering project that can induce significant impacts to both the diverted stream and the receiving channel. Large-scale water diversions where a relatively large flow is diverted across a divide and allowed to flow into a dry valley or into a relatively small stream, where it is left to develop a new larger channel, have been relatively rare. However, a number of projects were carried out in Canada during the early part of the 20th century. Following the Kemano project in 1954, where water from the Nechako River was diverted into the Lower Kemano River, there was a renewed interest in water diversions. For example, in the 1970s plans were made to divert a portion of the McGregor River (a tributary of the Fraser River) into the Parsnip River (a tributary of the Peace River) in order to increase power production from WAC Bennett Dam. As part of the assessment process,

Kellerhals et al. (1979) developed case histories from 19 projects in Canada in order to characterize the type and magnitude of long term impacts that occurred. These "morphological lessons" provided a basis for understanding the variability of effects and for assessing future project effects. Based on these studies, diversions were classified into three categories:

1. Bedrock-controlled diversion routes;
2. Steep diversions through unconsolidated materials;
3. Alluvial diversions, where the diversion flows are not significantly greater than the naturally occurring peak flows in the receiving channel.

Suggested methods for predicting project effects were described for each type of diversion.

A case history approach was also used by Church (1995) to document the long term effects of WAC Bennett Dam on the Peace River. The regulation of the Peace River in 1969 reduced the flood flows downstream of the dam by a factor of two to three times. This resulted in vegetation encroachment and suspended sediment deposition over parts of the formerly active channel zone, resulting in the formation of a new lower floodplain. As a consequence, many side channels are filling in and disappearing. At an intermediate time scale, the river is unable to move the sediment inflows from its major tributaries and this is gradually leading to a more irregular, stepped longitudinal profile, due to local aggradation near the tributary confluences. These observations are very different from the general experience on sand bed rivers, where channel degradation below dams is the norm.

Following earlier work by Schumm (1977), Kellerhals and Church (1989) compiled a list of the directions of change for a range of typical engineering projects, based on the case histories and experience on gravel bed rivers in western Canada. These projects illustrate imposed changes to discharge, bed material load and wash load, and the resulting response in terms of changes to width, depth, slope, sediment size, meander wave length and sinuosity.

Table 1 summarizes the expected channel response for the different types of projects. It was indicated that in most cases neither the rate nor the magnitude of the change is quantifiable on a general basis. Because of the imperfect knowledge of river behaviour and the complexity of imposed and response conditions, the overall assessment of morphological changes due to engineering works is in general a qualitative exercise. Because of this, case histories are of critical importance for improving our predictive ability.

Kellerhals and Church (1989) emphasized that river managers need to recognize the importance of documenting and publishing complete case histories of changes in large rivers and should attempt to encourage this activity wherever possible. This includes incorporating long term monitoring as part of approvals for major projects. Also, it is important to recognize the full complexity of the ecosystem that rivers represent, including its geophysical basis, and it is necessary to cooperate to ensure that the best experience is brought to bear on each aspect of the analysis of river systems.

Table 1: Direction of morphological change and relation to driving variables observed at selected case studies (reproduced from Kellerhals et al., 1979)

	Imposed changes			*Direction of resulting change*							
River reach	Q	q_{bm}	q_w	w	d	S	D_{50}	F	λ	P	M
Nechako River, below Cheslatta confl.	-		+	-	-	n.c.	-	+	n.c.	n.c.	+
Cheslatta River											
upper reach	**+**	-	-	**+**	**+**	+	**+**	+	+	-	-
lower reach	**+**	-	+	**+**	**+**	-	+	+	+	-	-
Kemano River											
initial 20 years	+	-	-	+	+	+	n.c.	+	?	-	-
final, upper reach	+	-	-	±?	+	-?	+?	-	?	-	-
final, lower reach	+	+	-	+	+	n.c.	n.c.?	+	+	-	n.c.?
Peace River											
below dam	-	-	-	-	-	n.c.	n.c.	+	n.c.	n.c.	±?
above Pine R. confl.	-	-	-	-	-	-	-	-?	n.c.	n.c.	-
below Pine R. confl.	-	+	+	-	-	+	-	+	n.c.	n.c.	-

Notes: ? Indicates that the direction of change is either not known or not well documented. Heavy symbols, e.g. **+** indicate a drastic change.

Where Q is channel-forming discharge, q_{bm} is bed material load, q_w is wash load, w is channel width, d is mean depth, S is slope, D_{50} is median sediment size, F is width/depth ratio, λ is meander wave length, P is sinuosity, and M is percent silt/clay in the bank materials. n.c. means "no change".

2.5 Re-assessing Assumptions About Sediment Yield in Western Canada

River engineering requires reliable long term field observations, including hydrological data. In the 1960s, Water Survey of Canada expanded the traditional hydrometric program to include measurements of suspended sediment load on many rivers. Church, Kellerhals and Day (1989) utilized this data to develop regional estimates of sediment yield in British Columbia. They found the pattern of sediment yield in British Columbia was very different from the commonly held view. Previously, most estimates of sediment yield were derived from measurements on low-lying agricultural lands and the sediment yield (defined as the sediment load per square kilometre) was found to decrease as the drainage area increased. However, the mountainous, glaciated valleys of British Columbia generally contain terraces of fluvial and glacial-fluvial outwash that often confine the rivers. These readily erodible deposits tend to cause the sediment yield to increase in the downstream direction, opposite to the accepted belief. This analysis fundamentally changed our understanding of sediment production and sediment supply in our watersheds. Without long term records at multiple stations along a river network, this type of understanding cannot be developed.

It is ironic that the sediment program in Canada was terminated in the 1990s as part of a cost saving policy. Today, many studies related to sedimentation in reservoirs or sedimentation processes are based on data that was collected over 30 years ago. It is often assumed by engineers that more sophisticated models can replace basic data and the need to make systematic observations. However, there is no evidence that this assumption is valid.

3. RELEVANCE OF KELLERHAL'S WORK TO THE PRESENT

Fluvial geomorphology has become a widely accepted practice in western Canada as a discipline and geoscientists often play a critical role in river engineering projects and in environmental impact assessments. It is also relatively common for civil engineers to take courses in fluvial geomorphology in university. As a result, river engineering often involves a multi-disciplinary team approach. The theme of much of the writing by Kellerhals and Church (1989) has, to some extent, been understood and accepted to a degree.

Progress has been made on many fronts in understanding sediment transport mechanics and morphodynamic modelling has also become a powerful tool for assessing project effects. However, the models are still limited by the physical understanding of three dimensional sediment transport processes and by the lack of long term data for properly validating their predictions (Cunge, 2013). This limitation is seldom addressed in most investigations.

There have been significant advances in field measurement and data collection that expand the tools for geomorphic assessments. These include:

- Multi-beam bathymetric surveys, which provide extremely high resolution, continuous coverage of the river bed allowing three dimensional representation of dunes and bar features.
- Acoustic Doppler Current Profiler (ADCP) systems, which allow continuous measurement of discharge. The back-scatter from the acoustic signal can be used to estimate suspended sediment concentrations in the water column, allowing for continuous monitoring of sediment transport.
- Acoustic detection of bed load movement, providing a means for determining initiation of bed load transport.
- LiDAR surveys, which provide a convenient, relatively low cost method for high resolution mapping of floodplains, bars and islands. GIS mapping software provides powerful tools for comparing changes over time to assess trends and to compute volumes of erosion and deposition.
- Drones, which provide a low cost means for surveying and monitoring channel changes over time.

These advances provide geomorphologists with a range of tools for expanding Kellerhals' interpretative methods into more quantitative fields of analysis.

4. CONCLUSIONS

Few engineers in consulting practice have been able to make fundamental advances in their discipline. Kellerhals' contribution included improving our understanding of gravel bed rivers and the role that geomorphology and geological history play in governing present river behaviour. He also taught engineers how interpretative methods could be used to improve our predictions about river behaviour and river response to engineering works. In this manner, he gave engineers a new way to view a river, not just as a conduit for carrying water, but as a dynamic, self-formed channel, often still responding to geological and climatic events that occurred thousands of years ago.

5 REFERENCES

Blench, T. 1969. *Mobile Bed Fluviology- a Regime Treatment for Canals and Rivers*, University of Alberta Press, Edmonton.

Bray, D. 1973. Regime Relations for Alberta Gravel Bed Rivers, *7th Canadian Hydrology Symposium, National Research Council of Canada*, 440-452.

Church, M., Kellerhals, R. and Day, T. 1989. Regional Clastic Sediment Yield in British Columbia. *Canadian Journal of Earth Sciences*, Vol. 26, 31-45.

Church, M. 1995. Geomorphic Response to River Flow Regulations: Case Studies and Time Scales. *Regulated Rivers: Research and Management,* Vol. 11, 3-22.

Church, M. 2015. *Channel Stability: Morphodynamics and the Morphology of Rivers, Chapter 12* in Rivers-Physical, Fluvial and Environmental Processes, ed. Rowinski, P. and Radecki-Paulik, A. Springer.

Cunge, J. 2013. Discussion, River Hydraulics, a View from Mid-Stream by Knight, D, *Journal of Hydraulic Research*, Vol. 51, 2-18.

Kellerhals. R. 1967. Stable Channels with Gravel-Paved Beds. *Journal of the Waterways and Harbors Division, American Society of Civil Engineers*, Vol. 1, 63-84.

Kellerhals, R., Neill, C.R. and Bray, D. 1972. *Hydraulic and Geomorphic Characteristics of Alberta*, Research Council of Alberta, Edmonton, Alberta, Report 72-1.

Kellerhals, R. and Church, M. 1990. *Hazard Management on Fans with Examples from British Columbia, in Alluvial Fans: A Field Approach*, ed. A. H. Rachocki and M. Church, John Wiley & Sons Ltd.

Kellerhals, R. and Church, M. 1986. *The Morphology of Large Rivers: Characterization and Management, in Proceedings of the International Large River Symposium*, ed. Dodge, D., Canadian Special Publication in Fisheries and Aquatic Science, 106, Ontario.

Kellerhals, R., Church, M. and Bray, D. 1976. Classification and Analysis of River Processes. *Journal of the Hydraulics Division, American Society of Civil Engineers*, HY7, 813-829.
Kellerhals, R., Church, M. and Davies, L.B. 1979. Morphological Effects of Interbasin Diversions. *Canadian Journal of Civil Engineering*, Vol 6, 18-31.
Kellerhals, R. 1973. Editor: *Proceedings of the First Canadian Hydraulics Conference*, University of Alberta, May 10-11, 1973, University of Alberta Water Resources Centre, Edmonton.
Kellerhals, R. and Gill, D. 1973. Observed and Potential Downstream Effects of Large Storage Projects in Northern Canada, *Proceedings of the International Commission of Large Dams*, 11th Congress, Madrid, Spain, 18 p.
Kellerhals, R. and Miles, M. 1996. Fluvial Geomorphology and Fish Habitat: Implications for River Restoration. *Ecohydraulics 2000*, Quebec.
Lacey, G. 1929. Stable Channels in Alluvium. *Proceedings of the Institution of Civil Engineers*, London, Vol. 229.
Lane, E. W. 1954. *The Importance of Fluvial Morphology in Hydraulic Engineering*. Hydraulic Lab Report No. 372, United States Department of the Interior, Bureau of Reclamation, Denver, Co.
Mollard, J.D. 1973. Air Photo Interpretation of Fluvial Features. In *Fluvial Processes and Sedimentation. Proc. Of Hydrology Symposium*, University of Alberta, Edmonton, 339-380.
Neill, C.R. and Galay, V. 1967. Systematic Evaluation of River Regime. *Journal of the Waterways and Harbours Division, American Society of Civil Engineers*, Vol 93, 25-53.
Parker, G. 1978. Self-formed River with Eroding Banks and Mobile Beds, Part 2: The Gravel Bed River. *Journal of Fluid Mechanics*, 89(1), 127-146.
Schumm, S. 1977. *The Fluvial System.* Wiley, New York.

Leadership in Sustainable Infrastructure
Leadership en Infrastructures Durables

Vancouver, Canada
May 31 – June 3, 2017/ *Mai 31 – Juin 3, 2017*

TUNNEL REPLACEMENT PROJECT: MORPHODYNAMIC MODELLING OF TRENCH MIGRATION

Robb, Daniel M.[1], Gellis, Matt S.[1,2], Vasquez, Jose A.[1] and Wang, Edwin C.[1]
[1] Northwest Hydraulic Consultants Ltd., Canada
[2] mgellis@nhcweb.com

Abstract: A highway tunnel passing under the main channel of the Lower Fraser River in British Columbia, Canada is nearing the end of its design life. Replacement with a new bridge has been proposed. The project includes decommissioning and removing the in-river segments of the tunnel. Preliminary plans do not propose backfilling after tunnel removal. This would result in a trench feature in the riverbed approximately 630 m long, 95 m wide, and 8 m deep. The tunnel is in a tidally-influenced reach of the river 5 km upstream from the sea. Because of the complex tidal conditions at the site, it was not feasible to assess the fate of the trench using analytical methods, especially the possibility that the trench could migrate upstream. In this paper, we use a three-dimensional hydrodynamic-morphodynamic model to investigate the alluvial channel response to the removal of the tunnel. The model used in the present study accounts for non-equilibrium effects; that is, the sediment concentration profile in the vertical direction does not adapt instantaneously to spatial and temporal changes in the flow. Non-equilibrium effects can influence the morphological evolution of a submerged trench. To validate the model, numerical results were first compared to existing experimental data from flume experiments on the morphodynamic development of dredged trenches. The numerical results agreed well with the experimental data over a range of initial trench dimensions. The model was then used to predict the morphological evolution of the trench resulting from the removal of the tunnel. The model results suggest that the trench will migrate downstream and will fill in within one or two freshet seasons.

1 INTRODUCTION

The highway tunnel analysed in this paper passes under the main channel of the sand-bedded Lower Fraser River between the municipalities of Richmond and Delta in British Columbia, Canada. During construction, each tunnel segment was floated into place and sunk into a shallow trench spanning the crossing. Since the tunnel is nearing the end of its useful life, replacement with a new bridge and removal of the in-river tunnel segments has been proposed. Decommissioning and removing the tunnel segments will result in a trench feature in the riverbed approximately 630 m long, 95 m wide, and 8 m deep. This study focuses on estimating the morphological evolution of the trench resulting from the removal of the tunnel.

The tunnel is in a tidally-influenced reach approximately 5 km upstream from where the river flows into the sea. The reach consists of deltaic distributary channels influenced by mixed semi-diurnal tides. Full flow reversal occurs in the winter months. Due to tidal effects, there was also the possibility that flow reversal would cause the trench to migrate upstream. To assess the fate of the trench, it is first important to correctly predict the hydraulic and morphological conditions in the study area. The complex tidal conditions present a challenge for many mathematical and numerical models. The problem is further complicated by the trench itself. The shape of the trench results in a relatively abrupt change in bed topography, which causes the sediment transport to be out of equilibrium with the flow conditions.

Early studies on trench migration proposed mathematical models for a few idealized cases (Galappatti and Vreugdenhil 1985; Lean 1980; Miles 1981). These studies focused on steady 1D or 2D approximate analytical solutions. For more complicated cases, involving flow reversal and tidal influence, these solutions have limited applicability. As a result, numerical models are often used for evaluating the hydraulic and morphological behaviour important for environmental assessment, navigation planning, and maintenance dredging in estuarine settings.

Experimental studies of trench migration have investigated bed evolution, velocity distribution and sediment concentration profiles (DHL 1980; Faraci et al. 2011; van Rijn 1986a; van Rijn 1986b). Many of these experimental studies produced datasets that have subsequently been used to validate numerical models. Of interest to this paper are sand-bed flume experiments described in van Rijn (1986a), in which suspended load was the dominant sediment transport mechanism. The bed evolution measurements from these experiments are used to validate the numerical model in this study.

Most river and coastal sediment transport models assume the bedload and suspended load are instantaneously in local equilibrium on each computational node (Sánchez and Wu 2011). That is, the vertical sediment concentration profile is assumed to adapt instantaneously to any spatial or temporal changes in the flow regime. As a result, the sediment flux between the bed and the water column is governed solely by the difference between the concentration of sediment in the water column and the equilibrium sediment concentration on the bed. These models are referred to as equilibrium or capacity transport models. Equilibrium models are usually adequate when modelling processes dominated by coarse bedload transport, but they may perform poorly when the deposition of suspended load is important. Assuming an instantaneous response of the sediment concentration profile to changes in the flow is invalid in many cases, for example, when there are changes in the flow conditions such as tidal variation, flow deceleration over the upstream edge of a dredged channel, or flow acceleration through a constriction. Because of inertial effects, it takes time for the velocity profile and the sediment concentration profile to adjust to the new flow conditions.

The time for the suspended sediment to adjust to changes in the fluid velocity is referred to as a lag effect (Pritchard and Hogg 2003). In tidal systems, lag effects can result in the net flux of sediment over a tidal cycle in a direction over which there is no net flux of fluid. Lag effects can, therefore, introduce a hysteresis in sediment concentrations over a tidal cycle. Furthermore, for equal flow velocities, lag effects result in higher sediment concentrations in a decelerating flow than in an accelerating flow (Humphries et al. 2012). This result is applicable to trench migration, where the flow decelerates over the upstream edge of the trench and accelerates over its downstream edge. Since the study area is tidally influenced and includes a trench, lag effects are likely to influence the morphological evolution of this physical system. Therefore, we selected a non-equilibrium transport model to predict the spatial and temporal lags between the flow and the sediment transport.

In this paper, we use a hydrodynamic-morphodynamic model to investigate the channel response to a trench in the riverbed, resulting from the removal of a tunnel. The model accounts for non-equilibrium effects, which can influence the morphological evolution of a trench. The objective of this study is to simulate the channel response to a trench using a numerical model. The rest of the paper is organized as follows. Section 2 describes the hydrodynamic and morphodynamic components of the numerical model, including the key assumptions and the applicability of the model to simulate trench migration. A validation test case, presented in Section 3, compares the numerical results to existing experimental data from flume experiments on the morphodynamic development of dredged trenches. The model is then used to predict the morphological evolution of the trench resulting from the removal of the highway tunnel. Finally, in Section 4, we discuss the main outcomes of the study and summarize the key findings.

2 NUMERICAL MODEL DESCRIPTION

The numerical results presented herein were computed using the Telemac hydro-informatic system (Hervouet 2000; Villaret et al. 2013). The Telemac system contains several modules including 2D and 3D hydrodynamic models and a sediment transport (morphodynamic) model. This approach allows for the coupling of a hydrodynamic model (Telemac-2D or Telemac-3D) with the morphodynamic model (Sisyphe). At every time step, the hydrodynamic model calculates the flow field and sends the main hydrodynamic variables: water depth, velocity, and bed shear stress to the morphodynamic model. The morphodynamic model uses these hydrodynamic variables to compute the sediment fluxes and the changes in bed elevation. The bed levels are updated and this information is sent back to the hydrodynamic model for the following time step. All Telemac system modules use an unstructured triangular discretization with finite-element or finite volume algorithms, described in Hervouet (2007). The main features of the hydrodynamic and morphodynamic models are presented in this section.

2.1 Hydrodynamic Model

Telemac-2D is based on the solution to the shallow-water equations, which assume that the vertical velocities are significantly smaller than the horizontal ones. As a result, the pressure field is found to be hydrostatic. Various options for horizontal turbulent diffusivity exist including depth-averaged k-ε and constant eddy viscosity models. The shallow-water approximation is valid for many river and coastal applications and allows for faster computation times compared to a 3D model. Telemac-3D is based on the solution to the Reynolds-averaged Navier-Stokes (RANS) equations. The model uses an unstructured triangular grid in the horizontal plane, which is replicated in the vertical direction to create several layers, discretizing the vertical dimension. The replicated layers do not need to be equally spaced, which enables refinement near the bed, allowing for better accuracy of the turbulence models and a better estimation of the bed shear stress (Villaret et al. 2013). Either hydrodynamic model (2D or 3D) can be coupled with the morphodynamic model.

2.2 Morphodynamic Model

The morphodynamic model, Sisyphe, simulates erosion, deposition, bedload and suspended load transport. The sediment transport rates are calculated using one of several classical semi-empirical formulae which generally involve decomposing the sediment transport rates into bedload and suspended load. The bedload transport rate is calculated as a function of the excess bed shear above a threshold value for movement, for example, Meyer-Peter and Muller (1948) or van Rijn (1984). The vertical suspended sediment profile can be computed in Telemac-3D by treating the suspended sediment as a passive scalar and solving the classical advection-diffusion equation with an extra advection term to represent the effect of particle settling due to gravity. Sisyphe uses the depth-averaged form of the advection-diffusion equation, where the source term is expressed in terms of net vertical sediment flux; that is, erosion minus deposition.

In the vertical direction, boundary conditions are needed at the free surface and at the interface between the bedload and the suspended load. At the free surface, the boundary condition is a zero sediment flux boundary through the free surface. The boundary condition at the bedload / suspended load interface is more challenging to define. Many models adopt the so-called *concentration boundary condition*, where the near-bed suspended load concentration is assumed to be at equilibrium. The other approach is to assume that the near-bed erosion rate (upward sediment flux) is at capacity under the given flow conditions. This approach is often called the *gradient boundary condition* (Wu 2008). Since it takes time for the suspended sediment load to adapt to spatial and temporal changes in the flow, the concentration boundary condition is only applicable to sediment transport conditions near equilibrium. However the near-bed erosion rate responds more readily to changes in the flow, so the gradient boundary condition is applicable to both equilibrium and non-equilibrium sediment transport.

In this study, we use a non-equilibrium model, applying the gradient boundary condition to simulate trench migration processes. To show the influence of non-equilibrium effects, we repeat a validation test case using the equilibrium, concentration boundary condition, approach. Non-equilibrium approaches are described further in Armanini and Di Silvio (1988), Galappatti and Vreugdenhil (1985), Miles (1981), and Wu (2008). The numerical results presented herein are based on the non-equilibrium approach proposed by Miles (1981) and included in recent versions of the Telemac system (version 6.3 or more recent). This approach has been recently applied to model trench migration processes in Tassi et al. (2011) and Knaapen and Kelly (2012).

3 NUMERICAL RESULTS

The numerical model described in Section 2 has been applied to two test cases. First, the migration of a trench in a laboratory was simulated and the results were compared to existing experimental data. The model was then applied to predict the morphological evolution of the trench resulting from the removal of the highway tunnel. The following section provides a description of each test and the corresponding numerical results.

3.1 Trench Migration Validation

As a first test case, the model was validated against flume data from experiments on the sediment transport processes associated with dredged trenches. The experiments are described in van Rijn (1986a). The purpose of selecting this test case was to assess the ability of the model to simulate trench migration.

3.1.1 Experimental Setup

Experiments were conducted at the Delft Hydraulics Laboratory in a straight rectangular flume 30 m long, 0.5 m wide, and 0.7 m high. The laboratory tests were carried out using sand of diameter, D_{50} (median diameter) of 0.16 mm and D_{90} (ninetieth percentile diameter) of 0.20 mm, supplied at a rate of 0.04 kg/s/m (total load) to maintain equilibrium conditions in the upstream section of the trench. During all tests, the mean flow velocity and water depth at the upstream section were $u_0 = 0.51$ m/s and $h_0 = 0.39$ m, respectively. The trench dimensions were varied using initial side slopes of 3H:1V, 7H:1V, and 10H:1V. Based on velocity and sediment concentration profile measurements, the equilibrium suspended sediment load in the upstream section was $q_s = 0.03$ kg/s/m (75% of total load), resulting in a bedload of $q_b = 0.01$ kg/s/m (25% of total load). From velocity profile measurements, the bed roughness (k_s) was found to be 0.025 m and based on suspended sediment samples, the settling velocity (w_s) of the suspended sediment was 0.013 m/s. The longitudinal bed profiles were measured at the beginning and end of the experiment. Further details on the experiment are provided in van Rijn (1986a).

3.1.2 Physical Description

In the upstream section of the channel, the steady uniform flow results in near equilibrium conditions for the sediment, as it is transported toward the trench. As the flow enters the trench, the velocity decreases due to the increase in water depth. This decrease in velocity results in a decrease in sediment transport capacity; therefore, particles from the bedload layer and some suspended particles will deposit in the trench. In the upstream (down-sloping) and middle regions of the trench, the dominant sediment process is the settling of sediment particles due to a decrease in transport capacity. The sediment trapped in the upstream and middle regions of the trench generate a deficit in sediment supply to the downstream region of the trench, which leads to erosion in this region.

The deposition in the upstream and middle regions of the trench and the erosion in the downstream region result in the trench migrating in the direction of flow. Figure 1 shows a schematic of this physical process for a trench oriented perpendicular to unidirectional flow. For bedload dominated conditions, the shape of the trench will remain relatively constant as it migrates downstream, whereas, under suspended load dominated conditions, the side slopes will smooth over time. Under tidal conditions, erosion may occur on both the upstream and downstream slopes.

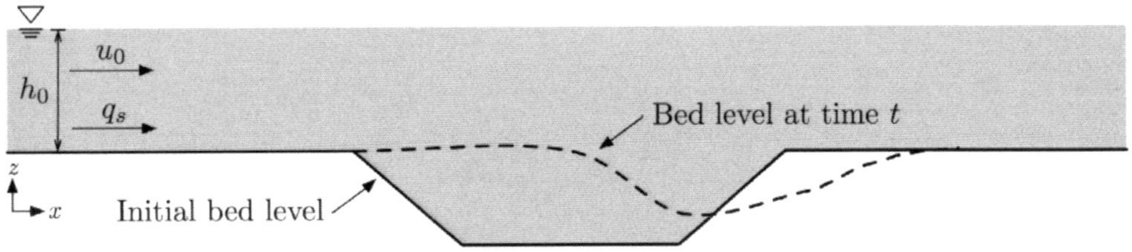

Figure 1: Problem definition sketch: trench migration

3.1.3 Model Setup

The coupled Telemac-2D/Sisyphe model was used to simulate trench migration. The model geometry, flow parameters, and sediment characteristics were selected to match the experimental setup in van Rijn (1986a). The computational domain was discretized into an unstructured triangular mesh with a resolution 0.1 m in the transverse direction and 0.2 m in the longitudinal direction, having 755 nodes and 1200 elements. A constant fluid and sediment discharge was imposed on the upstream boundary and a constant water level on the downstream boundary.

3.1.4 Model Results

Figure 2 shows the measured and computed bed levels after 15 hours for initial side slopes of 3H:1V, 7H:1V, and 10H:1V. The computed bed levels are in good agreement with the measured values. The maximum depth of the trench is predicted very well. Figure 2b shows that for initial side slopes of 7H:1V the computed results slightly under-predict the infilling of the trench at the upstream end, and the trench does not migrate as far downstream as in the laboratory experiments. For all three trench geometries, the computed results slightly under-predict the erosion at the downstream end of the trench as it migrates downstream.

To show the influence of non-equilibrium effects on the model results, Figure 3 compares computed bed levels using both the equilibrium and non-equilibrium approaches for the 3H:1V trench condition. The equilibrium approach results in a behaviour like that expected for a bedload dominated problem in which sediment deposits practically instantaneously when entering the trench. This behaviour forms an avalanching front at the upstream face of the trench that migrates downstream like the leeside of a dune or delta front entering a reservoir. Because all the sediment is deposited on the avalanching front, with no sediment depositing downstream of it, the bottom elevation of the trench downstream of the front remains unaltered. For the non-equilibrium case, suspended load deposits farther downstream inside the trench, causing the minimum bed elevation to gradually rise. The results demonstrate that to correctly model trench migration when suspended load is dominant (for example, in a fine sand river), a non-equilibrium model is needed. The equilibrium model results do not agree with the observations in terms of maximum depth of the trench, rate of infilling, or the general shape of the trench, as it migrates downstream.

Overall, the non-equilibrium model correctly predicts the migration and infilling of the trench over a range of trench geometries. The trench resulting from the removal of the tunnel is expected to have initial side slopes of approximately 4H:1V on the upstream slope and 5H:1V on the downstream slope. These slopes are similar to those modelled in the trench migration validation simulations presented herein. While these flume experiments are not necessarily representative of the field conditions in the study area, the validation shows that the model reproduces the laboratory simulations of an alluvial channel's response to an excavated trench. Also, in the laboratory experiments, 75% of the total load moves in suspension; in the field case this ratio is higher, making non-equilibrium conditions more relevant. Church and McLean (1994) found that, on average, 98.4% of the total sediment load in the Fraser River at Mission moved in suspension.

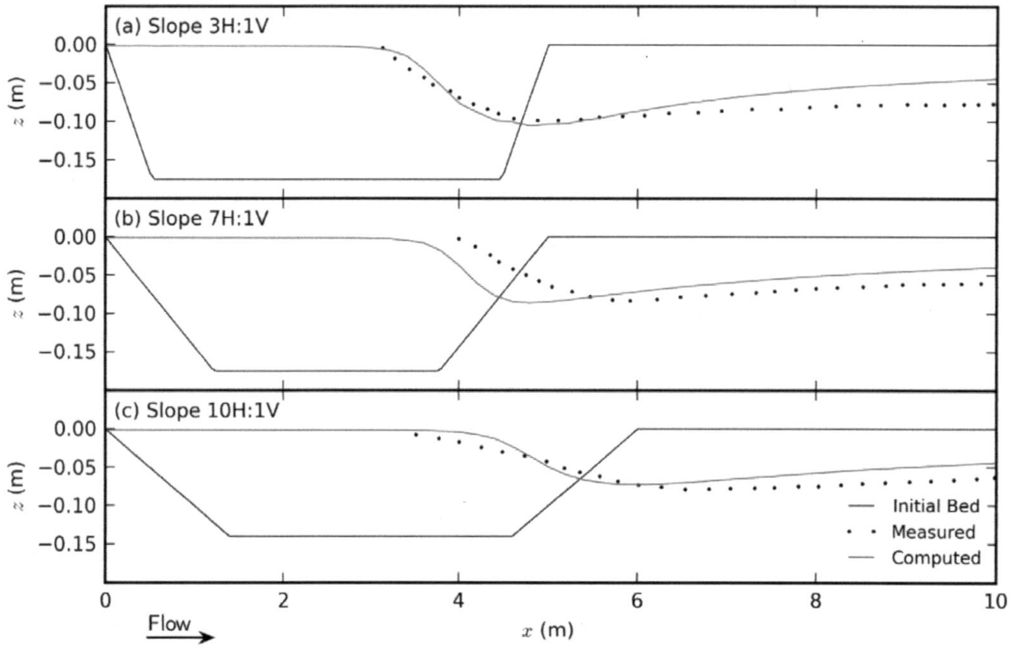

Figure 2: Computed and measured bed level profiles after 15 hours for a trench initially sloping at (a) 3H:1V, (b) 7H:1V, and (c) 10H:1V

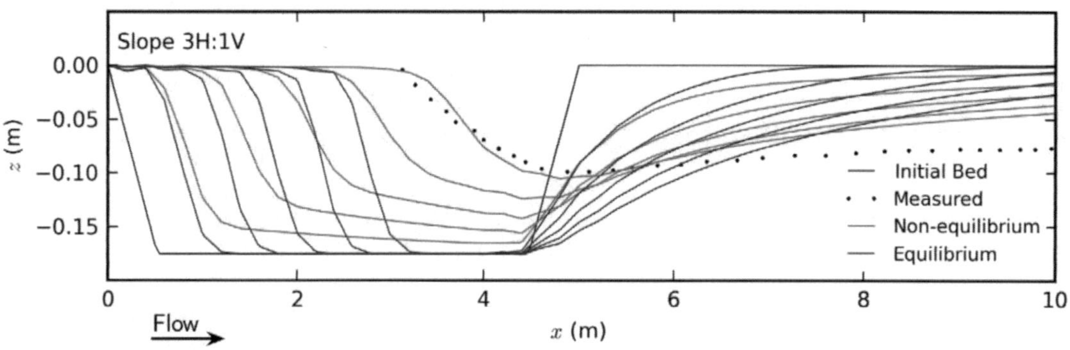

Figure 3: Comparison of equilibrium and non-equilibrium models when 75% of the sediment load is transported in suspension: computed and measured bed level profiles at $t = 3$ hr, 6 hr, 9 hr, 12 hr, and 15 hr for a trench initially sloping at 3H:1V

3.2 Tunnel Removal Simulation

3.2.1 Physical Description (hydraulic and morphological conditions)

The tunnel crosses the tidally-influenced main channel of the Fraser River between the municipalities of Richmond and Delta in British Columbia, about 5 km from the sea. The river's hydrologic regime is snowmelt-dominated, with the discharge typically rising in April, peaking between May and July (the freshet period) and then receding during the autumn and winter months. Near the mouth of the river, the mean tidal range is 3.1 m, with extremes of over 5.4 m (Villard and Church 2003). Full reversal of flow occurs in the study area during the winter months, when tidal amplitudes are large and freshwater discharges are small. During freshet, the flow remains tidally-influenced but does not fully reverse.

Suspended sediment load in the river averages 16.5 million tonnes/year, and ranges from 12.3 million to 31.0 million tonnes/year (McLean et al., 1999; McLean and Tassone, 1988; NHC, 2002; Tywoniuk, N., 1972). Fine sediments moving as washload remain in suspension without settling on the riverbed and therefore have little effect on sedimentation patterns. Of primary importance is the bed-material load, which includes the bedload and the suspended load capable of depositing in the river. These sediment fractions exert an influence on the river morphology. The bed-material load in the crossing reach is composed of sand with diameter larger than 0.18 mm; the estimated average annual bed-material load is 2.9 million tonnes/year, and ranges from 1.2 million to 8.9 million tonnes/year (NHC 2002).

A substantial amount of in-river infrastructure exists near the crossing. Of particular interest is a water main crossing about 600 m downstream of the tunnel, shown in Figure 4. Due to a relatively long history of industrial development, the river banks near the crossing are hardened by riprap and river training works, resulting in a narrower and deeper estuary than existed in the 19th century. Extensive dredging for navigation has also had a notable effect on the river. Bed degradation on the order of two to three metres occurred between 1951 and 1988 (NHC 2002). During the height of dredging activity in the 1970s and 1980s, about 15% of the total annual sediment removal was from the reaches adjacent to the tunnel, resulting in bed degradation of about 25 cm/year in the reach upstream of the tunnel.

3.2.2 Model Setup

A coupled Telemac-3D hydrodynamic model and Sisyphe morphodynamic model was used to evaluate changes in river hydraulics and morphology associated with the tunnel removal. The model mesh has on the order of 11,000 nodes, 21,000 elements and 10 levels in the vertical. The element lengths vary from 20 m near the upstream boundary to 5 m near the tunnel. Upstream and downstream boundary conditions were flow and water level, respectively.

We modelled a trench migration scenario to examine the channel response to the proposed tunnel removal. The validation case described in Section 3.1 supported the idea that the Telemac system could predict the non-equilibrium processes involved in trench migration. Existing conditions were modelled and validated against measured water levels and velocities. The model geometry was then altered to represent the proposed removal of tunnel segments, resulting in a trench 95 m wide at the top of slope, 630 m long and 8 m deep. Modelled initial bed elevations for existing and proposed conditions are shown in Figure 4. Sediment sizes in the model were based on measurements of the riverbed material. The model sediment had a D_{50} = 0.25 mm and D_{90} = 0.45 mm (McLaren and Ren 1995). The van Rijn (1993) formula was used to compute bedload and suspended load transport. The model was run for a 210 day period during low flow conditions, between 16 August 2012 and 13 March 2013. This was the most likely time frame for tunnel removal considering fisheries and constructability, and was expected to yield a conservative (i.e. long) estimate of the time required for the trench to fill in.

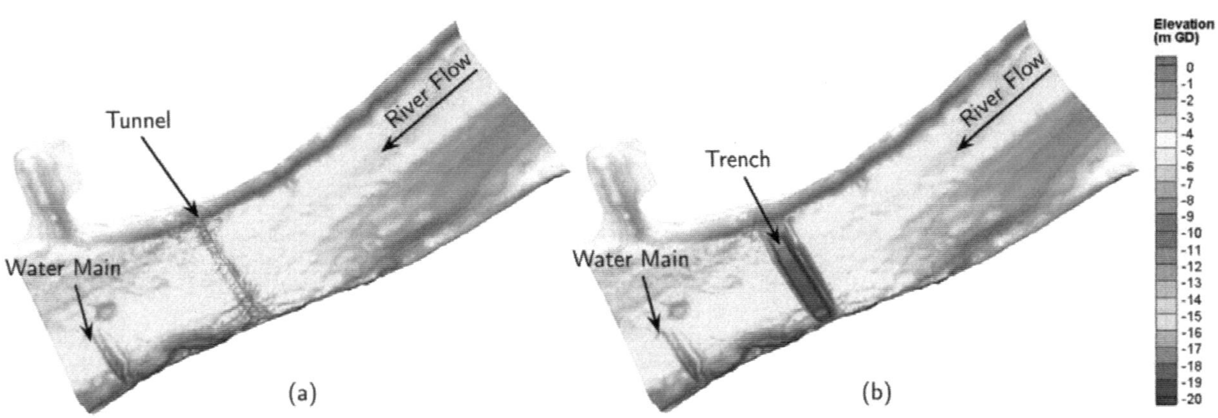

Figure 4: Initial bed elevations for (a) existing conditions and (b) proposed conditions (tunnel removal).

3.2.3 Model Results

The model results show flow decelerating and sediment depositing over the upstream region of the trench and erosion where flow accelerated over the downstream region of the trench. Figure 5 shows a time sequence of riverbed profiles along the navigation channel centreline. After 210 days of simulation, the trench has mostly filled in and we expect it to fill in completely by the end of the following freshet season. The model results support the hypothesis that the trench will migrate in the seaward direction and will have very little impact on upstream riverbed levels. The model predicts bed lowering of 1 to 2 m in the region between the tunnel and the water main. Over the same simulation period, a model of the existing conditions showed about 1 m of bed lowering; therefore, the incremental change in bed elevation due to the tunnel removal is roughly 1 m. The water main crossing is protected with riprap in the southern half of the channel, and bed elevations there remain unchanged throughout the model run. Figure 6 shows a time sequence of the trench evolution. While the trench is oriented perpendicular to the channel, the velocity distribution across the channel is asymmetric; therefore, the trench on the right side of the channel fills in faster than on the left side of the channel.

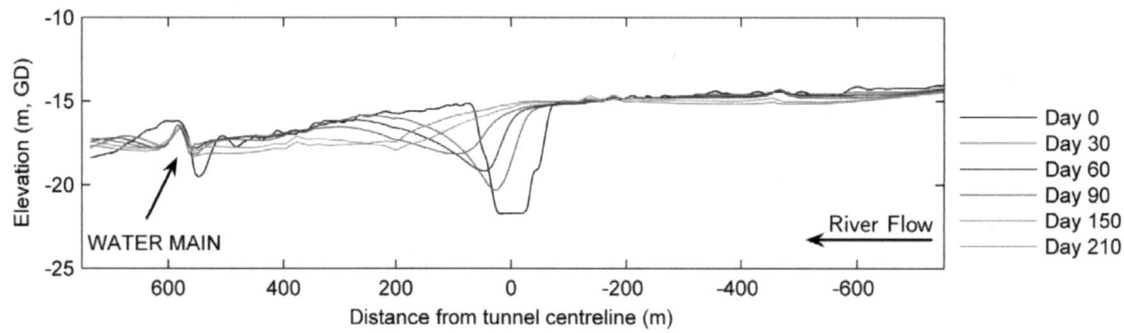

Figure 5: Riverbed profile along the centreline of the navigation channel from 750 m downstream to 750 m upstream of the tunnel with the tunnel removed.

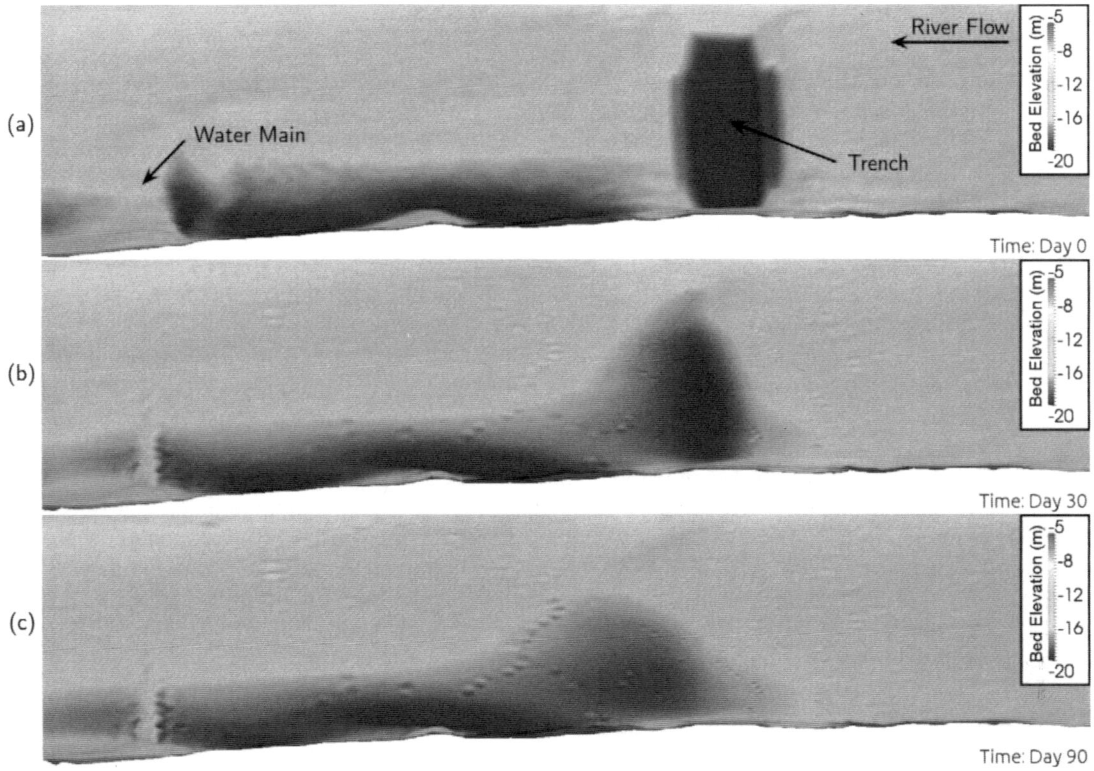

Figure 6: Three-dimensional view of trench migration at (a) $t = 0$ days, (b) $t = 30$ days, and (c) $t = 90$ days

4 SUMMARY AND CONCLUSIONS

In alluvial rivers, bed-material sediment is transported as bedload along the riverbed or as suspended load in the water column. Bedload sediment responds almost immediately to changes in flow conditions and sediment transport capacity; that is, the sediment remains in equilibrium with the flow. On the other hand, suspended sediment requires a finite distance to settle and deposit on the riverbed due to its small vertical fall velocity. Therefore, suspended sediment may not be in equilibrium with local flow conditions, causing lag effects. This distinction is important when studying sedimentation processes in sand bed rivers where suspended load is the main mode of sediment transport, as is the case in the Lower Fraser River.

The proposed removal of a highway tunnel passing under the Lower Fraser River would lead to the formation of a large trench where sediment would deposit. In contrast to other depositional features such as reservoirs or sand traps, which are fixed in placed by non-erodible material, trenches can migrate, potentially affecting nearby structures. Due to tidal effects, there was also the possibility that flow reversal would cause the trench to migrate upstream. To assess the trench dynamics, we used a numerical model with capabilities for unsteady flow and sediment transport with non-equilibrium effects.

We verified the numerical model by simulating the trench dynamics observed in flume experiments under suspended-load dominated conditions. We found that non-equilibrium effects play an important role in accurately reproducing the shape of the migrating trench, especially the gradual rising of the trench's bottom due to suspended load deposition.

Once verified, we applied the model to simulate the trench dynamics in the Lower Fraser River over a period of 210 days representing seasonal low flow conditions. Because tidal effects were not strong enough, the trench migrated downstream towards the sea. By the end of the simulation, the original trench was almost completely filled with sand. Due to its downstream migration, approximately 1 m of scour was predicted 600 m downstream at the location of a buried water main crossing.

References

Armanini, A. and Di Silvio, G. 1988. A one-dimensional model for the transport of a sediment mixture in non-equilibrium conditions. *Journal of Hydraulic Research*, **26**(3), 275-292.

Church, M. A., and McLean, D. G. (1994). Sedimentation in Lower Fraser River, British Columbia: Implications for Management. *The variability of large alluvial rivers* (pp. 221–241). American Society of Civil Engineers press.

Delft Hydraulics Laboratory (DHL). 1980. *Computation of Siltation in Dredged Trenches.* Report 1267-V/M 1570.

Faraci, C., Foti, E., Marini, A. and Scandura, P. 2011. Waves plus currents crossing at a right angle: sandpit case. *Journal of Waterway, Port, Coastal, and Ocean Engineering*, **138**(5), 339-361.

Galappatti, G. and Vreugdenhil, C. 1985. A depth-integrated model for suspended sediment transport. *Journal of Hydraulic Research*, **23**(4), 359-377.

Hervouet, J.-M. 2007. *Hydrodynamics of free surface flows: modelling with the finite element method*: John Wiley & Sons.

Hervouet, J.-M. 2000. TELEMAC modelling system: an overview. *Hydrological Processes*, **14**(13), 2209-2210.

Humphries, R., Venditti, J.G., Sklar, L.S. and Wooster, J.K. 2012. Experimental evidence for the effect of hydrographs on sediment pulse dynamics in gravel-bedded rivers. *Water Resources Research*, **48**(1).

Jurbin, T. 2006. Seismic upgrade for Massey Tunnel. *Tunnels & Tunnelling international*, (MAR).

Knaapen, M.A. and Kelly, D.M. 2012. Lag effects in morphodynamic modelling of engineering impacts. *33rd International Conference on Coastal Engineering (ICCE)*, Santander, Spain.

Lean, G. 1980. *Estimation of maintenance dredging for navigation channels.* Report by HR Wallingford Ltd.

McLean, D. G., Church, M. A., and Tassone, B. (1999). Sediment transport along lower Fraser River 1: Measurements and hydraulic computations. *Water Resources Research*, **35**(8), 2533–2548.

McLean, D. G., and Tassone, B. (1988). Budget of the Lower Fraser River. *Federal Inter-Agency Committee on Sedimentation,* 5th International Conference, Las Vegas, Nevada, pp. 16.

Meyer-Peter, E. and Müller, R. 1948. Formulas for bedload transport. *2nd International IAHR Congress*, Stockholm, Sweden.

NHC (2002). *Review of Lower Fraser River Sediment Budget.* Prepared by Northwest Hydraulic Consultants for Dredge Management Advisory Committee, Fraser River Estuary Management Program.

Miles, G. 1981. *Sediment transport models for estuaries.* Report by HR Wallingford Ltd.

Pritchard, D.T. and Hogg, A.J. 2003. Models of settling lag in coastal and estuarine settings. 3rd Symposium on River, *Coastal and Estuarine Morphodynamics*, Barcelona.

Sánchez, A. and Wu, W. 2011. A non-equilibrium sediment transport model for coastal inlets and navigation channels. *Journal of Coastal Research*, 39-48.

Tassi, P., Villaret, C., Huybrechts, N. and Hervouet, J. 2011. Numerical modelling of 2D and 3D suspended sediment transport in turbulent flows. *7th IAHR Symposium River Coastal and Estuarine Morphodynamics*, Beijing, China.

Tywoniuk, N. (1972). Sediment budget of the lower Fraser River. *American Society of Civil Engineering*, **Proc. 2**, 1105–1122.

van Rijn, L.C. 1984. Sediment transport, part I: bed load transport. *Journal of Hydraulic Engineering*, **110**(10), 1431-1456.

van Rijn, L.C. 1986a. Mathematical modeling of suspended sediment in nonuniform flows. *Journal of Hydraulic Engineering*, **112**(6), 433-455.

van Rijn, L.C. 1986b. Sedimentation of dredged channels by currents and waves. *Journal of Waterway, Port, Coastal, and Ocean Engineering*, **112**(5), 541-559.

Villard, P. and Church, M. 2003. Dunes and associated sand transport in a tidally influenced sand-bed channel: Fraser River, British Columbia. *Canadian Journal of Earth Sciences*, **40**(1), 115-130.

Villaret, C., Hervouet, J.-M., Kopmann, R., Merkel, U. and Davies, A.G. 2013. Morphodynamic modeling using the Telemac finite-element system. *Computers & Geosciences*, **53**, 105-113.

Wu, W. 2008. *Computational river dynamics*: CRC Press.

Leadership in Sustainable Infrastructure
Leadership en Infrastructures Durables

Vancouver, Canada
May 31 – June 3, 2017/ *Mai 31 – Juin 3, 2017*

WATER OSCILLATIONS IN VERTICAL TUBES FOR APPLICATION TO DRAINAGE SYSTEMS

Schulz, Harry E.[1,3], Zhu, David Z.[2] and Ma, Yiyi[2]
[1] School of Engineering at São Carlos, University of São Paulo, Brazil
[2] Faculty of Engineering, University of Alberta, Canada
[3] heschulz@sc.usp.br

Abstract: Vertical tubes are usual components of water distribution systems and of drainage systems (collecting tubes and manholes, for example). Drainage systems may be subjected to flooding events, which generate water columns with free surfaces in the vertical tubes. The water level corresponds, in these cases, to the hydraulic head where the vertical tube is located. Depending on the conditions of the flow in the system, pressure pulses may induce oscillations of the water level in the vertical tubes, eventually also generating phenomena like geysering. Oscillations may subject the buried structures to efforts not necessarily considered in the design phase. Eventual checking thus needs the periods of the oscillations. The amplitude of the oscillations may imply in spilling of water from the top section of the tube, or a momentary ejection above the outer ground level. The study of the conditions of possible spilling events is necessary to allow preventive measures. This study considers the dynamics of oscillating flows in vertical tubes. The governing equation for the flow is presented having the pressure imposed by the hydraulic head as the main impulsive factor of the movement. The resistive factors are taken into account through local and distributed losses. The equation is presented in normalized form, and predictions are compared with proper experimental data. Results obtained in adequate experimental devices, for different diameters of the vertical tube, were analyzed together. Conclusions about the different measured periods and the damping of fluctuations are presented.

Keywords: - Water oscillations in tubes, drainage systems, periods of oscillation, damping periodic flow, flooded manholes.

1 INTRODUCTION

The flooding of drainage systems in urban areas is becoming more common due to the accelerated increase of the cities, generating conditions not predicted in the design phase of these systems. This problem was already evidenced during the last two decades of the 20th century (Guo et al. 1990), but the frequency of flooding events is increasing since then (Lou et al. 2008, for example). Figures 1a and 1b sketch two conditions of possible flooding of urban drainage systems: the first showing only a flooded system of tubes, and the second showing a flood covering part of the surface of the urban area. The figure sketches vertical collectors, but also manholes may be considered. Implantation of new systems, extensions to existing systems, corrections of problematic sites, are procedures which design is based on existing or expected scenarios. If flooding is among these scenarios, the possibility of oscillations and geysering requires information of flow dynamics not only related to permanent and free surface cases. Computational tools allow the needed quantification, and calibration procedures are then followed.

Numerical codes for different transient situations in drainage systems may be found in the literature, some devoted to the transition from free to pressurized flows (Politano et al. 2005; Fuamba 2003, for example), others to the fluctuations in surge tanks due to valve operations (Wang et al. 2015, for example). Codes calibration considers generally simple flow conditions, in order to quickly verify the accuracy of obtained answers of, for example, periods of oscillation, pressure, and velocities. Vertical tubes may be of interest for the oscillation calculations, and also for its damping through the use of oscillation supressing devices, for protection of systems that may be subjected to some sudden blockage. In this case the tubes may be viewed as surge tanks, or, if capped, as air chambers, both viewed as a way of absorbing part of the energy of the flow oscillations.

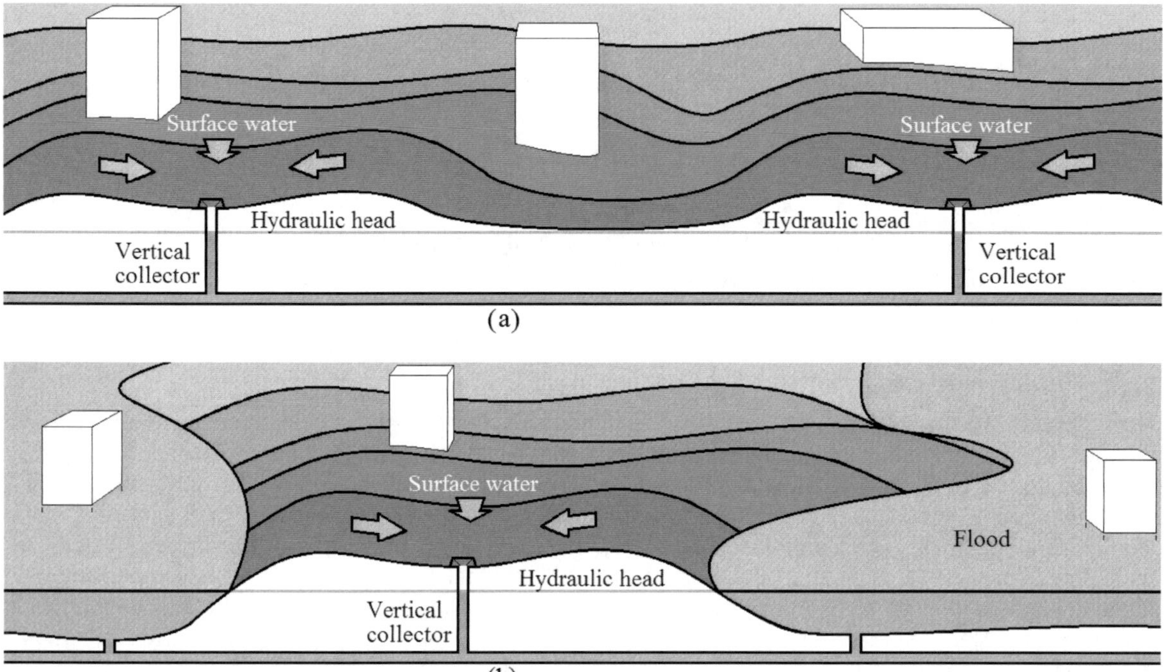

Figure 1: a) Flooded underground drainage system showing interconnected vertical collectors partially filled with water; b) Flood covering part of the urban surface area, and one vertical collector partially filled.

Complete drainage systems are simulated in the literature, but to provide a simple checking device, the vertical collector geometry was simplified in the present study as shown in Figure 2c.

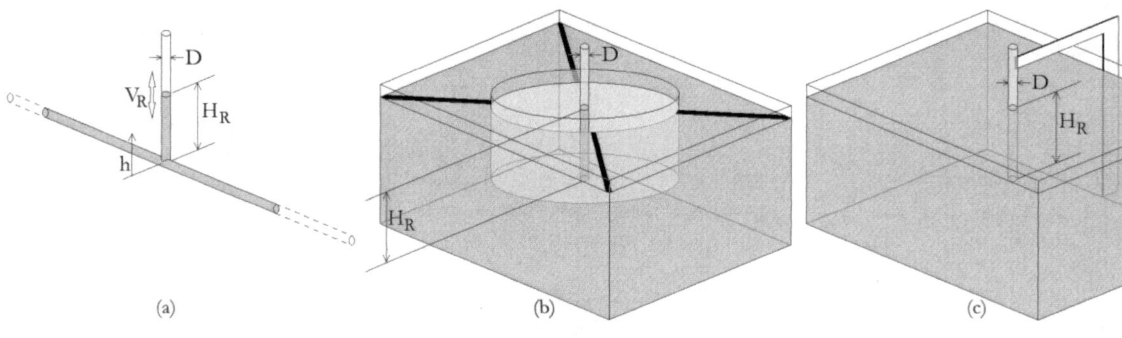

Figure 2 a) Real situation: vertical tube fixed to the horizontal system of tubes (shown in Figure 1); b) Possible laboratory setup: vertical column in a cylindrical empty reservoir which is fixed to the filled rectangular reservoir, c) Simplest setup of this study: only the vertical tube immersed in the rectangular reservoir. *D* is the diameter of the tube, H_R is the equilibrium position, V_R is the velocity, and *h* is the vertical axis that locates the water surface.

The oscillation in the vertical tube (collector, manhole, or tube for an oscillation supressing device) was analyzed here considering two governing differential equations solved analytically and numerically. The first is for the internal movement induced by an initial displacement from the equilibrium condition (H_R in Figure 2), considering concentrated and distributed losses in the vertical tube. The second is for the internal and external movements of the whole volume of water, for flows without losses. Comparisons with experimental data are presented. The experimental data were collected in devices properly built for this study. The aim of the study was to verify the applicability of the equations, in the sense of quantifying the periods of oscillation and the damping of the amplitudes in the geometry of vertical tubes.

2 MATHEMATICAL MODELS

The governing equation for the internal movement, induced by an imposed displacement from the equilibrium position (H_R), and considering local and distributed losses, is given by:

[1] $d^2y/dx^2 = (1/y - 1) + \theta_1(\theta_2 + \theta_3/y)(dy/dx)^2$

In this equation, $y = h/H_R$, $\tau = t\sqrt{(h/H_R)}$, $\theta_1 = \pm 1$, $\theta_2 = fH_R/2D$, $\theta_3 = \Sigma_i(K_i/2)$, where t is the time, D is the diameter of the vertical tube, f is the friction factor (for distributed losses), and K_i are local loss coefficients (i=1, 2, 3...). Equation 1 was obtained applying the principles of conservation of mass, momentum, and energy to the flow of the vertical tube, having atmospheric pressure at the water surface. As can be seen, θ_2 and θ_3 may be dependent on the flow itself, because the friction factor and the local loss coefficient are functions of the internal Reynolds number. In the present study θ_2 and θ_3 were taken as constants, which allow obtaining theoretical solutions, and also allow quickly verifying the adequacy of the equation in comparison to experimental data. Because θ_2 and θ_3 are multiplied by the square of the velocity, the constancy of the coefficients gives a turbulent characteristic to the quantified losses. This equation was integrated for natural values of $2\theta_3$ (that is, $2\theta_3$=1, 2, 3...), furnishing theoretical solutions of the vertical velocity for different values of the summed local losses. For example, for θ_3=1 and θ_2=1 the solution is given by the set of Eqs. [2] and [3]:

[2] $V_{Up} = \sqrt{C_{Up}y^{-2}e^{-2y} - \left(1 - \frac{1}{y}\right)^2}$

[3] $V_{Down} = \sqrt{e^{2y}y^2\left[-\frac{e^{-2y}}{y^2} + 4\frac{e^{-2y}}{y} + 8\ln y + 8\sum_{i=1}^{\infty}(-2y)^i/(i.i!) + C_{Down}\right]}$

V is the nondimensional velocity $V=V_R/\sqrt{(gH_R)}$, where g is the acceleration of the gravity, and V_R is the dimensional (real) velocity, as shown in Figure 2a. C_{Up} and C_{Down} are integration constants, and must be determined for each cycle "up/down" of the oscillating column. θ_2 depends on the parameters of its definition, being θ_2 = 1.0 used here as a convenient example. $2\theta_3$ > 2 produces more complicated equations. $2\theta_3$ = 2 and θ_2 = 1.0 were used for upwards and downwards movements, but different values may be used for each direction of movement. Upwards and downwards flows are stablished by setting θ_1 as -1 and +1, respectively. The theoretical solutions, Eqs. [2] and [3] may be used for calibration procedures. When intending to adjust θ_2 and θ_3 to experimental data, the direct use of Eq. [1] is recommended, because it allows using non-integer values for the two coefficients. As can be seen, although the global movement is a sequence of up/down cycles, it is not expressed as a damped sinusoidal function, with the intrinsic period of 2π, but by two different equations for the "up" and "down" directions. This is evident by the use of θ_1 = +1.0 or -1.0 in Eq. [1], and by the fact that each value of θ_1 leads to a different mathematical function for the movement. Although following different arguments and equations, similar conclusions are presented in the literature (Lorenceau et al., 2002; Masoodi et al., 2013, González-Santander and Martín, 2014, for example).

As mentioned, Eqs. [1], [2] and [3] consider the water inside the vertical tube, which is surely the relevant aspect of the flow. However, the moving water system also involves water outside the tube. Figure 3a shows a sketch of the water moving internally and externally of the vertical tube (up and down arrows). The external area A_E shows the limitation of the horizontal spreading of the water. We firstly considered two "ideal" conditions:

1. the outer superficial area has the same value of the internal superficial area ($A_E = A_t$). In this case, the situation is similar to a "U tube", which for a drainage system could be represented by the situation of Figure 1a. In this figure, if the right and left horizontal extensions of the drainage system are "closed" (momentously no flow), oscillations occur between the two vertical collectors, along a stretch of tube of constant area; and
2. the external area A_E tends to infinity. It corresponds to the sketch of Figure 1b, in which the oscillation occurs in the vertical collector, but it is not expected that the external level changes during the movement.

Figure 3b shows that the same experimental arrangement may be used to study much more complex situations, including long horizontal pipe systems by using coiled hoses. For the study of the vertical tubes only the arrangement of Figure 3a was used.

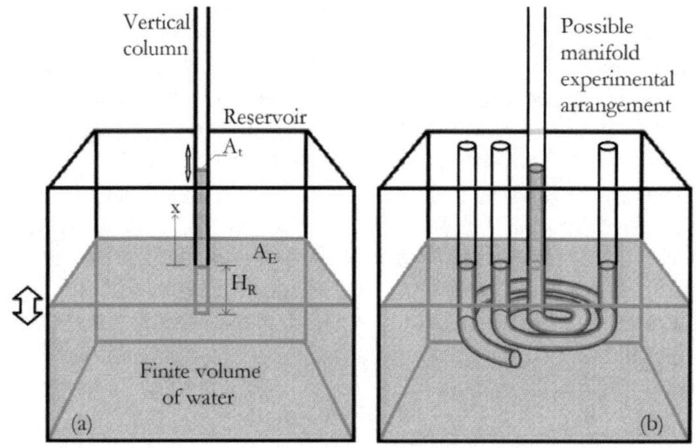

Figure 3: a) Whole moving volume of water. Origin of x at the equilibrium position, different of h of Figure 2. A_t is the transversal area of the tube, and A_E is the external superficial area; b) Example of possible complex arrangement allowed by this setup: manifold of vertical flows. The reference length may be more complex than only H_R.

The period of oscillation for the situation of Figure 3a is quantified here without considering shear losses (usual procedure in period evaluations). The governing Eq. [4] is used, and may be obtained by applying the energy conservation principle, or the balance of forces:

[4] $gx + \frac{\beta}{2}(dx/dt)^2 + (H_R + \beta x)d^2x/dt^2 = 0$, where $\beta = 1 - (A_t/A_E)$.

A_t is the transversal area of the tube, and A_E is the external superficial area. Equation [1] was integrated once to furnish the velocity of the flow in the tube using $dx/dt = 0$ at $x = x_0$, being x_0 the initial height of the water in the tube, leading to Eq. [5], presented here using nondimensional groups:

[5] $d\left(\frac{x}{x_0}\right)/d\sqrt{\frac{g}{H_R}}t = \pm\sqrt{[1-\left(\frac{x}{x_0}\right)^2]/[1+\sigma\left(\frac{x}{x_0}\right)]}$, where $\sigma = \frac{\beta x_0}{H_R}$.

If $\beta = 0$, the transversal area of the flow is constant, corresponding to a "U" tube, with the governing equation and its solution for $x = x_0$ at $t = 0$ presented as Eq. [6]:

[6] $d\left(\frac{x}{x_0}\right)/d\sqrt{\frac{g}{H_R}}t = \pm\sqrt{1-(x/x_0)^2}$, thus $\frac{x}{x_0} = \cos\left(\sqrt{\frac{g}{H_R}}t\right)$.

The period is $T = 2\pi\sqrt{(H_R/g)}$. In U tubes the total length (L) of the water volume is used, that is, $H_R = L/2$. For $\beta \rightarrow 1$, which implies $A_E \rightarrow \infty$, Eq. [5] may demand more efforts to be integrated. Considering, for simplicity, $H_R = x_0$, the governing equation and its solution are reduced, for $0 \le x \le x_0$, to the forms shown in Eqs. [7]:

$$[7]\ d\left(\frac{x}{x_0}\right)/d\sqrt{\frac{g}{H_R}}t = -\sqrt{1-(x/x_0)} \qquad \text{, thus} \qquad \frac{x}{x_0} = 1 - \frac{1}{2}\left(\sqrt{\frac{g}{H_R}}t\right)^2.$$

The time consumed moving along $0 \le x \le x_0$ is $t = 2\sqrt{(H_R/g)}$. The period for the surface of the water to return to the initial position is $T = 4t$, or $T = 8\sqrt{(H_R/g)}$ (remembering that $H_R = x_0$ in this derivation). As comparison tool, we may express $8 \approx 2.54\pi$. $\beta = 0$ leads to a sinusoidal function and to factor 2π for the fundamental period, while $\beta = 1$ conduces to a succession of parabolas and to the factor 2.54π for the period. The two functions are plotted in Figure 4a. Thus, for $H_R = x_0$ the periods of oscillation predicted by Eq. [5] are in the range $2\pi\sqrt{(H_R/g)} \le T \le 2.54\pi\sqrt{(H_R/g)}$, or, using the exact value, $2\pi\sqrt{(H_R/g)} \le T \le 8\sqrt{(H_R/g)}$. The possibility of different results considering or not the inertia of the water tank is mentioned in numerical studies of the literature, although not related to conceptual derivations as done here (Wang et al. 2015).

Equation [5] allows more detailed discussion. Considering the factor $\sigma = \beta x_0/H_R$, it tends to zero when $H_R \to \infty$, for any β and x_0. In other words, for long tubes (like the horizontal tubes in drainage systems), the period tends to $2\pi\sqrt{(H_R/g)}$ (in drainage systems $H_R = L/2$, being L the immersed length of the tubes, not only the hydraulic head). The condition $\beta = 1$ ($A_E \to \infty$) needs analyses of the effect of $\sigma = x_0/H_R$ on the period of vertical tubes. Figure 4b shows the evolution of x/x_0 for the first quarter of a period ($0 \le x \le x_0$) using different values of σ. Equation [6] implies $\sigma = 0$ and the period $2\pi\sqrt{(H_R/g)}$, so that any $\sigma > 0$ leads to larger periods. This occurs more evidently for the first periods. Along time, energy losses damp the amplitudes of the oscillations, producing smaller σ for each new period, probably inducing also a trend to $T = 2\pi\sqrt{(H_R/g)}$ for the larger times. Finally, the ideal and real conditions of flows for $\beta = 1$ are sketched in Figure 5, showing that in real cases the whole external volume of fluid does not dislocate with a homogeneous vertical velocity. A smaller amount of water dislocates vertically (closer to the tube), with a correspondent smaller mean A_E. As a conclusion, periods with value $T = 2\pi\sqrt{(H_R/g)}$ and larger are expected for vertical tubes, although smaller than the ideal period $T = 2.54\pi\sqrt{(H_R/g)}$.

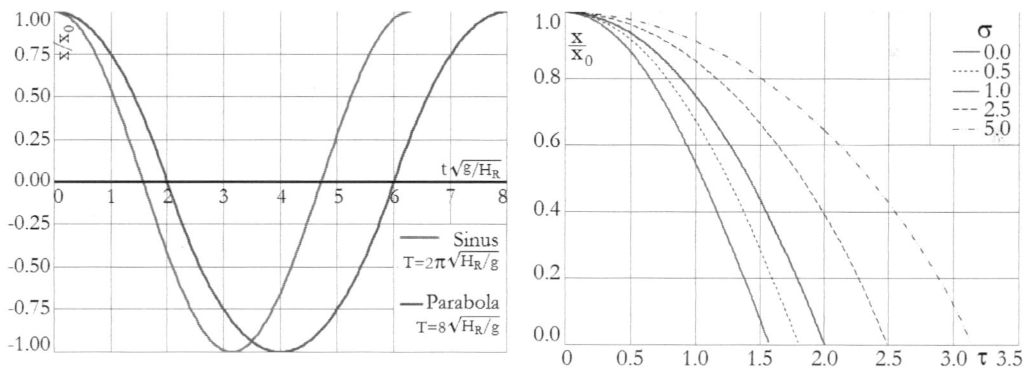

Figure 4: a) Shear free evolution of the water surface for $\beta = 0$ and $\beta = 1$ for vertical tubes; b) Shear free evolution for the first quarter of a period given by Eq. [5] for different σ. ($\tau = t\sqrt{(g/H_R)}$).

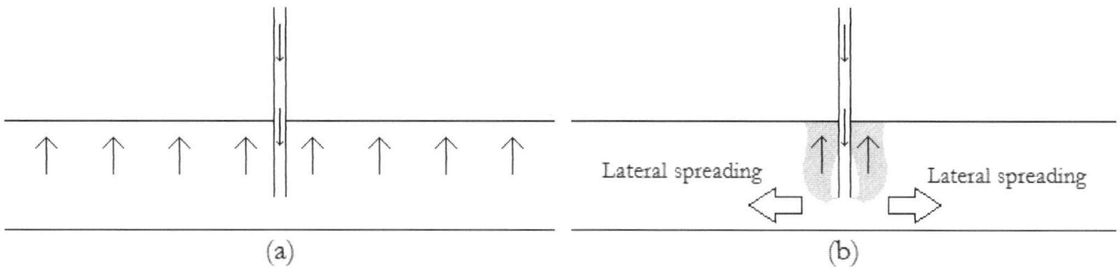

Figure 5: a) Ideal situation $\beta = 1$: fluid moves vertically with homogeneous velocity extending horizontally to infinity; b) Real situation: the fluid has a main vertical movement that does not extends to infinity.

3 EXPERIMENTAL DEVICES

Experiments were conducted in the Hydraulics Laboratory of the University of Alberta following the sketch of Figure 3. Two small scale devices were built for experiments conducted with tubes having diameter of 0.62 cm and 2.54 cm. A third larger scale setup was built for a vertical tube with diameter of 10.16 cm. The larger setup is shown in Figure 6a. A cast acrylic tube was used, with internal diameter of 4" (10.16 cm), and external diameter of 4½" (11.43 cm). The superficial area of the water tank was 1.769 m^2, having a square shape with sides of 1.33 m. The total height of the tank was about 91 cm, but experiments were also run for a depth of about 75 cm. The length of the acrylic tube was about 1.84 m.

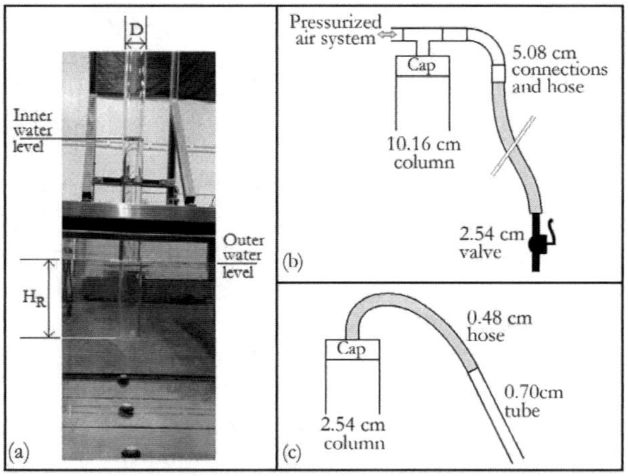

Figure 6: a) Experimental setup for column with $D = 10.16$ cm. b) Details of the control of the air flow for column with $D = 10.16$ cm. c) Details of the control of the air flow for column with $D = 2.54$ cm. The numbers in the figure are diameter values.

For the experiments with $D = 10.16$ cm, the upper section of the vertical column was covered with a cap fixed to a hose with a diameter of 5.08 cm (2"), using a "T" and a "curve" connections with a total length of 60 cm. The hose had a length of 4.11 m. A manual "open/close" valve was installed at the open end of the hose using a 10 cm long connection. Valve and connection had diameter of 2.54 cm (1"). The second opening of the "T" connection at the tube cap was used to inject air or to produce vacuum in the vertical column, for the adjustment of the inner water level. A sketch is shown in Figure 6b. Experiments were conducted in two ways: 1) by removing the cap and opening completely the end on the vertical tube, and 2) by opening the valve at the end of the hose. Thus free oscillations and damped oscillations were produced, respectively.

For the experiments with $D = 2.54$ cm, a plastic tube with length of 45.5 cm was used. Also for this diameter, the upper section of the vertical column was covered with a cap fixed to a hose. The hose was 30 cm long and had a diameter of 0.48 cm, being connected to a straight tube 21.5 cm long and with a diameter of 0.70 cm. The water in the vertical tube was adjusted by blowing in the 0.70 cm tube. A sketch is shown in Figure 6c. Also here experiments were conducted in two ways: 1) by removing the cap and opening completely the end on the vertical tube, and 2) by opening the end of the 0.70 cm tube. Once more, free oscillations and damped oscillations were produced, respectively.

For the experiments with $D = 0.62$ cm, a plastic tube 31.2 cm long was used. Considering the small diameter, only free oscillations were conducted in this tube. In this case capillary effects were present, thus involving forces not considered in the original formulation that also helped the damping of the oscillations.

The data were collected by filming the experiments with digital cameras, using speeds of 200, 120, 60, 40, and 30 fps. It was observed that higher speeds are more adequate (200, 120, and 60 fps) for the present experiments.

4 RESULTS

4.1 Periods of Oscillation

The theoretical calculations presented here in Eqs. [4] to [7] showed two possible "ideal" periods for shear free oscillations and $x_0 = H_R$: 2π and 8 (or 2π and $\sim 2.54\pi$), with a difference of 27%. As already mentioned, also numerical results of the literature show that different periods may be calculated for the same physical situation (Wang et al. 2015), depending on the adopted assumptions. Because such different results, the knowledge of the behavior of real data is necessary. The experiments were conducted using the vertical tubes with D = 10.16 cm, 2.54 cm, and 0.62 cm. Table 1 shows the experimental conditions and the measured periods, which were extracted from the post-processing of the obtained films. The post-processing involved frame by frame analyses for the location of the water surface, and the registering of the results. Corrections of refraction effects were performed.

Table 1: Experimental conditions and measured periods for vertical tubes

Free/Damped	Diameter (cm)	H_R (cm)	Measured Period (s)
Damped	10.16	22.5	1.008
Free	10.16	19.3	0.972
Damped	10.16	19.9	0.976
Free	10.16	19.0	0.970
Damped	10.16	33.6	1.287
Damped	10.16	33.6	1.279
Damped	10.16	41.2	1.400
Free	10.16	41.0	1.418
Damped	2.54	18.2	0.908
Free	2.54	18.2	0.878
Free	0.62	7.5	0.535
Free	0.62	5.8	0.500

The dots of Figure 7 show the measured periods of oscillation plotted against the measured $\sqrt{(H_R/g)}$. The lower and upper lines are the predictions of Eqs. [6] and [7], respectively. For smaller $\sqrt{(H_R/g)}$ the measured periods are closer to the line $T = 2\pi\sqrt{(H_R/g)}$, but they depart from this line for larger $\sqrt{(H_R/g)}$. The best fit of the data is $T = 2.17\pi\sqrt{(H_R/g)}$, with $R^2 = 0.99$, located between the two theoretical predictions.

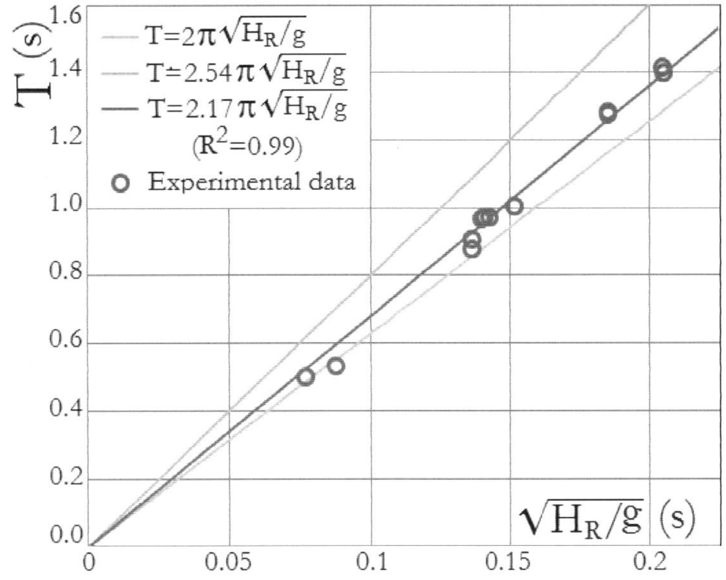

Figure 7: Theoretical and experimental results for the periods of oscillation in vertical tubes.

4.2 Damping of Oscillations

The calculation of oscillations in flooded scenarios allows verifying the distribution of additional efforts due to the water flow. But they may also suggest the use of devices that suppress the oscillations, intending to avoid induced damages. This study generated data for damped situations, for which a control of the air flow was applied. The top air pocket with the hose described in Figure 6 is the oscillation suppressing device used in the present study. In order to test Eq. [1], free oscillations experiments were firstly conducted by removing the cap of the vertical column and letting the air flow without restrictions. Experimental results and numerical calculations using Eq. [1] are shown in Figure 8 for D = 2.54 cm. Figure 8a shows the nondimensional water level plotted against the nondimensional time, and Figure 8b shows the nondimensional vertical velocity plotted against the nondimensional level.

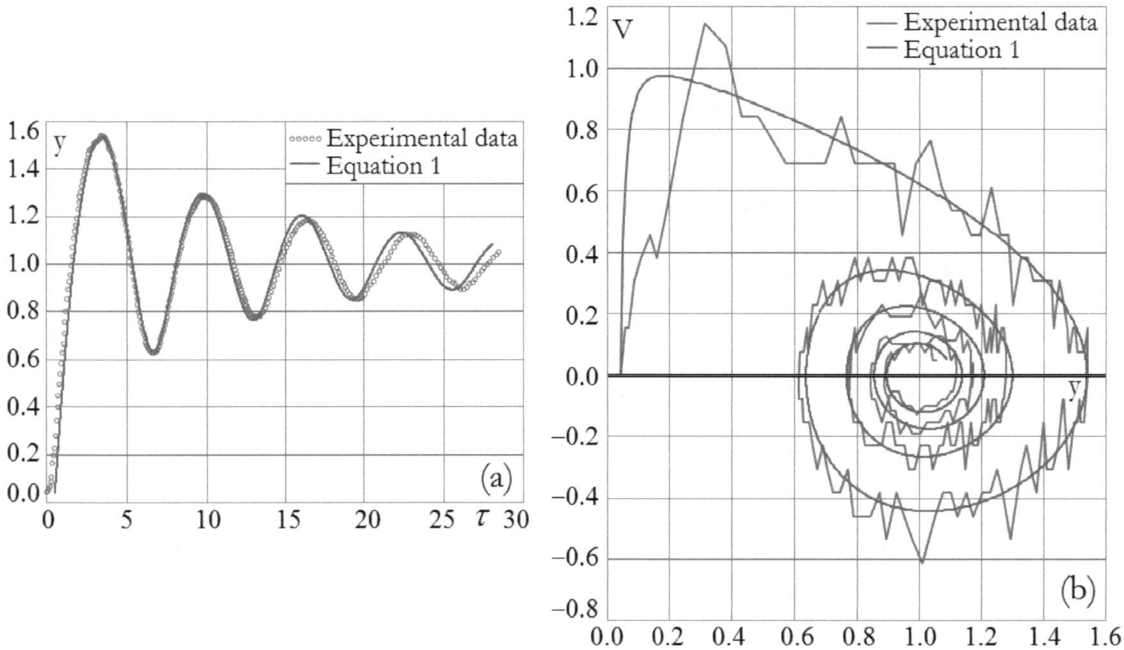

Figure 8: a) Measured and calculated water levels, D = 2.54 cm. Theoretical results with a shift t_0 = 0.4 (column uncapping lead to an experimental lag time); b) Measured and calculated velocities, D = 2.54 cm.

The calculations were performed using constant values of the coefficients θ_2 and θ_3 for each up/down movement, as follows: First and second up: θ_2 = 0.1, and θ_3 = 0.85; first and second down: θ_2 = 0.1, and θ_3 = 0.01, subsequent up and downs: θ_2 = 0.1, and θ_3 = 0.85. The adjustment to the observed movement is good, for both the water level and the velocity. The observed difference for higher times may be caused by the use of constant θ_2 and θ_3 for each up/down movement. These coefficients are functions of the movement itself (expressed through the Reynolds number, for example).

The damped oscillations shown in Figure 9 were produced by opening the valve and the straight tube in the setups for D = 10.16 cm and D = 2.54 cm, respectively. The air cap was maintained, so that pressurized air chambers were formed above the water level, together with high loss regions along the hoses. Figure 9a shows the evolution of the nondimensional water level of damped experiments for D = 10.16 cm and D = 2.54 cm, while Figure 9b shows the evolution of the nondimensional velocity for D = 10.16 cm. Numerical predictions would need Eq. [1] for the water flow and an additional equation for the air flow, not discussed in the present study.

A first evaluation of the suppression of oscillations was made based on the initial positive or negative pulse P_0, given as $P_0 = |y_0-1|$ (modulus was used, that is, positive values). It corresponds to a sudden lack or excess of pressure (in relation to the equilibrium level) applied at the basis of the vertical tube used as oscillation supressing device. The capacity to reduce initial pulses was analyzed through the subsequent $P_1 = y_1$-1 and $P_2 = y_2$-1 values, using the first maximum and minimum of the experimental y function. Table 2 shows values obtained for free and damped experiments with D = 10.16 cm and D = 2.54 cm.

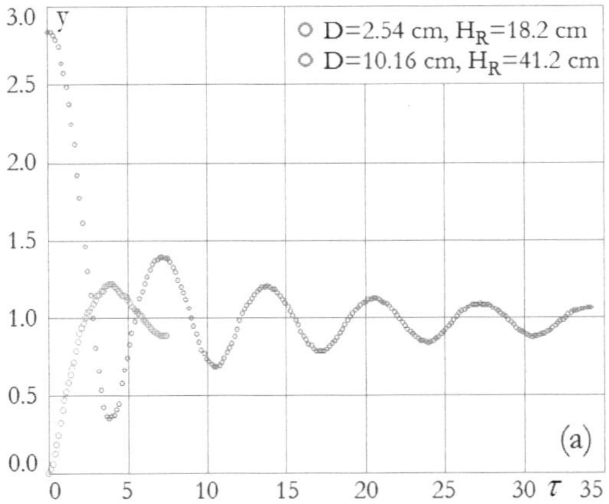

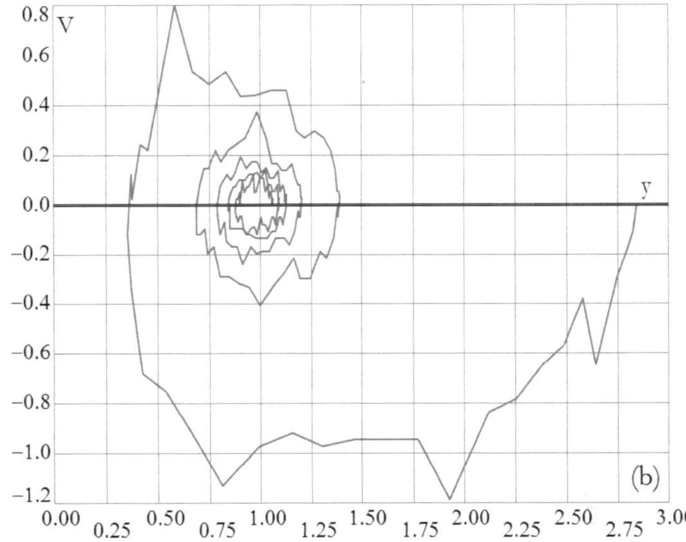

Figure 9: a) Experimental damped water levels; b) Experimental damped velocities for D = 10.16 cm.

The last column of Table 2 shows that the 10.16 cm vertical tube reduces to about 34.9% the initial pulse for one oscillation period (P_2) when working like an open surge tank (free oscillations). When using the air chamber and the hose, the reduction attains about 21.3% of the initial pulse. The same sequence for the 2.54 cm vertical column shows the results of 37.4% for free oscillations, and 11.6% for the air chamber with hose. The half period analysis (P_1) is possible for the 2.54 cm tube, showing the results of 54.2% for free oscillations, and 21.5% for air chamber with hose. The half period analysis is not possible for the 10.16 cm tube because the experimental procedures did not furnish the internal water levels when below the external water level. The results show that the experimental setup allows the study of suppression of oscillations, through the comparative analysis of free and damped oscillations, the last understood as being damped through an installed device.

Table 2: Reduction of oscillations for free and damped experiments (modulus of Pi were used)

Free/Damped	Diameter (cm)	H_R (cm)	P_0	P_1	P_2	P_1/P_0	P_2/P_0
Free	10.16	41.0	1.78	(*)	0.622	(*)	34.9%
Damped	10.16	41.2	1.84	0.648	0.391	35.2%	21.3%
Free	2.54	18.2	1.00	0.542	0.374	54.2%	37.4%
Damped	2.54	18.2	1.00	0.215	0.116	21.5%	11.6%

*P_1 not collected: internal water level not measured below the external water level.

5 CONCLUSIONS

Governing equations for the study of oscillating flows in vertical tubes were presented. A first equation considered the internal flow with energy losses, and a second equation considered the whole volume of water without energy losses.

Experimental setups were built in order to check with real data the theoretical predictions of the proposed equations. It was shown that the periods of oscillation in vertical tubes are preferentially larger than the solution $T = 2\pi\sqrt{(H_R/g)}$. The best fit equation for the present data was $T = 2.17\pi\sqrt{(H_R/g)}$, with a correlation coefficient $R^2 = 0.99$.

Numerical predictions for the evolution of the water level in the vertical tube are in good agreement with the measured results, a fact more evident for the first periods of oscillation. Differences observed for larger times are understood as a consequence of the use of constant friction factors and local loss coefficients in the numerical model (which in fact are dependent on the flow itself).

The use of the experimental devices to quantify observed damping in oscillation suppressing devices was tested. The initial condition was adjusted, and the subsequent maxima and minima were compared for free and damped oscillations, allowing the aimed direct comparison.

The obtained results show the adequacy of the methodology being followed in this research line.

Acknowledgements

The first author thanks the support of CAPES, which, through process BEX 5723/15-9, allowed the collaboration study with the University of Alberta. The authors thank Mr. Perry Fedun for the careful building of the larger experimental setup

References

Fuamba, M. 2003. Contribution on Transient Flow Modelling in Storm Sewers. *Journal of Hydraulic Research*, **40**(6): 685-693.

González-Santander, J.L. and Martín, G. 2015. New Analytical Approximations for the Liquid Rise in a Capillary Tube. *Fluid Dynamics Research*, **47**, 28p.

Guo, Q. and Song, C.C.S. 1990. Surging in Urban Storm Drainage Systems. *Journal of Hydraulic Engineering*, **116**(12): 1523-1537.

Lorenceau, É.; Quéré, D.; Jean-Yves, O. and Clanet, C. 2002. Gravitational Oscillations of a Liquid Column in a Pipe. *Physics of Fluids*, **14**(6), 1985-1992.

Lou, S.; Zhong, G.H.; Zhu, Y.D. and Liu, S.G. 2008. Investigation of Underground Traffic Facilities for Flood Control in Shanghai. *5th China-Japan Joint Seminar for the Graduate Students in Civil Engineering*, Tongji University, Shanghai, China, Proceedings, 136-141.

Masoodi, R.; Languri, E.; and Ostadhossein, A. 2013. Dynamics of Liquid Rise in a Vertical Capillary Tube. *Journal of Colloid and Interface Science*, **389**, 268-272.

Politano, M.; Odgaard, J. and Klecan, W. 2005. Numerical Simulation of Hydraulic Transients in Drainage Systems. *Mecánica Computacional*, **XXIV**: 297-310.

Wang, C.; Yang, J. and Nillson, H. 2015. Simulation of Water Level Fluctuations in a Hydraulic System Using a Coupled gas-Liquid Model. *Water*, **7**: 4446-4476.

AUTHOR INDEX

AUTHOR INDEX

Canadian Society for Civil Engineering (CSCE)
521-300 rue St-Sacrement
Montréal, QC
H2Y 1x4

ISBN 978-1-5108-7840-2

CURRAN ASSOCIATES INC.
proceedings.com

10th Mathematical Physics Meeting: School and Conference on Modern Mathematical Physics 2019

Belgrade, Serbia
9 – 14 September 2019